Numerical Computation in Science and Engineering

C. Pozrikidis

University of California, San Diego

New York *Oxford*
OXFORD UNIVERSITY PRESS
1998

Oxford University Press

Oxford New York
Athens Auckland Bangkok Bogotá Bombay
Buenos Aires Calcutta Cape Town Dar es Salaam
Delhi Florence Hong Kong Istanbul Karachi
Kuala Lumpur Madras Madrid Melbourne
Mexico City Nairobi Paris Singapore
Taipei Tokyo Toronto Warsaw

and associated companies in
Berlin Ibadan

Published by Oxford University Press, Inc.
198 Madison Avenue, New York, New York 10016

Library of Congress Cataloging-in-Publication Data
Pozrikidis, C.
Numerical computation in science and engineering / C. Pozrikidis.
p. cm.
Includes bibliographical references and index.
ISBN 0-19-511253-9 (cloth)
1. Numerical analysis. I. Title.
QA297.P69 1998
519.4—dc21 97-38185
CIP

Printing (last digit) 9 8 7 6 5 4 3 2 1

Printed in the United States of America
on acid-free paper

CONTENTS

PREFACE

After teaching and thinking about numerical computation for about fifteen years, I became convinced that there is a need for a relatively concise book on this subject that achieves an important yet challenging goal: *it allows the nonexpert, student or not, to read it in a relatively short period of time and get a good idea of what scientific and engineering numerical computation is and how it is applied.* The book should be rigorous enough so that the reader is not frustrated or misled, but practical enough, so that the gap between theory and practice is not wide. In a way, the book should be a manual of numerical computation, but not a primer, a recipe book, or a handbook. These thoughts motivated me to write this book whose intended features and presumed novelty hinge on its four goals:

- To discuss the fundamental and practical issues involved in a numerical computation in a *unified* manner, with a healthy but not excessive dosage of fundamentals.
- To concisely illustrate the practical implementation of the numerical algorithms, as well as expose the distinguishing features and relative merits of alternative approaches; and to demystify many seemingly esoteric numerical methods.
- To exemplify the need-to-develop numerical methods by discussing selected real-life examples drawn from the various branches of science and engineering.
- To address the important issues as they arise, acknowledging but not exploring problematic circumstances; these are treated best in research articles and specialized monographs.

Books on mature and diverse subjects such as numerical computation can easily run away in their volume. The present material was selected and presented according to the following guidelines:

- Numerical methods that can be explained without too much elaboration, and can be implemented within a few dozen lines of computer code, are discussed in detail.
- Numerical methods whose underlying theory requires long and elaborate explanations, and algorithms that are not likely to be used without resorting to numerical packages, are discussed at the level of first principles but are not explained in detail. Instead, references for further information are provided.
- The numerical methods discussed are *actually used* in practice. Knowledge of the various topics will prove useful at some point one way or another, but it does not have to be gained at once or in a strictly sequential manner.
- The use of long equations and specialized notation are avoided as much as possible, and the presentation relies on *schematic illustrations* accompanying the text but not being given the formal designation of a figure. The interplay between ideas and visual images is an important aspect of the discourse.

- Numerical examples are presented in cases where it would be difficult for the readers to produce their own. There is a fine line between *simple* and *simplistic*, and a Herculean effort was made not to cross it.

The book was written to be used as a text in *upper-level undergraduate and beginning graduate* courses in various disciplines of science and engineering. It should also be suitable for *self-study*, assuming no prior experience on numerical computation but a will to learn it. Prerequisites are basic knowledge of calculus and elementary computer programming. Several appendices are included for ready reference and to minimize distractions.

Each section is divided into several subsections, which are *learning entities*—the reader can pause between them—and each subsection is followed by theoretical and computational problems. Distributing the problems in this manner throughout a section is consistent with their intent, which is to *complement the theory* and to provide readers with an immediate opportunity to practice their skills, saving them from the frustration of trying to find them at a chapter's end; this chore is usually left for the instructor.

Explicit computer programs are not included in the text, but blueprints of numerical procedures are presented in the form of pseudocodes that can readily be translated into any computer language. The pseudocode commands are explained on page xi. The author strongly recommends that students write their own codes. If this is prohibited by lack of time due to extracurricular activities, the material should be useful for understanding the function of ready-made programs such as those accompanying this book.

The writing of a book is as painful as it is rewarding. *Madhu Goparakrishnan, Richard Charles, Pao Chau, Ian Do*, and *Jack Yee* read early drafts of chapters and offered me advice for improvement. *Audrey Hill* helped me with the presentation. *George Karniadakis, Jonathan Higdon, Bill Schowalter, George Triantafyllou, Yiannis Tsamopoulos* encouraged to pursue this project. *Bob Rogers* insightfully saw the need for this text. I sincerely thank them all.

The cover shows the interface of a liquid droplet extended under the influence of a shear flow; the color encodes the concentration of a surfactant. Computation and graphics by Xiaofan Li.

C. Pozrikidis

Kensington, San Diego
April 1997

Accompanying Software Package

The book comes with a bonus: appendix E gives instructions on how to access a public domain software library of FORTRAN 77 programs that have especially been written for this book and follow the notation of the text.

Numerical Methods Programs on the Internet

A variety of general and specialized numerical methods programs are available through the internet, many of them free of charge. Guided tours are available at the worldwide web sites:

http://gams.nist.gov
http://www.lahey.com

PseudoCode Language Commands

Pseudo derives from the Greek word ψεῦδος, which means *a lie.* But the pseudocodes do not lie; they simply emulate generic commands or computer instructions.

The pseducodes will be written in standard index notation, with a comma separating the indices.

Following are examples of key conventions:

$a = b + c$	Replace the value of the variable a with the value of the sum $b + c$
$a = bc$	Replace the value of the variable a with the value of the product bc
$a \leftarrow a + c$	Replace the value of the variable a with the value of the sum $a + c$

Do $i = 8, 27$
  $u_i = v_i + w_i$
End Do

Structured ascending Do loop. The index i takes the integer values: $8, 9, \ldots, 27$

Do $i = 8, 64, 4$
  $u_i = v_i + w_i$
End Do

Structured ascending Do loop. The index i takes the integer values: $8, 12, 16, \ldots, 64$

Do $i = 20, 10, -2$
  $u_i = v_i + w_i$
End Do

Structured descending Do loop. The index i takes the values: 20, 18, 16, 14, 12, 10

Do $i = 2, 10, -2$
  $u_i = v_i + w_i$
End Do

Illegal Do loop

DO $i = 14, 1$
  $u_i = v_i + w_i$
End Do

Illegal Do loop

If$(i = j)$Then
  $u_i = A_{i,j}$
End If

Structured If loop. Operation is executed only if the statement in the parantheses is true

Chapter 1

Numerical Computation

1.1 Analytical and Numerical Computation

The term *computation* usually brings to mind the image of a computing machine that generates strings or tables of numbers, or the image of a roomful of engineers, scientists, or accountants, as seen in a PBS or BBC documentary, patiently sitting at their desks and carrying out thousands of additions and multiplications. The latter was actually realized in 1794 when the French government ordered the manual compilation of an enormous set of mathematical tables: Among them, the logarithms of the natural numbers from 1 to 200,000, accurate to the nineteenth decimal place (Hayes 1988, p. 1). The project was never completed. This picture, however, is not entirely accurate, and it is useful to make a distinction between *analytical* and *numerical* computation at the outset.

Two ways of computing the solution of the zeroth-order Bessel ordinary differential equation

$$t^2\frac{d^2x}{dt^2} + t\frac{dx}{dt} + t^2x = 0 \qquad (1.1.1)$$

for the function $x(t)$, subject to appropriate conditions or constraints, are as follows:

1. By finding a certain polynomial that represents or approximates the desired function, that is, by deriving explicit expressions for the coefficients of the polynomial, followed by *numerical evaluation.*
2. By generating tables of numbers that display values of the function $x(t)$ at different values of the independent variable t.

The first computation has an analytical character, while the second computation has a numerical nature. The preferable approach depends on the particular application that necessitated the solution, the desired accuracy, and the relative difficulty of the two methods. In practice, the complexity of the analytical treatment often makes numerical computation desirable if not inevitable.

To make the distinction between analytical and numerical computation more clear, consider the quadratic algebraic equation $ax^2+bx+c = 0$, where a, b, and c are three known coefficients. Analytical computation yields the well-known solution in closed analytic form,

$$x = \frac{-b \pm \sqrt{b^2 - 4ac}}{2a} = -\frac{2c}{b \pm \sqrt{b^2 - 4ac}} \tag{1.1.2}$$

Numerical computation, on the other hand, produces a three-parameter family of tables that list the two roots as functions of a, b, and c. Clearly, in this case the analytical approach is desirable. But the evaluation of the right-hand side of equation (1.1.1) is expedited with the assistance of a calculator whose function relies on numerical computation. It may then be argued that numerical computation is unavoidable at a certain level.

The necessity for numerical computation becomes more evident by considering the cubic equation $ax^3 + bx^2 + cx + d = 0$, whose analytical solution is available but involved (e.g., Lanczos 1988). An analytical solution for polynomial equations of higher order is not generally feasible, and one must rely on numerical computation in order to generate numerical solutions to the particular problem at hand, working on a case-by-case basis.

As a final argument in favor of numerical computation, is that it can be used to obtain quick answers: To find the limit of the function $f(x) = x \ln x$, as x tends to zero, without computing derivatives, we can simply evaluate it at a small value of x with the help of a calculator.

Ready-made Computer Programs

With the availability of ready-made *algebraic* or *symbolic manipulation* computer programs, a properly equipped general-purpose computer is able to perform both tasks of analytical and numerical computation with efficiency and at a low cost. Developing programs for analytical computation involves the networking of logical operations whose implementation requires expertise from specialized disciplines of mathematics and computer science. Given the reliability of such programs, an understanding of the precise manner in which they work is not essential in general-purpose scientific and engineering computation.

The same could be argued about programs that carry out numerical computation: Why not use a ready-made package? This is a dangerous temptation: *A poor understanding of how a numerical solution is computed may lead down a dangerous path between failure and misinterpretation.* Numerical computation, as straightforward as it might appear at first sight, can be subtle and capricious. For example, within the scope of analytical computation, the two expressions in equation (1.1.2) are equivalent. But, as we will see in Section 1.3, under certain conditions, their numerical evaluation can produce wildly different answers.

A story told by the prominent English physicist Eddington and quoted by Hamming (1962, p. 709) is relevant to this discussion. A person went fishing with a net of a certain size, observed the minimum size of the fish he caught, and erroneously concluded that this was the minimum size of all the fish in the ocean. His mistake was that he did not understand how fishing was done. And he was too careless to verify the validity of his conclusions by using a net of different size.

Preparing the Numerical Problem

It is sometimes tempting to instruct the computer to tackle a certain numerical problem in its raw form. The importance of preceding the numerical computations with some analytical work that simplifies the numerical problem, however, cannot be overemphasized. In some cases, one may even find that carrying out a numerical computation is not necessary after all, and a simple numerical evaluation of an analytical expression will do the job.

As a general rule, the more effort and time one spends preparing and refining the statement of the computational problem and designing the strategy of solution, also called an *algorithm* (Section 1.3), the more accurate, reliable, and efficient the numerical solution will be. Such preparation also yields dividends when debugging a computer program and facilitates the extension of the program to a more general framework.

Consider, for example, computing the Cartesian coordinates of the center $\boldsymbol{x}_c = (x_c, y_c, z_c)$ and radius a of a sphere that passes through four specified points in space. If the four points do not lie in the same plane, the sphere is unique, but if the points do lie in the same plane, the sphere either does not exist or else it is not unique.

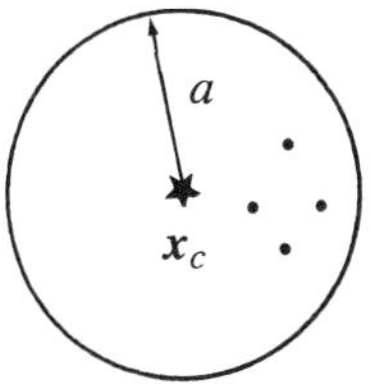

The distance between a point $\boldsymbol{x} = (x, y, z)$ that lies on the sphere and the center of the sphere is, of course, equal to the radius a. Using the Pythagorean theorem, we find

$$|\boldsymbol{x} - \boldsymbol{x}_c|^2 = (x - x_c)^2 + (y - y_c)^2 + (z - z_c)^2 = a^2 \tag{1.1.3}$$

Applying equation (1.1.3) for the four points that lie on the sphere, and expanding the squares on the right-hand side, we obtain four *quadratic* equations for the four unknowns x_c, y_c, z_c, a. Solving this primary problem requires a substantial amount of work.

The solution of the nonlinear equations can be circumvented by subtracting the fourth equation from the first three equations, thereby obtaining a reduced system of three *linear* equations for x_c, y_c, z_c,

$$2(x_i - x_4)x_c + 2(y_i - y_4)y_c + 2(z_i - z_4)z_c = x_i^2 + y_i^2 + z_i^2 - x_4^2 - y_4^2 - z_4^2 \tag{1.1.4}$$

for $i = 1, 2, 3$. The linear system can be solved readily by elementary methods such as Cramer's rule taught in high school. Physically, the preceding three equations describe the intersection of the three planes that are perpendicular to the midpoints of three straight segments; each segment passes through the fourth point and one of the first three points. Once the center of the sphere has been computed, the results are substituted into equation (1.1.3), and the resulting expression is evaluated at one of the four points to produce the radius a.

As a second example, we consider the *Fibonacci sequence* of integers, defined by the recursive relation

$$a_k = a_{k-1} + a_{k-2} \qquad \text{for } k > 2 \tag{1.1.5}$$

with

$$a_1 = 1, \qquad a_2 = 1$$

which has been argued to model the structure and growth of certain biological patterns. As more members of the sequence emerge, the ratio a_{k-1}/a_k tends to limiting value r, called the *golden ratio.* One way to obtain r is to compute a large number of members of the sequence and to observe the behavior of the ratio a_{k-1}/a_k. Another way is to divide the recursive formula (1.1.5) by a_{k-1}, set $a_k/a_{k-1} = 1/r$ and $a_{k-2}/a_{k-1} = r$, and thus obtain the quadratic equation $1/r = 1 + r$ or $r^2 + r - 1 = 0$. The positive root of this quadratic equation yields the golden ratio $r = \frac{1}{2}(5^{1/2} - 1)$.

PROBLEMS

1.1.1 *Points on a sphere.*

(*a*) Compute the center and radius of the sphere that passes through the four points (0, 0, 0), (1, 0, 0), (1, 1, 0), (1, 1, 0.90). (*b*) Consider a hypersphere in the N-dimensional space. How many points do you need to compute its center and radius, and is the solution unique? Discuss possible pathologies of the numerical computation associated with a nonunique or a nonexistent solution.

1.1.2 *Convergence of the Fibonacci series.*

Compute the golden ratio r accurate to the eighth decimal place by monitoring the behavior of the ratios a_{k-1}/a_k of the Fibonacci series. How many terms in the series do you require? Plot the difference $a_{k-1}/a_k - r$ against k on a log–log scale, and discuss its behavior.

Murphy's Law

The origin of Murphy's law is quoted in the lovely *Macintosh Bible* (1994, Peachpit Press), and its consequences are expounded by Matthews (1997). In 1949, Captain Edward A. Murphy was the director of a NASA laboratory at the Edwards United States Air Force Base. When an important experiment was grossly miswired, Captain Murphy caustically remarked: "*If there are two or more ways of doing something, and some of them can lead to catastrophe, then someone will do it.*" His comment was later generalized to become the universal Murphy's law, one version of which is: "*If something can go wrong, it will.*"

The editors of the *Macintosh Bible* are careful to point out that Murphy's law is not the result of some kind of malevolence on behalf of the cosmos, but a reflection of the complexity of the real world. Numerical computation presents one with a multitude of opportunities to confirm the pervasiveness and universality of some version of Murphy's law.

1.2 Hardware and Software

Since the fabrication of the first all-electronic computer at the University of Pennsylvania in 1946, named the Electrical Numerical Integrator and Automatic Computer, ENIAC™, computers have changed a lot, and certainly for the better. But certain consequences of this progress still endure: The accepted psychological term *cyberphobia* describes the fear of, and aversion for, computers; the reader is assumed not to suffer from it.

A rudimentary knowledge of the various computer components, their functions, and the way in which information is processed and stored in them are necessary for recognizing the potentialities and pragmatic limitations of small- and large-scale numerical computation. A detailed discussion of a computer's operation is outside the scope of this book, but a brief description may serve to illuminate some dark spots.

The computer's physical components, called hardware, and programmed instructions, called software, can be described as follows.

- A central processor and assisting coprocessors perform numerical computations and make logical decisions, collectively called *operations*. For example, a math coprocessor specializes in mathematical computations. Processors in the *Program-Control Unit*, PCU, interprete and prioritize instructions, and processors in the *Arithmetic-Logic Unit*, ALU, execute these instructions.

  A processor can be rated in terms of its *clock frequency* or *clock rate*, that is, the frequency of an internal vibrating crystal; a decent processor vibrates at 200 MHz. The clock rate is the pulse time of the computer circuitry, representing the highest attainable rate at which a particular computer can process information, and can be compared with the blood pulse of a living being. The clock rate is a meaningful measure of a processor's efficiency, however, only when processors of a particular brand name are compared; other considerations are also important.

- The *Random-Access Memory*, RAM, is used by the processor and coprocessors to store temporary information, run the operating system to be discussed shortly, and execute the various instructions.

  Physically, the RAM consists of a large collection of silicon chips called *Very Large Scale Integrated Circuits*, VLSIC, and each chip contains tens of thousands of microchip switches called *flip-flops*. The collective on–off positions of the switches provide us with a means for transmitting and executing instructions.

  A RAM unit is subdivided into a collection of cells that are called *memory positions* and are identified by numbers called *memory addresses*. The qualifier *random-access* signifies that all memory positions can be accessed at approximately the same length of time. The capacity of a computer in terms of RAM size is limited by the maximum allowable address number, which is specified by the manufacturer and is measured in terms of available number of memory address bits, as will be discussed in Section 1.3. Each memory address can store a fundamental piece of information called a *word*, to be defined precisely in Section 1.3. The information within a memory position can be erased, replaced, or recalled at any time. An important point is that *erasing does not occur by default, and it is important to initialize all variables at the beginning of a computation; otherwise the day-old values may be used.*

  There are different types of RAM, distinguished by their specialization and access time. When the computer is turned off, the information stored in most of the RAM disappears. The Preserved RAM, PRAM, runs on a battery, and its contents are preserved even after the main power has been switched off. A relatively small amount of high-speed memory, called the *cache memory*, is situated between the processors and the main RAM. All information received or transmitted by the processors passes through the cache.

  There is a certain amount of permanent information that is used by the operating system on a routine basis, pertaining, in particular, to the way in which information is transmitted and received. This information is stored in a special memory bank called the Read-Only Memory, ROM. The access time of the ROM is much shorter than that of the RAM; for example, 60–80 nanoseconds versus 20 milliseconds. The Macintosh™ toolbox is housed in a ROM.

- The *Operating System*, OS, is a set of instructions that tell the computer how to organize itself and communicate with its environment. Examples are the UNIX™ system and its many variations, the VMS™ system, the DOS™ system, and the Macintosh OS™. The operating system is a program, like any other, written in a low-level language, as will be discussed later in this section. When a computer boots up, it actually loads the OS.

The heart of the OS is its *kernel.* For example, the UNIX™ kernel is a very small portion of the UNIX™ OS or its variants that allows it to run on many types of computers, from Personal Computers using LINUX™, to CRAY™ supercomputers using UNICOS™.

- The *storage device* or *secondary memory* serves as a storehouse for the operating system, accepting and storing in its partitions data files, application files, and computer language programs. A storage device can be a Hard Disk, HD, a removable disk, also called a floppy disk (it used to flop, but not any more), or an optical Compact Disk with Read-Only Memory, CD-ROM. At the present time, a floppy disk can hold 0.36–1.4 Mbytes, whereas a CD-ROM can hold 600 Mbytes; 1 Mbyte is equal to 10^6 bytes; the byte will be difined Section 1.3. Before a disk can be used it must be formatted and then partitioned into blocks of desirable size.

  The distinction between a storage device and the RAM is nominal but not fundamental. Information that is stored in the RAM could also have been stored in a storage device. The access time of a storage device, however, is typically many orders of magnitude higher than that of the RAM.

  Many operating systems allow the storage device to act as an extension of their RAM by *swapping* pages of overflow information in a dynamic manner, in a process called *paging,* thereby creating a *virtual memory.* If set too low, the swap page size—typically a block of the RAM or sometimes an auxiliary RAM—can deter the computer's performance. In the worst-case scenario, the system will spend all of its time importing and exporting portions of an executable code without actually doing anything useful. *Absolute machines,* such as the CRAY™ supercomputers, do not have virtual memory.

- The *keyboard* and a *mouse* or a *trackball* are used to enter information in an interactive manner; these are the *standard input devices.*

- The *standard output* is the *video terminal* that displays different sorts of information. The video terminal is accompanied by an electronic video card that instructs the terminal how to behave. The video card is fabricated on a printed circuit that is installed in the box that houses the computer.

  Each screen character is defined by a number of illuminated dots called *pixels* in a pattern that depends on the selected *font.* A video terminal equipped with a *Video Graphics Adapter,* VGA, can accommodate an array of 640×480 pixels, a high-resolution graphics monitor can accommodate an array of 1000×1000 pixels; but these numbers are changing by the day. When a personal computer operates in the *character mode,* the video screen can display 25 or so lines of 80 characters each.

  When a computer operates in the *graphics mode,* the screen can display drawings that are formed by groups of neighboring pixels. In *bit-mapped computer graphics,* each pixel is controlled independently by one or more bits of information (Section 1.3). The OS of the Macintosh™ is distinguished by the default use of bit-mapped computer graphics. The process of finding the pixels that define a character or a drawing is called *rasterization* or *scan conversion.*

- An *Electronic-Control Unit,* ECU, initiates, supervises, and coordinates the activities of the remaining units; this unit resides on the *system bus.*

- The processors and the RAM are usually housed in the same unit, called the *Central Processing Unit,* CPU. Sometimes the Electronic Control Unit is also part of the CPU.

PROJECT

1.2.1 *Deciphering an advertisement.*

Decipher and explain the jargon of a computer advertisement in a contemporary magazine or newspaper.

Multiprocessors and Massively Parallel Computers

A class of advanced but accessible computers contain many independent arithmetic-logic units that are housed in different processors. These computers are called *multiprocessors* or *parallel computers. Massively parallel computers* contain a few hundred or even several thousand processors. For example, the CM 5 computer of the *Thinking Machines Corporation*™ has 16,384 processors.

Parallel computers distribute the job of executing a set of instructions among their various processors, with the benefits of enhanced efficiency due to synergism. The gain in efficiency is usually quantified in terms of the *speedup*. It is rumored that the largest parallel computer in the United States is owned by the discount retail store chain WALMART™, and its main purpose is to update the prices in an interactive manner with the objective of emptying the pockets of the customers at the fastest possible rate. It is revealing to note that the research budget of the leading parallel business computer manufacturer—typically 10% of the profits—would exceed *all* of the scientific parallel processor manufacturers' total operating cost budgets combined.

Finer computer classifications are based on the particular way in which the processors are supervised by the control unit, the way in which the RAM is distributed and its contents are accessed and exchanged by the processors, and whether the processors work in a *synchronous* or *asynchronous* mode. For example, the processors of a Single-Instruction–Multiple-Processes computer, SIMP, are supervised by the same unit; all processors execute the same task but on different sets of data. In contrast, the processors of a Multiple-Instructions–Multiple-Processes computer, MIMP, work on independent tasks with generally different sets of data. The exchange of information between the processor RAMs requires careful design in order to avoid a high level of trafficking and bottle-necking (e.g., Wilson 1995).

A cluster of identical or similar but stand-alone single-processor computers can be instructed to tackle a particular task in a synergistic manner that makes them behave like multiprocessors. For example, the *Parallel Virtual Machine*, PVM™, is a communication program that links a group of computers through a *Local Access Network*, LAN (e.g., Fosdick et al. 1995).

Vector-pipelining Computers

A parallel computer should be distinguished from a *vector-pipelining computer* such as the CRAY™ C-90. The latter contains a relatively small number of highly specialized, expensive but sophisticated processors that execute mathematical operations with vectors in an expedient fashion. In contrast, parallel computers use generic and inexpensive processors. The computer architecture of a vector-pipelining computer is similar to that of an assembly line that manufactures a continuous stream of multicomponent products, such as automobiles, working in sequential stages.

Computer Taxonomy

Today's computers can be classified roughly into *massively parallel computers, parallel computers, supercomputers, mainframes, high-performance workstations*, and *desktop* or *laptop personal computers*. The distinguishing features of each category, however, are evolving, and there is significant overlap (e.g., Fosdick et al. 1995).

Computers at the workstation level and above are designed for *time-sharing* and *multitasking*; that is, they can work on different tasks simultaneously according to a certain priority protocol that is enforced by the operating system. But some high-end personal computers can also do multitasking.

Many workstations use the *Reduced Instruction Set Computer*, RISC™, architecture, featuring a small set of simple OS instructions, pipelined instruction execution, and extensive use of the cache memory. The goal is to approach the theoretical limit of one instruction per clock cycle, but it takes a higher number of instructions to complete a certain job.

Desktops

To make a computer user-friendly, personal and desk computers run application programs that make the video screen resemble a desktop, that is, the surface of a desk. These programs implement the *Graphical User Interface*, GUI. One can then move around icons of files, choose options, and select parameters simply by using the mouse to move the cursor, emulating the motion of a finger or hand (*icon* derives from the Greek word εἰκόνα, which means *painting*). Examples are the x11™ program and its variations that run on workstations. A desktop facility is built into the Macintosh OS™ and does not have to be run as a separate application. Certain advanced GUIs allow the computer to run simultaneously a number of desktops, which are selected by the cursor from a menu, and thus avoid cluttering the video screen.

Computer Programming

How does one instruct a computer to carry out a computation or execute another operation? The OS can respond only to a particular set of instructions that constitute the *machine language*. Even more challenging, these instructions are expressed by numerical codes written in the binary system, to be discussed in Section 1.3. Each instruction consists of an operation code and an accompanying set of parameters or arguments. For example, the binary equivalent of instruction 67099098095 may request addition—designated by the first two digits 67—of the content of the memory positioned at the address 099 and that of the memory positioned at the address 098, placing the sum in the memory positioned at the address 095.

A machine language is difficult to master. *Symbolic languages* allow the use of familiar words instead of operation codes and refer to operations by symbolic terms such as ADD or SCRATCH. An instruction in a symbolic language, also called an *assembly language*, is translated into the corresponding instruction in the machine language by a translation program called the *assembler*. The instructions of an assembly language refer to the actual *loading* and *fetching* of particular variables to, or from, memory locations. This is the lowest level of coding that is meaningful to contemporary humans.

An assembly language is still inconvenient, let alone notoriously difficult to debug. These concerns have motivated the development of *high-level* algorithmic languages that employ English words and standard mathematical notation. Examples are the BASIC™ language (Beginner's All-purpose Symbolic Instruction Code) developed in the mid-1960s and still surviving (e.g., Catlin 1992), the FORTRAN™ (FORmula TRANslator) language developed in the mid-1950s and thriving (e.g., Ellis 1990, Borse 1991), the Pascal™ language (named after the noted scientist Blaise Pascal who was also the inventor of an early computing machine that could carry out additions and subtractions) introduced in 1971, the C™ language developed in the mid-1970s (e.g., Schildt 1987), and their variations. The C™ language allows the manipulation of bits, bytes, and memory addresses, to be discussed in Section 1.3, and is considered a middle-level language, half a step above the assembler. The ranking of these computer languages in terms of efficiency and ease of learning can be the object of a heated debate.

To write a set of instructions in a middle- or high-level language, one must first create a number of files or documents that contain the main program and subroutines using a *text editor*, such as the *vi* editor that comes with UNIX™. These are the *source codes*. Second, one must *compile* the program and subroutines with the aid of the *compiler* to create the *object codes*. These are the translations of the computer language documents into the machine language, which is understood by the CPU. Third, one must *link* the object codes with computer library or OS codes that are called by the program, thereby producing an *executable code*. Library files may contain mathematical functions as well as graphics tools that allow a program to run on its own exclusive space on a desktop. Some compilers have their own linkers; others use a linker that is supplied by the operating system. Finally, one simply loads the executable code into the memory and lets it run; the presence of the compiler is not required.

While these are the general rules, exceptions do occur. For example, a program written in the standard version of the BASIC™ language is compiled or, more accurately, *interpreted* line by line while it is executed; the presence of the compiler or interpreter is necessary as the program runs. Executable codes run much faster than interpreted programs.

Since different computers have different machine languages, an executable file that has been compiled on a certain computer will not necessarily run on another, unless *binary compatibility* has been established. In order to decrease the size of the executable code before loading, some computers use Compile/Link/Load systems with *run-time loaded libraries*. But using them can cause compatibility problems between similar but slightly different systems.

The parameters of a computation are usually contained in a separate file or must be entered from the keyboard as the program runs. Certain *object-oriented* languages allow the input to be entered through screen windows. Examples are the VISUAL BASIC™ language for PCs, and the THINK PASCAL™ language for the Macintosh. A computer program and its input parameters define an *object* that can be duplicated or embedded in a more general design.

Computer Performance

The performance of a computer in scientific or engineering computation can be rated in terms of the FLOPS, that is, the number of *floating point operations per second.* A FLOP is a scalar addition, subtraction, or multiplication. On some computers, these take a comparable amount of time. Previously, and often today, a multiplication was much slower than an addition or a subtraction, and the latter were not counted as a FLOP. On most systems, division is slower than multiplication by a factor that can be as high as 3 and 5: it is faster to multiply a number by 0.25 than to divide it by 4.

A respectable high-performance workstation can perform 10 to 100 MFLOPS. Today's fastest computers can perform up to 10^9 FLOPS corresponding to approximately 1 to 3 nanoseconds per FLOP.

Another measure of a computer's efficiency is the MIPS, which expresses the ability of the computer to respond to millions of instructions per second. Ideally, the MIPS of a workstation with RISC™ architecture is equal to the clock rate measured in megahertz (MHz).

The *Central Processor Unit* time, CPU time, is the most important measure of a computer's efficiency in the context of numerical computation. This is the time required by the CPU to receive, generate, and deposit numbers into the RAM. If the computer were dedicated to a single user, and the input/output (I/O) were not excessive, the CPU time would be very close to the *wall clock time*, that is, real time that elapses from the beginning to the end of the computation.

The CPU time necessary to accomplish a certain task can be a very sensitive function of the particular method or algorithm chosen to solve a problem, as will be discussed in Section 1.4. The CPU times of two different methods that solve the same problem can differ by a factor that is as high as a few thousand or even higher.

In designing a numerical method to be implemented on a particular computer system, it is important to have a knowledge of the time required for the execution of the various mathematical or logical operations. An example is the evaluation of a *transcendental function*, defined as a function that cannot be expressed as the sum, the difference, the product, the ratio of two polynomials, or the root of a polynomial. These time requirements can be found by running various tests where a particular operation is repeated many times.

Plotting, Graphics Visualization, and Animation

Numerical computation produces arrays or matrices of numbers, which are then placed in properly formatted output files. Preparing graphs of various sorts of the functions that are represented by these numbers is an essential aspect of scientific and engineering data processing. Interpretation and visualization tools include *scatter plots, line graphs, histograms, pie charts, vector field plots,* and *contour plots.*

A video terminal can display graphs provided that it is equipped with an appropriate graphics control card and it works on the *graphics mode,* as discussed previously in this section. At the present time, all Macintosh™ computers, nearly all personal computer video terminals, and all workstation terminals are manufactured to support color graphics. Plotting and graphics display software packages are available at low or no-cost when they reside in a public domain. Many advanced graphics packages can perform animation; that is, they can plot successive curves or surfaces that describe the motion of objects.

Plotters, whose operation is based on the mechanical motion of pens, and printers of various sorts are used to produce copies on paper or transparencies, called *hard copies.* To instruct a plotter or a printer to make a graph, one must prepare a file that contains commands in the printer's language, such as *PostScript*™. The instructions are interpreted by an interpreter that has been built into the printer so that the PostScript™ files are portable across many computer platforms.

Like a typewriter, a *dot-matrix impact printer* imprints an array of densely spaced small dots on a surface through an ink ribbon. A collection of dots forms a figure or character.

An *ink-jet printer* prints a continuous stream of electrically charged tiny ink droplets whose motion is guided by an electrical field, to form a desirable pattern. In one method, the drops are formed by the disintegration of an ink jet that has been ejected from an ink pool, stimulated by electrical pulses (e.g., Olfe, 1995). A relatively inexpensive ink-jet printer can print a matrix of 720 dots per inch, *dpi,* which is adequate for most applications. The publication industry operates printers with resolution up to 2500 dpi.

A *laser printer* traces an image on the photoconductive coating of a drum with a laser beam, so that the precharged coating is locally discharged. A plastic powder, appropriately called the *toner,* is distributed over the drum and sticks to the charged portion of the surface. The drum presses against the printed surface, the loose toner it transferred onto it, and the toner is then melted to form a permanent image. Laser printers with resolution of 600 dpi are readily available at a reasonable cost. The laser-printer technology is similar to, and was borrowed from, that in electronic photocopying.

It is not surprising then that dot-matrix and ink-jet printers operate at room temperature, whereas laser-printers operate at elevated temperatures.

PROBLEM

1.2.2 *Visualizing the temperature field due to an impulsive point source of heat.*

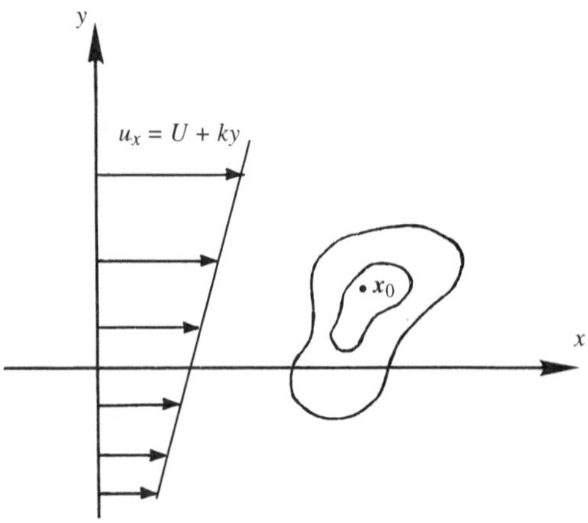

Consider a two-dimensional point source of heat located at the point $\boldsymbol{x}_0$ and activated impulsively at the time instant t_0, in the presence of a fluid flow in the xy plane as shown in the diagram. This may be accomplished, for example, by sending an electrical signal through a wire. The velocity components along the x and y axes are given, respectively, by

$$u_x = U + ky, \qquad u_y = 0 \tag{1.2.1}$$

where k is the constant shear rate and U is a constant velocity. When $k = 0$, we obtain uniform flow along the x axis with velocity U.

The induced temperature field G, which is the *Green's function* of the problem, satisfies the two-dimensional singularly forced convection–diffusion equation

$$\frac{\partial G}{\partial t} + (U + ky)\frac{\partial G}{\partial x} = D\left(\frac{\partial^2 G}{\partial x^2} + \frac{\partial^2 G}{\partial y^2}\right) + \delta(x - x_0)\,\delta(x - y_0)\,\delta(t - t_0) \tag{1.2.2}$$

where δ is the one-dimensional delta function, and D is the thermal diffusivity, assumed to be constant. The solution of equation (1.2.2), subject to the condition that G tends to zero at infinity, is

$$G(\mathbf{x}, \hat{t}; \mathbf{x}_0) = \frac{1}{4\pi D\hat{t}}\frac{1}{(1 + \frac{1}{12}k^2\hat{t}^2)^{1/2}} \exp\left(-\frac{(\hat{x} - \frac{1}{2}k\hat{t}\,\hat{y})^2}{4\,(1 + \frac{1}{12}k^2\hat{t}^2)D\hat{t}} - \frac{\hat{y}^2}{4D\,\hat{t}}\right) \tag{1.2.3}$$

where

$$\hat{x} = x - x_0 - (U + ky_0)\,\hat{t}, \qquad \hat{y} = y - y_0 \qquad \hat{t} = t - t_0 \tag{1.2.4}$$

(a) Consider uniform convection corresponding to $k = 0$ but $U \neq 0$, and prepare *isoscalar contour plots* of the Green's function at a proper sequence of times. That is, plot the contours over which G is constant (*iso* derives from the Greek word ἴσος, which means *equal*). Discuss the behavior of your results in physical terms.
(b) Repeat part (a) for convection in a symmetric shear flow corresponding to $k \neq 0$ and $U = -ky_0$.

1.3 Computer Arithmetic and Round-off Error

The word *arithmetic* derives from the Greek word ἀριθμός, which means *number*. Most computers work with the binary or base-two system of numbers that uses the two digits 0 and 1, instead of the ten digits 0 to 9 of the familiar decimal or base-ten system that we humans use in everyday life. The number 2 is the *radix* of the binary system, and the number 10 is the *radix* of the decimal system. Computers are two-fingered devices; humans are ten-fingered creatures.

A number in the binary system is denoted as

$$(b_k b_{k-1} \cdots b_0 \odot b_{-1} \cdots b_{-l})_2 \tag{1.3.1}$$

where k and l are two integers, and where the *b*inary dig*its* or *bits*, b_i, take the values 0 or 1. The corresponding decimal value is equal to

$$b_k \times 2^k + b_{k-1} \times 2^{k-1} + \cdots + b_0 \times 2^0 + b_{-1} \times 2^{-1} + \cdots + b_{-l} \times 2^{-l} \tag{1.3.2}$$

The same number in the decimal system is denoted as

$$(d_m\, d_{m-1} \cdots d_0 \odot d_{-1} \cdots d_{-n})_{10} \tag{1.3.3}$$

where m and n are two integers, and the decimal digits d_i take values from 0 to 9. The subscript 10 is, of course, omitted by convention in everyday exchange. The implied decimal value is

$$d_m \times 10^m + d_{m-1} \times 10^{m-1} + \cdots + d_0 \times 10^0 + d_{-1} \times 10^{-1} + \cdots + d_{-n} \times 10^{-n} \tag{1.3.4}$$

which is identical to that computed from expression (1.3.2).

Since the bits can be represented by the on–off positions of the electrical and forthcoming electron switches that are built in the microchips, the binary system is suitable for developing a computer architecture. This convenience, however, comes at the expense of economy: A binary string is much longer than a decimal string, and some architects prefer to economize by using the hexadecimal system, employing sixteen figures, as discussed in Problem 1.3.2.

Decimal to Binary Conversion

To express the decimal number 6.28125 in the binary system, we first consider the integral part, 6, and compute the ratios

$$6/2 = 3 + \mathbf{0}/2, \qquad 3/2 = 1 + \mathbf{1}/2, \qquad 1/2 = 0 + \mathbf{1}/2 \tag{1.3.5}$$

We stop when the integral part has become equal to zero. Taking the boldface figures expressing the remainders in reverse order, we find

$$(6)_{10} = (110)_2 = \mathbf{1} \times 2^2 + \mathbf{1} \times 2^1 + \mathbf{0} \times 2^0 \tag{1.3.6}$$

Next, we consider the decimal part and compute the products

$$\begin{gathered} 0.28125 \times 2 = \mathbf{0}.5625, \qquad 0.5625 \times 2 = \mathbf{1}.125, \qquad 0.125 \times 2 = \mathbf{0}.250, \\ 0.250 \times 2 = \mathbf{0}.500, \qquad 0.500 \times 2 = \mathbf{1}.000 \end{gathered} \tag{1.3.7}$$

We stop when the decimal part has become equal to zero. Taking the integer boldface figures in forward order, we find

$$(0.28125)_{10} = (.01001)_2 \tag{1.3.8}$$

Combining the preceding results, we write

$$\begin{gathered} (6.28125)_{10} = (110.01001)_2 \\ = \mathbf{1} \times 2^2 + \mathbf{1} \times 2^1 + \mathbf{0} \times 2^0 + \mathbf{0} \times 2^{-1} + \mathbf{1} \times 2^{-2} + \mathbf{0} \times 2^{-3} + \mathbf{0} \times 2^{-4} + \mathbf{1} \times 2^{-5} \end{gathered} \tag{1.3.9}$$

Eight *bits* and one *binary point* are necessary to represent this number. In FORTRAN, extracting the remainder can be done by using the intrinsic functions MOD and INT (Appendix C).

A set of eight bits is one *byte*. The largest number that can be represented with one byte is

$$(11111111)_2 \tag{1.3.10}$$

The decimal-number equivalent is

$$\mathbf{1} \times 2^7 + \mathbf{1} \times 2^6 + \mathbf{1} \times 2^5 + \mathbf{1} \times 2^4 + \mathbf{1} \times 2^3 + \mathbf{0} \times 2^2 + \mathbf{0} \times 2^1 + \mathbf{1} \times 2^0 = (255)_{10} \tag{1.3.11}$$

Memory addresses

Computer operating systems are designed to work with memory addresses whose maximum number is expressed by the number of available bits. A 24-bit or 3-byte addressing system can accommodate approximately up to $33.6 \cdot 10^6$ positions, which limits the maximum size of the RAM to approximately 33.6 Mbytes.

Binary to Decimal Conversion

To compute the decimal number corresponding to a certain binary number, we simply use expression (1.3.2). The evaluation of the powers of 2 requires $k + l$ multiplications; their subsequent multiplication with the binary digits requires an equal number of multiplications; and the final evaluation of the decimal number requires $k + l$ additions: A total of $2(k + l)$ multiplications and $k + l$ additions.

The computational cost may be reduced substantially by expressing the sum in the equivalent form

$$\begin{aligned} &[\cdots\{(b_k 2 + b_{k-1})\,2 + b_{k-3}\}\,2 + \cdots + b_1]\,2 + b_0 \\ &+ [\cdots\{(b_{-l}0.5 + b_{-l+1})\,0.5 + b_{-l+2}\}\,0.5 + \cdots + b_{-1}]\,0.5 \end{aligned} \tag{1.3.12}$$

and then carrying out the computations according to *Horner's algorithm.*

First, set

$$a_k = b_k \tag{1.3.13}$$

and compute the sequence

$$a_{k-1} = 2a_k + b_{k-1}, \quad \cdots, \quad a_i = 2a_{i+1} + b_i, \quad \cdots, \quad a_0 = 2a_1 + b_0 \tag{1.3.14}$$

Second, set

$$c_{-l} = b_{-l} \tag{1.3.15}$$

and compute the sequence

$$c_{-l+1} = 0.5c_{-l} + b_{-l+1}, \ \cdots, \ c_i = 0.5c_{i-1} + b_i, \ \cdots, \ c_{-1} = 0.5c_{-2} + b_{-1}, \ c_0 = 0.5c_{-1} \tag{1.3.16}$$

The required number d is equal to $a_0 + c_0$. The computations can be programmed in terms of two Do loops according to Algorithm 1.3.1.

Computing the decimal number in this manner requires a reduced number of $k + l$ multiplications and an equal number of additions. When the cost of addition is much less than the number of multiplications, using Horner's algorithm reduces the execution time by nearly a factor of 2.

ALGORITHM 1.3.1 Horner's algorithm for converting the binary number $(b_k b_{k-1} \cdots b_0 \odot b_{-1} \cdots b_{-l})_2$ to the decimal number d.

$a = b_k$
Do $i = k - 1, 0, -1$
$a \leftarrow 2a + b_i$
END DO

$c = b_{-l}$
Do $i = -l + 1, -1$
$c \leftarrow 0.5c + b_i$
END DO

$d = a + 0.5c$

PROBLEMS

1.3.1 *Binary system.*

Express the numbers 152 and 7.777 in the binary system. How many bits do you require in each case?

1.3.2 *Hexadecimal system.*

Some computers work with the hexadecimal system with the benefit of shorter strings. To express the number 11.28125 in the hexadecimal system that uses the sixteen symbols, 0 to 9 and A to F, we work as follows. For the integral part, we write

$$11/16 = 0 + \boldsymbol{11}/16$$

Thus $(11)_{10} = (B)_{16}$. For the decimal part, we write

$$0.28125 \times 16 = \mathbf{4}.500, \qquad 0.50 \times 16 = \mathbf{8}.000$$

Thus $(11.28125)_{10} = (B.48)_{16}$.

What is the year of birth of your favorite person in ancient Egyptian history in the hexadecimal system? And why is he or she your favorite person?

Character Representation

A computer represents characters—including the letters of the English alphabet A–Z and special symbols such as % and @—with binary strings corresponding to integers. In certain early computers, these were labeled by a special code that designates that the string is actually a character, not a number. In others, the distinction is made by keeping track of the memory addresses.

According to the American Standard Code for Information Interchange convention, ASCII, 128 characters are represented by the numerical values 0 to 127. The uppercase letters of the English alphabet, A to Z, correspond to the successive integers 65 to 90. The lowercase letters are assigned another set of numbers. Computers with an *extended set of characters* have an additional set of 128 characters including, for example, Greek letters, various mathematical symbols, musical notes, and sounds.

Round-off Error

An arbitrary number that has a *finite* number of digits in the decimal system generally requires an *infinite* number of bits in the binary system. An ideal computing machine would be able to register the number and carry out additions and multiplications with infinite precision, yielding the exact result to all figures. Regrettably, one must work with nonideal machines that work with only a finite number of bits and thus generate *round-off error.*

Some computers *round* a real number to the closest number that they can describe, with an equal probability of positive or negative error. Others simply *chop off* the extra digits in a guillotine-like fashion.

Word length

The number of bytes assigned to a variable, called the *word length*, is controlled by the programmer under the options of *single precision, double precision*, or *extended precision* of the programming language or application program. A variable can be declared as an *integer*, in which case the binary point is fixed at the end of the word length, or as a *real*, in which case the binary point floats across the word length. Operations between integers do not generate round-off error.

Floating-point representation

The floating-point representation allows us to describe a broad band of real numbers, as well as to carry out meaningful mathematical operations between numbers with disparate magnitudes. In a computer architecture based on the binary system, a real variable is usually expressed and stored in the floating-point representation:

$$\pm s 2^e \tag{1.3.17}$$

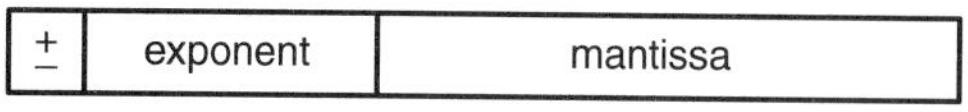

where s is a real number called the *mantissa* or *significand*, and e is an integer called the *exponent*. This representation requires one bit for the sign, a set of bytes for the significand, and another set of bytes for the mantissa.

When the exponent takes a value that is higher than the maximum value or lower than the minimum value that can be described with the available number of bits, the operating system protests by sending the message of *system overflow* or *system underflow*. Division by a virtual zero may activate the NaN, *Not-a-Number* warning; but one should be careful: some computers, designed primarily for business applications, suppress this warning.

There are many combinations of s and e that produce the same number, and the normalized one is the one that leaves as many zeros as possible at the end of the binary string. This means that the first digit of the significand is always nonzero and thus equal to 1. Certain computer systems exploit this convention to avoid storing this bit and gain one binary digit of accuracy. We then say that the mantissa has a *hidden bit*. The exponent is usually stored after it has been shifted by a certain number called the *bias*, so that it has become positive.

Single precision

Single precision usually employs 32-bit or 4-byte word lengths. A real number in the range between 1 and 2 can be resolved only up to the eighth decimal place, and the machine accuracy is 10^{-8}. The mantissa is usually described by 23 bits, and the biased exponent is described by 8 bits. System overflow or underflow typically occurs when $|e| > 127$, and the maximum and minimum positive numbers that can be described by the computer are $1.701 \cdot 10^{38}$ and $5.877 \cdot 10^{-39}$.

Double precision

Double precision usually employs 60- or 64-bit word lengths. A real number in the range between 1 and 2 can be resolved up to the 14th or 16th decimal place. The mantissa is usually described by 52 bits, and the biased exponent is described by 11 bits. System overflow or underflow typically occurs when $|e| > 1023$, and the maximum and minimum positive numbers that can be described by the computer are $8.988 \cdot 10^{307}$ and $1.123 \cdot 10^{-308}$.

The CRAY™ computers use 64-bit word lengths for single precision, and 128-bit word lengths for double precision.

Extended precision

It is sometimes necessary to use *extended precision* that works with 128-bit word lengths and allows us to describe a number up to the 20th significant figure. This level of resolution may appear excessive, but it turns out to be necessary in solving certain highly sensitive, nearly ill-posed problems.

PROBLEM

1.3.3 ***Find the unit round-off error of your computer.***

The *unit round-off error* is the smaller positive number that, when added to 1.0, is found to be larger than 1.0 by the computer. Write a program that deduces the unit round-off error of your computer in single precision.

Additions and Multiplications in the Binary System

Rules for adding, subtracting, multiplying, and dividing two numbers in the binary system have been developed and implemented in electrical circuitries that reside on chips. To add two binary numbers, we add the corresponding digits according to the four basic rules:

$$\begin{aligned}
(0)_2 + (0)_2 &= (0)_2 \\
(0)_2 + (1)_2 &= (1)_2 \\
(1)_2 + (0)_2 &= (1)_2 \\
(1)_2 + (1)_2 &= (0)_2, \text{ and carry } (1)_2 \text{ to the left}
\end{aligned} \tag{1.3.18}$$

To subtract a binary number from another, we subtract the corresponding digits according to the four basic rules:

$$\begin{aligned}
(0)_2 - (0)_2 &= (0)_2 \\
(1)_2 - (1)_2 &= (0)_2 \\
(1)_2 - (0)_2 &= (1)_2 \\
(0)_2 - (1)_2 &= (1)_2, \text{ with } (1)_2 \text{ borrowed from the left}
\end{aligned} \tag{1.3.19}$$

Multiplying a number by $2 = (10)_2$ shifts the binary point by one position to the right. Multiplying it by $0.5 = (0.1)_2$, which amounts to dividing it by 2, shifts the binary point by one position to the left. These observations suggest the following rules of binary digit multiplication:

$$\begin{aligned}
(0)_2 \times (0)_2 &= (0)_2 \\
(0)_2 \times (1)_2 &= (0)_2 \\
(1)_2 \times (0)_2 &= (0)_2 \\
(1)_2 \times (1)_2 &= (1)_2
\end{aligned} \tag{1.3.20}$$

In Chapter 4, we shall see that, surprising though it may seem, division can actually be carried out in terms of a sequence of multiplications (Problem 4.4.3).

In the computer hardware, addition and subtraction are implemented by straightforward combinations of electrical signals. Multiplication is a more complex operation involving multiplication by a single digit using the rules described in the preceding paragraph, column shifting, and addition of the various subtotal quantities. To expedite the process, these operations are carried out on special electronic processors, and this avoids the tardy execution of machine-coded instructions.

Evaluation of Transcendental Functions

Any respectable calculator is able to produce the value of a selection of transcendental functions at the stroke of a button. Examples are the exponential function, the logarithmic function, various trigonometric functions, and even Bessel functions.

Evaluating one of these functions may then appear to be a fundamental operation, similar to number addition or multiplication. In fact, a calculator computes a transcendental function by carrying out a sequence of additions and multiplications, according to instructions that have been printed on special-purpose microchips by the manufacturer. These instructions have been optimized so as to accomplish the desired task in the shortest possible amount of time and with adequate accuracy, as will be discussed in forthcoming sections.

Mathematical libraries that evaluate transcendental functions are necessary companions of computer language compilers. But, in writing a computer program, one should bear in mind that these evaluations can be costly. Fortunately, many scientific library function routines have been written and optimized in assembly codes, which expedites their execution.

PROBLEM

1.3.4 ***Evaluation of an exponential.***

Compute the number $e^{4.7}$ accurate to the third decimal place using your calculator, making only additions, multiplications, and divisions. Raising a number to an integral power is considered a multiplication and may be done using the calculator. *Hint*: You may begin by writing $e^{4.7} = (e^{4.7/5})^5$, and then evaluating the inner exponential in terms of its Maclaurin expansion.

Propagation of the Round-off Error

When two real numbers are added or subtracted in their floating-point system representation, the significant digits of the number with the smaller exponent are shifted in order to align the decimal point, and this causes the loss of significant digits; floating-point normalization of the resulting number incurs additional losses. Consequently, arithmetic operations between real variables exacerbate the magnitude of the round-off error. Unless integers only are involved, identities that are precise in *exact arithmetic* become approximate in *computer arithmetic.*

The accumulation of round-off error during a computation may range from negligible, to observable, to significant, to disastrous. Depending on the nature of the problem and the sequence of the computations, the round-off error may amplify, become comparable to, or even exceed the magnitude of the actual variables.

There is a class of physically ill-posed problems where the amplification of the round-off error is so violent that even using extended precision cannot prevent the corruption of the results after a certain stage. The potentially disasterous effect of the round-off error has motivated the development of sophisticated algorithms and clever tricks, to be discussed in subsequent chapters, and is the *raison d'être* (*reason to be* in French) for many branches of numerical analysis.

One might argue that the random nature of the round-off error will keep its magnitude small after a sufficiently large number of operations, but the probability of this happening is minuscule. Statistical mechanics says that there is a finite probability that all molecules comprising a brick will move in the same direction at a particular time instant, in which case the brick will fly into the universe; experience shows that this rarely happens. (It is occasionally reported in the press that such events are actually witnessed, but the reliability of such reports has been questioned.)

The behavior of the round-off error is similar to that of a walker executing random steps along the axis of real numbers, as will be discussed in Section 1.4. The probability that the walker comes back to its initial position becomes slimmer as the walker takes more steps. In the case of numerical computation, biases in preprogrammed computer operations makes this probability even slimmer.

To see the effect of the round-off error, consider the polynomial

$$P(x) = (x-10)^4 + 0.2(x-10)^3 + 0.05(x-10)^2 - 0.005(x-10) + 0.001 \quad (1.3.21)$$

discussed by Dahlquist and Björck (1974, p. 56). Note that the magnitude of successive coefficients multiplying the powers of the monomial $x - 10$ drops by one order. Carrying out the binomial expansions, and keeping the six leading figures of each number at each stage, we obtain the rounded form

$$P^R(x) = x^4 - 39.8000x^3 + 594.050x^2 - 3941.00x + 9805.05 \quad (1.3.22)$$

Evaluating $P(10.11)$ keeping only the three leading figures at every stage yields an answer in the range of 0.0015 ± 10^{-4}, which includes the exact value. A similar computation for $P^R(10.11)$ keeping eight figures at every stage yields the grossly inaccurate range $0.0481 \pm (\frac{1}{2})\ 10^{-4}$.

Control of the round-off error

In certain simple cases, the damaging effect of the round-off error can be predicted and thus minimized or controlled.

As a general rule, *one should avoid subtracting two nearly equal numbers.* Consider, for example, computing the roots of the quadratic equation $ax^2 + bx + c = 0$, and assume that b is positive and the magnitude of b^2 is much larger than that of $4ac$. Using the expression in the middle of equation (1.1.2) for the root corresponding to the negative sign, and the expression on the right-hand side for the root corresponding to the positive sign, minimizes the damage of the round-off error.

Another rule can be stated as follows: in computing the sum of a sequence of numbers, one should start summing up the numbers with the smaller magnitudes first, and the largest magnitudes last.

It also goes without saying that, numerical instabilities to be discussed in Section 1.4 aside, the fewer the number of operations, the smaller the resulting round-off error. This is another reason why the Horner algorithm described previously in this section is so attractive: not only does it reduce the execution time, but it also reduces the magnitude of the round-off error.

PROBLEMS

1.3.5 *Computing* π.

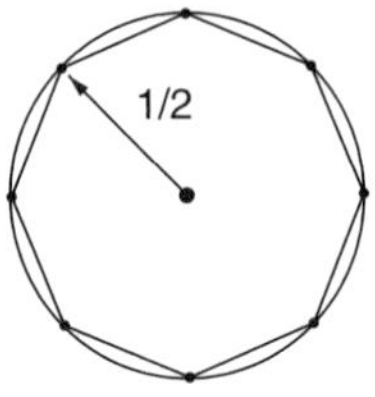

The great Greek thinker Αρχιμήδης noted that the number π can be approximated with the perimeter of a regular n-sided polygon that is inscribed within a circle whose radius is equal to $\frac{1}{2}$. The approximation improves as n becomes larger.

Consider a family of polygons parametrized by the index m; $m = 2$ corresponds to a square with four sides; $m = 3$ corresponds to a regular polygon with eight sides; each time m increases by one unit, the number of sides is doubled, yielding a polygon with 2^m sides whose perimeter is equal to p_m.

(*a*) Using elements of geometry and trigonometry, show that

$$p_{m+1} = 2^{m+1/2}\left[1 - \left(1 - \frac{p_m^2}{2^{2m}}\right)^{1/2}\right]^{1/2} \tag{1.3.23}$$

(*b*) Compute 40 members of the sequence p_m in single and then in double precision beginning with $p_2 = 2\sqrt{2}$, and discuss the deleterious effects of the round-off error (e.g., Borse 1991, p. 639).

1.3.6 ***Reduction of the round-off error.***

Indicate the proper way of computing the quantity

$$\sqrt{x^2 + a^2} - |a| \tag{1.3.24}$$

when the magnitude of x is small compared to that of a. Then verify the effectiveness of your method by means of a numerical example.

The Importance of Being Nice

An analogy can be made between the propagation of the round-off error and the following events. A person walks into a store to buy a pack of gum and is rude to the cashier. This ruins the cashier's day. Later, as the cashier drives home, still agitated by the incident, he almost has an accident with an oncoming taxi cab. The taxi driver and her passenger are shaken up and develop bad moods. The taxi driver delivers the passenger to the airport. As she boards the plane, the passenger makes an impolite comment to the pilot, who then becomes agitated and neglects to check the engine oil pressure before take-off. In the middle of the flight, oil starts leaking. The lives of passengers and crew are at risk, simply because the gum-buyer was rude to the cashier, a seemingly minor event.

1.4 *Algorithms*

An algorithm prescribes a sequence of procedures that accomplish a desired task by making *logical decisions*, by *repeating a sequence of computations*, or by doing both. A recursive formula for computing a new number in terms of old ones, such as that shown in equation (1.1.5), represents a simple algorithm whose execution illustrates the benefits of using a computer or a programmable calculator. Certain algorithms provide us with systematic ways of *eliminating events* and *narrowing down possibilities*. Others provide us with craftily devised methods of *producing a sequence of approximations* to a solution.

The origin of the word *algorithm* is discussed by Knuth (1968, p.1). After much debate and numerous linguistics speculations, it was decided that the term was coined after the author of a famous Persian textbook of mathematics. The author's name was *Abu Ja'far Mohammed ibn Mûsâ al-Khowârizm*; the qualifier *al-Khowârizm* means native of *Khowârizm*, which is the small contemporary city of *Khiva* in Uzbekistan.

Algorithm design

It seems impossible to give a complete set of rules for the design of algorithms without vagueness and unwarranted generalization. We will compromise by stating that the design of an efficient algorithm requires creativity and imagination; practice helps improve both.

A key to a successful, unambiguous, and efficient numerical computation using a computer or even a hand calculator is:

> *Plan the sequence of computations the way you would have done them by hand, and then instruct the computer to do them.*

The best way for a human is usually the best way for the computer, although exceptions may occur when dealing with problems of large size. These exceptions, however, are mostly attributed to the way in which the computer stores and retrieves information in its memory bank, which may be different from the way it is done in the human brain.

For example, the fastest way of computing the sum of a sequence of numbers is by computing the sequence of partial sums: Initialize the sum to have the value of zero, and then add to it successive terms of the sequence, monitoring only the incremented sum. In a computer language, this can be done using a single Do loop as taught in elementary courses of computer programming.

Algorithms and Mathematical Logic

It was mentioned earlier that a certain class of algorithms are implementations of logical procedures for assessing the truthfulness of statements and the validity of conjectures. In certain cases, the making of logical decisions can be reduced to the making of arithmetic comparisons.

For example, a simple way to assess whether a positive integer m is odd or even, without calling intrinsic computer language functions, is to compute the quantity $(-1)^m$, which is equal to -1 if m is odd, or 1 if m is even.

Euclid's algorithm for the greatest common divisor

A more subtle example takes us all the way back to ancient Greece. The ingenious Greek geometrist Εὐκλείδης developed an algorithm that finds the greatest common divisor of two positive integers m and n, call it k (e.g., Knuth 1969, p. 293). The greatest common divisor is the largest integer that evenly divides both m and n. The instructions are shown in Algorithm 1.4.1.

A wealth of algorithms are discussed in the comprehensive trilogy of monographs by Knuth (1969, 1973) and in more recent monographs of computer science (e.g., Cormen et al. 1990).

ALGORITHM 1.4.1 Euclid's algorithm for computing the greatest common divisor k of two positive integers m and n.

```
1     IF (n = m) THEN
         k = n
         STOP
      END IF
                                   If necessary, interchange n and m to make m > n
      IF (n > m) THEN
         save = m
         m = n
         n = save
      END IF
```

$k = m - n$
$m = n$
$n = k$
Go To 1

PROBLEMS

1.4.1 *Prime numbers.*

Devise an algorithm that allows you to assess whether an integer m is prime; an even integer that is higher than 2 cannot be a prime. A prime number has only two divisors: unity and itself. The set of prime numbers provides us with a basis that can be used to build the set of all integers by means of multiplication.

1.4.2 *A computer game.*

A robot moves inside a rectangular room by executing unit steps in four directions: toward the North, the South, the West, or the East. At each instant in time, the robot is located at the center of a unit cell, as illustrated in the diagram. The room contains an obstacle whose area has been digitized so that its perimeter appears to be a polygon.

Write a program that instructs the robot to move toward an object and then to move around its perimeter. The robot should be able to find the object from any initial position, and then go around it once. Run the program for a triangular, a rectangular, and a cross-shaped object; verify that it works, and plot the path that the robot has described.

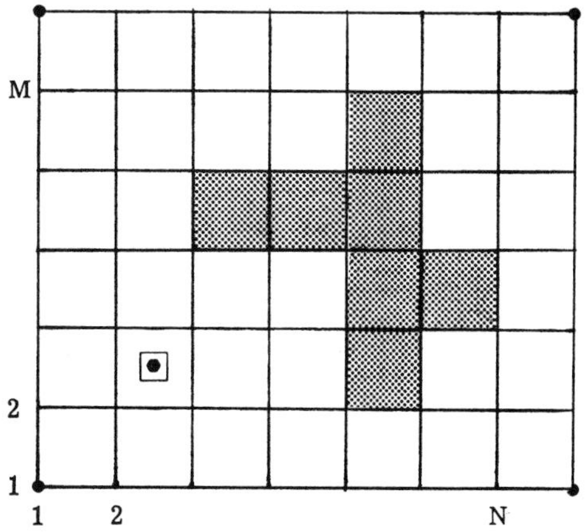

Indexing and Sorting

Consider a set of N real numbers that occupy the entries of the vector $\boldsymbol{x}$ representing, for example, the income of taxpayers in a certain group. We want to label these numbers in descending order using the index m_i, so that the index of the largest number is equal to 1, and the index of the smallest number is equal to N; that is, we want to find the richest person in the group. After the numbers have been indexed, they can be printed in a desirable order.

A simple indexing algorithm can be designed on the observation that m_i is equal to the number of times that $x_i - x_j$ is nonnegative, for $j = 1, \ldots, N$. The instructions are:

```
Do i = 1, N
  m_i = 1
  Do j = 1, N
    If (x_i < x_j) m_i ← m_i + 1
  End Do
End Do
```

(1.4.1)

Bubble sort

The bubble-sort algorithm arranges a group of N numbers x_i in descending order. At the end, the maximum number is placed at the last entry of the vector $\boldsymbol{x}$, and the minimum number is placed at the top. A simple modification does the converse.

The sorting is done in $N-1$ sweeps. In the first sweep, we find the maximum of all numbers by switching, if appropriate, the order of numbers in successive pairs, and put it at the bottom; in the second sweep, we find the maximum of the top $N-1$ numbers and put it at the penultimate entry; and the process continues until we reach the top. Each time a number is sorted, the smaller numbers bubble up to the top, and the procedure stops when there is no more bubbling. The algorithm is:

```
  k = N − 1         Number of points to be compared
1 ISTOP = 1         read " I stop"; This is a flag: ISTOP = 1 means no bubbling
  Do i = 1, k

    IF (x_i > x_{i+1}) THEN
      save = x_i
      x_i = x_{i+1}
      x_{i+1} = save
      ISTOP = 0
    END IF

  END DO

  k = k − 1

  IF (ISTOP = 0) GO TO 1
```
(1.4.2)

Faster than bubble sort

The preceding two algorithms requires N^2 operations; when N is large, the performance is slow. There are much more efficient methods of indexing and sorting that reduce the operation count down to the order of $N \log_2 N$. Faster algorithms are discussed in the texts of Knuth (1973), Cormen et al. (1990), and Press et al. (1992, Chapter 8).

Sorting and searching through a list of names

How can we sort a list of names in alphabetical order? This can be done by assigning to the letters of the alphabet appropriate numerical values, such as the ASCII values discussed in Section 1.3. The sorting is then done by comparing pairs of names, character by character, and switching them accordingly. Certain computer languages, including TRUE BASIC™ and FORTRAN™, provide intrinsic functions that perform these comparisons by inequality checks (e.g., Catlin 1992, p. 273; Ellis 1990, p. 303).

Binary search

Searching through a *sorted* list of numbers or names with the objective of finding a particular entry and attached information can be done efficiently using the *binary search method.* We compare the target number or name with the number or name in the middle of the list. If the comparison places the target at the upper half of the list, we discard the bottom half and repeat the process with the upper half. Otherwise we do the converse.

In Section 4.2, we shall see that the binary search method has a cousin who specializes in the finding of the roots of nonlinear algebraic equations.

PROBLEMS

1.4.3 *Indexing and sorting a set of numbers.*

Write a program that (*a*) indexes a set of N real numbers x_i in ascending or descending order, and (*b*) sorts the numbers, that is, produces a two-column table of i, x_i, where the x_i are arranged in ascending or descending order.

1.4.4 *Binary search.*

Write a program that implements the method of binary search for a sorted list of real numbers. The program should return the position of the number in the list that is closest to a specified number.

Algorithms and Numerical Efficiency

Several types of large-scale numerical computation have become feasible only thanks to the availability of ultrafast ingenious algorithms. Computations that once would have required scores of years, today can be done within several minutes on an inexpensive desktop computer or even a calculator.

For example, the Fourier transform of a large set of data was practically impossible before the invention of the Fast Fourier Transform method, FFT, to be discussed in Section 8.9. Other less dramatic examples include the computation of the interaction of a large number of vortices in a fluid by the multipole fast summation method (e.g., Pozrikidis 1997, Chapter 9).

Nested multiplication and Horner's algorithm

Nested multiplication provides us with a characteristic example where the careful design of an algorithm rewards us with considerable savings. Consider the evaluation of the Nth degree polynomial

$$P(x) = a_1 x^N + a_2 x^{N-1} + a_3 x^{N-2} + \cdots + a_N x + a_{N+1} \tag{1.4.3}$$

where a_i, $i = 1, \ldots, N+1$ is a collection of $N+1$ real or complex coefficients. The sequential evaluation of the monomials requires N multiplications, their subsequent multiplication with the coefficients requires N additional multiplications, and the final evaluation of $P(x)$ requires N more additions: A total of $2N$ multiplications and N additions.

The efficiency of the evaluation can be improved notably by rewriting the polynomial in the nested form

$$P(x) = [\cdots\{(a_1 x + a_2)x + a_3\}x \cdots + a_N]x + a_{N+1} \tag{1.4.4}$$

which provides us with a basis for *Horner's algorithm* discussed earlier in Section 1.3 in the context of binary to decimal conversion. Set

$$b_1 = a_1$$

and compute the sequence

$$b_2 = xb_1 + a_2, \qquad b_3 = xb_2 + a_3, \cdots, \qquad b_{N+1} = xb_N + a_{N+1}$$

then $P(x) = b_{N+1}$. The evaluation of $P(x)$ requires N multiplications and an equal number of additions. The pertinent pseudocode is:

```
p = a_1
Do i = 2, N + 1
   p ← xp + a_i
END Do
P(x) = p
```

(1.4.5)

PROBLEMS

1.4.5 ***Computing a power of a number with the least amount of effort.***

Devise an algorithm that yields the power a^m, where m is a natural number and a is a positive real number, with the least number of multiplications. You can assume that m is given in its decimal as well as in its binary form. *Hint*: See Algorithm 2.2.1, Section 2.2.

1.4.6 ***Computing the combinatorial.***

Lay out on a carpet n wooden turtles, and pick up groups of m of them; $m \leq n$. The number of all possible combinations is equal to the combinatorial

$$\binom{n}{m} \equiv \frac{n!}{m!\,(n-m)!} \tag{1.4.6}$$

where ! indicates the factorial; for example, $n! = 1 \cdot 2 \cdot 3 \cdots n$. Show that an equivalent way of computing the combinatorial is

$$\binom{n}{m} = \prod_{k=1}^{l} \frac{n-k+1}{k} \tag{1.4.7}$$

where l is the minimum of m and $n - m$. Develop and implement a pertinent algorithm.

From Easy to Difficult Problems

Algorithms range from simple-minded, to involved, to convoluted. To solve a certain difficult problem, it is sometimes expedient to build the solution in successive stages of increasing difficulty: Use an algorithm to solve a sequence of simpler problems that lead to the difficult problem. The accumulation of information and expertise along the way is the cornerstone of this approach.

To illustrate the method, consider the construction of the generalized *von Koch self-similar line* defined as follows:

- We begin by dividing the x axis into an infinite sequence of equal segments with length L, as shown in Figure 1.4.1(a). This division provides us with the *zeroth*-order shape corresponding to the index $m = 0$.

- To obtain the *first-order* shape corresponding to $m = 1$, in the middle of each horizontal segment we introduce a triangular isosceles protrusion, as shown in the second panel of Figure 1.4.1(a), and adjust the width of the protrusion so that all sides of the emerging infinite periodic line have equal length $s_1 = \alpha L$, where α is a specified coefficient. When $\alpha = \frac{1}{3}$, the line collapses to the x axis, whereas when $\alpha = 1.0$, the protrusions become flat plates with infinitesimal thickness.

- After the first stage, for $m > 1$, each time m is increased by one unit, one asperity is added in the middle of each segment, and its width is adjusted so that the lengths of all sides of the emerging periodic line have the common value s_m, as shown in Figure 1.4.1(a). Elementary geometry shows that

$$s_m = \tfrac{1}{6}(1 + 2\alpha)s_{m-1} \qquad (1.4.8)$$

As α tends to the value of $\frac{1}{4}$, the height of the protrusions tends to vanish, whereas as α tends to the value of 1.0, the protrusions tend to become flat plates with infinitesimal thickness. Note, however, that the construction of the first-order shape requires that $\alpha > \frac{1}{3}$.

The number of vertices over one period is equal to $3 \cdot 4^{m-1} + 1$, and the length of the line over one period is equal to

$$L_m = L3\alpha[\tfrac{2}{3}(1 + 2\alpha)]^{m-1} \qquad (1.4.9)$$

Thus, as m tends to infinity, the length of the line becomes infinite; this is a distinctive property of a *fractal* line. Another peculiar feature of a fractal line is that it does not have a uniquely defined tangential line at any point.

Other fractal lines arise as limits of different sequences of self-similar shapes. Some of those lines have been shown to be reasonable representations of certain natural shapes including the edges of cloud patterns and the surfaces of natural rocks.

Generating the mth level periodic von Koch line in a direct manner by computing the positions of its vertices in a single pass requires a considerable amount of analytical work. The construction can be simplified by successively generating all lines from the zeroth up to the mth order, computing the coordinates of the vertices of a new line in terms of those of the previous line, as shown in FORTRAN Program 1.4.1. Two shapes computed by this program are displayed in Figure 1.4.1(b, c).

PROGRAM 1.4.1 A FORTRAN program that generates the vertices of the generalized von Koch line. Results are displayed in Figure 1.4.1.

```
Program von_Koch

c-----------------------------------
c Computes a periodic generalized von Koch line
c
c m        :   order of fractal line
c L        :   period
c alpha    :   aspect ratio of spikes
c mp       :   number of vertices over one period
c ifr      :   vertex index 1 for a sharp corner
c                           2 for a blunt corner
```

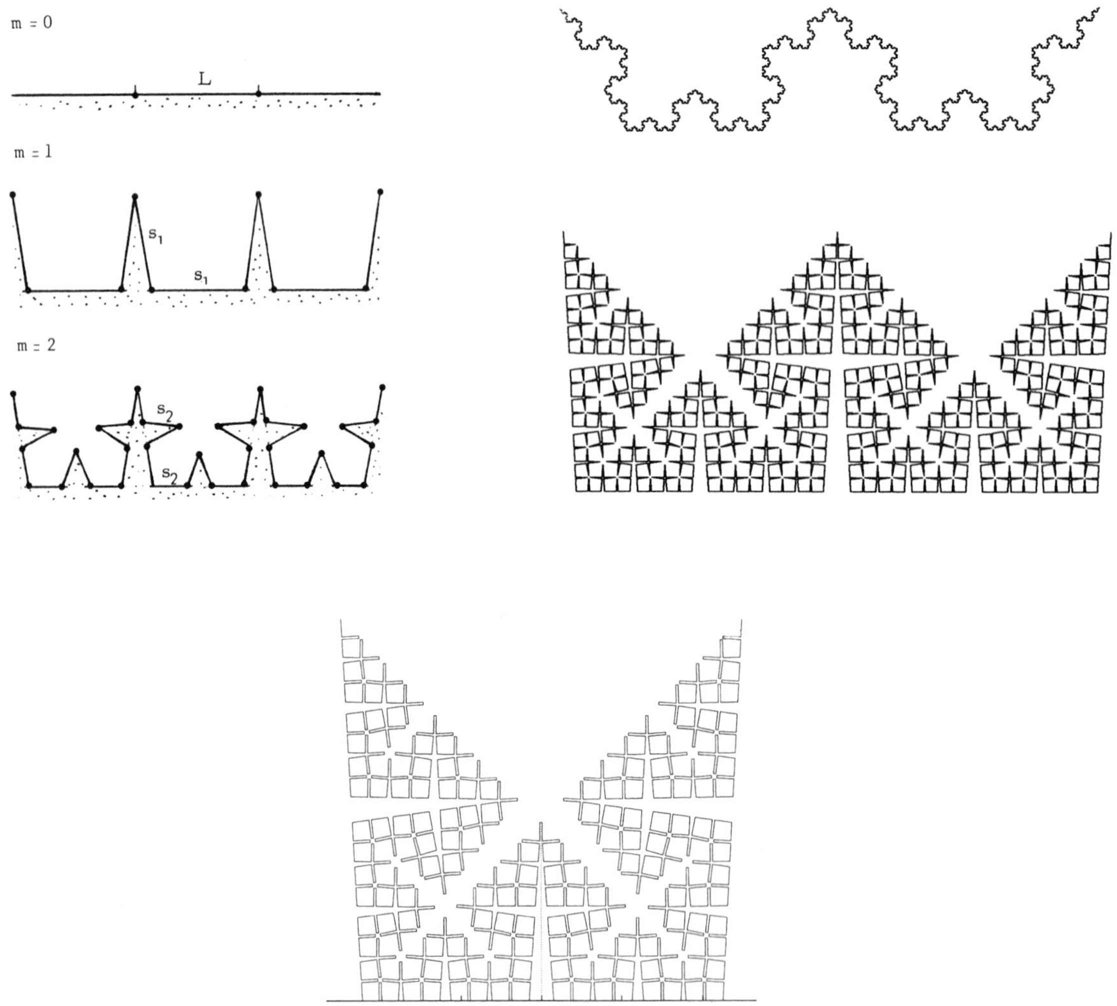

Figure 1.4.1 The generalized von Koch line. (*a*) Parental division, $m = 0$; first generation, $m = 1$; second generation, $m = 2$. The von Koch line for $m = 5$ with (*b*) $\alpha = 0.50$ and (*c*) $\alpha = 0.90$. (*d*) A von Koch line where the triangles are replaced by rectangles.

Program 1.4.1 (*Continued*)

```
c   xfr,yfr          :      coordinates of vertices
c- - - - - - - - - - - - - - - - - - - - - - - - - - - - - - - - - -

        Implicit double precision (a–h,o-z)
        Double precision L
```

```
      Dimension xfr (4001),yfr(4001),ifr(4001)

      common/Intg/ifr
      common/Real/xfr,yfr

      write  (6,*)  "Enter period L"
      write  (6,*)  "- - - - - - - -  "
      read   (5,*)  L
      write  (6,*)  "Enter m; should be an integer"
      write  (6,*)  "- - - - - - - - - - - - - - - - - - "
      read   (5,*)  m
 98   write  (6,*)  "Enter alpha; 1/3 < alpha < 1.0"
      write  (6,*)  "- - - - - - - - - - - - - - - - - - "
      read   (5,*)  alpha

      If(alpha.le.0.3333333.or.alpha.ge.1.0) Then
        write (6,*) "Please try again "
        Go to 98
      End If

c- - -

      CALL von_koch_f (L,m,alpha,mp)

c- - -
c Print vertices in file PLOTDAT
c- - -

      Open (1,file="PLOTDAT")
      write (1,100) mp
      Do 1 i = i,mp
        write (1,100) i,xfr(i),yfr(i),ifr(i)
        write (6,100) i,xfr(i),yfr(i),ifr(i)
 1    Continue
      Close (1)

 100  Format (1x,i3,2(1x,f10.5),1x,i3)

      Stop
      End

c= = = = = = = = = = = = = = = = = = = = = = = = =

      SUBROUTINE von_koch_f (L,m,alpha,mp)
c- - - - - - - - - -
c  Computes the vertices of the periodic von Koch line
c  xaux, yaux     :      auxiliary points
c- - - - - - - - - -
```

Program 1.4.1 (*Continued*)

```
      Implicit double precision (a-h,o-z)
      Double precision L

      Dimension xfr (4001),yfr (4001),ifr (4001)
      Dimension xaux(4001),yaux(4001),iaux(4001)

      common/Intg/ifr
      common/Real/xfr,yfr

        three    =  3.0
        sr3      =  sqrt(three)

c- - - - - - - - - - - - -
c  the m = 1 line first
c- - - - - - - - - - - - -

        height   =  sqrt( 3.0*alpha**2+2.0*alpha-1.0 )

        mp       =  4
        xfr(4)   =  L
        yfr(4)   =  L*height
        ifr(4)   =  1
        xfr(3)   =  L*alpha
        yfr(3)   =  0.
        ifr(3)   =  2
        xfr(2)   =  -xfr(3)
        yfr(2)   =  yfr(3)
        ifr(2)   =  2
        xfr(1)   =  -xfr(4)
        yfr(1)   =  yfr(4)
        ifr(1)   =  1

c- - - - - - - - - - - - - - -
      IF(m.eq.1) Go to 99
c- - - - - - - - - - - - - - -

      Do 1 i = 2,m

c- - - - - - - - - - - - - - -
c  Save the vertices
c  of the previous shape
c- - - - - - - - - - - - - - -

        kp = mp
        Do 2 j = 1,kp
          xaux(j)  =  xfr(j)
          yaux(j)  =  yfr(j)
          iaux(j)  =  ifr(j)
 2      Continue
```

Program 1.4.1 (*Continued*)

```
      mp = 3*4**(i-1)+1

      xfr(mp) = xaux(kp)
      yfr(mp) = yaux(kp)
      ifr(mp)  = iaux(kp)

      j     =   mp

      Do 1 l = 1,kp-1

      k     =   kp-l
      xa    =   xaux(k)
      ya    =   yaux(k)
      xb    =   xaux(k+1)
      yb    =   yaux(k+1)

      xc    =   0.50*(xa+xb)
      yc    =   0.50*(ya+yb)
      dx    =   xb-xa
      dy    =   yb-ya
      ds    =   sqrt (dx**2+dy*2)
      height = sqrt(3.0*(4.0*alpha-1.0)) / 6.0

      j = j-1
      xfr(j)   =   xc + (1.0-alpha)*dx/3.0
      yfr(j)   =   yc + (1.0-alpha)*dy/3.0
      ifr(j)   =   2
      j = j-1
      xfr(j)   =   xc - dy * height
      yfr(j)   =   yc + dx * height
      ifr(j)   =   1
      j = j-1
      xfr(j)   =   xc - (1.0-alpha)*dx/3.0
      yfr(j)   =   yc - (1.0-alpha)*dy/3.0
      ifr(j)   =   2
      j = j-1
      xfr(j)   =   xa
      yfr(j)   =   ya
      ifr(j)   =   ifr(k)

 1    Continue

 99   Return
      End
```

PROBLEMS

1.4.7 *A von Koch line with rectangular protrusions.*

Modify Program 1.4.1 into a program that computes the vertices of a modified von Koch line over one period, as shown in Figure 1.4.1(*d*). This is similar to the von Koch line except that, after the von Koch lines have been constructed, the triangular protrusions are replaced by rectangular projections. Plot families of shapes for α = 0.35, 0.50, 0.90 and m = 1, 2, 3.

1.4.8 *Random von Koch line.*

Modify Program 1.4.1 into a program that generates the random von Koch line. This is similar to the regular von Koch line described in the text, but the orientation of each protrusion is chosen randomly as the protrusion is generated, with an equal probability that it points up or down. Plot families of shapes for α = 0.35, 0.50, 0.90 and m = 1, 2, 3. For the choice of the random orientations, see text around statements (1.5.1).

1.4.9 *A fractal tree.*

Write a program that generates a fractal tree by replicating an initial pattern on a smaller scale, as exempified in the diagram. Produce various kinds of trees by varying the parameters of your algorithm.

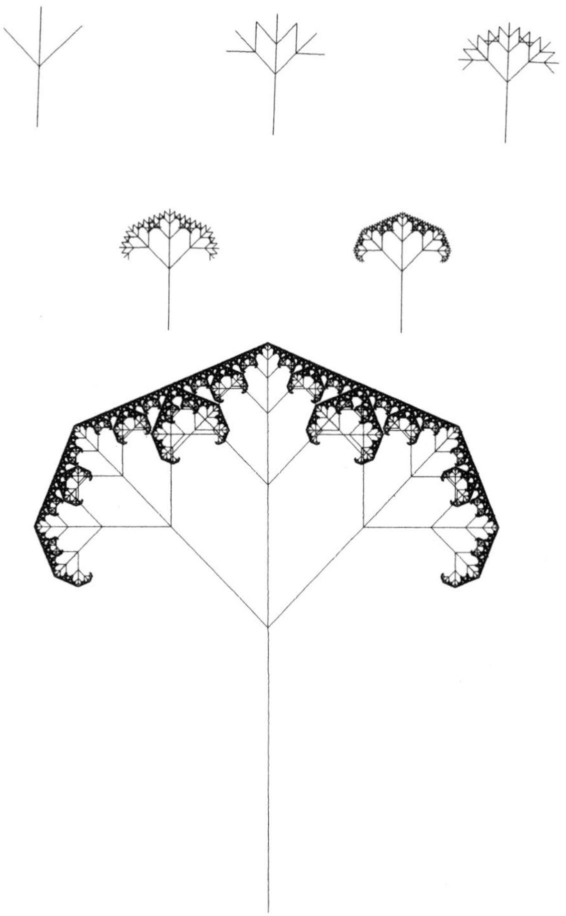

Algorithms and Numerical Stability

The common goal of many algorithms is to produce a sequence of numbers using a properly designed recursive formula; new numbers emerge from old ones. An ill-mannered class of such algorithms, classified as *numerically unstable*, foster the growth of an accidental or of the inevitable round-off error.

A dramatic demonstration of the amplification of the round-off error due to a poorly behaving algorithm is presented by Kincaid and Cheney (1996, p. 69). Consider the sequence of numbers x_i computed by the algorithm

$$
\begin{aligned}
x_0 &= 1 \\
x_1 &= \tfrac{1}{3} \\
x_i &= \tfrac{13}{3} x_{i-1} - \tfrac{4}{3} x_{i-2}, \qquad \text{for } i \geq 2
\end{aligned}
\tag{1.4.10}
$$

It is not difficult to show by induction that exact arithmetic yields the equivalent definitive formula

$$x_i = (\frac{1}{3})^i \tag{1.4.11}$$

Computing successive members of the sequence using algorithm (1.4.10) with 32-bit computer arithmetic, that is, using single precision, we obtain

$$x_0 = 1.000\,000, \quad \ldots, \quad x_6 = 0.001\,385, \quad \ldots, \quad x_{10} = 0.002\,588, \quad \ldots, \quad x_{15} = 3.657\,493, \quad \ldots$$

Exact-arithmetic results based on formula (1.4.11) yield the values

$$x_0 = 1.000\,000, \quad \ldots, \quad x_6 = 0.001\,372, \quad \ldots, \quad x_{10} = 0.000\,017, \quad \ldots, \quad x_{15} = 0.000\,000, \quad \ldots$$

The deleterious effect of the round-off error is stunning.

What is the reason for these discrepancies? Successive substitution of the expressions generated by the algorithm (1.4.10) into their descendants produces an expression for x_i in terms of x_0 and x_1, whose leading-order term is

$$x_i = (\tfrac{13}{2})^{i-1} x_1 + \ldots \tag{1.4.12}$$

We can see now that a small error in the exact value of x_1 due to finite computer precision will be multiplied by the growing factor $(\frac{13}{2})^{i-1}$, which is on the order of 10^9 when $i = 15$. It is then inevitable that the numerical error will dominate and eventually overtake the exact value whose magnitude is decreasing rapidly with increasing i.

The design of numerically stable algorithms is a central objective of numerical analysis.

1.5 Computer Simulation

The term *computer simulation* describes the numerical solution of a set of algebraic or differential equations that govern the behavior of a natural, technological, or abstract mathematical system, possibly with drastic approximations. For example, computer simulation of a liquid involves the numerical solution of the equations that govern the motion of the individual molecules with specified atomic properties, taking into consideration intermolecular interactions and other extraneous constraints. In a sense, computer simulation is the natural extension of mathematical modeling.

Experiment by computer simulation has complemented, and in some cases replaced, physical experimentation, with the benefits of expediency and reduced monetary cost. For example, the design of an aircraft today is based, to a large extent, on computational testing. The performance of a proposed model is evaluated by solving the equations that govern the flow around the aircraft, as well as the equations that govern the motion of the aircraft's structural components. This is the direct analog of physical experimentation where the performance of a pilot-model is tested in a wind tunnel.

An important advantage of computational experimentation is that it provides us with a perfectly or nearly controlled environment that is free of unwanted noise and perturbations. Furthermore, computer simulation can illustrate how nature would have behaved in a one-, two-, or four-dimensional world. A risk is excessive idealization.

Dynamic and Monte Carlo Simulation

Dynamic simulation is used to study the *evolution* of a system from a specified initial state. An example is the evolution of a suspension of macromolecules in a polymeric solution. The results allow us to assess the time-average and statistical properties of the simulated system, as well as analyze the dynamics of the microstructure.

Monte Carlo simulation, on the other hand, investigates the statistical properties of a large number of *instantaneous* configurations corresponding to the same macroscopic conditions. An example is the study of the total energy of a collection of interacting molecules whose instantaneous position and velocity are consistent with certain probability distributions.

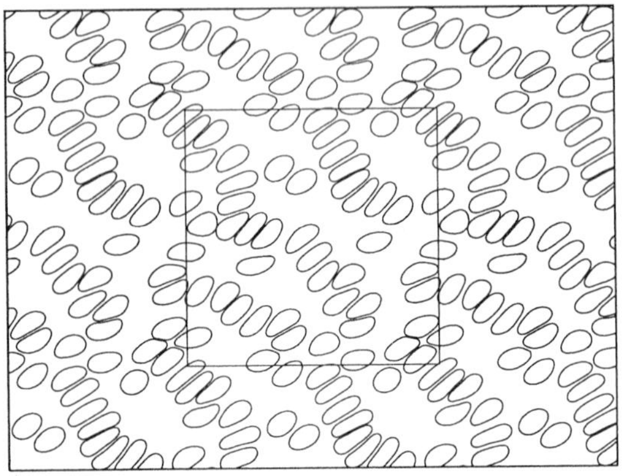

Over the past few decades, computer simulation has become an invaluable tool of scientific and engineering research. Simulation of turbulent motion has given us insights into the nature of high-speed flow (e.g., Rogallo and Moin 1984). Simulations of particulate solids and suspensions of rigid and deformation particles of various sorts have elucidated the behavior of complex solids and fluids, such as blood. The diagram shows the instantaneous structure of a suspension of liquid drops in shear flow, computed by numerical simulation (Li et al. 1996). Finally, simulation of molecular motion continues to provide us with information on the properties of solids, gases, and liquids, and the physics of their transitions (Allen and Tildeslay 1987).

In the remainder of this section, we shall discuss several characteristic examples.

PROBLEM

1.5.1 ***Simulation.***

Describe a natural, technological, or abstract mathematical system of your choice that is amenable to dynamic or Monte Carlo simulation.

A Monte Carlo Method of Computing π

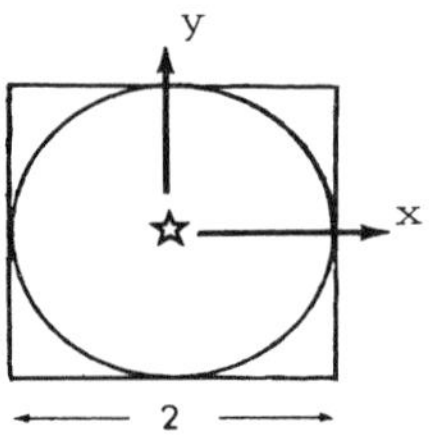

Consider a circular disk of unit radius centered at the origin, representing a target, and inscribe it within a square box of side length equal to 2. Suppose that N people throw a dart at the target, in a random manner and with no particular bias. As N tends to infinity, the probability $P(N)$ that a dart land inside the disk is equal to the ratio of the area of the disk to the area of the square, which is known to be $\pi/4$. Accordingly, π can be computed as $\pi = 4P(\infty)$.

To implement the computational counterpart of the process just described, we work in the context of Monte Carlo simulation. Specifically, we ask the computer to generate N pairs of random numbers with *uniform probability density distribution* lying between -1 and 1, and regard these numbers as the (x, y) coordinates of the dart. The probability $P(N)$ is equal to the ratio of the number of points whose distance from the origin is less than 1, to the total number of trials N.

Random numbers with uniform pdf

Most computer operating systems provide subroutines for generating sequences of random or nearly random numbers r with uniform probability density function ranging between 0 and 1, called *uniform deviates*. The associated series of numbers $1 - r$ is called the *antithetic series*.

For example, a simple way of generating a sequence of N uniform deviates in single or double precision, with a SUN Microsystems OS, is by setting, respectively,

$$r = r\_\text{lcran\_}() \qquad \text{or} \qquad r = d\_\text{lcran\_}() \tag{1.5.1}$$

sequentially N times. The acronym *lcran* derives from the terminology *linear congruential pseudorandom generator*. Improved and more reliable methods of producing uniform deviates are discussed by Knuth (1969, Vol. II) and Press et al. (1992, Chapter 7).

To obtain a series of random numbers x with uniform probability density function varying between the limits a and b, we simply introduce the linear transformation $x = a(1 - r) + br$. As r varies between 0 and 1, x varies between a and b.

PROBLEM

1.5.2 *Computing* π.

Write a computer program that computes the number π using the Monte Carlo method described in the text, and discuss the convergence of the results with respect to N. Specifically, prepare a plot of $P(N) - \frac{1}{4}\pi$ as a function of N on a log–log scale; examine and discuss the shape and slope of your graph.

Buffon's Needle Problem

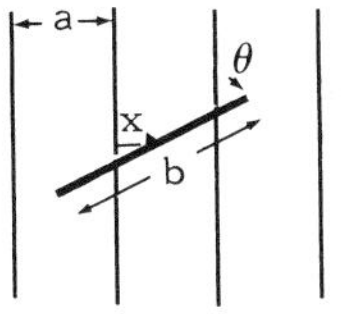

Consider an infinite number of evenly spaced parallel lines drawn on a flat horizontal surface, separated by distance a representing, for example, a ruled paper pad. Pick up a needle of length b, and let it fall on the surface many times, each time in a random fashion. We want to compute the probability P that the needle crosses one of the parallel lines. Clearly, P will be a function of the ratio a/b.

To formulate the associated computational problem, we note that if x is the distance of the midpoint of the needle from the closest line, and θ is the angle that the needle forms with the parallel array, then the needle will cross that line as long as $x \le \frac{1}{2}b \sin\theta$, or

$$c \equiv x - \tfrac{1}{2}b \sin\theta \le 0 \tag{1.5.2}$$

where x varies between 0 and $\frac{1}{2}a$, and θ varies between 0 and π. Accordingly, we generate a sequence of N pairs of unrelated random numbers (x, θ) with uniform probability distribution and values that fall within their respective ranges of definition, and we compute the quantity c defined in equation (1.5.2). If c is negative M times, then $P(a/b)$ is equal to the limit of the ratio M/N as the number of trials N tends to infinity.

Buffon observed that, as long as $b \le a$, $P(a/b) = 2b/(\pi a)$. Thus the value of P computed by the Monte Carlo method just described may be applied to estimate π.

PROBLEM

1.5.3 ***Buffon's needle problem.***

Write a computer program that produces $P(a/b)$ for $a/b = 0.10, 0.50, 1.0, 2.0$ and the associated estimate for π using the method described in the text. In each case, discuss the convergence of the results with respect to N as described in Problem 1.5.2.

Assessing Whether a Point Lies Inside a Planar Region

It will be misleading to give the impression that computer experimentation does not require ingenuity and skill. Let us return to the dart problem, and replace the disk with a region of some other geometry. An important part of the algorithm is the component that allows us to assess whether a particular point with Cartesian coordinates $\boldsymbol{x}_0 = (x_0, y_0)$ lies inside or outside the region. Can this be done in a manner that is free of logical complexities and does not consume an exorbitant amount of time?

Semi-analytic approach

If the boundary of the region can be described by an equation of the form $f(x, y) = 0$, where $f(x, y)$ is a continuous function, then the sign of the function $f(x, y)$ will change as the boundary of the region is crossed. Accordingly, a decision on whether a point lies inside or outside the region may be made simply by sign inspection.

For example, for a disk of radius a, we define $f(x, y) = x^2 + y^2 - a^2$. If $f(x_0, y_0) > 0$, then the point (x_0, y_0) lies outside the disk, whereas if $f(x_0, y_0) < 0$ the point lies inside the disk. A similar inequality may be derived for the ellipse.

Polygonal regions

The problem becomes more challenging by considering a polygonal area that is bounded by a closed polygon with M vertices $\boldsymbol{x}_i$, $i = 1, \ldots, M$, arranged in the counterclockwise sense as shown in Figure 1.5.1. In this case, a continuous function $f(x, y)$ cannot be found, and the semi-analytic method described in the preceding subsection is no longer effective. Fortunately, there are a number of alternatives (Milgram 1989).

In one method, we identify the xy plane with the complex z plane and introduce the angle $\theta = \text{Arg}(z - z_0)$ that is subtended between the x axis and the line that passes through the points z and z_0, where Arg( ) signifies the argument of the enclosed complex variable. As the point z traces the boundary of a closed region and comes back to its initial point of departure, θ changes by an amount $\Delta\theta$ that is

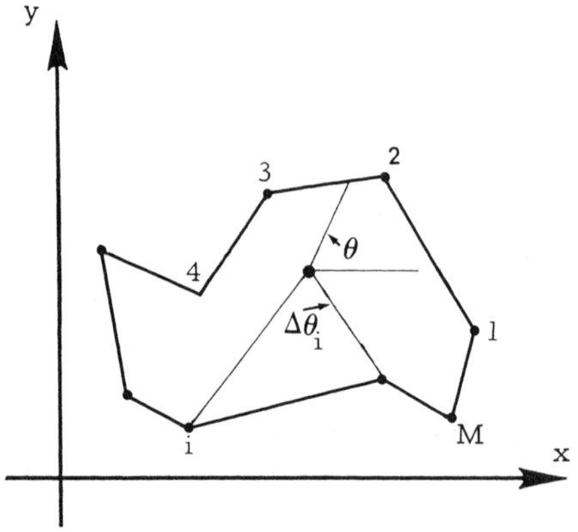

Figure 1.5.1 A closed polygon with M vertices $\boldsymbol{x}_i, i = 1, \ldots, M$, numbered in the counterclockwise sense. We want to develop an algorithm that allows us to assess whether a point lies inside or outside the area enclosed by the polygon.

equal to 0 when z_0 is located outside the region, or 2π when z_0 is located inside the region. In the world of fluid mechanics, we say that the circulation around a simple closed loop that encloses a *point vortex* of unit strength located at z_0 is equal to 2π, whereas the circulation around a loop that does not enclose the point vortex is equal to zero. This observation provides us with a simple criterion for assessing whether or not a point is located inside or outside the region.

To implement the method, we consider the polygon shown in Figure 1.5.1 and note that the total change of the argument $\Delta\theta$ is equal to the sum of the incremental changes across the M sides $\Delta\theta_i$, $i = 1, \ldots, M$. Exploiting the geometrical interpretation of the cross product of the vectors $\boldsymbol{x}_i - \boldsymbol{x}_0$ and $\boldsymbol{x}_{i+1} - \boldsymbol{x}_0$, we write

$$|\boldsymbol{x}_{i+1} - \boldsymbol{x}_0||\boldsymbol{x}_i - \boldsymbol{x}_0| \sin\Delta\theta_i = (x_i - x_0)(y_{i+1} - x_0) - (y_i - y_0)(x_{i+1} - x_0) \tag{1.5.3}$$

Solving for $\sin\Delta\theta_i$ and inverting the resulting trigonometric equation yields the value of $\Delta\theta_i$, where $-\pi \le \Delta\theta_i < \pi$.

Other less general but more efficient methods for triangular shapes will be discussed in Section 6.9. Alternative methods for rectangles are reviewed by Milgram (1989).

To assess whether a point lies inside an arbitrary region that is enclosed by a nonpolygonal boundary, we can approximate the boundary of the region with a polygonal line that passes through a sequence of marker points, and then use the preceding algorithm to obtain an estimate.

PROBLEMS

1.5.4 *Inside or outside a loop?*

Consider a closed polygonal line in the xy plane described by a set of M points as shown in Figure 1.5.1. Write a program that tells you whether a certain point $\boldsymbol{x}_0$ lies in the interior or exterior of the polygon by responding with "*inside*" or "*outside.*" Test the reliability of the program by considering several test cases of your choice.

1.5.5 *Overlapping elliptical disks.*

Develop an algorithm that allows you to assess whether two elliptical disks in a plane, specified by their centers, orientations, and axes, overlap.

Assessing Whether a Point Lies Inside a Certain Volume

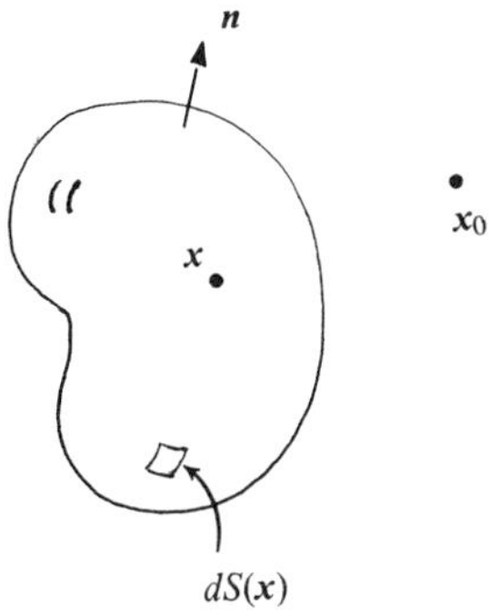

Next, we consider a method for assessing whether a specified point lies within a certain volume of the three-dimensional space. When the boundary of the volume can be described in closed form in terms of a continuous function as $f(x, y, z) = 0$, we can use the semi-analytic method for the planar region discussed in the preceding subsection.

Consider now a volume whose boundary cannot be described by the equation above. Unfortunately, the point-vortex method for the polygonal area described in the preceding subsection does not have a counterpart. To this end, mathematical training proves an invaluable arsenal. Using the theory of Green's functions and the divergence theorem, we find that the surface integral computed over the boundary of any finite volume, is equal to 4π when the point $\boldsymbol{x}_0 = (x_0, y_0, z_0)$ lies inside the region, or zero when the point $\boldsymbol{x}_0$ lies outside the region; $\boldsymbol{n}$ is the unit vector normal to the boundary, dS is the differential surface area, and the integration point $\boldsymbol{x}$ is located within the boundary.

$$I = \int_{Boundary} \frac{(x - x_0)\, n_x(\boldsymbol{x}) + (y - y_0)\, n_y(\boldsymbol{x}) + (z - z_0)\, n_z(\boldsymbol{x})}{\left((x - x_0)^2 + (y - y_0)^2 + (z - z_0)^2\right)^{3/2}}\, dS(\boldsymbol{x}) \tag{1.5.4}$$

These observations provide us with a criterion for assessing whether a point is located inside or outside the volume.

In practice, the integral on the right-hand side of equation (1.5.4) can be computed from knowledge of the position of a set of marker points distributed over the boundary using the numerical methods discussed in Chapter 7.

PROBLEMS

1.5.6 *Green's functions.*

The counterpart of equation (1.5.4) in two dimensions is the line integral

$$I = \oint_{Boundary} \frac{(x - x_0)\, n_x(\mathbf{x}) + (y - y_0)\, n_y(\mathbf{x})}{(x - x_0)^2 + (y - y_0)^2}\, dl(\mathbf{x}) \tag{1.5.5}$$

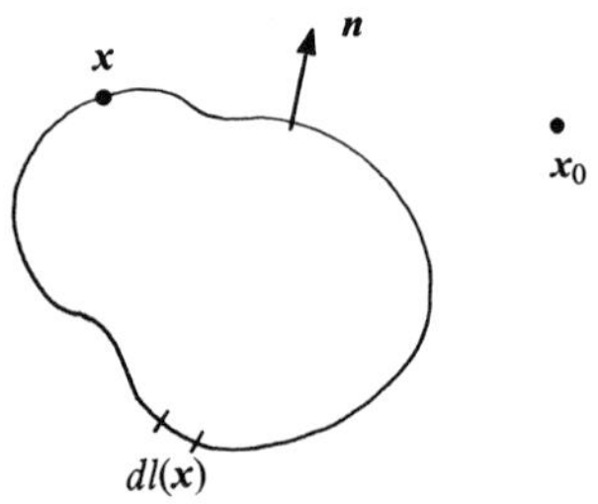

computed along the closed boundary of an arbitrary region in the xy plane. This integral is equal to 2π when the point $\mathbf{x}_0 = (x_0, y_0)$ lies inside the region, or zero when the point $\mathbf{x}_0$ lies outside the region; $\mathbf{n}$ is the unit vector normal to the boundary pointing outward from the region, dl is the differential arc length, and the point $\mathbf{x}$ traces the boundary.

To compute the integral on the right-hand side of equation (1.5.5) along the perimeter of a polygon, we describe the ith side as

$$y = f_i(x) = y_i + (x - x_i)\alpha_i$$

where $\alpha_i = (y_{i+1} - y_i)/(x_{i+1} - x_i)$ is the slope. The corresponding components of the normal vector are given by

$$n_x = (y_{i+1} - y_i)/|\mathbf{x}_{i+1} - \mathbf{x}_i|, \qquad n_y = -(x_{i+1} - x_i)/|\mathbf{x}_{i+1} - \mathbf{x}_i|$$

and the differential arc length can be expressed as

$$dl = \beta\, (1 + \alpha_i^2)^{1/2}\, dx$$

where $\beta = (x_{i+1} - x_i)/|x_{i+1} - x_i| = \pm 1$. Substituting these expressions into equation (1.5.5) and simplifying, we find

$$\begin{aligned} I &= \sum_{i=1}^{M} \int_{x_i}^{x_i+1} \frac{\alpha_i (x - x_0) - (y - y_0)}{(x - x_0)^2 + (y - y_0)^2}\, dx \\ &= \sum_{i=1}^{M} [y_0 - f_i(x_0)] \int_{x_i}^{x_i+1} \frac{dx}{(x - x_0)^2 + [f_i(x) - y_0]^2} \end{aligned} \tag{1.5.6}$$

Perform the integration by analytical means, and show that this method is identical to the point-vortex method described previously in this section.

1.5.7 *Inside a polyhedron?*

Consider a polyhedral boundary, and investigate whether the integral in equation (1.5.4) over each side can be computed by analytical means.

Modeling Stochastic and Random Processes

In most cases, computers are used to solve *deterministic* problems that are governed by well-established algebraic or differential equations, with well-defined initial and boundary conditions. But numerical simulation also allows us to study the behavior of *stochastic* or *random* systems that emulate the seemingly random nature of the physical world. For example, *computational point particles* that execute random walks model the motion of molecules or small fluid parcels and reproduce the physical process of diffusion. Similarly, lines and surfaces with random irregularities, possibly of fractal nature, are used to model rough walls.

Random walkers with discrete steps

Consider L point particles initially sitting at the origin of the x axis. At the origin of time, each particle starts making random steps to the left with probability q, and to the right with probability $1-q$. After a step has been executed, each particle has moved by one unit of length. After N steps have been executed, the particles have spread out from the origin and occupy discrete positions along the x axis corresponding to positive or negative integral displacements, with the ith notch hosting $k_i(N)$ particles, as shown in the diagram. Since the particles move by one unit of length, $k_i(N) = 0$ for $i > N$ or $i < -N$.

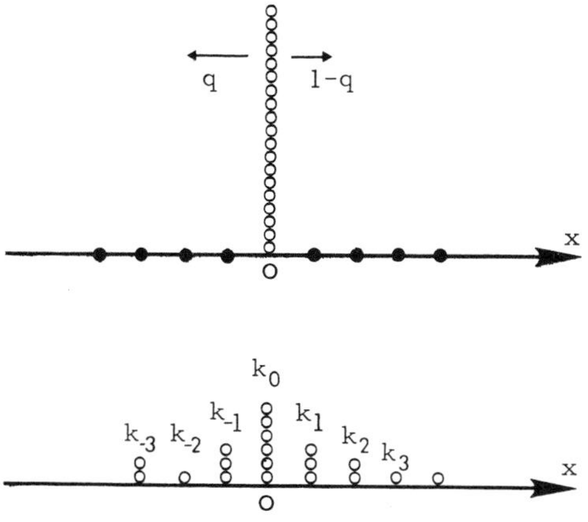

The motion of the walkers is relevant to a variety of physical processes that can be analyzed by methods of statistical mechanics (e.g., Chandrasekhar 1943). To study the statistics of the motion, we introduce the *discrete probability function*

$$p_i(N) \equiv \frac{k_i(N)}{L} \tag{1.5.7}$$

The *mean value* of the particle distribution is given by

$$m(N) = \sum_{i=-N}^{N} i\,p_i(N) \tag{1.5.8}$$

and the *variance* is given by

$$s^2(N) = \sum_{i=-N}^{N} [i - m(N)]^2 p_i(N) = -m(N)^2 + \sum_{i=-N}^{N} i^2 p_i(N) \tag{1.5.9}$$

where s is the *standard deviation.*

The problem is ideally suited for numerical simulation. In the numerical method, we assess the direction of motion of each particle by using a random-number generator that produces random numbers with a uniform probability density function ranging between 0 and 1, as discussed previously in this section. If a random number is less than q, a particle moves to the left; if it is higher than q, it moves to the right.

PROBLEMS

1.5.8 ***Random walkers with discrete steps in one dimension.***

(*a*) Compute and plot $p_i(N)$ for $L = 1000$, $q = 0.25$ at several values of N, and discuss its shape as the number of steps N becomes larger. Discuss the dependence of the mean value and variance on N.
(*b*) Repeat part (*a*) with $q = 0.50$.

1.5.9 ***Random walkers with discrete steps in two dimensions.***

Discuss the generalization of Problem 1.5.8 to random walkers moving in *two* dimensions, and repeat parts (*a*) and (*b*).

Random walkers with continuous steps

A physically more appealing problem considers walkers that execute random motion with a continuous range of displacements d after each step. For travel in one dimension, the positive or negative travel distance d is a stochastic variable with a certain probability density function $q(d)$ ranging between two specified limits.

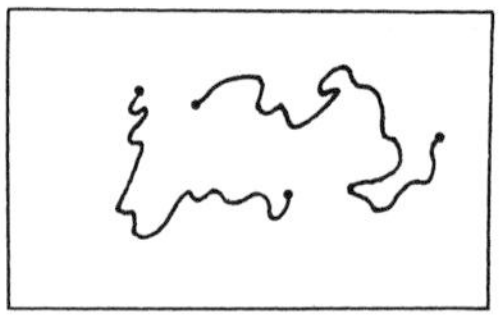

Physically, the walkers may be identified with the molecules of a diffusing chemical species, and the motion may be attributed to the thermal energy of the fluid; the higher the temperature, the more vigorous the motion. Thus we can imagine that the walkers move with randomly varying velocities, while their position is illuminated by a strobe-light at time intervals of equal length. On a more observable level, the walkers may be identified with small particles, with dimensions less than about 10 microns, suspended in an ambient liquid. When observed through a microscope, these particles are seen to execute *Brownian motion*, moving under the influence of the forces imparted to them by the erratically moving molecules of the ambient liquid. This motion was first noted by the Dutch physician Jan Ingenhausz in 1785 and was subsequently documented by the British botanist Robert Brown in 1928 (Probstein 1994, p. 116; Klafter et al. 1996).

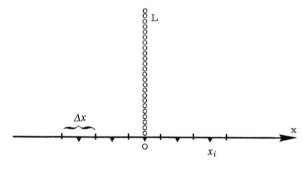

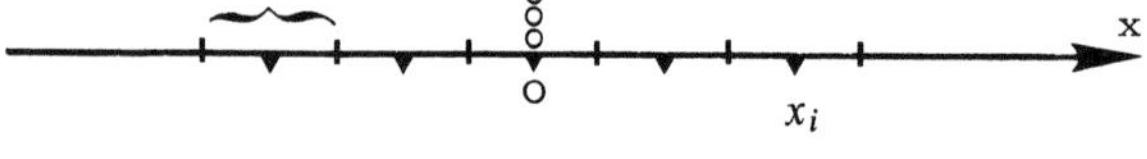

Consider L such walkers initially sitting at the origin of the x axis. After N steps have been executed, the walkers have spread out from the origin to random positions. To study the statistics of the motion, we divide the x axis into evenly spaced intervals of size Δx centered at the points $x_i = i\Delta x$, where i is an integer. The probability of finding a particle within the ith interval is given by the *discrete probability function* defined in equation (1.5.7), where $k_i(N)$ is the number of particles residing within that interval.

It is useful to introduce the staircase-like *probability density function*

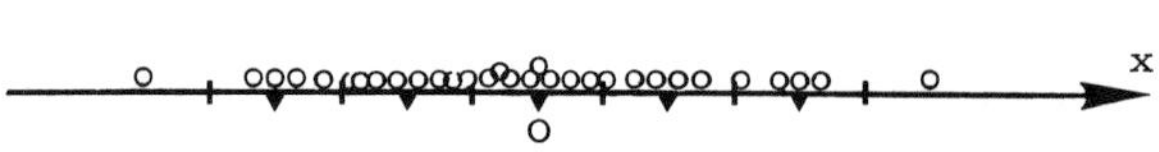

$$g(x, N) \equiv \frac{p_i(N)}{\Delta x} \qquad \text{for } x_i - \tfrac{1}{2}\Delta x < x < x_i + \tfrac{1}{2}\Delta x \tag{1.5.10}$$

In the limit as Δx tends to zero and the number of walkers L tends to infinity, $g(x, N)$ tends to a continuous function called the *continuous probability density function.*

Of particular interest are the mean value and variance of the walker distribution given by

$$m(N) = \sum_{i=-\infty}^{\infty} x_i p_i(N) \cong \int_{-\infty}^{\infty} x g(x, N)\, dx \tag{1.5.11}$$

and

$$s^2(N) = \sum_{i=-\infty}^{\infty} [x_i - m(N)]^2 p_i(N) \cong \int_{-\infty}^{\infty} [x - m(N)]^2 g(x, N)\, dx \tag{1.5.12}$$

where s is the standard deviation.

To carry out the numerical simulations, we require a sequence of displacements d that conform with a specified probability density function $q(d)$ taking nonzero values over the interval $[a, b]$. This can be obtained on the basis of the following theorem.

Let r be a stochastic variable with uniform probability density function taking values in the range between 0 and 1, and define the cumulative distribution

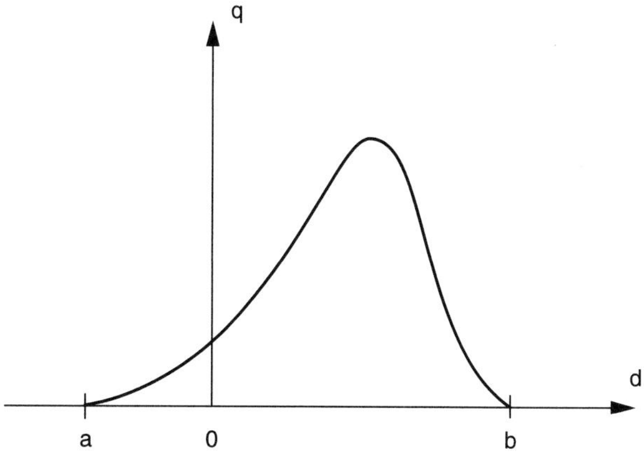

$$F(d) \equiv \int_a^d q(u)\, du \tag{5.1.13}$$

where $F(a) = 0$ and $F(b) = 1$. Then the d series arises from an r series by solving either one of the following equations for d:

$$F(d) = r, \qquad F(d) = 1 - r \tag{1.5.14}$$

For example, in the contrived case where $q(d)$ has the uniform value $1/(b - a)$, we obtain $F(d) = (d - a)/(b - a)$, and use the first of equations (1.5.14) to find $d = r(b - a) + a$. A more interesting example will be discussed in Problem 1.5.10.

Gaussian displacement distribution

A Gaussian or normal distribution with mean value equal to α and standard deviation equal to σ is described by

$$q(d) = \frac{1}{\sigma\sqrt{2\pi}} \exp\left[-\tfrac{1}{2}\left(\frac{d - \alpha}{\sigma}\right)^2\right] \tag{1.5.15}$$

where d ranges from $-\infty$ to $+\infty$. The application of the aforementioned theorem results in an analytically intractable algebraic equation $F(d) = r$, which is also impractical to invert using numerical methods.

Alternatively, the random steps d can be obtained with much less effort by exploiting the implications of the *central-limit theorem.* Select a group of ν random numbers $r_{i,1}$, $i = 1, \ldots, \nu$, with a uniform probability distribution, ranging in the interval [0, 1], and compute the number

$$x_1 = \alpha + \sigma \left(-0.5\, \nu + \sum_{i=1}^{\nu} r_{i,1} \right) \tag{1.5.16}$$

Then select another group $r_{i,2}$, $i = 1, 2, \ldots, \nu$, and compute the corresponding number x_2. Repeat this many times. The central-limit theorem guarantees that as ν tends to infinity, the random variable x_i will obey a Gaussian distribution with mean value equal to α and standard deviation equal to σ. In practice, setting $\nu = 12$ yields a satisfactory Gaussian-like behavior.

PROBLEMS

1.5.10 ***Random walkers with exponentially distributed step sizes.***

Write a program that produces a time series for d corresponding to

$$q(d) = \tfrac{1}{2}\mu \, \exp(-\mu|d|) \tag{1.5.17}$$

where μ is a constant, and verify that the series observes the required probability distribution.

1.5.11 ***Random walkers in one dimension.***

Compute and plot the probability density function $g(x, N)$ for $L = 1000$ walkers moving to the left or right with the Gaussian probability density function given in equation (1.5.15), for a series of values of N. Discuss the dependence of the mean value and variance on N, α, and σ.

1.5.12 ***Random walkers in two dimensions.***

Discuss the generalization of Problem 1.5.11 to random walkers moving in two dimensions, and perform several series of dynamical simulations. To obtain two sequences of normally distributed random variables d_1 and d_2 with mean values α_1, α_2 and standard deviations σ_1, σ_2, introduce two random variables r_1 and r_2 with uniform probability density functions ranging between 0 and 1, and use the Box–Muller transformation (e.g., Dahlquist and Björck 1974, p. 453)

$$\begin{aligned} d_1 &= \alpha_1 + \sigma_1(-2 \ln r_1)^{1/2} \cos(2\pi r_2) \\ d_2 &= \alpha_2 + \sigma_2(-2 \ln r_1)^{1/2} \sin(2\pi r_2) \end{aligned} \tag{1.5.18}$$

Numerical Evidence

Is it possible to prove a mathematical theorem by numerical computation? The pure mathematician will revolt, but the physical scientist and the practicing engineer will rejoice. The computer can exhaustively examine a large number of possibilities in a relatively short period of time and thus confirm or dismiss a conjecture by elimination. The proof of the four-color theorem is an example. By examining a large number of possibilities, one can show that only four colors are needed to paint all countries of a continent so that two neighboring countries do not have the same color, provided that one country is not completely enclosed by another. Varga (1993) discusses other examples.

Perhaps more importantly, numerical computation may be used to collect evidence for a suspected behavior. The discipline of nonlinear science sprung from the discovery that the solution of a certain class of nonlinear differential equations is so sensitive to the initial condition that the round-off error dominates the results after some time, and the effect of the initial state is lost. Since exact solutions to these nonlinear equations are not available, the theory must be built on numerical evidence that is collected by careful numerical experimentation.

PROBLEMS

1.5.13 *Scrutinizing the behavior of numerical sequences.*

(*a*) Compute the sequence of numbers

$$x^{(k)} = \left(1 + \frac{1}{k}\right)^k \tag{1.5.19}$$

for $k = 0, 1, 2, 3, \ldots$. Can you identify the limit?
(*b*) Select a positive number c, and another positive number $x^{(0)}$, and compute the sequence of numbers $x^{(1)}, x^{(2)}, x^{(3)}, \ldots$ using the formula

$$x^{(k+1)} = x^{(k)} - \tfrac{1}{3}\frac{x^{(k)^3} - c}{x^{(k)^2}} \tag{1.5.20}$$

for $k = 0, 1, 2, 3, \ldots$. Does this sequence converge to a limit, that is, do the $x^{(k)}$ tend to become identical? If yes, can you identify a relationship between the limit and the number c? Does the behavior of the sequence depend on the choice of $x^{(0)}$ and c?

1.5.14 *Computing functions in terms of Fourier series.*

(*a*) Plot the *periodic* function

$$f(x) = \sum_{k=-N}^{N} \cos(2\pi k x) \tag{1.5.21}$$

over one period, $0 < x < 1$, for $N = 10, 20, 30, 40$, and discuss the behavior as N becomes larger.
(*b*) Repeat part (*a*) for the function

$$f(x) = \sum_{k=-N}^{N} e^{-k} \cos(2\pi k x) \tag{1.5.22}$$

What seems to be the key difference between the two cases? Can you derive a closed-form expression for the function displayed in equation (1.5.22) in the limit as N tends to infinity?

1.5.15 *Sturges's law.*

The *Macintosh Bible* (1994, Peachpit Press) discusses Sturges's law, one variation of which is: "*Ninety percent of science-fiction writing is rubbish, but then again ninety percent of everything is rubbish.*" Discuss whether and how Sturges's law applies to numerical simulation.

1.6 Systematic Error and its Reduction

The results of a typical numerical computation provide us with an approximation to the exact solution of a particular problem under consideration. Examples are the value of a function computed from a truncated Taylor series expansion, the value of a definite integral of a function computed by approximating the function with a polynomial, and the value of the first derivative of a function whose values are known only at discrete points, computed using a finite-difference method.

Absolute and Relative Error

The difference between the numerical value b and the exact value a is the *absolute numerical error*

$$e = b - a \tag{1.6.1}$$

Some authors define $e = a - b$, but definition (1.6.1) is more intuitive: When the numerical value is larger than the exact value, the error is positive.

When the exact value a is either very small or very large, it is more appropriate to work with the *relative numerical error*, which expresses the relative deviation of the numerical from the exact value, defined as

$$e_R = e/a \tag{1.6.2}$$

Multiplying the right-hand side by 100 yields the *percent error.*

Systematic Error

The numerical error consists of the round-off error discussed in Section 1.3, and a systematic error due to the various approximations involved in the fabrication of the numerical method.

Consider, for example, computing the first derivative of a function $f(x)$ at a certain point x_0. By definition,

$$f'(x_0) \equiv \lim_{h\to 0} \frac{f(x_0+h) - f(x_0)}{h} \tag{1.6.3}$$

The finite-difference formula

$$b(h) \equiv \frac{f(x_0+h) - f(x_0)}{h} \tag{1.6.4}$$

provides us with an approximation to the exact value of $f'(x_0)$ for any finite value of h, with an associated numerical error $e(h) \equiv b(h) - f'(x_0)$. If the magnitude of all variables on the right-hand side of equation (1.6.4) is higher than the computer's unit round-off error, then $e(h)$ will be virtually equal to the systematic numerical error; the round-off error will make a negligible contribution.

Order of a method

In a typical application, the magnitude of the systematic error depends on the value of a numerical parameter h, which is a *control variable* of the numerical method. A computation with a large value of h is a *rough* or a *crude* one, whereas a computation with a small value of h is a *refined* one. Of course, the classifications *large* and *small* depend on the circumstances.

In the vast majority of numerical methods, the systematic error is expressible in the form of an asymptotic series as

$$e(h) = c_1 h^{p1} + c_2 h^{p2} + \cdots \tag{1.6.5}$$

where the *positive* integer or real exponents p_i have been arranged in an ascending order of magnitude, $p_1 < p_2 < \ldots$; c_i are constants. The value of p_1, in particular, defines the *order of the numerical error*, sometimes called the *order of the numerical method.*

Although the preceding power-law behavior is the norm, there are circumstances where the dependence of the systematic error on h is more involved. Two examples are

$$e(h) = c_1 h \ln h + \cdots, \qquad e(h) = c_1 \frac{1}{\left(\frac{L}{h}\right)!} + \cdots \tag{1.6.6}$$

In the second example, L is a constant and h is an integral subdivision of L so that L/h is a positive integer.

For all types of behavior displayed in equations (1.6.5) and (1.6.6), the magnitude of the systematic error decreases as h is made smaller and the computation is refined. But there are extraordinary circumstances where this does not happen and the preceding scalings break down in a catastrophic way: The seemingly numerical error does not vanish as h is made smaller. These exceptions will not concern us in this book, but it is important to be aware of their existence.

Consider, for example, measuring the arc length of a line using a straight ruler of unit length h. The measured length of the line is a function of h, call it $L(h)$. Normally, we should expect that as h tends to zero, $L(h)$ will tend to the exact value $L(0)$, with the corresponding error possibly exhibiting the power-law behavior shown in equation (1.6.5). But if the line happens to have an endless sequence of self-similar indentations with increasingly smaller size, which classify it as a fractal (Section 1.3), then as h tends to vanish, $L(h)$ will keep increasing without a bound.

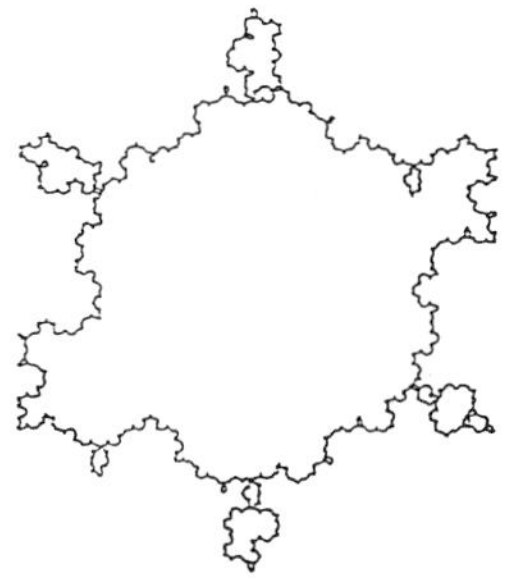

Assessing the Order of the Error

In the remainder of this chapter, we shall assume that the error obeys the power-law relation shown in equation (1.6.5). How can we identify the exponents p_i?

Error analysis

In certain simple cases, this can be done without much difficulty by performing an *error analysis of the numerical method.* Examples will be discussed in the ensuing chapters. Unfortunately, performing an error analysis is often either impractical or prohibitively involved to carry out.

Benchmarking

A quick way of assessing the order of the error is by numerical benchmarking. Consider, for example, formula (1.6.4), and select

$$f(x) = \ln x, \qquad x_0 = 1$$

Now, perform two computations, the first one with $h_1 = \varepsilon$, and the second one with $h_2 = \frac{1}{2}\varepsilon$, where ε is a small number. Setting ε equal to 0.10, we find readily with the aid of a calculator

$$b(0.10) = 0.953, \qquad b(0.05) = 0.976$$

Knowledge of the exact value $f'(x_0) = 1$, which is obtained by elementary analytical methods, allows us to compute the associated errors

$$e(0.10) = -0.047, \qquad e(0.05) = -0.024$$

We observe that decreasing h by a factor of 2 reduces the error roughly by a factor of 2, and this indicates that $p_1 = 1$. If the error were reduced by a factor of 4, we would have concluded that $p_1 = 2$. And if the error were reduced by a factor of r, then $p_1 = \ln_2 r$.

Observing the limit of a geometrical series

In the preceding example, we estimated the order of the error from a knowledge of the exact value, but the availability of the latter is not imperative.

Let us perform a series of m computations with values of h that differ by a certain positive constant factor $q > 1$, forming the geometric sequence

$$h_1 = \varepsilon, \quad h_2 = \varepsilon/q, \quad h_3 = \varepsilon/q^2, \quad \ldots, \quad h_m = \varepsilon/q^{m-1} \tag{1.6.7}$$

We denote the corresponding values of the numerical solution by

$$b_1, b_2, \ldots, b_m \tag{1.6.8}$$

where b_m is the most accurate value. Using equation (1.6.5), we write

$$y_k \equiv \frac{b_{k+1} - b_k}{b_{k+2} - b_{k+1}} = \frac{e_{k+1} - e_k}{e_{k+2} - ek + 1} \cong \frac{\left(\frac{\varepsilon}{q^k}\right)^{p_1} - \left(\frac{\varepsilon}{q^{k-1}}\right)^{p_1}}{\left(\frac{\varepsilon}{q^{k+1}}\right)^{p_1} - \left(\frac{\varepsilon}{q^k}\right)^{p_1}} = q^{p_1} \tag{1.6.9}$$

Thus

$$p_1 \cong \frac{\ln y_k}{\ln q} = \ln_q y_k \tag{1.6.10}$$

The accuracy of this estimate improves as we use more advanced triplets in the sequence (1.6.8).

For example, the numerical computation of the definite integral

$$a = \int_0^1 x^{3/2}\, dx \tag{1.6.11}$$

using Simpson's $\frac{1}{3}$ rule with step size h (see Section 7.2) gives the following table of values

k	h_k	b_k
1	0.500	0.4023689
2	0.250	0.4004319
3	0.125	0.4000772

which corresponds to $\varepsilon = 0.500$, $q = 2$. Using the left-hand side of equation (1.6.9), we find $y_1 = 5.46$, which indicates that the order of the method is $p_1 = \ln 5.46/\ln 2 = 2.45$.

PROBLEM

1.6.1 ***Assessing the order of the error.***

Replace the right-hand side of equation (1.6.4) with the expression $[f(x_0 + h) - f(x_0 - h)]/(2h)$, and assess the order of the systematic error using the two methods discussed in the text.

Error Reduction by Richardson Extrapolation

One way to improve the accuracy of a numerical computation is to use a very small value for the numerical parameter h. This works fine, provided that h is kept above the level where the numerical error becomes comparable to the round-off error. If not, the results will be rubbish. On a more pragmatic level, the computational expense typically increases dramatically as the value of h is made smaller, placing realistic limitations on the performance of the numerical method. For instance, in many problems the computational expense, quantified in some sense, scales with $1/h^2$.

Fortunately, when at least the first of the exponents p_1 is known, the results of a sequence of relatively crude computations may be used to estimate the magnitude of the error and hence improve the accuracy of the results by a method that is known as the *deferred approach to the limit* or *Richardson extrapolation.*

To illustrate the method, we use the definition of the error in equation (1.6.1) and recast equation (1.6.5) into the form

$$b(h) = a + c_1 h^{p1} + c_2 h^{p2} + \cdots \tag{1.6.12}$$

where $b(h)$ is the result of the numerical computation with parameter size h, and a is the exact value. Having computed two numerical values $b(h_1)$ and $b(h_2)$, we retain two terms on the right-hand side of equation (1.6.12) and, assuming that p_1 is known, we obtain a system of two linear algebraic equations for a and c_1. The value of a obtained in this manner will be a much better approximation to the exact value than either one of $b(h_1)$ or $b(h_2)$, and the associated error will be on the order of h_1^{p2} or h_2^{p2}.

When, in particular, $h_1 = \varepsilon$, and $h_2 = \varepsilon/q$, where q is a certain positive factor that is greater than unity, the solution of the linear system yields the improved estimate

$$a = \frac{q^{p_1} b(\varepsilon/q) - b(\varepsilon)}{q^{p_1} - 1} \tag{1.6.13}$$

For the popular choice $q = 2$, we obtain

$$
\boxed{
\begin{aligned}
a &= 2b(\varepsilon/2) - b(\varepsilon), \quad &\text{for} \quad p_1 = 1 \\
a &= \tfrac{1}{3}[4\, b(\varepsilon/2) - b(\varepsilon)], \quad &\text{for} \quad p_1 = 2 \\
a &= \tfrac{1}{7}[8\, b(\varepsilon/2) - b(\varepsilon)], \quad &\text{for} \quad p_1 = 3 \\
a &= \tfrac{1}{15}[16\, b(\varepsilon/2) - b(\varepsilon)], \quad &\text{for} \quad p_1 = 4
\end{aligned}
}
\tag{1.6.14}
$$

The right-hand sides of these equations express weighted averages of the numerical values, where the more accurate values are given higher weights.

For the numerical example discussed in the preceding subsection concerning the derivative of the logarithmic function, corresponding to $p_1 = 1$, we use the first of equations (1.6.14) and find $a = 2b(0.10) - b(0.05) = 2 \times 0.976 - 0.953 = 0.999$, which is much closer to the exact value of 1.0.

Similarly, having computed three numerical values $b(h_1)$, $b(h_2)$, and $b(h_3)$, we retain three terms on the right-hand side of equation (1.6.12) and, assuming that the values of both p_1 and p_2 are known, we form a system of three linear algebraic equations for a, c_1, and c_2. The value of a obtained in this manner will be a much better approximation to the exact value than either one of $b(h_1)$, $b(h_2)$, or $b(h_3)$, and the associated error will be on the order of h_1^{p3}, h_2^{p3}, or h_3^{p3}.

The generalization of extrapolation from a series of m numerical results is now evident: The problem is reduced to solving a system of m linear equations for m unknown constants. Fortunately, practice has shown that the solution of this system is not sensitive to the round-off error or any other type of error that perturbs the numerical results by a small amount.

Extrapolation from a geometric sequence of h

The solution of the aforementioned linear system can be expedited by performing a series of m computations with values of h that form the geometric sequence (1.6.7), and corresponding numerical values shown in sequence (1.6.8). A much improved value of a emerges by successively filling out the columns of an $m \times m$ lower triangular matrix $\mathbf{A}$.

We begin by setting the elements of the first column equal to the computed values

$$
A_{1,1} = b_1, \quad A_{2,1} = b_2, \quad \ldots, \quad A_{m,1} = b_m \tag{1.6.15}
$$

and then fill in the rest of the elements working as follows:

$$
\begin{bmatrix}
A_{1,1} & & & \\
A_{2,1} & A_{2,2} = (q^{p1} A_{2,1} - A_{1,1})/(q^{p1} - 1) & & \\
A_{3,1} & A_{3,2} = (q^{p1} A_{3,1} - A_{2,1})/(q^{p1} - 1) & A_{3,3} = (q^{p2} A_{3,2} - A_{2,2})/(q^{p2} - 1) & \\
A_{4,1} & A_{4,2} = (q^{p1} A_{4,1} - A_{3,1})/(q^{p1} - 1) & A_{4,3} = (q^{p2} A_{4,2} - A_{3,2})/(q^{p2} - 1) & \ldots \\
\ldots & \ldots & \ldots &
\end{bmatrix}
\tag{1.6.16}
$$

The bottom value $A_{m,m}$ is a much improved approximation to the exact value a, and the associated error is on the order of $\varepsilon^{p_{m+1}}$ (Dahlquist and Björck 1974, p. 271).

The computations may be programmed according to the algorithm:

$$\begin{array}{lcl} \text{Do } k & = & 2, m \\ \omega & = & q^{pk-1} \\ \text{Do } i & = & k, m \\ A_{i,k} & = & (\omega A_{i,k-1} - A_{i-1,k-1})/(\omega - 1) \\ \text{END DO} & & \\ \text{END DO} & & \end{array} \qquad (1.6.17)$$

PROBLEMS

1.6.2 ***Improvement of Taylor series expansion.***

The exponents of the error associated with the formula given in Problem 1.6.1 are known to be $p_1 = 2$, $p_2, = 4$, $p_6 = 6, \ldots$. Using this knowledge, compute $f'(x_0)$ for $f(x) = \ln x$ and $x_0 = 1$, with $\varepsilon = 0.10$, $q = 3.0$, and $m = 3$.

1.6.3 ***Improvement of accuracy by extrapolation.***

An involved system of differential equations describing the propagation of flame in a porous medium has been solved numerically, by discretizing the porous medium into N elements. It is known that the accuracy of the solution increases as N becomes larger, and the numerical solution produces the exact value in the limit as N tends to infinity. Part of the solution includes the temperature at point A, denoted by T. The calculations gave the following pairs of (N, T): (10, 0.6162), (20, 0.4118), (30, 0.3609), (40, 0.3545). Based on these data, estimate as well as you can the exact value of T. *Hint*: Define $h = 1/N$.

1.7 Iterations, Numerical Sequences, and their Convergence

Many numerical algorithms produce a sequence of approximations $x^{(k)}$, $k = 0, 1, \ldots$, to an exact value a, using an iterative or recursive method, with associated error

$$e^{(k)} = x^{(k)} - a \qquad (1.7.1)$$

An example was given in Problem 1.5.13, and further examples will be presented in subsequent chapters.

The iterations are said to *converge* when the magnitude of $e^{(k)}$ keeps decreasing during the iterations; otherwise they either *diverge* or *stall* at some value.

Rate of Convergence

An error analysis of the algorithm that generates the numerical sequence typically produces a relationship between the next error $e^{(k+1)}$ and the current and previous errors $e^{(k)}, e^{(k-1)}, \ldots, e^{(0)}$; specific examples will be discussed in Section 4.3. In many cases, this relation involves only the next and the current errors, $e^{(k+1)}$ and $e^{(k)}$, and has the power-law form:

$$e^{(k+1)} = c e^{(k)^m} \qquad (1.7.2)$$

where m is a positive real number, and c is a coefficient. The value of m defines the *rate of convergence of the iterative method.*

When $m = 1$, the rate of convergence is linear: The error behaves in a *geometric manner*, and the sequence converges as long as $|c| < 1$.

When $m = 2$, the rate of convergence is quadratic: The sequence converges for any value of c provided that $|e^{(0)}|$ is sufficiently small.

Acceleration of Convergence

Knowledge of the value of m can be exploited to improve the rate of convergence by a method that is known as the *Aitken extrapolation*; the idea was actually first conceived by Thiele in 1909 (see Bodewig 1959, p. 160). This is done by applying equation (1.7.2) for two successive approximations, obtaining

$$\begin{aligned} x^{(k+1)} - a &= c(x^{(k)} - a)^m \\ x^{(k)} - a &= c(x^{(k-1)} - a)^m \end{aligned} \tag{1.7.3}$$

which provides us with a system of two nonlinear algebraic equations for the two unknowns a and c. The value of c is of limited interest, but the value of a provides us with a much better approximation to the limit than $x^{(k+1)}$.

It is important to note that *extrapolation cannot make a divergent sequence converge*; when trading in mathematics one cannot get something out of nothing. The extrapolation simply makes a convergent sequence converge even faster. When the sequence diverges, it makes it diverge even faster.

Linear Convergence

When $m = 1$, equations (1.7.3) can be combined to yield a single linear equation for a, whose solution is

$$\begin{aligned} a^{(k)} &= x^{(k+1)} - \frac{\left(x^{(k+1)} - x^{(k)}\right)^2}{x^{(k+1)} - 2x^{(k)} + x^{(k-1)}} \\ &= \frac{x^{(k-1)}x^{(k+1)} - x^{(k)2}}{x^{(k+1)} - 2x^{(k)} + x^{(k-1)}} \end{aligned} \tag{1.7.4}$$

If the primary sequence $x^{(k)}$ converges, the extrapolated sequence $a^{(k)}$ will converge at an even faster rate. The procedure involves computing the $a^{(k)}$ sequence either simultaneously or after the $x^{(k)}$ has been produced.

In *Steffensen's modification*, we consider three successive values $x^{(k)}$, obtain $a^{(k)}$ according to equation (1.7.4), and then repeat the procedure with three new triplets $x^{(k)}$, where the first member is set equal to $a^{(k)}$ while the other two are produced using the numerical method.

Time series with exponential decay

As an application, we consider a *time series*, that is, a sequence of values $x^{(k)}$ corresponding to evenly spaced time instants $t^{(k)}$ separated by the time interval Δt, and assume that the series contains the values of the exponentially decaying function

$$x = a + d\ \exp(-\beta t) \tag{1.7.5}$$

where a and d are two real constants, and β is a positive rate of decay. Our objective is to estimate the value of a.

To assess the rate of convergence of the time series, we write

$$\begin{aligned} e^{(k)} &\equiv x^{(k)} - a = d\ \exp(-\beta t^{(k)}) \\ e^{(k+1)} &\equiv x^{(k+1)} - a = d\ \exp[-\beta(t^{(k)} + \Delta t)] \end{aligned} \tag{1.7.6}$$

and set the ratio of the right-hand sides equal to the ratio of the left-hand sides to obtain relation (1.7.2) with

$$m = 1, \qquad c = \exp(-\beta \Delta t) \tag{1.7.7}$$

Thus *an exponentially decaying function yields a linearly converging time series*, and this explains why the Aitken extrapolation is sometimes called *exponential extrapolation*. The use of formula (1.7.4) is therefore justified.

As an example, we select $a = 1, d = 1, \beta = 1, t^{(0)} = 0$, and $\Delta t = 1$ and generate three terms in the time series, $x^{(0)} = 2, x^{(1)} = 1.3679, x^{(2)} = 1.1353$. Evaluating the right-hand side of equation (1.7.4) for $k = 1$ yields $a^{(1)} = 1$, which is identical to the exact value! Δt does not even have to be small.

Aitken extrapolation theorem

A more general statement can be made about the Aitken extrapolation of a linearly converging sequence. Let the error of the converging sequence $x^{(k)}$ behave like

$$e^{(k+1)} = (c + \delta^{(k)})e^{(k)} \tag{1.7.8}$$

where $|c| < 1$, and the sequence $\delta^{(k)}$ tends to zero as k becomes larger. Note that this is a slight generalization of the behavior shown in equation (1.7.2) with $m = 1$. Then the sequence $a^{(k)}$ computed from equation (1.7.4) converges faster to the limit than $x^{(k)}$ (Problem 1.7.2).

Performing the Richardson extrapolation

It is illuminating to make a connection between the Aitken extrapolation and the Richardson extrapolation discussed in Section 1.6. Let us assume that the systematic numerical error is describable by equation (1.6.5), and let us perform a series of computations with the values of h shown in equations (1.6.7), to obtain the corresponding values b_i. Next, we consider the sequence $x^{(k)} \equiv b_k$ and compute the associated error

$$e^{(k)} \equiv x^{(k)} - a \cong c_1 \left(\frac{\varepsilon}{q^{k-1}} \right)^{p_1} \tag{1.7.9}$$

which clearly satisfies the linear-convergence law

$$e^{(k+1)} = q^{-p_1} e(k) \tag{1.7.10}$$

It is reassuring to observe that producing a from the first of equations (1.7.3) with $c = q^{-p_1}$ *is equivalent to performing the first stage of Richardson extrapolation*, as discussed in Section 1.6.

PROBLEMS

1.7.1 *A sequence with linear rate of convergence.*

Set $x^{(0)} = 0$ and compute $x^{(1)}, x^{(2)}, \ldots$ using the formula $x^{(k+1)} = 1 - 0.2\, x^{(k)^3}$. Verify *a posteriori* that the convergence is linear, apply the Aitken extrapolation formula to estimate the limit, and discuss the benefits of the extrapolation.

1.7.2 *General statement of Aitken extrapolation.*

With reference to the general form of the Aitken extrapolation involving equation (1.7.8), show that, as k becomes larger, the ratio $(a^{(k)} - a)/(x^{(k)} - a)$ tends to vanish; a is the common limit of the sequences $x^{(k)}$ and $a^{(k)}$.

1.7.3 *Summing a series.*

Summation of infinite series arises in several contexts including the approximate evaluation of functions using Taylor expansions, and the computation of field functions induced by a periodic arrangement of singularities such as point charges. In one application, the objective is to compute the infinite sum

$$S = \sum_{i=1}^{\infty} \alpha_i \tag{1.7.11}$$

It is known that as i tends to infinity, the terms α_i are one-signed and decay like i^{-2}; that is, a log–log plot of $|\alpha_i|$ against i produces a straight line with slope equal to -2.
Consider the sequence

$$x^{(k)} = -\tfrac{1}{2}\alpha_{M_k} + \sum_{i=1}^{M_k} \alpha_i \tag{1.7.12}$$

for $k = 1, 2, \ldots$, where $M_k = Np^{k-1}$ and N, p are two arbitrary positive integers. Show that the use of formula (1.7.4) is appropriate, and S may be computed in an expedient fashion as the limit of the sequence $a^{(k)}$. Confirm your arguments by a numerical example of your choice. *Hint*: Use the trapezoidal rule of integration discussed in Section 7.2 to show that the error $e^{(k)} = S - x^{(k)}$ obeys (1.7.2) with $m = 1$.

References

Allen, M. P., and Tildeslay, D. J., 1987, *Computer Simulation of Liquids.* Oxford University Press.
Bodewig, E., 1959, *Matrix Calculus.* North-Holland.
Borse, G. J., 1991, *FORTRAN 77 and Numerical Methods for Engineers.* PWS-Kent.
Catlin, A., 1992, *Standard Basic Programming with True BASIC.* Prentice-Hall.
Chandrasekhar, S., 1943, Stochastic problems in physics and astronomy. *Rev. Modern Phys.* **15**, 1–89.
Cormen, T. H., Leiserson, C. E., and Rivest, R. L., 1990, *Introduction to Algorithms.* McGraw-Hill.
Dahlquist, G., and Björck, Å., 1974, *Numerical Methods.* Prentice-Hall.
Ellis, T. M. R., 1990, *FORTRAN 77 Programming.* Addison-Wesley.
Fosdick, L. D., Jessup, E. R., Schauble, C. J. C., and Domik, G., 1995, *An Introduction to High-Performance Scientific Computing.* MIT Press.
Hayes, J. P., 1988, *Computer Architecture and Organization.* McGraw-Hill.
Hamming, R. W., 1962, *Numerical Methods for Scientists and Engineers.* Reprinted by Dover, 1986.
Kincaid, D., and Cheney, W., 1996, *Numerical Analysis.* Brooks/Cole.
Klafter, J., Shlesinger, M. F., and Zumofen, G., 1996, Beyond Brownian motion. *Phys. Today* **Feb.**, 33–39.
Knuth, D. E., 1968, *The Art of Scientific Computing. Vol. I: Fundamental Algorithms.* Addison-Wesley.
Knuth, D. E., 1969, *The Art of Scientific Computing. Vol. II: Seminumerical Algorithms.* Addison-Wesley.

KNUTH, D. E., 1973, *The Art of Scientific Computing. Vol. III: Sorting and Searching.* Addison-Wesley.
LANCZOS, C., 1988, *Applied Analysis.* Dover.
LI, X., CHARLES, R., and POZRIKIDIS, C., 1996, Shear flow of suspensions of liquid drops. *J. Fluid Mech.* **320**, 395–416.
MATTHEWS, R. A., 1997, The science of Murphy's Law. *Sci. Am.* **Apr.**, 88–91.
MILGRAM, M. S., 1989, Does a point lie inside a polygon? *J. Comp. Phys.* **84**, 134–44.
OLFE, D. B., 1995, *Computer Graphics for Design.* Prentice-Hall.
POZRIKIDIS, C., 1997, *Introduction to Theoretical and Computational Fluid Dynamics.* Oxford University Press.
PRESS, W. H., FLANNERY, B. P., TEUKOLSKY, S. A., and VETTERLING, W. T., 1992, *Numerical Recipes* 2nd ed. Cambridge University Press.
PROBSTEIN, R. F., 1994, *Physicochemical Hydrodynamics.* Wiley.
ROGALLO, R. S., and MOIN, P., 1984, Numerical simulation of turbulent flows. *Annu. Rev. Fluid Mech.* **16**, 99–137.
SCHILDT, H., 1987, *C: The Complete Reference.* McGraw-Hill.
VARGA, R. S., 1993, *Scientific Computation on Mathematical Problems and Conjectures.* SIAM Press.
WILSON, G. V., 1995, *Practical Parallel Programming.* MIT Press.

Chapter 2

Numerical Matrix Algebra and Matrix Calculus

Matrix operations are engines underneath the hood of a broad range of numerical methods to be discussed in subsequent chapters, and their mastering is a fundamental prerequisite of numerical computation.

As with everything else in life, there are two ways of introducing the set of rules that govern arithmetic matrix operations, concisely called *linear algebra.* The first is on a need-to-know basis: Define concepts just before you need them. The second is to treat linear algebra as a distinct introductory topic, such as number theory or differential calculus.

We prefer the second approach for several reasons: It organizes the development of the various concepts; it reveals the main body of linear algebra pulling aside its branches; and it serves as a prelude to advanced numerical computation by means of the relatively new but well-established field of *numerical linear algebra.*

2.1 Matrix Algebra

A matrix is a collection of real or complex numbers that have been arranged in a table in a certain order. A *two-dimensional* matrix $\mathbf{A}$ with M *rows* and N *columns* looks like

$$\mathbf{A} = \begin{bmatrix} A_{1,1} & A_{1,2} & \cdots & \cdots & A_{1,N-1} & A_{1,N} \\ A_{2,1} & A_{2,2} & \cdots & \cdots & A_{2,N-1} & A_{2,N} \\ \cdots & \cdots & \cdots & \cdots & \cdots & \cdots \\ A_{M-1,1} & A_{M-1,2} & \cdots & \cdots & A_{M-1,N-1} & A_{M-1,N} \\ A_{M,1} & A_{M,2} & \cdots & \cdots & A_{M,N-1} & A_{M,N} \end{bmatrix} \tag{2.1.1}$$

The dimensions of this matrix are $M \times N$. A 1×1 matrix is simply the real or complex number $A_{1,1}$.

The individual entries of a two-dimensional matrix $\mathbf{A}$ with M rows and N columns are denoted as $A_{i,j}$, where the *subscripts* or *indices* i and j take values in the ranges

$$i = 1, \ldots, M, \qquad j = 1, \ldots, N$$

Note that *the first index designates the number of the host row*, and *the second index designates the number of the host column.*

In order to prevent ambiguities, we have inserted a comma between the two indices. This practice is not followed in texts of theoretical and applied linear algebra. Our reason for doing it, at the expense of longer notation, is to facilitate the translation of equations into computer language instructions where a comma is always inserted between the two indices.

A Matrix Contains Vectors

The matrix displayed in equation (2.1.1) may be regarded either as a collection of M row vectors or as a collection of N column vectors: *A matrix is a device that allows us to handle small or large collections of vectors in an orderly and methodological fashion.*

Storage in Computer Memory

A computer stores the elements of a two-dimensional matrix in its memory bank in a manner that depends on the selected computer language. FORTRAN instructs the computer to store the elements in sequentially numbered addresses by *columns*, whereas Pascal and C instruct storage by *rows*. Accordingly, when programming in FORTRAN, it is expedient to recall the elements in a column-wise rather than a row-wise fashion.

Suppose, for example, that we want to compute the sum of all elements. The following FORTRAN program produces the sum faster than that when the two Do statements are interchanged:

```
Sum = 0
Do j = 1, M
Do i = 1, N
  Sum = Sum + A(i,j)
END Do
END Do
```

(2.1.2)

Tensors

Consider a matrix whose elements are certain physical quantities such as the three components of the velocity of a flying insect; the values of these quantities depend on the coordinate system in which the physical system is described. If the values in one system are related to those in another system by specific relationships that involve the relative positions of the two coordinate systems, as described in standard texts of tensor analysis (e.g., Synge and Schild 1978), then the matrix is a tensor.

Consider, for example, a 1×1 matrix with a single element $A_{1,1}$. If the value of this element remains unchanged when a coordinate system is replaced by another one, then the matrix is a tensor. This is certainly true when $A_{1,1}$ represents the temperature of a star, which is independent of the position from which the star is observed. But if the surface of the star is painted with different colors, then the color observed through a telescope will depend on the location of the telescope, and the 1×1 matrix containing the color function will not be a tensor.

The tensor-like nature of a matrix is not relevant in the context of matrix algebra and matrix calculus where the physical consequences of transformations between coordinate systems are not considered.

PROBLEM

2.1.1 ***A tensor or not a tensor?***

Discuss a 1×1 matrix that is a tensor, and a 1×1 matrix that is not a tensor.

Transpose of a Matrix

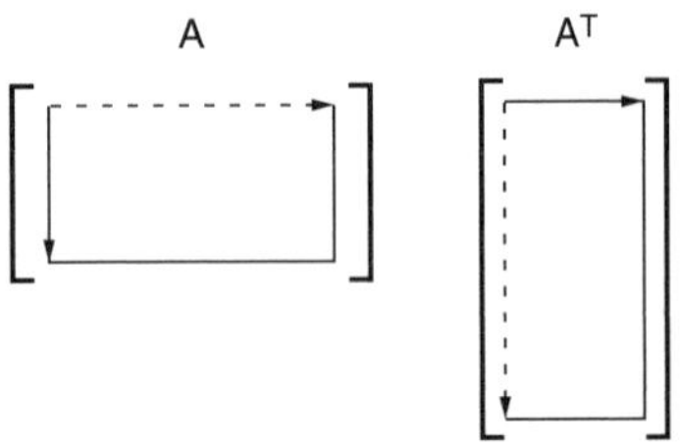

To every matrix $\boldsymbol{A}$ there corresponds another matrix $\boldsymbol{B}$, called the transpose of $\boldsymbol{A}$ and denoted by $\boldsymbol{B} = \boldsymbol{A}^T$, constructed by turning the rows of $\boldsymbol{A}$ into columns, whereupon the columns become rows. Thus, if $\boldsymbol{A}$ is an $M \times N$ matrix, $\boldsymbol{A}^T$ will be an $N \times M$ matrix. If $\boldsymbol{A}$ is a slender horizontal matrix, $\boldsymbol{A}^T$ will be a slender vertical matrix and vice versa, as illustrated in the diagram.

The transpose of the matrix displayed in equation (2.1.1) is

$$\boldsymbol{B} = \boldsymbol{A}^T = \begin{bmatrix} A_{1,1} & A_{2,1} & \cdots & A_{M-1,1} & A_{M,1} \\ A_{1,2} & A_{2,2} & \cdots & A_{M-1,2} & A_{M,2} \\ \cdots & \cdots & \cdots & \cdots & \cdots \\ \cdots & \cdots & \cdots & \cdots & \cdots \\ A_{1,N-1} & A_{2,N-1} & \cdots & A_{M-1,N-1} & A_{M,N-1} \\ A_{1,N} & A_{2,N} & \cdots & A_{M-1,N} & A_{M,N} \end{bmatrix} \tag{2.1.3}$$

Formally, we write

$$B_{i,j} = A^T_{i,j} = A_{j,i} \tag{2.1.4}$$

where $i = 1, \ldots, N$ and $j = 1, \ldots, M$.

The transpose of the matrix $\boldsymbol{A}^T$ is identical to the original matrix $\boldsymbol{A}$. Although this is not difficult to accept, its proof using index notation demonstrates the power of this approach. We simply define $\boldsymbol{B} = \boldsymbol{A}^T$, introduce the intermediate matrix $\boldsymbol{C} = \boldsymbol{B}^T$, and write

$$C_{i,j} = B^T_{i,j} = B_{j,i} = A^T_{j,i} = A_{i,j} \tag{2.1.5}$$

The use of intermediate matrices simplifies the proof of many matrix identities to be discussed in subsequent sections and should be considered a first resort.

Why introduce the transpose? We shall see in later sections that a matrix and its transpose share several important features: Studying the properties of the latter yields important information about those of the former and vice versa. Other more technical reasons will become evident in subsequent sections.

Adjoint of a Matrix

Taking the complex conjugate of a complex matrix $\boldsymbol{A}$, that is, taking the complex conjugate of all elements, and then forming its transpose, produces the new matrix

$$\boldsymbol{A}^A \equiv \boldsymbol{A}^{*T} \tag{2.1.6}$$

which is called the adjoint of $\boldsymbol{A}$; an asterisk designates the complex conjugate. If all elements of $\boldsymbol{A}$ are real, then $\boldsymbol{A}^A = \boldsymbol{A}^T$. The adjoint of a matrix is important in the theoretical analysis of matrices regarded as engines that drive linear mappings, to be discussed in subsequent sections.

An example of a 2×2 matrix and its adjoint is

$$\boldsymbol{A} = \begin{bmatrix} 1+i & i \\ 1+2i & i \end{bmatrix} \qquad \boldsymbol{A}^A = \begin{bmatrix} 1-i & 1-2i \\ -i & -i \end{bmatrix} \tag{2.1.7}$$

where i is the imaginary unit.

Important note on terminology

In classical books of linear algebra, the adjoint of a matrix is called the *Hermitian* and is denoted by $\boldsymbol{A}^H$. Our terminology is consistent with modern conventions of functional analysis concerning linear operators and mappings, as will be discussed at the end of Section 2.2.

Vectors are Slender Matrices

A matrix $\boldsymbol{u}$ with dimensions $M \times 1$ is an M-dimensional *column vector*. For convenience, we designate the components of $\boldsymbol{u}$ with a single index as u_i, where $i = 1, \ldots, M$.

A matrix $\boldsymbol{v}$ with dimensions $1 \times N$ is an N-dimensional *row vector*. For convenience again, we designate the components of this vector using a single index as v_i, where $i = 1, \ldots, N$.

$$\boldsymbol{u} = \begin{bmatrix} u_1 \\ u_2 \\ \vdots \\ u_{M-1} \\ u_M \end{bmatrix} \qquad \boldsymbol{v} = [v_1, v_2, \ldots, v_N]$$

The transpose of a column vector is a row vector and vice versa. It is a standard convention of numerical analysis, and one that we shall observe throughout this book, to regard a vector $\boldsymbol{u}$ *by default* as a column vector, in which case $\boldsymbol{u}^T$ is a row vector. Thus, unless otherwise specified, every time we say *a vector*, we really mean *a column vector.*

$$\boldsymbol{u}^T = [u_1, u_2, \ldots, u_M] \qquad \boldsymbol{v}^T = \begin{bmatrix} v_1 \\ v_2 \\ \vdots \\ v_{N-1} \\ v_N \end{bmatrix}$$

Vectors representing functions

The information encapsulated in a vector may have diverse physical or mathematical meanings. We shall see, in particular, that a vector is routinely used to represent a function in some approximate sense.

For example, the N elements of an N-dimensional vector may represent the values of a function $f(x)$ at N selected values of the independent variable x, or contain the first N coefficients of its Fourier components. In both cases, the representation improves as N becomes larger and the function is described in more detail. Taking the limit as N tends to infinity leads us to the interpretation of a function as a point in an infinite-dimensional vector space.

It is then not surprising that the study of vectors is relevant to, and provides us with insights into, the study of functions. The rule of thumb is: Study the behavior of vectors, and then take the limit as N tends to infinity to recover the behavior of functions (e.g., Kolmogorov and Fomin 1970). While this extension is legitimate in most cases, exceptions do occur when dealing with difficult functions or toilsome mathematical operations.

Matrix Algebra and Matrix Calculus

In order to make the apparatus of matrices useful in practice, we must build an appropriate algebra and develop the apparatus of calculus. We begin with algebra and continue with differential calculus.

Equality

Two matrices $\boldsymbol{A}$ and $\boldsymbol{B}$ are equal provided that they have identical dimensions, and all corresponding elements are equal,

$$A_{i,j} = B_{i,j} \tag{2.1.8}$$

It then follows that equality is *transitive*; that is, if $\boldsymbol{A} = \boldsymbol{B}$ and $\boldsymbol{B} = \boldsymbol{C}$, then $\boldsymbol{A} = \boldsymbol{C}$.

Addition and subtraction

Before attempting to add or subtract two matrices $\boldsymbol{A}$ and $\boldsymbol{B}$, we must make sure that they have matching dimensions, say, $M \times N$. The sum of the two matrices is another matrix $\boldsymbol{S} = \boldsymbol{A} + \boldsymbol{B}$ with the same dimensions $M \times N$, whose elements are found by adding the corresponding elements of $\boldsymbol{A}$ and $\boldsymbol{B}$; that is,

$$S_{i,j} = A_{i,j} + B_{i,j} \tag{2.1.9}$$

It is then evident that $\boldsymbol{S} = \boldsymbol{A} + \boldsymbol{B} = \boldsymbol{B} + \boldsymbol{A}$; that is, the order of addition does not matter, and the *matrices commute with respect to addition.*

Similarly, the difference between the two matrices $\boldsymbol{A}$ and $\boldsymbol{B}$ with matching dimensions $M \times N$ is a new matrix $\boldsymbol{D} = \boldsymbol{A} - \boldsymbol{B}$ with the same dimensions, whose elements are given by

$$D_{i,j} = A_{i,j} - B_{i,j} \tag{2.1.10}$$

The null matrix

If $\boldsymbol{A} \pm \boldsymbol{B} = \boldsymbol{A}$, then the matrix $\boldsymbol{B}$ must be filled up with zeros; that is, it must be the *null matrix* denoted as $\boldsymbol{0}$. Undoubtedly, $\boldsymbol{A} - \boldsymbol{A} = \boldsymbol{0}$.

Multiplication by a scalar

We can multiply a matrix $\boldsymbol{A}$ by a real or complex number α, and the result is the new matrix $\boldsymbol{B} = \alpha \boldsymbol{A}$ whose elements are computed by multiplying the corresponding elements of $\boldsymbol{A}$ by α; that is,

$$B_{i,j} = \alpha\, A_{i,j} \tag{2.1.11}$$

If $\alpha = 0$, then $\boldsymbol{B}$ is the null matrix $\boldsymbol{0}$. If α is a natural number equal to l, then $\boldsymbol{B} = \boldsymbol{A} + \boldsymbol{A} + \ldots + \boldsymbol{A}$, where the addition is repeated l times.

PROBLEM

2.1.2 *Transpose of a linear combination of matrices.*

Prove the identity

$$(\alpha \boldsymbol{A} \pm \beta \boldsymbol{B})^T = \alpha \boldsymbol{A}^T \pm \beta \boldsymbol{B}^T \tag{2.1.12}$$

where α and β are two arbitrary scalars, and $\boldsymbol{A}$, $\boldsymbol{B}$ are two arbitrary matrices with matching dimensions.

Inner vector product

If $\boldsymbol{v}$ and $\boldsymbol{u}$ are two *vertical* N-dimensional vectors, the scalar

$$s = \boldsymbol{v}^T \boldsymbol{u} = \boldsymbol{u}^T \boldsymbol{v} = u_1 v_1 + u_2 v_2 + \ldots + u_N v_N \tag{2.1.13}$$

is their *inner* or *dot product.* Identifying $\boldsymbol{v}$ with $\boldsymbol{u}$ provides us with the square of the *length of the vector* $|\boldsymbol{u}|$, sometimes called the *Euclidean norm,*

inner product of two vectors

$$\boldsymbol{v}^T \boldsymbol{u} = \begin{bmatrix} v_1, & v_2, & \cdots, & v_N \end{bmatrix} \begin{bmatrix} u_1 \\ u_2 \\ \vdots \\ u_N \end{bmatrix} = s$$

$$|\boldsymbol{u}|^2 = \boldsymbol{u}^T \boldsymbol{u} = u_1^2 + u_2^2 + \ldots + u_N^2 \tag{2.1.14}$$

PROBLEM

2.1.3 ***Geometrical interpretation of the inner product.***

Show that the inner product of two two- or three-dimensional vectors $\boldsymbol{v}$ and $\boldsymbol{u}$ can be placed in the form $\boldsymbol{v}^T \boldsymbol{u} = |\boldsymbol{u}||\boldsymbol{v}| \cos \theta$, where θ is the angle subtended between them.

Matrix-Matrix multiplication

Multiplication of two matrices $\boldsymbol{A}$ and $\boldsymbol{B}$, arranged in this particular order, is intended to provide us with information on the relative directions of the vectors that comprise their columns and rows. Specifically, multiplication is defined in terms of the inner products of the horizontal vectors that comprise the rows of $\boldsymbol{A}$ and the vertical vectors that comprise the columns of $\boldsymbol{B}$.

To be able to compute the aforementioned inner products, matrices $\boldsymbol{A}$ and $\boldsymbol{B}$ must have, respectively, dimensions $M \times N$ and $N \times L$; if they do, then they are *conformable.* The product of $\boldsymbol{A}$ and $\boldsymbol{B}$ is another matrix $\boldsymbol{C} = \boldsymbol{A}\boldsymbol{B}$ with dimensions $M \times L$, where $C_{i,j}$ is the inner product of the vectors represented by the ith row of $\boldsymbol{A}$ and the jth column of $\boldsymbol{B}$. Thus, by definition,

$$C_{i,j} = \sum_{k=1}^{N} A_{i,k} B_{k,j} \tag{2.1.15}$$

The process of multiplication is illustrated in Figure 2.1.1.

The matrix $\boldsymbol{C}$ may be regarded as the result of *premultiplication* of $\boldsymbol{B}$ by $\boldsymbol{A}$, or *postmultiplication* of $\boldsymbol{A}$ by $\boldsymbol{B}$.

For example, if

$$\boldsymbol{A} = \begin{bmatrix} 1 & 2 \\ 1 & 3 \end{bmatrix}, \qquad \boldsymbol{B} = \begin{bmatrix} -1 & 1 & 0 \\ 0 & 3 & 2 \end{bmatrix} \tag{2.1.16}$$

then

$$\boldsymbol{C} = \boldsymbol{A}\boldsymbol{B} = \begin{bmatrix} -1 & 7 & 4 \\ -1 & 10 & 6 \end{bmatrix} \tag{2.1.17}$$

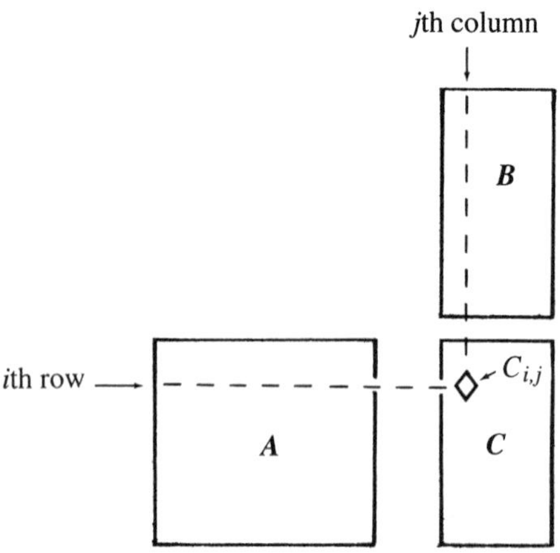

Figure 2.1.1 The product of a matrix $\boldsymbol{A}$ with dimensions $M \times N$, with another matrix $\boldsymbol{B}$ with dimensions $N \times K$, is a new matrix $\boldsymbol{C}$ with dimensions $M \times K$, so that $C_{i,j}$ is the inner product of the vectors represented by the ith row of $\boldsymbol{A}$ and the jth column of $\boldsymbol{B}$.

It is important to note that the existence of the product $\boldsymbol{AB}$ does not guarantee the existence of the product $\boldsymbol{BA}$. The matrices involved in the second product may not even be conformable. If $\boldsymbol{A}$ and $\boldsymbol{B}$ are square matrices with matching dimensions, both $\boldsymbol{AB}$ and $\boldsymbol{BA}$ are defined but are not generally equal. The two matrices $\boldsymbol{AB}$ and $\boldsymbol{BA}$, however, do share certain properties, as will be discussed in later sections. Thus the commutative property of number multiplication does not generally carry over to matrices in a general form. If $\boldsymbol{AB} = \boldsymbol{BA}$, the matrices $\boldsymbol{A}$ and $\boldsymbol{B}$ are said to *commute* with respect to multiplication, or simply commute.

Furthermore, unlike in number multiplication, if $\boldsymbol{AB} = \boldsymbol{0}$, where $\boldsymbol{0}$ is the null matrix, it is not necessarily true that either $\boldsymbol{A}$ or $\boldsymbol{B}$ must be null (Problem 2.1.5).

Additional properties of matrix–matrix multiplication will be discussed after we have introduced the repeated-index summation convention in the next subsection.

PROBLEMS

2.1.4 ***Rows and columns of the product.***

Show that the columns of the matrix $\boldsymbol{AB}$ are linear combinations of the columns of $\boldsymbol{A}$, and its rows are linear combinations of the rows of $\boldsymbol{B}$. What are the coefficients in these linear combinations?

2.1.5 ***The product of two nonnull matrices can be the null matrix.***

Devise two nonnull 2×2 matrices whose product is the null matrix $\boldsymbol{0}$.

Repeated-index summation convention

To avoid using the cumbersome summation symbol, we introduce the *Einstein summation convention,* which requires that *if the same index appears twice in an index array or across index arrays involved in a product, then summation is implied over that index over its range.* Under this convention, the right-hand side of equation (2.1.15) may be simply written as $A_{i,k} B_{k,j}$.

Properties of Matrix–Matrix multiplication

The rules of matrix multiplication work out so that

$$(\boldsymbol{AB})^T = \boldsymbol{B}^T\boldsymbol{A}^T \tag{2.1.18}$$

If the dimensions of $\boldsymbol{A}$ are $M \times N$ and the dimensions of $\boldsymbol{B}$ are $N \times L$, then both sides of this equation represent a matrix with dimensions $L \times N$. To prove this identity, we denote the left-hand side as $\boldsymbol{C}$ and the right-hand side as $\boldsymbol{D}$. Then, by definition, $C_{i,j} = (A_{i,k}B_{k,j})^T = A_{j,k}B_{k,i}$ and $D_{i,j} = B^T_{i,k}A^T_{k,j} = B_{k,i}A_{j,k}$, which completes the proof. The use of intermediate matrices has simplified once again a seemingly difficult proof.

Multiplication satisfies (a) the *associative* property

$$\boldsymbol{A}\,(\boldsymbol{BC}) = (\boldsymbol{AB})\,\boldsymbol{C} \tag{2.1.19}$$

which allows us to denote the product of three arbitrary matrices $\boldsymbol{A}$, $\boldsymbol{B}$, and $\boldsymbol{C}$ simply as $\boldsymbol{ABC}$, and (b) the *distributive* property

$$\boldsymbol{A}(\boldsymbol{B} + \boldsymbol{C}) = \boldsymbol{AB} + \boldsymbol{AC} \tag{2.1.20}$$

These properties can be proved readily using index notation and working with intermediate matrices as we have done in proving identity (2.1.18).

Matrix–Vector multiplication

We can premultiply a *column vector* $\boldsymbol{u}$ with dimensions $N \times 1$ by a matrix $\boldsymbol{A}$ with dimensions $M \times N$, and the result will be the new *column vector* $\boldsymbol{v} = \boldsymbol{Au}$ with dimensions $M \times 1$ whose components are given by $v_i = A_{i,j}u_j$. The vector $\boldsymbol{v}$ is a linear combination of the column vectors of $\boldsymbol{A}$; the coefficients of the linear combination are the corresponding entries of $\boldsymbol{u}$.

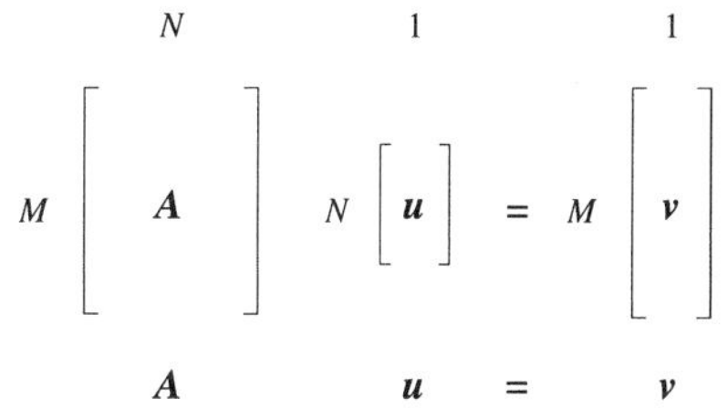

Physically, the matrix $\boldsymbol{A}$ represents a process, and the column vector $\boldsymbol{u}$ represents a set of physical or initial conditions. Multiplying the matrix by the vector produces a new vector $\boldsymbol{v}$ that expresses the action of the process on the physical or initial conditions. For instance, the vector $\boldsymbol{u}$ may contain the operating conditions of a petroleum refinery, and the vector $\boldsymbol{v}$ may contain the outlet conditions, such as the concentrations of a collection of chemical species in a mixture. In another example, the vector $\boldsymbol{u}$ may contain the wind velocity at a sequence of elevations in the atmosphere at a particular time, and the vector $\boldsymbol{v}$ may contain the corresponding velocities at a subsequent time.

As a third example, we consider the sequential rotation of an object about the x_1, x_2, and x_3 Cartesian axes, with respective angles equal to φ_1, φ_2, and φ_3; the first rotation is shown in the diagram. The Cartesian coordinates of a particular point on the object after rotation, contained in the column vector $\boldsymbol{y}$, are related to those before rotation, contained in the column vector $\boldsymbol{x}$, by the relation $\boldsymbol{y} = \boldsymbol{Ax}$ where $\boldsymbol{A} = \boldsymbol{R}^{(1)}\boldsymbol{R}^{(2)}\boldsymbol{R}^{(3)}$, and $\boldsymbol{R}^{(1)}, \boldsymbol{R}^{(2)}, \boldsymbol{R}^{(3)}$ are three *rotation matrices* defined as

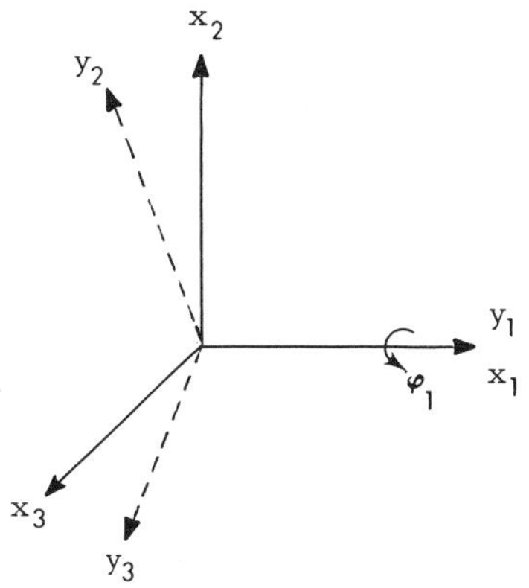

$$
\boldsymbol{R}^{(1)} = \begin{bmatrix} 1 & 0 & 0 \\ 0 & \cos\varphi_1 & -\sin\varphi_1 \\ 0 & \sin\varphi_1 & \cos\varphi_1 \end{bmatrix}, \qquad \boldsymbol{R}^{(2)} = \begin{bmatrix} \cos\varphi_2 & 0 & \sin\varphi_2 \\ 0 & 1 & 0 \\ -\sin\varphi_2 & 0 & \cos\varphi_2 \end{bmatrix}
$$

$$
\boldsymbol{R}^{(3)} = \begin{bmatrix} \cos\varphi_3 & -\sin\varphi_3 & 0 \\ \sin\varphi_3 & \cos\varphi_3 & 0 \\ 0 & 0 & 1 \end{bmatrix} \tag{2.1.21}
$$

One may readily verify that $\boldsymbol{R}^{(m)}(\varphi_m = \alpha)\boldsymbol{R}^{(m)}(\varphi_m = \beta) = \boldsymbol{R}^{(m)}(\varphi_m = \alpha + \beta)$, in agreement with geometrical intuition regarding sequential rotation.

$$
\underset{\boldsymbol{v}^T}{[\, v_1, v_2, \ldots, v_N \,]} \underset{\boldsymbol{A}}{\begin{bmatrix} A_{1,1} & A_{1,2} & \cdots & A_{1,M} \\ A_{2,1} & \cdots & \cdots & \cdots \\ \cdots & \cdots & \cdots & \cdots \\ \cdots & \cdots & \cdots & \cdots \\ \cdots & \cdots & \cdots & \cdots \\ A_{N,1} & \cdots & \cdots & A_{N,M} \end{bmatrix}} = \underset{\boldsymbol{u}^T}{[\, u_1, u_2, \ldots, u_M \,]}
$$

We can also postmultiply any *row vector* $\boldsymbol{v}^T$ with dimensions $1 \times N$ by any matrix $\boldsymbol{A}$ with dimensions $N \times M$, and this produces the new row vector $\boldsymbol{u}^T = \boldsymbol{v}^T\boldsymbol{A}$ with dimensions $1 \times M$, whose components are given by $u_j = v_i A_{i,j}$. The row vector $\boldsymbol{u}^T$ is a linear combination of the row vectors of $\boldsymbol{A}$; the coefficients multiplying the row vectors are the entries of $\boldsymbol{v}$. The physical interpretation of this operation is similar to that of column-vector multiplication discussed in a previous paragraph.

PROBLEMS

2.1.6 *Computer graphics: rotation of an object.*

Consider a unit cube with one vertex at the origin and three sides along the three Cartesian axes. Rotate the cube by an angle of 25° about the x_3 axis, and then by an angle of 25° about the x_2 axis. Plot the sides of the cube before and after the two rotations.

2.1.7 *Physical action of the product of two matrices.*

Discuss the physical interpretation of the product of two matrices, where each matrix represents a distinct physical process.

2.1.8 *Successive mappings.*

Given an N-dimensional vector $\boldsymbol{v}^{(0)}$ and a square matrix $\boldsymbol{A}$ of size $N \times N$, let us compute a sequence of vectors by the *successive mappings*

$$
\boldsymbol{v}^{(k+1)} = \frac{1}{|\boldsymbol{A}\boldsymbol{v}^{(k)}|}\boldsymbol{A}\boldsymbol{v}^{(k)} \tag{2.1.22}
$$

where $|\boldsymbol{w}|$ designates the length of the vector $\boldsymbol{w}$. In practice, this mapping is used to generate numerical solutions to the differential equations of mathematical physics using a variety of numerical methods. What happens to the sequence of vectors $(\boldsymbol{v}^{(0)}, \boldsymbol{v}^{(1)}, \boldsymbol{v}^{(2)}, \ldots)$ as we continue the mappings? Does this sequence of vectors tend to a certain limiting vector?

To develop insights by computational experimentation, consider the matrix

$$A = \begin{bmatrix} 0.5 & 0.4 & 0.3 & 0.2 & 0.1 \\ 0.1 & 0.2 & 0.7 & 0.1 & 0.8 \\ 0.1 & 0.2 & 0.0 & 0.1 & 0.0 \\ 0.2 & 0.1 & 0.0 & 0.1 & 0.0 \\ 0.1 & 0.0 & 0.0 & 0.5 & 0.1 \end{bmatrix} \tag{2.1.23}$$

Compute three sequences $(\boldsymbol{v}^{(0)}, \ldots, \boldsymbol{v}^{(50)})$ corresponding to the three different starting column vectors $\boldsymbol{v}^{(0)} = (1, 0, 0, 0, 0)^T$, $(0, 1, 0, 0, 0)^T$, and $(0, 0, 1, 0, 0)^T$. Do you find that any of these sequences tend to a limiting vector? If yes, are the limiting vectors identical, that is, do they depend on the choice of $\boldsymbol{v}^{(0)}$?

Double-dot matrix product

The double-dot product of two matrices $\boldsymbol{A}$ and $\boldsymbol{B}$ with matching dimensions is a scalar quantity defined as

$$s = \boldsymbol{A} : \boldsymbol{B} \equiv A_{i,j} B_{i,j} \tag{2.1.24}$$

Summation is implied over the repeated indices i and j. The use of the double-dot product facilitates the notation in theoretical manipulations involving matrix operations.

Gram–Schmidt Orthogonalization

An important problem in the theory and practice of numerical computation concerns the construction of a set of N *orthogonal* M-dimensional vectors

$$\boldsymbol{u}^{(i)}, i = 1, \ldots, N \tag{2.1.25}$$

from an arbitrary set of N linearly independent M-dimensional vectors

$$\boldsymbol{v}^{(i)}, i = 1, \ldots, N \tag{2.1.26}$$

For simplicity, we shall assume that all vectors are real. This construction plays an important role in the theory of orthogonal polynomials discussed in Appendix B.

According to the Gram–Schmidt orthogonalization process, the orthogonal set (2.1.25) is constructed by working as follows:

1. We begin by setting $\boldsymbol{u}^{(1)} = \boldsymbol{v}^{(1)}$.
2. We require that $\boldsymbol{u}^{(2)}$ lie in the plane of $\boldsymbol{u}^{(1)}$ and $\boldsymbol{v}^{(2)}$ and is orthogonal to $\boldsymbol{u}^{(1)}$.
3. We require that $\boldsymbol{u}^{(3)}$ lie in the space of $\boldsymbol{u}^{(1)}$, $\boldsymbol{u}^{(2)}$, and $\boldsymbol{v}^{(3)}$ and is orthogonal to $\boldsymbol{u}^{(1)}$ and $\boldsymbol{u}^{(2)}$.

   ⋮

We continue in this manner until the desired set is complete, following the steps of Algorithm 2.1.1.

One may readily verify that, with the definitions shown in Algorithm 2.1.1, $\boldsymbol{u}^{(i)T}\boldsymbol{u}^{(j)} = 0$ when $i \neq j$.

A straightforward modification of the basic Gram–Schmidt algorithm produces a set of *orthonormal* vectors

$$w^{(i)}, i = 1, \ldots, N \tag{2.1.27}$$

whose lengths are equal to unity, according to Algorithm 2.1.2. That is, $w^{(i)T}w^{(j)} = 0$ when $i \neq j$, and $w^{(i)T}w^{(j)} = 1$ when $i = j$.

ALGORITHM 2.1.1 Gram–Schmidt algorithm for generating a set of M-dimensional *orthogonal* vectors $u^{(i)}, i = 1, \ldots, N$, from an arbitrary set of linearly independent M-dimensional vectors $v^{(i)}, i = 1, \ldots, N$.

$$u^{(1)} = v^{(1)}$$

Do $j = 2, N$

$$u^{(j)} = v^{(j)} - \sum_{i=1}^{j-1} \alpha_{i,j} u^{(i)}$$

where $\alpha_{i,j} = \dfrac{u^{(i)^T} v^{(j)}}{u^{(i)^T} u^{(i)}}$

END DO

ALGORITHM 2.1.2 Gram–Schmidt algorithm for generating a set of M-dimensional *orthonormal* vectors $w^{(i)}, i = 1, \ldots, N$, from an arbitrary set of linearly independent M-dimensional vectors $v^{(i)}, i = 1, \ldots, N$.

$$w^{(1)} = v^{(1)}, \beta_1 = |v^{(1)}|, w^{(1)} = \frac{v^{(1)}}{\beta_1}$$

Do $j = 2, N$

$$w^{(j)} = v^{(j)} - \sum_{i=1}^{j-1} (w^{(i)^T} v^{(j)}) w^{(i)}$$

$$\beta_j = |w^{(j)}|$$

$$w^{(j)} \leftarrow \frac{w^{(j)}}{\beta_j}$$

END DO

PROBLEM

2.1.9 *Orthogonalization of the columns of a matrix.*

Identify the vectors $v^{(i)}$ with the five vectors comprising the columns of the matrix given in equation (2.1.23). Use the Gram–Schmidt method to generate (*a*) an orthogonal and (*b*) an orthonormal set.

Matrix Functions and Their Derivatives

Let us assume that the elements of a matrix A are functions of the independent variable t. In order to signify this dependence, we write $A(t)$. If we change t by a small amount dt, the elements of A will change by the corresponding differential amount $dA_{i,j}$. The derivative dA/dt is a new matrix B whose elements are given by $B_{i,j} = dA_{i,j}/dt$. Partial derivatives of matrices whose elements are functions of more than one variable are defined in a similar fashion.

2.2 *Square Matrices*

When the vertical and horizontal dimensions of an $M \times N$ matrix are equal, $M = N$, we obtain a square matrix that can fit into a square box. A 1×1 square matrix is simply a number.

The diagonal line containing the elements $A_{1,1}, A_{2,2}, \ldots, A_{N,N}$ is called the *principal diagonal* line or simply the *diagonal*. The diagonal line located immediately above the diagonal is called the *first superdiagonal*, and the diagonal line located immediately below the diagonal is called the *first subdiagonal*; the qualifier *first* is usually omitted. The diagonal line passing through the elements $A_{1,N}$ and $A_{N,1}$ is called the *back-diagonal*.

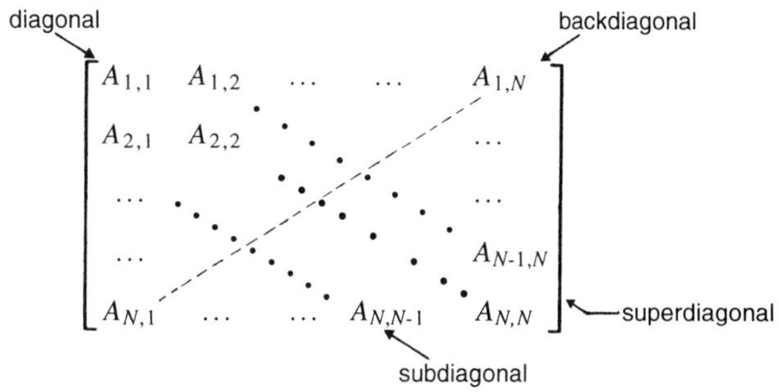

We can check for equality and compute the sum, the difference, and the product of any two square matrices $\boldsymbol{A}$ and $\boldsymbol{B}$ with same dimensions. Recall, however, that the product $\boldsymbol{AB}$ will not necessarily be equal to $\boldsymbol{BA}$. If it is, the matrices *commute with respect to multiplication*, or simply commute.

Powers of a Matrix

We can multiply a square matrix $\boldsymbol{A}$ by itself p times—the associative property of multiplication guarantees that the order of the pairwise multiplications is immaterial—and the result is a new square matrix denoted by $\boldsymbol{A}^p$. As a consequence of this definition,

$$\boldsymbol{A}^p \boldsymbol{A}^q = \boldsymbol{A}^{p+q} \tag{2.2.1}$$

where p and q are natural numbers. Using identity (2.1.18), we find

$$(\boldsymbol{A}^p)^T = (\boldsymbol{A}^T)^p \tag{2.2.2}$$

which ensures that we can either do the multiplications first and then take the transpose, or we can take the transpose first and then do the multiplications.

The most economical method of computing the matrix $\boldsymbol{B} = \boldsymbol{A}^p$ relies on the binary representation of the integer exponent p. We express p in the binary form $b_k\, 2^k + \ldots + b_0\, 2^0$, where the binary digits b_i take the value 0 or 1, and then write

$$\boldsymbol{A}^p = \boldsymbol{A}^{b_k 2^k + \ldots + b_2 2^2 + b_1 2^1 + b_0 2^0} = (\boldsymbol{A}^{2^k})^{b_k} \ldots (\boldsymbol{A}^{2^2})^{b_2} (\boldsymbol{A}^{2^1})^{b_1} (\boldsymbol{A}^{2^0})^{b_0} \tag{2.2.3}$$

The ith factor on the right-hand side of equation (2.2.3) is equal to $\boldsymbol{I}$, if $b_i = 0$, or $\boldsymbol{A}^{2^i}$, if $b_i = 1$. Accordingly, we successively square the matrix $\boldsymbol{A}$ a number of k times, while multiplying with each other only the powers whose corresponding binary digit b_i is equal to unity, thereby forming $\boldsymbol{A}^p$.

For example, when $p = 9$, we set $b_0 = 1, b_1 = 0, b_2 = 0, b_3 = 1$, compute $\boldsymbol{A}^2, \boldsymbol{A}^4, \boldsymbol{A}^8$, and then multiply $\boldsymbol{A}^9 = \boldsymbol{A}\boldsymbol{A}^8$.

The general algorithm is (Golub and van Loan 1989, p. 552):

```
    C = A                         C is an auxiliary matrix used to square A

    j = 0         Will begin multiplying when the first binary digit is equal to 1
1   IF (b_j = 0) THEN
       C ← C^2
       j ← j + 1
       GO TO 1
    END IF
                                                                                  (2.2.4)
    B = C                                          Begin the multiplications

    DO i = j + 1, k
    C ← C^2
       IF (b_i ≠ 0) B ← BC
    END DO
    A^p = B
```

As in the case of numbers, we can allow p to have a real positive value. When p has the positive fractional value r/q, where r and q are two natural numbers—that is, p is a rational number—the requisite matrix $\boldsymbol{B} = \boldsymbol{A}^{r/q}$ satisfies the equation $\boldsymbol{B}^q = \boldsymbol{A}^r$. When p is an irrational number, which is not expressible as the ratio of two integers, the matrix $\boldsymbol{A}^p$ can be interpreted as the limit of a sequence of matrices $\boldsymbol{A}^{r/q}$, arising as the rational number r/q tends to the irrational number p. Methods of computing roots and fractional powers of a matrix will be discussed at the end of this chapter.

Symmetric and Skew-symmetric Matrices

$$\begin{bmatrix} A_{1,1} & A_{1,2} & \cdots & \cdots & A_{1,N} \\ A_{2,1} & A_{2,2} & \cdots & A_{i,j} & \cdots \\ \cdots & \cdots & \cdots & \cdots & \cdots \\ \cdots & A_{j,i} & \cdots & \cdots & A_{N-1,N} \\ A_{N,1} & \cdots & \cdots & A_{N,N-1} & A_{N,N} \end{bmatrix}$$

Symmetric matrix:

$A_{i,j} = A_{j,i}$

$\boldsymbol{A}^T = \boldsymbol{A}$

If the elements of a square matrix have a reflective type of symmetry with respect to the diagonal line, that is, $A_{i,j} = A_{j,i}$, then $\boldsymbol{A} = \boldsymbol{A}^T$ and $\boldsymbol{A}$ is *symmetric*.

It can be shown readily using index notation that all three matrices $\boldsymbol{B}+\boldsymbol{B}^T$, $\boldsymbol{B}\boldsymbol{B}^T$, and $\boldsymbol{B}^T\boldsymbol{B}$ are symmetric independently of the structure of the matrix $\boldsymbol{B}$.

Antisymmetric or skew-symmetric matrix:

$A_{i,j} = -A_{j,i}$

$\boldsymbol{A}^T = -\boldsymbol{A}$

If, in contrast, $A_{i,j} = -A_{j,i}$, then $\boldsymbol{A} = -\boldsymbol{A}^T$, and $\boldsymbol{A}$ is *antisymmetric* or *skew-symmetric*. The diagonal elements of such a matrix must necessarily vanish.

It can be shown readily using index notation that the matrix $\boldsymbol{B}-\boldsymbol{B}^T$ is skew-symmetric independently of the structure of the matrix $\boldsymbol{B}$.

PROBLEMS

2.2.1 ***Product of symmetric matrices.***

Let $\boldsymbol{A}$ and $\boldsymbol{B}$ be two symmetric $N \times N$ matrices. Show that a necessary and sufficient condition for the matrix $\boldsymbol{AB}$ to also be symmetric is that $\boldsymbol{A}$ and $\boldsymbol{B}$ commute.

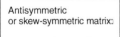

2.2.2 ***Any matrix can be split into a symmetric and a skew-symmetric component.***

An arbitrary matrix $\boldsymbol{A}$ can be decomposed into a symmetric and a skew-symmetric part given, respectively, by $\frac{1}{2}(\boldsymbol{A}+\boldsymbol{A}^T)$ and $\frac{1}{2}(\boldsymbol{A}-\boldsymbol{A}^T)$. Show that this decomposition is unique.

2.2.3 ***Rigid-body rotation.***

The velocity of a fluid at a point $\boldsymbol{x}$ is given by $\boldsymbol{u} = \boldsymbol{A}\boldsymbol{x}$, where $\boldsymbol{A}$ is a 3×3 matrix. Show that when $\boldsymbol{A}$ is a skew-symmetric matrix with constant elements, then the fluid executes rigid-body rotation about the origin. Compute the angular velocity of rotation in terms of $\boldsymbol{A}$ (e.g., Pozrikidis 1997, Chapter 1).

Hermitian or Self-adjoint Matrices

Consider a complex matrix $\boldsymbol{A}$ and decompose it into its real and imaginary parts, writing

$$\boldsymbol{A} = \boldsymbol{A}_R + I\boldsymbol{A}_I, \tag{2.2.5}$$

where $\boldsymbol{A}_R$ and $\boldsymbol{A}_I$ are two real matrices, and I is the imaginary unit. If $\boldsymbol{A}_R$ is symmetric and $\boldsymbol{A}_I$ is antisymmetric, then $\boldsymbol{A}$ is called *Hermitian* or *self-adjoint*. This is another way of saying that a Hermitian matrix is equal to its adjoint, that is, $\boldsymbol{A} = \boldsymbol{A}^{*T} = \boldsymbol{A}^A$, where an asterisk designates the complex conjugate. The diagonal elements of a Hermitian matrix must thus be real.

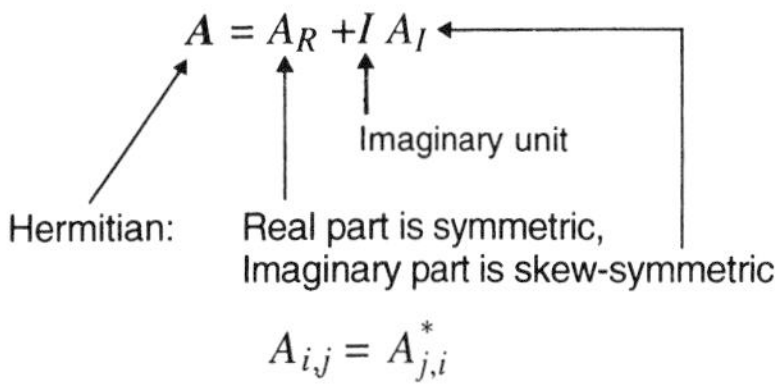

It can readily be seen that (*a*) any real and symmetric matrix is Hermitian, and (*b*) the matrices $\boldsymbol{B}^A\boldsymbol{B}$ and $\boldsymbol{B}\boldsymbol{B}^A$ are Hermitian independently of the structure of the generally complex matrix $\boldsymbol{B}$.

The physical significance of a Hermitian matrix will be discussed at the end of this section, where the terminology *self-adjoint* will be explained. In subsequent sections, we shall see that Hermitian matrices enjoy certain distinctive properties that make them desirable in several classes of numerical computation.

Matrices with a Special Structure

Square matrices with a special structure interest us for two reasons: They occur frequently in scientific and engineering applications, as will be seen in forthcoming sections, and they provide us with devices for carrying out various matrix alterations.

Diagonal matrices

A matrix $\boldsymbol{D}$ whose elements are equal to zero, except for the N diagonal elements that are not necessarily equal to zero, is called *diagonal.*

Postmultiplying an arbitrary matrix $\boldsymbol{A}$ by $\boldsymbol{D}$ produces a new matrix whose ith column is equal to the ith column of $\boldsymbol{A}$ multiplied by the corresponding element $D_{i,i}$; summation is not implied over i. Premultiplication has a similar effect on the rows.

Diagonal matrix:

$$\boldsymbol{D} = \begin{bmatrix} D_{1,1} & 0 & \dots & \dots & 0 \\ 0 & D_{2,2} & 0 & \dots & \dots \\ \dots & 0 & \ddots & 0 & \dots \\ \dots & \dots & 0 & D_{N-1,N-1} & 0 \\ 0 & \dots & \dots & 0 & D_{N,N} \end{bmatrix}$$

Back-diagonal matrices

Back-diagonal matrix:

$$Y = \begin{bmatrix} 0 & \cdots & \cdots & 0 & Y_{1,N} \\ \cdots & \cdots & 0 & Y_{2,N-1} & 0 \\ \cdots & 0 & \cdot^{\cdot^{\cdot}} & 0 & \cdots \\ 0 & Y_{N-1,2} & 0 & \cdots & \cdots \\ Y_{N,1} & 0 & \cdots & \cdots & 0 \end{bmatrix}$$

A matrix $\boldsymbol{Y}$ whose elements are equal to zero, except for the elements along the back-diagonal $Y_{1,N}, Y_{2,N-1}, \ldots, Y_{N,1}$ that are not necessarily equal to zero, is called *back-diagonal.*

The transpose of $\boldsymbol{Y}$ is another back-diagonal matrix whose elements run in reverse order with respect to the elements of $\boldsymbol{Y}$. Postmultiplying or premultiplying an arbitrary matrix $\boldsymbol{A}$ by $\boldsymbol{Y}$ produces a new matrix that is constructed by multiplying the ith column or row of $\boldsymbol{A}$ by the corresponding element $Y_{i,i}$, for all values of i, and then reversing the order of the columns or rows.

Super- and subdiagonal matrices

A matrix $\boldsymbol{S}$ whose elements are equal to zero, except for the elements along the superdiagonal line $S_{1,2}, S_{2,3}, \ldots, S_{N-1,N}$ that are not necessarily equal to zero, is called *superdiagonal.*

Superdiagonal matrix:

$$S = \begin{bmatrix} 0 & S_{1,2} & 0 & \cdots & 0 \\ 0 & 0 & S_{2,3} & 0 & \cdots \\ \cdots & 0 & 0 & \ddots & 0 \\ \cdots & \cdots & 0 & 0 & S_{N-1,N} \\ 0 & \cdots & \cdots & 0 & 0 \end{bmatrix}$$

Subdiagonal matrix:

$$Z = \begin{bmatrix} 0 & 0 & \cdots & \cdots & 0 \\ Z_{2,1} & 0 & 0 & \cdots & \cdots \\ \cdots & \ddots & 0 & 0 & \cdots \\ \cdots & \cdots & Z_{N-1,N-2} & 0 & 0 \\ 0 & \cdots & \cdots & Z_{N,N-1} & 0 \end{bmatrix}$$

Correspondingly, a matrix $\boldsymbol{Z}$ whose elements are equal to zero, except for the elements along the subdiagonal line $Z_{2,1}, Z_{3,2}, \ldots, Z_{N,N-1}$ that are not necessarily equal to zero, is called *subdiagonal.*

The transpose of $\boldsymbol{S}$ is a $\boldsymbol{Z}$, and the transpose of $\boldsymbol{Z}$ is an $\boldsymbol{S}$. All elements of the matrices $\boldsymbol{S}^p$ and $\boldsymbol{Z}^p$, where p is an integer, are equal to zero, except for those along the pth superdiagonal or subdiagonal line. Consequently, $\boldsymbol{S}^p = \boldsymbol{0}$ and $\boldsymbol{Z}^p = \boldsymbol{0}$ for $p \geq N$.

PROBLEM

2.2.4 ***Action of superdiagonal and subdiagonal matrices.***

Describe the result of postmultiplying or premultiplying a matrix by a superdiagonal or a subdiagonal matrix.

Tridiagonal and banded matrices

Tridiagonal matrix:

$$T = \begin{bmatrix} T_{1,1} & T_{1,2} & 0 & \cdots & 0 \\ T_{2,1} & T_{2,2} & T_{2,3} & 0 & \cdots \\ 0 & T_{3,2} & T_{3,3} & T_{3,4} & 0 \\ \cdots & 0 & \ddots & \ddots & \ddots \\ 0 & \cdots & 0 & T_{N,N-1} & T_{N,N} \end{bmatrix}$$

All elements of a *tridiagonal* matrix $\boldsymbol{T}$ are equal to zero, except for the elements along the diagonal, the superdiagonal, and the subdiagonal line that are not necessarily equal to zero. The number of nonzero elements is thus equal to $N + 2(N-1) = 3N - 2$. The transpose of a tridiagonal matrix is another tridiagonal matrix.

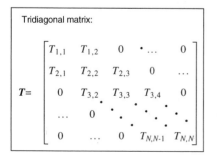

Tridiagonal matrices belong to the broader family of banded matrices. All elements of these matrices are equal to zero, with the possible exception of the elements along the diagonal line, the k diagonal lines above the diagonal, and the k diagonal lines below the diagonal. The number $l = k + 1$ is called the *semi-bandwidth* of the matrix, and the number $p = 2k + 1$ is called the *bandwidth*. A tridiagonal matrix corresponds to $k = 1, l = 2, p = 3$.

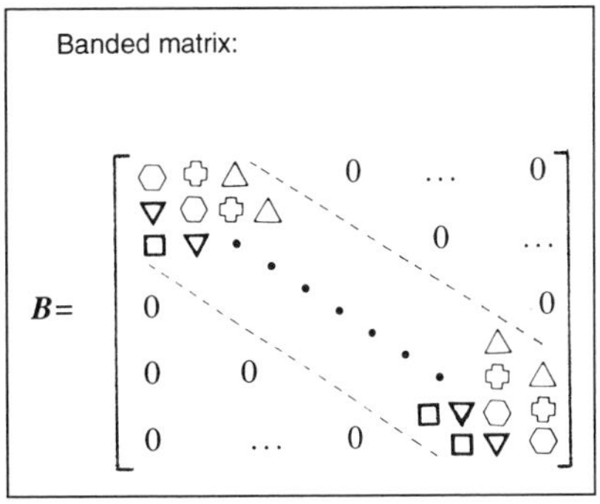

Triangular matrices

All elements of a *lower triangular matrix* $\boldsymbol{L}$ residing above the diagonal line are equal to zero, but the elements below and along the diagonal are not necessarily equal to zero.

Lower triangular matrix:

$$\boldsymbol{L} = \begin{bmatrix} L_{1,1} & 0 & \cdots & \cdots & 0 \\ L_{2,1} & L_{2,2} & 0 & \cdots & \cdots \\ \cdots & \cdots & \ddots & 0 & \cdots \\ \cdots & \cdots & \cdots & L_{N-1,N-1} & 0 \\ L_{N,1} & L_{N,2} & L_{N,3} & \cdots & L_{N,N} \end{bmatrix}$$

Upper triangular matrix:

$$\boldsymbol{U} = \begin{bmatrix} U_{1,1} & U_{1,2} & U_{1,3} & \cdots & U_{1,N} \\ 0 & U_{2,2} & U_{2,3} & \cdots & U_{2,N} \\ \cdots & 0 & \ddots & \cdots & \cdots \\ \cdots & \cdots & 0 & U_{N-1,N-1} & U_{N-1,N} \\ 0 & \cdots & \cdots & 0 & U_{N,N} \end{bmatrix}$$

Correspondingly, all elements of an *upper triangular matrix* $\boldsymbol{U}$ residing below the diagonal are equal to zero, but the elements above and along the diagonal are not necessarily equal to zero.

The number of nonzero elements of a triangular matrix is thus equal to $1+2+\ldots+N = \frac{1}{2}N(N+1)$. The transpose of $\boldsymbol{L}$ or $\boldsymbol{U}$ is, respectively, a $\boldsymbol{U}$ and an $\boldsymbol{L}$.

The product of two lower or upper triangular matrices is another lower triangular or upper triangular matrix. But the products $\boldsymbol{LU}$ and $\boldsymbol{UL}$ do not necessarily have a particular structure.

If the diagonal elements of a triangular matrix vanish, then the matrix is *strictly triangular*. If $\boldsymbol{M}$ is a strictly upper triangular matrix, then all elements of the matrix $\boldsymbol{M}^p$, where p is a positive integer, are equal to zero, except for the elements along and above the pth superdiagonal; similarly for a strictly lower triangular matrix. As a result, when $p \geq N, \boldsymbol{M}^p = \boldsymbol{0}$. Similar observations reveal that if $\boldsymbol{M}^{(1)}, \boldsymbol{M}^{(2)}, \ldots, \boldsymbol{M}^{(p)}$ is a collection of p strictly lower or upper triangular matrices, then $\boldsymbol{M}^{(1)}\boldsymbol{M}^{(2)} \ldots \boldsymbol{M}^{(p)} = \boldsymbol{0}$.

PROBLEM

2.2.5 *Can an LU be an L or a U and vice versa?*

Show that if the matrix $\boldsymbol{A} = \boldsymbol{LU}$ is upper triangular, then $\boldsymbol{L}$ must necessarily be diagonal; whereas if $\boldsymbol{A}$ is lower triangular, $\boldsymbol{U}$ must be diagonal.

Unit diagonal matrices

Setting all diagonal elements of a diagonal matrix equal to unity produces the *unit* or *identity matrix* $\boldsymbol{I}$. The elements of this matrix are represented by Kronecker's delta $\delta_{i,j}$ which is equal to 0, if $i \neq j$, or 1, if $i = j$.

The rules of matrix multiplication ensure that $\boldsymbol{AI} = \boldsymbol{A}$ and $\boldsymbol{IA} = \boldsymbol{A}$ for any matrix $\boldsymbol{A}$ whose dimensions match those of $\boldsymbol{I}$. Hence $A_{i,j}\delta_{j,k} = A_{i,k}$.

Unit matrix:

$$\boldsymbol{I} = \begin{bmatrix} 1 & 0 & \cdots & 0 & 0 \\ 0 & 1 & 0 & \cdots & 0 \\ \cdots & 0 & \ddots & 0 & \cdots \\ 0 & \cdots & 0 & 1 & 0 \\ 0 & \cdots & \cdots & 0 & 1 \end{bmatrix}$$

Unit back-diagonal matrix:

$$\boldsymbol{J} = \begin{bmatrix} 0 & \cdots & \cdots & 0 & 1 \\ \cdots & \cdots & 0 & 1 & 0 \\ \cdots & 0 & \iddots & 0 & \cdots \\ 0 & 1 & 0 & \cdots & \cdots \\ 1 & 0 & \cdots & \cdots & 0 \end{bmatrix}$$

Setting all nonzero elements of a back-diagonal matrix equal to unity yields the *unit backdiagonal* matrix $\boldsymbol{J}$, sometimes called an *exchange* matrix. Postmultiplying or premultiplying a matrix $\boldsymbol{A}$ by $\boldsymbol{J}$ reverses the order of the columns or rows; the double multiplication $\boldsymbol{JAJ}$ does both. Setting $\boldsymbol{A} = \boldsymbol{I}$ yields $\boldsymbol{J}^2 = \boldsymbol{I}$.

The *unit superdiagonal* and *unit subdiagonal* matrices $\boldsymbol{K}$ and $\boldsymbol{K}^T$, also called *shift* matrices, are defined in a similar manner: The nonzero elements are all equal to unity. Premultiplying a matrix by $\boldsymbol{K}$ produces another matrix with zeros in the bottom row, all other rows raised to the ones above them, and the top row missing. Premultiplying a matrix by $\boldsymbol{K}^T$ depresses the rows, causing the bottom row to disappear. Postmultiplication has a corresponding effect on the columns. All elements of the matrices $\boldsymbol{K}^p$, where p is a positive integer, are equal to zero, except for those along the pth superdiagonal line that are equal to unity. Consequently, if $p \geq N$, $\boldsymbol{K}^p = \boldsymbol{0}$.

PROBLEMS

2.2.6 *Elementary triangular matrices.*

(*a*) Consider an $\boldsymbol{L}$ matrix that is equal to the identity matrix $\boldsymbol{I}$, expect that the kth column below the diagonal is filled with nonzero elements; this is an elementary matrix. Describe the result of premultiplying or postmultiplying an arbitrary matrix $\boldsymbol{A}$ by $\boldsymbol{L}$. (*b*) Repeat part (*a*) for an elementary $\boldsymbol{U}$ matrix.

2.2.7 *Unit imaginary matrix.*

(*a*) Consider a 2×2 back-diagonal matrix $\boldsymbol{Z}$ with elements $Z_{1,2} = 1$ and $Z_{2,1} = -1$. Compute successive powers of this matrix, and show that they behave similarly to the imaginary unit i, in the sense that $i^2 = -1$, $i^3 = -i, \ldots$. (*b*) The results of part (*a*) suggest that if a and b are two real numbers, we can define the complex matrix $a\boldsymbol{I} + b\boldsymbol{Z}$ whose rules of multiplication parallel those of complex numbers. Verify that this is indeed true.

2.2.8 *Action of unit super- and subdiagonal matrices.*

Describe the action of a unit super- or subdiagonal matrix concerning multiplication.

Positive-definite Matrices

Previously in this section, we saw that premultiplying a matrix $\boldsymbol{A}$ by a column vector $\boldsymbol{x}$, producing the new column vector $\boldsymbol{y} = \boldsymbol{Ax}$, can be interpreted as operating on a certain set of physical or initial conditions producing a new set of conditions. To quantify the degree to which the operation has changed the original conditions, we take the inner product of the new vector $\boldsymbol{y}$ and the old vector $\boldsymbol{x}$ and obtain the scalar

$$s = \boldsymbol{x}^T\boldsymbol{y} = \boldsymbol{x}^T(\boldsymbol{Ax}) = x_i A_{i,j} x_j \tag{2.2.6}$$

Real matrices

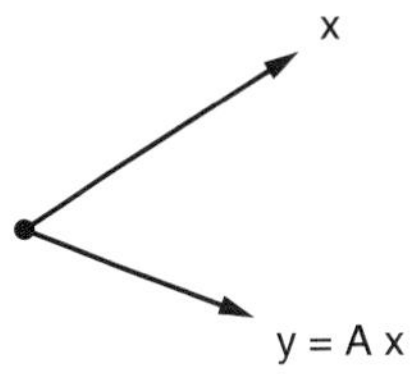

Let us assume that $\boldsymbol{A}$ and $\boldsymbol{x}$ are both real. If the physical process represented by the matrix $\boldsymbol{A}$ is weak, the lengths and orientations of the input and output vectors $\boldsymbol{x}$ and $\boldsymbol{y}$ will be nearly identical, and the magnitude of s will be nearly equal to the square of the length of the vector $\boldsymbol{x}$. But if the physical process is strong, the magnitude of s could be small compared to the square of the length of the vector $\boldsymbol{x}$. Under extreme conditions, s may even vanish or be negative. When $s = 0$, the operation has made the input and output vectors orthogonal.

If the scalar s defined in equation (2.2.6) is positive for any real vector $\boldsymbol{x}$, then the matrix $\boldsymbol{A}$ is *positive-definite.* Any diagonal matrix with positive diagonal elements is positive-definite.

A satisfying geometrical interpretation of a real and positive-definite matrix $\boldsymbol{A}$ emerges by considering the scalar functional expressed by the quadratic form $f(\boldsymbol{x}) = \boldsymbol{x}^T\boldsymbol{A}\boldsymbol{x}$, where $\boldsymbol{x}$ is a real vector. In the present context, a functional is a factory that receives vectors and produces numbers. Observing that $\partial f/\partial x_i$ vanishes when $\boldsymbol{x} = \boldsymbol{0}$ nominates the null point as a candidate for a maximum, a minimum, or a saddle point. If the matrix $\boldsymbol{A}$ is positive-definite, the null point is guaranteed to be a minimum.

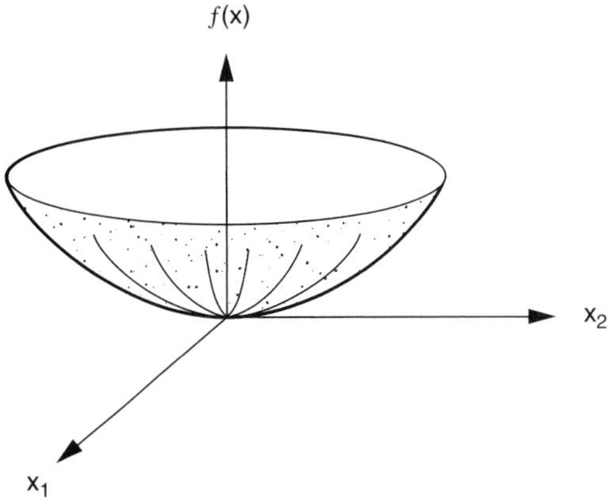

Although it is not possible to just look at a matrix and tell whether or not it is positive-definite, we can perform several tests:

- Identifying the input vector $\boldsymbol{x}$ with the vectors $(1, 0, 0, \ldots, 0)$, $(0, 1, 0, \ldots, 0)$, $\ldots$, $(0, 0, \ldots, 0, 1)$, shows that the diagonal elements of a positive-definite matrix are all positive.

- The determinant of a matrix is a scalar to be defined in Section 2.3. If a matrix $\boldsymbol{A}$ is positive definite, then its determinant is positive. More generally, a matrix $\boldsymbol{A}$ is positive-definite if and only if the determinants of the N submatrices

$$[A_{1,1}], \quad \begin{bmatrix} A_{1,1} & A_{1,2} \\ A_{2,1} & A_{2,2} \end{bmatrix}, \quad \begin{bmatrix} A_{1,1} & A_{1,2} & A_{1,3} \\ A_{2,1} & A_{2,2} & A_{2,3} \\ A_{3,1} & A_{3,2} & A_{3,3} \end{bmatrix}, \quad \ldots \tag{2.2.7}$$

  are all positive (e.g., Nobel and Daniel 1988, p. 413).

- The element with the maximum magnitude of a symmetric positive-definite matrix must lie on the diagonal.

Complex Hermitian matrices

Consider now a complex *Hermitian* square matrix $\boldsymbol{A}$. If the real number $s = \boldsymbol{x}^{T*}\boldsymbol{y} = \boldsymbol{x}^{T*}\boldsymbol{A}\boldsymbol{x}$ is positive for any *complex* vector $\boldsymbol{x}$, then $\boldsymbol{A}$ is positive-definite. The requirement for $\boldsymbol{A}$ to be Hermitian is dictated by the necessity for s to be real.

The concept of a non-Hermitian complex positive-definite matrix has not been invented.

PROBLEM

2.2.9 ***Positive-definite matrices.***

Show that the matrix $\boldsymbol{A} = \boldsymbol{B}^A \boldsymbol{B}$, where $\boldsymbol{B}$ is an arbitrary square complex matrix, is Hermitian and positive-definite.

Trace of a Matrix

The trace of a square matrix $\boldsymbol{A}$, sometimes called the *spur*, is the sum of the diagonal elements, that is,

$$\text{Tr}(\boldsymbol{A}) = A_{i,i} = A_{1,1} + A_{2,2} + \cdots + A_{N,N} \tag{2.2.8}$$

If $\boldsymbol{A}$ and $\boldsymbol{B}$ are two arbitrary matrices with matching dimensions, then

$$\text{Tr}(\boldsymbol{A} + \boldsymbol{B}) = \text{Tr}(\boldsymbol{A}) + \text{Tr}(\boldsymbol{B}) \tag{2.2.9}$$

and

$$\text{Tr}(\boldsymbol{AB}) = \text{Tr}(\boldsymbol{BA}) \tag{2.2.10}$$

PROBLEM

2.2.10 ***Double-dot product in terms of the trace.***

Consider two matrices $\boldsymbol{A}$ and $\boldsymbol{B}$ with matching dimensions. Using index notation, show that

$$\boldsymbol{A} : \boldsymbol{B} = \text{Tr}(\boldsymbol{AB}^T) = \text{Tr}(\boldsymbol{BA}^T) \tag{2.2.11}$$

The double-dot product was defined in equation (2.1.24).

Significance of the Adjoint

In Section 2.1, we introduced the adjoint of a matrix as the transpose of its complex conjugate, $\boldsymbol{A}^A = \boldsymbol{A}^{*T}$. We can now explain this seemingly esoteric terminology working from a broader framework.

First, we return to the interpretation of a matrix as an *operator* that premultiplies a vector to produce another vector. Second, we consider a certain complex vector $\boldsymbol{x}$ and premultiply it by $\boldsymbol{A}$ to form the new vector $\boldsymbol{Ax}$. Third, we compute the inner product of the new vector with the complex conjugate of another vector $\boldsymbol{y}$, to form the scalar

$$s = \boldsymbol{y}^{*^T} \boldsymbol{Ax} \tag{2.2.12}$$

By straightforward algebraic manipulations we obtain

$$s = y_i^* A_{i,j} x_j = y_i^* A_{j,i}^T x_j = x_j (A_{j,i}^{*T} y_i)^* = x_j (A_{j,i}^A y_i)^* \tag{2.2.13}$$

that is,

$$s = \boldsymbol{x}^T (\boldsymbol{A}^A \boldsymbol{y})^* \tag{2.2.14}$$

Comparing equation (2.2.12) to equation (2.2.14) shows that

$$y^{*T} A x = x^T (A^A y)^* \tag{2.2.15}$$

To this end, we consider two complex vectors u and v, define the inner product operator

$$\langle u, v \rangle \equiv u^{*T} v \tag{2.2.16}$$

and rewrite equation (2.2.15) in the symbolic form

$$\langle y, Ax \rangle = \langle A^A y, x \rangle \tag{2.2.17}$$

This equation defines the adjoint of a linear operator acting on the linear vector space of x and y.

If the matrix A is Hermitian, $A = A^A$, the operator expressed by A is *self-adjoint*. In this case, we can either operate on x and take the inner product with y, or we can operate on y and take the inner product with x. The result will be the same.

2.3 *Inverse of a Matrix, Cofactors, and the Determinant*

We have defined matrix addition, subtraction, and multiplication. How about division? Given two square matrices A and B with the same dimensions, we want to find another matrix C so that $AC = B$. In the case of 1×1 matrices, $A = [a]$, $B = [b]$, $C = [c]$, we readily find $c = b/a$. To compute C for matrices of larger size, we must work in two steps: First, we compute a matrix D so that $AD = I$, where I is the identity matrix; and second, we set $C = DB$.

Matrix Inverse

Given a square matrix A, we thus want to know if it is possible to find another matrix D with the same dimensions, so that $AD = I$. The matrix D is called the *inverse of* A and is denoted by A^{-1}. If A^{-1} exists, then we can be sure that its inverse also exists and is equal to A; that is,

$$AA^{-1} = A^{-1}A = I \tag{2.3.1}$$

This is another way of saying that the matrices A and A^{-1} commute. Identity (2.3.1) can be proved by assuming the existence of a matrix E such that $EA = I$, and then writing $E = EI = EAA^{-1} = IA^{-1} = A^{-1}$.

A matrix that has an inverse is called *nonsingular* or *invertible*, and a matrix that does not have an inverse is called *singular* or *noninvertible*. Using the definition (2.3.1), one can show that *if the inverse exists, then it is unique.*

PROBLEM

2.3.1 ***Property of the identity matrix.***

Let A be a nonsingular matrix. Show that the only matrix B with the property $AB = A$ is the unit matrix $B = I$.

Can We Tell Just By Looking?

Is it possible simply to look at a matrix and tell whether it has an inverse? Unfortunately, the answer is negative, in general. But if the matrix is *diagonally dominant,* that is, if the magnitude of each diagonal element is larger than the sum of the magnitudes of the off-diagonal elements in the corresponding column *or* row, then its inverse is guaranteed to exist. The proof will be discussed in Problem 2.9.7.

Properties of the Inverse

Using the rules of matrix multiplication, we find that the set of the row vectors of $\boldsymbol{A}$, call them $\boldsymbol{r}^{(i)T}$, is orthonormal to the set of the column vectors of $\boldsymbol{A}^{-1}$, call them $\boldsymbol{c}^{(i)}$, and vice versa; that is, $\boldsymbol{r}^{(i)T}\boldsymbol{c}^{(j)} = \delta_{i,j}$.

As a result of this property, if we interchange the *rows or columns* of $\boldsymbol{A}$ to obtain a modified matrix $\boldsymbol{A}^{MOD}$, we must also interchange the corresponding *columns or rows* of $\boldsymbol{A}^{-1}$ to obtain the inverse of the modified matrix $(\boldsymbol{A}^{MOD})^{-1}$.

A number of identities stemming from the definition of the inverse are listed in Table 2.3.1. The first of these identities arises by taking the transpose of all sides of equation (2.3.1), and using the property (2.1.18).

Woodbury formula

The Woodbury formula relates the inverse of a real $N \times N$ matrix $\boldsymbol{A}$ to the inverse of the perturbed matrix $\boldsymbol{A} + \boldsymbol{U}\boldsymbol{V}^T$, where $\boldsymbol{U}$ and $\boldsymbol{V}$ are two real $N \times K$ matrices with $K < N$:

$$(\boldsymbol{A} + \boldsymbol{U}\boldsymbol{V}^T)^{-1} = \boldsymbol{A}^{-1} - \boldsymbol{A}^{-1}\boldsymbol{U}(\boldsymbol{I} + \boldsymbol{V}^T\boldsymbol{A}^{-1}\boldsymbol{U})^{-1}\,\boldsymbol{V}^T\boldsymbol{A}^{-1} \tag{2.3.2}$$

$\boldsymbol{I}$ is the $K \times K$ identity matrix (e.g., Householder 1964, p. 123).

Sherman–Morrison formula

In the particular case where $K = 1$, the matrices $\boldsymbol{U}$ and $\boldsymbol{V}$ are N-dimensional column vectors $\boldsymbol{u}$ and $\boldsymbol{v}$, the quantity $\boldsymbol{I} + \boldsymbol{v}^T\boldsymbol{A}^{-1}\boldsymbol{u} = 1 + s$ is a scalar, and the Woodbury formula reduces to the Sherman–Morrison formula (e.g., Bodewig 1959, p. 39):

$$(\boldsymbol{A} + \boldsymbol{u}\boldsymbol{v}^T)^{-1} = \boldsymbol{A}^{-1} - \frac{1}{1+s}\boldsymbol{A}^{-1}\boldsymbol{u}\boldsymbol{v}^T\boldsymbol{A}^{-1} \tag{2.3.3}$$

Generalizations of the preceding two formulas to complex matrices are discussed by Householder (1964, p. 123). In practice, these formulas suggest clever ways of computing the inverse of a mildly difficult matrix

Table 2.3.1 Identities involving the inverses of two generally unrelated square nonsingular matrices A and B; α is a scalar.

$(\boldsymbol{A}^{-1})^T = (\boldsymbol{A}^T)^{-1}$
$(\alpha\boldsymbol{A})^{-1} = \alpha^{-1}\boldsymbol{A}^{-1}$
$(\boldsymbol{A}\boldsymbol{B})^{-1} = \boldsymbol{B}^{-1}\boldsymbol{A}^{-1}$
$\boldsymbol{A}(\boldsymbol{A}^{-1} + \boldsymbol{B}^{-1})\boldsymbol{B} = \boldsymbol{B}(\boldsymbol{A}^{-1} + \boldsymbol{B}^{-1})\boldsymbol{A} = \boldsymbol{A} + \boldsymbol{B}$
$(\boldsymbol{A}^{-1} + \boldsymbol{B}^{-1})^{-1} = \boldsymbol{A}(\boldsymbol{A} + \boldsymbol{B})^{-1}\boldsymbol{B} = \boldsymbol{B}(\boldsymbol{A} + \boldsymbol{B})^{-1}\boldsymbol{A}$

that is slightly perturbed with respect to a matrix $\boldsymbol{A}$ whose inverse is easy to compute. An example will be discussed at the end of Section 3.3.

PROBLEM

2.3.2 ***Inverse of an elementary triangular matrix.***

Show that the inverse of an elementary lower triangular matrix, defined in Problem 2.2.6, is given by $\boldsymbol{L}^{-1} = 2\boldsymbol{I} - \boldsymbol{L}$. What is the inverse of an elementary upper triangular matrix?

Inverse of a Diagonal Matrix

The inverse of a diagonal matrix $\boldsymbol{D}$ is easy to compute. One may readily verify that $\boldsymbol{D}^{-1}$ is another diagonal matrix with diagonal elements equal to $D^{-1}_{i,i} = 1/D_{i,i}$, where summation is *not* implied over the repeated index i. It is then evident that if one of the diagonal elements of $\boldsymbol{D}$ is equal to zero, the corresponding element of $\boldsymbol{D}^{-1}$ will have an infinite value: $\boldsymbol{D}$ will be singular, and $\boldsymbol{D}^{-1}$ will not be defined.

Inverse of a diagonal matrix

$$\boldsymbol{D}^{-1} = \begin{bmatrix} 1/D_{1,1} & 0 & \dots & 0 & 0 \\ 0 & 1/D_{2,2} & 0 & \dots & 0 \\ \dots & 0 & \ddots & 0 & \dots \\ 0 & \dots & 0 & 1/D_{N-1,N-1} & 0 \\ 0 & \dots & \dots & 0 & 1/D_{N,N} \end{bmatrix}$$

Minors, Cofactors, and the Determinant

To compute the inverse of an arbitrary matrix, we regard its N^2 elements as unknowns and form a linear system of N^2 linear equations by demanding the satisfaction of equation (2.3.1). The system may have a unique solution or no solution. To distinguish between these two possibilities, we consider the detailed algebraic manipulations involved in the actual solution, and this leads us to the determinant, the minors, and the cofactors.

The determinant of a matrix $\boldsymbol{A}$ is a scalar quantity, denoted as Det($\boldsymbol{A}$), which is defined in an implicit fashion:

- First, we define the determinant of a 1×1 matrix to be the number comprising the matrix.
- Next, to every element of a matrix with higher dimensions, $A_{i,j}$, we assign a number called its *minor*, denoted as $M_{i,j}$. This is defined as the determinant of the matrix that arises by (a) replacing the elements in the ith row and jth column of $\boldsymbol{A}$ with blank spaces, and (b) compressing the matrix to eliminate the blank spaces, as shown in the diagram. The result is a reduced matrix with dimensions $(N-1) \times (N-1)$. The corresponding cofactor is defined as

*j*th column ↓

*i*th row →

$$\begin{bmatrix} A_{1,1} & \dots & \dots & A_{1,j} & \dots & A_{1,N} \\ \dots & \dots & \dots & \dots & \dots & \dots \\ A_{i,1} & \dots & A_{i,i} & \dots & \dots & A_{i,N} \\ \dots & \dots & \dots & A_{j,j} & \dots & \dots \\ \dots & \dots & \dots & \dots & \dots & \dots \\ A_{N,1} & \dots & \dots & A_{N,j} & \dots & A_{N,N} \end{bmatrix}$$

$$C_{i,j} = (-1)^{i+j} M_{i,j} \qquad (2.3.4)$$

- The determinant of $\boldsymbol{A}$ is finally defined in terms of the *Laplace expansion*:

$$\text{Det}(\boldsymbol{A}) = \sum_{j=1}^{N} A_{i,j} C_{i,j} = \sum_{i=1}^{N} A_{i,k} C_{i,k} \tag{2.3.5}$$

where the values of i and k are arbitrary; we have temporarily given up the repeated-index summation convention. In words, the determinant is equal to the sum of the product of the elements of the matrix with the associated cofactors, where the sum is computed across an *arbitrary column* or *row*.

We have thus reduced the computation of the determinant of the original $N \times N$ matrix $\boldsymbol{A}$ to the computation of the determinants of a collection of N^2 matrices of size $(N-1) \times (N-1)$ that define the N^2 minors. The process is repeated N times, and the final result is an expression for the determinant in terms of the determinants of a collection of 1×1 matrices, which have been defined at the outset.

As a simple example, we compute the determinant of a 2×2 matrix by performing the Laplace expansion with respect to the first column:

$$\text{Det}\begin{bmatrix} A_{1,1} & A_{1,2} \\ A_{2,1} & A_{2,2} \end{bmatrix} = A_{1,1}C_{1,1} + A_{2,1}C_{2,1}$$

$$= A_{1,1}M_{1,1} - A_{2,1}M_{2,1} = A_{1,1}A_{2,2} - A_{2,1}A_{1,2} \tag{2.3.6}$$

PROBLEMS

2.3.3 *Vandermonde matrix.*

Vandermonde matrix:

$$\boldsymbol{V} = \begin{bmatrix} 1 & 1 & 1 & \dots & 1 \\ a_1 & a_2 & a_3 & \dots & a_N \\ a_1^2 & a_2^2 & a_3^2 & \dots & a_N^2 \\ \dots & \dots & \dots & \dots & \dots \\ a_1^{N\text{-}1} & a_2^{N\text{-}1} & a_3^{N\text{-}1} & \dots & a_N^{N\text{-}1} \end{bmatrix}$$

Select a set of N numbers A_j, $j = 1, \dots, N$, and compute the Vandermonde matrix $V_{i,j} = a_j^{i-1}$, where $i, j = 1, \dots, N$. This matrix arises in the approximation of functions using the least-squares method to be discussed in Section 8.2. Show that $\text{Det}(\boldsymbol{V})$ is equal to the product of all factors $(a_m - a_n)$, where $1 \le n < m \le N$.

2.3.4 *Zero of a determinant.*
(a) Is it possible that by changing the value of a single element of a matrix one can make the matrix singular?
(b) Consider a $N \times N$ matrix $\boldsymbol{A}$, maintain all of its elements constant, and set the elements in the ith row equal to the variable x. Discuss whether there is one or more values of x for which $\text{Det}(\boldsymbol{A}) = 0$. Present a schematic illustration of the determinant as a function of x.
(c) Repeat part (b), but this time assume that all elements of the ith row *and* all elements of the jth column are equal to x.

Properties of the Determinant

Using the preceding definitions, we find that the determinant obeys the relations shown in Table 2.3.2. The last inequality, in particular, can be verified by identifying the matrices $\boldsymbol{A}$ and $\boldsymbol{B}$ with the $N \times N$ identity matrix, in which case the left-hand side is equal to 2^N, whereas the right-hand side is equal to 2. Equality is possible only when $N = 1$.

Further properties of a matrix concerning its determinant are:

- The determinant of a matrix will not change if we add to any row or column, respectively, any other row or column multiplied by an arbitrary scalar constant. Consequently, if any two rows or columns

Table 2.3.2 Identities involving the determinants of two generally unrelated $N \times N$ square matrices $\boldsymbol{A}$ and $\boldsymbol{B}$; α is a scalar.

$\mathrm{Det}(\alpha\,\boldsymbol{A}) = \alpha^N\,\mathrm{Det}(\boldsymbol{A})$
$\mathrm{Det}(\boldsymbol{A}^T) = \mathrm{Det}(\boldsymbol{A})$
$\mathrm{Det}(\boldsymbol{A}^{-1}) = 1/\mathrm{Det}(\boldsymbol{A})$
$\mathrm{Det}(\boldsymbol{A}\boldsymbol{B}) = \mathrm{Det}(\boldsymbol{A})\,\mathrm{Det}(\boldsymbol{B})$
$\mathrm{Det}(\boldsymbol{A}^n) = [\,\mathrm{Det}(\boldsymbol{A})\,]^n$
$\mathrm{Det}(\boldsymbol{A}+\boldsymbol{B}) \neq \mathrm{Det}(\boldsymbol{A}) + \mathrm{Det}(\boldsymbol{B})$, in general

of a matrix are identical, or at least one row or column can be expressed as a linear combination of the others—which means that the columns of rows are linearly dependent—the determinant will be equal to zero.

- If the columns are linearly dependent, then the rows must also be linearly dependent, and vice versa. Conversely, if the determinant of a matrix is *not* equal to zero, then the column vectors and row vectors are guaranteed to be linearly independent.
- The sign of the determinant reverses when we interchange two columns or two rows; an even number of interchanges leaves the sign unaffected.

PROBLEM

2.3.5 *Properties of the determinant.*

(*a*) Verify all identities listed in Table 2.3.2 for 2×2 matrices.
(*b*) Show that the determinant of a 2×2 matrix $\boldsymbol{A}$ can be computed as

$$\mathrm{Det}(\boldsymbol{A}) = \tfrac{1}{2}[\mathrm{Tr}^2(\boldsymbol{A}) - \mathrm{Tr}(\boldsymbol{A}^2)] \tag{2.3.7}$$

Confirm this identity with an example.
(*c*) Generalize equation (2.3.7) to matrices of larger size (see Section 5.5).
(*d*) How will the value of the determinant change if we multiply k rows by the scalar α and l rows by the scalar β?

Inverse in Terms of the Cofactors

Having established rules for evaluating the matrix of cofactors $\boldsymbol{C}$ and the determinant, we proceed to compute the inverse of a matrix $\boldsymbol{A}$ as

$$\boldsymbol{A}^{-1} = \frac{1}{\mathrm{Det}(\boldsymbol{A})}\boldsymbol{C}^T \tag{2.3.8}$$

This relation makes it clear that if $\mathrm{Det}(\boldsymbol{A}) = 0$, then some or even all of the elements of the matrix represented by the right-hand side will have *infinite* values: The inverse of $\boldsymbol{A}$ will not exist, and $\boldsymbol{A}$ will be *singular*. Thus, *if the determinant of a matrix vanishes, the matrix does not have an inverse.*

The only 1×1 matrix that does not have an inverse is the zero matrix [0], but there are many nonnull matrices in higher dimensions that do not have an inverse.

Equation (2.3.8) can be used to demonstrate that the inverse of a lower triangular, an upper triangular, or a tridiagonal matrix, $\boldsymbol{L}$, $\boldsymbol{U}$, or $\boldsymbol{T}$, is, respectively, another $\boldsymbol{L}$, $\boldsymbol{U}$, or $\boldsymbol{T}$, and the inverse of a symmetric matrix is also symmetric.

The Matrix Adjoint

The transpose of the matrix of cofactors, $\boldsymbol{C}^T$, is called the *matrix adjoint* of $\boldsymbol{A}$ and is denoted by Adj($\boldsymbol{A}$). Thus, by definition,

$$\boldsymbol{A}^{-1} \equiv \frac{1}{\text{Det}(\boldsymbol{A})}\text{Adj}(\boldsymbol{A}) \tag{2.3.9}$$

The matrix adjoint should be distinguished from the adjoint of a matrix $\boldsymbol{A}^A \equiv \boldsymbol{A}^{*T}$, where an asterisk designates the complex conjugate. In classical texts of liner algebra, Adj($\boldsymbol{A}$) is coined the adjoint, but this terminology contradicts well-established definitions in the theory of *linear operators* regarding the *adjoint of linear maps.*

Orthogonal Matrices

A real matrix $\boldsymbol{A}$ whose inverse is equal to its transpose is called orthogonal. Since $\boldsymbol{A}^{-1} = \boldsymbol{A}^T$, the inverse of an orthogonal matrix is extremely easy to find.

Taking advantage of the identity $\boldsymbol{A}\boldsymbol{A}^T = \boldsymbol{A}^T\boldsymbol{A} = \boldsymbol{I}$, we find that, if $\boldsymbol{r}^{(i)}$ and $\boldsymbol{c}^{(j)}$ are the vectors comprising the rows and columns of $\boldsymbol{A}$, then

$$\boldsymbol{c}^{(i)T}\boldsymbol{c}^{(j)} = \delta_{i,j}, \qquad \boldsymbol{r}^{(i)T}\boldsymbol{r}^{(j)} = \delta_{i,j} \tag{2.3.10}$$

that is, the two sets of vectors are *orthonormal.* The former implies the latter and *vice versa.*

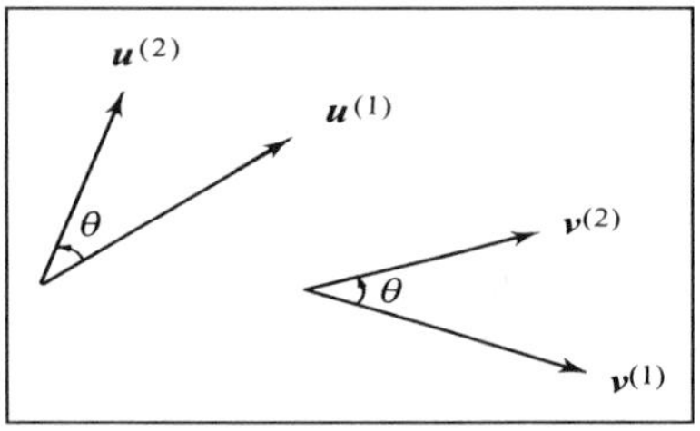

Using the fourth identity in Table 2.3.2, we find that Det($\boldsymbol{A}\boldsymbol{A}^T$) $=$ Det$(\boldsymbol{A})^2 = 1$, which requires that Det($\boldsymbol{A}$) $= \pm 1$. For example, each rotation matrix defined in equations (2.1.21) is orthogonal, and its determinant is equal to 1.

When an orthogonal matrix $\boldsymbol{A}$ premultiplies a vector $\boldsymbol{u}$ to yield the new vector $\boldsymbol{v} = \boldsymbol{A}\boldsymbol{u}$, it leaves its length unaffected. This can be seen immediately by writing

$$\boldsymbol{v}^T\boldsymbol{v} = (\boldsymbol{A}\boldsymbol{u})^T\boldsymbol{v} = \boldsymbol{u}^T\boldsymbol{A}^T\boldsymbol{A}\boldsymbol{u} = \boldsymbol{u}^T\boldsymbol{u} \tag{2.3.11}$$

Working in the same vein, one can show that the inner product between the transformed vectors $\boldsymbol{v}^{(1)} = \boldsymbol{A}\boldsymbol{u}^{(1)}$ and $\boldsymbol{v}^{(2)} = \boldsymbol{A}\boldsymbol{u}^{(2)}$ is identical to that between the original vectors $\boldsymbol{u}^{(1)}$ and $\boldsymbol{u}^{(2)}$: *Orthogonal transformations do not alter the relative positions of two vectors.*

PROBLEM

2.3.6 ***A triangular orthogonal matrix must be diagonal.***

Show that if a lower or upper triangular matrix is orthogonal, then it must necessarily be diagonal.

Unitary Matrices

A unitary matrix is the counterpart of an orthogonal matrix in the complex space: A complex matrix $\boldsymbol{A}$ whose inverse is equal to its adjoint, that is, $\boldsymbol{A}^{-1} = \boldsymbol{A}^A$, is *unitary.* A real orthogonal matrix is evidently unitary.

Since $AA^A = A^A A = I$, if $r^{(i)}$ and $c^{(i)}$ are the vectors comprising the rows and columns of A, then

$$c^{(i)*T} c^{(j)} = \delta_{i,j}, \qquad r^{(i)*T} r^{(j)} = \delta_{i,j} \tag{2.3.12}$$

The first of these relations implies the second and vice versa.

An ambassador of the province whose residents are the 2×2 unitary matrices is

$$A = \begin{bmatrix} \cos\theta & I \sin\theta \\ -I \sin\theta & -\cos\theta \end{bmatrix} \tag{2.3.13}$$

where I is the imaginary unit, and θ has an arbitrary real value.

When a unitary matrix A premultiplies a complex vector u to yield the new complex vector $v = Au$, it leaves its norm unaffected. This can be seen immediately by writing

$$v^T v^* = (Au)^T v^* = u^T A^T A^* u^* = u^T (A^A A)^* u^* = u^T u^* \tag{2.3.14}$$

Plane Rotation

There is a family of orthogonal matrices, called *plane rotation matrices*, used to change the coordinates of a point in the N-dimensional space subject to rotation of a pair of Cartesian axes in their plane.

Let u be the coordinates of a point in a certain Cartesian system, and v be the corresponding coordinates in a Cartesian system that arises by rotating the ith and jth axes of the first system in their plane about the origin, by an angle θ. The vectors u and v are related by $v = P^{(i,j)} u$, where $P^{(i,j)}$ is a plane rotation matrix given by

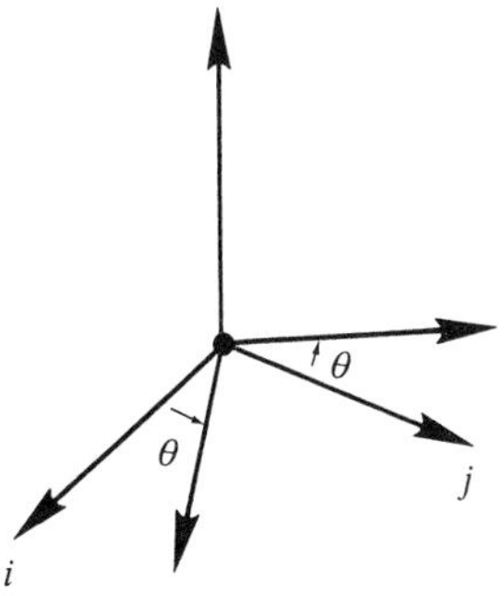

$$P^{(i,j)} = \begin{array}{c} \\ \\ \\ \\ i \\ \\ j \\ \\ \\ \\ \end{array} \begin{bmatrix} & & & i & & j & & & \\ 1 & 0 & \ldots & 0 & \ldots & 0 & \ldots & 0 & 0 \\ 0 & 1 & \ldots & 0 & \ldots & 0 & \ldots & 0 & 0 \\ \ldots & \ldots & \ldots & \ldots & \ldots & \ldots & \ldots & \ldots & \ldots \\ 0 & 0 & 0 & \cos\theta & \ldots & \sin\theta & 0 & 0 & 0 \\ \ldots & \ldots & \ldots & \ldots & \ldots & \ldots & \ldots & \ldots & \ldots \\ 0 & 0 & 0 & -\sin\theta & \ldots & \cos\theta & 0 & 0 & 0 \\ \ldots & \ldots & \ldots & \ldots & \ldots & \ldots & \ldots & \ldots & \ldots \\ 0 & 0 & \ldots & 0 & \ldots & 0 & \ldots & 1 & 0 \\ 0 & 0 & \ldots & 0 & \ldots & 0 & \ldots & 0 & 1 \end{bmatrix} \tag{2.3.15}$$

Note that $\mathrm{Det}(P^{(i,j)}) = 1$.

It is instructive to observe that the 3×3 version of this matrix differs from the rotation matrices given in equations (2.1.21) only by the sign of the off-diagonal elements. Equation (2.3.15) expresses the change in coordinates due to the *rotation of the axes*, whereas equations (2.1.21) express the change in coordinates due to the *rotation of a point about fixed axes.*

One important property of the plane rotation matrices is that premultiplying a certain matrix by $P^{(i,j)}$ changes only rows i and j, whereas postmultiplying it by $P^{(i,j)}$ changes only the corresponding columns.

Plane rotation matrices will be used in Chapter 5 to transform an arbitrary matrix to a simpler form, such as the diagonal form.

PROBLEM

2.3.7 ***Zeroing a component of a vector by rotation of the axes.***

Let $\boldsymbol{u}$ be an N-dimensional vector. Show that if $\theta = -\arctan(u_i/u_j)$, then the ith component of the vector $\boldsymbol{v} = \boldsymbol{P}^{(i,j)}\boldsymbol{u}$ is equal to zero for any value of j. Which value would you choose and why?

Reflection and Householder Matrices

Another important class of symmetric and orthogonal, or more generally Hermitian and unitary, matrices that enjoy extensive usage in the theory and practice of numerical computation are the Householder matrices with elements

$$H_{i,j} = \delta_{i,j} - 2w_i w_j^* \tag{2.3.16}$$

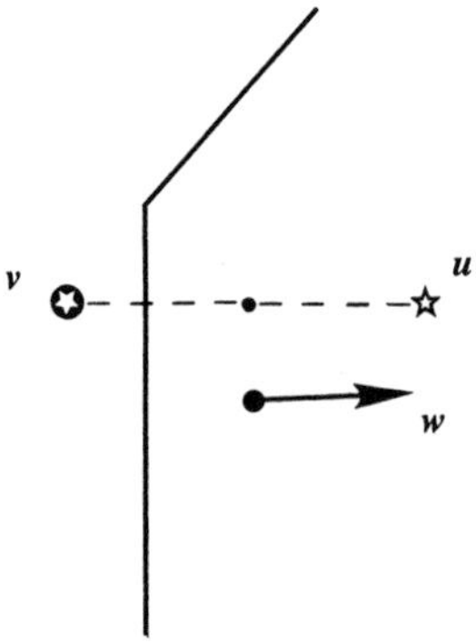

where $\boldsymbol{w}$ is an arbitrary N-dimensional complex vector whose norm is equal to unity, that is, $\boldsymbol{w}^T\boldsymbol{w}^* = 1$. One may readily verify that, subject to this constraint, $\boldsymbol{H}^T\boldsymbol{H}^* = \boldsymbol{I}$, that is, $\boldsymbol{H}^{-1} = \boldsymbol{H}^A$. Furthermore, it is clear that $\boldsymbol{H}$ is Hermitian, that is, $\boldsymbol{H} = \boldsymbol{H}^A \equiv \boldsymbol{H}^{T*}$, and $\boldsymbol{H}^{-1} = \boldsymbol{H}$.

To investigate the properties of the Householder matrices, let us set $\boldsymbol{w} = (1, 0)^T$, yielding $H_{1,1} = -1$, $H_{1,2} = 0$, $H_{2,1} = 0$, $H_{2,2} = 1$. If $\boldsymbol{u} = (u_1, u_2)^T$ is an arbitrary two-dimensional vector, then $\boldsymbol{v} = \boldsymbol{H}\boldsymbol{u} = (-u_1, u_2)$. This example, along with the observation that $\boldsymbol{H}\boldsymbol{w} = -\boldsymbol{w}$, reveals that when a vector $\boldsymbol{u}$ is premultiplied by a Householder matrix to produce the vector $\boldsymbol{v} = \boldsymbol{H}\boldsymbol{u}$, the corresponding point in the N-dimensional space is reflected with respect to the plane that passes through the origin and is perpendicular to the unit vector represented by the vector $\boldsymbol{w}$.

PROBLEMS

2.3.8 ***Reflection.***

Demonstrate the reflective action of the Householder matrices by expressing $\boldsymbol{u}$ as a linear combination of N orthogonal vectors, one of which is the vector $\boldsymbol{w}$.

2.3.9 ***Plane of reflection.***

Consider two vectors $\boldsymbol{u}$ and $\boldsymbol{v}$ with equal norms, and compute the unit vector $\boldsymbol{w}$ so that $\boldsymbol{v} = \boldsymbol{H}\boldsymbol{u}$.

Matrix Functions

As a last topic, we consider a matrix function of the independent variable t. If t changes by the small amount dt, the determinant of $\boldsymbol{A}$ changes by the differential amount $d\ \mathrm{Det}\boldsymbol{A}(t)$, and its inverse changes by the differential amount $d\boldsymbol{A}^{-1}(t)$.

Taking the derivative of both sides of equation (2.3.1), and then expanding it out into two terms, we find that the rate of change of the inverse is

$$\frac{d\boldsymbol{A}^{-1}}{dt} = -\boldsymbol{A}^{-1}\frac{d\boldsymbol{A}}{dt}\boldsymbol{A}^{-1} \tag{2.3.17}$$

Analogous expressions can be derived for higher-order derivatives.

The rate of change of the determinant may be computed conveniently from the expression

$$\frac{1}{\text{Det}(A)} \frac{d\text{Det}(A)}{dt} = \frac{dA}{dt} : A^{-1^T} = \text{Tr}\left[A^{-1}\frac{dA}{dt}\right] \tag{2.3.18}$$

(Bodewig 1959, p. 41). The proof requires a pen full of ink and plenty of scrap paper.

PROBLEM

2.3.10 *Rate of change of the determinant.*

Verify the validity of equations (2.3.17) and (2.3.18) for a 2×2 diagonal matrix function whose first and second diagonal elements are, respectively, equal to t and t^2.

2.4 *Computation of the Determinant*

Having established rules for evaluating the determinant, we turn to discussing its practical computation considering matrices with increasingly more general structure.

Diagonal and Triangular Matrices

Expressing the determinant of a *diagonal* or *triangular* matrix in terms of the cofactors, we find that it is equal to the *product of the diagonal elements*. Thus, if one of the diagonal elements is equal to zero, the matrix will be singular.

Tridiagonal Matrices

We consider next a tridiagonal matrix T and denote its nonzero elements as

$$T_{i,i-1} = c_i, \qquad T_{i,i} = a_i, \qquad T_{i,i+1} = b_i \tag{2.4.1}$$

as shown in the diagram, where summation is *not* implied over i: A tridiagonal matrix can be described and stored in terms of the three vectors a, b, c.

Let P_i be the determinant of the upper-left-corner diagonal $i \times i$ block of T. Expanding P_i in terms of the cofactors down the last column, we find

$$T = \begin{bmatrix} a_1 & b_1 & 0 & \cdots & \cdots & \cdots & \cdots & \cdots \\ c_2 & a_2 & b_2 & 0 & \cdots & \cdots & \cdots & \cdots \\ 0 & c_3 & a_3 & b_3 & 0 & \cdots & \cdots & \cdots \\ \cdots & 0 & \ddots & \ddots & \ddots & 0 & \cdots & \cdots \\ \cdots & \cdots & 0 & c_{i-1} & a_{i-1} & b_{i-1} & 0 & \cdots \\ \cdots & \cdots & \cdots & 0 & c_i & a_i & b_i & 0 \\ \cdots & \cdots & \cdots & \cdots & 0 & c_{i+1} & a_{i+1} & b_{i+1} \\ \cdots & \cdots & \cdots & \cdots & \cdots & \cdots & \cdots & \cdots \end{bmatrix}$$

*i*th column $\downarrow$

*i*th row $\longrightarrow$

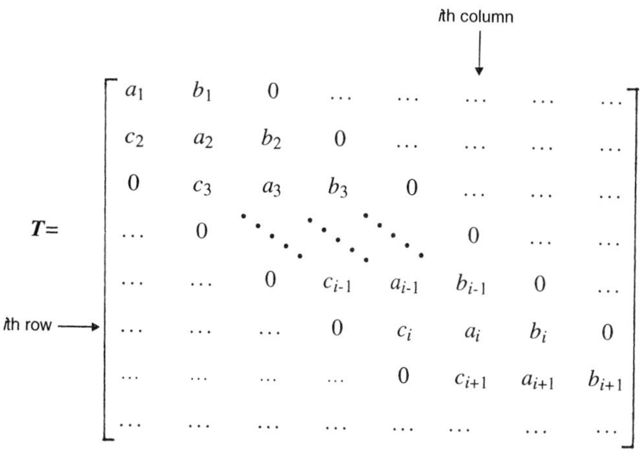

$$P_i = a_i P_{i-1} - b_{i-1} \, \mathrm{Det}(\boldsymbol{Q}) \tag{2.4.2}$$

$$\boldsymbol{Q} = \begin{bmatrix} a_1 & b_1 & 0 & \cdots & \cdots & \cdots \\ c_2 & a_2 & b_2 & 0 & \cdots & \cdots \\ 0 & c_3 & a_3 & b_3 & 0 & \cdots \\ \cdots & 0 & \ddots & \ddots & \ddots & 0 \\ \cdots & \cdots & 0 & c_{i-2} & a_{i-2} & 0 \\ \cdots & \cdots & \cdots & 0 & 0 & c_i \end{bmatrix}$$

where the matrix $\boldsymbol{Q}$ is shown on the right. We note that $\mathrm{Det}(\boldsymbol{Q}) = c_i P_{i-2}$ and work recursively to derive the algorithm algorithm

$$\boxed{\begin{array}{l} P_0 = 1 \\ P_1 = a_1 \\ \text{Do } i = 2, N \\ P_i = a_i P_{i-1} - b_{i-1} c_i P_{i-2} \\ \text{End Do} \\ \mathrm{Det}(\boldsymbol{T}) = P_N \end{array}} \tag{2.4.3}$$

Tridiagonal matrices with equal determinants

Consider a tridiagonal matrix $\boldsymbol{T}$ whose elements are given in equations (2.4.1), and introduce another tridiagonal matrix $\boldsymbol{T}'$ with elements

$$T'_{i,i-1} = c'_i, \qquad T'_{i,i} = a'_i, \qquad T'_{i,i+1} = b'_i \tag{2.4.4}$$

which are related to the elements of $\boldsymbol{T}$ by

$$\begin{aligned} a'_i &= a_i \\ b'_{i-1} c'_i &= b_{i-1} c_i \end{aligned} \tag{2.4.5}$$

Note that the diagonal elements and products of corresponding sub- and superdiagonal elements are equal. Cursory inspection of the third equation in the Do loop of algorithm (2.4.3) reveals that

$$\mathrm{Det}(\boldsymbol{T}') = \mathrm{Det}(\boldsymbol{T}) \tag{2.4.6}$$

This seemingly fortuitous agreement is only the tip of an iceberg. A more fundamental relation between the two matrices will be discussed at the end of Section 2.9 in the context of eigenvalues.

Determinants of special classes of tridiagonal matrices in terms of orthogonal polynomials

Appendix B discusses the properties orthogonal polynomials. This is an excellent time to digress and take a look. Equation (B.9), in particular, relates the determinant of families of $N \times N$ tridiagonal matrices to the values of Nth degree orthogonal polynomials that belong to a corresponding class.

For example, the determinant of an $N \times N$ tridiagonal matrix with constant elements along the three diagonals given by $a_i = 2\mu$, $b_i = \nu$, and $c_i = 1/\nu$, where μ and ν are two arbitrary constants, is equal to $S_N(\mu)$, where S_N is the Nth degree Chebyshev polynomial of the second kind discussed in Table B.4 of Appendix B.

PROBLEM

2.4.1 ***Tridiagonal matrices and Legendre polynomials.***

Invent a one-parameter family of $N \times N$ tridiagonal matrices whose determinants are equal to the Nth degree Legendre polynomial L_N evaluated at some point. The Legendre polynomials are discussed in Table B.2 of Appendix B.

Nearly Triangular (Hessenberg) Matrices

A similar algorithm can be devised for *Hessenberg* matrices. These are lower or upper triangular matrices, but the elements along the first super- or subdiagonal are not necessarily equal to zero.

Considering first an *upper* Hessenberg matrix $\mathbf{A}$, we express it in the form

Upper Hessenberg matrix:

$$
A = \begin{bmatrix}
A_{1,1} & A_{1,2} & A_{1,3} & \cdots & \cdots & A_{1,N} \\
A_{2,1} & A_{2,2} & A_{2,3} & \cdots & \cdots & A_{2,N} \\
0 & A_{3,2} & A_{3,3} & \cdots & \cdots & \cdots \\
0 & 0 & \ddots & \cdots & \cdots & \cdots \\
\cdots & 0 & 0 & A_{N-1,N-2} & A_{N-1,N-1} & A_{N-1,N} \\
0 & \cdots & 0 & 0 & A_{N,N-1} & A_{N,N}
\end{bmatrix}
$$

$$
\mathbf{A} = (\mathbf{U} - \mathbf{K}^T)\mathbf{D} \tag{2.4.7}
$$

where:

- $\mathbf{K}^T$ is the unit subdiagonal matrix.
- $\mathbf{D}$ is a diagonal matrix with elements

$$
D_{i,i} = -A_{i+1,i} \tag{2.4.8}
$$

for $i = 1, \ldots, N-1$, where summation is not implied over i. The last element $D_{N,N}$ has an arbitrary value, which, for simplicity, we set equal to unity.

- $\mathbf{U}$ is an upper triangular matrix, with elements

$$
U_{i,j} = \frac{A_{i,j}}{D_{j,j}} \tag{2.4.9}
$$

for $j = i, \ldots, N$, where summation is not implied over j. We have assumed that none of the diagonal elements $D_{j,j}$ is equal to zero.

$$H = \begin{bmatrix} U_{1,1} & U_{1,2} & U_{1,3} & \cdots & \cdots & U_{1,N} \\ -1 & U_{2,2} & U_{2,3} & \cdots & \cdots & U_{2,N} \\ 0 & -1 & U_{3,3} & \cdots & \cdots & \cdots \\ 0 & 0 & \ddots & \cdots & \cdots & \cdots \\ \cdots & 0 & 0 & -1 & U_{N-1,N-1} & U_{N-1,N} \\ 0 & \cdots & 0 & 0 & -1 & U_{N,N} \end{bmatrix}$$

Next, we consider the Hessenberg matrix $\boldsymbol{H} \equiv \boldsymbol{U} - \boldsymbol{K}^T$ depicted in the diagram. Denoting the determinant of the upper-left-corner diagonal $i \times i$ block of $\boldsymbol{H}$ by P_i, where $P_N = \text{Det}(\boldsymbol{H})$, and expanding it in terms of the cofactors down the last column, we obtain

$$P_1 = U_{1,1}, \quad P_2 = U_{1,2} + U_{2,2}P_1, \quad \ldots \tag{2.4.10}$$

which can be encoded in the algorithm

$$\boxed{\begin{array}{l} P_0 = 1 \\ P_1 = U_{1,1} \\ \text{Do } i = 2, N \\ \qquad P_i = \sum_{j=1}^{i} U_{j,i} P_{j-1} \\ \text{END DO} \\ \text{Det}(\boldsymbol{H}) = P_N \end{array}} \tag{2.4.11}$$

We recall that $\boldsymbol{A} = \boldsymbol{HD}$, and use the fourth identity of Table 2.3.2 to set $\text{Det}(\boldsymbol{A}) = \text{Det}(\boldsymbol{H})\text{Det}(\boldsymbol{D})$, where $\text{Det}(\boldsymbol{D})$ is equal to the product of the diagonal elements of $\boldsymbol{D}$.

If $\boldsymbol{A}$ is a lower Hessenberg matrix, we take its transpose and repeat the preceding procedure. Taking the transpose does not affect the value of the determinant.

PROBLEM

2.4.2 ***Performance when a subdiagonal element is close to zero.***

Write a computer program that computes the determinant of an upper Hessenberg matrix using the method described in the text. Then consider an upper Hessenberg matrix of your choice, hold all of its elements constant, but start decreasing the magnitude of one of the subdiagonal elements. Discuss the performance of the numerical method as the magnitude of this element tends to zero.

Arbitrary Matrices

The rules for computing the determinant of an arbitrary matrix in terms of the cofactors discussed in Section 2.3 typically require an exorbitant amount of CPU time, even for matrices of moderate size. Regrettably, the primary form is not designated for numerical computation, and we must develop alternative approaches.

An expedient method of computing the determinant of an arbitrary matrix $\boldsymbol{A}$ proceeds by (a) decomposing it into the product of a lower triangular matrix $\boldsymbol{L}$ and an upper triangular matrix $\boldsymbol{U}$, $\boldsymbol{A} = \boldsymbol{LU}$, and ($b$) using the fourth identity of Table 2.3.2 to write

$$\mathrm{Det}(\boldsymbol{A}) = \mathrm{Det}(\boldsymbol{L})\mathrm{Det}(\boldsymbol{U}) \tag{2.4.12}$$

The right-hand side is equal to the product of the diagonal elements of $\boldsymbol{L}$ and those of $\boldsymbol{U}$. Algorithms for carrying out the ***LU decomposition*** will be discussed in Section 2.6 and then again in Chapter 3.

2.5 *Computation of the Inverse*

Computing the inverse of an arbitrary matrix $\boldsymbol{A}$ on the basis of the right-hand side of equation (2.3.8) requires a prohibitively large amount of computational effort, even for matrices of moderate size. We must look for alternatives.

Computing A^{-1} Column-by-Column

The standard way of computing the inverse of an $N \times N$ matrix $\boldsymbol{A}$ is by solving N systems of linear equations for its N columns. Let us denote the jth column of $\boldsymbol{A}^{-1}$ by $\boldsymbol{c}^{(j)}$, where $j = 1, \ldots, N$. The definition $\boldsymbol{A}\boldsymbol{A}^{-1} = \boldsymbol{I}$, requires that $\boldsymbol{c}^{(j)}$ satisfy the linear system of equations

$$\boldsymbol{A}\boldsymbol{c}^{(j)} = \boldsymbol{e}^{(j)} \tag{2.5.1}$$

where all components of the vector $\boldsymbol{e}^{(j)}$ are equal to zero, except for the jth component that is equal to unity: $\boldsymbol{e}^{(j)}$ is the unit vector of the N-dimensional space pointing in the jth direction.

We have thus reduced the task of computing the inverse of an $N \times N$ matrix to the task of solving N systems of linear equations for $\boldsymbol{c}^{(j)}$, which can be done using a variety of methods discussed in Chapter 3.

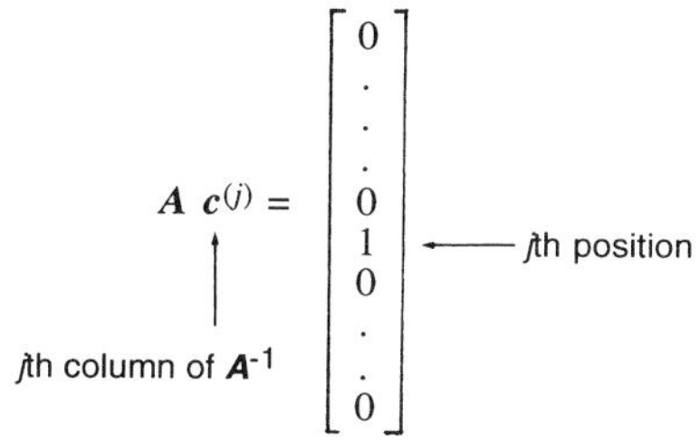

Computing A^{-1} from the *LU* Decomposition

An alternative method of computing the inverse of an arbitrary matrix $\boldsymbol{A}$ is based on its $\boldsymbol{LU}$ decomposition to be discussed in Section 2.6. If $\boldsymbol{A} = \boldsymbol{L}\boldsymbol{U}$, where $\boldsymbol{L}$ is a lower triangular matrix and $\boldsymbol{U}$ is an upper triangular matrix, then

$$\boldsymbol{A}^{-1} = \boldsymbol{U}^{-1}\boldsymbol{L}^{-1} \tag{2.5.2}$$

The two inverses $\boldsymbol{L}^{-1}$ and $\boldsymbol{U}^{-1}$ may be computed efficiently as follows.

Inverse of a Lower Triangular Matrix

Let us assume that none of the diagonal elements of the lower triangular matrix $\boldsymbol{L}$ is equal to zero, that is, $\boldsymbol{L}$ is nonsingular, and decompose it into a diagonal matrix $\boldsymbol{D}$ and a strictly lower triangular matrix $\boldsymbol{L}'$ (with zeros on the diagonal), so that

$$\boldsymbol{L} = \boldsymbol{D} + \boldsymbol{L}' \tag{2.5.3}$$

We know that $\boldsymbol{L}^{-1}$ is lower triangular and express it in the form

$$\boldsymbol{L}^{-1} = \boldsymbol{D}^{-1} + \boldsymbol{L}'' \tag{2.5.4}$$

where $\boldsymbol{L}''$ is strictly lower triangular. The diagonal elements of $\boldsymbol{D}^{-1}$ are the inverses of the diagonal elements of $\boldsymbol{D}$. Requiring that

$$(\boldsymbol{D} + \boldsymbol{L}')(\boldsymbol{D}^{-1} + \boldsymbol{L}'') = \boldsymbol{I} \tag{2.5.5}$$

and rearranging, we obtain the seemingly cumbersome expression

$$\boldsymbol{L}'' = -\boldsymbol{D}^{-1}\boldsymbol{L}'(\boldsymbol{D}^{-1} + \boldsymbol{L}'') \tag{2.5.6}$$

Inspection of the right-hand side, however, shows that the elements of $\boldsymbol{L}''$ can be determined successively one-by-one, for $\boldsymbol{L}'\boldsymbol{L}''$ is one element behind $\boldsymbol{L}''$ down each column at every stage.

Setting $\boldsymbol{L}^{-1} = \boldsymbol{B}$, we obtain the simple algorithm

Do $i = 1, N$
$B_{i,i} = 1/L_{i,i}$
END DO

Do $j = 1, N - 1$
Do $i = j + 1, N$
$$B_{i,j} = -\frac{1}{L_{i,i}} \sum_{m=j}^{i-1} L_{i,m} B_{m,j}$$
END DO
END DO
$\boldsymbol{L}^{-1} = \boldsymbol{B}$

(2.5.7)

Inverse of an Upper Triangular Matrix

The inverse of an upper triangular matrix $\boldsymbol{U}$ is another upper triangular matrix $\boldsymbol{B}$. Working as in the preceding subsection, we find that the columns of $\boldsymbol{B}$ may be computed successively according to the algorithm

Do $i = 1, N$
$B_{i,i} = 1/U_{i,i}$
END DO

Do $j = 2, N$
Do $i = j - 1, 1, -1$
$$B_{i,j} = -\frac{1}{U_{i,i}} \sum_{m=i+1}^{j} U_{i,m} B_{m,j}$$
END DO
END DO
$\boldsymbol{U}^{-1} = \boldsymbol{B}$

(2.5.8)

PROBLEM

2.5.1 ***Inverse of triangular matrices.***

Write a routine that computes the inverse of an upper triangular or a lower triangular matrix. In Problem 2.6.1, you will be asked to use this routine to compute the inverse of an arbitrary matrix.

Computing A^{-1} from the QR Decomposition

Yet another method of computing the inverse of an arbitrary matrix is based on the $\boldsymbol{QR}$ decomposition to be discussed in Section 2.7. If $\boldsymbol{A} = \boldsymbol{QR}$, where $\boldsymbol{R}$ is an upper (right) triangular matrix and $\boldsymbol{Q}$ is an orthogonal matrix, then

$$\boldsymbol{A}^{-1} = \boldsymbol{R}^{-1}\boldsymbol{Q}^T \qquad (2.5.9)$$

The inverse $\boldsymbol{R}^{-1}$ may be computed efficiently using the method described in the preceding subsection.

Improving an Approximation to the Inverse

Suppose that we have somehow obtained an approximation $\boldsymbol{B}$ to the inverse of an arbitrary matrix $\boldsymbol{A}$. If the approximation is good, then the product $\boldsymbol{AB}$ will be almost equal to $\boldsymbol{I}$; that is,

$$\boldsymbol{AB} = \boldsymbol{I} - \boldsymbol{E} \qquad (2.5.10)$$

where the elements of the error matrix $\boldsymbol{E}$ are small compared to unity. Premultiplying both sides of equation (2.5.10) by $\boldsymbol{A}^{-1}$, and rearranging the resulting expression, we find

$$\boldsymbol{A}^{-1} = \boldsymbol{B} + \boldsymbol{A}^{-1}\boldsymbol{E} \qquad (2.5.11)$$

Replacing $\boldsymbol{A}^{-1}$ on the right-hand side with the whole of the right-hand side, we find

$$\boldsymbol{A}^{-1} = \boldsymbol{B} + \boldsymbol{BE} + \boldsymbol{A}^{-1}\boldsymbol{E}^2 \qquad (2.5.12)$$

Repeating this substitution many times, we derive the infinite expansion

$$\boldsymbol{A}^{-1} = \boldsymbol{B}\,(\boldsymbol{I} + \boldsymbol{E} + \boldsymbol{E}^2 + \cdots) \qquad (2.5.13)$$

Since the elements of the matrix $\boldsymbol{E}$ are small compared to unity, the magnitudes of the elements of its power $\boldsymbol{E}^m$ will keep reducing as m becomes larger. The emerging algorithm involves (a) computing $\boldsymbol{E} = \boldsymbol{I} - \boldsymbol{AB}$, and ($b$) obtaining $\boldsymbol{A}^{-1}$ on the basis of equation (2.5.13).

Rearranging equation (2.5.10), we find

$$\boldsymbol{A}^{-1} = \boldsymbol{B}\,(\boldsymbol{I} - \boldsymbol{E})^{-1} \qquad (2.5.14)$$

We now compare the right-hand side of this equation with the right-hand side of equation (2.5.13), and this allows us to write

$$(\boldsymbol{I} - \boldsymbol{E})^{-1} = \boldsymbol{I} + \boldsymbol{E} + \boldsymbol{E}^2 + \cdots \qquad (2.5.15)$$

which is analogous to the well-known scalar expansion $1/(1-\varepsilon) = 1+\varepsilon+\varepsilon^2+\dots$. The analogy between scalar and matrix calculus is uncanny.

Inverse of a Matrix in Terms of a Polynomial Expansion

We are about to do something bold. We shall approximate the inverse of a matrix $\boldsymbol{A}$ with the inverse of the diagonal matrix $\boldsymbol{D}$ that arises by replacing the off-diagonal elements of $\boldsymbol{A}$ with zeros; in the notation of the preceding subsection, $\boldsymbol{B} = \boldsymbol{D}^{-1}$. Note that the computation of $\boldsymbol{B}$ involves a simple inversion of the diagonal elements of $\boldsymbol{D}$, which requires that none of the diagonal elements of $\boldsymbol{A}$ vanish. With this simple choice, the error matrix $\boldsymbol{E}$ defined in equation (2.5.10) is given by

$$\boldsymbol{E} \equiv \boldsymbol{I} - \boldsymbol{AB} = \boldsymbol{I} - \boldsymbol{AD}^{-1} = (\boldsymbol{D} - \boldsymbol{A})\,\boldsymbol{D}^{-1} \tag{2.5.16}$$

Cursory inspection reveals that the diagonal elements of $\boldsymbol{E}$ vanish,

$$E_{i,i} = 0, \tag{2.5.17}$$

and the off-diagonal elements are given by

$$D_{i,j} = -\frac{A_{i,j}}{A_{j,j}} \qquad \text{for } i \neq j \tag{2.5.18}$$

where summation is not implied over the repeated index j.

If the matrix $\boldsymbol{A}$ happens to be *lower or upper* triangular, then the matrix $\boldsymbol{E}$ is strictly lower or strictly upper triangular, which means that $\boldsymbol{E}^m = \boldsymbol{0}$ for $m \geq N$, yielding the *finite* expansion

$$\boldsymbol{A}^{-1} = \boldsymbol{D}^{-1}(\boldsymbol{I} + \boldsymbol{E} + \boldsymbol{E}^2 + \cdots + \boldsymbol{E}^{N-1}) \tag{2.5.19}$$

Equations (2.5.15) and (2.5.19) seemingly provide us with an attractive method of computing the inverse of an arbitrary matrix. Unfortunately, the sum on the right-hand side of the first equation converges only when the spectral radius of the matrix $\boldsymbol{E}$ is less than unity. Even then, computing the powers of $\boldsymbol{E}$ can be demanding, let alone likely to support the growth of the round-off error. Regrettably, these difficulties render this seemingly powerful method inefficient for practical computation.

2.6 LU and LDU Decompositions

In the preceding two sections, we made two references to the decomposition of a matrix $\boldsymbol{A}$ into the product of a lower triangular matrix $\boldsymbol{L}$ and an upper triangular matrix $\boldsymbol{U}$, so that $\boldsymbol{A} = \boldsymbol{LU}$. One reference was in regard to the computation of Det($\boldsymbol{A}$), and the second reference was in regard to the computation of $\boldsymbol{A}^{-1}$. In this section, we address the theoretical issue of whether this decomposition is feasible and unique, and the practical issue of how it can be achieved at a low cost.

How Many Decompositions?

The total number of unknown scalar elements of the requisite matrices $\boldsymbol{L}$ and $\boldsymbol{U}$ is equal to $K =$

$2 \cdot \frac{1}{2} \cdot N(N+1) = N^2 + N$. Setting the elements of the product $\boldsymbol{LU}$ equal to the corresponding elements of $\boldsymbol{A}$, we obtain a system of N^2 quadratic algebraic equations:

$$A_{i,j} = \sum_{m=1}^{\text{Min}(i,j)} L_{i,m} U_{m,j} \qquad (2.6.1)$$

It is clear that the decomposition carries $K - N^2 = N$ degrees of freedom: N elements of $\boldsymbol{L}$, $\boldsymbol{U}$, or both can be assigned arbitrary values, and the rest of them must be computed to satisfy equation (2.6.1).

We can arrive at this conclusion in a more formal way, by considering two particular decompositions $\boldsymbol{A} = \boldsymbol{L}_1\boldsymbol{U}_1 = \boldsymbol{L}_2\boldsymbol{U}_2$. A straightforward manipulation yields $\boldsymbol{U}_1 = \boldsymbol{L}_3\boldsymbol{U}_2$, where $\boldsymbol{L}_3 = \boldsymbol{L}_1^{-1}\boldsymbol{L}_2$ is a lower triangular matrix. But the product $\boldsymbol{L}_3\boldsymbol{U}_2$ can be an upper triangular matrix $\boldsymbol{U}_1$ only if $\boldsymbol{L}_3$ is a diagonal matrix $\boldsymbol{D}$, in which case $\boldsymbol{L}_2 = \boldsymbol{L}_1\boldsymbol{D}$ and $\boldsymbol{U}_2 = \boldsymbol{D}^{-1}\boldsymbol{U}_1$. The nonzero elements of $\boldsymbol{D}$ represent the aforementioned N degrees of freedom.

Doolittle Decomposition

In a standard method of $\boldsymbol{LU}$ decomposition, named after Doolittle, we stipulate that the diagonal elements of $\boldsymbol{L}$ be equal to unity. With this constraint, however, it is not certain that the decomposition will be feasible. For instance, if $A_{1,1} = 0$ the decomposition cannot be achieved.

When the decomposition exists, the matrices $\boldsymbol{L}$ and $\boldsymbol{U}$ can be found most efficiently using the Doolittle Algorithm 2.6.1, according to which same-numbered rows of $\boldsymbol{U}$ and columns of $\boldsymbol{L}$ are computed one after another in N passes. At the kth pass, we work in two stages:

Doolittle decomposition

$$\boldsymbol{L} = \begin{bmatrix} 1 & 0 & \dots & \dots & 0 \\ L_{2,1} & 1 & 0 & \dots & \dots \\ \dots & \dots & \ddots & 0 & \dots \\ \dots & \dots & \dots & 1 & 0 \\ L_{N,1} & L_{N,2} & L_{N,3} & \dots & 1 \end{bmatrix}$$

- First, we consider equations (2.6.1) corresponding to the elements in the kth *row* of $\boldsymbol{A}$, $A_{k,j}$ for $j = k, \dots, N$, and solve for the corresponding elements of $\boldsymbol{U}$.
- Second, we consider the equations corresponding to the elements in the kth *column* of $\boldsymbol{A}$, $A_{i,k}$ for $i = k, \dots, N$, and solve for the corresponding elements of $\boldsymbol{L}$.

ALGORITHM 2.6.1 Doolittle $\boldsymbol{LU}$ decomposition.

Do $i = 1, N$	
$\quad L_{i,i} = 1$	Diagonals are set equal to unity
$\quad U_{1,i} = A_{1,i}$	First row of $\boldsymbol{U}$ is first row of $\boldsymbol{A}$
$\quad L_{i,1} = A_{i,1}/U_{1,1}$	First column of $\boldsymbol{L}$ is easy to compute
END DO	
Do $k = 2, N$	Second and subsequent rows and columns
$\quad$ Do $j = k, N$	

$$U_{k,j} = A_{k,j} - \sum_{m=1}^{k-1} L_{k,m} U_{m,j}$$ Compute kth row of $\boldsymbol{U}$

END DO

Row pivoting of A and L is done here

Do $i = k+1, N$

$$L_{i,k} = \frac{1}{U_{k,k}} \left(A_{i,k} - \sum_{m=1}^{k-1} L_{i,m} U_{m,k} \right)$$ Compute kth column of $\boldsymbol{L}$

END DO

END DO

To economize the memory storage, we can place the elements of $\boldsymbol{L}$ and $\boldsymbol{U}$ at the corresponding positions of $\boldsymbol{A}$ as soon as the former are available. At the end, $\boldsymbol{L}$ and $\boldsymbol{U}$ will reside, respectively, at the lower and upper triangular blocks of $\boldsymbol{A}$. This practice endows the method with the qualities of a *compact scheme.* In developing and implementing a compact scheme, care must be taken so that variables that are to be used in subsequent operations are not overwritten.

A slight variation of the method just described produces successive *columns* of $\boldsymbol{U}$ and $\boldsymbol{L}$ in a sequential fashion. At the kth pass, we consider equations (2.6.1) corresponding to the elements in the kth column of $\boldsymbol{A}$, $A_{j,k}$, for $j = 1, \ldots, N$, and solve for the corresponding elements of $\boldsymbol{U}$; the computation of $\boldsymbol{L}$ remains unchanged. With reference to Algorithm 2.6.1, the only necessary modification is the replacement of the inner loop, Do $j = k, N$, with the loop

Do $j = 2, k$

$$U_{j,k} = A_{j,k} - \sum_{m=1}^{j-1} L_{j,m} U_{m,k}$$ Compute kth column of $\boldsymbol{U}$ (2.6.2)

END DO

A different way of carrying out the Doolittle decomposition is provided by the method of Gauss elimination discussed in Section 3.5 in the context of systems of linear equations.

LDU decomposition

After the decomposition has been completed, we can put the diagonal elements of $\boldsymbol{U}$ at the diagonal entries of a diagonal matrix $\boldsymbol{D}$ and write $\boldsymbol{U} = \boldsymbol{D}\boldsymbol{U}'$, where $\boldsymbol{U}'$ is an upper diagonal matrix with every diagonal element equal to 1, thereby obtaining the decompostion $\boldsymbol{A} = \boldsymbol{L}\boldsymbol{D}\boldsymbol{U}'$. The kth row of $\boldsymbol{U}'$ is equal to the kth row of $\boldsymbol{U}$ divided by $U_{k,k}$.

This seems like an academic exercise, but there is an important exception: When the matrix $\boldsymbol{A}$ is symmetric, the preceding manipulations suggest a way of avoiding the explicit computation of $\boldsymbol{L}$ and thus reducing the computational cost by a substantial factor. Specifically, requiring that $\boldsymbol{L}\boldsymbol{D}\boldsymbol{U}'$ be symmetric demands that the matrix $\boldsymbol{U}'$ be the transpose of $\boldsymbol{L}$ and vice versa; that is, $\boldsymbol{A} = \boldsymbol{L}\boldsymbol{D}\boldsymbol{L}^T = \boldsymbol{U}'^T\boldsymbol{D}\boldsymbol{U}'$. With reference to Algorithm 2.6.1, after we have computed the kth row of $\boldsymbol{U}$, we simply divide it by $U_{k,k}$ to produce the kth row of $\boldsymbol{U}$, which is identical to the kth column of $\boldsymbol{L}$.

Tridiagonal matrices

Consider now a tridiagonal matrix $\mathbf{A}$ and denote its nonzero elements as shown in equations (2.4.1). Using the formulas of Algorithm 2.6.1, we find that both matrices $\mathbf{L}$ and $\mathbf{U}$ are *bidiagonal*, and the superdiagonal line of $\mathbf{U}$ is identical to that of $\mathbf{A}$. Denoting the nonzero elements as

Doolittle decomposition of a tridiagonal matrix

$$\begin{bmatrix} a_1 & b_1 & 0 & \dots & \dots \\ c_2 & a_2 & b_2 & 0 & \dots \\ 0 & c_3 & a_3 & b_3 & 0 \\ 0 & 0 & \cdot & \cdot & \cdot \end{bmatrix} = \begin{bmatrix} 1 & 0 & 0 & \dots & \dots \\ d_2 & 1 & 0 & \dots & \dots \\ 0 & d_3 & 1 & 0 & \dots \\ 0 & 0 & \cdot & \cdot & \cdot \end{bmatrix} \begin{bmatrix} e_1 & b_1 & 0 & \dots & \dots \\ 0 & e_2 & b_2 & 0 & \dots \\ 0 & 0 & e_3 & b_3 & 0 \\ 0 & 0 & 0 & \cdot & \cdot \end{bmatrix}$$

$$L_{i,i} = 1, \qquad L_{i,i-1} = d_i, \qquad U_{i,i} = e_i, \qquad U_{i,i+1} = b_i \tag{2.6.3}$$

we find that the general Doolittle algorithm simplifies to

$$\boxed{\begin{array}{l} e_1 = a_1 \\ d_2 = c_2/e_1 \\ \text{Do } i = 2, N-1 \\ \quad e_i = a_i - d_i b_{i-1} \\ \quad d_{i+1} = c_{i+1}/e_i \\ \text{END DO} \\ e_N = a_N - d_N b_{N-1} \end{array}} \tag{2.6.4}$$

PROBLEMS

2.6.1 *Doolittle algorithm.*

(*a*) Write a computer program called *lu_d* that performs the $\mathbf{LU}$ decomposition of an arbitrary matrix according to the Doolittle algorithm.
(*b*) Run the program to decompose the matrix $\mathbf{A}$ given in equation (2.1.23), and compute Det($\mathbf{A}$).
(*c*) Combine the program with the subroutine of Problem 2.5.1 to compute the inverse of the matrix $\mathbf{A}$ given in equation (2.1.23).

2.6.2 *Hilbert matrix.*

The *Hilbert matrix* $\mathbf{H}$ with elements

$$H_{i,j} = \frac{1}{i+j-1} \tag{2.6.5}$$

arises in the approximation of functions with polynomials using the least-squares method, to be discussed in Section 8.2. Explicitly, the $N \times N$ Hilbert matrix is given by

$$\mathbf{H}^{(N)} = \begin{bmatrix} 1 & \frac{1}{2} & \frac{1}{3} & \dots & 1/N \\ \frac{1}{2} & \frac{1}{3} & \frac{1}{4} & \dots & 1/(N+1) \\ \dots & \dots & \dots & \dots & \dots \\ 1/N & 1/(N+1) & \dots & \dots & 1/(2N-1) \end{bmatrix} \tag{2.6.6}$$

Use the program of Problem 2.6.1 to compute the determinants of $\boldsymbol{H}^{(1)}, \boldsymbol{H}^{(2)}, \boldsymbol{H}^{(3)}, \boldsymbol{H}^{(4)}, \boldsymbol{H}^{(5)}$. Based on the numerical results, assess the limiting value of the determinant as N tends to infinity.

Crout Decomposition

Crout decomposition

$$U = \begin{bmatrix} 1 & U_{1,2} & U_{1,3} & \cdots & U_{1,N} \\ 0 & 1 & U_{2,3} & \cdots & U_{2,N} \\ \cdots & 0 & \ddots & \cdots & \cdots \\ \cdots & \cdots & 0 & 1 & 0 \\ 0 & \cdots & \cdots & 0 & 1 \end{bmatrix}$$

The Crout decomposition emerges by stipulating that the diagonal elements of the matrix $\boldsymbol{U}$ are equal to unity. At the kth stage, we consider equations (2.6.1) corresponding to the elements in the kth *column* of $\boldsymbol{A}$, $A_{i,k}$ for $i = k, \ldots, N$, and solve for the corresponding elements of $\boldsymbol{L}$; we then consider the equations corresponding to the elements in the kth *row* of $\boldsymbol{A}$, $A_{k,j}$ for $j = k, \ldots, N$, and solve for the corresponding elements of $\boldsymbol{U}$. The computations are done according to Algorithm 2.6.2.

The Doolittle decomposition derives from the Crout decomposition, and vice versa, by a simple matrix transformation. If $\boldsymbol{L}^C$ and $\boldsymbol{U}^C$ are the Crout matrices, and if $\boldsymbol{D}$ is the diagonal component of $\boldsymbol{L}^C$, then the Doolittle matrices are

$$\boldsymbol{L}^D = \boldsymbol{L}^C \boldsymbol{D}^{-1}, \qquad \boldsymbol{U}^D = \boldsymbol{D}\boldsymbol{U}^C \tag{2.6.7}$$

where the diagonal elements of $\boldsymbol{D}^{-1}$ are the inverses of corresponding elements of $\boldsymbol{D}$. A generalization of this transformation is discussed in Problem 2.6.5.

PROBLEMS

2.6.3 ***Crout decomposition.***

Repeat Problem 2.6.1 with the Crout decomposition. The computer program will now be called *lu_c*.

2.6.4 ***Crout decomposition of a tridiagonal matrix.***

Develop an algorithm, similar to algorithm (2.6.4), for the Crout decomposition of a tridiagonal matrix, and confirm its reliability with an example.

2.6.5 ***General form of the LU decomposition.***

Develop a method that, given the Crout matrices $\boldsymbol{L}^C$ and $\boldsymbol{U}^C$, produces an $\boldsymbol{LU}$ decomposition, where N diagonal elements of $\boldsymbol{L}$ or $\boldsymbol{U}$ have specified values.

ALGORITHM 2.6.2 Crout $\boldsymbol{LU}$ decomposition.

Do $i = 1, N$	
$\quad U_{i,i} = 1$	Diagonals are set equal to unity
$\quad L_{i,1} = A_{i,1}$	First column of $\boldsymbol{L}$ is first column of $\boldsymbol{A}$
$\quad U_{1,i} = A_{1,i}/L_{1,1}$	First row of $\boldsymbol{U}$ is easy to compute
End Do	

Do $k = 2, N$

Do $i = k, N$

$$L_{i,k} = A_{i,k} - \sum_{m=1}^{k-1} L_{i,m} U_{m,k}$$ Compute kth column of $\boldsymbol{L}$

End Do

Row pivoting of A and L is done here

Do $j = k + 1, N$

$$U_{k,j} = \frac{1}{L_{k,k}} \left(A_{k,j} - \sum_{m=1}^{k-1} L_{k,m} U_{m,j} \right)$$ Compute kth column of $\boldsymbol{U}$

End Do

End Do

Pivoting and Gauss Elimination

It is possible that, in the course of the computations, one of the denominators involved in the equations of the preceding algorithms will be equal to zero; this means that the decomposition cannot be carried out. Furthermore, because one of the denominators may be very small, the decomposition may exist in principle, but the results may be notably sensitive to the round-off error. To circumvent these difficulties, we re-order the rows or columns of the matrix $\boldsymbol{A}$ in a dynamic manner in the course of the computations, in a process that is called *pivoting*.

We defer the discussion of pivoting to Section 3.5, where we explain it in the context of solving linear systems of equations by Gauss elimination. *Row pivoting* during the $\boldsymbol{LU}$ decomposition, in particular, is performed as discussed in Section 3.5. The pivoting module is placed at the marked places in Algorithms 2.6.1 and 2.6.2. Pivoting is not necessary for *symmetric and positive-definite matrices*, or more generally *diagonally dominant* matrices (a matrix is diagonally dominant when the norm of each diagonal element is greater than the sum of the norm of the rest of the elements in the corresponding row).

It is important to keep in mind that the determinant and inverse of the modified matrix that arises by interchanging the rows or columns of $\boldsymbol{A}$ are related to those of the original matrix as described on page 72.

One advantage of the *explicit* Doolittle or Crout $\boldsymbol{LU}$ decomposition methods discussed in this section, compared with the Gauss elimination method, concerns the behavior of the round-off error. Certain computers are built to accumulate, by default, inner vector products in double precision and, if instructed, produce the result in single precision. This practice saves CPU time required in the normal operation with double-length memory blocks. Unlike explicit decomposition methods, the Gauss elimination method does not involve operations in the form of inner vector products. Thus the explicit methods are more accurate when accumulation in double precision is available.

Cholesky's Algorithms

When the matrix $\boldsymbol{A}$ is symmetric, comparing the number of unknowns to the number of equations shows that it is permissible to set $\boldsymbol{L}$ equal to the transpose of $\boldsymbol{U}$, in which case $\boldsymbol{A} = \boldsymbol{L}\boldsymbol{L}^T$. When, in addition, $\boldsymbol{A}$ is real and positive-definite, the elements of $\boldsymbol{L}$ are all real. The columns of $\boldsymbol{L}$ may then be computed successively according to Cholesky's Algorithm 2.6.3, sometimes attributed to Banachiewicz.

Cholesky decomposition

$$A = \begin{bmatrix} L_{1,1} & 0 & 0 & \cdots & 0 \\ L_{2,1} & L_{2,2} & 0 & \cdots & \cdots \\ \cdots & \cdots & \ddots & 0 & 0 \\ \cdots & \cdots & \cdots & L_{N\text{-}1,N\text{-}1} & 0 \\ L_{N,1} & L_{N,2} & \cdots & \cdots & L_{N,N} \end{bmatrix} \begin{bmatrix} L_{1,1} & L_{2,1} & \cdots & \cdots & L_{N,1} \\ 0 & L_{2,2} & \cdots & \cdots & L_{N,2} \\ 0 & 0 & \ddots & \cdots & \cdots \\ \cdots & \cdots & 0 & L_{N\text{-}1,N\text{-}1} & \cdots \\ 0 & \cdots & 0 & 0 & L_{N,N} \end{bmatrix}$$

Another version of Cholesky's algorithm produces the elements of $\boldsymbol{L}$ row-by-row, as shown in Algorithm 2.6.4.

PROBLEM

2.6.6 *Cholesky decomposition of a tridiagonal matrix.*

Develop an algorithm, similar to Algorithm (2.6.3), for the Cholesky decomposition of a tridiagonal matrix, and test its reliability with an example.

ALGORITHM 2.6.3 Cholesky's algorithm for the $\boldsymbol{LL}^T$ decomposition by column.

$L_{1,1} = A_{1,1}^{1/2}$

Do $i = 2, N$

$$L_{i,1} = \frac{A_{i,1}}{L_{1,1}}$$

End Do

Do $j = 2, N$

$$L_{j,j} = \left(A_{j,j} - \sum_{m=1}^{j-1} L_{j,m}^2 \right)^{1/2}$$

Do $i = j + 1, N$

$$L_{i,j} = \frac{1}{L_{j,j}} \left(A_{i,j} - \sum_{m=1}^{j-1} L_{i,m} L_{j,m} \right)$$

End Do

End Do

ALGORITHM 2.6.4 Cholesky's algorithm for the $\boldsymbol{LL}^T$ decomposition by row.

$L_{1,1} = A_{1,1}^{1/2}$

Do $i = 2, N$

Do $j = 1, i - 1$

$$L_{i,j} = \frac{1}{L_{j,j}} \left(A_{i,j} - \sum_{m=1}^{j-1} L_{i,m} L_{j,m} \right)$$

END DO

$$L_{i,i} = \left(A_{i,i} - \sum_{m=1}^{i-1} L_{i,m}^2 \right)^{1/2}$$

END DO

2.7 *QR* Decomposition

We want to know if it is possible to express a matrix $\boldsymbol{A}$ as the product of (*a*) an orthogonal—or more generally unitary—matrix $\boldsymbol{Q}$, and (*b*) an upper triangular matrix $\boldsymbol{R}$, so that

$$\boldsymbol{A} = \boldsymbol{QR} \tag{2.7.1}$$

The familiar designation of an upper triangular matrix as $\boldsymbol{U}$ is temporarily abandoned to follow standard convention.

The motivation for this decomposition is threefold: The $\boldsymbol{QR}$ decomposition provides us with a nifty way of computing the inverse of $\boldsymbol{A}$ according to equation (2.5.9); it allows us to solve difficult systems of linear equations, as will be discussed in Section 3.3; and it may be used to compute the eigenvalues of large matrices, as will be discussed in Section 5.6.

The *QR* Decomposition Is not Unique

To assess how much flexibility we have, we first investigate whether the $\boldsymbol{QR}$ decomposition is unique. For simplicity, we confine our attention to real matrices. Let us assume that there are two sets of matrices $\boldsymbol{Q}$ and $\boldsymbol{R}$ so that

$$\boldsymbol{A} = \boldsymbol{Q}_1\boldsymbol{R}_1 = \boldsymbol{Q}_2\boldsymbol{R}_2 \tag{2.7.2}$$

which can be rearranged to yield

$$\boldsymbol{R}_1\boldsymbol{R}_2^{-1} = \boldsymbol{Q}_1^T\boldsymbol{Q}_2 \tag{2.7.3}$$

The left-hand side is an upper triangular matrix, whereas the right-hand side is an orthogonal matrix. But the only real upper triangular matrix that is also orthogonal is the diagonal matrix $\boldsymbol{I}^{1/2}$, where the square roots of the diagonal elements are equal to ± 1. We thus find $\boldsymbol{R}_1 = \boldsymbol{I}^{1/2}\boldsymbol{R}_2$, which shows that the signs of the elements of $\boldsymbol{R}$ in each row may be switched independently and arbitrarily, providing us with N degrees of freedom and allowing for 2^N possible combinations.

Next, we discuss methods of computing particular pairs of matrices $\boldsymbol{Q}$ and $\boldsymbol{R}$ beginning with a method that has already been developed.

QR Decomposition by Gram–Schmidt Orthogonalization

Let us identify the collection of vectors $\boldsymbol{v}^{(i)}$, $i = 1, \ldots, N$, with the columns of an $N \times N$ matrix $\boldsymbol{A}$, and use the Gram–Schmidt orthogonalization process described in Algorithm 2.1.1 to compute the collection of *orthogonal* vectors $\mathbf{u}^{(i)}$, $i = 1, \ldots, N$. After we are done, we put the orthogonal vectors at the columns of a matrix $\mathbf{Q}'$ as shown in the diagram. Inspection of the Gram–Schmidt algorithm shows that

$$\mathbf{A} = \mathbf{Q}'\mathbf{R}' \tag{2.7.4}$$

where $\mathbf{R}'$ is an upper $N \times N$ triangular matrix with diagonal elements equal to unity and upper triangular elements equal to the coefficients $\alpha_{i,j}$ defined in Algorithm 2.1.1; $R'_{i,j} = \alpha_{i,j}$. The matrix $\mathbf{Q}'$, however, is not generally orthonormal: The lengths of the vectors represented by its rows are not equal to unity.

$$\mathbf{A} = \underbrace{\begin{bmatrix} \uparrow & \uparrow & & \uparrow & \uparrow \\ \mathbf{u}^{(1)} & \mathbf{u}^{(2)} & \ldots & \mathbf{u}^{(N-1)} & \mathbf{u}^{(N)} \\ \downarrow & \downarrow & & \downarrow & \downarrow \end{bmatrix}}_{\mathbf{Q}'} \underbrace{\begin{bmatrix} 1 & \alpha_{1,2} & \alpha_{1,3} & \ldots & \alpha_{1,N} \\ 0 & 1 & \alpha_{2,3} & \ldots & \alpha_{2,N} \\ 0 & 0 & \ddots & \ldots & \ldots \\ \ldots & \ldots & 0 & 1 & \ldots \\ 0 & \ldots & 0 & 0 & 1 \end{bmatrix}}_{\mathbf{R}'}$$

To obtain the $\mathbf{QR}$ decomposition, we introduce the diagonal matrix $\mathbf{D}$ whose ith diagonal element is equal to the norm of $\mathbf{u}^{(i)}$, that is,

$$D_{i,i} \equiv (\mathbf{u}^{(i)T}\mathbf{u}^{(i)})^{1/2} \tag{2.7.5}$$

where summation over i is not implied. It may readily be seen by inspection that the matrices

$$\mathbf{Q} = \mathbf{Q}'\mathbf{D}^{-1}, \qquad \mathbf{R} = \mathbf{D}\mathbf{R}' \tag{2.7.6}$$

satisfy equation (2.7.1) and are consistent with its underlying definitions. The columns of the orthogonal matrix $\mathbf{Q}$ are the vectors $\mathbf{w}^{(i)}$, $i = 1, \ldots, N$, produced by the modified Gram–Schmidt Algorithm 2.1.2.

QR Decomposition by Reflection

There is a more expedient method of carrying out a $\mathbf{QR}$ decomposition, based on a seemingly simple observation: Let $\mathbf{A}$ be an arbitrary real matrix, and suppose that we have found an orthogonal matrix $\mathbf{P}$ so that

$$\mathbf{PA} = \mathbf{R} \tag{2.7.7}$$

Then, we can be sure that

$$\mathbf{Q} = \mathbf{P}^T \tag{2.7.8}$$

The matrix $\mathbf{P}$ can be computed as the product of $N - 1$ real Householder matrices introduced in Section 2.3,

$$H_{i,j}^{(m)} = \delta_{i,j} - 2w_i^{(m)}w_j^{(m)}, \qquad m = 1, \ldots, N-1 \tag{2.7.9}$$

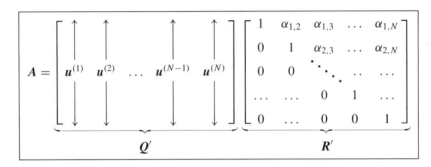

which are designed so as to make the elements of successive columns of $\boldsymbol{A}$ below the diagonal vanish. Then,

$$\boldsymbol{P} = \boldsymbol{H}^{(N-1)}\boldsymbol{H}^{(N-2)} \ldots \boldsymbol{H}^{(2)}\boldsymbol{H}^{(1)} \tag{2.7.10}$$

and $\boldsymbol{Q} = \boldsymbol{P}^T = \boldsymbol{H}^{(1)}\boldsymbol{H}^{(2)} \ldots \boldsymbol{H}^{(N-2)}\boldsymbol{H}^{(N-1)}$.

Projection to a hyperplane

To illustrate how the matrix $\boldsymbol{H}^{(m)}$ can be built, we introduce the real vector

$$\boldsymbol{w}^{(m)} = (0, \ldots, 0, w_m, \ldots, w_N)^T \tag{2.7.11}$$

select an arbitrary vector $\boldsymbol{u}$, and compute the new vector

$$\boldsymbol{v} = \boldsymbol{H}^{(m)}\boldsymbol{u} \tag{2.7.12}$$

We want to adjust the $N - m + 1$ elements $w_m, \ldots, w_N$ so that the last $N - m$ elements of $\boldsymbol{v}$ vanish, that is,

$$\boldsymbol{v} = (v_1, v_2, \ldots, v_m, 0, 0, \ldots, 0)^T \tag{2.7.13}$$

subject to the constraint that the length of $\boldsymbol{w}$ is equal to unity. The number of unknowns is equal to the number of equations, and the statement of the problem is complete.

To compute $w_m, \ldots, w_N$, we recast equation (2.7.12) into the more explicit form

$$\boldsymbol{u} - 2\boldsymbol{w}^{(m)}(\boldsymbol{u}^T\boldsymbol{w}^{(m)}) = \boldsymbol{v} \tag{2.7.14}$$

and take the inner product of both sides with $\boldsymbol{w}^{(m)}$ to find

$$\boldsymbol{u}^T\boldsymbol{w}^{(m)} = -w_m v_m \tag{2.7.15}$$

where summation over m is *not* implied in this and in the subsequent equations. Substituting the right-hand side of equation (2.7.15) in place of the term enclosed by the parentheses on the left-hand side of equation (2.7.14), we find

$$\boldsymbol{u} + 2w_m v_m \boldsymbol{w}^{(m)} = \boldsymbol{v} \tag{2.7.16}$$

which reveals that

$$\begin{aligned} v_1 &= u_1 \\ v_2 &= u_2 \\ &\ldots \\ v_{m-1} &= u_{m-1} \end{aligned} \tag{1.7.17}$$

but

$$v_m \neq u_m \tag{2.7.18}$$

Since $\boldsymbol{H}^{(m)}$ is orthogonal, the lengths of the vectors $\boldsymbol{u}$ and $\boldsymbol{v}$ are equal, and this requires that

$$v_m^2 = \sum_{i=m}^{N} u_i^2 \equiv s^2 \tag{2.7.19}$$

With v_m being a known, we return to equation (2.7.16) and obtain the desired equations:

$$\begin{aligned} w_m &= \left[\tfrac{1}{2}\left(1 \pm \frac{u_m}{s}\right)\right]^{1/2} \\ w_i &= \pm\frac{u_i}{2sw_m} \qquad \text{for } i = m+1, \ldots, N \end{aligned} \tag{2.7.20}$$

Since $|u_m| < s$, the right-hand side of the first of these equations is real. To reduce the possibility of large round-off error due to division by a large number in the computation of w_i, we select the plus or minus sign in the first of equations (2.7.20) so as to send w_m as far away from zero as possible. If it should happen that $w_m = 0$, then $\boldsymbol{w} = \boldsymbol{0}$ and $\boldsymbol{H}^{(m)} = \boldsymbol{I}$, in which case there is no need to carry out the projection.

In practice, there is no need to compute the square root on the right-hand side of the first of equations (2.7.20), for either it will cancel itself, or it will be squared within each term during the computation of $\boldsymbol{H}^{(m)}$.

Building the reflection matrices

The manner in which the projection matrices $\boldsymbol{P}^{(m)}$ arise is now evident. They are found by applying formulas (2.7.19) and (2.7.20) $N-1$ times, each time identifying the vector $\boldsymbol{u}$ with successive columns of the updated matrix $\boldsymbol{A}^{(m)} = \boldsymbol{H}^{(m)}\boldsymbol{H}^{(m-1)}\ldots\boldsymbol{H}^{(2)}\boldsymbol{H}^{(1)}\boldsymbol{A}$, where $m = 1, 2, \ldots, N-1$. This algorithm generates a sequence of matrices, obtained most effectively by first computing the vector

$$\boldsymbol{c}^T = \boldsymbol{w}^{(m)T}\boldsymbol{A}^{(m-1)} \tag{2.7.21}$$

and then setting

$$\boldsymbol{A}^{(m)} = \boldsymbol{A}^{(m-1)} - 2\boldsymbol{w}^{(m)}\boldsymbol{c}^T \tag{2.7.22}$$

where $\boldsymbol{A}^{(N-1)} = \boldsymbol{R}$. The sequence of computations is shown in Algorithm 2.7.1.

ALGORITHM 2.7.1 ***QR*** decomposition of a matrix $\boldsymbol{A}$ by the Householder reflection method.

```
Q = I
Do m = 1, N − 1
  Do j = m, N
  u(j) = A(j, m)
  End Do
```

$$s = \left(\sum_{i=m}^{N} u_i^2\right)^{1/2}$$

$$w(m) = \begin{cases} \left[\frac{1}{2}\left(1 + \frac{u(m)}{s}\right)\right]^{1/2} & \textit{if } u(m) > 0 \\ \left[\frac{1}{2}\left(1 - \frac{u(m)}{s}\right)\right]^{1/2} & \textit{if } u(m) < 0 \end{cases}$$

Do $i = m + 1, N$

$$w(i) = \pm \frac{u(i)}{2s\,w(m)}$$

End Do

$$\mathbf{c}^T = \mathbf{w}^T \mathbf{A}$$

$$\mathbf{A} \leftarrow \mathbf{A} - 2\mathbf{w}\mathbf{c}^T$$

$$\mathbf{Q} \leftarrow \mathbf{Q}(\mathbf{I} - 2\mathbf{w}\mathbf{w}^T)$$

End Do

QR Decomposition by Rotation

A third way to accomplish the $\boldsymbol{QR}$ decomposition proceeds by postmultiplying the matrix $\boldsymbol{A}$ with a sequence of plane rotation matrices $\boldsymbol{P}^{(i,j)}$ defined in equation (2.3.15), using appropriate values for θ, so that $\boldsymbol{A}$ eventually reduces to the triangular matrix $\boldsymbol{R}$ (see Problem 2.3.7). The method targets the annihilation of successive elements of successive *columns* below the diagonal. Various improvements can be made to avoid the explicit computation of $\boldsymbol{P}^{(i,j)}$, resulting in Algorithm 2.7.2 (Cullen 1994, p. 131).

The method requires a higher number of operations than the Householder reflection method but is simpler in both conception and implementation.

PROBLEM

2.7.1 *QR decomposition by three methods.*

Carry out the $\boldsymbol{QR}$ decomposition of the matrix shown in equation (2.1.23) using (*a*) the Gram–Schmidt orthogonalization method, (*b*) the Householder reflection-matrix method, and (*c*) the rotation-matrix method. Verify the consistency of your results, and discuss the relative efficiency of the three approaches.

ALGORITHM 2.7.2 $\boldsymbol{QR}$ decomposition of an $N \times N$ matrix $\boldsymbol{A}$ using the plane-rotation method; $\boldsymbol{I}$ is the $N \times N$ identity matrix.

$$\mathbf{B} = [\mathbf{A}|\mathbf{I}]$$

Do $j = 1, N - 1$

Do $i = j + 1, N$

$$p = \sqrt{B_{j,j}^2 + B_{i,j}^2}$$

$$c = B_{j,j}/p$$

$$s = B_{i,j}/p$$

Do $k = j, 2N$

$$t_1 = cB_{j,k} + sB_{i,k}$$

$$t_2 = -s B_{j,k} + c B_{i,k}$$
$$B_{j,k} = t_1$$
$$B_{i,k} = t_2$$

END DO

END DO

END DO

$[\boldsymbol{R}|\boldsymbol{Q}^T] = \boldsymbol{B}$

2.8 Systems of Linear Algebraic Equations

Given an N-dimensional column vector $\boldsymbol{b}$ and an $N \times N$ matrix $\boldsymbol{A}$, we want to find another N-dimensional column vector $\boldsymbol{x}$ so that

$$\boldsymbol{A}\boldsymbol{x} = \boldsymbol{b} \tag{2.8.1}$$

In physical terms, we want to find a set of physical or initial conditions represented by the vector $\boldsymbol{x}$, so that when we operate on them by a process represented by the matrix $\boldsymbol{A}$, we obtain a desired set of physical or final conditions represented by the vector $\boldsymbol{b}$.

The vector equation (2.8.1) encompasses a system of N linear equations for the N scalar components of $\boldsymbol{x}$. The ways in which such systems arise in practice will be discussed in Chapter 3; an example is given in Problem 2.8.1.

PROBLEM

2.8.1 ***Wheatstone bridge.***

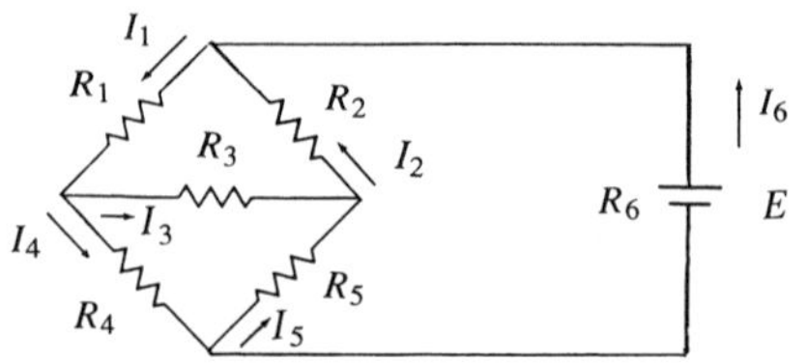

Consider the electrical network depicted in the diagram. The electrical current is driven by a voltage E imposed across the poles of a battery in branch 6. The battery can be regarded as an electron pump, analogous to a fluid pump. Given the electrical resistances R_i of branches 1, 2, 4, 5, 6, we want to compute the currents I_i.

Kirchhoff's first law requires that the algebraic sum of the currents approaching any node vanishes. Kirchhoff's second law requires that the voltage around a closed circuit that does not contain an electron pump vanishes. Ohm's law requires that the electrical current through a circuit that does not contain an electron pump is related to the corresponding voltage drop E_i by $E_i = R_i I_i$.

(*a*) Using Kirchhoff's laws and Ohm's law, derive the system of linear equations

$$\begin{bmatrix} R_1 + R_2 + R_3 & -R_3 & -R_2 \\ -R_3 & R_3 + R_4 + R_5 & -R_5 \\ -R_2 & -R_6 & R_2 + R_5 + R_6 \end{bmatrix} \begin{bmatrix} I_1 \\ I_4 \\ I_6 \end{bmatrix} = \begin{bmatrix} 0 \\ 0 \\ E \end{bmatrix} \tag{2.8.2}$$

The coefficient matrix on the left-hand side is the *resistance matrix* of the bridge; its inverse is the *capacitance matrix*. The symmetry of these matrices is a consequence of the physical symmetry of the circuit with respect to the third branch.

(*b*) Discuss whether it is possible that the resistance matrix may become singular.

Cramer's Rule

When the matrix $\boldsymbol{A}$ is nonsingular, the unknown vector $\boldsymbol{x}$ can be found in a straightforward but not necessarily efficient manner by first computing $\boldsymbol{A}^{-1}$, and then setting

$$\boldsymbol{x} = \boldsymbol{A}^{-1}\boldsymbol{b} \tag{2.8.3}$$

Using the expression for $\boldsymbol{A}^{-1}$ in terms of the matrix of cofactors $\boldsymbol{C}$ given in equation (2.3.8), we obtain *Cramer's rule*:

$$x_i = \frac{1}{\mathrm{Det}(\boldsymbol{A})} b_j C_{j,i} = \frac{\mathrm{Det}(\boldsymbol{A}^{(i)})}{\mathrm{Det}(\boldsymbol{A})} \tag{2.8.4}$$

where the matrix $\boldsymbol{A}^{(i)}$ arises by replacing the ith column of $\boldsymbol{A}$ with the column vector $\boldsymbol{b}$.

Unfortunately, this method of computing $\boldsymbol{x}$ requires an exorbitant amount of computational effort even for systems of moderate size. In practice, the solution is found using the algorithms discussed in Chapter 3.

PROBLEM

2.8.2 *Solution of an underdetermined homogeneous system.*

Let $\boldsymbol{A}$ be an $N \times (N+1)$ matrix, and $\boldsymbol{x}$ be an $(N+1)$-dimensional vector. Show that a nontrivial solution of the homogeneous equation $\boldsymbol{A}\boldsymbol{x} = \boldsymbol{0}$ is given by

$$x_i = (-1)^i \,\mathrm{Det}\left(\boldsymbol{A}^{(i)}\right) \tag{2.8.5}$$

where $\boldsymbol{A}^{(i)}$ is the $N \times N$ matrix that arises by rejecting the ith column of $\boldsymbol{A}$ and compressing the resulting matrix.

Singular Matrices and Linear Mapping

What happens when the matrix $\boldsymbol{A}$ is singular and its inverse does not exist? To gain insights into this problematic case, we regard the matrix $\boldsymbol{A}$ as a factory that receives an input vector $\boldsymbol{x}$ and produces a new vector $\boldsymbol{y}$, running the engine

$$\boldsymbol{y} = \boldsymbol{A}\boldsymbol{x} \tag{2.8.6}$$

as illustrated in Figure 2.8.1(a). This point of view establishes a *linear mapping* between the space of the input vectors $\boldsymbol{x}$ and the space of the output vectors $\boldsymbol{y}$, called, respectively, the *domain of definition* and the *image or range* of the mapping. The vector $\boldsymbol{y}$ is the *image* or projection of the vector $\boldsymbol{x}$ subject to the linear mapping associated with the matrix $\boldsymbol{A}$; in this light, $\boldsymbol{A}$ is a projection matrix.

The rules of matrix–vector multiplication show that the vector $\boldsymbol{y}$ is a linear combination of the column vectors of $\boldsymbol{A}$; the factory has a limited line of products. In more formal terms, *the range of the mapping is the space of vectors that can be expressed as linear combinations of column vectors of the matrix* $\boldsymbol{A}$.

When the matrix $\boldsymbol{A}$ is nonsingular, the column vectors of $\boldsymbol{A}$ are linearly independent: Any N-dimensional vector can be expressed as a linear combination of them, and the aforementioned mapping is one-to-one. For every $\boldsymbol{x}$ there is a unique corresponding $\boldsymbol{y} = \boldsymbol{A}\boldsymbol{x}$, and for every $\boldsymbol{y}$ there is a unique corresponding $\boldsymbol{x}$ that is equal to $\boldsymbol{A}^{-1}\boldsymbol{y}$.

Consider now a nonsingular matrix, and modify one or several of its elements so as to make its determinant nearly equal to zero, that is, so as to make the matrix nearly singular. In this limit, the column vectors tend

(a) (b) (c)

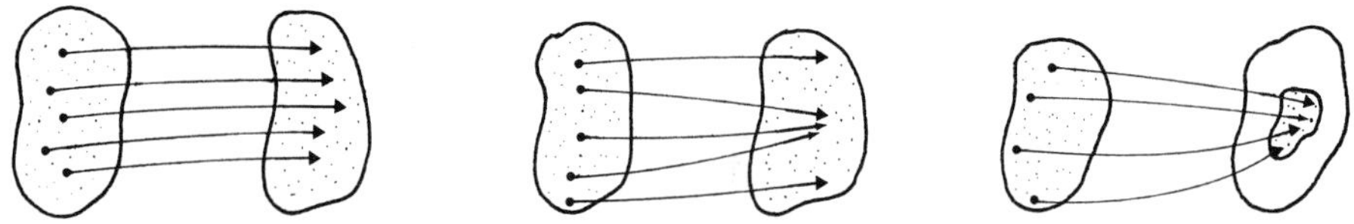

Figure 2.8.1 Mapping the N-dimensional vector $\boldsymbol{x}$ to the N-dimensional vector $\boldsymbol{y}$ by means of an $N \times N$ matrix $\boldsymbol{A}$, so that $\boldsymbol{y} = \boldsymbol{A}\boldsymbol{x}$. The matrix $\boldsymbol{A}$ is (a) nonsingular, (b) nearly singular, and (c) singular. In the third case, the range of the mapping is manifold whose dimension, called the rank of matrix $\boldsymbol{A}$, is less than N.

to become linearly dependent, and the range of the mapping is dilute everywhere except near an isolated island where it is dense, as illustrated in Figure 2.8.1(b). When the determinant of $\boldsymbol{A}$ is exactly equal to zero, the range of the mapping is a *manifold* whose dimension is lower than N, as illustrated in Figure 2.8.1(c).

Matrix rank

The minimum number of linearly independent vectors that are necessary to describe an arbitrary vector that lies in the image manifold of a matrix $\boldsymbol{A}$ as a linear combination of them, is the matrix *rank*. When Rank($\boldsymbol{A}$) $= 0$, the manifold is the null point $\boldsymbol{0}$; when Rank($\boldsymbol{A}$) $= 1$, the manifold is a line passing through the null point; when Rank($\boldsymbol{A}$) $= 2$, the manifold is a plane containing the null point; when Rank($\boldsymbol{A}$) $= N$, the manifold is the whole N-dimensional space.

Solution of a singular system

If the matrix $\boldsymbol{A}$ is singular and the vector $\boldsymbol{b}$ happens to reside within the image manifold of $\boldsymbol{A}$, then equation (2.8.1) will have an infinity of solutions. But if $\boldsymbol{b}$ does not reside within the manifold, the equation will not have a solution.

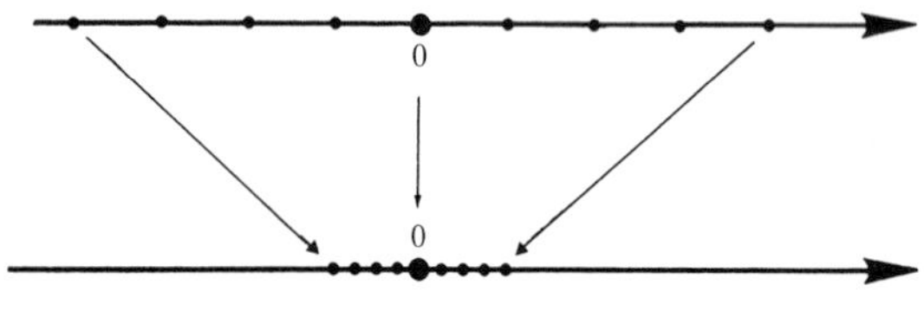

Consider, for example, the 1×1 matrix $[a]$. As long as $a \neq 0$, the domain of definition and the range of the linear mapping expressed by the equation $y = ax$ are the whole real axis. As the value of the coefficient a approaches zero, the images of an array of evenly spaced points on the real axis tend to concentrate in the vicinity of the origin, and the mapping tends to become singular. When $a = 0$, all images fall at the origin $y = 0$, and the image manifold is comprised of the single point $y = 0$: The rank of the matrix is equal to zero.

The image manifold of a singular 2×2 matrix can be either the null point or a line passing through the null point. In the first case, all elements of the matrix are equal to zero. In the second case, one of the matrix columns or rows is a multiple of the other: The columns and rows are linearly dependent; one implies the other.

Further properties of linear mapping mediated by matrices will be discussed in Section 2.11.

Homogeneous Systems and the Null Space

We are now in a position to answer the question: Does the homogeneous equation

$$\boldsymbol{A}\boldsymbol{x} = \boldsymbol{0} \tag{2.8.7}$$

have a solution apart from the obvious solution $\boldsymbol{x} = \boldsymbol{0}$? The answer is affirmative, provided that the rank of the matrix $\boldsymbol{A}$ is less than N; that is, $\text{Det}(\boldsymbol{A}) = 0$. In particular, if $\text{Rank}(\boldsymbol{A}) = N - m$, where m is the deficiency of the range, then there is an m-parameter family of such solutions. If $m = 0$, the homogeneous equation does not have a solution.

The space of vectors $\boldsymbol{x}$ that satisfy the homogeneous equation (2.8.7) is called the *null space* or the *kernel* of the matrix $\boldsymbol{A}$. The dimension of the null space $\text{Dim}(Nu(\boldsymbol{A}))$ is the maximum number of linearly independent vectors that reside in it.

Sylvester's law of nullity identifies the deficiency of the range m with $\text{Dim}(Nu(\boldsymbol{A}))$, stating that

$$\text{Rank}(\boldsymbol{A}) + \text{Dim}(Nu(\boldsymbol{A})) = N \tag{2.8.8}$$

Thus to assess the rank of a matrix, all we need to do is compute the dimension of its null space.

PROBLEM

2.8.3 *Image manifold of* 2×2 *matrices.*

Derive and discuss the image manifold of a 2×2 singular matrix with rank 1 of your choice.

2.9 *Eigenvalues and Eigenvectors*

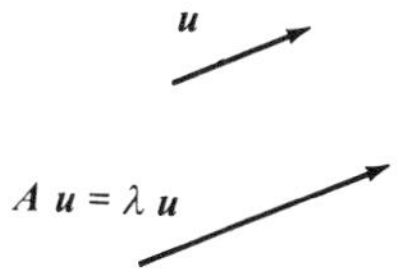

We recall the interpretation of a matrix $\boldsymbol{A}$ as a physical or mathematical process that operates on a particular set of conditions to yield a new set of conditions, and ask: Is it possible to find a set of conditions expressed by the vector $\boldsymbol{u}$, so that when these are modified by the process producing the new conditions $\boldsymbol{Au}$, they maintain their character? This means that the vector $\boldsymbol{u}$ will retain its orientation but possibly shrink or expand by the scalar factor λ.

If the answer is affirmative, the vector $\boldsymbol{u}$ with the aforementioned property is an eigenvector of the matrix $\boldsymbol{A}$, and λ is the corresponding eigenvalue (*eigen* in German means *characteristic*). By definition, we then have

$$\boldsymbol{Au} = \lambda \boldsymbol{u} \tag{2.9.1}$$

which may be restated as

$$(\boldsymbol{A} - \lambda \boldsymbol{I})\boldsymbol{u} = \boldsymbol{0} \tag{2.9.2}$$

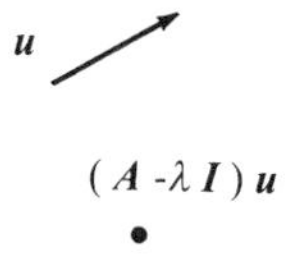

with the understanding that $\boldsymbol{u}$ *is not the null vector.* Demanding that the homogeneous equation (2.9.2) have a nontrivial solution for $\boldsymbol{u}$ requires that the matrix $\boldsymbol{A} - \lambda \boldsymbol{I}$ be singular: By definition, an eigenvector falls into the *null space* of the singular matrix $\boldsymbol{A} - \lambda \boldsymbol{I}$.

If $\boldsymbol{u}$ is an eigenvector corresponding to a certain eigenvalue, then $\alpha \boldsymbol{u}$ is also an eigenvector corresponding to the same eigenvalue, for any value of the coefficient α. Eigenvectors that derive from one another by multiplication with a scalar constant, however, are not considered distinct.

Although seemingly esoteric, eigenvalues and eigenvectors play an important role in the analysis of a broad class of numerical methods. Furthermore, eigenvalues arise naturally in the analysis and modeling of several classes of physical systems and engineering processes including stability analysis and vibration mechanics, as will be discussed in Chapters 5 and 10. Roughly speaking, the eigenvectors of a matrix provide us with a natural framework for describing the process that underlies the matrix.

Real or Complex Numbers?

A real or complex matrix may have real or complex eigenvalues and associated eigenvectors. Later in this section, we shall see that a *real* matrix has real and pairs of complex conjugate eigenvalues, whereas a *real and symmetric* matrix, and more generally a *Hermitian complex* matrix, has only real eigenvalues. The eigenvalues of a *tridiagonal* matrix will be discussed at the end of this section.

If a matrix is *real*, an eigenvector corresponding to a real eigenvalue is necessarily real, and an eigenvector corresponding to a complex eigenvalue is necessarily complex. If the matrix is *complex*, an eigenvector corresponding to a real eigenvalue is necessarily complex.

PROBLEMS

2.9.1 ***Generalized eigenvalue problems.***

(*a*) Suppose that the value of the scalar μ is such that $(\boldsymbol{A} - \mu \boldsymbol{J})\boldsymbol{u} = \boldsymbol{0}$, where $\boldsymbol{J}$ is the unit back-diagonal matrix and $\boldsymbol{u}$ is a vector. Can you find a relation between μ and an eigenvalue of a matrix that is related to $\boldsymbol{A}$ in some way?
(*b*) Let $\boldsymbol{A}$ and $\boldsymbol{B}$ be two arbitrary matrices, and suppose that the value of the scalar ν is such that $(\boldsymbol{A} - \nu \boldsymbol{B})\boldsymbol{u} = \boldsymbol{0}$ where $\boldsymbol{u}$ is a vector. Can you find a relation between ν and an eigenvalue of a matrix that is related to $\boldsymbol{A}$ and $\boldsymbol{B}$ in some way?

2.9.2 ***Eigenvalues of a product.***

Let $\boldsymbol{A}$ and $\boldsymbol{B}$ be two arbitrary square matrices. Show that $\boldsymbol{AB}$ and $\boldsymbol{BA}$ have the same eigenvalues, and find the relation between the corresponding eigenvectors.

2.9.3 ***Eigenvalues of a positive-definite matrix.***

Show that the eigenvalues of a positive-definite matrix are all positive.

Characteristic Polynomial

Expressing the determinant of the $N \times N$ matrix $\boldsymbol{A} - \lambda \boldsymbol{I}$ in terms of the cofactors, we obtain an Nth degree polynomial with respect to λ, called the *characteristic polynomial* of the matrix $\boldsymbol{A}$, defined as

$$P(\lambda) = \text{Det}(\boldsymbol{A} - \lambda \boldsymbol{I}) \tag{2.9.3}$$

Note that $P(0) = \text{Det}(\boldsymbol{A})$. More generally, we find

$$P(\lambda) = (-\lambda)^N + (-\lambda)^{N-1}\,\text{Tr}(\boldsymbol{A}) + (-\lambda)^{N-2} \sum_{\substack{i,j=1 \\ i<j}}^{N} (A_{i,i}A_{j,j} - A_{i,j}A_{j,i}) + \cdots + \text{Det}(\boldsymbol{A}) \tag{2.9.4}$$

Algorithms for the programmable computation of the coefficients of the characteristic polynomial will be discussed in Section 5.4.

If the matrix $\boldsymbol{A}$ is *diagonal* or *triangular*, we obtain the simpler expansion

$$P(\lambda) = (A_{1,1} - \lambda)(A_{2,2} - \lambda)\ldots(A_{N,N} - \lambda) \tag{2.9.5}$$

Computing an eigenvalue is equivalent to finding a value of λ that renders the matrix $\mathbf{A} - \lambda \mathbf{I}$ singular. Thus, by definition, *an eigenvalue of a matrix is also a root of its characteristic polynomial*, satisfying

$$P(\lambda) = 0 \tag{2.9.6}$$

Since an Nth degree polynomial has precisely N roots (see Section 4.1), an $N \times N$ matrix $\mathbf{A}$ is guaranteed to have exactly N real or complex eigenvalues $\lambda_1, \lambda_2, \ldots, \lambda_N$, some of which may be repeated. If a particular eigenvalue λ is repeated m times, that is,

$$P(\lambda) = 0, \quad P'(\lambda) = 0, \quad \ldots, \quad P^{(m-1)}(\lambda) = 0 \tag{2.9.7}$$

where $P^{(k)}$ signifies the kth derivative, its *algebraic* multiplicity is equal to m.

Equation (2.9.5) shows that the eigenvalues of a diagonal or triangular matrix are equal to the diagonal elements; a repeated element yields a multiple eigenvalue. Thus the identity matrix has a single eigenvalue equal to unity with algebraic multiplicity equal to N.

We note that the coefficients of the characteristic polynomial of a real matrix are real, and this requires that the eigenvalues must either be real or else come in pairs of complex conjugates.

Spectrum and Spectral Radius

The set of all eigenvalues of a matrix is called the *spectrum of the eigenvalues*, and the maximum of the norm of all real and complex eigenvalues is called the *spectral radius*. We shall see in subsequent sections and chapters that the spectral radius serves as a diagnostic of certain important properties of the matrix concerning, in particular, the action of linear mappings.

Relationship Between the Eigenvalues, the Trace, and the Determinant

The characteristic polynomial defined in equation (2.9.3) may be placed in an alternative form involving its roots, λ_i, as

$$P(\lambda) = (\lambda_1 - \lambda)(\lambda_2 - \lambda)\ldots(\lambda_N - \lambda) \tag{2.9.8}$$

If the matrix is diagonal or triangular, this is equivalent to the expansion shown in equation (2.9.5). Expanding out the product on the right-hand side of equation (2.9.8), we find

$$P(\lambda) = (-\lambda)^N + (-\lambda)^{N-1}\sum_i^N \lambda_i + \cdots + \prod_i^N \lambda_i \tag{2.9.9}$$

Comparing the right-hand sides of equations (2.9.9) and (2.9.4) shows that

$$\sum_i^N \lambda_i = \mathrm{Tr}(\mathbf{A}) \tag{2.9.10}$$

and

$$\prod_i^N \lambda_i = \mathrm{Det}(\mathbf{A}) \tag{2.9.11}$$

Equation (2.9.11), in particular, reveals that *if one of the eigenvalues is equal to zero, the determinant will vanish and the matrix will be singular.*

PROBLEM

2.9.4 ***Shifting an eigenvalue.***

Show that if λ is an eigenvalue of the matrix $\boldsymbol{A}$, then $\lambda - \alpha$ is an eigenvalue of the matrix $\boldsymbol{A} - \alpha \boldsymbol{I}$.

2.9.5 ***A singular matrix.***

Consider two vectors $\boldsymbol{w}$ and $\boldsymbol{u}$ whose inner product is equal to unity, $\boldsymbol{w}^T \boldsymbol{u} = 1$. Show that the matrix $\boldsymbol{A}$ with elements $A_{i,j} = \delta_{i,j} - w_i u_j$ is singular. *Hint*: Show that one eigenvalue is equal to zero.

Eigenvalues and Eigenvectors of a Power, the Inverse, and a Function of a Matrix

Premultiplying both sides of equation (2.9.1) by $\boldsymbol{A}$, we find

$$\boldsymbol{A}(\boldsymbol{A}\boldsymbol{u}) = \lambda(\boldsymbol{A}\boldsymbol{u}) = \lambda^2 \boldsymbol{u} \tag{2.9.12}$$

Identifying the left-hand side with $\boldsymbol{A}^2\boldsymbol{u}$ shows that λ^2 is an eigenvalue of the matrix $\boldsymbol{A}^2$, and the corresponding eigenvector is $\boldsymbol{u}$. Working in a similar manner, we find that λ^p is an eigenvalue of the matrix $\boldsymbol{A}^p$, and the corresponding eigenvector is $\boldsymbol{u}$, where p is a positive real number.

Similar arguments reveal that $1/\lambda$ is an eigenvalue of the inverse matrix $\boldsymbol{A}^{-1}$, and the corresponding eigenvector is $\boldsymbol{u}$.

PROBLEM

2.9.6 ***Eigenvalues of a polynomial of a matrix.***

Assume that λ is an eigenvalue of the matrix $\boldsymbol{A}$ corresponding to the eigenvector $\boldsymbol{u}$.

(*a*) Let $Q(x)$ be an arbitrary polynomial. Show that $Q(\lambda)$ is an eigenvalue of the matrix $\boldsymbol{Q}(\boldsymbol{A})$, with corresponding eigenvector $\boldsymbol{u}$.

(*b*) Let $Q(x)$ and $R(x)$ be two generally unrelated polynomials. Show that $Q(\lambda)/R(\lambda)$ is an eigenvalue of the matrix $[\boldsymbol{R}(\boldsymbol{A})]^{-1}\boldsymbol{Q}(\boldsymbol{A})$, with corresponding eigenvector $\boldsymbol{u}$.

Eigenvalues of Real and Symmetric, Hermitian, and Positive-definite Matrices

Hermitian matrices, and hence real and symmetric matrices, have real eigenvalues. To show this, we take the complex conjugate and then the transpose of both sides of equation (2.9.1) to obtain $\boldsymbol{u}^{*T}\boldsymbol{A}^{*T} = \lambda^*\boldsymbol{u}^{*T}$, which is equivalent to $\boldsymbol{u}^{*T}\boldsymbol{A} = \lambda^*\boldsymbol{u}^{*T}$; an asterisk signifies the complex conjugate. Taking the inner product of both sides with $\boldsymbol{u}$ yields $\boldsymbol{u}^{*T}\boldsymbol{A}\boldsymbol{u} = \lambda^*\boldsymbol{u}^{*T}\boldsymbol{u}$ or $\lambda\boldsymbol{u}^{*T}\boldsymbol{u} = \lambda^*\boldsymbol{u}^{*T}\boldsymbol{u}$, which requires that $\lambda = \lambda^*$ or λ be real.

If the Hermitian matrix $\boldsymbol{A}$ is positive-definite, then the scalar $\boldsymbol{u}^{*T}\boldsymbol{A}\boldsymbol{u} = \lambda\boldsymbol{u}^{*T}\boldsymbol{u}$ and thus λ are real and positive. We shall see later in this section that, conversely, if all eigenvalues of a Hermitian matrix are positive, then the matrix is guaranteed to be positive-definite.

Approximate Location of Eigenvalues

Before attempting to compute the eigenvalues of a matrix using numerical methods, it is helpful, and sometimes imperative, to have a good idea about their location.

Gerschgorin's first theorem guarantees that the eigenvalues of an $N \times N$ matrix reside within a disk that is centered at the origin of the complex plane. The radius of the disk is equal to the minimum of (a) the maximum of the sums of the norms of the elements in each column, and (b) the maximum of the sums of the norms of the elements in each row.

Gerschgorin's circle theorem locates the eigenvalues of an arbitrary $N \times N$ matrix inside the union of N disks in the complex plane. The ith disk is centered at the diagonal element $A_{i,i}$, and the corresponding radius is equal to the minimum of

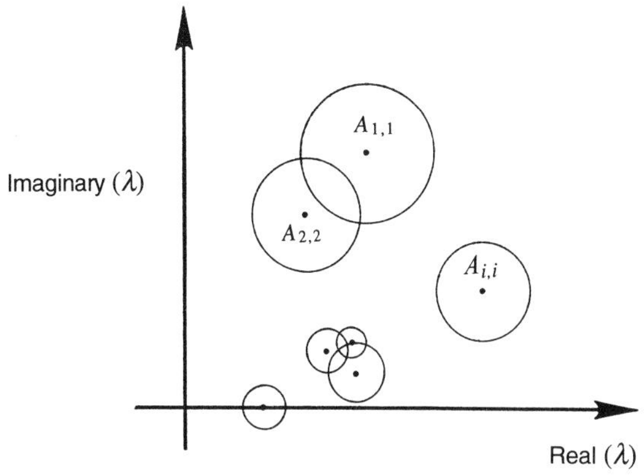

$$\sum_{\substack{j=1 \\ i \neq j}}^{N} |A_{i,j}|, \qquad \sum_{\substack{j=1 \\ i \neq j}}^{N} |A_{j,i}| \tag{2.9.13}$$

If the disks do not overlap, then each one of them contains one eigenvalue. If m disks overlap forming a connected set, then there are precisely m eigenvalues within their union.

Other theorems for locating eigenvalues are discussed by Bodewig (1959, p. 67) and Wilkinson (1965).

PROBLEMS

2.9.7 ***A diagonally dominant matrix cannot be singular.***

Show that a diagonally dominant matrix cannot be singular. A matrix is diagonally dominant when the norm of each diagonal element is greater than the maximum of the sums of the norms of the rest of the elements in the corresponding column or row.

2.9.8 ***Eigenvalues and eigenvectors of the unit back-diagonal matrix.***

Compute all eigenvalues and eigenvectors of the unit back-diagonal matrix, and state their multiplicity.

Eigenvectors

After the eigenvalues have been computed, the eigenvectors can be found by solving the homogeneous system of linear equations (2.9.2). For each eigenvalue with algebraic multiplicity equal to one, there will be a multiplicity of solutions reflecting the arbitrary length of the corresponding eigenvector. But this degree of freedom can be removed by stipulating, for example, that the magnitude of one component of the eigenvector, or the length of the eigenvector, takes a preset value. Certain important theorems concerning the eigenvectors are the following:

- The eigenvectors corresponding to distinct eigenvalues are *linearly independent.* A sketch of the proof is as follows: Assume that one eigenvector depends on the others; express it as a linear combination of them, and premultiply this expansion by the matrix $\boldsymbol{A}$. Using equation (2.9.1), and comparing the resulting equation with the original expansion of the eigenvector, we find that the eigenvector must be the null vector, which is a contradiction.

- If a matrix has N distinct eigenvalues, then it is guaranteed to have N linearly independent eigenvectors that form a basis of the N-dimensional vector space. Any vector may be expressed as a linear combination of the eigenvectors.
- If a matrix does not have N distinct eigenvalues, and this assumes that one or more eigenvalues are multiple, it is not certain that there will be N linearly independent eigenvectors. The number of eigenvectors k corresponding to a particular eigenvalue of algebraic multiplicity m is called the *geometric multiplicity* of the eigenvalue; $k \le m$. Since equation (2.9.2) has at least one family of solutions, $k \ge 1$.
- *Hermitian* matrices are guaranteed to have N linearly independent and mutually orthogonal eigenvectors, even though they may have multiple eigenvalues; in this case $k = m$. The proof relies on the existence of the Schur normal form to be discussed later in this section.

Can we find two matrices that have the same set of linearly independent eigenvectors? The answer is affirmative. It can be shown, in particular, that if two Hermitian matrices commute with respect to multiplication, then they share eigenvectors.

Principal Vectors

If the number of linearly independent eigenvectors L is less than N, then the set of eigenvectors is deficient by the dimension $N - L$ and *cannot* form a basis for the N-dimensional vector space. One way of completing the basis is to introduce the *generalized eigenvectors* or *principal vectors* associated with each multiple eigenvalue λ of algebraic multiplicity m and geometric multiplicity k, where $k < m$ (Bodewig 1959, p. 91). These satisfy the equation

$$(\boldsymbol{A} - \lambda \boldsymbol{I})^r \boldsymbol{u} = \boldsymbol{0} \tag{2.9.14}$$

where r is an integer, and thus fall within the null space of the matrix $(\boldsymbol{A} - \lambda \boldsymbol{I})^r$. The smallest value of r for which equation (2.9.14) holds is called the *grade* of $\boldsymbol{u}$; when $r = 1$, $\boldsymbol{u}$ is an eigenvector.

Wielandt's representation theorem ensures that the eigenvectors and principal vectors corresponding to the various eigenvalues of a matrix form a basis. Any vector can be expressed as a linear combination of them (e.g., Bodewig 1959, p. 92).

PROBLEM

2.9.9 ***A deficient set of eigenvectors.***

Consider a 2×2 matrix $\boldsymbol{A}$ with elements $A_{1,1} = a$, $A_{1,2} = b$, $A_{2,1} = 0$, and $A_{2,2} = a$. Show that this matrix has a single eigenvalue $\lambda = a$ with algebraic and geometric multiplicity equal to 2 and 1, respectively. Then compute the eigenvector and principal vector, and show that they form a basis for the two-dimensional space.

Eigenvalues and Eigenvectors of the Transpose of a Matrix

The determinant and therefore the characteristic polynomial and the set of eigenvalues of a matrix $\boldsymbol{A}$ and its transpose $\boldsymbol{A}^T$ are equal. Unless the matrix is symmetric, however, the corresponding eigenvectors will not necessarily point in parallel directions.

The eigenvector $\boldsymbol{v}$ of $\boldsymbol{A}^T$ corresponding to the eigenvalue λ_1 is orthogonal to the eigenvector $\boldsymbol{u}$ of $\boldsymbol{A}$ corresponding to a different eigenvalue λ_2. To show this, we take the transpose of both sides of equation $\boldsymbol{A}\boldsymbol{u} = \lambda_2 \boldsymbol{u}$, to find $\boldsymbol{u}^T \boldsymbol{A}^T = \lambda_2 \boldsymbol{u}^T$. Taking the inner product of both sides of the last equation with $\boldsymbol{v}$, we obtain $\boldsymbol{u}^T\boldsymbol{A}^T\boldsymbol{v} = \lambda_2\boldsymbol{u}^T\boldsymbol{v}$. Finally, we use the definition $\boldsymbol{A}^T\boldsymbol{v} = \lambda_1\boldsymbol{v}$ to find $(\lambda_2 - \lambda_1)\boldsymbol{u}^T\boldsymbol{v} = 0$, which proves the assertion. A similar proof shows that $\boldsymbol{v}$ is orthogonal to the principal vectors of $\boldsymbol{A}$ corresponding to a multiple eigenvalue λ_2.

The number of linearly independent eigenvectors of a matrix and its transpose, corresponding to a particular multiple eigenvalue, are identical. Thus *a matrix and its transpose have identical eigenvalues and the same number of linearly independent eigenvectors.*

Let us assume that a matrix $\boldsymbol{A}$ has N eigenvectors $\boldsymbol{u}^{(i)}$, and put them at successive columns of the matrix $\boldsymbol{U}$. Furthermore, let us put the corresponding eigenvectors $\boldsymbol{v}^{(i)}$ of $\boldsymbol{A}^T$ at successive columns of the matrix $\boldsymbol{V}$, and normalize them so that a corresponding pair of $\boldsymbol{u}$ and $\boldsymbol{v}$ satisfies the condition $\boldsymbol{v}^{(i)T}\boldsymbol{u}^{(i)} = 1$. Subject to these definitions, $\boldsymbol{V}^T\boldsymbol{U} = \boldsymbol{U}^T\boldsymbol{V} = \boldsymbol{I}$, or

$$\boldsymbol{U} = \begin{bmatrix} \uparrow & \uparrow & \uparrow & \uparrow & \uparrow \\ \boldsymbol{u}^{(1)} & \boldsymbol{u}^{(2)} & \cdots & \boldsymbol{u}^{(N-1)} & \boldsymbol{u}^{(N)} \\ \downarrow & \downarrow & \downarrow & \downarrow & \downarrow \end{bmatrix}$$

$$\boldsymbol{U}^{-1} = \boldsymbol{V}^T, \qquad \boldsymbol{V}^{-1} = \boldsymbol{U}^T \tag{2.9.15}$$

The collection $\boldsymbol{u}^{(i)}$ and the collection $\boldsymbol{v}^{(i)}$ provide us with a pair of bi-orthonormal sets.

PROBLEMS

2.9.10 ***Numerical exercise.***

Consider a 2×2 matrix $\boldsymbol{A}$ with elements $A_{1,1} = I$, $A_{1,2} = 2$, $A_{2,1} = 0$, and $A_{2,2} = 1$, where I is the imaginary unit. Compute the matrices $\boldsymbol{U}$ and $\boldsymbol{V}$, and verify the validity of equations (2.9.15).

2.9.11 ***Eigenvectors for the generalized eigenvalue problem.***

Let the vector $\boldsymbol{u}$ satisfy the equation $\boldsymbol{A}\boldsymbol{u} = \lambda_1\boldsymbol{B}\boldsymbol{u}$, and the vector $\boldsymbol{v}$ satisfy the equation $\boldsymbol{v}^T\boldsymbol{A} = \lambda_2\boldsymbol{v}^T\boldsymbol{B}$, where $\boldsymbol{A}$ and $\boldsymbol{B}$ are two arbitrary square matrices. Show that if $\lambda_1 \neq \lambda_2$, then $\boldsymbol{v}^T\boldsymbol{B}\boldsymbol{u} = 0$.

Eigenvectors of a Symmetric Matrix

One corollary of the preceding results is that *two eigenvectors of a symmetric matrix corresponding to two different eigenvalues are orthogonal.* Equations (2.9.15) show that $\boldsymbol{U} = \boldsymbol{V}$ and $\boldsymbol{U}^{-1} = \boldsymbol{U}^T$, which means that *the matrix of eigenvectors $\boldsymbol{U}$ is orthogonal.*

The eigenvalues and therefore the eigenvectors of a real and symmetric matrix are real: *An $N \times N$ real symmetric matrix has N real eigenvalues and N real and orthogonal eigenvectors.*

PROBLEM

2.9.12 ***Numerical exercise.***

Consider a 2×2 matrix $\boldsymbol{A}$ with elements $A_{1,1} = 5$, $A_{1,2} = 2$, $A_{2,1} = 2$, and $A_{2,2} = 3$. Compute the matrix $\boldsymbol{U}$ and verify that it is orthogonal.

Normal Matrices

What is the most general class of matrices with the property that eigenvectors corresponding to distinct eigenvalues are orthogonal? This is the class of normal matrices containing, as a subset, real and symmetric matrices. By definition, a normal matrix $\boldsymbol{A}$ commutes with its adjoint, that is, $\boldsymbol{A}\boldsymbol{A}^A = \boldsymbol{A}^A\boldsymbol{A}$. Hermitian and orthogonal matrices are examples of normal matrices.

Eigenvalues and Eigenvectors of the Adjoint

The determinant of a matrix is equal to the complex conjugate of the determinant of its adjoint. Thus its characteristic polynomial and therefore its eigenvalues are the complex conjugates of those of the adjoint. Although the associated eigenvectors are not generally related in a simple way, an eigenvector of the adjoint $\boldsymbol{A}^A$ corresponding to an eigenvalue μ_1, call it $\boldsymbol{w}$, is orthogonal to the complex conjugate of an eigenvector of $\boldsymbol{A}$ corresponding to an eigenvalue λ_2, call it $\boldsymbol{u}^*$, where $\mu_1 \neq \lambda_2^*$.

Assume that $\boldsymbol{A}^A$ has N eigenvectors $\boldsymbol{w}^{(i)}$, put them at the columns of the matrix $\boldsymbol{W}$ and normalize them so that the eigenvectors $\boldsymbol{w}^{(i)}$ and $\boldsymbol{u}^{(i)*}$ corresponding to complex conjugate eigenvalues satisfy the condition $\boldsymbol{w}^{(i)T}\boldsymbol{u}^{(i)*} = 1$. Then $\boldsymbol{W}^A\boldsymbol{U} = \boldsymbol{U}^A\boldsymbol{W} = \boldsymbol{I}$, which reveals that

$$\boldsymbol{U}^{-1} = \boldsymbol{W}^A, \qquad \boldsymbol{W}^{-1} = \boldsymbol{U}^A \tag{2.9.16}$$

If the matrix $\boldsymbol{A}$ is Hermitian, then $\boldsymbol{U} = \boldsymbol{W}$ and $\boldsymbol{U}^{-1} = \boldsymbol{U}^A$.

PROBLEM

2.9.13 ***Numerical exercise.***

Consider the matrix described in Problem 2.9.10. Compute the matrices $\boldsymbol{U}$ and $\boldsymbol{W}$, and verify that $\boldsymbol{U}^{-1} = \boldsymbol{W}^A$.

Positive-definite Hermitian Matrices

We saw that all eigenvalues of a Hermitian matrix are real. If, in addition, all eigenvalues are positive, then the matrix is positive-definite. To show this, we consider an arbitrary vector $\boldsymbol{x}$ and express it as a linear combination of the eigenvectors of a Hermitian matrix $\boldsymbol{A}$, as $\boldsymbol{x} = c_1\boldsymbol{u}^{(1)} + \ldots + c_N\boldsymbol{u}^{(N)}$. Using the results of the preceding sections, we find $\boldsymbol{x}^{*T}\boldsymbol{A}\boldsymbol{x} = c_1^*c_1\lambda_1\boldsymbol{u}^{(1)*T}\boldsymbol{u}^{(1)} + \ldots + c_N^*c_N\lambda_N\boldsymbol{u}^{(N)*T}\boldsymbol{u}^{(N)}$, which shows that so long as all eigenvalues are positive $\boldsymbol{x}^{*T}\boldsymbol{A}\boldsymbol{x}$ is guaranteed to be positive.

Similarity Transformations

Let us consider a nonsingular matrix $\boldsymbol{P}$, and compute its inverse $\boldsymbol{P}^{-1}$. A *similarity transformation* of the matrix $\boldsymbol{A}$, mediated by the matrix $\boldsymbol{P}$, produces the new matrix $\boldsymbol{B}$:

$$\boldsymbol{B} = \boldsymbol{P}^{-1}\boldsymbol{A}\boldsymbol{P} \tag{2.9.17}$$

We shall see that the similar matrices $\boldsymbol{B}$ and $\boldsymbol{A}$ are related in theoretically significant and practically useful ways.

Premultiplying both sides of equation (2.9.1) by $\boldsymbol{P}^{-1}$, we obtain $\boldsymbol{P}^{-1}\boldsymbol{A}\boldsymbol{u} = \lambda\boldsymbol{P}^{-1}\boldsymbol{u}$, which can be restated as $\boldsymbol{P}^{-1}\boldsymbol{A}\boldsymbol{P}\boldsymbol{P}^{-1}\boldsymbol{u} = \lambda\boldsymbol{P}^{-1}\boldsymbol{u}$ or $\boldsymbol{B}\boldsymbol{P}^{-1}\boldsymbol{u} = \lambda\boldsymbol{P}^{-1}\boldsymbol{u}$. The last equation demonstrates that λ is an eigenvalue of $\boldsymbol{B}$ with associated eigenvector $\boldsymbol{P}^{-1}\boldsymbol{u}$. Thus the matrices $\boldsymbol{A}$ and $\boldsymbol{B}$ share eigenvalues, but not necessarily eigenvectors: *Similarity transformations preserve the eigenvalues.*

Identities (2.9.10) and (2.9.11) require that

$$\begin{aligned} \mathrm{Tr}(\boldsymbol{B}) &= \mathrm{Tr}(\boldsymbol{A}) \\ \mathrm{Det}(\boldsymbol{B}) &= Det(\boldsymbol{A}) \end{aligned} \tag{2.9.18}$$

Thus the trace and the determinant of a matrix remain invariant under similarity transformations. It is impossible to transform a singular matrix into a nonsingular matrix by a similarity transformation, and vice versa. Two similar matrices have the same rank.

PROBLEM

2.9.14 ***Similarity transformation by decomposition.***

Assume that $\mathbf{A} = \mathbf{CD}$, where $\mathbf{A}$, $\mathbf{C}$, and $\mathbf{D}$ are three square matrices. Show that $\mathbf{B} = \mathbf{DC}$ is similar to $\mathbf{A}$.

How Many Matrices to a Set of Eigenvalues?

It is clear that an $N \times N$ square matrix has a unique Nth degree characteristic polynomial, but the correspondence is not one-to-one: A certain Nth degree polynomial is the characteristic polynomial of an infinite family of $N \times N$ matrices related by similarity transformations. This means that the set of eigenvalues cannot be used to fingerprint a matrix; we shall see shortly that neither can the set of eigenvectors.

For example, one may readily verify that the polynomial

$$P(x) = (-1)^N (x^N + a_1 x^{N-1} + \cdots + a_N) \tag{2.9.19}$$

is the characteristic polynomial of the so-called *companion or Frobenius matrix*

$$\mathbf{F} = \begin{bmatrix} 0 & 1 & 0 & \dots & 0 \\ 0 & 0 & 1 & \dots & 0 \\ \dots & \dots & \dots & \dots & \dots \\ 0 & 0 & 0 & \dots & 1 \\ -a_N & -a_{N-1} & -a_{N-2} & \dots & -a_1 \end{bmatrix} \tag{2.9.20}$$

The components of the eigenvector $\mathbf{u}$ corresponding to a particular eigenvalue λ of this matrix are given by $u_i = \lambda^{i-1}$, where $i = 1, \dots, N$.

PROBLEM

2.9.15 ***Other companion matrices.***

Derive three additional companion matrices for the polynomial (2.9.19), with the polynomial coefficients deployed along the first column, the first row, or the last column.

Computation of Eigenvalues

Equation (2.9.5) shows that the eigenvalues of a *diagonal* or *triangular* matrix are equal to their diagonal elements. Unfortunately, the eigenvalues of a *bidiagonal* or *tridiagonal* matrix whose elements do not exhibit special patterns cannot be identified at a glance, and their computation requires detailed consideration. Some mild exceptions will be discussed in Section 5.3.

Computing the eigenvalues of a matrix with an arbitrary structure is an arduous task. The most general method of computing all eigenvalues of a matrix proceeds by transforming the matrix using a sequence of similarity transformations, until it obtains a simple form; for instance, until it becomes tridiagonal, triangular, or diagonal. To this end, the feasibility of transforming a certain matrix into a similar matrix with a simple form must be considered.

Schur normal form

It can be shown by mathematical induction that, given a complex matrix $\mathbf{A}$, there exists a unique unitary matrix $\mathbf{U}$ so that the matrix $\mathbf{R} = \mathbf{U}^A \mathbf{A} \mathbf{U}$, which is similar to $\mathbf{A}$, has the upper triangular form, called the *Schur normal form* (e.g., Atkinson 1989, p. 474). The diagonal elements of $\mathbf{R}$ are the eigenvalues of $\mathbf{A}$.

Matrix Diagonalization

Next, we address the issue of diagonalization: Given a certain matrix $\boldsymbol{A}$, we want to know if we can find a nonsingular matrix $\boldsymbol{P}$, so that its similar matrix $\boldsymbol{B} = \boldsymbol{P}^{-1}\boldsymbol{A}\boldsymbol{P}$ is diagonal.

If the matrix $\boldsymbol{A}$ has N distinct eigenvectors, the existence of $\boldsymbol{P}$ can be established by construction. We apply equation (2.9.1) for all eigenvectors and put the resulting system of equations in a compact form to obtain the matrix equation,

$$\boldsymbol{A}\boldsymbol{U} = \boldsymbol{U}\boldsymbol{\Lambda} \tag{2.9.21}$$

$$\Lambda = \begin{bmatrix} \lambda_1 & 0 & 0 & \dots & 0 \\ 0 & \lambda_2 & 0 & \dots & \dots \\ 0 & 0 & \ddots & 0 & 0 \\ \dots & \dots & 0 & \lambda_{N-1} & 0 \\ 0 & \dots & 0 & 0 & \lambda_N \end{bmatrix}$$

where $\boldsymbol{\Lambda}$ is a diagonal matrix with the N possibly multiple eigenvalues of $\boldsymbol{A}$ residing along the diagonal. The columns of the matrix $\boldsymbol{U}$ contain the corresponding eigenvectors. Premultiplying equation (2.9.21) by $\boldsymbol{U}^{-1}$, and remembering that $\boldsymbol{U}^{-1} = \boldsymbol{V}^T$ where the columns of the matrix $\boldsymbol{V}$ contain the corresponding eigenvectors of $\boldsymbol{A}^T$ as explained immediately before equation (2.9.15), we find

$$\boldsymbol{\Lambda} = \boldsymbol{U}^{-1}\boldsymbol{A}\boldsymbol{U} = \boldsymbol{V}^T\boldsymbol{A}\boldsymbol{U} \tag{2.9.22}$$

Identifying $\boldsymbol{P}$ with the matrix of eigenvectors $\boldsymbol{U}$ provides us with the desired similarity transformation that renders $\boldsymbol{A}$ diagonal.

Conversely, if we can find a matrix $\boldsymbol{P}$ so that its similar matrix $\boldsymbol{B} = \boldsymbol{P}^{-1}\boldsymbol{A}\boldsymbol{P}$ is diagonal, then we can be certain that $\boldsymbol{B} = \boldsymbol{\Lambda}$ and $\boldsymbol{P} = \boldsymbol{U}$.

Spectral expansion

Postmultiplying equation (2.9.21) by $\boldsymbol{U}^{-1}$, we obtain an expression for $\boldsymbol{A}$ in terms of its eigenvalues, its eigenvectors, and the eigenvectors of its transpose,

$$\boldsymbol{A} = \boldsymbol{U}\boldsymbol{\Lambda}\boldsymbol{U}^{-1} = \boldsymbol{U}\boldsymbol{\Lambda}\boldsymbol{V}^T \tag{2.9.23}$$

Expressing the right-hand side of equation (2.9.23) in terms of products of the eigenvectors of $\boldsymbol{A}$ and its transpose, we obtain the *spectral expansion*

$$\boldsymbol{A} = \sum_{m=1}^{N} \lambda_m \boldsymbol{u}^{(m)} \boldsymbol{v}^{(m)^T} \tag{2.9.24}$$

This is analogous to the expansion of a vector in terms of the eigenvectors.

Jordan Canonical Form

If the $N \times N$ matrix $\boldsymbol{A}$ does not have N linearly independent eigenvectors, then it cannot be diagonalized. But we still want to know whether it is possible to find a nonsingular matrix $\boldsymbol{P}$, so that the matrix $\boldsymbol{B} = \boldsymbol{P}^{-1}\boldsymbol{A}\boldsymbol{P}$ is as nearly diagonal as we can make it.

It can be shown that any matrix can be transformed into a similar matrix that has a nearly diagonal form, called the *Jordan canonical form.* In this form, all elements of the matrix are equal to zero, except for the elements within a sequence of square blocks $J_l(\lambda), l = 1, 2, \ldots, p$, called the *Jordan blocks*, deployed along the diagonal (e.g., Bodewig 1959, p. 88). The relationship between p and N will be described shortly. The blocks are upper bidiagonal matrices with superdiagonal elements equal to 1 and diagonal elements equal to the corresponding eigenvalue λ. Each block corresponds to a single or multiple eigenvalue.

Jordan canonical form:

$$
\begin{bmatrix}
\mathbf{J}_1 & \mathbf{0} & \mathbf{0} & \ldots & \mathbf{0} \\
\mathbf{0} & \mathbf{J}_2 & \mathbf{0} & \ldots & \ldots \\
\mathbf{0} & \mathbf{0} & \ddots & \mathbf{0} & \mathbf{0} \\
\ldots & \ldots & \mathbf{0} & \mathbf{J}_{p-1} & \mathbf{0} \\
\mathbf{0} & \ldots & \mathbf{0} & \mathbf{0} & \mathbf{J}_p
\end{bmatrix}
\qquad
\mathbf{J}_l =
\begin{bmatrix}
\lambda_k & 1 & 0 & \ldots & 0 \\
0 & \lambda_k & 1 & \ldots & \ldots \\
0 & 0 & \ddots & 1 & 0 \\
\ldots & \ldots & 0 & \lambda_k & 1 \\
0 & \ldots & 0 & 0 & \lambda_k
\end{bmatrix}
$$

To understand the structure of the blocks, we note the following:

- If all N eigenvalues of the matrix are distinct, there are N diagonal Jordan blocks, each one of them of size 1×1, containing the eigenvalues. This is the standard diagonal form.
- If there are m distinct eigenvectors corresponding to a certain eigenvalue of multiplicity m, then, corresponding to this eigenvalue, there are m identical blocks of size 1×1 containing this eigenvalue. This is still the standard diagonal form.
- If there are k eigenvectors corresponding to a certain eigenvalue of multiplicity m, with $k < m$,then, corresponding to this eigenvalue, there are $k - 1$ identical blocks of size 1×1 containing this eigenvalue, and another Jordan block of size $(m-k+1) \times (m-k+1)$ with the eigenvalue placed along the diagonal. The Jordan canonical form of a matrix with $N - 2$ distinct eigenvalues is shown in Figure 2.9.1.

The Jordan canonical form derives from the transformation (2.9.22), where the columns of $\mathbf{U}$ contain the eigenvectors and principal vectors of $\mathbf{A}$ (Bodewig 1959, p. 92).

Eigenvalues of Tridiagonal Matrices

We return to investigating the eigenvalues of a tridiagonal matrix $\mathbf{T}$. For simplicity, we designate the nonzero elements of $\mathbf{T}$ as shown in equations (2.4.1).

Figure 2.9.1 Jordan canonical form of an $N \times N$ matrix with $N-2$ distinct eigenvalues, one eigenvalue λ_1 with algebraic multiplicity equal to 2 and geometric multiplicity equal to 1, and another eigenvalue λ_2 with algebraic multiplicity and geometric multiplicity both equal to 2.

$$
\begin{bmatrix}
\lambda_1 & 1 & 0 & \ldots & \ldots & \ldots & \ldots \\
0 & \lambda_1 & 0 & 0 & \ldots & \ldots & \ldots \\
0 & 0 & \lambda_2 & 0 & 0 & \ldots & \ldots \\
\ldots & 0 & 0 & \lambda_2 & 0 & 0 & \ldots \\
\ldots & \ldots & 0 & 0 & \lambda_3 & 0 & \ldots \\
\ldots & \ldots & \ldots & 0 & 0 & \lambda_4 & \ldots \\
\ldots & \ldots & \ldots & \ldots & \ldots & \ldots & \ldots
\end{bmatrix}
$$

Let us consider a nonsingular diagonal transformation matrix $\boldsymbol{P}$ with diagonal elements equal to $d_1, \ldots, d_N$; $\boldsymbol{P}^{-1}$ is another nonsingular diagonal matrix with diagonal elements equal to $1/d_1, \ldots, 1/d_N$. According to our previous discussion, the matrix $\boldsymbol{T}' = \boldsymbol{P}^{-1}\boldsymbol{T}\boldsymbol{P}$ is similar to $\boldsymbol{T}$. Denoting the nonzero elements of the transformed tridiagonal matrix $\boldsymbol{T}'$ as shown in equations (2.4.1) but with a prime, we find

$$a_i' = a_i \tag{2.9.25}$$

and

$$b_i' c_{i+1}' = b_i c_{i+1} \tag{2.9.26}$$

for $i = 1, \ldots, N-1$. Two tridiagonal matrices whose elements satisfy these relations are similar, and therefore have the same eigenvalues and equal determinants.

Assigning to d_1, in particular, an arbitrary value, and computing the rest of the diagonal elements of $\boldsymbol{D}$ from the relation

$$d_{i+1} = d_i \sqrt{\frac{c_{i+1}}{b_i}} \tag{2.9.27}$$

renders the matrix $\boldsymbol{T}'$ symmetric, with

$$b_i' = c_{i+1}' = \sqrt{b_i c_{i+1}} \tag{2.9.28}$$

If the diagonal elements of $\boldsymbol{T}$ are real, and if all products $b_i c_{i+1}$ are real and positive, then $\boldsymbol{T}'$ will be real and symmetric. Our previous discussion guarantees that the eigenvalues of $\boldsymbol{T}'$ and thus of $\boldsymbol{T}$ will be *real.*

We conclude that the eigenvalues of a tridiagonal matrix with real diagonal elements and real and positive products $b_i c_{i+1}$ are real. This class of matrices includes real and symmetric matrices, as well as Hermitian matrices as special categories.

2.10 *Wielandt's Deflation*

Certain computational procedures for solving systems of linear equations $\boldsymbol{A}\boldsymbol{x} = \boldsymbol{b}$, or for computing the eigenvalues of a matrix $\boldsymbol{A}$, require generating a *singular* matrix $\boldsymbol{B}$ that shares the eigenvalues of $\boldsymbol{A}$, except for one eigenvalue λ_1 that has been replaced by zero. Specific applications will be presented in Sections 3.8 and 5.5. The process of shifting one of the eigenvalues to zero while preserving the rest of them is called *single eigenvalue deflation or eigenvalue spectrum deflation.*

Single Deflation with an Eigenvector

Let λ_1 be a *single* eigenvalue of $\boldsymbol{A}$ corresponding to the eigenvector $\boldsymbol{u}^{(1)}$, and introduce the vector $\boldsymbol{w}$ that satisfies the constraint $\boldsymbol{w}^T\boldsymbol{u}^{(1)} = 1$ but is arbitrary otherwise. *Wielandt's theorem* states that the matrix $\boldsymbol{B}$ with elements

$$B_{i,j} = A_{i,j} - \lambda_1 u_i^{(1)} w_j \tag{2.10.1}$$

shares the eigenvalues of $\boldsymbol{A}$, with the exception of the eigenvalue λ_1 that has been replaced by zero. The eigenvectors of $\boldsymbol{B}$ corresponding to the unaltered eigenvalues are generally different from those of $\boldsymbol{A}$. An extension of the theorem to *multiple* eigenvalues will be discussed at the end of this section.

Wielandt's theorem can be proved in two stages. First, we note that $\mathbf{B}\mathbf{u}^{(1)} = \mathbf{0}$, and this shows that $\mathbf{u}^{(1)}$ falls within the null space of $\mathbf{B}$, as discussed in Section 2.8. Second, we assume that $\mathbf{v}^{(2)}$ is an eigenvector of $\mathbf{A}^T$ corresponding to a different eigenvalue λ_2, and recall that $\mathbf{v}^{(2)}$ is orthogonal to $\mathbf{u}^{(1)}$ to find

$$\mathbf{B}^T\mathbf{v}^{(2)} = \mathbf{A}^T\mathbf{v}^{(2)} - \lambda_1 \mathbf{w}(\mathbf{u}^{(1)T}\mathbf{v}^{(2)}) = \mathbf{A}^T\mathbf{v}^{(2)} = \lambda_2\mathbf{v}^{(2)} \tag{2.10.2}$$

This shows that λ_2 is an eigenvalue of $\mathbf{B}^T$ with corresponding eigenvalue equal to $\mathbf{v}^{(2)}$. Since, however, the eigenvalues of a matrix and its transpose are equal, λ_2 is also an eigenvalue of $\mathbf{B}$.

As an example, the matrix

$$\mathbf{A} = \begin{bmatrix} 1 & 9 \\ \frac{1}{4} & 1 \end{bmatrix} \tag{2.10.3}$$

has two eigenvalues $\lambda_1 = \frac{5}{2}$ and $\lambda_2 = -\frac{1}{2}$. The eigenvector corresponding to λ_1 is $\mathbf{u}^{(1)} = (6, 1)^T$. Choosing $\mathbf{w} = (0, 1)^T$, we find

$$\mathbf{B} = \begin{bmatrix} 1 & 9 \\ \frac{1}{4} & 1 \end{bmatrix} - \frac{5}{2}\begin{bmatrix} 0 & 6 \\ 0 & 1 \end{bmatrix} = \begin{bmatrix} 1 & -6 \\ \frac{1}{4} & -\frac{3}{2} \end{bmatrix} \tag{2.10.4}$$

whose eigevalues are $\lambda_{B_1} = 0$ and $\lambda_{B_2} = -\frac{1}{2}$.

Deflation of the spectrum

Identifying the vector $\mathbf{w}$ with the eigenvector $\mathbf{v}^{(1)}$ of $\mathbf{A}^T$ corresponding to the eigenvalue λ_1, reduces the deflating term $\lambda_1 u_i^{(1)} w_j$ on the right-hand side of equation (2.10.1) to the corresponding term in the spectral expansion shown in equation (2.9.24). It is not surprising then that subtracting off this term annihilates the corresponding eigenvalue.

Single Deflation with an Eigenvector of the Transpose

Let λ_1 be a single eigenvalue of $\mathbf{A}^T$ corresponding to the eigenvector $\mathbf{v}^{(1)}$, and introduce the vector $\mathbf{q}$ that satisfies $\mathbf{q}^T\mathbf{v}^{(1)} = 1$ but is arbitrary otherwise. Repeating the preceding steps, we find that the matrix

$$B_{i,j} = A_{i,j} - \lambda_1 q_i v_j^{(1)} \tag{2.10.5}$$

shares the eigenvalues of $\mathbf{A}$, with the exception of the eigenvalue λ_1 that has been replaced by zero. The eigenvectors of $\mathbf{B}$ corresponding to the unaltered eigenvalues are parallel to those of $\mathbf{A}$.

For example, for the matrix $\mathbf{A}$ defined in equation (2.10.3), $\mathbf{v}^{(1)} = (1, 6)^T$. Choosing $\mathbf{q} = (1, 0)^T$, we find

$$\mathbf{B} = \begin{bmatrix} 1 & 9 \\ \frac{1}{4} & 1 \end{bmatrix} - \frac{5}{2}\begin{bmatrix} 1 & 6 \\ 0 & 0 \end{bmatrix} = \begin{bmatrix} -\frac{3}{2} & -6 \\ \frac{1}{4} & 1 \end{bmatrix} \tag{2.10.6}$$

whose eigevalues are $\lambda_{B_1} = 0$ and $\lambda_{B_2} = -\frac{1}{2}$.

Identifying $\boldsymbol{q}$ with the eigenvector $\boldsymbol{u}^{(1)}$ of $\boldsymbol{A}$ corresponding to the eigenvalue λ_1, reduces the deflating term $\lambda_1 q_i v_j^{(1)}$ to the corresponding term in the spectral expansion shown in equation (2.9.24), written for $\boldsymbol{A}^T$. It is then not surprising that subtracting off this term annihilates the corresponding eigenvalue.

PROBLEM

2.10.1 ***Deflation of the Hilbert matrix.***

Remove the eigenvalue with the maximum norm of the 2×2 Hilbert matrix defined in equation (2.6.5) using (a) its eigenvector and (b) the eigenvector of its transpose.

Sequential Deflations

The single deflation just described may be repeated in a sequential fashion a number of times, to successively shift further eigenvalues to zero. Each deflation requires computing one eigenvalue and the corresponding eigenvector of the matrix or its transpose.

Simultaneous Double Deflation

It is possible to combine two single deflations and carry out a simultaneous *double deflation.* Let λ_1 be a single eigenvalue of $\boldsymbol{A}$ corresponding to the eigenvector $\boldsymbol{u}^{(1)}$, and λ_2 be a different single eigenvalue of $\boldsymbol{A}^T$ corresponding to the eigenvector $\boldsymbol{v}^{(2)}$, and introduce the vectors $\boldsymbol{w}$ and $\boldsymbol{q}$ that satisfy $\boldsymbol{w}^T\boldsymbol{u}^{(1)} = 1$ and $\boldsymbol{q}^T\boldsymbol{v}^{(1)} = 1$, but are arbitrary otherwise. It can be shown, working as previously, that the matrix $\boldsymbol{B}$ with elements

$$B_{i,j} = A_{i,j} - \lambda_1 u_i^{(1)} w_j - \lambda_2 q_i v_j^{(1)} \tag{2.10.7}$$

shares eigenvalues of matrix $\boldsymbol{A}$, with the exception of the eigenvalues λ_1 and λ_2 that have been replaced by zeros. The eigenvectors of $\boldsymbol{B}$ are generally different from those of $\boldsymbol{A}$.

For example, for the matrix $\boldsymbol{A}$ defined in equation (2.10.3), $\lambda_1 = \frac{5}{2}$, $\lambda_2 = -\frac{1}{2}$, $\boldsymbol{u}^{(1)} = (6, 1)^T$, and $\boldsymbol{v}^{(2)} = (1, -6)^T$. Accordingly, we choose $\boldsymbol{w} = (0, 1)^T$ and $\boldsymbol{q} = (1, 0)^T$ and find

$$\boldsymbol{B} = \begin{bmatrix} 1 & 9 \\ \frac{1}{4} & 1 \end{bmatrix} - \frac{5}{2}\begin{bmatrix} 0 & 6 \\ 0 & 1 \end{bmatrix} + \frac{1}{2}\begin{bmatrix} 1 & -6 \\ 0 & 0 \end{bmatrix} = \begin{bmatrix} \frac{3}{2} & -9 \\ \frac{1}{4} & -\frac{3}{2} \end{bmatrix} \tag{2.10.8}$$

which has a double eigenvalue equal to zero.

PROBLEM

2.10.2 ***Double deflation of the Hilbert matrix.***

Perform the double deflation of the 2×2 Hilbert matrix defined in equation (2.6.5).

Eigenvalues with Higher Multiplicity

The preceding results can be generalized to account for multiple eigenvalues as follows:

- Let λ_1 be an eigenvalue of $\boldsymbol{A}$ with algebraic multiplicity equal to m corresponding to the eigenvectors $\boldsymbol{u}^{(i)}, i = 1, \ldots, m$, and λ_2 be a different eigenvalue of $\boldsymbol{A}^T$ with algebraic multiplicity equal to l corresponding to the eigenvectors $\boldsymbol{v}^{(i)}, i = 1, \ldots, l$.

- Let us then introduce the set of vectors $\boldsymbol{w}^{(i)}$, $i = 1, \ldots, m$, that are orthonormal to $\boldsymbol{u}^{(i)}$, that is,

$$\boldsymbol{u}^{(i)T}\boldsymbol{w}^{(j)} = \delta_{i,j} \tag{2.10.9}$$

and the set of vectors $\boldsymbol{q}^{(i)}$, $i = 1, \ldots, l$, that are orthonormal to $\boldsymbol{v}^{(i)}$, that is,

$$\boldsymbol{v}^{(i)T}\boldsymbol{q}^{(j)} = \delta_{i,j} \tag{2.10.10}$$

The general statement of Wielandt's theorem asserts that:

- The matrix

$$B_{i,j} = A_{i,j} - \lambda_1 \sum_{n=1}^{m} u_i^{(n)} w_j^{(n)} \tag{2.10.11}$$

shares the eigenvalues of $\boldsymbol{A}$ with the exception of the eigenvalue λ_1 that has been replaced by zero.

- The matrix

$$B_{i,j} = A_{i,j} - \lambda_2 \sum_{n=1}^{l} q_i^{(n)} v_j^{(n)} \tag{2.10.12}$$

shares the eigenvalues of $\boldsymbol{A}$ with the exception of the eigenvalue λ_2 that has been replaced by zero.

- The matrix

$$B_{i,j} = A_{i,j} - \lambda_1 \sum_{n=1}^{m} u_i^{(n)} w_j^{(n)} - \lambda_2 \sum_{n=1}^{l} q_i^{(n)} v_j^{(n)} \tag{2.10.13}$$

shares the eigenvalues of $\boldsymbol{A}$ with the exception of λ_1 and λ_2; both have been replaced by zero.

- The eigenvectors of $\boldsymbol{B}$ are generally different than those of $\boldsymbol{A}$.

2.11 *Successive Linear Mappings*

In Sections 2.3 and 2.8, we interpreted a matrix $\boldsymbol{A}$ as a device that associates a certain *input* vector $\boldsymbol{x}$ with another *output* vector

$$\boldsymbol{y} = \boldsymbol{A}\boldsymbol{x} \tag{2.11.1}$$

This association has the formal identity of a *linear mapping*, also called a *linear transformation*, *linear projection*, or *linear operator*.

A general linear mapping of a vector space, expressed by the function $\boldsymbol{y} = \boldsymbol{g}(\boldsymbol{x})$, has the distinctive property

$$\boldsymbol{g}(\boldsymbol{x}_1 + \boldsymbol{x}_2) = \boldsymbol{g}(\boldsymbol{x}_1) + \boldsymbol{g}(\boldsymbol{x}_2) \tag{2.11.2}$$

where $\boldsymbol{x}_1$ and $\boldsymbol{x}_2$ are two arbitrary vectors. This property is certainly satisfied when $\boldsymbol{g}(\boldsymbol{x}) = \boldsymbol{A}\boldsymbol{x}$.

PROBLEM

2.11.1 ***Linear mapping or not?***

Examine whether the following mappings are linear: (*a*) $\boldsymbol{g}(\boldsymbol{x}) = \alpha\boldsymbol{x}$, where α is a scalar constant; (*b*) $\boldsymbol{g}(\boldsymbol{x}) = \boldsymbol{A}\boldsymbol{x} + \boldsymbol{b}$, where $\boldsymbol{b}$ is a constant vector.

Null Space and Eigenvalues

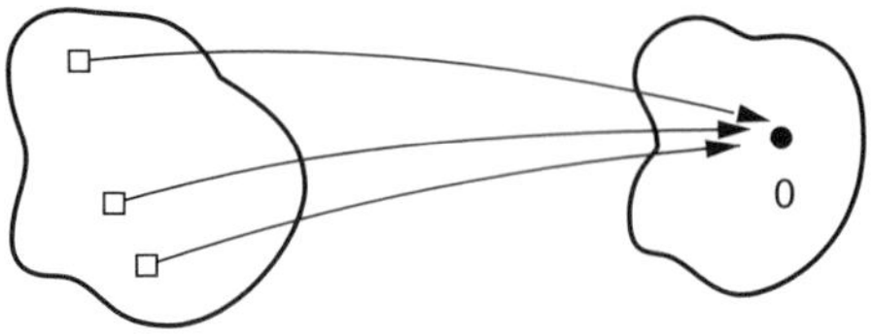

The space of the vectors $\boldsymbol{x}$ whose image is the null vector $\boldsymbol{0}$ is the *null space* or *kernel* of the linear mapping. Each vector in the null space is an eigenvector of $\boldsymbol{A}$ corresponding to the eigenvalue $\lambda = 0$. Consequently, if all eigenvalues of $\boldsymbol{A}$ are nonzero, that is, the matrix is nonsingular, then the null space is empty.

Fixed Points

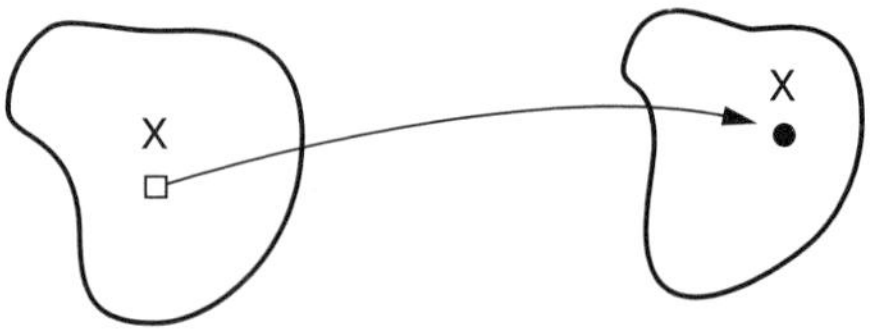

A vector $\boldsymbol{X}$ that is mapped onto itself, that is, $\boldsymbol{X} = \boldsymbol{A}\boldsymbol{X}$, is a *fixed point* of the linear mapping (2.11.1). The null point $\boldsymbol{X} = \boldsymbol{0}$ is an obvious, albeit trivial, fixed point of any matrix $\boldsymbol{A}$; any point is a fixed point of the identity matrix $\boldsymbol{I}$; when $\alpha \neq 1$, the mapping associated with the matrix $\alpha\boldsymbol{I}$ does not have a nontrivial fixed point. These observations make it clear that the number of fixed points depends on the structure of the matrix $\boldsymbol{A}$.

By definition, a fixed point of a particular matrix $\boldsymbol{A}$ falls into the null space of the linear mapping associated with the matrix $\boldsymbol{A} - \boldsymbol{I}$, which is nonempty only when $\text{Det}(\boldsymbol{A} - \boldsymbol{I}) = 0$. When this condition is fulfilled, at least one eigenvalue of $\boldsymbol{A} - \boldsymbol{I}$ will be equal to zero, and one eigenvalue of $\boldsymbol{A}$ will be equal to unity. A matrix can have a nontrivial fixed point only when its spectral radius is equal to or larger than unity.

PROBLEM

2.11.2 ***Fixed points.***

(*a*) Consider a continuous function $f(x)$ defined over the semi-closed interval $a \leq x < b$, and assume that its range is included in that interval. Explain, on geometrical grounds, why the function must have at least one fixed point that satisfies $X = f(X)$.
(*b*) Generalize the statement in part (*a*) to higher dimensions.

Image Manifold of a Singular Matrix

Consider now the image manifold of a singular matrix discussed previously in Section 2.8. Since $\text{Det}(\boldsymbol{A}) = 0$, at least one eigenvalue λ of $\boldsymbol{A}$ must be equal to zero. Let $\boldsymbol{v}$ be a corresponding eigenvector of $\boldsymbol{A}^T$, $\boldsymbol{x}$ be an arbitrary vector, and define $\boldsymbol{y} = \boldsymbol{A}\boldsymbol{x}$. Then

$$\boldsymbol{v}^T\boldsymbol{y} = \boldsymbol{v}^T\boldsymbol{A}\boldsymbol{x} = \lambda\boldsymbol{v}^T\boldsymbol{x} = 0 \tag{2.11.3}$$

that is, the image of $\boldsymbol{x}$ is orthogonal to $\boldsymbol{v}$.

We thus find that the image of the linear mapping associated with a singular matrix $\boldsymbol{A}$ is orthogonal to the space of the eigenvectors $\boldsymbol{v}$. Phrased differently, the space of eigenvectors $\boldsymbol{v}$ is the orthogonal complement of the image of the linear mapping associated with the singular matrix $\boldsymbol{A}$.

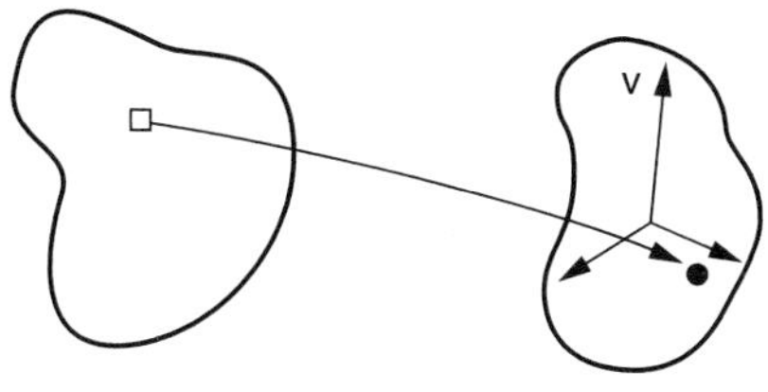

Existence of Solutions of Linear Systems with Singular Matrices

Based on the preceding results, we can establish conditions for the existence of a solution of the equation $\boldsymbol{A}\boldsymbol{x} = \boldsymbol{b}$ when $\text{Det}(\boldsymbol{A}) = 0$. A solution will exist only when the *solvability condition*

$$\boldsymbol{v}^T \boldsymbol{b} = 0 \tag{2.11.4}$$

is fulfilled.

We note that condition (2.11.4) is satisfied in a trivial way when $\boldsymbol{b} = \boldsymbol{0}$, and this shows that the homogeneous equation $\boldsymbol{A}\boldsymbol{x} = \boldsymbol{0}$, where $\text{Det}(\boldsymbol{A}) = 0$, has at least one family of nontrivial solutions parametrized by appropriate constants. Conversely, a necessary and sufficient condition for the homogeneous equation $\boldsymbol{A}\boldsymbol{x} = \boldsymbol{0}$ to have a nontrivial solution is that $\text{Det}(\boldsymbol{A}) = 0$.

Behavior of a Vector Subject to Successive Mappings

Let us select a certain N-dimensional vector $\boldsymbol{x}^{(0)}$ and compute a sequence of images using the linear mapping associated with the $N \times N$ matrix $\boldsymbol{A}$, setting

$$\boldsymbol{x}^{(k+1)} = \boldsymbol{A}\boldsymbol{x}^{(k)} \tag{2.11.5}$$

We want to know what happens to the images as we continue the projections. In Section 3.7 we shall see that understanding the behavior of this sequence is a prerequisite for the design of algorithms that solve systems of linear equations using iterative methods. For simplicity, we assume that $\boldsymbol{A}$ has N eigenvectors.

To find the answer to the problem just posed, we express the initial vector $\boldsymbol{x}^{(0)}$ in terms of the N eigenvectors of $\boldsymbol{A}$ as

$$\boldsymbol{x}^{(0)} = \sum_{i=1}^{N} c_i \boldsymbol{u}^{(i)} \tag{2.11.6}$$

where c_i are constant coefficients. Using the distinguishing property of the eigenvectors stated in equation (2.9.1), we write

$$\boldsymbol{x}^{(k)} = \sum_{i=1}^{N} c_i \lambda_i^k \boldsymbol{u}^{(i)} = \rho^k \sum_{i=1}^{N} c_i \left(\frac{\lambda_i}{\rho}\right)^k \boldsymbol{u}^{(i)} \tag{2.11.7}$$

where λ_1 is an eigenvalue with the maximum norm, and $\rho = |\lambda_1|$ is the spectral radius of $\boldsymbol{A}$. The magnitudes of the ratios within the sum on the right-hand side are less than unity or, at most, equal to unity. As k tends to infinity, the powers of the ratios whose magnitudes are less than unity diminish, whereas the powers of the ratios whose magnitudes are exactly equal to unity stay constant.

If, in particular, λ_1 is a single eigenvalue with corresponding eigenvector $\boldsymbol{u}^{(1)}$, and if there is no other eigenvalue with the same norm, then

$$\boldsymbol{x}^{(k)} \to c_1 \lambda_1^k \boldsymbol{u}^{(1)} \tag{2.11.8}$$

which shows that

$$\frac{|\boldsymbol{x}^{(k+1)}|}{|\boldsymbol{x}^{(k)}|} \cong \rho \tag{2.11.9}$$

Assuming that c_1 does not vanish due to a fortuitous selection of $\boldsymbol{x}^{(0)}$, we conclude:

- If ρ is larger than unity, the length of $\boldsymbol{x}^{(k)}$ will keep increasing during the mappings.
- If ρ is less than unity, the length of $\boldsymbol{x}^{(k)}$ will keep decreasing during the mappings.
- In both cases, $\boldsymbol{x}^{(k)}$ will tend to become parallel to the eigenvector $\boldsymbol{u}^{(1)}$ corresponding to the eigenvalue with the maximum norm.

PROBLEMS

2.11.3 ***Successive mappings.***

Discuss the behavior of $\boldsymbol{x}^{(k)}$ when (a) the coefficient c_1 vanishes and (b) the algebraic multiplicity of the eigenvalue with the maximum norm is higher than one.

2.11.4 ***Linear mapping with offset.***

(a) Consider the scalar sequence produced by the mapping $x^{(k+1)} = ax^{(k)} + b$, and show that

$$x^{(k)} = a^k x^{(0)} + \frac{1 - a^k}{1 - a} b \tag{2.11.10}$$

(b) Discuss the asymptotic behavior of the images $\boldsymbol{x}^{(k+1)} = \boldsymbol{A}\boldsymbol{x}^{(k)} + \boldsymbol{b}$, where $\boldsymbol{b}$ is a constant vector.

2.12 *Functions of Matrices*

A function $\boldsymbol{f}$ of a square matrix $\boldsymbol{A}$ is a device that associates the matrix $\boldsymbol{A}$ with another square matrix $\boldsymbol{B}$ of the same dimensions. Symbolically, we write

$$\boldsymbol{B} = \boldsymbol{f}(\boldsymbol{A}) \tag{2.12.1}$$

where $\boldsymbol{f}$ is a two-index operator.

Functions of Matrices from Functions of Scalars

Matrix functions $\boldsymbol{f}(\boldsymbol{A})$ can be produced from scalar functions $f(x)$ by expanding $f(x)$ in its Maclaurin or Taylor series (Appendix A), replacing x with $\boldsymbol{A}$, and computing the designated powers of $\boldsymbol{A}$.

When the matrix $\boldsymbol{A}$ is diagonal, we find

$$B_{i,j} = 0 \qquad \text{if } i \neq j$$
$$B_{i,i} = f_{i,i}(\boldsymbol{A}) = f(A_{i,i}) \tag{2.12.2}$$

where summation is not implied over the repeated index i. Thus $\boldsymbol{B}$ is another diagonal matrix with nonzero elements equal to the scalar functions of the corresponding elements of $\boldsymbol{A}$.

For example, $\boldsymbol{exp}(2\boldsymbol{I})$ is a diagonal matrix with all diagonal elements equal to e^2.

Computation of Matrix Functions

The Maclaurin-series expansion method may present difficulties associated with possible lack of convergence and high computational cost. An infallible, but not necessarily expedient, alternative approach derives from the diagonal form shown in equation (2.9.23); one condition is that the matrix has N distinct eigenvectors. We note that

$$\boldsymbol{A}^p = \boldsymbol{U}\boldsymbol{\Lambda}^p\boldsymbol{U}^{-1} \tag{2.12.3}$$

where p is a positive integer, expand $\boldsymbol{f}(\boldsymbol{A})$ in its Maclaurin series, and find

$$\boldsymbol{f}(\boldsymbol{A}) = \boldsymbol{U}\boldsymbol{f}(\boldsymbol{\Lambda})\boldsymbol{U}^{-1} \tag{2.12.4}$$

where the columns of $\boldsymbol{U}$ contain the eigenvectors of $\boldsymbol{A}$ corresponding to the eigenvalues $\lambda_1, \lambda_2, \ldots, \lambda_N$, and where $f(\boldsymbol{\Lambda})$ is a diagonal matrix with diagonal elements equal to $f(\lambda_1), f(\lambda_2), \ldots, f(\lambda_N)$. This method, however, requires the availability of all eigenvalues and eigenvectors of $\boldsymbol{A}$, whose computation may be a demanding task.

More efficient methods of computing and approximating matrix functions are discussed by Golub and van Loan (1989, Chapter 11).

PROBLEM

2.12.1 ***Square root of a matrix.***

Compute all real and complex matrices $\boldsymbol{B}$ that satisfy the equation $\boldsymbol{B}^2 = \boldsymbol{A}$, and are thus the square roots of a 2×2 matrix $\boldsymbol{A}$ with $A_{1,1} = 3$, $A_{1,2} = 2$, $A_{2,1} = 2$, $A_{2,2} = 2$.

The Hamilton–Cayley Theorem

Identifying the scalar function $f(x)$ with the characteristic polynomial of the $N \times N$ matrix $\boldsymbol{A}$, that is, setting $f(x) = P(x)$, noting that, by definition, $\boldsymbol{P}(\boldsymbol{\Lambda}) = \boldsymbol{0}$, and referring to equation (2.12.4), we obtain the mathematical statement of the *Hamilton–Cayley theorem* asserting that a matrix $\boldsymbol{A}$ satisfies the equation

$$\boldsymbol{P}(\boldsymbol{A}) = \boldsymbol{0} \tag{2.12.5}$$

This equation allows us to express the integral powers of a matrix, $\boldsymbol{A}^m$ with $m \geq N$, in terms of the N matrices $\boldsymbol{I}, \boldsymbol{A}, \boldsymbol{A}^2, \ldots, \boldsymbol{A}^{N-1}$. And by extension, it allows us to reduce any polynomial function of a matrix with degree $m \geq N$ to an equivalent polynomial of degree $N - 1$.

PROBLEM

2.12.2 ***Confirmation of the Hamilton–Cayley theorem.***

Derive the characteristic polynomial and confirm the validity of the Hamilton–Cayley theorem for a 3×3 matrix of your choice. Then consider a fifth degree polynomial of this matrix and reduce it to an equivalent second degree polynomial.

The Exponential Function

The exponential function of a matrix arises naturally during the analytical solution of systems of ordinary linear differential equations with constant coefficients, to be discussed in Section 9.2. According to the preceding discussion,

$$\boldsymbol{exp}(\boldsymbol{A}) = \boldsymbol{I} + \boldsymbol{A} + \tfrac{1}{2}\boldsymbol{A}^2 + \tfrac{1}{6}\boldsymbol{A}^3 + \cdots \tag{2.12.6}$$

Not all rules of number exponentiation carry over to matrix exponentiation. For example, the identities

$$[\boldsymbol{exp}(\boldsymbol{A})]^p = \boldsymbol{exp}(p\boldsymbol{A}) \tag{2.12.7}$$

where p is a natural number, and

$$\boldsymbol{exp}(\boldsymbol{A}^T) = [\boldsymbol{exp}(\boldsymbol{A})]^T \tag{2.12.8}$$

hold for any matrix $\boldsymbol{A}$, but

$$\boldsymbol{exp}(\boldsymbol{A} + \boldsymbol{B}) \neq \boldsymbol{exp}(\boldsymbol{A})\, \boldsymbol{exp}(\boldsymbol{B}) \tag{2.12.9}$$

Equality holds only when the matrices $\boldsymbol{A}$ and $\boldsymbol{B}$ commute. This can be seen readily by expressing both sides of equation (2.12.9) in Maclaurin series and carrying out the multiplications.

Setting, in particular, $\boldsymbol{B} = -\boldsymbol{A}$, and noting that $\boldsymbol{exp}(\boldsymbol{0}) = \boldsymbol{I}$, we find

$$\boldsymbol{exp}(-\boldsymbol{A}) = [\boldsymbol{exp}(\boldsymbol{A})]^{-1} \tag{2.12.10}$$

which extends the validity of equation (2.12.7) to $p = -1$.

Another useful identity is

$$\frac{d e^{t\boldsymbol{A}}}{dt} = \boldsymbol{A} e^{t\boldsymbol{A}} \tag{2.12.11}$$

where t is a scalar.

References

Atkinson, K. E., 1989, *An Introduction to Numerical Analysis.* Wiley.
Bodewig, E., 1959, *Matrix Calculus.* North-Holland.
Cullen, C. G., 1994, *An Introduction to Numerical Linear Algebra.* PWS-Kent.

Golub, G. H., and van Loan, C. F., 1989, *Matrix Computation.* The Johns Hopkins University Press.
Householder, A. S., 1964, *The Theory of Matrices in Numerical Analysis.* Dover.
Kolmogorov, A. N., and Fomin, S. V., 1970, *Introductory Real Analysis.* Dover.
Noble, B., and Daniel, J. W., 1988, *Applied Linear Algebra.* Prentice-Hall.
Pozrikidis, C., 1997, *Introduction to Theoretical and Computational Fluid Dynamics.* Oxford University Press.
Press, W. H, Flannery, B. P., Teukolsky, S. A., and Vetterling, W. T., 1992, *Numerical Recipes,* 2nd ed. Cambridge University Press.
Synge, J. L., and Schild, A., 1978, *Tensor Calculus.* Dover.
Wilkinson, J. H., 1965, *The Algebraic Eigenvalue Problem.* Oxford University Press.

Chapter 3

Linear Algebraic Equations

3.1 Significance and Applications

It is not an exaggeration to say that solving the vast majority of problems in science and engineering is reduced, one way or another, to computing the solution of systems of linear algebraic equations using numerical methods. In the final stage, we look for a vector $\boldsymbol{x}$ that, when multiplied by the matrix $\boldsymbol{A}$, yields the known vector $\boldsymbol{b}$, that is,

$$\boldsymbol{A}\boldsymbol{x} = \boldsymbol{b} \tag{3.1.1}$$

The quantities $\boldsymbol{A}$, $\boldsymbol{x}$, and $\boldsymbol{b}$ may have a broad range of interpretations depending on the particular physical or mathematical context underlying equation (3.1.1).

The following examples, drawn from the areas of structural mechanics, mathematical physics, and chemical engineering, illustrate the manner in which problems of this nature typically arise in practice. Additional examples will be presented in subsequent chapters.

Forces on the Elements of a Crane Truss

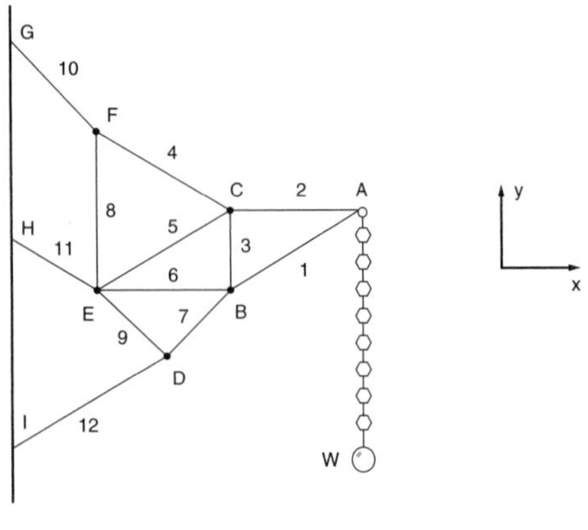

Consider the truss of a crane lifting a demolition ball with weight $W = 800$ kg in a designated hard-hat area, as depicted in the diagram. We want to compute the tensile or compressive forces exerted on the 12 linear elements, F_i, $i = 1, 2, \ldots, 12$. A positive force signifies element tension, and a negative force signifies element compression.

Proceeding with the mathematical formulation, we write force balances in the directions of the x and y axes at the five joints A–F and thus obtain a system of 12 equations for the 12 unknowns. The force balances require that the sums of the x or y components of the forces exerted at a joint must add up to zero. Measuring all forces in kilograms, and dropping the units in the balance equations, we obtain the following:

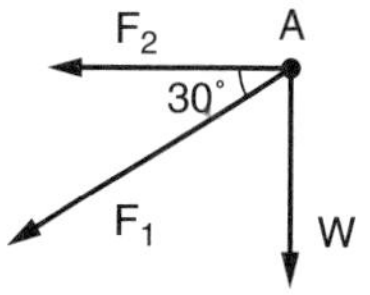

Joint A:	x balance	$-F_1 \cos(30°) - F_2 = 0$
	y balance	$-800 - F_1 \cos(60°) = 0$
Joint B:	x balance	$F_1 \cos(30°) - F_6 - F_7 \cos(45°) = 0$
	y balance	$F_3 + F_1 \cos(60°) - F_7 \cos(45°) = 0$
Joint C:	x balance	$F_2 - F_4 \cos(30°) - F_5 \cos(30°) = 0$
	y balance	$-F_3 - F_5 \cos(60°) + F_4 \cos(60°) = 0$
Joint D:	x balance	$F_7 \cos(45°) - F_9 \cos(45°) - F_{12} \cos(30°) = 0$
	y balance	$F_7 \cos(45°) + F_9 \cos(45°) - F_{12} \cos(60°) = 0$
Joint E:	x balance	$F_6 + F_5 \cos(30°) + F_9 \cos(45°) - F_{11} \cos(30°) = 0$
	y balance	$F_5 \cos(60°) - F_9 \cos(45°) + F_8 + F_{11} \cos(60°) = 0$
Joint F:	x balance	$F_4 \cos(30°) - F_{10} \cos(45°) = 0$
	y balance	$-F_8 - F_4 \cos(60°) + F_{10} \cos(45°) = 0$

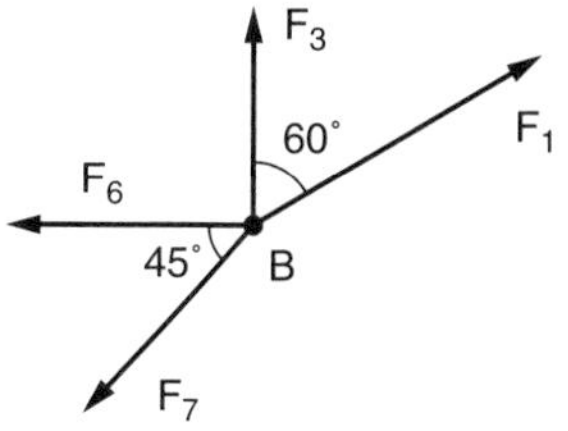

There are three more joints, G, H, and I, but the corresponding force balances would simply say that elements 10, 11, and 12 pull the vertical wall in their respective directions with the developing forces.

The preceding 12 equations may be placed in the standard form of the linear system shown in equation (3.1.1):

$$\begin{bmatrix} \cos(30°) & 1 & 0 & 0 & 0 & 0 & 0 & 0 & 0 & 0 & 0 & 0 \\ \cos(60°) & 0 & 0 & 0 & 0 & 0 & 0 & 0 & 0 & 0 & 0 & 0 \\ \cos(30°) & 0 & 0 & 0 & 0 & -1 & -\cos(45°) & 0 & 0 & 0 & 0 & 0 \\ \cos(60°) & 0 & 1 & 0 & 0 & 0 & -\cos(45°) & 0 & 0 & 0 & 0 & 0 \\ 0 & 1 & 0 & -\cos(30°) & -\cos(30°) & 0 & 0 & 0 & 0 & 0 & 0 & 0 \\ 0 & 0 & -1 & \cos(60°) & -\cos(60°) & 0 & 0 & 0 & 0 & 0 & 0 & 0 \\ 0 & 0 & 0 & 0 & 0 & 0 & \cos(45°) & 0 & -\cos(45°) & 0 & 0 & -\cos(30°) \\ 0 & 0 & 0 & 0 & 0 & 0 & \cos(45°) & 0 & \cos(45°) & 0 & 0 & -\cos(60°) \\ 0 & 0 & 0 & 0 & \cos(30°) & 1 & 0 & 0 & \cos(45°) & 0 & -\cos(30°) & 0 \\ 0 & 0 & 0 & 0 & \cos(60°) & 0 & 0 & 1 & -\cos(45°) & 0 & \cos(60°) & 0 \\ 0 & 0 & 0 & \cos(30°) & 0 & 0 & 0 & 0 & 0 & -\cos(45°) & 0 & 0 \\ 0 & 0 & 0 & -\cos(60°) & 0 & 0 & 0 & -1 & 0 & \cos(45°) & 0 & 0 \end{bmatrix} \begin{bmatrix} F_1 \\ F_2 \\ F_3 \\ F_4 \\ F_5 \\ F_6 \\ F_7 \\ F_8 \\ F_9 \\ F_{10} \\ F_{11} \\ F_{12} \end{bmatrix} = \begin{bmatrix} 0 \\ -800 \\ 0 \\ 0 \\ 0 \\ 0 \\ 0 \\ 0 \\ 0 \\ 0 \\ 0 \\ 0 \end{bmatrix} \tag{3.1.2}$$

We note that the matrix coefficient on the left-hand side does not have a particular structure although it is noticeably sparse. Systems of this nature are solved most efficiently using custom-made iterative methods, to be discussed in Section 3.4.

PROBLEM

3.1.1 ***A statically determinate, simple plane truss.***

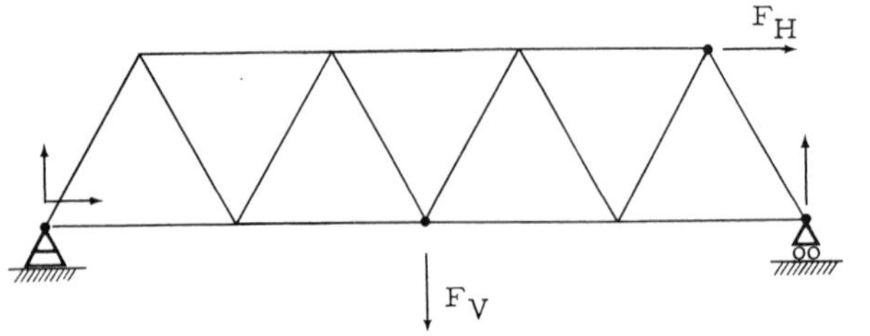

The number of joints N_J and the number of elements N_E of a statically determinate, simple plane truss with a pivot and a roller satisfy the equation $N_E = 2N_J - 3$ (e.g., Beer and Johnston 1988, Chapter 6). A pivot can support both a horizontal and a vertical load, whereas a roller can support only a vertical load.

Assess whether the simple truss shown in the diagram is statically determinate, and derive a system of linear equations for the forces exerted on its elements. The truss is loaded with a horizontal force F_H and a vertical load F_V.

Temperature Distribution in a Plate

In the second example, we illustrate how systems of linear equations arise from the discretization of partial differential equations of mathematical physics using *finite-difference* methods. Additional examples and similar methodologies leading to similar systems of linear equations will be discussed in Chapter 11.

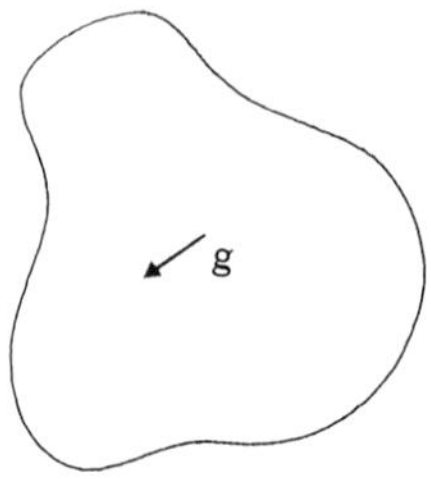

Consider the steady-state temperature distribution in a two-dimensional domain such as a thermally conducting plate. Heat conservation requires that the distribution of the temperature, denoted by f, satisfy the Poisson equation in two dimensions

$$\nabla^2 f \equiv \frac{\partial^2 f}{\partial x^2} + \frac{\partial^2 f}{\partial y^2} = g(x, y) \tag{3.1.3}$$

where ∇^2 is the two-dimensional Laplacian operator, and $g(x, y)$ is a known function of position expressing homogeneous heat loss or production. A positive value of $g(x, y)$ signifies loss due, for example, to radiation, and a negative value signifies production due, for example, to a chemical reaction. To complete the definition of the problem, we must also specify one scalar boundary condition around the plate perimeter.

Rectangular plate

To simplify the numerical implementation, we consider a plate with a rectangular shape confined between $a \le x \le b$ and $c \le y \le d$, as shown in Figure 3.1.1, and specify the temperature distribution along the left and the two horizontal sides and the flux distribution along the fourth side. These stipulations provide us with the *Dirichlet boundary conditions*

$$f = w(y) \quad \text{at } x = a, \qquad f = z(x) \quad \text{at } y = c, \qquad f = v(x) \quad \text{at } y = d \tag{3.1.4}$$

and the *Neumann boundary condition*

$$\frac{\partial f}{\partial x} = q(y) \qquad \text{at } x = b \tag{3.1.5}$$

By definition, a Dirichlet boundary condition provides us with the value of a function at a boundary, whereas a Neumann boundary condition provides us with the value of the normal derivative; a *mixed* or *Robin* boundary condition specifies a linear combination.

Finite-difference grid

We proceed to develop an entry-level *finite-difference method*; a more complete discussion will be given in Section 11.4.

We begin by dividing the x interval (a, b) into N evenly spaced subintervals that are separated by the spacing $\Delta x = (b - a)/N$, and draw the grid lines $x = x_i$, where

$$x_i = a + (i - 1)\Delta x$$

$i = 1, 2, \ldots, N + 1$, as shown in Figure 3.1.1. Similarly, we divide the y interval (c, d) into M evenly spaced subintervals that are separated by the spacing $\Delta y = (d - c)/M$, and draw the grid lines $y = y_j$, where

$$y_j = c + (j - 1)\,\Delta y$$

$j = 1, \ldots, M + 1$. The intersections between the vertical and the horizontal grid lines define the *grid points* or *nodes*. For simplicity, we denote the value of the function f at the (i, j) node as

$$f_{i,j} \equiv f(x_i, y_j)$$

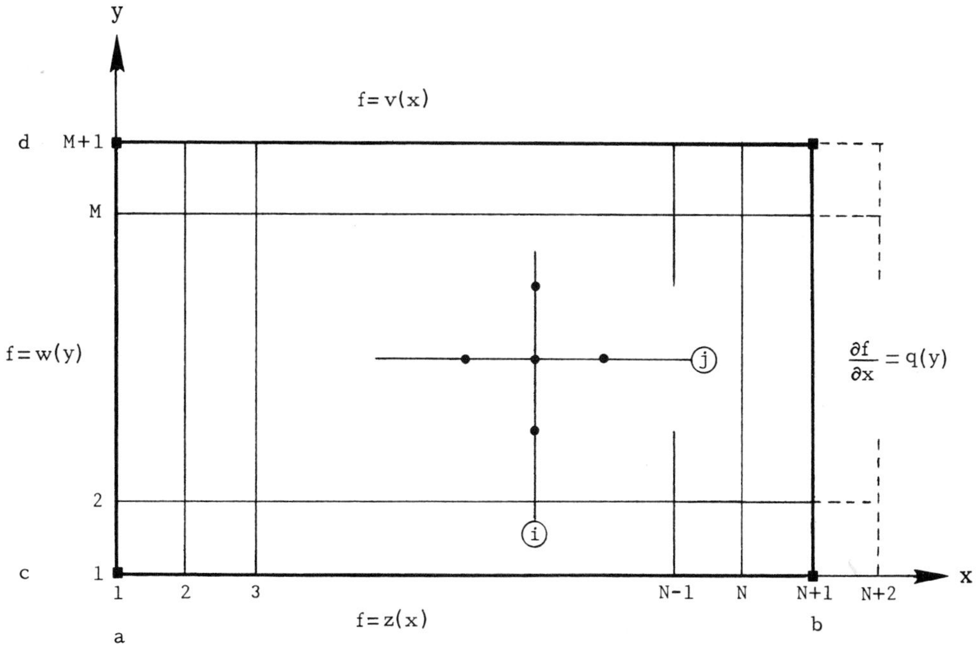

Figure 3.1.1 A finite-difference grid for solving the Poisson equation in a rectangular dmoain.

The Dirichlet boundary conditions (3.1.4) require

$$f_{1,j} = w(y_j), \qquad f_{i,1} = z(x_i), \qquad f_{i,M+1} = v(x_i) \tag{3.1.6}$$

Our objective is to compute the unknown values $f_{i,j}$ at the grid points $i = 2, \ldots, N+1$, $j = 2, \ldots, M$, a total of $K = N(M-1)$ unknowns.

Finite-difference equations

To build a system of linear equations involving the aforementioned unknowns, we apply the differential equation (3.1.3) at the (i, j) grid point and approximate the second partial derivatives using central differences, setting

$$\begin{aligned}(\partial^2 f/\partial x^2)_{i,j} &\cong (f_{i+1,j} - 2f_{i,j} + f_{i-1,j})/\Delta x^2 \\ (\partial^2 f/\partial y^2)_{i,j} &\cong (f_{i,j+1} - 2f_{i,j} + f_{i,j-1})/\Delta y^2\end{aligned} \tag{3.1.7}$$

In Section 6.11, we shall see that these approximations introduce a numerical error that is comparable to the size of Δx^2 or Δy^2. These approximations allow us to convert the partial differential equation (3.1.3) applied at the (i, j) grid point to the *finite-difference equation*

$$\frac{f_{i+1,j} - 2f_{i,j} + f_{i-1,j}}{\Delta x^2} + \frac{f_{i,j+1} - 2f_{i,j} + f_{i,j-1}}{\Delta y^2} = g_{i,j} \tag{3.1.8}$$

which can be rearranged to yield

$$f_{i+1,j} - 2(1+\beta) f_{i,j} + f_{i-1,j} + \beta f_{i,j+1} + \beta f_{i,j-1} = \Delta x^2 g_{i,j} \tag{3.1.9}$$

for $i = 2, \ldots, N$ and $j = 2, \ldots, M$, where

$$\beta = \left(\frac{\Delta x}{\Delta y}\right)^2 \tag{3.1.10}$$

Note that the left-hand side of equation (3.1.9) is a weighted average of the values of the unknown function at a group of five neighboring nodes.

Equation (3.1.9) may certainly be applied at the $L = (N-1)(M-1)$ internal grid points corresponding to $i = 2, \ldots, N$ and $j = 2, \ldots, M$, to provide us with L linear equations, but not at the boundary grid points, since some of the grid points involved will then lie outside the solution domain. We must somehow produce $K - L = N(M-1) - (N-1)(M-1) = M - 1$ additional equations.

Neumann boundary condition

The missing equations must involve the Neumann boundary condition at the right boundary of the plate, at $x = b$, which have not yet been employed. One way to implement this boundary condition with an error that is comparable to the size of Δx^2, which matches the error due to the discretization of the differential equation, is to extend the domain of solution beyond the physical boundary $x = b$ and introduce the *phantom* grid line $x_{N+2} = b + \Delta x$ and associated values $f_{N+2,j}$. In Section 6.11, we shall see that approximating $(\partial f/\partial x)_{N+1,j}$ with the central-difference formula $(f_{N+2,j} - f_{N,j})/(2\,\Delta x)$ introduces a numerical error that is comparable to the size of Δx^2. Adopting this approximation, we replace equation (3.1.5) with the finite-difference equation $(f_{N+2,j} - f_{N,j})/(2\,\Delta x) = q_j$, which can be rearranged to give

$$f_{N+2,j} - f_{N,j} = 2\,\Delta x\, q_j \tag{3.1.11}$$

Having introduced the phantom nodes, we may apply equation (3.1.9) at the boundary grid points with $i = N + 1$ and $j = 2, \ldots, M$. Using equation (3.1.11) to express $f_{N+2,j}$ in terms of $f_{N,j}$, we find

$$-2(1+\beta) f_{N+1,j} + 2 f_{N,j} + \beta f_{N+1,j+1} + \beta f_{N+1,j-1} = \Delta x^2\, g_{N+1,j} - 2\Delta x q_j \tag{3.1.12}$$

The linear system

Equations (3.1.9) and (3.1.12) provide us with the desired system of $N(M-1)$ linear algebraic equations for the unknowns. To obtain the standard form shown in equation (3.1.1), we place the values $f_{i,j}$, $i = 2, \ldots, N+1$, $j = 2, \ldots, M$, at the entries of a long $N(M-1)$-dimensional vector $\boldsymbol{S}$, row-by-row starting from the bottom, as shown in Table 3.1.1. Collecting the aforementioned equations also row-by-row starting from the bottom, we form the linear system $\boldsymbol{AS} = \boldsymbol{b}$, where the *block tridiagonal* coefficient matrix $\boldsymbol{A}$ and constant vector $\boldsymbol{b}$ are shown in Table 3.1.1. In practice, systems of this nature are solved using custom-made iterative methods to be discussed in Sections 3.7 and 10.5. Two comments are worth stating:

- There is nothing special about the row-by-row compilation, and a column-by-column complitation will work fine too (Problem 3.1.3). Both the row-by-row and the column-by-column compilations are *lexicographic.*
- In Section 11.4, we shall describe a programmable method for constructing $\boldsymbol{A}$ and $\boldsymbol{b}$ so as to circumvent the mundane task of manual bookkeeping.

PROBLEMS

3.1.2 *Explicit form of the linear system.*

Display the complete linear system $\boldsymbol{AS} = \boldsymbol{b}$ for the problem discussed in the text with $N = 2$ and $M = 2$.

3.1.3 *Column-by-column.*

Derive the counterpart of the linear system shown in Table 3.1.1, with the unknowns $f_{i,j}$, $i = 2, \ldots, N+1$, $j = 2, \ldots, M$, placed at successive entries of a long $N(M-1)$-dimensional vector column-by-column, and the finite-difference equations also collected column-by-column, in both cases starting from southwest.

3.1.4 *Neumann boundary conditions all around: A singular system.*

Derive a system of $(N+1)(M+1)$ linear equations arising from the finite-difference discretization of Poisson's equation with Neumann boundary conditions that specify the normal derivative of the unknown function f around *all* boundaries of a rectangular domain. Then show that the coefficient matrix is *singular,* and discuss the physical reason.

3.1.5 *Helmholtz equation.*

Work out the finite-difference formulation of the problem discussed in the text with $g = \alpha f$, where α is a constant. In this case, the Poisson equation reduces to the Helmholtz equation.

Table 3.1.1 System of linear equations arising from the finite-difference discretization of the Poisson equation describing heat conduction in a rectangular plate depicted in Figure 3.1.1. $\boldsymbol{T}$ and $\boldsymbol{D}$ are $N \times N$ matrices.

$$\begin{bmatrix} \boldsymbol{T} & \boldsymbol{D} & \boldsymbol{0} & \cdots & \cdots & \cdots & \boldsymbol{0} \\ \boldsymbol{D} & \boldsymbol{T} & \boldsymbol{D} & \boldsymbol{0} & \cdots & \cdots & \cdots \\ \boldsymbol{0} & \boldsymbol{D} & \boldsymbol{T} & \boldsymbol{D} & \boldsymbol{0} & \cdots & \cdots \\ \cdots & \boldsymbol{0} & & & & \boldsymbol{0} & \cdots \\ \cdots & \cdots & \boldsymbol{0} & \boldsymbol{D} & \boldsymbol{T} & \boldsymbol{D} & \boldsymbol{0} \\ \cdots & \cdots & \cdots & \boldsymbol{0} & \boldsymbol{D} & \boldsymbol{T} & \boldsymbol{D} \\ \boldsymbol{0} & \cdots & \cdots & \cdots & \boldsymbol{0} & \boldsymbol{D} & \boldsymbol{T} \end{bmatrix} \begin{bmatrix} \boldsymbol{f}^{(2)} \\ \boldsymbol{f}^{(3)} \\ \cdots \\ \cdots \\ \cdots \\ \boldsymbol{f}^{(M-1)} \\ \boldsymbol{f}^{(M)} \end{bmatrix} = \begin{bmatrix} \boldsymbol{b}^{(2)} \\ \boldsymbol{b}^{(3)} \\ \cdots \\ \cdots \\ \cdots \\ \boldsymbol{b}^{(M-1)} \\ \boldsymbol{b}^{(M)} \end{bmatrix}$$

$$\boldsymbol{T} = \begin{bmatrix} -2(1+\beta) & 1 & 0 & \cdots & \cdots & 0 \\ 1 & -2(1+\beta) & 1 & 0 & \cdots & 0 \\ 0 & 1 & -2(1+\beta) & 1 & 0 & \cdots \\ \cdots & 0 & & \ddots & & 0 \\ \cdots & \cdots & 0 & 1 & -2(1+\beta) & 1 \\ 0 & \cdots & \cdots & 0 & 2 & -2(1+\beta) \end{bmatrix}$$

$$\boldsymbol{D} = \begin{bmatrix} \beta & 0 & \cdots & \cdots & 0 \\ 0 & \beta & 0 & \cdots & \cdots \\ \cdots & 0 & \ddots & 0 & \cdots \\ \cdots & \cdots & 0 & \beta & 0 \\ 0 & \cdots & \cdots & 0 & \beta \end{bmatrix}$$

$$\boldsymbol{f}^{(j)} = (f_{2,j}, f_{3,j}, \ldots, f_{N+1,j})^T$$

$$\boldsymbol{b}^{(2)} = \begin{bmatrix} -w(y_2) - \beta z(x_2) + \Delta x^2 g_{2,2} \\ -\beta z(x_3) + \Delta x^2 g_{3,2} \\ \cdots \\ -\beta z(x_N) + \Delta x^2 g_{N,2} \\ -2\Delta x q(y_2) - \beta z(x_{N+1}) + \Delta x^2 g_{N+1,2} \end{bmatrix}, \quad \boldsymbol{b}^{(3)} = \begin{bmatrix} -w(y_3) + \Delta x^2 g_{2,3} \\ \Delta x^2 g_{3,3} \\ \cdots \\ \Delta x^2 g_{N,3} \\ -2\Delta x q(y_3) + \Delta x^2 g_{N+1,3} \end{bmatrix}$$

$$\boldsymbol{b}^{(j)} = \begin{bmatrix} -w(y_i) + \Delta x^2 g_{2,j} \\ \Delta x^2 g_{3,j} \\ \cdots \\ \Delta x^2 g_{N,j} \\ -2\Delta x q(y_i) + \Delta x^2 g_{N+1,j} \end{bmatrix} \quad \text{for } j = 3, \ldots, M-1, \quad \boldsymbol{b}^{(M)} = \begin{bmatrix} -w(y_M) - \beta v(x_2) + \Delta x^2 g_{2,M} \\ -\beta v(x_3) + \Delta x^2 g_{3,M} \\ \cdots \\ -\beta v(x_N) + \Delta x^2 g_{N,M} \\ -2\Delta x q(y_M) - \beta v(x_{N+1}) + \Delta x^2 g_{N+1,M} \end{bmatrix}$$

Mass Balance Around a Dual Distillation Column

A solution containing 30% benzene denoted by B, 45% toluene denoted by T, and 25% xylene denoted by X, where the percentages are per weight, is fed into the chemical process shown in the diagram. The overhead product from the first column contains 91.40% B, 8.30% T, and 0.30% X. The overhead product from the second column contains 4.25% B, 91.60% T, and 4.15% X. One-third of the bottom product of the second column is recycled into the first column.

Manufacturing specifications require that the benzene in the first column overhead product be 20 times that of the second column overhead product, and 2 times that of the second column bottom product. We

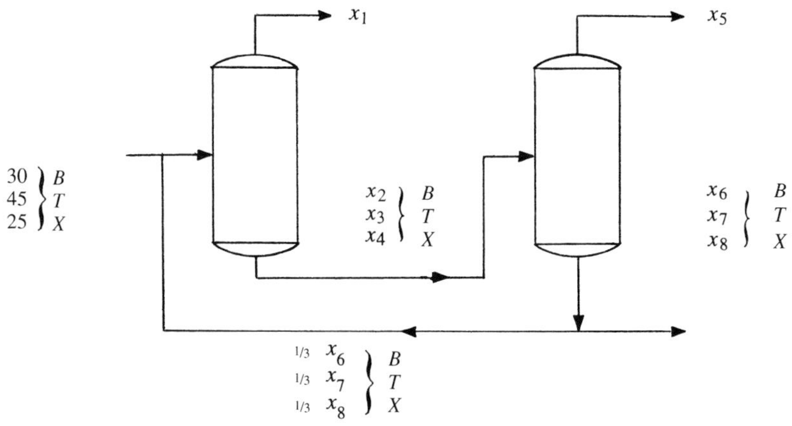

want to find how much benzene is recovered in the first and second column overhead product and in the second column bottom product.

We base our calculations on 100 kg of feed and introduce the following variables, all measured in kilograms:

x_1 : total *first-column overhead* product

x_2 : B in the *first-column bottom* product

x_3 : T in the *first-column bottom* product

x_4 : X in the *first-column bottom* product

x_5 : total *second-column overhead* product

x_6 : B in the *second-column bottom* product

x_7 : T in the *second-column bottom* product

x_8 : X in the *second-column exit* bottom product

We have 8 unknowns $x_i, i = 1, \dots, 8$, and require 8 equations, which arise as follows:

Species balance around the first column:

$$\begin{aligned} 30 + \tfrac{1}{3}x_6 &= 0.914\, x_1 + x_2 \\ 45 + \tfrac{1}{3}x_7 &= 0.083\, x_1 + x_3 \\ 25 + \tfrac{1}{3}x_8 &= 0.003\, x_1 + x_4 \end{aligned} \tag{3.1.13}$$

Species balance around the second column:

$$\begin{aligned} x_2 &= 0.0425\, x_5 + x_6 \\ x_3 &= 0.9160\, x_5 + x_7 \\ x_4 &= 0.0415\, x_5 + x_8 \end{aligned} \tag{3.1.14}$$

Additional requirements:

$$0.914x_1 = (20)(0.0425)\, x_5 = 2x_6 \tag{3.1.15}$$

The preceding eight equations may be placed in the standard form of a linear system,

$$\begin{bmatrix} 0.9140 & 1 & 0 & 0 & 0 & -\frac{1}{3} & 0 & 0 \\ 0.0830 & 0 & 1 & 0 & 0 & 0 & -\frac{1}{3} & 0 \\ 0.0030 & 0 & 0 & 1 & 0 & 0 & 0 & -\frac{1}{3} \\ 0 & 1 & 0 & 0 & -0.0425 & -1 & 0 & 0 \\ 0 & 0 & 1 & 0 & -0.9160 & 0 & -1 & 0 \\ 0 & 0 & 0 & 1 & -0.0415 & 0 & 0 & -1 \\ 0.9140 & 0 & 0 & 0 & -0.8500 & 0 & 0 & 0 \\ 0.9140 & 0 & 0 & 0 & 0 & -2 & 0 & 0 \end{bmatrix} \begin{bmatrix} x_1 \\ x_2 \\ x_3 \\ x_4 \\ x_5 \\ x_6 \\ x_7 \\ x_8 \end{bmatrix} = \begin{bmatrix} 30 \\ 45 \\ 25 \\ 0 \\ 0 \\ 0 \\ 0 \\ 0 \end{bmatrix} \tag{3.1.16}$$

A sparse structure of the coefficient matrix is noticeable.

PROBLEM

3.1.6 ***Voltages across the nodes of an electrical network.***

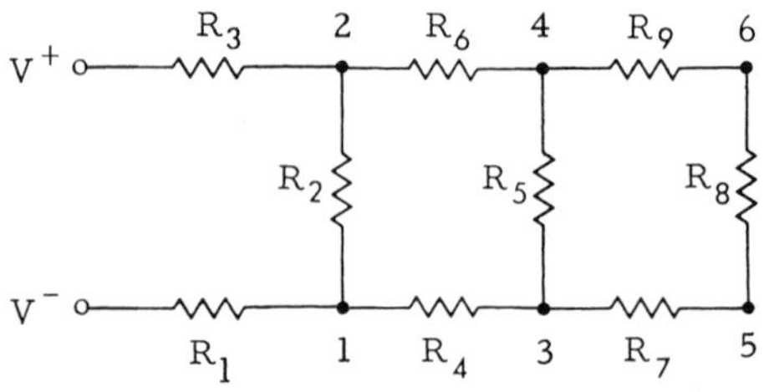

Use Ohm's law to derive a system of linear equations for the voltages at the six nodes of the electrical network shown in the diagram. Discuss the structure of the coefficient matrix with reference to the geometry of the network.

3.2 Diagonal and Triangular Systems

We begin developing methods for solving the linear system $\boldsymbol{Ax} = \boldsymbol{b}$ by addressing three cases where the coefficient matrix $\boldsymbol{A}$ has a conveniently simple form.

Diagonal Systems

$$\begin{bmatrix} A_{1,1} & 0 & \dots & \dots & 0 \\ 0 & A_{2,2} & 0 & \dots & \dots \\ \dots & 0 & \ddots & 0 & \dots \\ \dots & \dots & 0 & A_{N\text{-}1,N\text{-}1} & 0 \\ 0 & \dots & \dots & 0 & A_{N,N} \end{bmatrix} \boldsymbol{x} = \boldsymbol{b}$$

When $\boldsymbol{A}$ is diagonal, each scalar equation of the linear system contains only one scalar unknown, that is, the equations are *decoupled.* The vector $\boldsymbol{x}$ may then be computed using the simple algorithm:

$$\begin{array}{|l|} \hline \text{Do } i = 1, N \\ \quad x_i = \dfrac{b_i}{A_{i,i}} \\ \text{END DO} \\ \hline \end{array} \tag{3.2.1}$$

If $A_{i,i} = 0$ for some value of i, then the corresponding variable x_i takes an infinite value, and the system does not have a solution. In that case, $\text{Det}(\boldsymbol{A}) = 0$ according to our discussion in Section 2.4. But if it also happens that $b_i = 0$, then the ith equation is satisfied for any value of x_i, and the system has an infinite number of solutions.

Lower Triangular Systems and Forward Substitution

When the matrix $\mathbf{A}$ is lower triangular, the first equation contains only the first unknown, the second equation contains the first and the second unknowns, the ith equation contains the first i unknowns, and the last equation contains all unknowns.

$$\begin{bmatrix} A_{1,1} & 0 & 0 & \dots & 0 \\ A_{2,1} & A_{2,2} & 0 & 0 & \dots \\ \dots & \dots & \ddots & 0 & 0 \\ \dots & \dots & \dots & A_{N\text{-}1,N\text{-}1} & 0 \\ A_{N,1} & A_{N,2} & A_{N,3} & \dots & A_{N,N} \end{bmatrix} \mathbf{x} = \mathbf{b}$$

Correspondingly, we solve the first equation for the first unknown, which then becomes known, solve the second equation for the second unknown, and so forth, according to the *forward substitution* algorithm:

$$x_1 = \frac{b_1}{A_{1,1}}$$

Do $i = 2, N$

$$x_i = \frac{1}{A_{i,i}} \left(b_i - \sum_{j=1}^{i-1} A_{i,j}\, x_j \right) \tag{3.2.2}$$

End Do

The number of required operations is on the order of N^2.

If the ith diagonal element of $\mathbf{A}$ vanishes, $A_{i,i} = 0$, the corresponding unknown x_i takes an infinite value, signaling that the system does not have a solution. But if the corresponding numerator on the right-hand side of the equation inside the Do loop also happens to vanish, the system will have an infinite number of solutions, with the unknown x_i taking an arbitrary value.

Upper Triangular Systems and Backward Substitution

When the matrix $\mathbf{A}$ is upper triangular, we work in a similar manner but in reverse order according to the *backward substitution* algorithm:

$$\begin{bmatrix} A_{1,1} & A_{1,2} & A_{1,3} & \dots & A_{1,N} \\ 0 & A_{2,2} & A_{2,3} & \dots & A_{2,N} \\ 0 & 0 & \ddots & \dots & \dots \\ \dots & 0 & 0 & A_{N\text{-}1,N\text{-}1} & A_{N\text{-}1,N} \\ 0 & \dots & 0 & 0 & A_{N,N} \end{bmatrix} \mathbf{x} = \mathbf{b}$$

$$x_N = \frac{b_N}{A_{N,N}}$$

Do $i = N - 1, 1, -1$

$$x_i = \frac{1}{A_{i,i}} \left(b_i - \sum_{j=i+1}^{N} A_{i,j}\, x_j \right) \tag{3.2.3}$$

End Do

The number of required operations is on the order of N^2.

Note that, in this case, the last unknown is computed first. The comments made at the end of the previous subsection regarding the existence and uniqueness of solution are also applicable to this case.

PROBLEM

3.2.1 ***Backward substitution.***

Write a subroutine called BACK_SUB that solves an upper triangular system of equations, and confirm its reliability by solving a system of your choice. Discuss the dependence of the required CPU time on the number of equations. This subroutine will be used later as part of a more general program that solves an arbitrary system of equations by the method of Gauss elimination.

Block-diagonal and Block-triangular Systems

Simple modifications of the preceding algorithms allow us to solve block-diagonal and block-triangular systems, where the scalar elements of the matrix $\boldsymbol{A}$ are replaced by square matrix blocks of appropriate sizes, and the components of $\boldsymbol{x}$ and $\boldsymbol{b}$ are replaced by smaller vectors with corresponding dimensions. The key idea is to *replace number division by matrix multiplication with the inverse.*

Block-diagonal systems

For example, an algorithm for solving the block-diagonal system,

$$\begin{bmatrix} \boldsymbol{A}^{(1,1)} & \boldsymbol{0} & \boldsymbol{0} & \cdots & \boldsymbol{0} \\ \boldsymbol{0} & \boldsymbol{A}^{(2,2)} & \boldsymbol{0} & \boldsymbol{0} & \cdots \\ \boldsymbol{0} & \boldsymbol{0} & \ddots & \boldsymbol{0} & \boldsymbol{0} \\ \cdots & \boldsymbol{0} & \boldsymbol{0} & \boldsymbol{A}^{(N-1,N-1)} & \boldsymbol{0} \\ \boldsymbol{0} & \cdots & \boldsymbol{0} & \boldsymbol{0} & \boldsymbol{A}^{(N,N)} \end{bmatrix} \begin{bmatrix} \boldsymbol{x}^{(1)} \\ \boldsymbol{x}^{(2)} \\ \vdots \\ \boldsymbol{x}^{(N)} \end{bmatrix} = \begin{bmatrix} \boldsymbol{b}^{(1)} \\ \boldsymbol{b}^{(2)} \\ \vdots \\ \boldsymbol{b}^{(N)} \end{bmatrix}$$

where the diagonal blocks $\boldsymbol{A}^{(i,i)}$ may have different dimensions matching the lengths of the subvectors $\boldsymbol{x}^{(i)}$, is

$$\begin{array}{|l|} \hline \text{Do } i = 1, N \\ \boldsymbol{x}^{(i)} = \boldsymbol{A}^{(i,i)^{-1}} \boldsymbol{b}^{(i)} \\ \text{END DO} \\ \hline \end{array} \qquad (3.2.4)$$

The equation in the Do loop is only symbolic; in practice, the subvectors $\boldsymbol{x}^{(i)}$ are found by solving the individual linear systems $\boldsymbol{A}^{(i,i)}\boldsymbol{x}^{(i)} = \boldsymbol{b}^{(i)}$ using one of the methods reviewed in Section 3.3. For example, if $\boldsymbol{A}^{(i,i)}$ is upper triangular, $\boldsymbol{x}^{(i)}$ is computed by the method of backward substitution.

PROBLEM

3.2.2 ***Block-triangular systems.***

Design algorithms for solving block lower-triangular and upper-triangular systems.

3.3 General Procedures and Overview

One way to compute the solution of the arbitrary system $Ax = b$ is by *Cramer's rule* discussed in Section 2.8. Unfortunately, the computations can be extremely cumbersome and may demand an excessive amount of CPU time even for systems of moderate size. Because of these difficulties, we must search for alternative, more efficient methods. Many such methods have been invented over the years. Some are pertinent to particular classes of systems, while others have a more general applicability.

Before embarking to discuss specific methods and their implementation, we pause to present a general overview and introduce certain general procedures.

Systems with Complex Numbers

Nearly all numerical methods to be discussed in the remainder of this chapter are capable of handling complex systems where the coefficient matrix A, the constant vector b, or both, are complex. But since programming in complex variables is often undesirable, it might be preferable to convert a complex system into an equivalent real system at the outset.

Considering the complex system $Ax = b$, we decompose A, x, and b into their real and imaginary constituents, writing

$$A = A_R + IA_I, \qquad x = x_R + Ix_I, \quad b = b_R + Ib_I \tag{3.3.1}$$

where I is the imaginary unit. We then introduce the real vectors

$$y = (x_R, x_I)^T, \qquad c = (b_R, b_I)^T \tag{3.3.2}$$

and decompose Ax into its real and imaginary parts to derive the real system

$$Ey = c \tag{3.3.3}$$

where E is the real matrix:

$$E = \begin{bmatrix} A_R & -A_I \\ A_I & A_R \end{bmatrix} \tag{3.3.4}$$

Not surprisingly, an $N \times N$ system of complex equations is equivalent to a system of real equations with twice its size.

PROBLEM

3.3.1 ***An electrical circuit with inductors and capacitors.***

The complex resistance (impedance) of an *inductor* in an alternating current, AC, with frequency ω, is equal to $I\omega L$, where L is the *inductance*. The impedance of a *capacitor* is equal to $-I/\omega C$, where C is the *capacitance*. Using Kirchhoff's law and Ohm's law, derive a system of complex equations governing the complex currents in the three branches of the circuit shown in the diagram. Then display and discuss the structure of the matrix E defined in equation (3.3.4).

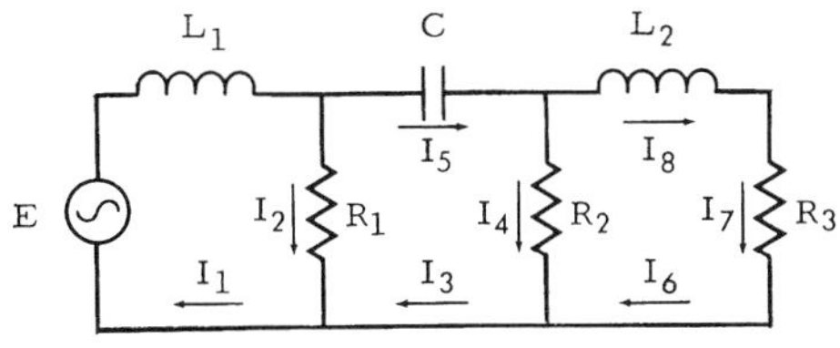

Diagonal, Triangular, Tridiagonal, Pentadiagonal, Sparse, and Arbitrary Systems

Efficient algorithms for solving systems of equations with *diagonal* and *triangular* coefficient matrices were discussed in Section 3.2. Systems of equations whose coefficient matrices are *tridiagonal, pentadiagonal,* or *sparse* may be solved with comparable efficiency using the algorithms to be discussed in Section 3.4; a sparse matrix contains many zeros. These carefully designed algorithms allow us to handle systems with large sizes, containing a few thousand unknowns, with only modest computational resources.

But the efficiency of the aforementioned algorithms should not be misleading. Computing the solution of *arbitrary* systems whose size is higher than about 200 × 200 requires a substantial amount of CPU time.

LU Decomposition and Gauss Elimination

It is a pleasant surprise to realize that a method for solving an *arbitrary* system has already been developed. The method works in three stages:

- First, we decompose the matrix $\boldsymbol{A}$ into the product $\boldsymbol{LU}$ as discussed in Section 2.6. In practice, the $\boldsymbol{LU}$ decomposition is often carried out indirectly by the method of *Gauss elimination* to be discussed in Section 3.5.

Once the $\boldsymbol{L}$ and $\boldsymbol{U}$ matrices have been obtained, the solution of the linear system $\boldsymbol{Ax} = \boldsymbol{b}$ or, equivalently, $\boldsymbol{LUx} = \boldsymbol{b}$ is found in two stages; we define $\boldsymbol{y} = \boldsymbol{Ux}$, and then:

- Solve the system $\boldsymbol{Ly} = \boldsymbol{b}$ for $\boldsymbol{y}$ using forward substitution.
- Solve the system $\boldsymbol{Ux} = \boldsymbol{y}$ for $\boldsymbol{x}$ using backward substitution.

Thus combining an $\boldsymbol{LU}$ decomposition algorithm, or in the case of real symmetric and positive definite matrices the Cholesky decomposition, with the forward or backward substitution algorithms enables us to solve any arbitrary system.

PROBLEM

3.3.2 Solution by the LU decomposition.

Perform the $\boldsymbol{LU}$ decomposition of the coefficient matrix shown in equation (3.1.2), and produce the solution using the forward and backward substitution algorithms.

QR Decomposition

In Section 2.7, we discussed three ways of decomposing an arbitrary matrix $\boldsymbol{A}$ into the product $\boldsymbol{QR}$, where $\boldsymbol{Q}$ is an orthogonal matrix and $\boldsymbol{R}$ is an upper triangular matrix: Gram–Schmidt orthogonalization, sequential reflection, and sequential rotation. Once the matrices $\boldsymbol{Q}$ and $\boldsymbol{R}$ have been obtained, the solution of the linear system $\boldsymbol{Ax} = \boldsymbol{b}$ or, equivalently, $\boldsymbol{QRx} = \boldsymbol{b}$ can be found by solving the triangular system of equations

$$\boldsymbol{Rx} = \boldsymbol{d} \tag{3.3.5}$$

using the method of backward substitution, where $\boldsymbol{d} = \boldsymbol{Q}^T \boldsymbol{b}$.

The method of $\boldsymbol{QR}$ decomposition is particularly effective for systems with a nearly singular coefficient matrix $\boldsymbol{A}$, for which the method of $\boldsymbol{LU}$ decomposition is likely to promote the growth of round-off error. The singular-value decomposition method described next is even more powerful.

Singular-Value Decomposition, *SVD*

A square matrix $\boldsymbol{A}$ can be decomposed into the product of three square matrices, as

$$\boldsymbol{A} = \boldsymbol{V}\boldsymbol{M}\boldsymbol{U}^A \tag{3.3.6}$$

where the superscript A designates the adjoint, that is, the transpose of the complex conjugate, and where:

- $\boldsymbol{V}$ and $\boldsymbol{U}$ are two *unitary* matrices: $\boldsymbol{V}^{-1} = \boldsymbol{V}^A$ and $\boldsymbol{U}^{-1} = \boldsymbol{U}^A$.
- $\boldsymbol{M}$ is a *diagonal* matrix with real and nonnegative diagonal elements, called the *singular values* of $\boldsymbol{A}$. The decomposition becomes unique by specifying that the singular values are arranged in a certain specified order; for example, in the order of ascending or descending magnitude.

Having accomplished the SVD of the matrix $\boldsymbol{A}$, we compute the solution of the system $\boldsymbol{A}\boldsymbol{x} = \boldsymbol{b}$ or, equivalently, $\boldsymbol{V}\boldsymbol{M}\boldsymbol{U}^A\boldsymbol{x} = \boldsymbol{b}$ simply by setting

$$\boldsymbol{x} = \boldsymbol{U}\boldsymbol{M}^{-1}\boldsymbol{V}^A\boldsymbol{b} \tag{3.3.7}$$

Note that $\boldsymbol{M}^{-1}$ arises by taking the inverse of the diagonal elements of $\boldsymbol{M}$.

To demonstrate the feasibility of the singular-value decomposition, as well as illustrate a method of achieving it, we note that the matrix $\boldsymbol{A}^A\boldsymbol{A}$ is both *Hermitian* and *positive-definite* and, as such, it has real and nonnegative eigenvalues (see Problem 2.2.9 and Section 2.9). Applying equation (2.9.22) with $\boldsymbol{A}^A\boldsymbol{A}$ in place of $\boldsymbol{A}$, and noting that $\boldsymbol{U}^{-1} = \boldsymbol{U}^A$, we find

$$\boldsymbol{\Lambda} = \boldsymbol{U}^A\boldsymbol{A}^A\boldsymbol{A}\boldsymbol{U} = (\boldsymbol{A}\boldsymbol{U})^A\boldsymbol{A}\boldsymbol{U} \tag{3.3.8}$$

We proceed by introducing the real diagonal matrix

$$\boldsymbol{M} = \boldsymbol{\Lambda}^{\frac{1}{2}} \tag{3.3.9}$$

whose diagonal elements are the singular values of $\boldsymbol{A}$, and express equation (3.3.8) in the form

$$\boldsymbol{M}^2 = (\boldsymbol{A}\boldsymbol{U})^A\boldsymbol{A}\boldsymbol{U} \tag{3.3.10}$$

Rearranging, we find

$$\boldsymbol{M} = (\boldsymbol{A}\boldsymbol{U}\boldsymbol{M}^{-1})^A\boldsymbol{A}\boldsymbol{U} \tag{3.3.11}$$

and then

$$\boldsymbol{M}\boldsymbol{U}^A = (\boldsymbol{A}\boldsymbol{U}\boldsymbol{M}^{-1})^A\boldsymbol{A} \tag{3.3.12}$$

The last equation reduces to equation (3.3.6) provided that $\boldsymbol{V}^{-1} = (\boldsymbol{AUM}^{-1})^A$ or, since $\boldsymbol{V}$ is alleged to be unitary,

$$\boldsymbol{V} = \boldsymbol{AUM}^{-1} \tag{3.3.13}$$

It remains to verify that $\boldsymbol{V}$ is unitary indeed, but this is evident from equations (3.3.8), (3.3.9), and the definition (3.3.13):

$$\begin{aligned} \boldsymbol{V}^A\boldsymbol{V} &= (\boldsymbol{AUM}^{-1})^A\boldsymbol{AUM}^{-1} \\ &= \boldsymbol{M}^{-1}(\boldsymbol{AU})^A\boldsymbol{AUM}^{-1} = \boldsymbol{M}^{-1}\boldsymbol{\Lambda}\boldsymbol{M}^{-1} = \boldsymbol{I} \end{aligned} \tag{3.3.14}$$

The efficiency of the SVD method hinges on our ability to compute the eigenvalues and eigenvectors of a Hermitian and positive-definite matrix in an economical fashion. Numerical methods that produce these eigenvalues will be discussed in Chapter 5.

PROBLEM

3.3.3 *SVD of a nonsquare matrix.*

Show that equation (3.3.6) holds even when $\boldsymbol{A}$ is a $K \times N$ rectangular matrix. In that case, $\boldsymbol{V}$ and $\boldsymbol{U}$ are square matrices with respective dimensions equal to $K \times K$, and $N \times N$, and $\boldsymbol{M}$ is a rectangular $K \times N$ matrix with real and nonnegative diagonal elements $M_{i,i}$.

Iterative Methods

The $\boldsymbol{LU}$, $\boldsymbol{QR}$, and SVD methods are all *direct.* This means that the computations are terminated after a finite number of operations whose count is a polynomial function of the system size N. At that point, if the computer were able to carry out the computations with infinite precision, the solution would have been exact.

Pragmatic or essential difficulties are encountered when the size of a system is large, containing more than a few hundred equations. Such large systems arise routinely in solving various partial differential equations of mathematical physics using numerical methods. The frequent occurrence of large systems has motivated the development of *iterative methods* to be discussed in Sections 3.6 to 3.9. Iterative methods generate successive approximations to the solution based on algorithms that employ matrix–vector multiplications.

A typical iterative method proceeds by selecting a certain initial vector $\boldsymbol{x}^{(0)}$ and then computing the sequence of vectors

$$\boldsymbol{x}^{(k+1)} = \boldsymbol{Px}^{(k)} + \boldsymbol{c} \tag{3.3.15}$$

where $\boldsymbol{P}$ is an appropriate *projection* or *iteration* matrix, and $\boldsymbol{c}$ is a constant vector. $\boldsymbol{P}$ and $\boldsymbol{c}$ are designed craftily in terms of the coefficient matrix $\boldsymbol{A}$ and constant vector $\boldsymbol{b}$, so that when the sequence $\boldsymbol{x}^{(k)}$ converges, the limit will coincide with the solution of the linear system $\boldsymbol{Ax} = \boldsymbol{b}$. The *Jacobi* method, the *Gauss–Seidel* method, and the SOR method, to be discussed in Section 3.7, provide us with particular ways of designing $\boldsymbol{P}$ and $\boldsymbol{c}$.

Iterative methods are especially effective for *sparse* systems, where the coefficient and iteration matrices $\boldsymbol{A}$ and $\boldsymbol{P}$ contain many zeros: Idle multiplications by the zeros can readily be bypassed. In contrast, the triangular factors of a sparse matrix are not necessarily sparse, and the direct methods do not generally benefit from the pronounced uncoupling of the unknowns.

One important feature of iterative methods is that they are amenable to vector and parallel computation: The matrix–vector multiplication $\boldsymbol{P}\boldsymbol{x}^{(k)}$ is broken up into several blocks, and multiplication of portions of $\boldsymbol{x}^{(k)}$ with corresponding blocks of $\boldsymbol{P}$ are assigned to different processors.

$$\begin{bmatrix} \boldsymbol{P}^{(1,1)} & \boldsymbol{P}^{(1,2)} & \cdots \\ \boldsymbol{P}^{(2,1)} & \boldsymbol{P}^{(2,2)} & \cdots \\ \cdots & \cdots & \cdots \end{bmatrix} \begin{bmatrix} \boldsymbol{x}^{(k,1)} \\ \boldsymbol{x}^{(k,2)} \\ \cdots \end{bmatrix}$$

On the down side, iterative methods require an infinite number of repetitive computations before they produce the exact solution, even on an ideal computer that has infinite precision and no round-off error. In practice, however, if the method is designed well, the numerical error becomes tolerably small after only a small or moderate number of iterations, much less than the dimension of the unknown vector. If this is not the case, then the iterative method is abandoned in favor of a direct method. The design and performance of iterative methods will be discussed in Section 3.6.

Iterative Solution Improvement

Once the solution of a linear system has been found, call it $\boldsymbol{x}^{(0)}$, it should be validated by ensuring that the magnitude of the residual vector

$$\boldsymbol{r}^{(0)} = \boldsymbol{A}\boldsymbol{x}^{(0)} - \boldsymbol{b} \tag{3.3.16}$$

is sufficiently small, hopefully comparable to the computer's round-off error. This practice is in line with the Arabic saying: *trust in Allah, but always tie your camel.* Unfortunately, because of the accumulation of the round-off error, the magnitude of $\boldsymbol{r}^{(0)}$ may be higher than it can be tolerated.

To rectify this deficiency, we make *a posteriori* improvements. The exact solution $\boldsymbol{X}$ satisfies the equation $\boldsymbol{0} = \boldsymbol{A}\boldsymbol{X} - \boldsymbol{b}$; subtracting corresponding sides from equation (3.3.16), we find

$$\boldsymbol{A}\boldsymbol{e} = \boldsymbol{r}^{(0)} \tag{3.3.17}$$

where $\boldsymbol{e} = \boldsymbol{x}^{(0)} - \boldsymbol{X}$ is the error. The three basic steps are:

- Solve equation (3.3.17) for $\boldsymbol{e}$.
- Compute the improved solution $\boldsymbol{x}^{(1)} = \boldsymbol{x}^{(0)} - \boldsymbol{e}$.
- Repeat if necessary.

In practice, one or two improvements are usually what it takes to obtain a high-quality solution.

Provided that the linear systems for $\boldsymbol{x}^{(0)}$ and $\boldsymbol{e}$ are solved by the method of $\boldsymbol{LU}$ or $\boldsymbol{QR}$ decomposition, the computation of the correction $\boldsymbol{e}$ requires a reduced amount of effort: The decomposition of the matrix $\boldsymbol{A}$ is already available from the very first step.

Singular Systems with an Infinity of Solutions

If it happens that one equation of an $N \times N$ linear system can be expressed as a linear combination of the remaining $N - 1$ linearly independent equations, then $\text{Rank}(\boldsymbol{A}) = N - 1$, the matrix $\boldsymbol{A}$ is singular, and the system $\boldsymbol{A}\boldsymbol{x} = \boldsymbol{b}$ has an infinite number of solutions that can be parametrized by a constant. Contrary to conventional wisdom, in certain applications it is meaningful to obtain one of these many solutions.

For example, a method of computing the flow of an incompressible fluid involves solving a Poisson equation for the pressure with Neumann boundary conditions around all boundaries (e.g., Pozrikidis 1997). The numerical implementation leads to a singular system of linear equations that has an infinite number of solutions. The physical explanation is that the pressure in an incompressible fluid may be set to an arbitrary level without any consequences on the flow.

To compute a solution of such a singular system, we simply assign to one scalar unknown an arbitrary value, discard one equation, and solve the remaining $N-1$ equations for the remaining $N-1$ unknowns. Round-off error aside, the solution obtained in this manner is guaranteed to satisfy the discarded equation.

Forcing a singular system to have a solution

Because of the various numerical approximations involved in deriving systems of linear algebraic equations from the discretization of differential equations, the expected linear dependence of one equation on the remaining equations might be inexact. As a consequence, the matrix $\boldsymbol{A}$ may be singular, as required, but the system $\boldsymbol{A}\boldsymbol{x} = \boldsymbol{b}$ may not have a solution.

A rather violent, but effective, method of extracting a solution is to perturb the components of the constant vector $\boldsymbol{b}$ by a small amount, so as to shift it into the image manifold of $\boldsymbol{A}$, working as follows:

1. Set the last unknown equal to zero, $x_N = 0$, solve the first $N-1$ equations of the system $\boldsymbol{A}\boldsymbol{x} = \boldsymbol{b}$ for the first $N-1$ unknowns, call the solution $\boldsymbol{x}^{(1)}$, and compute the residual of the last equation $R^{(1)} = A_{N,i}x_i^{(1)} - b_N$; summation over i is implied.
2. Introduce an arbitrary or suitably designed vector $\boldsymbol{c}$, set the last unknown equal to zero, $x_N = 0$, solve the first $N-1$ equations of the linear system $\boldsymbol{A}\boldsymbol{x} = \boldsymbol{c}$ for the first $N-1$ unknowns, call the solution $\boldsymbol{x}^{(2)}$, and compute the residual of the last equation $R^{(2)} = A_{N,i}x_i^{(2)} - c_N$.
3. Set $\varepsilon = -R^{(1)}/R^{(2)}$ and compute the final solution $\boldsymbol{x} = \boldsymbol{x}^{(1)} + \varepsilon\boldsymbol{x}^{(2)}$.

It can be shown by direct substitution that the vector $\boldsymbol{x} = \boldsymbol{x}^{(1)} + \varepsilon\boldsymbol{x}^{(2)}$ satisfies all N equations of the modified linear system $\boldsymbol{A}\boldsymbol{x} = \boldsymbol{b} + \varepsilon\boldsymbol{c}$, which is, hopefully, only slightly perturbed with respect to the original system $\boldsymbol{A}\boldsymbol{x} = \boldsymbol{b}$.

Sensitivity and the Condition Number

Let us assume that $\boldsymbol{X}$ is the unique solution of a certain linear system $\boldsymbol{A}\boldsymbol{x} = \boldsymbol{b}$. Let us now modify the constant vector $\boldsymbol{b}$ by the small amount $\Delta\boldsymbol{b}$, keeping $\boldsymbol{A}$ fixed; the solution of the linear system will be changed from $\boldsymbol{X}$ to $\boldsymbol{X} + \Delta\boldsymbol{X}$. We want to know what the relative change in $\boldsymbol{X}$ is with respect to the relative change in $\boldsymbol{b}$; that is, we want to assess the sensitivity of the solution to a small perturbation of the right-hand side. Clearly, this question is related to the importance of the round-off error represented by $\Delta\boldsymbol{b}$.

As a preliminary, we introduce a norm for the vectors $\boldsymbol{x}$ and $\boldsymbol{b}$ and a corresponding norm for the matrix $\boldsymbol{A}$, denoted by $||\boldsymbol{x}||$, $||\boldsymbol{b}||$, and $||\boldsymbol{A}||$. A general discussion of vector and matrix norms can be found in texts of linear algebra and numerical analysis (e.g., Atkinson 1989, pp. 480–490).

For example, the *maximum of the norm of the components of a vector* is an appropriate norm for $\boldsymbol{x}$ and $\boldsymbol{b}$ denoted as $||\ ||_\infty$; the corresponding norm for $\boldsymbol{A}$ is the *maximum of the sum of the norms of the elements in each row*. Another choice is the *Euclidean length of a vector* for the norm of $\boldsymbol{x}$ and $\boldsymbol{b}$ denoted as $||\ ||_2$; the *square root of the spectral radius of the Hermitian and positive-definite matrix* $\boldsymbol{A}^A\boldsymbol{A}$ is the corresponding norm for $\boldsymbol{A}$; $\boldsymbol{A}^A$ is the adjoint of $\boldsymbol{A}$.

Using the properties of vectors and matrix norms, it can be shown that the fractional increase $||\Delta\boldsymbol{X}||\ /\ ||\boldsymbol{X}||$ can be located anywhere in the range

$$\frac{1}{\mathrm{Cond}(\boldsymbol{A})}\frac{||\Delta\boldsymbol{b}||}{||\boldsymbol{b}||} \le \frac{||\Delta\boldsymbol{X}||}{||\boldsymbol{X}||} \le \mathrm{Cond}(\boldsymbol{A})\frac{||\Delta\boldsymbol{b}||}{||\boldsymbol{b}||} \tag{3.3.18}$$

where Cond(A) is a positive scalar constant called the *condition number*, defined as

$$\mathrm{Cond}(\boldsymbol{A}) \equiv ||\boldsymbol{A}||\,||\boldsymbol{A}^{-1}|| \tag{3.3.19}$$

(e.g., Atkinson 1989, pp. 529–540). For example, $\mathrm{Cond}_2(\boldsymbol{A}) = ||\boldsymbol{A}||_2\,||\boldsymbol{A}^{-1}||_2 = (|\lambda_1|/|\lambda_N|)^{\frac{1}{2}}$, where λ_1 and λ_N are, respectively, the eigenvalues of $\boldsymbol{A}\boldsymbol{A}^T$ with the maximum and minimum norm.

The value of the condition number clearly depends on the selected matrix norm but is always larger than unity. In fact, it can be shown that $\mathrm{Cond}(\boldsymbol{A}) \geq |\lambda_1|/|\lambda_N|$, where λ_1 and λ_N are, respectively, the eigenvalues of $\boldsymbol{A}$ with the maximum and minimum norm. The right-hand side of the last inequality defines a lower bound that is sometimes used in place of the condition number.

Equation (3.3.18) suggests that the range of the relative variation $||\Delta\boldsymbol{X}||\,/\,||\boldsymbol{X}||$ becomes broader as the condition number becomes larger. Thus *a large condition number typically indicates that the solution is sensitive to small variations on the right-hand side.*

Similar results emerge when we consider the relative variation of the solution with respect to perturbation in both $\boldsymbol{A}$ and $\boldsymbol{b}$. Specific expressions are derived and discussed in texts on numerical analysis (e.g., Wilkinson 1965, pp. 209–216; Golub and van Loan 1989, Chapter 4).

Nearly Easy Problems

The numerical treatment of a certain class of physical problems results in systems of linear equations $\boldsymbol{A}\boldsymbol{x} = \boldsymbol{b}$, where the matrix $\boldsymbol{A}$ is *almost* but *not exactly* diagonal, triangular, tridiagonal, or pentadiagonal. A small number of nonzero elements prevent us from using efficient special-purpose methods that are applicable to these desirable forms. An example of a nearly tridiagonal system will be discussed in Problem 3.4.3.

In some cases, the matrix $\boldsymbol{A}$ can be placed into the *Sherman–Morrison form* $\boldsymbol{A} = \boldsymbol{B} + \boldsymbol{u}\boldsymbol{v}^T$, and more explicitly,

$$A_{i,j} = B_{i,j} + u_i v_j \tag{3.3.20}$$

where the matrix $\boldsymbol{B}$ is diagonal, triangular, tridiagonal, or pentadiagonal; $\boldsymbol{u}$ and $\boldsymbol{v}$ are two vectors with appropriate dimensions.

For example, $\boldsymbol{A}$ may be tridiagonal but, in addition,

$$A = \begin{bmatrix} A_{1,1} & A_{1,2} & 0 & \cdots & 0 & \alpha \\ A_{2,1} & A_{2,2} & A_{2,3} & 0 & \cdots & 0 \\ 0 & A_{3,2} & A_{3,3} & \ddots & 0 & \cdots \\ \cdots & 0 & \ddots & \ddots & \ddots & \cdots \\ 0 & \cdots & 0 & A_{N-1,N-2} & A_{N-1,N-1} & A_{N-1,N} \\ \beta & 0 & \cdots & 0 & A_{N,N-1} & A_{N,N} \end{bmatrix}$$

$$A_{1,N} = \alpha \neq 0, \qquad A_{N,1} = \beta \neq 0$$

In this case, the matrix $\boldsymbol{B}$ defined in the last equation is tridiagonal, and we may put

$$\boldsymbol{u}^T = (\gamma, 0, \ldots, 0, \beta)$$

$$\boldsymbol{v}^T = (1, 0, \ldots, 0, \alpha/\gamma)$$

where γ is an arbitrary constant (Press et al. 1992, p. 67).

When the decomposition (3.3.20) is feasible, we can produce the solution by solving two easy problems as follows. First, we decompose, $\boldsymbol{x} = \boldsymbol{y} + \boldsymbol{z}$, and require that

$$(\boldsymbol{B} + \boldsymbol{u}\boldsymbol{v}^T)(\boldsymbol{y} + \boldsymbol{z}) = \boldsymbol{b} \tag{3.3.21}$$

Stipulating that

$$\boldsymbol{B}\boldsymbol{y} = \boldsymbol{b} \tag{3.3.22}$$

and defining

$$\boldsymbol{w} \equiv -\frac{1}{\boldsymbol{v}^T(\boldsymbol{v} + \boldsymbol{z})}\boldsymbol{z} \tag{3.3.23}$$

we find

$$\boldsymbol{B}\boldsymbol{w} = \boldsymbol{u} \tag{3.3.24}$$

The method proceeds by solving equation (3.3.22) for $\boldsymbol{y}$ and equation (3.3.24) for $\boldsymbol{w}$, which can be done with little effort. To recover $\boldsymbol{z}$, we take the inner product of both sides of equation (3.3.23) with $\boldsymbol{v}$, solve for $\boldsymbol{v}^T\boldsymbol{z}$, substitute the result back into equation (3.3.23), and finally solve the resulting equation for $\boldsymbol{z}$ to find

$$\boldsymbol{z} = -\frac{\boldsymbol{v}^T\boldsymbol{y}}{1 + \boldsymbol{v}^T\boldsymbol{w}}\boldsymbol{w} \tag{3.3.25}$$

The foundation of this procedure may be traced back to the Sherman–Morrison formula (2.3.3).

As an example, we consider a 2×2 system with

$$\boldsymbol{A} = \begin{bmatrix} 2 & 1 \\ 4 & 3 \end{bmatrix}, \qquad \boldsymbol{b} = \begin{bmatrix} 3 \\ 7 \end{bmatrix} \tag{3.3.26}$$

and define

$$\boldsymbol{B} = \begin{bmatrix} -2 & 0 \\ 0 & 2 \end{bmatrix}, \qquad \boldsymbol{u} = \begin{bmatrix} 1 \\ 1 \end{bmatrix}, \qquad \boldsymbol{v} = \begin{bmatrix} 4 \\ 1 \end{bmatrix} \tag{3.3.27}$$

The solution of the two easy problems is

$$\boldsymbol{y} = \tfrac{1}{2}\begin{bmatrix} -3 \\ 7 \end{bmatrix}, \qquad \boldsymbol{w} = \tfrac{1}{2}\begin{bmatrix} -1 \\ 1 \end{bmatrix} \tag{3.3.28}$$

and the straightfoward evaluation of the right-hand side of equation (3.3.25) gives $\boldsymbol{z}^T = \frac{5}{2}(1, -1)$. Finally, $\boldsymbol{x}^T = \boldsymbol{y}^T + \boldsymbol{z}^T = (1, 1)$, which obviously solves the system $\boldsymbol{A}\boldsymbol{x} = \boldsymbol{b}$.

PROBLEM

3.3.4 *Two easy diagonal problems.*

Find the solution of a 3×3 system $\boldsymbol{A}\boldsymbol{x} = \boldsymbol{b}$ with

$$A = \begin{bmatrix} 1 & 0 & 2 \\ 0 & 2 & 0 \\ -1 & 0 & 1 \end{bmatrix}, \qquad b = \begin{bmatrix} 3 \\ 4 \\ 0 \end{bmatrix} \tag{3.3.29}$$

using the method described in the text, where the matrix $\boldsymbol{B}$ is chosen to be diagonal.

3.4 Tridiagonal, Pentadiagonal, and Sparse Systems

Systems of linear equations with tridiagonal, pentadiagonal, and more general sparse coefficient matrices arise invariably in the numerical solution of differential equations using finite-difference and related methods. One example was presented in Table 3.1.1, a second example follows, and further examples will be discussed in Chapters 10 and 11.

Consider the steady-state distribution of temperature, denoted by $f(x)$, along a pipe extending along the x axis between the points $x = a$ and $x = b$, with boundary conditions $f(a) = c$ and $f(b) = d$. The pipe carries a reacting fluid that generates heat, and the rate of heat production is proportional to the temperature. In an appropriate dimensionless form, $f(x)$ satisfies the second-order linear ordinary differential equation

$$\frac{d^2 f}{dx^2} + 4f = 0 \tag{3.4.1}$$

Although it is possible to obtain the solution exactly by analytical means (Problem 3.4.2(*a*)), we proceed to compute it for the purpose of illustration using a *finite-difference method.*

We begin by discretizing the solution domain into N evenly spaced intervals that are separated by the nodes $x_i = a + (i-1)\,\Delta x$, where $i = 1, \ldots, N+1$, $\Delta x = (b-a)/N$, and apply the differential equation (3.4.1) at the ith node. Approximating the second derivative with the second-order central-difference formula, we set $(d^2 f/dx^2)_{x_i} = (f_{i+1} - 2f_i + f_{i-1})\,/\Delta x^2$, where we have denoted $f_i \equiv f(x_i)$ (see Table 6.11.1), and obtain the finite-difference equation

$$-f_{i-1} + (2 - 4\Delta x^2) f_i - f_{i+1} = 0 \tag{3.4.2}$$

Applying equation (3.4.2) for $i = 2, \ldots, N$, and requiring the boundary conditions $f_1 = c$ and $f_{N+1} = d$, we derive the tridiagonal system of equations:

$$\begin{bmatrix} 2-4\Delta x^2 & -1 & 0 & 0 & \ldots & 0 & 0 & 0 \\ -1 & 2-4\Delta x^2 & -1 & 0 & \ldots & 0 & 0 & 0 \\ 0 & -1 & 2-4\Delta x^2 & -1 & \ldots & 0 & 0 & \\ \ldots & \ldots & \ldots & \ldots & \ldots & \ldots & \ldots & \ldots \\ 0 & 0 & 0 & 0 & \ldots & -1 & 2-4\Delta x^2 & -1 \\ 0 & 0 & 0 & 0 & \ldots & 0 & -1 & 2-4\Delta x^2 \end{bmatrix} \begin{bmatrix} f_2 \\ f_3 \\ f_4 \\ \ldots \\ f_{N-1} \\ f_N \end{bmatrix} = \begin{bmatrix} c \\ 0 \\ 0 \\ \ldots \\ 0 \\ d \end{bmatrix} \tag{3.4.3}$$

for the $N-1$ unknowns f_i, $i = 2, \ldots, N$.

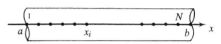

If we had approximated the second derivative $(d^2 f/dx^2)_{xi}$ with a more accurate formula involving five instead of three grid points, we would have obtained a pentadiagonal coefficient matrix with five nonzero diagonals.

PROBLEM

3.4.1 ***Finite-difference solution of a linear* ODE *with nonconstant coefficients.***

Develop a finite-difference method that results in a tridiagonal system of linear algebraic equations for solving the differential equation

$$\frac{d^2 f}{dx^2} + 4(1 + \cos x) f = 0 \tag{3.4.4}$$

over the same domain, and with boundary conditions identical to those described in the text for equation (3.4.1). Discuss the differences between the derived system and the system shown in equation (3.4.3).

Thomas Algorithm for Tridiagonal Systems

According to the method of $\boldsymbol{LU}$ decomposition described in Section 3.3, the linear system $\boldsymbol{Ax} = \boldsymbol{b}$ is solved in three stages: Decompose $\boldsymbol{A} = \boldsymbol{LU}$; solve the system $\boldsymbol{Ly} = \boldsymbol{b}$ for $\boldsymbol{y}$; and solve the system $\boldsymbol{Ux} = \boldsymbol{y}$ for $\boldsymbol{x}$. If the matrix $\boldsymbol{A}$ is tridiagonal, the matrices $\boldsymbol{L}$ and $\boldsymbol{U}$ are *bidiagonal*, and the computation of the upper triangular matrix $\boldsymbol{U}$ and vector $\boldsymbol{y}$ can be significantly economized working as follows.

Consider the linear system

$$\boldsymbol{Tx} = \boldsymbol{s} \tag{3.4.5}$$

where $\boldsymbol{T}$ is an $N \times N$ tridiagonal matrix, and $\boldsymbol{s}$ is an N-dimensional vector. The Thomas algorithm produces the equivalent *upper bidiagonal* system

$$\boldsymbol{Ux} = \boldsymbol{y} \tag{3.4.6}$$

$$\begin{bmatrix} 1 & d_1 & \cdots & \cdots & 0 \\ 0 & 1 & d_2 & \cdots & \cdots \\ \cdots & 0 & \ddots & 0 & \cdots \\ \cdots & \cdots & 0 & 1 & d_{N-1} \\ 0 & \cdots & \cdots & 0 & 1 \end{bmatrix} \boldsymbol{x} = \boldsymbol{y}$$

where the diagonal elements of $\boldsymbol{U}$ are equal to unity, $U_{i,i} = 1$, and summation is not implied over i.

For convenience, and to save computer memory space, we denote the nonzero elements of $\boldsymbol{T}$ as shown in equations (2.4.1), and the superdiagonal elements of $\boldsymbol{U}$ by $d_i = U_{i,i+1}$, for $i = 1, \ldots, N-1$. The reduced bidiagonal system (3.4.6) is shown in the diagram.

To develop the algorithm, we assume that we have managed to put the ith equation into the bidiagonal form

$$x_i + d_i x_{i+1} = y_i \tag{3.4.7}$$

Solving for x_i, substituting the resulting expression into the $i+1$ equation of the original system (3.4.5), and recasting the emerging expression into the corresponding form $x_{i+1} + d_{i+1} x_{i+2} = y_{i+1}$, we obtain the relation shown in the first Do loop of the Thomas Algorithm 3.4.1.

Once the matrix $\boldsymbol{U}$ and constant vector $\boldsymbol{y}$ have been found, the solution is computed by applying the simplified backward substitution algorithm shown in the second part of Algorithm 3.4.1.

ALGORITHM 3.4.1 Thomas algorithm for solving a system of equations $\boldsymbol{T}\boldsymbol{x} = \boldsymbol{s}$ with a tridiagonal matrix $\boldsymbol{T}$, where $T_{i,i-1} = c_i$, $T_{i,i} = a_i$, $T_{i,i+1} = b_i$. The first part reduces the original system to the upper bidiagonal system $\boldsymbol{U}\boldsymbol{x} = \boldsymbol{y}$, where $U_{i,i} = 1$ and $U_{i,i+1} = d_i$, and the second part performs the back-substitution.

$$\begin{bmatrix} d_1 \\ y_1 \end{bmatrix} = \frac{1}{a_1}\begin{bmatrix} b_1 \\ s_1 \end{bmatrix}$$

Do $i = 1, N-1$

$$\begin{bmatrix} d_{i+1} \\ y_{i+1} \end{bmatrix} = \frac{1}{a_{i+1} - c_{i+1} d_i}\begin{bmatrix} b_{i+1} \\ s_{i+1} - c_{i+1} y_i \end{bmatrix}$$

End Do

$x_N = y_N$

Do $i = N-1, 1, -1$

$x_i = y_i - d_i x_{i+1}$

End Do

The Thomas algorithm is a special case of the Gauss elimination algorithm for arbitrary systems to be discussed in Section 3.5. Its important advantage is that it circumvents a lot of idle multiplications by zeros that would otherwise have to be bypassed with costly If statement executions.

Performance

It is possible that the first part of the Thomas algorithm may fail even though the system may have a unique solution, for the denominator in the Do loop may fortuitously become equal to zero. Fortunately, this rarely occurs in practice. It can be shown, in particular, that when the system is diagonally dominant, that is, $|a_i| > |b_i| + |c_i|$ for all i, the algorithm will *not* fail.

The backward-substitution part of the Thomas algorithm produces the values of successive unknowns in terms of a *linear mapping with a shift*: x_{i+1} is multiplied by $-d_i$, and then shifted by y_i to produce x_i. According to our discussion at the end of Section 1.4, the magnitudes of d_i determine the behavior of the round-off error. If, for example, the elements d_i are constant and equal to d, the algorithm will be stable when $|d| < 1$, and unstable when $|d| > 1$. In the second case, the numerical error will grow and may ultimately overtake the solution during the execution of the Do loop.

Block-tridiagonal systems

The Thomas algorithm may be extended to block-tridiagonal systems where the nonzero elements become square matrices with appropriate dimensions, as discussed at the end of Section 3.2. The only required modification is that division by a scalar coefficient be replaced by multiplication with the inverse of the corresponding matrix.

For example, the first two equations of Algorithm 3.4.1 become

$$\boldsymbol{d}_1 = \boldsymbol{a}_1^{-1}\boldsymbol{b}_1, \qquad \boldsymbol{y}_1 = \boldsymbol{a}_1^{-1}\boldsymbol{s}_1 \tag{3.4.8}$$

The ith column of the matrix $\boldsymbol{d}_1$, denoted as $\boldsymbol{d}_{1,i}$, arises by solving the linear system $\boldsymbol{a}_1\boldsymbol{d}_{1,i} = \boldsymbol{b}_{1,i}$, where $\boldsymbol{b}_{1,i}$ is the ith column of $\boldsymbol{b}$. The vector $\boldsymbol{y}_1$ arises by solving the linear system $\boldsymbol{a}_1\boldsymbol{y}_1 = \boldsymbol{s}_1$. Thus each pass requires solving a set of systems of equations with the *same coefficient matrix* but *multiple right-hand sides.* Using the ***LU*** or the ***QR*** decomposition method or grouping the systems into a unified form, as will be discussed in Section 3.5, will prevent repetitive computations.

PROBLEMS

3.4.2 *Finite-difference solution of ODEs.*

(*a*) Compute the solution of equation (3.4.1) with the stated boundary conditions analytically and in closed form.
(*b*) Solve the tridiagonal systems of equations (3.4.3) for $a = 0, b = \pi/4, c = 0, d = 5$, and $N = 4, 8, 16, 32, 64$. Compare the numerical results with the exact solution derived in part (*a*), and discuss the efficiency of the numerical method.
(*c*) Repeat part (*b*) with the same values of the parameters, except that $b = \pi/2$, and discuss the behavior of the numerical solution with reference to the exact solution.

This computation will show that *it is possible to obtain a spurious numerical solution, even though an exact solution may not exist.* The error introduced by the various numerical approximations alters the problem described by the differential equation and produces a modified problem described by the difference equations. In this case, as the numerical error is reduced, the results of the numerical computation diverge to infinity, reflecting the lack of an exact solution. Ensuring the convergence of a numerical solution with respect to the numerical parameters, in this case N, guarantees a reliable scientific computation.

3.4.3 *A periodic solution.*

Consider a smooth periodic function $f(x)$ with period L, that is, $f(x) = f(x+L)$, satisfying the differential equation

$$\frac{d^2 f}{dx^2} + 4f = \exp(\sin kx) \tag{3.4.9}$$

where $k = 2\pi/L$ is the wave number. Discretize one period of the x axis into N intervals, and develop a finite-difference method that produces a *nearly tridiagonal* system of difference equations. Compute the solution using the tricky method described at the end of Section 3.3 for $k = 1$ and $N = 4, 8, 16, 32, 64$, and discuss the accuracy of the results.

Modified Thomas Algorithm for Pentadiagonal Systems

An extension of the Thomas algorithm provides us with an efficient method for converting the pentadiagonal system

$$\boldsymbol{P}\boldsymbol{x} = \boldsymbol{s} \tag{3.4.10}$$

to the *quadradiagonal* system

$$\boldsymbol{Q}\boldsymbol{x} = \boldsymbol{s}' \tag{3.4.11}$$

and then to the *upper tridiagonal* system

$$\boldsymbol{T}\boldsymbol{x} = \boldsymbol{s}'' \tag{3.4.12}$$

where the diagonal elements of $\boldsymbol{T}$ are equal to unity, as shown in the diagram. System (3.4.12) is finally solved using a simplified version of backward substitution.

To save computer memory space, we represent the nonzero elements of $\boldsymbol{P}$, $\boldsymbol{Q}$, and $\boldsymbol{T}$ with one-index variables as

$$\boldsymbol{T} = \begin{bmatrix} 1 & b_1'' & c_1'' & 0 & \cdots & 0 \\ 0 & 1 & b_2'' & c_2'' & 0 & \cdots \\ 0 & 0 & 1 & \ddots & \ddots & 0 \\ \cdots & 0 & 0 & \ddots & b_{N-2}'' & c_{N-2}'' \\ \cdots & \cdots & 0 & 0 & 1 & b_{N-1}'' \\ 0 & \cdots & \cdots & 0 & 0 & 1 \end{bmatrix}$$

$$\begin{aligned} &P_{i,i-2} = e_i, \quad P_{i,i-1} = d_i, \quad P_{i,i} = a_i, \quad P_{i,i+1} = b_i, \quad P_{i,i+2} = c_i \\ &Q_{i,i-1} = d_i', \quad Q_{i,i} = a_i', \quad Q_{i,i+1} = b_i', \quad Q_{i,i+2} = c_i' \\ &T_{i,i} = 1 \quad T_{i,i+1} = b_i'', \quad T_{,i+2} = c_i'' \end{aligned} \tag{3.4.13}$$

Working as previously for tridiagonal systems, we develop a method in three parts. The first part produces the system (3.4.11) according to the algorithm:

$$\begin{aligned} &a_1' = a_1, \quad b_1' = b_1, \quad c_1' = c_1, \quad s_1' = s_1 \\ &d_2' = d_2, \quad a_2' = a_2, \quad b_2' = b_2, \quad s_2' = s_2 \\ &\text{Do } i = 2, N-1 \\ &\quad c_i' = c_i \\ &\quad \begin{bmatrix} a_{i+1}' \\ b_{i+1}' \\ d_{i+1}' \\ s_{i+1}' \end{bmatrix} = \begin{bmatrix} a_{i+1} \\ b_{i+1} \\ d_{i+1} \\ s_{i+1} \end{bmatrix} - \frac{e_{i+1}}{d_i'} \begin{bmatrix} b_i' \\ c_i' \\ a_i' \\ s_i' \end{bmatrix} \\ &\text{End Do} \end{aligned} \tag{3.4.14}$$

The second part produces the system (3.4.12) according to the algorithm:

$$\begin{aligned} &\begin{bmatrix} b_1'' \\ c_1'' \\ s_1'' \end{bmatrix} = \frac{1}{a_1'} \begin{bmatrix} b_1' \\ c_1' \\ s_1' \end{bmatrix} \\ &\text{Do } = 1, N-1 \\ &\begin{bmatrix} b_{i+1}'' \\ c_{i+1}'' \\ s_{i+1}'' \end{bmatrix} = \frac{1}{a_{i+1}' - d_{i+1}' b_i''} \begin{bmatrix} b_{i+1}' - d_{i+1}' c_i'' \\ c_{i+1}' \\ s_{i+1}' - d_{i+1}' s_i' \end{bmatrix} \\ &\text{End Do} \end{aligned} \tag{3.4.15}$$

And the third part provides us with the desired solution $\boldsymbol{x}$ based on the modified backward-substitution algorithm:

$$
\boxed{\begin{aligned}
&x_N = s''_N \\
&x_{N-1} = s''_{N-1} - b''_{N-1} x_N \\
&\text{Do } i = N-2, 1, -1 \\
&x_i = s''_i - b''_i x_{i+1} - c''_i x_{i+2} \\
&\text{END DO}
\end{aligned}} \tag{3.4.16}
$$

The modified Thomas algorithm just described is a special version of the Gauss elimination algorithm, which is capable of solving arbitrary systems, to be discussed in Section 3.5. Its paramount advantage is that it circumvents many idle multiplications by zeros and thus reduces the memory storage.

PROBLEM

3.4.4 ***A pentadiagonal system from the finite-difference solution of an ODE.***

(*a*) Develop a finite-difference method for solving equation (3.4.1) with the boundary conditions stated in the text, resulting in a pentadiagonal system of linear equations. The second derivative should be approximated with the fourth-order central finite-difference formula involving five points at the nodes $i = 3, \ldots, N-2$, and with the second-order forward or backward finite-difference formula involving four points at the nodes $i = 2$ and $N-1$. The finite-difference formulas are given in Table 6.11.1.
(*b*) Repeat parts (*b*) and (*c*) of Problem 3.4.2 using the modified Thomas algorithm discussed in the text for $N = 4, 8, 16, 32, 64$, and discuss the efficiency of the method.

Cyclic Reduction

Having established general-purpose methods for tridiagonal and pentadiagonal systems, we turn to considering the nifty but specialized method of cyclic reduction. In practice, this method is used to solve tridiagonal systems that arise from finite-difference or related discretizations of ordinary and partial differential equations, including the Laplace and Poisson equations. An example was discussed at the beginning of this section, and more elaborate examples will be discussed in Section 11.4.

Modified back-substitution

As a first preliminary, we make a simple observation: Consider the $N \times N$ tridiagonal system of equations $\boldsymbol{T}\boldsymbol{x} = \boldsymbol{s}$, where the elements of the tridiagonal matrix $\boldsymbol{T}$ are denoted as shown in equations (2.4.1). If we knew the value of the last unknown x_N, then we would be able to compute the rest of the unknowns using the modified back-substitution algorithm:

$$
\boxed{\begin{aligned}
&x_{N-1} = \frac{s_N - a_N x_N}{c_N} \\
&\text{Do } i = N-1, 2, -1 \\
&x_{i-1} = \frac{s_i - a_i x_i - b_i x_{i+1}}{c_i} \\
&\text{END DO}
\end{aligned}} \tag{3.4.17}
$$

The computed values of x_1 and x_2 are guaranteed to satisfy the first equation $a_1 x_1 + b_1 x_2 = s_1$.

PROBLEM

3.4.5 ***Trial and error.***

Explain why the following procedure produces a solution of the tridiagonal system $\boldsymbol{Tx} = \boldsymbol{s}$:

1. Set $x_N = 0$, execute algorithm (3.4.17), and call the computed vector $\boldsymbol{x}^{(1)}$. Then compute the residual $a_1 x_1^{(1)} + b_1 x_2^{(1)} - s_1 \equiv r_1$.
2. Set $x_N = 1$, execute algorithm (3.4.17), and call the computed vector $\boldsymbol{x}^{(2)}$. Then compute the residual $a_1 x_1^{(2)} + b_1 x_2^{(2)} - s_1 \equiv r_2$.
3. Set $x_N = r_1/(r_1 - r_2)$, and execute algorithm (3.4.17), to obtain the desired solution $\boldsymbol{x}$.

Discuss circumstances under which this procedure will fail.

Expanded tridiagonal system

As a second preliminary, we consider an *expanded* $N \times N$ *tridiagonal system of equations*, where the coefficient matrix has the form:

$$\begin{bmatrix} \clubsuit & \clubsuit & 0 & 0 & 0 & 0 & 0 & 0 & 0 & 0 & \cdots \\ 0 & \blacklozenge & 0 & \blacklozenge & 0 & 0 & 0 & 0 & 0 & 0 & \cdots \\ 0 & \clubsuit & \clubsuit & \clubsuit & 0 & 0 & 0 & 0 & 0 & 0 & \cdots \\ 0 & \underline{\blacklozenge} & 0 & \blacklozenge & 0 & \underline{\blacklozenge} & 0 & \underline{0} & 0 & 0 & \cdots \\ 0 & 0 & 0 & \clubsuit & \clubsuit & \clubsuit & 0 & 0 & 0 & 0 & \cdots \\ 0 & \underline{0} & 0 & \underline{\blacklozenge} & 0 & \blacklozenge & 0 & \underline{\blacklozenge} & 0 & \underline{0} & \cdots \\ 0 & 0 & 0 & 0 & 0 & \clubsuit & \clubsuit & \clubsuit & 0 & 0 & \cdots \\ \cdots & \cdots & \cdots & \cdots & \cdots & \cdots & \cdots & \cdots & \cdots & \cdots & \cdots \end{bmatrix} \tag{3.4.18}$$

An $\blacklozenge$ or $\clubsuit$, underlined or not, designates a nonzero value. Note that every other element along the first and second sub- and superdiagonal lines is equal to zero. To this end, we make two important observations:

- The even-numbered equations contain only even-numbered unknowns; collecting them produces a *tridiagonal* system for the even-numbered unknowns with corresponding coefficients $\blacklozenge$.
- If we knew the values of the even-numbered unknowns, the odd-numbered equations would provide us with a *diagonal* system of equations for the odd-numbered unknowns with corresponding coefficients $\clubsuit$.

Reduction of a tridiagonal system to an expanded tridiagonal system

Turning next to the tridiagonal system (3.4.5), denoted as TS, we seek to reduce it to an equivalent expanded tridiagonal system, denoted as ETS, where the coefficient matrix has the structure of the matrix (3.4.18). Our objective is to produce a descendant tridiagonal system for the even-numbered unknowns.

The reduction is straightforward: Denoting the nonzero elements of the tridiagonal matrix $\boldsymbol{T}$ as shown in equations (2.4.1), we simply set

$$[i\text{th equation of ETS}] = [i\text{th equation of TS}] - (c_i/a_{i-1})x[(i-1)\text{ equation of TS}]$$
$$-(b_i/a_{i+1}) \times [(i+1)\text{ equation of TS}] \tag{3.4.19}$$

for $i = 2, 4, \ldots$, with the understanding that the last term is omitted when $i = N$. When N is even, the descendant tridiagonal system contains the $N/2$ even-numbered unknowns $x_2, x_4, \ldots, x_{N/2}$.

Cyclic reduction

When $N/2$ is even, we can repeat the process just described once more, thereby producing a second-generation descendant tridiagonal system for the $N/4$ unknowns $x_4, x_8, \ldots, x_{N/4}$. And if $N = 2^p$, where p is a positive integer, we can repeat the reduction p times. At the end, we will have one equation for the sole unknown x_N.

Effectively, the pth reduction shifts selected off-diagonal elements of the coefficient matrix by 2^{p-1} lines to the left or the right normal to the diagonal, thereby decoupling the unknowns. For example, after the second reduction has been completed, the elements $\underline{\blacklozenge}$ of matrix (3.4.18) will have been replaced by zeros, while the underlined zeros will have assumed finite values. Straightforward modifications can be made to handle systems where N is not a power of 2.

For example, the first reduction of the 4×4 system,

$$\begin{aligned} x_1 + 2x_2 \qquad\qquad &= 3 \\ x_1 + 3x_2 - 2x_3 \qquad &= 2 \\ 5x_2 - 2x_3 - 2x_4 &= 1 \\ x_3 - 2x_4 &= -1 \end{aligned} \tag{3.4.20}$$

yields the 2×2 system,

$$\begin{aligned} (3-2-5)x_2 + \qquad 2x_4 &= 2-3-1 \\ \tfrac{5}{2}x_2 + (-2-1)x_4 &= -1 + \tfrac{1}{2} \end{aligned}$$

or

$$\begin{aligned} -4x_2 + 2x_4 &= -2 \\ \tfrac{5}{2}x_2 - 3x_4 &= -\tfrac{1}{2} \end{aligned}$$

A further reduction eliminates x_2 from the last equation, producing $x_4 = 1$. Applying the back-substitution algorithm (3.4.17) to the system (3.4.20), we find $x_3 = 1, x_2 = 1, x_1 = 1$.

In practice, the method of cyclic reduction is particularly attractive when the elements of the original tridiagonal system are constant along each diagonal, that is, $a_i = a, b_i = b, c_i = c$. In that case, since the new elements of the descendant tridiagonal system are equal, they need to be computed only once;

an exception is the bottom diagonal element that differs from the rest of the elements. The sequence of computations is shown in Algorithm 3.4.2. An operation count shows that the computational cost scales with $p = \ln_2 N$, which allows for tremendous saving when N is large. Other implementations of the method are discussed by Boisvert (1991).

The method of cyclic reduction is a cousin of the powerful method of fast Fourier transform, FFT, which is used to interpolate or approximate a function from knowledge of its values at a large number of data points, as will be discussed in Section 8.7.

ALGORITHM 3.4.2 Cyclic reduction algorithm for solving the $N \times N$ tridiagonal linear system $\boldsymbol{T}\boldsymbol{x} = \boldsymbol{s}$ with constant diagonal elements a, b, c, where $N = 2^p$.

```
α = a                                    Greek symbols are coefficients
β = b                                              of descendant systems
γ = c
α_L = a                               α_L is the last diagonal element
Nmax = N                                    Size of descendant system
Do i = 1, Nmax
σ_i = s_i                      Right-hand side of the descendant system
END DO

Do m = 1, p                                           Cyclic reduction
  α_T = α − 2βγ/α                                    Temporary storage
  β_T = −β²/α
  γ_T = −γ²/α
  α_L ← α_L − βγ/α
    Do i = 2, Nmax − 2, 2
    θ_{i/2} = −(γ/α)σ_{i−1} + σ_i − (β/α)σ_{i+1}     Temporary storage
    END DO
      θ_{Nmax/2} = −(γ/α)σ_{Nmax−1} + σ_{Nmax}        Temporary storage
  α = α_T
  β = β_T
  γ = γ_T
  Nmax ← Nmax/2
  Do i = 1, Nmax
    σ_i = θ_i
  END DO
END DO
x_N = σ_1/α_L
```

The backward-substitution algorithm (3.4.17) follows.

PROBLEM

3.4.6 ***Numerical exercise with a calculator.***

Apply the method of cyclic reduction to solve an 8×8 tridiagonal system of your choice with constant elements along the three diagonals. Exploit the constancy of the coefficients along the three diagonals to reduce the number of operations.

Sparse Systems

Tridiagonal and pentadiagonal systems are representative of sparse systems, that is, systems whose coefficient matrix contains many zeros. Sparse systems composed of block matrices arise in a wide variety of applications; some patterns are illustrated in Table 3.4.1.

The Thomas algorithm and its generalization to pentadiagonal systems are designed to circumvent idle multiplications by zero, and thus reduce both the computational cost and storage requirements. Similar

Table 3.4.1 **Sample sparse matrices with various patterns, after Tewarson (1973) and Press et al. (1992).**

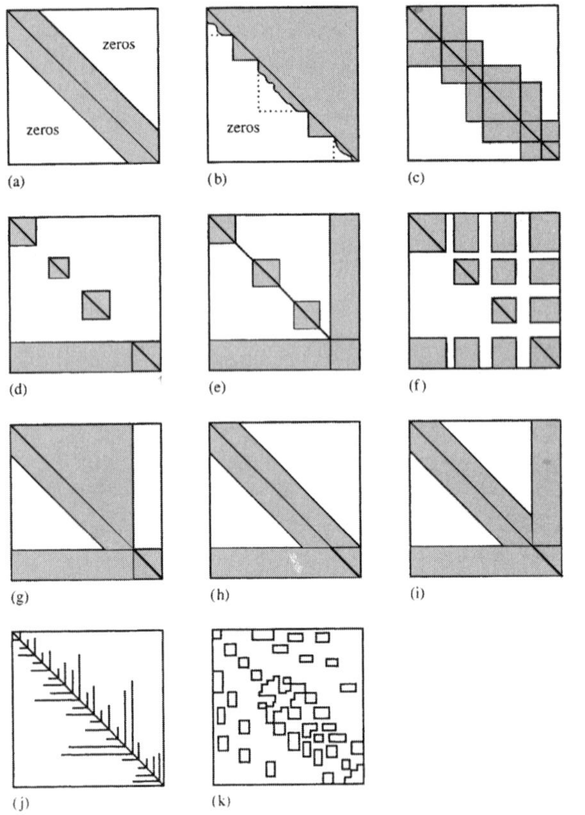

(a) Band diagonal, (b) block triangular, (c) block tridiagonal, (d) singly bordered block diagonal, (e) doubly bordered block diagonal, (f) singly bordered block triangular, (g) bordered band triangular, (h) and (i) singly and doubly bordered band diagonal, (j) and (k) other.

algorithms applicable to other families of sparse systems can be developed by exercising intuition and imagination. Reviews and practical advice on implementation are given by Tewarson (1973), George and Liu (1981), Pissanetzky (1984), and Press et al. (1992, pp. 63–82).

3.5 Gauss Elimination and Related Methods

Gauss elimination is the simplest and most popular *direct* method for solving systems of equations of moderate size. It works well for matrices that are not nearly singular; that is, their condition number defined in Section 3.3 is small or moderate, but not too large. Nearly singular systems are best treated by the methods of $\boldsymbol{QR}$ and singular-value decomposition discussed in Section 3.3.

Gauss elimination transforms the system $\boldsymbol{A}\boldsymbol{x} = \boldsymbol{b}$ to the upper triangular system $\boldsymbol{U}\boldsymbol{x} = \boldsymbol{y}$, and simultaneously produces a lower diagonal matrix $\boldsymbol{L}$ so that $\boldsymbol{L}\boldsymbol{U} = \boldsymbol{A}$, where the diagonal elements of $\boldsymbol{L}$ are all equal to unity. Thus Gauss elimination performs the Doolittle $\boldsymbol{L}\boldsymbol{U}$ decomposition of the coefficient matrix $\boldsymbol{A}$ and can be used for that purpose without a reference to a system of equations. When the option of row pivoting, to be discussed shortly, is enabled, the method performs the Doolittle decomposition of a matrix that arises by interchanging pairs of rows of $\boldsymbol{A}$. In the final step, the vector $\boldsymbol{x}$ arises by backward substitution.

Strategy

The basic idea is to solve the first equation for the first unknown x_1, and use the resulting expression to eliminate x_1 from all subsequent equations. We then retain the first equation as is and replace the subsequent equations with their descendants that do not contain x_1.

In the second stage, we solve the second equation for the second unknown x_2, and use the resulting expression to eliminate x_2 from all subsequent equations. We then retain the first and second equations and replace all subsequent equations with their descendants that do not contain x_1 or x_2.

Continuing this process, we arrive at the last equation, which contains only the last unknown. At that point, the reduced system of equations has the targeted upper triangular form $\boldsymbol{U}\boldsymbol{x} = \boldsymbol{y}$ and may thus be solved using backward substitution.

Immediately before the mth equation is solved for the mth unknown, where $m = 1, \ldots, N-1$, the linear system has the form

$$\begin{bmatrix}
A^{(m)}_{1,1} & A^{(m)}_{1,2} & A^{(m)}_{1,3} & A^{(m)}_{1,4} & \cdots & \cdots & \cdots & A^{(m)}_{1,N-1} & A^{(m)}_{1,N} \\
0 & A^{(m)}_{2,2} & A^{(m)}_{2,3} & A^{(m)}_{2,4} & \cdots & \cdots & \cdots & A^{(m)}_{2,N-1} & A^{(m)}_{2,N} \\
0 & 0 & A^{(m)}_{3,3} & A^{(m)}_{3,4} & \cdots & \cdots & \cdots & A^{(m)}_{3,N-1} & A^{(m)}_{3,N} \\
\cdots & \cdots & \cdots & \cdots & \cdots & \cdots & \cdots & \cdots & \cdots \\
0 & 0 & 0 & A^{(m)}_{m-1,m-1} & A^{(m)}_{m-1,m} & A^{(m)}_{m-1,m+1} & \cdots & A^{(m)}_{m-1,N-1} & A^{(m)}_{m-1,N} \\
0 & 0 & 0 & 0 & A^{(m)}_{m,m} & A^{(m)}_{m,m+1} & \cdots & A^{(m)}_{m,N-1} & A^{(m)}_{m,N} \\
0 & 0 & 0 & 0 & A^{(m)}_{m+1,m} & A^{(m)}_{m+1,m+1} & \cdots & A^{(m)}_{m+1,N-1} & A^{(m)}_{m+1,N} \\
\cdots & \cdots & \cdots & \cdots & \cdots & \cdots & \cdots & \cdots & \cdots \\
0 & 0 & 0 & 0 & A^{(m)}_{N,m} & A^{(m)}_{N,m+1} & \cdots & A^{(m)}_{N,N-1} & A^{(m)}_{N,N}
\end{bmatrix} \boldsymbol{x} = \boldsymbol{b}^{(m)} \quad (3.5.1)$$

where $\boldsymbol{A}^{(m)}$ is an intermediate coefficient matrix, and $\boldsymbol{b}^{(m)}$ is an intermediate constant vector. The first row of $\boldsymbol{A}^{(m)}$ is identical to the first row of $\boldsymbol{A}$, but subsequent rows are different. Similarly, the first element of $\boldsymbol{b}^{(m)}$ is equal to the first element of $\boldsymbol{b}$, but the subsequent elements are different.

Pivoting and Scaling

An apparent difficulty arises when the coefficient $A_{m,m}^{(m)}$ is nearly or exactly equal to zero, for then, we can no longer solve the mth equation for x_m, as required. But the failure of the method *does not* mean that the system does not have a solution. To circumvent this stumbling block, we simply rearrange the equations or relabel the unknowns, so as to bring the mth unknown to the mth equation, using the celebrated method of *pivoting*. If there is no way we can make this happen, then the system has either no solution or an infinite number of solutions.

Before discussing how pivoting is actually implemented, we pause to make a few important observations:

- Pivoting is not necessary for systems with *diagonally dominant* coefficient matrices. The norm of each diagonal element of a diagonally dominant matrix is larger than the sum of the norms of the elements in the corresponding row.
- Pivoting is not necessary for systems with *symmetric and positive-definite* coefficient matrices.
- It appears that pivoting should be enabled only when the magnitude of $A_{m,m}^{(m)}$ is smaller than a pre-established threshold. In practice, however, we want to make $|A_{m,m}^{(m)}|$ as large as possible in order to reduce the round-off error. Thus pivoting should be enabled even when $|A_{m,m}^{(m)}|$ is not necessarily small.
- Pivoting prohibits the parallelization of the computations.

Row pivoting and scaling

In the method of *row pivoting*, we switch the mth equation of the system (3.5.1) with the subsequent kth equation, $k > m$, which is chosen so that $|A_{k,m}^{(m)}|$ is the maximum norm of the element in the mth column below the diagonal, $|A_{i,m}^{(m)}|$ for $i \geq m$.

If it happens that $|A_{i,m}^{(m)}| = 0$ for all $i \geq m$, the original system of equations does not have a unique solution: The rank of $\mathbf{A}$ is less than N, and the system has either no solution, or an infinite number of solutions.

One cannot help thinking that row pivoting on its own is somewhat arbitrary: One can always multiply both sides of the mth equation by a large factor so as to make the magnitude of $A_{m,m}^{(m)}$ dominant, and thereby prevent row pivoting. This trick may work in some cases, but it usually disguises instead of curing intrinsic problems with problematic systems. One way to eliminate these ambiguities is to scale the equations at the outset. This can be done by dividing each equation either with the maximum norm of its coefficients, so that the value of one coefficient becomes equal to ± 1, or with the length of the vector represented by each row.

Column pivoting

In the method of *column pivoting*, we switch the mth column of the coefficient matrix of the system (3.5.1) with the subsequent kth column, $k > m$, which is chosen so that $|A_{m,k}^{(m)}|$ is the maximum of $|A_{m,j}^{(m)}|$ for $j \geq m$.

If it happens that $|A_{m,j}^{(m)}| = 0$ for all $j \geq m$, then the original system of equations does not have a unique solution: The rank of $\mathbf{A}$ is less than N, and the system has either no solution or an infinite number of solutions.

Interchanging the columns of the coefficient matrix is equivalent to interchanging the positions or switching the labels of the corresponding unknowns. To obtain a consistent solution at the end, we must keep track of the original and final positions of the interchanged unknowns. The additional work puts the method of column pivoting at a disadvantage.

Complete pivoting

In the method of *complete pivoting*, we find the element with the maximum norm of the $m \times m$ southeast lower diagonal block of the coefficient matrix of the system (3.5.1); this block is bordered from the left by the mth column and from the top by the mth row. We then switch either the mth equation or the mth column with those corresponding to the maximum element, as described in the preceding two subsections. Complete pivoting, however, is not necessary in practice; row pivoting works well and is the standard choice.

Algorithm of Gauss Elimination with Row Pivoting

To implement the method of Gauss elimination with row pivoting, we proceed along the following steps:

Setting up

Form the $N \times (N+1)$ partitioned *augmented matrix* $\boldsymbol{C}^{(1)} \equiv [\boldsymbol{A}|\boldsymbol{b}]$, and introduce the $N \times N$ matrix $\boldsymbol{L}$ whose elements are initialized to zero.

First Pass

1. Find the location of the element with the maximum norm in the first column of $\boldsymbol{C}^{(1)}$. That is, search for the maximum norm of the elements $|C_{i,1}^{(1)}|$, $i = 1, \ldots, N$. Assume that this is equal to $|C_{k,1}^{(1)}|$, corresponding to the kth row.
2. Interchange the first row with the kth row of $\boldsymbol{C}^{(1)}$; repeat for $\boldsymbol{L}$. If $k = 1$, skip this step.
3. Compute the first column of $\boldsymbol{L}$ by setting $L_{i,1} = C_{i,1}^{(1)}/C_{1,1}^{(1)}$ for $i = 1, \ldots, N$.
4. Subtract from the ith row of $\boldsymbol{C}^{(1)}$ the first row multiplied by $L_{i,1}$ for $i = 2, \ldots, N$, to obtain the new augmented matrix $\boldsymbol{C}^{(2)} \equiv [\boldsymbol{A}^{(2)}|\boldsymbol{b}^{(2)}]$.

Second Pass

5. Find the location of the element with the maximum norm in the second column of $\boldsymbol{C}^{(2)}$ below the diagonal. That is, search for the maximum norm of the elements $|C_{i,2}^{(2)}|$, $i = 2, \ldots, N$. Assume that this is equal to $|C_{k,2}^{(2)}|$, corresponding to the kth row.
6. Interchange the second row with the kth row of $\boldsymbol{C}^{(2)}$; repeat for $\boldsymbol{L}$. If $k = 2$, skip this step.
7. Compute the second column of $\boldsymbol{L}$, setting $L_{i,2} = C_{i,2}^{(2)}/C_{2,2}^{(2)}$ for $i = 2, \ldots, N$.
8. Subtract from the ith row of $\boldsymbol{C}^{(2)}$ the second row multiplied by $L_{i,2}$ for $i = 3, \ldots, N$, to obtain the new augmented matrix $\boldsymbol{C}^{(3)} \equiv [\boldsymbol{A}^{(3)}|\boldsymbol{b}^{(3)}]$.

...

mth Pass

$4m-3$. Find the location of the element with the maximum norm in the mth column of $\boldsymbol{C}^{(m-1)}$ below the diagonal. That is, search for the maximum norm of the elements $|C_{i,m}^{(m)}|$, $i = m, \ldots, N$. Assume that this is equal to $|C_{k,m}^{(m)}|$, corresponding to the kth row.

$4m-2$. Interchange the mth row with the kth row of $\boldsymbol{C}^{(m)}$; repeat for $\boldsymbol{L}$. If $k = m$, skip this step.

$4m-1$. Compute the mth column of $\boldsymbol{L}$, setting $L_{i,m} = C_{i,m}^{(m)}/C_{m,m}^{(m)}$ for $i = m, \ldots, N$.

$4m$. Subtract from the ith row of $\boldsymbol{C}^{(m)}$ the mth row multiplied by $L_{i,m}$ for $i = m+1, \ldots, N$, to obtain the new augmented matrix $\boldsymbol{C}^{(m+1)} \equiv [\boldsymbol{A}^{(m+1)} | \boldsymbol{b}^{(m+1)}]$.

$(N-1)$ Pass

At the end of the $(N-1)$ pass, corresponding to $m = N-1$, the augmented matrix $\boldsymbol{C}^{(N)}$ has the form

$$\boldsymbol{C}^{(N)} \equiv [\boldsymbol{A}^{(N)} | \boldsymbol{b}^{(N)}] \tag{3.5.2}$$

where $\boldsymbol{A}^{(N)} \equiv \boldsymbol{U}$ is an upper triangular matrix.

Backward substitution

Finally, solve the upper triangular system $\boldsymbol{Ux} = \boldsymbol{b}^{(N)}$ using the backward substitution algorithm described in Section 3.2, to obtain the solution of the original system of equations $\boldsymbol{Ax} = \boldsymbol{b}$.

The procedure up to the point where backward substitution begins is displayed in Algorithm 3.5.1. A simple modification, with the objective of memory economization, places the matrix $\boldsymbol{L}$ at the bottom part of $\boldsymbol{C}$.

Detailed inspection shows that the number of required multiplications is on the order of $N^3/3$. Thus when N is on the order of a few hundred or higher, the required CPU time is substantial.

ALGORITHM 3.5.1 Algorithm for Gauss elimination with row pivoting, also performing the $\boldsymbol{LU}$ decomposition.

- Define the $N \times (N+1)$ augmented matrix $\boldsymbol{C}^{(1)} \equiv [\boldsymbol{A}|\boldsymbol{b}]$.
- Introduce the matrix $\boldsymbol{L}$ and initialize its elements to 0.
- Set $Icount = 0$ — Read: $I\ count$; Will count row interchanges due to pivoting

```
Do m = 1, N − 1                               m counts elimination passes

  ipv = m                                     Assess if pivoting is required
  pivot = |C_{m,m}|                           Find pivotal element
    Do j = m + 1, N
```

```
        IF (|C_{j,m}| > pivot) THEN
        ipv = j
        pivot = |C_{j,m}|
        END IF
      END DO

    IF (pivot < ε) STOP                     The system is singular within a specified tolerance ε
        IF (ipv ≠ m) THEN                   If pivoting is required, interchange rows
          DO j = m, N + 1                   of augmented matrix
            save = C_{m,j}
            C_{m,j} = C_{ipv,j}
            C_{ipv,j} = save
          END DO
          DO j = 1, m − 1                   Repeat for L
            save = L_{m,j}
            L_{m,j} = L_{ipv,j}
            L_{ipv,j} = save
          END DO
            Icount ← Icount + 1
          END IF

        DO i = m + 1, N                     Proceed with the elimination
        L_{i,m} = C_{i,m}/C_{m,m}           For a symmetric system substitute: L_{i,m} = C_{m,i}/C_{m,m}
        C_{i,m} = 0
            DO j = m + 1, N + 1             For a symmetric system substitute: DO j = i, N + 1
            C_{i,j} ← C_{i,j} − L_{i,m}C_{m,j}
            END DO
        END DO
END DO                                      Closes the DO m loop
DO i = 1, N                                 Compute the matrices L and U
    L_{i,i} = 1.0
    DO j = i, N
    U_{i,j} = C_{i,j}
    END DO
END DO
```

Backward substitution follows after this stage.

LU Decomposition via Gauss Elimination

It can be shown, by straightforward algebraic manipulations, that the matrices $\boldsymbol{L}$ and $\boldsymbol{U}$ provide us with the $\boldsymbol{LU}$ decomposition of the matrix $\boldsymbol{A}$, that is,

$$\boldsymbol{LU} = \boldsymbol{A}^{Mod} \tag{3.5.3}$$

The matrix $\boldsymbol{A}^{Mod}$ is identical to $\boldsymbol{A}$, except that the rows may have been interchanged due to pivoting. If pivoting is not done, $\boldsymbol{A}^{Mod} = \boldsymbol{A}$.

The determinant of $\boldsymbol{A}$ follows from

$$\pm \text{ Det}(\boldsymbol{A}) = \text{Det}(\boldsymbol{A}^{Mod}) = \text{Det}(\boldsymbol{L}) \text{ Det}(\boldsymbol{U}) = U_{1,1}U_{2,2}\dots U_{N,N} \tag{3.5.4}$$

where the plus sign applies when an even number of row interchanges have been made because of pivoting, and the minus sign otherwise.

Preconditioning and Permutation Matrices

Each step in the Gauss elimination process may be effected by premultiplying the partially reduced system of equations with appropriate matrices. Thus the algorithm effectively premultiplies both sides of the original equation $\boldsymbol{Ax} = \boldsymbol{b}$ with a sequence of matrices so as to *precondition* it, that is, so as to bring it to the desired upper triangular form.

Pivoting and permutation matrices

Pivoting can be effected by premultiplying the intermediate systems by appropriate permutation matrices. These matrices arise by interchanging one or more pairs of rows or columns of the identity matrix $\boldsymbol{I}$.

All elements of a permutation matrix $\boldsymbol{P}$ are equal to zero, except for one element in each column and one element in each row that is equal to 1. Thus $P_{i,j} = 1$ when $j = k_i$, and $P_{i,j} = 0$ when $j \neq k_i$, where the collection of the natural numbers k_i is a permutation of the collection of the sequential natural numbers $1, 2, \dots, N$. Premultiplying a matrix $\boldsymbol{A}$ by $\boldsymbol{P}$ produces a new matrix $\boldsymbol{B}$, where the k_ith row of $\boldsymbol{A}$ is the ith row of $\boldsymbol{B}$.

$$\boldsymbol{P} = \begin{bmatrix} 1 & 0 & 0 & 0 & 0 & 0 & 0 \\ 0 & 1 & 0 & 0 & 0 & 0 & 0 \\ 0 & 0 & 0 & 0 & 0 & 1 & 0 \\ 0 & 0 & 0 & 1 & 0 & 0 & 0 \\ 0 & 0 & 0 & 0 & 1 & 0 & 0 \\ 0 & 0 & 1 & 0 & 0 & 0 & 0 \\ 0 & 0 & 0 & 0 & 0 & 0 & 1 \end{bmatrix}$$

Interchanging the ith and jth rows of the augmented matrix can be effected by premultiplying it with a permutation matrix that arises by interchanging the ith and jth rows of $\boldsymbol{I}$. An example of such a permutation matrix with $N = 7, i = 3$, and $j = 6$ is shown in the diagram.

The success of the Gauss elimination method with row pivoting relies on the following theorem: If $\boldsymbol{A}$ is a nonsingular matrix, then there exists a permutation matrix $\boldsymbol{P}$ so that $\boldsymbol{PA}$ has a specified $\boldsymbol{LU}$ decomposition (e.g., Noble and Daniel 1988, p. 120).

PROBLEMS

3.5.1 ***The permutation matrices are orthogonal.***

Show that the permutation matrices are orthogonal and therefore nonsingular.

3.5.2 ***Cramer's rule and Gauss elimination for a singular system.***

Consider the following system of three equations with three unknowns $x + 2y + z = 3, 2x - 3y - z = 1, 3x - y = 4$.

(*a*) Compute the solution using Cramer's rule. Show all steps, and discuss the significance of your results regarding the existence and uniqueness of solution.

(*b*) Compute the solution using Gauss elimination with row pivoting, just like a computer would have done it, and discuss the significance of your results regarding the existence and uniqueness of solution.

3.5.3 ***Gauss elimination with row pivoting.***

(*a*) Write a computer program that produces the solution of a system of linear algebraic equations using the method of Gauss elimination with *row pivoting*. The program should compute and store the $\boldsymbol{L}$ and $\boldsymbol{U}$ matrices, verify that the product $\boldsymbol{LU}$ is equal to $\boldsymbol{A}$ except that the rows may have been switched due to pivoting, and compute Det($\boldsymbol{A}$).

Furthermore, the program should verify that the solution has been found accurately, by computing the residual vector $\boldsymbol{r} = \boldsymbol{Ax} - \boldsymbol{b}$ and making sure that it contains small numbers, on the order of the round-off error. Test the reliability of the program by solving a system of your choice.

(*b*) Use the program to solve systems (3.1.2) and (3.1.16).

(*c*) Use the program to compute the determinant of the Hilbert matrix defined in equation (2.6.5) for $N = 1, 2, 3, 4, 5$. Based on the numerical results, estimate the behavior of the determinant as N becomes larger.

3.5.4 ***Poisson equation with Neumann boundary conditions all around.***

Consider heat conduction in a rectangular plate with Neumann boundary conditions around the edges, discussed in Problem 3.1.4. With reference to the notation of Section 3.1, select a certain source function g and Neumann boundary conditions of your choice, and follow the discussion of Section 3.3 to produce one solution of the singular system of equations.

Multiple Right-hand Sides

A simple modification of the basic Gauss elimination algorithm allows us to solve, at once, multiple systems of equations with the same coefficient matrix but different right-hand sides, $\boldsymbol{Ax}^{(j)} = \boldsymbol{b}^{(j)}$, $j = 1, \ldots, L$, in a single run.

The standard method reduces the primary system $\boldsymbol{Ax}^{(j)} = \boldsymbol{b}^{(j)}$ to the modified system $\boldsymbol{Ux}^{(j)} = \boldsymbol{c}^{(j)}$, whose solution is then found by backward substitution. The computation of the vectors $\boldsymbol{c}^{(j)}$ may be done simultaneously working with the $N \times (N + L)$ augmented matrix

$$\boldsymbol{C}^{(1)} \equiv [\boldsymbol{A}|\boldsymbol{b}^{(1)}|\boldsymbol{b}^{(2)}|\ldots|\boldsymbol{b}^{(L)}] \tag{3.5.5}$$

At the end of the $N - 1$ pass, the augmented matrix $\boldsymbol{C}^{(N)}$ will have the partially block-triangular form $\boldsymbol{B}^{(N)} \equiv [\boldsymbol{A}^{(N)}|\,\boldsymbol{D}^{(N)}]$, where $\boldsymbol{A}^{(N)} \equiv \boldsymbol{U}$ is an upper triangular matrix, and the N columns of $\boldsymbol{D}^{(N)}$ contain the $\boldsymbol{c}^{(j)}$.

Computation of the Inverse

In Section 2.5, we argued that the jth column of the inverse of a matrix $\boldsymbol{A}$, denoted by $\boldsymbol{x}^{(j)}$, where $j = 1, \ldots, N$, satisfies the linear system $\boldsymbol{Ax}^{(j)} = \boldsymbol{e}^{(j)}$, where all components of the vector $\boldsymbol{e}^{(j)}$ are equal to 0 except for the jth component that is equal to 1. The computation of the vectors $\boldsymbol{x}^{(j)}$ may be done in a compact manner, working with the $N \times 2N$ augmented matrix

$$\boldsymbol{C}^{(1)} \equiv [\boldsymbol{A}|\boldsymbol{I}] \tag{3.5.6}$$

where $\boldsymbol{I}$ is the $N \times N$ identity matrix. At the end of the $N - 1$ pass, the augmented matrix $\boldsymbol{C}^{(N)}$ will have the form $\boldsymbol{C}^{(N)} \equiv [\boldsymbol{A}^{(N)}|\,\boldsymbol{D}^{(N)}]$, where $\boldsymbol{A}^{(N)} \equiv \boldsymbol{U}$ is an upper triangular matrix, and the N columns of $\boldsymbol{D}^{(N)}$ contain the $\boldsymbol{c}^{(j)}$, where $\boldsymbol{Ux}^{(j)} = \boldsymbol{c}^{(j)}$. The algorithm concludes with N backward substitutions.

PROBLEM

3.5.5 *Computation of the inverse.*

(*a*) Modify the program described in Problem 3.5.3 into a program that computes the inverse of a matrix.
(*b*) Use the program of part (*a*) to compute the inverse of the matrix

$$\boldsymbol{A} = \begin{bmatrix} 0.5 & 0.4 & 0.3 & 0.2 & 0.1 \\ 0.1 & 0.2 & 0.7 & 0.1 & 0.8 \\ 0.1 & 0.2 & 0.0 & 0.1 & 0.0 \\ 0.2 & 0.1 & 0.0 & 0.1 & 0.0 \\ 0.1 & 0.0 & 0.0 & 0.5 & 0.1 \end{bmatrix} \tag{3.5.7}$$

(*c*) Use the program of part (*a*) to compute the inverse of the Hilbert matrix $\boldsymbol{H}$ defined in equation (2.6.5) for $N = 1, 2, 3, 4, 5$. Compare your results with the exact solution (e.g., Atkinson 1989, p. 534):

$$H^{-1}_{i,j} = (-1)^{i+j} \frac{(N+i-1)!\,(N+j-1)!}{[(i-1)!\,(j-1)!\,]^2 (N-i)!\,(N-j)!\,(i+j-1)} \tag{3.5.8}$$

Report and explain difficulties with round-off error.
(*d*) Consider a matrix $\boldsymbol{A}$ whose elements are functions of the scalar variable t. The rate of change of the determinant of $\boldsymbol{A}$ with respect to t was given in equation (2.3.18). Approximating the derivatives with centered finite differences, we obtain the approximate discrete form

$$\mathrm{Det}(\boldsymbol{A}(t+\varepsilon)) - \mathrm{Det}(\boldsymbol{A}(t-\varepsilon)) = \mathrm{Det}(\boldsymbol{A}(t))\,[\boldsymbol{A}(t+\varepsilon) - \boldsymbol{A}(t-\varepsilon)] : \boldsymbol{A}^{-1^T}(t) \tag{3.5.9}$$

where ε is a positive or negative number with small magnitude.
Consider the $N \times N$ matrix $\boldsymbol{A}$ with elements $A_{i,j} = t(i-1) + e^{jt}$, and use the program of part (*a*) to verify identity (3.5.9) at two different values of t of your choice.

Economization for Symmetric Systems

If the original matrix $\boldsymbol{A}$ is symmetric *and pivoting is not enabled*, at the risk of having to divide by zero or promote the growth of the round-off error, the lower square diagonal $(N-m+1) \times (N-m+1)$ block of the matrix displayed in equation (3.5.1) will remain symmetric at every stage, for any value of m. This property can be exploited for storing, and working only with the upper or lower triangular part of $\boldsymbol{A}$ and its descendants.

Specifically, for any value of m, we work only with the elements of the augmented matrix that lie on or above the diagonal, and set the values of the elements below the diagonal equal to their symmetric counterparts, as required. This modification reduces the number of operations by nearly a factor of 2. When $\boldsymbol{A}$ is positive-definite, pivoting is not necessary, and this modification can be implemented without a risk. The necessary changes are pointed out in the right margin of Algorithm 3.5.1.

Gauss–Jordan Elimination

The Gauss–Jordan method is a cousin of the Gauss elimination method. The algorithm proceeds along the following steps:

- Divide the first equation by $A^{(1)}_{1,1}$, solve it for the first unknown x_1, and use the resulting expression to eliminate x_1 from *all* subsequent equations.

- Divide the second equation by $A^{(2)}_{2,2}$, solve it for the second unknown x_2, and use the resulting expression to eliminate x_2 from *the first and all subsequent* equations.
- Divide the third equation by $A^{(3)}_{3,3}$, solve it for the third unknown x_3, and use the resulting expression to eliminate x_3 from *the first, the second, and all subsequent* equations.
- Continue in this manner, until the last equation contains only the last unknown. At that point, the modified system of equations will be diagonal, with all diagonal elements of the coefficient matrix equal to unity. Consequently, the solution will be displayed on the right-hand side.

To implement the method, we form the augmented matrix $\boldsymbol{C}^{(1)} \equiv [\boldsymbol{A}|\boldsymbol{b}]$ and proceed as in Gauss elimination with straightforward modifications. At the end of the $N-1$ pass, corresponding to $m = N-1$, $\boldsymbol{C}^{(N)}$ will have the form $\boldsymbol{B}^{(N)} \equiv [\boldsymbol{I}|\boldsymbol{b}^{(N)}]$, where $\boldsymbol{x} = \boldsymbol{b}^{(N)}$ is the required solution. Row pivoting is implemented as in Gauss elimination.

It would appear that the Gauss–Jordan elimination is competitive with, if not preferable over, the Gauss elimination. Counting the number of operations, however, shows that the Gauss–Jordan method is slower than the Gauss method roughly by a factor of 3. In spite of this deficiency, the robustness of the Gauss–Jordan method makes it attractive for systems of small and moderate size.

When solving groups of L systems of linear equations using the formulation of multiple right-hand side discussed earlier in this section, where L is comparable to N, the efficiency of the Gauss–Jordan elimination is comparable, or even superior, to that of the Gauss elimination.

Computation of the inverse

To compute the inverse of a matrix, we begin with the $N \times 2N$ augmented matrix $\boldsymbol{C}^{(1)} \equiv [\boldsymbol{A}|\boldsymbol{I}]$. At the end of the $N-1$ pass, $\boldsymbol{C}^{(N)}$ will have the form $\boldsymbol{B}^{(N)} \equiv [\boldsymbol{I}|\boldsymbol{D}^{(N)}]$, where $\boldsymbol{D}^{(N)} = \boldsymbol{A}^{-1}$ except that its columns may have been switched due to pivoting. Switching back the columns can be done by keeping a record of the positions of the original and reduced rows.

PROBLEM

3.4.6 *Program for Gauss–Jordan elimination with column pivoting.*

(*a*) Write a computer program that produces the solution of a system of linear algebraic equations using the method of Gauss–Jordan elimination with *row pivoting*. Test the reliability of the program by solving a system of linear equations of your choice.
(*b*) Use the program of part (a) to solve systems (3.1.2) and (3.1.16).

3.6 Iterative Methods

In Section 3.3, we outlined the general principles of iterative methods and briefly mentioned their advantages and disadvantages. We noted, in particular, that iterative methods are especially but not exclusively suited for *sparse systems with large dimensions*. In this section, we become more specific about their development and implementation.

Fixed-point Iterations for one Equation

It is instructive to begin by considering one scalar equation containing one single scalar unknown x,

$$ax = b \tag{3.6.1}$$

Applying either the forward or the backward substitution algorithm, we obtain the well-known solution in a single step, $x = b/a$.

For reasons that will soon become apparent, we now recast equation (3.6.1) into the equivalent form $(a + n)x = nx + b$, where n is a real or complex number other than $-a$. Rearranging the last equation, we obtain

$$x = px + c \tag{3.6.2}$$

where we have defined

$$p = n/(a + n), \qquad c = b/(a + n) \tag{3.6.3}$$

We propose to compute the solution in the following iterative manner:

1. Select a starting value $x^{(0)}$.
2. Compute a sequence of numbers using the recursive formula:

$$x^{(k+1)} = px^{(k)} + c \tag{3.6.4}$$

Note that this formula stems directly from equation (3.6.2). To produce $x^{(k+1)}$, we multiply $x^{(k)}$ by the coefficient p, and then shift it by c.

If it happens that the sequence (3.6.4) converges to a limit denoted by X, then, by definition, this limit will satisfy the equation

$$X = pX + c \tag{3.6.5}$$

and therefore the original equation (3.6.1); that is, X is the desired solution.

The number X is the *fixed point* of the mapping expressed by equation (3.6.4), corresponding to the linear iteration function

$$f(x) = px + c \tag{3.6.6}$$

that is, X is the point where $X = f(X)$. Accordingly, the iterative procedure just described falls into the general category of *fixed-point iterations*; other names include *successive substitutions, Picard iterations, successive approximations*, and *one-point iterations.* Two-point iterations produce $x^{(k+1)}$ in terms of $x^{(k)}$ and $x^{(k-1)}$.

Convergence for One Equation

How can we assess whether the sequence computed from equation (3.6.4) will tend to a limit? One way is to use the results of Problem 2.11.4, finding

$$x^{(k)} = p^k x^{(0)} + \frac{1-p^k}{1-p} c \tag{3.6.7}$$

and then examine the behavior of $x^{(k)}$ as k becomes larger. An equivalent way is to subtract from both sides of equation (3.6.4) corresponding sides of equation (3.6.5), and thus derive an evolution equation for the error $e^{(k)} = x^{(k)} - X$,

$$e^{(k+1)} = pe^{(k)} \tag{3.6.8}$$

for $k = 0, 1, 2, \ldots$. Either way, it is evident that the sequence will converge to X provided that

$$|p| < 1 \tag{3.6.9}$$

that is, provided that the generally complex number p falls within a disk of unit radius centered at the origin of the complex plane. The satisfaction of this inequality is the sole prerequisite for the success of the iterative method. If the computations converge, they will do so for any initial error $e^{(0)}$ and hence for any arbitrary initial guess $x^{(0)}$.

The optimal value of p and thus of n is clearly equal to zero! In that case, the exact solution arises after a single step, which seems to trivialize this approach. But, in the case of a single equation presently considered, the shift $c = b/(a+n)$ is easy to calculate. When we extend this method to a system of equations, the scalars will become matrices, and the matrix counterpart of n will be selected so as to facilitate the computation of the vector counterpart of c, in a way that circumvents the computation of a matrix inverse.

Fixed-point Iterations for a System of Equations

Taking the preceding ideas one step further, we consider the system of linear equations

$$\mathbf{A}\mathbf{x} = \mathbf{b} \tag{3.6.10}$$

and rewrite it in the equivalent form $(\mathbf{A} + \mathbf{N})\,\mathbf{x} = \mathbf{N}\mathbf{x} + \mathbf{b}$, where $\mathbf{N}$ is some matrix, and then

$$\mathbf{M}\mathbf{x} = \mathbf{N}\mathbf{x} + \mathbf{b} \tag{3.6.11}$$

where we have defined

$$\mathbf{M} = \mathbf{A} + \mathbf{N} \tag{3.6.12}$$

We say that $\mathbf{A} = \mathbf{M} - \mathbf{N}$ is a splitting of the matrix $\mathbf{A}$. Premultiplying both sides of equation (3.6.11) by $\mathbf{M}^{-1}$, we obtain

$$\mathbf{x} = \mathbf{P}\mathbf{x} + \mathbf{c} \tag{3.6.13}$$

where

$$\mathbf{P} = \mathbf{M}^{-1}\mathbf{N}, \qquad \mathbf{c} = \mathbf{M}^{-1}\mathbf{b} \tag{3.6.14}$$

Note that the explicit computation of $\boldsymbol{P}$ and $\boldsymbol{c}$ requires the availability of $\boldsymbol{M}^{-1}$. The method of one-point iterations proceeds as follows:

1. Select a starting vector $\boldsymbol{x}^{(0)}$.
2. Compute a sequence of vectors using the recursive formula:

$$\boldsymbol{M}\boldsymbol{x}^{(k+1)} = \boldsymbol{N}\boldsymbol{x}^{(k)} + \boldsymbol{b} \tag{3.6.15}$$

which originates from equation (3.6.11). More explicitly,

$$\boldsymbol{x}^{(k+1)} = \boldsymbol{P}\boldsymbol{x}^{(k)} + \boldsymbol{c} \tag{3.6.16}$$

where $\boldsymbol{P}$ and $\boldsymbol{c}$ are defined in equations (3.6.14). The matrix $\boldsymbol{P}$ projects $\boldsymbol{x}^{(k)}$ onto its range, and the projection is shifted by $\boldsymbol{c}$ to produce $\boldsymbol{x}^{(k+1)}$.

Convergence of the Method for a System of Equations

The fixed point $\boldsymbol{X}$ of the linear mapping defined by equations (3.6.15) and (3.6.16) satisfies the equations

$$\boldsymbol{M}\boldsymbol{X} = \boldsymbol{N}\boldsymbol{X} + \boldsymbol{b}, \qquad \boldsymbol{X} = \boldsymbol{P}\boldsymbol{X} + \boldsymbol{c} \tag{3.6.17}$$

and therefore equation (3.6.10). Working as in the case of a single equation, we find that the error vector $\boldsymbol{e}^{(k)} = \boldsymbol{x}^{(k)} - \boldsymbol{X}$ evolves according to

$$\boldsymbol{e}^{(k+1)} = \boldsymbol{P}\boldsymbol{e}^{(k)} \tag{3.6.18}$$

Our discussion in Section 2.11—and, in particular, equation (2.11.9)—shows that, when k is sufficiently large,

$$\frac{|\boldsymbol{e}^{(k+1)}|}{|\boldsymbol{e}^{(k)}|} \cong \rho \tag{3.6.19}$$

where ρ is the spectral radius of $\boldsymbol{P}$.

Accordingly, if ρ is less than unity, that is, the norm of each eigenvalue of $\boldsymbol{P}$ is less than unity, the norms of the vectors $\boldsymbol{e}^{(k)}$ will keep decreasing during the iterations, and the sequence $\boldsymbol{x}^{(k)}$ will converge to the fixed point $\boldsymbol{X}$. If ρ is larger than unity, the iterations will diverge, and if ρ is equal to unity, the iterations will stall.

Clearly, the optimal choice for $\boldsymbol{N}$ is the null matrix $\boldsymbol{0}$: The spectral radius of $\boldsymbol{P}$ is equal to zero, and the solution is obtained in a single step. But with this selection, computing the shift $\boldsymbol{c}$ from the second of equations (3.6.14) requires solving the equation $\boldsymbol{M}\boldsymbol{c} = \boldsymbol{b}$, which is what we are trying to avoid in the first place.

The matrix $\boldsymbol{M}$ should be simple enough so that equation (3.6.15) can be solved for $\boldsymbol{x}^{(k+1)}$ using an efficient method. For example, $\boldsymbol{M}$ should be diagonal, triangular, or tridiagonal. In addition, $\boldsymbol{M}$ should be such that the spectral radius of $\boldsymbol{P}$ is less than unity. The interplay between these two requirements yields different types of iterative methods.

Aitken extrapolation

Equation (3.6.18) shows that the rate of convergence of the method of successive substitutions is linear. This knowledge allows us to use the method of Aitken extrapolation to the individual components of the iteration vector $\mathbf{x}^{(k)}$, as discussed in Section 1.7, in order to accelerate the approach toward the fixed point. But remember that, when the iterations diverge, the Aitken extrapolation will make them diverge even faster.

PROBLEM

3.6.1 ***Explicit formula for the sequence in terms of the initial vector.***

Derive an explicit expression for $\mathbf{x}^{(k+1)}$ in terms of $\mathbf{x}^{(0)}$, $\mathbf{P}$, $\mathbf{c}$, and k. Based on this expression, assess the behavior of $\mathbf{x}^{(k+1)}$ as k becomes larger.

A Sufficient Condition for Convergence

The spectral radius of the iteration matrix $\mathbf{P}$ may be difficult to compute. As a compromise, we use Gerschgorin's first theorem, discussed in Section 2.9, and find that a *sufficient but not necessary* condition for the successive iterations to converge is that the minimum of (a) the sums of the norms of the elements in each column of $\mathbf{P}$ and (b) the sums of the norms of the elements in each row of $\mathbf{P}$ be less than unity.

3.7 *Jacobi, Gauss–Seidel, and the SOR Method*

Having laid out the theoretical foundation of iterative methods, we proceed to develop specific mapping functions and discuss their performance.

Jacobi's Method

In this method, the iteration matrix $\mathbf{P}$ and shift $\mathbf{c}$ defined in equations (3.6.14) are constructed simply by solving the individual scalar equations of the original system $\mathbf{A}\mathbf{x} = \mathbf{b}$ for the diagonal unknowns, obtaining

$$\boxed{x_i^{(k+1)} = \frac{1}{A_{i,i}}\left(b_i - \sum_{\substack{j=1\\ j\neq i}}^{N} A_{i,j}x_j^{(k)}\right)} \tag{3.7.1}$$

where summation over i is *not* implied in the denominator. With reference to equation (3.6.12), the matrix $\mathbf{M}$ is the diagonal component of $\mathbf{A}$, and the matrix $\mathbf{N}$ is the negative of $\mathbf{A}$ with the diagonal elements set equal to zero. The projection matrix is thus given by

$$P_{i,j} = \begin{cases} -\dfrac{A_{i,j}}{A_{i,i}} & \text{when } i \neq j \\ 0 & \text{when } i = j \end{cases} \tag{3.7.2}$$

and the shift is given by

$$c_i = -b_i / A_{i,i} \tag{3.7.3}$$

where summation is *not* implied over the repeated index i in both equations.

The computer programming is done in a straightforward manner on the basis of equation (3.7.1). It is important to note that the current and next approximations $\boldsymbol{x}^{(k)}$ and $\boldsymbol{x}^{(k+1)}$ must be stored as two different vectors.

For example, for the 2 × 2 system,

$$\begin{aligned} 5x_1 + 3x_2 &= 8 \\ -x_1 + 4x_2 &= 3 \end{aligned} \tag{3.7.4}$$

we obtain

$$\begin{aligned} x_1^{(k+1)} &= -\tfrac{3}{5}x_2^{(k)} + \tfrac{8}{5} \\ x_2^{(k+1)} &= \tfrac{1}{4}x_1^{(k)} + \tfrac{3}{4} \end{aligned} \tag{3.7.5}$$

Selecting $x_1^{(0)} = 0$ and $x_2^{(0)} = 0$, we compute $x_1^{(1)} = \frac{8}{5}$ and $x_2^{(1)} = \frac{3}{4}$. Continuing the substitutions, we ultimately arrive at the fixed point $X_1 = 1$ and $X_2 = 1$, which is the exact solution of the system (3.7.4).

Convergence

The sequence $\boldsymbol{x}^{(k)}$ will converge to the fixed point provided that the spectral radius of $\boldsymbol{P}$ is less than unity, but this is generally difficult to estimate. As a compromise, and use the convergence criteria discussed at the end of Section 3.5 and find that a *sufficient* but not necessary condition for the successive substitutions to converge is that the matrix $\boldsymbol{A}$ or its transpose be *diagonally dominant*; that is,

$$|A_{i,i}| > \sum_{j=1}^{N} |A_{i,j}| \qquad \text{or} \qquad |A_{i,i}| > \sum_{j=1}^{N} |A_{j,i}| \tag{3.7.6}$$

for all values of i, where summation over i is *not* implied on the left-hand sides (Problem 3.7.1). This condition is clearly satisfied for the system (3.7.4).

If $\boldsymbol{A}$ is not diagonally dominant, then rearranging the equations or relabeling the unknowns can make it so, although an efficient and programmable way of making these rearrangments is elusive.

PROBLEM

3.7.1 *Sufficient criterion for convergence.*

Derive criterion (3.7.6) by requiring that the spectral radius of the projection matrix be less than unity, and using Gerschgorin's first theorem discussed in Section 2.9.

Gauss–Seidel Method

The Gauss–Seidel method emerges by a simple modification of Jacobi's method. Its distinguishing feature is that the individual components of $\boldsymbol{x}^{(k+1)}$ replace the corresponding components of $\boldsymbol{x}^{(k)}$ as soon as the former are available. For example, with reference the equations (3.7.5), we select $x_1^{(0)} = 0$ and $x_2^{(0)} = 0$, and compute $x_1^{(1)} = \frac{8}{5}$ and $x_2^{(1)} = (\frac{1}{4})(\frac{8}{5}) + \frac{3}{4} = \frac{23}{20}$.

The programming of the method is somewhat simpler than that for Jacobi's method; there is no need to store $\boldsymbol{x}^{(k)}$ and $\boldsymbol{x}^{(k+1)}$ as two distinct vectors.

Effective projection matrix

The effective algorithm of the Gauss–Seidel method is

$$x_i^{(k+1)} = \frac{1}{A_{i,i}} \left(b_i - \sum_{j=1}^{i-1} A_{i,j} x_j^{(k+1)} - \sum_{j=i+1}^{N} A_{i,j} x_j^{(k)} \right) \tag{3.7.7}$$

The associated projection matrix becomes evident by recasting equation (3.7.7) into the form

$$(\boldsymbol{D} + \boldsymbol{L})\boldsymbol{x}^{(k+1)} = -\boldsymbol{U}\boldsymbol{x}^{(k)} + \boldsymbol{b} \tag{3.7.8}$$

where $\boldsymbol{D}$ is a diagonal matrix, $\boldsymbol{L}$ is a lower triangular matrix with zeros on the diagonal, and $\boldsymbol{U}$ is an upper triangular matrix with zeros on the diagonal containing, respectively, the diagonal, strictly lower triangular, and strictly upper triangular components of $\boldsymbol{A}$.

The projection matrix and shift are thus given by

$$\boldsymbol{P} = -(\boldsymbol{D} + \boldsymbol{L})^{-1}\boldsymbol{U}, \qquad \boldsymbol{c} = (\boldsymbol{D} + \boldsymbol{L})^{-1}\boldsymbol{b} \tag{3.7.9}$$

Convergence

In general, when Jacobi's method converges, the Gauss–Seidel method will also converge at an even faster rate. A *sufficient* but not necessary condition for the successive substitutions to converge is that the matrix $\boldsymbol{A}$ be symmetric and positive definite (e.g., Varga 1962, Hageman and Young 1981).

Symmetric Gauss–Seidel Method

Clearly, there is a bias in the updating of the unknowns: The last unknown is served with the most recent information, whereas the first unknown uses the old values. To prevent this unfair treatment, we can reverse the order of the updating after a complete pass. This modification results in the symmetric Gauss–Seidel method.

PROBLEMS

3.7.2 ***Jacobi's and Gauss–Seidel methods for solving a linear system.***

Find the solution of the following system of equations accurate to the third decimal place using Jacobi's method: $x + 9y = 10$, $8x - y = 7$. Repeat with the Gauss–Seidel method, and compare the rates of convergence.

3.7.3 ***Programming Jacobi's and the Gauss–Seidel method.***

(*a*) Write a computer program that solves a system of linear algebraic equations using Jacobi's method, with an option to perform the Gauss–Seidel iterations. The computations should be terminated when the magnitudes of the individual scalar corrections are smaller than a specified threshold.
(*b*) Use the program of part (*a*) to solve the following system of linear equations, accurate to the fifth decimal place, using both methods:

$$\begin{aligned} 0.2x_1 + 0.20x_2 + 12.0x_3 + \quad x_4 &= 1.0 \\ 0.3x_1 + 15.0x_2 + \quad x_3 + 0.5x_4 &= 9.0 \\ 0.1x_1 + 0.40x_2 + \ 0.2x_3 + \ 19x_4 &= 5.0 \\ 10x_1 + 0.50x_2 + \ 0.2x_3 + \quad x_4 &= 0.0 \end{aligned} \tag{3.7.10}$$

Discuss differences in the rates of convergence.

Successive Over-relaxation

We return to Jacobi's algorithm expressed by equation (3.7.1) and restate it in the residual-correction form

$$\boldsymbol{x}^{(k+1)} = \boldsymbol{x}^{(k)} + \boldsymbol{r}^{(k)} \tag{3.7.11}$$

where

$$\boldsymbol{r}^{(k)} = (\boldsymbol{P} - \boldsymbol{I})\,\boldsymbol{x}^{(k)} + \boldsymbol{c} \tag{3.7.12}$$

is the residual or correction. If all components of $\boldsymbol{r}^{(k)}$ have the same sign through the iterations, the convergence will be accelerated by overcorrecting; whereas if the signs fluctuate, the convergence will be accelerated by undercorrecting.

These considerations motivate modifying the corrections, by introducing the relaxation parameter ω, and setting

$$\boldsymbol{x}^{(k+1)} = \boldsymbol{x}^{(k)} + \omega\boldsymbol{r}^{(k)} \tag{3.7.13}$$

The associated algorithm is

$$\boldsymbol{x}^{(k+1)} = (1-\omega)x_i^{(k)} + \frac{\omega}{A_{i,i}}\left(b_i - \sum_{\substack{j=1 \\ j\neq i}}^{N} A_{i,j}x_j^{(k)}\right) \tag{3.7.14}$$

Updating the components of $\boldsymbol{x}^{(k)}$ as soon as their new values are available, in the spirit of the Gauss–Seidel method, yields the successive over-relaxation method, SOR. In practice, the method is implemented on the basis of equation (3.7.14), where the new values replace the old ones, as soon as they are available.

Effective projection matrix and optimal value of ω

The effective algorithm of the SOR method is

$$\boxed{\boldsymbol{x}^{(k+1)} = (1-\omega)x_i^{(k)} + \frac{\omega}{A_{i,i}}\left(b_i - \sum_{j=1}^{i-1} A_{i,j}x_j^{(k+1)} - \sum_{j=i+1}^{N} A_{i,j}x_j^{(k)}\right)} \tag{3.7.15}$$

Switching to vector notation and rearranging, we find

$$(\boldsymbol{D} + \omega \boldsymbol{L})\boldsymbol{x}^{(k+1)} = \left[(1 - \omega)\boldsymbol{D} - \omega \boldsymbol{U}\right]\boldsymbol{x}^{(k)} + \omega \boldsymbol{b} \tag{3.7.16}$$

which shows that the projection matrix is given by

$$\boldsymbol{P} = (\boldsymbol{D} + \omega \boldsymbol{L})^{-1}\left((1 - \omega)\boldsymbol{D} - \omega \boldsymbol{U}\right) \tag{3.7.17}$$

Ideally, ω should take the value that minimizes the spectral radius of $\boldsymbol{P}$ but, unfortunately, this is generally unknown and must be found by numerical experimentation. Methods of computing the optimal value during the iterations are discussed by Hageman and Young (1981, Chapter 9).

It is certainly useful to have an idea about the range of variation of ω. It can be shown that a necessary condition for the successive substitutions to converge is that $0 < \omega < 2$. To see this, we set the product of the eigenvalues of $\boldsymbol{P}$ equal to its determinant and recall that the determinant of a triangular matrix is equal to the product of the diagonal elements, to obtain

$$\prod_{i=1}^{N} \lambda_1 = \text{Det}(\boldsymbol{P}) = \text{Det}\left((\boldsymbol{D} + \omega \boldsymbol{L})^{-1}\right)\text{Det}\left((1 - \omega)\boldsymbol{D} - \omega \boldsymbol{U}\right) = (1 - \omega)^N \tag{3.7.18}$$

If $|1 - \omega|$ were larger than unity, then the norm of at least one of the eigenvalues and thus the spectral radius of $\boldsymbol{P}$ would have to be larger than unity, which is certainly unacceptable.

S-SOR

The symmetric SOR method, S-SOR, is implemented as the symmetric Gauss–Seidel method, with the objective of treating the first and the last unknown equatably and without prejudice.

PROBLEM

3.7.4 *Understanding the* SOR *method.*

Consider the system of two linear equations:

$$\begin{bmatrix} 1 & -a \\ -a & 1 \end{bmatrix} \boldsymbol{x} = \boldsymbol{b} \tag{3.7.19}$$

where a is a real number, $\boldsymbol{b}$ is a unknown two-dimensional vector, and $\boldsymbol{x}$ is an unknown two-dimensional vector.
(*a*) Show that the solution coincides with the fixed point of the mapping

$$\begin{bmatrix} 1 & 0 \\ -\omega a & 1 \end{bmatrix} \boldsymbol{x}^{(k+1)} = \begin{bmatrix} 1 - \omega & \omega a \\ 0 & 1 - \omega \end{bmatrix} \boldsymbol{x}^{(k)} + \omega \boldsymbol{b} \tag{3.7.20}$$

where ω is an arbitrary parameter, and derive the effective projection matrix in an explicit form.
(*b*) Set $\omega = 1$, and find the range of a for which the fixed-point iterations converge.
(*c*) Set $a = 0.5$, and compute the spectral radius of the projection matrix for $\omega = 0.8, 0.9, 1.0, 1.1, 1.2, 1.3$. For which of these values of ω is the spectral radius minimum?
(*d*) Perform the iterations for $a = 0.50, \boldsymbol{b} = (1, 2)^T, \boldsymbol{x}^{(0)} = (0, 0)^T$, for all values of ω given in part (*c*), until the solution has been computed accurate to the fifth significant figure. Discuss the number of necessary iterations for the various values of ω.

3.8 *Acceleration of Iterative Methods by Deflation*

The rate of convergence of the iterative methods discussed in Section 3.7 is linear: Each time we carry out one iteration, the magnitude of the error vector is reduced by a constant factor that is roughly equal to the spectral radius of the projection matrix, as indicated by equation (3.6.19). Clearly, the rate of convergence will be improved by removing the eigenvalue with the maximum norm or, more precisely, by shifting this eigenvalue to zero while leaving the other eigenvalues unchanged. The Wielandt deflation method discussed in Section 2.10 is suited ideally for this purpose.

Single Deflation with the Eigenvector of the Transpose of the Iteration Matrix

Let λ_1 be a single eigenvalue of the projection matrix $\boldsymbol{P}$ with the maximum norm and $\boldsymbol{v}^{(1)}$ be the corresponding eigenvector of $\boldsymbol{P}^T$, and let us introduce the vector $\boldsymbol{q}$ subject to the restriction that $\boldsymbol{q}^T\boldsymbol{v}^{(1)} = 1$. Wielandt's theorem, discussed in Section 2.10, guarantees that the matrix

$$Q_{i,j} = P_{i,j} - \lambda_1 q_i v_j^{(1)} \tag{3.8.1}$$

shares the eigenvalues of $\boldsymbol{P}$, with the exception of the eigenvalue λ_1 that has been replaced by zero. Clearly, the spectral radius of $\boldsymbol{Q}$ is less than that of $\boldsymbol{P}$.

To make this result useful, we consider replacing the equation

$$\boldsymbol{x} = \boldsymbol{P}\boldsymbol{x} + \boldsymbol{c} \tag{3.8.2}$$

which provides us with a basis for the iterative procedure as discussed in the preceding two sections, with the modified equation

$$\boldsymbol{x} = \boldsymbol{Q}\boldsymbol{x} + \boldsymbol{h} \tag{3.8.3}$$

where $\boldsymbol{h}$ is a constant vector yet to be determined, and carry out the associated fixed-point iterations. Specifically, we select a starting vector $\boldsymbol{x}^{(0)}$ and compute the sequence

$$\boldsymbol{x}^{(k+1)} = \boldsymbol{Q}\boldsymbol{x}^{(k)} + \boldsymbol{h} \tag{3.8.4}$$

in the hope that it will lead us to a fixed point.

Replacing the master equation $\boldsymbol{x} = \boldsymbol{P}\boldsymbol{x} + \boldsymbol{c}$ with the master equation $\boldsymbol{x} = \boldsymbol{Q}\boldsymbol{x} + \boldsymbol{h}$ is legitimate only if the fixed points of the mappings $\boldsymbol{g}_1(\boldsymbol{x}) = \boldsymbol{P}\boldsymbol{x} + \boldsymbol{c}$ and $\boldsymbol{g}_2(\boldsymbol{x}) = \boldsymbol{Q}\boldsymbol{x} + \boldsymbol{h}$ coincide or else are related in a simple way. If $\boldsymbol{X}$ is the fixed point of $\boldsymbol{g}_1(\boldsymbol{x})$, that is, $\boldsymbol{X} = \boldsymbol{P}\boldsymbol{X} + \boldsymbol{c}$, then

$$\begin{aligned} \boldsymbol{g}_2(\boldsymbol{X}) &= \boldsymbol{Q}\boldsymbol{X} + \boldsymbol{h} = \boldsymbol{P}\boldsymbol{X} - \lambda_1 \boldsymbol{q}\left(\boldsymbol{X}^T\boldsymbol{v}^{(1)}\right) + \boldsymbol{h} \\ &= \boldsymbol{X} - \boldsymbol{c} - \lambda_1 \boldsymbol{q}\left(\boldsymbol{X}^T\boldsymbol{v}^{(1)}\right) + \boldsymbol{h} \end{aligned} \tag{3.8.5}$$

which shows that $\boldsymbol{X}$ will also be the fixed point of $\boldsymbol{Q}\boldsymbol{X} + \boldsymbol{h}$ provided that we set

$$\boldsymbol{h} = \boldsymbol{c} + \lambda_1 \boldsymbol{q}\left(\boldsymbol{X}^T\boldsymbol{v}^{(1)}\right) \tag{3.8.6}$$

But since $\boldsymbol{X}$ is *a priori* unknown, the usefulness of this approach is not clear. Not to panic: Taking the inner product of both sides of the equation $\boldsymbol{X} = \boldsymbol{P}\boldsymbol{X} + \boldsymbol{c}$ with $\boldsymbol{v}^{(1)}$, and rearranging, we find

$$X^T v^{(1)} = \frac{1}{1-\lambda_1} c^T v^{(1)} \tag{3.8.7}$$

Combining the preceding two equations, we find

$$h = c + q \frac{\lambda_1}{1-\lambda_1} \left(c^T v^{(1)}\right) = \left(I + \frac{\lambda_1}{1-\lambda_1} q v^{(1)^T}\right) c \tag{3.8.8}$$

A difficulty arises when $\lambda_1 = 1$. The denominator on the right-hand side of equation (3.8.8) vanishes, and h is not defined. Rewriting equation (3.8.2) as $(I - P)x = c$, however, shows that at least one eigenvalue, and thus the determinant of the coefficient matrix $I - P$, is equal to zero. In that case, according to our discussion in Section 2.11, an infinite number of solutions will exist only if $c^T v^{(1)} = 0$, which is precisely the condition for the last term of equations (3.8.8) to be equal to zero.

As an example, we consider the mapping $x^{(k+1)} = Px^{(k)} + c$, where c is a two-dimensional vector and P is the 2×2 projection matrix:

$$P = \begin{bmatrix} 1 & 9 \\ \frac{1}{4} & 1 \end{bmatrix} \tag{3.8.9}$$

The fixed point provides us with the solution of the linear system $(I - P)\, x = c$, where I is the identity matrix. The eigenvalues of P are readily found to be $\frac{5}{2}$ and $-\frac{1}{2}$. We want to remove the largest eigenvalue $\lambda_1 = \frac{5}{2}$, so that the spectral radius of the projection matrix becomes less than unity. Following the preceding discussion, we introduce the corresponding eigenvector of P^T, $v^{(1)} = (1, 6)^T$. Selecting $q^T = (1, 0)$, we find the new projection matrix

$$Q = \begin{bmatrix} 1 - \frac{5}{2} & 9 - \frac{30}{2} \\ \frac{1}{4} & 1 \end{bmatrix} = \begin{bmatrix} -\frac{3}{2} & -6 \\ \frac{1}{4} & 1 \end{bmatrix} \tag{3.8.10}$$

whose eigenvalues are equal to 0 and $-\frac{1}{2}$. The new shift h, computed from equation (3.8.8), is equal to $c - \frac{5}{3}(c_1 + 6c_2, 0)^T$.

Sequential deflation

The single deflation just described may be repeated in a sequential manner to remove further eigenvalues of the projection matrix. A second deflation is effected by replacing equation (3.8.3) with a new equation that is constructed with reference to the maximum eigenvalue of Q and the associated eigenvector of its transpose. In principle, the process may continue until all eigenvalues have been shifted to zero, whereupon the iteration matrix is filled up with zeros.

PROBLEM

3.8.1 ***Single deflation with the eigenvector of the transpose of the projection matrix.***

Consider the mapping

$$x^{(k+1)} = \begin{bmatrix} 0.50 & 0.10 \\ 0.08 & 0.30 \end{bmatrix} x^{(k)} + \begin{bmatrix} 0.2 \\ 0.1 \end{bmatrix} \tag{3.8.11}$$

Perform a single deflation to reduce the spectral radius of the projection matrix, and compute the common fixed point of the original and deflated mappings. Compare and discuss the rates of convergence.

Single Deflation with the Eigenvector of the Projection Matrix

Let λ_1 be a single eigenvalue of the projection matrix $\boldsymbol{P}$ with the maximum norm, corresponding to the eigenvector $\boldsymbol{u}^{(1)}$, and introduce the vector $\boldsymbol{w}$, with the only restriction that $\boldsymbol{w}^T\boldsymbol{u}^{(1)} = 1$. Wielandt's theorem, discussed in Section 2.10, guarantees that the matrix

$$R_{i,j} = P_{i,j} - \lambda_1 u_i^{(1)} w_j \tag{3.8.12}$$

shares the eigenvalues of $\boldsymbol{P}$, with the exception of the eigenvalue λ_1 that has been replaced by zero.

Working as in the previous subsection, we introduce the equation

$$\boldsymbol{x} = \boldsymbol{R}\boldsymbol{x} + \boldsymbol{c} \tag{3.8.13}$$

and inquire whether the fixed point $\boldsymbol{X}_R$ of the associated mapping $\boldsymbol{g}_3(\boldsymbol{x}) = \boldsymbol{R}\boldsymbol{x} + \boldsymbol{c}$ can be related in a simple way to the fixed point $\boldsymbol{X}$ of the original mapping $\boldsymbol{g}_1(\boldsymbol{x}) = \boldsymbol{P}\boldsymbol{x} + \boldsymbol{c}$. We proceed by computing

$$\boldsymbol{g}_3(\boldsymbol{X}) = \boldsymbol{R}\boldsymbol{X} + \boldsymbol{c} = \boldsymbol{P}\boldsymbol{X} - \lambda_1\boldsymbol{u}^{(1)}(\boldsymbol{w}^T\boldsymbol{X}) + \boldsymbol{c} = \boldsymbol{X} - \lambda_1\boldsymbol{u}^{(1)}(\boldsymbol{w}^T\boldsymbol{X}) \tag{3.8.14}$$

and note that $\boldsymbol{R}\boldsymbol{u}^{(1)} = \boldsymbol{0}$ to obtain

$$\boldsymbol{X}_R = \boldsymbol{X} - \lambda_1\boldsymbol{u}^{(1)}(\boldsymbol{w}^T\boldsymbol{X}) \tag{3.8.15}$$

What we really want to do is to solve equation (3.8.15) for $\boldsymbol{X}$ in terms of $\boldsymbol{X}_R$. For this purpose, we take the inner product of both sides with $\boldsymbol{w}$ and rearrange to find $\boldsymbol{w}^T\boldsymbol{X}_R = (1 - \lambda_1)(\boldsymbol{w}^T\boldsymbol{X})$. Substituting the result back into equation (3.8.15) yeilds

$$\boldsymbol{X} = \boldsymbol{X}_R + \frac{\lambda_1}{1-\lambda_1}\boldsymbol{u}^{(1)}(\boldsymbol{w}^T\boldsymbol{X}_R) \tag{3.8.16}$$

The numerical procedure involves two stages:

1. Carry out the iterations $\boldsymbol{x}^{(k+1)} = \boldsymbol{R}\boldsymbol{x}^{(k)} + \boldsymbol{c}$ until the fixed point $\boldsymbol{X}_R$ has been obtained.
2. Recover the fixed point $\boldsymbol{X}$ using equation (3.8.16).

When $\lambda_1 = 1$, the right-hand side of equation (3.8.16) is not defined. Tracing back our steps, we find that an infinite number of solutions for $\boldsymbol{X}$ will exist only when $\boldsymbol{w}^T\boldsymbol{X}_R = 0$. Clearly, if $\boldsymbol{X}$ is a fixed point, then $\boldsymbol{X} + \alpha\boldsymbol{u}^{(1)}$ is also a fixed point for any value of the scalar α.

As an example, the eigenvector of the projection matrix $\boldsymbol{P}$ shown in equation (3.8.9), corresponding to the large eigenvalue $\lambda_1 = \frac{5}{2}$, is $\boldsymbol{u}^{(1)} = (6, 1)^T$. Selecting $\boldsymbol{w} = (0, 1)^T$, we find the new projection matrix

$$\boldsymbol{R} = \begin{bmatrix} 1 & 9 - \frac{30}{2} \\ \frac{1}{4} & 1 - \frac{5}{2} \end{bmatrix} = \begin{bmatrix} 1 & -6 \\ \frac{1}{4} & -\frac{3}{2} \end{bmatrix} \tag{3.8.17}$$

whose eigenvalues are equal to 0 and $-\frac{1}{2}$. The computation of fixed points will be requested in Problem 3.8.2.

The single deflation just described may be repeated in a sequential manner to remove further eigenvalues as described in the preceding subsection.

PROBLEM

3.8.2 ***Single deflation with the eigenvector of the projection matrix.***

(a) Consider the numerical example discussed in the text, and compute the fixed points $\boldsymbol{X}_R$ and $\boldsymbol{X}$ for $\boldsymbol{c} = (1, 1)^T$.
(b) Consider the mapping given in equation (3.8.11). Perform a single deflation with an eigenvector of the projection matrix to reduce its spectral radius, and compute the fixed point by iterating the deflated mapping.

Simultaneous Double Deflation

The preceding discussion can be generalized in two ways: We can remove eigenvalues of multiplicity higher than one; and we can implement simultaneous double deflation of two eigenvalues.

Let λ_1 be an eigenvalue of $\boldsymbol{P}$ with multiplicity m corresponding to the generalized eigenvectors $\boldsymbol{u}^{(i)}, i = 1, \ldots, m$, and λ_2 be a different eigenvalue of $\boldsymbol{P}^T$ with multiplicity l corresponding to the generalized eigenvectors $\boldsymbol{v}^{(i)}, i = 1, \ldots, l$, and introduce the set of vectors $\boldsymbol{w}^{(i)}, i = 1, \ldots, m$, that is orthonormal to the set $\boldsymbol{u}^{(i)}$, that is,

$$\boldsymbol{u}^{(i)T}\boldsymbol{w}^{(j)} = \delta_{i,j} \tag{3.8.18}$$

and the set of vectors $\boldsymbol{q}^{(i)}, i = 1, \ldots, l$, that is orthonormal to the set $\boldsymbol{v}^{(i)}$, that is,

$$\boldsymbol{v}^{(i)T}\boldsymbol{q}^{(j)} = \delta_{i,j} \tag{3.8.19}$$

The fixed point $\boldsymbol{X}$ of the mapping $\boldsymbol{g}_1(\boldsymbol{x}) = \boldsymbol{P}\boldsymbol{x} + \boldsymbol{c}$ is found in two steps:

1. Compute, by one-point iterations, the fixed point $\boldsymbol{X}_R$ of the mapping $\boldsymbol{x} = \boldsymbol{R}\boldsymbol{x} + \boldsymbol{h}$, where

$$R_{i,j} = P_{i,j} - \lambda_1 \sum_{n=1}^{m} u_i^{(n)} w_j^{(n)} - \lambda_2 \sum_{n=1}^{l} q_i^{(n)} v_j^{(n)} \tag{3.8.20}$$

and

$$\boldsymbol{h} = \boldsymbol{c} + \frac{\lambda_2}{1-\lambda_2} \sum_{n=1}^{l} \boldsymbol{q}^{(n)} \left(\boldsymbol{c}^T \boldsymbol{v}^{(n)}\right) \tag{3.8.21}$$

2. Recover the desired fixed point $\boldsymbol{X}$ from the equation

$$\boldsymbol{X} = \boldsymbol{X}_R + \frac{\lambda_1}{1-\lambda_1} \sum_{n=1}^{m} \boldsymbol{u}^{(n)} \left(\boldsymbol{X}_R^T \boldsymbol{w}^{(n)}\right) \tag{3.8.22}$$

This method may also be applied when $\lambda_1 = \lambda_2$, in which case λ_1 is a multiple eigenvalue with algebraic multiplicity $m + l$. In that case, the two sums in equation (3.8.20) must contain linearly independent eigenvectors.

PROBLEM

3.8.3 ***Simultaneous double deflation.***

Perform the double deflation of the projection matrix shown in equations (3.8.11), and compute the associated fixed point.

3.9 Minimization and Conjugate-gradients Methods

A different class of iterative methods arises by recasting the problem of solving a system of linear equations into the problem of finding a vector that minimizes a scalar function of its components, called a functional. These methods are somewhat more general than the ones discussed in the preceding sections, in the sense that they can be extended to systems of nonlinear equations, as will be discussed in Chapter 4. Their main advantage is that, provided certain mild conditions are met, convergence is guaranteed.

The Minimization Problem

When the real matrix $\boldsymbol{A}$ is *symmetric and positive-definite,* computing the solution $\boldsymbol{X}$ of the linear system $\boldsymbol{Ax} = \boldsymbol{b}$ is equivalent to finding the vector $\boldsymbol{x}$ that minimizes the quadratic form or functional

$$f(\boldsymbol{x}) = \tfrac{1}{2}\boldsymbol{x}^T\boldsymbol{A}\boldsymbol{x} - \boldsymbol{b}^T\boldsymbol{x} \tag{3.9.1}$$

The equivalence is evident for a single equation, $ax = b$, where $f(x) = x(\frac{1}{2}ax - b)$; the minimum value of $f(x)$ is attained at the point $x = X$, where $\partial f/\partial x = aX - b = 0$. To demonstrate the equivalence for a higher number of equations, we compute the gradient of $f(\boldsymbol{x})$

$$\nabla f = \frac{\partial f}{\partial x_i} = \tfrac{1}{2}A_{k,i}x_k + \tfrac{1}{2}A_{i,k}x_k - b_i = A_{i,k}x_k - b_i \tag{3.9.2}$$

and note that it vanishes when $\boldsymbol{x} = \boldsymbol{X}$, where $\boldsymbol{AX} = \boldsymbol{b}$. Thus $\boldsymbol{X}$ is a minimum, a maximum, or a saddle point of $f(\boldsymbol{x})$. To see which one it is, we expand $f(\boldsymbol{x})$ in a Taylor series about the point $\boldsymbol{X}$, observe that all but the constant and quadratic terms vanish, and obtain the *exact* relation

$$f(\boldsymbol{x}) = -\tfrac{1}{2}\boldsymbol{b}^T\boldsymbol{X} + \tfrac{1}{2}\widehat{\boldsymbol{x}}^T\boldsymbol{A}\widehat{\boldsymbol{x}} \tag{3.9.3}$$

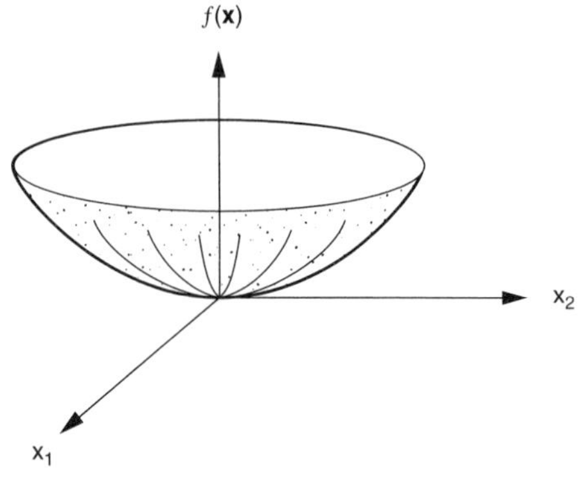

where $\widehat{\boldsymbol{x}} = \boldsymbol{x} - \boldsymbol{X}$. Because the matrix $\boldsymbol{A}$ is positive definite, the second term on the right-hand side is positive, and this guarantees that $f(\boldsymbol{x})$ reaches a minimum value when $\boldsymbol{x} = \boldsymbol{X}$.

In summary, we have reduced the problem of computing the solution of the equation $\boldsymbol{Ax} = \boldsymbol{b}$ to the problem of finding the minimum of a quadratic form. For a certain class of matrices $\boldsymbol{A}$, the minimization problem can be solved with a much lesser number of operations than that required by the direct or iterative methods discussed in the preceding sections.

What if the matrix $\boldsymbol{A}$ is not symmetric and positive-definite? In that case, we must make a conversion prior to seeking minimization.

One option is to premultiply both sides of the equation $\boldsymbol{Ax} = \boldsymbol{b}$ by the transpose of $\boldsymbol{A}$, and thereby obtain the preconditioned system $\boldsymbol{Cx} = \boldsymbol{d}$, where the new coefficient matrix $\boldsymbol{C} = \boldsymbol{A}^T\boldsymbol{A}$ is symmetric and positive-definite, and $\boldsymbol{d} = \boldsymbol{A}^T\boldsymbol{b}$ is a new right-hand side. Other options are reviewed by Broyden (1996) and discussed at the end of this section.

Steepest-descent Search

The steepest-descent method is the oldest and most straightforward minimization method. The algorithm proceeds by making an initial guess $\boldsymbol{x}^{(0)}$ and then improving it by proceeding in the direction where the functional $f(\boldsymbol{x})$ appears to change most rapidly. The steepest-descent direction is aligned with the vector $\boldsymbol{r} = -\nabla f$ evaluated at $\boldsymbol{x}^{(0)}$, which will be denoted by $\boldsymbol{r}^{(0)}$. In the case of one equation $ax = b$, we move toward infinity if $\partial f/\partial x$ is negative, and toward minus infinity when $\partial f/\partial x$ is positive. But by how long a distance shall we move?

To answer this question, we note that, as we travel in the direction of $\boldsymbol{r}^{(0)}$, the vector $\boldsymbol{x}$ is described by $\boldsymbol{x} = \boldsymbol{x}^{(0)} + \alpha\boldsymbol{r}^{(0)}$, where α is a scalar parameter, and the value of the functional is given by

$$f\left(\boldsymbol{x}^{(0)}, \alpha\right) = \tfrac{1}{2}\left(\boldsymbol{x}^{(0)} + \alpha\boldsymbol{r}^{(0)}\right)^T \boldsymbol{A}\left(\boldsymbol{x}^{(0)} + \alpha\boldsymbol{r}^{(0)}\right) - \boldsymbol{b}^T\left(\boldsymbol{x}^{(0)} + \alpha\boldsymbol{r}^{(0)}\right) \tag{3.9.4}$$

This is an algebraic quadratic function of α. We want to stop traveling when $f(\boldsymbol{x})$ has reached a minimum; that is, at the point where $\partial f/\partial\alpha = 0$. Taking the partial derivative of the right-hand side of equation (3.9.4) with respect to α, and setting it equal to zero, provides us with the desired value of α and produces the steepest-descent algorithm:

Choose the starting point $\boldsymbol{x}^{(0)}$

Do $k = 1, 2, \ldots,$ *Satisfaction*

$$\boldsymbol{r}^{(k-1)} = \boldsymbol{b} - \boldsymbol{Ax}^{(k-1)}$$

$$\alpha_k = \frac{\boldsymbol{r}^{(k-1)^T}\boldsymbol{r}^{(k-1)}}{\boldsymbol{r}^{(k-1)^T}\boldsymbol{Ar}^{(k-1)}} \tag{3.9.5}$$

$$\boldsymbol{x}^{(k)} = \boldsymbol{x}^{(k-1)} + \alpha_k\boldsymbol{r}^{(k-1)}$$

END DO

It is instructive to test the performance of the method for a single equation $ax = b$. We readily find $r^{(0)} = b - ax^{(0)}$ and $\alpha_1 = 1/a$, which produces the exact solution $x^{(1)} = b/a$ in a single step!

Unfortunately, this amazing performance does not hold for two or a higher number of equations. While the method is guaranteed to convergence, the rate of convergence is often prohibitively slow. When, in particular, the graph of the functional has narrow valleys, successive approximations bounce off opposite sides slowly approaching the bottom. A typical path is illustrated in the diagram for $N = 2$; the closed curves are isoscalar contours of the quadratic form.

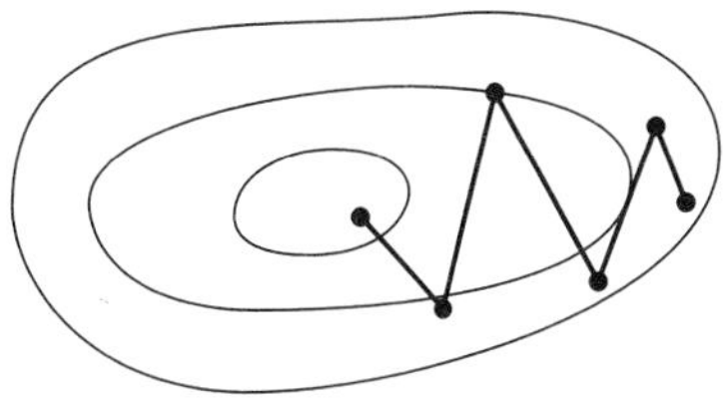

PROBLEM

3.9.1 ***Steepest descent for two equations.***

Discuss the performance of the steepest-descent method for two equations, and support your arguments by numerical examples.

Directional Search

We are off to searching for better and faster minimization methods. Our general strategy is to select a set of search directions expressed by a collection of vectors

$$\boldsymbol{p}^{(1)}, \boldsymbol{p}^{(2)}, \boldsymbol{p}^{(3)}, \ldots \tag{3.9.6}$$

and then advance the solution according to the algorithm

$$\boldsymbol{x}^{(k)} = \boldsymbol{x}^{(k-1)} + \alpha_k \boldsymbol{p}^{(k)} \tag{3.9.7}$$

for $k = 1, 2, \ldots$, where α_k are appropriate coefficients. Working as in the preceding subsection with the objective of minimizing the quadratic form at every step, we find that the optimal value of α_k is given by

$$\alpha_k = \frac{\boldsymbol{p}^{(k)^T} \boldsymbol{r}^{(k-1)}}{\boldsymbol{p}^{(k)^T} \boldsymbol{A} \boldsymbol{p}^{(k)}} \tag{3.9.8}$$

where $\boldsymbol{r}^{(k)} = \boldsymbol{b} - \boldsymbol{A}\boldsymbol{x}^{(k)}$ is the residual.

There are many degrees of freedom concerning the selection of the search directions. In the method of steepest descent, we stipulated that $\boldsymbol{p}^{(k)} = \boldsymbol{r}^{(k-1)}$. But, as we saw, when the graph of the functional has narrow valleys, the search directions tend to align, and successive points bounce off opposite sides, oscillating back and forth.

To this end, the issue of optimal search directions must be carefully considered. Fortunately, we have available several good choices.

Method of Conjugate Gradients

Without loss of generality, we begin the search from the origin; that is, we set $\boldsymbol{x}^{(0)} = \boldsymbol{0}$. Our goal is to compute the search directions so that the exact solution $\boldsymbol{X}$ is found exactly after N steps; we shall see shortly that this is an ambitious yet achievable goal.

The repetitive application of equation (3.9.7) yields

$$\boldsymbol{X} = \alpha_1 \boldsymbol{p}^{(1)} + \ldots + \alpha_N \boldsymbol{p}^{(N)} \tag{3.9.9}$$

The distinguishing feature of the method of conjugate gradients is that the N search directions are required to be $\boldsymbol{A}$-conjugate with each other, which means that

$$\boldsymbol{p}^{(i)^T} \boldsymbol{A} \boldsymbol{p}^{(j)} = 0 \qquad \text{for } i \neq j \tag{3.9.10}$$

We shall see shortly this constraint arises in a natural manner during the minimization of the quadratic form with respect to the search directions at every stage.

It is important to note that requiring the conjugation constraint (3.9.10) does *not* determine the search directions; it only imposes restrictions among them.

Equivalence to minimization

Assuming that equation (3.9.10) is fulfilled, we take the inner product of both sides of equation (3.9.9) with the vector $\boldsymbol{A}\boldsymbol{p}^{(k)}$ and find

$$\alpha_k = \frac{\boldsymbol{p}^{(k)^T}\boldsymbol{A}\boldsymbol{X}}{\boldsymbol{p}^{(k)^T}\boldsymbol{A}\boldsymbol{p}^{(k)}} = \frac{\boldsymbol{p}^{(k)^T}\boldsymbol{b}}{\boldsymbol{p}^{(k)^T}\boldsymbol{A}\boldsymbol{p}^{(k)}} \tag{3.9.11}$$

which appears to somewhat different from the formula shown in equation (3.9.8); we shall show, however, that, in fact, the two formulas are equivalent.

As a preliminary, we introduce the residual at the kth step

$$\boldsymbol{r}^{(k)} \equiv \boldsymbol{b} - \boldsymbol{A}\boldsymbol{x}^{(k)} = \boldsymbol{b} - \sum_{i=1}^{k} \alpha_i \boldsymbol{A}\boldsymbol{p}^{(i)} \tag{3.9.12}$$

Taking the inner product of both sides with the vector $\boldsymbol{p}^{(i)}$, where $i \le k$, and using equations (3.9.10) and (3.9.11), we obtain

$$\boldsymbol{p}^{(i)^T}\boldsymbol{r}^{(k)} = 0 \qquad \text{for } i \le k \tag{3.9.13}$$

which shows that *any vector that is expressible as a linear combination of* $\boldsymbol{p}^{(i)}$ *with* $i \le k$ *is orthogonal to the residual* $\boldsymbol{r}^{(k)}$. We proceed by expressing equation (3.9.12) in the form

$$\boldsymbol{r}^{(k)} = \boldsymbol{r}^{(k-1)} - \alpha_k \boldsymbol{A}\boldsymbol{p}^{(k)} \tag{3.9.14}$$

take the inner product of both sides with $\boldsymbol{p}^{(k)}$, note that $\boldsymbol{p}^{(k)^T}\boldsymbol{r}^{(k)} = 0$, and thus find

$$\boldsymbol{p}^{(k)^T}\boldsymbol{r}^{(k-1)} = \alpha_k \boldsymbol{p}^{(k)^T}\boldsymbol{A}\boldsymbol{p}^{(k)} \tag{3.9.15}$$

which provides us with an alternative expression for α_k. But this expression is identical to that given previously in equation (3.9.8). Thus *the projection method of computing* α_k *based on equation (3.9.11) is equivalent to the minimization method based on equation (3.9.8).*

Proof of termination after N steps

How can we be sure that the exact solution will be found after N steps? All we have to do is ensure that of all vectors that can be written as linear combinations of $\boldsymbol{p}^{(i)}$, with $i \le k$, the vector

$$\boldsymbol{x}^{(k)} = \alpha_1 \boldsymbol{p}^{(1)} + \ldots + \alpha_k \boldsymbol{p}^{(k)} \tag{3.9.16}$$

with the coefficients computed from equation (3.9.8) or (3.9.11) minimizes the quadratic form (3.9.1). To prove this assertion, we write $\boldsymbol{x}^{(k)} + \boldsymbol{d}$, where $\boldsymbol{d}$ is expressible as a linear combination of the $\boldsymbol{p}^{(i)}$, with $i \le k$, and compute

$$f\left(\boldsymbol{x}^{(k)}+\boldsymbol{d}\right)=\tfrac{1}{2}\boldsymbol{x}^{(k)^T}\boldsymbol{A}\boldsymbol{x}^{(k)}-\boldsymbol{b}^T\boldsymbol{x}^{(k)}+\tfrac{1}{2}\boldsymbol{d}^T\boldsymbol{A}\boldsymbol{d}+\boldsymbol{d}^T\left(\boldsymbol{A}\boldsymbol{x}^{(k)}-\boldsymbol{b}\right) \qquad (3.9.17)$$

The last term vanishes because of equation (3.9.13), and the penultimate term is nonnegative because $\boldsymbol{A}$ is positive-definite, suggesting that f reaches its minimum value when $\boldsymbol{d}=\boldsymbol{0}$.

Choice of search directions

The issue of how to compute the search directions is still pending and leaves us with several possibilities. In the original method of conjugate gradients developed by Hestenes and Stiefel (1952), the direction $\boldsymbol{p}^{(k)}$ is aligned as much as possible with $\boldsymbol{r}^{(k-1)}$ subject to the $\boldsymbol{A}$-conjugation constraint. The objective is to make the magnitude of the numerator in equation (3.9.8) as large as possible, and thereby move toward the minimum of the functional in the fastest possible way. In this case, it can be shown that the residual vectors $\boldsymbol{r}^{(k)}$ form an orthogonal set, that is,

$$\boldsymbol{r}^{(i)^T}\boldsymbol{r}^{(k)}=0 \qquad \text{for } i\le k \qquad (3.9.18)$$

If the matrix $\boldsymbol{A}$ is nonsingular, the directions are guaranteed to be linearly independent. The sequence of computations is shown in Algorithm 3.9.1 (e.g., Golub and van Loan 1989, p. 523).

ALGORITHM 3.9.1 A conjugate-gradients algorithm for solving the linear system $\boldsymbol{A}\boldsymbol{x}=\boldsymbol{b}$ with a *real symmetric* and *positive-definite* coefficient matrix $\boldsymbol{A}$.

$\boldsymbol{x}^{(0)}=\boldsymbol{0}$

Do $k=1,N$

  If $k=1$

    $\boldsymbol{p}^{(1)}=\boldsymbol{r}^{(0)}=\boldsymbol{b}$

  Else If $k>1$

    $\beta_k=\dfrac{\boldsymbol{r}^{(k-1)^T}\boldsymbol{r}^{(k-1)}}{\boldsymbol{r}^{(k-2)^T}\boldsymbol{r}^{(k-2)}}$

    $\boldsymbol{p}^{(k)}=\boldsymbol{r}^{(k-1)}+\beta_k\,\boldsymbol{p}^{(k-1)}$

  End If

  $\alpha_k=\dfrac{\boldsymbol{r}^{(k-1)^T}\boldsymbol{r}^{(k-1)}}{\boldsymbol{p}^{(k)^T}\boldsymbol{A}\boldsymbol{p}^{(k)}}$

  $\boldsymbol{x}^{(k)}=\boldsymbol{x}^{(k-1)}+\alpha_k\,\boldsymbol{p}^{(k)}$

  $\boldsymbol{r}^{(k)}=\boldsymbol{r}^{(k-1)}-\alpha_k\,\boldsymbol{A}\boldsymbol{p}^{(k)}$

End Do

Convergence and preconditioning

The conjugate-gradients method produces a sequence of vectors that approximate the solution of a linear system with increasing accuracy. But the reduction in error may be uneven through the iterations; some steps may improve the solution a little, and others a lot.

In practice, we want to make a number of steps that are substantially less than the size of the system, and yet obtain a reasonable approximation to the solution. If the large corrections are made at the beginning, we are fortunate; but if they are made at the end, we are unfortunate. The first scenario occurs when the condition number of the matrix is sufficiently small.

To reduce the condition number, we precondition the linear system $\boldsymbol{A}\boldsymbol{x} = \boldsymbol{b}$ before we solve it. This is done by multiplying both sides by the inverse of a preconditioning matrix $\boldsymbol{C}$, obtaining the system

$$\boldsymbol{A}^P \boldsymbol{y} = \boldsymbol{b}^P \tag{3.9.19}$$

where

$$\boldsymbol{A}^P = \boldsymbol{C}^{-1}\boldsymbol{A}\boldsymbol{C}^{-1}, \qquad \boldsymbol{y} = \boldsymbol{C}\boldsymbol{x}, \qquad \boldsymbol{b}^P = \boldsymbol{C}^{-1}\boldsymbol{b} \tag{3.9.20}$$

The algorithm involves solving the system (3.9.19) by the conjugate-gradients method and then recovering $\boldsymbol{x}$ by inverting the second of equations (3.9.20), setting $\boldsymbol{x} = \boldsymbol{C}^{-1}\boldsymbol{y}$. Note that the matrix $\boldsymbol{C}$ does not need to be computed explicitly.

Nonsymmetric systems

The method of conjugate gradients can be extended in several ways to handle nonsymmetric systems $\boldsymbol{A}\boldsymbol{x} = \boldsymbol{b}$. In one extension, the method just described is applied to the extended symmetric system

$$\begin{bmatrix} \boldsymbol{0} & \boldsymbol{A} \\ \boldsymbol{A}^T & \boldsymbol{0} \end{bmatrix} \begin{bmatrix} \boldsymbol{y} \\ \boldsymbol{x} \end{bmatrix} = \begin{bmatrix} \boldsymbol{b} \\ \boldsymbol{0} \end{bmatrix} \tag{3.9.21}$$

which consists of the two decoupled systems $\boldsymbol{A}\boldsymbol{x} = \boldsymbol{b}$ and $\boldsymbol{A}^T\boldsymbol{y} = \boldsymbol{0}$; the solution of the second system is irrelevant. The resulting bicongugate-gradients method is described in Algorithm 3.9.2.

Advanced methods

Advanced versions of the conjugate-gradients method that include sophisticated preconditioning are discussed by Khosla and Rubin (1981) and Golub and van Loan (1989, pp. 527–535). Other related methods have been developed by Eisenstat et al. (1983), Saad and Schultz (1986), Freund and Nachtigal (1991), and van den Vorst (1992). A comprehensive review has been presented by Broyden (1996). The theory of iterative methods is the subject of a recent monograph by Axelsson (1996).

ALGORITHM 3.9.2 A biconjugate-gradients algorithm for solving the general linear system $\boldsymbol{A}\boldsymbol{x} = \boldsymbol{b}$.

$\boldsymbol{x}^{(0)} = \boldsymbol{0}$

Do $k = 1, N$

  If $k = 1$

    $\boldsymbol{p}^{(1)} = \boldsymbol{r}^{(0)} = \boldsymbol{b}$

    $\overline{\boldsymbol{p}}^{(1)} = \overline{\boldsymbol{r}}^{(0)}$ (a specified vector)

  Else If $k > 1$

$$\beta_k = \frac{\bar{\boldsymbol{r}}^{(k-1)^T}\boldsymbol{r}^{(k-1)}}{\bar{\boldsymbol{r}}^{(k-2)^T}\boldsymbol{r}^{(k-2)}}$$

$$\boldsymbol{p}^{(k)} = \boldsymbol{r}^{(k-1)} + \beta_k\,\boldsymbol{p}^{(k-1)}$$

$$\bar{\boldsymbol{p}}^{(k)} = \bar{\boldsymbol{r}}^{(k-1)} + \beta_k\,\bar{\boldsymbol{p}}^{(k-1)}$$

END IF

$$\alpha_k = \frac{\bar{\boldsymbol{r}}^{(k-1)^T}\boldsymbol{r}^{(k-1)}}{\bar{\boldsymbol{p}}^{(k)^T}\boldsymbol{A}\boldsymbol{p}^{(k)}}$$

$$\boldsymbol{x}^{(k)} = \boldsymbol{x}^{(k-1)} + \alpha_k\,\boldsymbol{p}^{(k)}$$

$$\boldsymbol{r}^{(k)} = \boldsymbol{r}^{(k-1)} - \alpha_k\,\boldsymbol{A}\boldsymbol{p}^{(k)}$$

$$\bar{\boldsymbol{r}}^{(k)} = \bar{\boldsymbol{r}}^{(k-1)} - \alpha_k\,\boldsymbol{A}^T\bar{\boldsymbol{p}}^{(k)}$$

END DO

3.10 Overdetermined Systems

When a real system $\boldsymbol{Dx} = \boldsymbol{e}$ is overdetermined, which means that the number of equations is higher than the number of unknowns, we precondition it by multiplying both sides by $\boldsymbol{D}^T$, and thus obtain the associated *normal* system $\boldsymbol{D}^T\boldsymbol{Dx} = \boldsymbol{D}^T\boldsymbol{e}$ whose coefficient matrix is square, symmetric, and positive definite. The solution of the normal system will satisfy the original system in the least-squares sense. This and other ways of handling overdetermined systems are discussed by Atkinson (1989, pp. 633–645).

References

Atkinson, K. E., 1989, *An Introduction to Numerical Analysis.* Wiley.

AXELSSON, O., 1996, *Iterative Solution Methods.* Cambridge University Press.

BEER, F. P., and JOHNSTON, E. R., 1988, *Vector Mechanics for Engineers: Statics and Dynamics.* McGraw-Hill.

BOISVERT, R. F., 1991, Algorithms for special tridiagonal systems. *SIAM J. Sci. Statist. Comput.* **12**, 423–42.

BROYDEN, G. G., 1996, A new taxonomy of conjugate gradient methods. *Comput. Math. Applic.* **31**, 7–17.

EISENSTAT, S. C., ELMAN, H. C., and SCHULTZ, M. H., 1983, Variational iterative methods for nonsymmetric systems of linear equations. *SIAM J. Numer. Anal.* **20**, 345–57.

FREUND, R. W., and NACHTIGAL, N. M., 1991, QMR: a quasi-minimal residual method for non-Hermitian linear systems. *Numer. Math.* **60**, 315–39.

GEORGE, A., and LIU, J., 1981, *Computer Solution of Large Sparse Positive Definite Systems.* Prentice-Hall.

GOLUB, G. H., and VAN LOAN, C. F., 1989, *Matrix Computations.* The Johns Hopkins University Press.

HAGEMAN, L. A. and YOUNG, D. M., 1981, *Applied Iterative Methods.* Academic Press.

HESTENES, M., and STIEFEL, E. 1952, Methods of conjugate gradients for solving linear systems. *J. Res. Natl. Bur. Stand.* **49**, 409–36.

KHOSLA, P. K., and RUBIN, S. G., 1981, A conjugate gradient iterative method. *Comput. Fluids* **9**, 109–21.

NOBLE, B., and DANIEL, J. W., 1988, *Applied Linear Algebra.* Prentice-Hall.

PISSANETZKY, S., 1984, *Sparse Matrix Technology.* Academic Press.

POZRIKIDIS, C., 1997, *Introduction to Theoretical and Computational Fluid Dynamics.* Oxford University Press.

PRESS, W. H, FLANNERY, B. P., TEUKOLSKY, S. A., and VETTERLING, W. T., 1992, *Numerical Recipes*, 2nd ed. Cambridge University Press.

SAAD, Y., and SCHULTZ, M. H., 1986, GMRES: a generalized minimal residual algorithm for solving nonsymmetric linear systems. *SIAM J. Sci. Statist. Comput.* **7**, 856–69.

TEWARSON, R. P., 1973, *Sparse Matrices.* Academic Press.

VAN DEN VORST, H. A., 1992, BI-CGSTAB: a fast and smoothly converging variant of BI-CG for the solution of non-symmetric linear systems. *SIAM J. Sci. Statist. Comput.* **13**, 631–44.

VARGA, R. S., 1962, *Matrix Iterative Analysis.* Prentice-Hall.

WILKINSON, J. H., 1965, *The Algebraic Eigenvalue Problem.* Oxford University Press.

Chapter 4

Nonlinear Algebraic Equations

4.1 Mathematical and Physical Context

Consider a collection of N real or complex variables $x_1, x_2, \ldots, x_N$. We want to find one or more sets of values of these variables, denoted by $x_1 = X_1, x_2 = X_2, \ldots, x_N = X_N$, that satisfy a given system of N nonlinear algebraic equations

$$f_i(x_1, x_2, \ldots, x_N) = 0 \tag{4.1.1}$$

where $i = 1, \ldots, N$.

Note that the number of unknowns is equal to the number of equations, which means that the problem is neither under- nor overdetermined. Overdetermined systems where the number of equations is greater than the number of unknowns do not generally have a solution; and underdetermined systems generally have continuous families of solutions.

One Equation

When $N = 1$, we obtain a single equation $f_1(x_1) = 0$ that is usually written in a simpler form, without the indices, as

$$f(x) = 0 \tag{4.1.2}$$

An example is the quadratic equation corresponding to $f(x) = ax^2 + bx + c$, where a, b, and c are three constants; in this simple case, the solution can be found analytically in closed form and is displayed in equation (1.1.2). More generally, $f(x)$ can be a polynomial or a transcendental function; the latter is a function that cannot be expressed as the sum, the difference, the product, the ratio of two polynomials, or the root of one polynomial.

Two Equations

When $N = 2$, we obtain a system of two equations whose general form is

$$f_1(x_1, x_2) = 0, \qquad f_2(x_1, x_2) = 0 \tag{4.1.3}$$

One example is the system

$$\begin{aligned} f_1(x_1, x_2) &= x_1 \exp(x_1 + x_2) - 2 = 0 \\ f_2(x_1, x_2) &= x_1 x_2 - 0.10 \exp(-x_2) = 0 \end{aligned} \tag{4.1.4}$$

Vector Form

To simplify the notation, we introduce the vector of independent variables

$$\boldsymbol{x} = (x_1, x_2, \ldots, x_N)^T \tag{4.1.5}$$

and the vector function

$$\boldsymbol{f} = (f_1, f_2, \ldots, f_N)^T \tag{4.1.6}$$

and rewrite the system (4.1.1) in the compact vector form

$$\boldsymbol{f}(\boldsymbol{x}) = \boldsymbol{0} \tag{4.1.7}$$

A solution of this vector equation can be viewed as a point in the N-dimensional real or complex space, where the coordinates of this point are represented by the vector

$$\boldsymbol{X} = (X_1, X_2, \ldots, X_N) \tag{4.1.8}$$

Alternatively, we say that the point $\boldsymbol{X}$ is a *zero* of the function $\boldsymbol{f}(\boldsymbol{x})$, or a *root* of the vectorial equation (4.1.7). In practice, the terms *zero* and *root* are used interchangeably and without discrimination.

Nonlinear Mapping

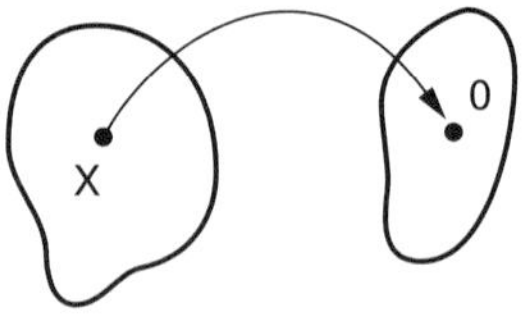

The left-hand side of equation (4.1.7) defines a nonlinear mapping that associates a given input vector $\boldsymbol{x}$ with another image vector $\boldsymbol{f}(\boldsymbol{x})$. The problem of solving equation (4.1.7) may then be restated as follows: Find one or more vector points, denoted by $\boldsymbol{X}$, that are mapped to the null vector point $\boldsymbol{0}$.

Elementary properties of linear mappings were discussed in Sections 2.8 and 2.11. Nonlinear mappings are much more difficult to analyze. In practice, their properties are often deduced by numerical experimentation as will be discussed in Section 4.3.

PROBLEM

4.1.1 ***Range of a mapping.***

Find the range of the mapping of (*a*) the linear function $f(x) = ax + b$, where a and b are real coefficients and x is a real variable spanning the whole real axis; (*b*) the quadratic function $f(x) = ax^2 + bx + c$, where a, b, and c are real coefficients and x is a real variable spanning the whole real axis; and (*c*) the two functions

$f_1(x_1, x_2)$ and $f_2(x_1, x_2)$ involved in the system (4.1.4), where x_1 and x_2 are real variables taking values over the whole real plane.

Equivalence to Minimization

For reasons that will soon become apparent, we introduce a continuous scalar function of N independent variables

$$H(x_1, x_2, \ldots, x_N) \tag{4.1.9}$$

concisely denoted as $H(\boldsymbol{x})$, called a *functional*, under the stipulation that the equation $\partial H/\partial x_i = f_i$ implies the equation $f_i = 0$ and vice versa. In vector notation, we require that

$$\nabla H = \boldsymbol{f} \tag{4.1.10}$$

where ∇ is the gradient operator. Rudimentary calculus shows that the set of values of the independent variables $x_1, x_2, \ldots, x_N$ for which H assumes a local minimum, or a maximum, or exhibits a saddle-point behavior, satisfies equation (4.1.7).

More specifically, we introduce a functional whose local *minima* are reached at the roots of the function $\boldsymbol{f}(\boldsymbol{x})$. There are many choices of such functionals; one example is the least-squares quadratic form

$$H(\boldsymbol{x}) = |\boldsymbol{f}(\boldsymbol{x})|^2 \equiv f_1^2 + \ldots + f_N^2 \tag{4.1.11}$$

The problem of solving a system of nonlinear algebraic equations is thus reduced to the problem of minimizing a functional without any constraints. Numerical methods for achieving *unconstrained function minimization* are discussed in monographs on optimization (e.g., Gill et al. 1981). A generalization of the method of conjugate gradients, discussed in Section 3.8, to nonlinear systems falls into this category.

The problem recasting just described is designated when the numerical solution of the nonlinear algebraic system is so sensitive, that it cannot be obtained by the methods discussed in the ensuing sections of this chapter, or the number of scalar equations is large.

Existence, Uniqueness, and Computation of a Solution

When the vector function $\boldsymbol{f}$ is a *linear* function of $\boldsymbol{x}$, we can express it in the familiar form

$$\boldsymbol{f}(\boldsymbol{x}) = \boldsymbol{A}\boldsymbol{x} - \boldsymbol{b} \tag{4.1.12}$$

where $\boldsymbol{A}$ is a constant coefficient matrix, and $\boldsymbol{b}$ is a constant vector. The corresponding linear system is

$$\boldsymbol{A}\boldsymbol{x} - \boldsymbol{b} = \boldsymbol{0} \qquad \text{or} \qquad \boldsymbol{A}\boldsymbol{x} = \boldsymbol{b} \tag{4.1.13}$$

The theoretical issue of existence and uniqueness of solution, and the practical issue of how to compute a solution were discussed in Chapter 3.

A general method of computing the solution of a *nonlinear* system, where the components of $\boldsymbol{f}$ are nonlinear functions of the components of $\boldsymbol{x}$, is not available, and we must resort to approximate iterative methods; these will be the object of our discussion in subsequent sections. None of these methods is capable of producing the exact solution of an *arbitrary* system in a *finite* number of steps, even in the theoretical absence of round-off error. That is, *direct general-purpose methods for solving nonlinear systems of equations are not available.*

Perhaps more disturbingly, a general theory pertaining to the existence of a solution of a nonlinear system is elusive, and we must carry out investigations for each given system on a case-by-case basis. The best general result we have is the *fundamental theorem of algebra,* a consequence of which is: *An Nth degree polynomial has exactly N roots residing in the complex plane* (e.g., Householder 1970, p. 2; Lanczos 1988, p. 5).

The Importance of Function Evaluations

In many problems, the vector function $\boldsymbol{f}(\boldsymbol{x})$ is given explicitly in closed form or in terms of transcendental functions as, for example, in equations (4.1.4), and its evaluation is a matter of a simple computation. But there are circumstances where the evaluation of $\boldsymbol{f}(\boldsymbol{x})$ requires a substantial amount of work.

For example, in the discussion of differential-equation boundary-value problems in Chapter 10, we shall see that the evaluation of $\boldsymbol{f}(\boldsymbol{x})$ may require solving a system of ordinary differential equations using numerical methods. In other applications, the evaluation of $\boldsymbol{f}(\boldsymbol{x})$ may require laboratory experiments or physical observations under specific conditions. When the required amount of work is substantial, the number of function evaluations, that is, the number of evaluations of the vector function $\boldsymbol{f}(\boldsymbol{x})$, becomes an important parameter of the numerical method.

The evaluation of the partial derivatives of the scalar components of $\boldsymbol{f}(\boldsymbol{x})$, required in certain numerical methods, may be even more demanding, to the extent that it must be avoided at all cost. These considerations have motivated the development of many of the more sophisticated methods to be discussed in the following sections.

Applications

Systems of nonlinear algebraic equations arise routinely in a broad range of theoretical and practical applications. Their frequent occurrence is a reflection of the predominantly nonlinear nature of the physical laws. Before proceeding to develop specific numerical methods, we pause to discuss several characteristic examples.

The Redlich–Kwong equation of state in thermodynamics

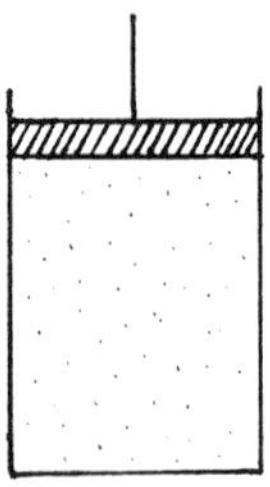

An equation of state relates the volume v occupied by 1 gmol of a gas to the instantaneous pressure p and to the Kelvin absolute temperature T of the gas. There are a number of equations of state with varying degrees of accuracy and sophistication; the simplest is the ideal gas law $pv = RT$. The coefficient R is a universal constant equal to 1.987 cal/(gmol · K) = 82.06 atm · cc/(gmol · K).

More accurate predictions are provided by the Redlich–Kwong equation of state

$$p = \frac{RT}{v-b} - \frac{a}{v(v+b)\sqrt{T}} \tag{4.1.14}$$

where a and b are two physical constants related to the *critical temperature* T_c and *critical pressure* P_c by the equations

$$a = 0.4278\frac{R^2 T_c^{2.5}}{P_c}, \qquad b = 0.0867\frac{RT_c}{P_c} \tag{4.1.15}$$

(e.g., Prausnitz 1969, p. 152). When $a = b = 0$, the Redlich–Kwong equation of state reduces to the ideal gas law. The critical temperature is defined as the minimum temperature above which a substance or a mixture cannot be in the liquid phase, no matter how high the pressure. The pressure that would suffice to liquify a fluid at the critical temperature is the critical pressure (e.g., Moore 1972, p. 21).

A chemical engineer was put in charge of storing a certain amount of hydrogen in a tank. To design the tank, she must have available a table that gives the molar volume in units of cc / gmol, at various pressures and temperatures. Evidently, this job requires solving the nonlinear algebraic equation (4.1.14) for v for a range of specified values of p. The reader will be asked to complete this project in one of the problems of the following sections.

Reversible chemical reactions

Benzene, C_2H_6, denoted by B, is dehydrogenated to yield *diphenyl*, $C_{12}H_{10}$, denoted by D, which then reacts further with benzene to yield *triphenyl* $C_{18}H_{14}$, denoted by T, and hydrogen, H_2, denoted by H, according the following pair of reversible reactions

$$2\ C_2H_6 \leftrightharpoons C_{12}H_{10} + H_2 \tag{4.1.16}$$

$$C_6H_6 + C_{12}H_{10} \rightleftharpoons C_{18}H_{14} + H_2 \tag{4.1.17}$$

The rates of these reactions are known to follow their stoichiometry, as will be shown below. At a total pressure of 1 atm and elevated temperatures so that all compounds are in the gas phase, experimental measurements have shown that the rate of decomposition of benzene due to reaction (4.1.16) is

$$r_1 = -\frac{dc_B}{dt} = 14.96 \times 10^6 \exp\left(-\frac{15,200}{T}\right)\left(p_B^2 - \frac{p_D p_H}{k_1}\right) \tag{4.1.18}$$

and the corresponding rate of decomposition due to reaction (4.1.17) is

$$r_2 = -\frac{dc_B}{dt} = 8.67 \times 10^6 \exp\left(-\frac{15,200}{T}\right)\left(p_B p_D - \frac{p_T p_H}{k_2}\right) \tag{4.1.19}$$

where:

- c_B is the concentration of benzene measured in lb-mol/ft^3,
- T is the absolute temperature,
- p_B, p_T, p_D, and p_H are the partial pressures of the subscribed species,
- k_1 and k_2 are two dimensionless equilibrium constants.

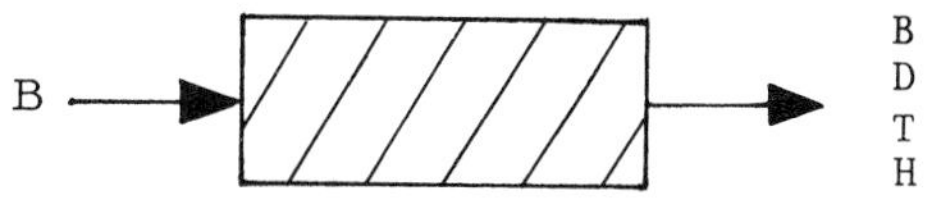

In a refinery plant, benzene is supplied into an isothermal tubular flow reactor under atmospheric pressure at 1400 °F. Because of the aforementioned reactions, a certain amount of benzene decomposes to yield diphenyl, triphenyl, and hydrogen. At the exit of the reactor, the mixture has reached chemical equilibrium, that is, $r_1 = r_2 = 0$. Engineer Justine Case was asked to compute the mole fraction of each substance after equilibrium has been established. The theoretical division of the company has provided the values $k_1 = 0.312$ and $k_2 = 0.480$.

To set up the mathematical formulation of the problem, we assume that a fraction of x_1 moles of benzene reacts according to reaction (4.1.16), and another fraction of x_2 reacts according to reaction (4.1.17). If a moles of benzene *enter* the reactor, then $a(1 - x_1 - x_2)$ moles of benzene, $a(\frac{1}{2}x_1 - x_2)$ moles of diphenyl, ax_2 moles of triphenyl, and $a(\frac{1}{2}x_1 + x_2)$ moles of hydrogen will *exit*. The total number of exiting moles is thus equal to a. Note that the reacting mixture does not expand due to reaction.

Setting the partial pressure of each compound at the exit equal to the corresponding mole fraction, we obtain

$$p_B = 1 - x_1 - x_2, \qquad p_D = \tfrac{1}{2}x_1 - x_2, \qquad p_T = x_2, \qquad p_H = \tfrac{1}{2}x_1 + x_2 \tag{4.1.20}$$

Substituting these expressions into the right-hand sides of equations (4.1.18) and (4.1.19), and setting the left-hand sides equal to zero, provides us with a system of two nonlinear equations for the unknown fractions x_1 and x_2:

$$\begin{aligned}(\tfrac{1}{2}x_1 - x_2)(\tfrac{1}{2}x_1 + x_2) &= 0.312(1 - x_1 - x_2)^2 \\ x_2(\tfrac{1}{2}x_1 + x_2) &= 0.480(1 - x_1 - x_2)(\tfrac{1}{2}x_1 - x_2)\end{aligned} \tag{4.1.21}$$

Computing the solution and substituting the results back into equations (4.1.11) completes the job.

Flash vaporization

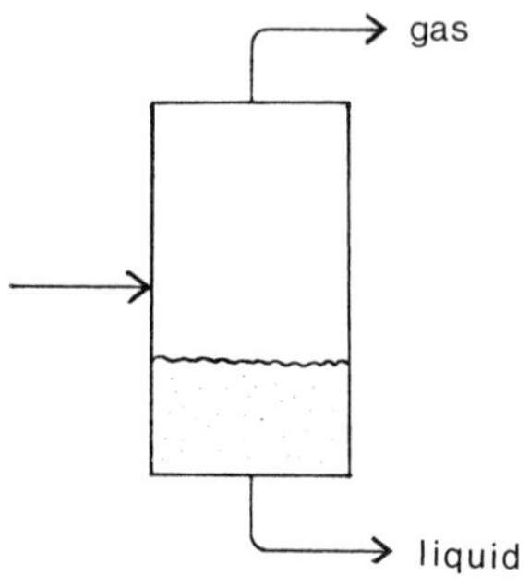

We end this section with yet another example from the multifaceted field of chemical engineering. A liquid containing 50 mol % of *benzene*, denoted by B, 25 mol % of *toluene*, denoted by T, and 25 mol % of *o-xylene*, denoted by O, is supplied into a flash evaporator operating at 1 atm or 760 mm Hg, and 100 °C (Treybal 1968, p. 304). We want to compute the fraction of the total moles that end up in the gas and in the liquid phase, as well as the compositions of the gas and the liquid phases.

It is reasonable to assume that the gas in the evaporator is in equilibrium with the liquid. Furthermore, for lack of a better alternative, we assume that the equilibrium molar fractions are given by *Henry's law*, relating the mole fraction of the ith species in the gas phase $f_{G,i}$ to the corresponding mole fraction in the liquid phase $f_{L,i}$ at equilibrium, by means of the equation

$$f_{G,i} = m_i f_{L,i} \tag{4.1.22}$$

where m_i are constants. Raoult's law says that

$$m_i = \frac{P_i}{P_{Tot}} \tag{4.1.23}$$

where P_i is the vapor pressure of the ith species, and P_{Tot} is the total pressure of the mixture. The vapor pressures of the benzene, toluene, and *o*-xylene at 100 °C are, respectively, equal to 1370, 550, and 200 mm Hg.

Proceeding with the mathematical formulation, we denote the feed flow rate by F, the liquid flow rate by L, and the gas flow rate by G, all measured in mol/min. The mole fractions of the feed stream are

$$f_{F,B} = 0.50, \qquad f_{F,T} = 0.25, \qquad f_{F,O} = 0.25 \tag{4.1.24}$$

Next, we introduce the unknown fractions of total moles that end up in the liquid and in the gas phase,

$$x_1 = L/F, \qquad x_2 = G/F \tag{4.1.25}$$

and the unknown mole fractions of the species in the gas and liquid phases,

$$\begin{aligned} x_3 &= f_{G,B} & x_4 &= f_{G,T} & x_5 &= f_{G,O} \\ x_6 &= f_{L,B} & x_7 &= f_{L,T} & x_8 &= f_{L,O} \end{aligned} \tag{4.1.26}$$

We have a total of eight unknowns and require an equal number of equations.

Two equations arise by requiring that the sum of the mole fractions of all species add up to unity:

$$x_3 + x_4 + x_5 = 1 \tag{4.1.27}$$

and

$$x_6 + x_7 + x_8 = 1 \tag{4.1.28}$$

Three additional equations arise by performing mass balances. An overall mole balance gives $L + G = F$. Dividing both sides by F, we obtain

$$x_1 + x_2 = 1 \tag{4.1.29}$$

A mole balance for benzene gives $F f_{F,B} = L f_{L,B} + G f_{G,B}$. Dividing both sides by F, we obtain

$$0.50 = x_1 x_6 + x_2 x_3 \tag{4.1.30}$$

Working similarly with the toluene, we find

$$0.25 = x_1 x_7 + x_2 x_4 \tag{4.1.31}$$

A mole balance for *o*-xylene is implicit in equations (4.1.27)–(4.1.31); that is, it can be derived from them and does not have to be written down explicitly.

Equations (4.1.22) and (4.1.23) provide us with three additional equations,

$$x_3 = \frac{1370}{760} x_6, \qquad x_4 = \frac{550}{760} x_7, \qquad x_5 = \frac{200}{760} x_8 \tag{4.1.32}$$

We have obtained the desired system of eight linear and nonlinear equations. The solution must be found using a numerical method.

PROBLEM

4.1.2 ***Reduction to two equations.***

Reduce the system of equations (4.1.27)–(4.1.32) to a system of two nonlinear equations in two unknowns.

4.2 *Bracketing Methods*

Bracketing describes the policy of chopping up a particular search interval into smaller pieces in a certain methodological manner, with the objective of confining a suspected root into a subinterval of an increasingly smaller size. A simple bracketing method involves graphing followed by visual inspection. This works fine for one or two equations. But have you ever succeeded in observing the hyperspace? Bracketing methods are used predominantly to solve one equation, and we confine our discussion to this case.

Bisection

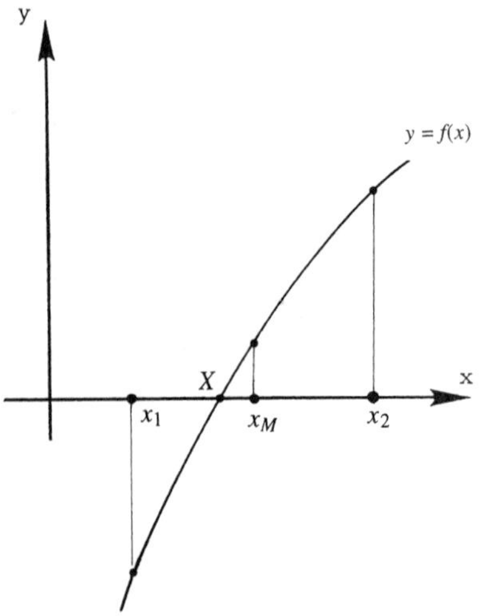

A simple but dependable bracketing method of computing a solution of a single equation $f(x) = 0$, called the *bisection* method or more descriptively the method of *halving-the-interval*, proceeds according to the following steps:

1. Find two values x_1 and x_2 for which $f(x_1)$ and $f(x_2)$ have different signs. If the function $f(x)$ does not have any singularities within the interval (x_1, x_2), then the desired root X, where $f(X) = 0$, is guaranteed to lie between x_1 and x_2.

2. Compute $x_M = \frac{1}{2}(x_1 + x_2)$ and evaluate $f(x_M)$. The value x_M is our current estimate of the root. In order to minimize the round-off error, it is better to compute $d = x_2 - x_1$ and then set $x_M = x_1 + \frac{1}{2}d$.

3. If $f(x_1)$ and $f(x_M)$ have different signs, replace x_2 by x_M, whereas if $f(x_2)$ and $f(x_M)$ have different signs, replace x_1 by x_M and return to step 2. If $f(x_M) = 0$ within a specified tolerance, stop the computation: x_M is the required root.

As the iterations continue, x_M is guaranteed to approximate a root with increasing accuracy, with one exception: If the function $f(x)$ happens to become infinite at some point within the interval (x_1, x_2), the algorithm may produce the location of the singularity instead of a root, even though $f(x_1)$ and $f(x_2)$ have different signs. This occurs, for example, for the function $f(x) = g(x)/(x - 1)$ with $x_1 = 0.50$ and $x_2 = 1.5$, provided that the function $g(x)$ does not vanish at $x = 1$.

In practice, however, physical intuition or mathematical hindsight usually allow us to detect the presence of a singularity and eliminate it from the equation at the outset. In the aforementioned example, we simply apply the bisection method to the nonsingular function $g(x) = (x - 1)f(x)$.

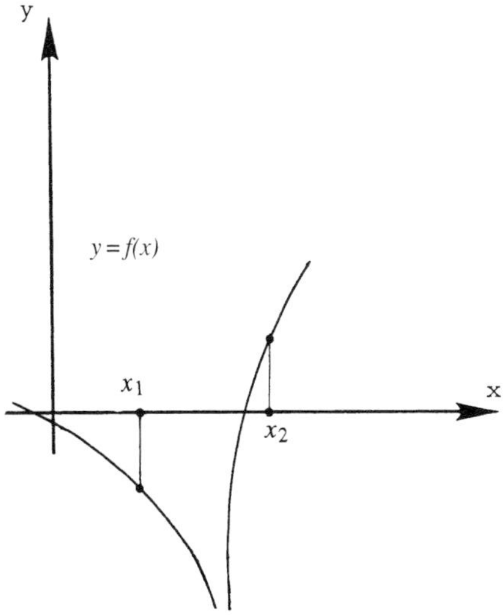

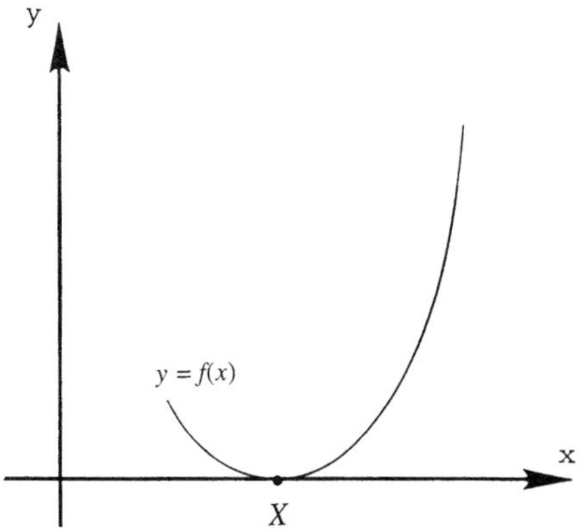

Another weak point of the bisection method is that it may not be able to produce *all* of the roots. For example, the method cannot capture a root X of multiplicity equal to two, for which $f(X) = 0$ and $f'(X) = 0$ but $f''(X) \neq 0$.

Convergence

Cursory inspection of the bisection algorithm shows that, after one iteration has been carried out, the interval of uncertainty has been reduced by a factor of 2. This corresponds to a gain of one binary digit per step, or to one decimal figure per 3.3 steps, independent of the nature of the function $f(x)$. After k iterations have been carried out, the magnitude of the error $e^{(k)} \equiv x^{(k)} - X$ has been confined in the range

$$|e^{(k)}| \leq \frac{1}{2^k}|b - a| \tag{4.2.1}$$

The inequality tends to become an equality as the root X approaches the ends of the search interval a or b. Although the bisection method will not fail to converge, the rate of convergence is linear and thus slow (see Section 4.3). Faster methods are desirable.

The bisection method is a first cousin of the method of *binary search* described in Section 1.4. The latter is used when values of the function $f(x)$ are known only at discrete values of the independent variable x. Every time you make a plane reservation, the computer of the travel agent is ordered to perform a logarithmic search.

PROBLEMS

4.2.1 *The bisection/regula-falsi method.*

In an advanced implementation of the bisection method, the point x_M is placed at the intersection of the line that passes through the points $(x_1, f(x_1))$, $(x_2, f(x_2))$, and the x axis. Develop the pertinent algorithm in the form of a pseudocode, and discuss the convergence of the method in comparison with that of the standard bisection method.

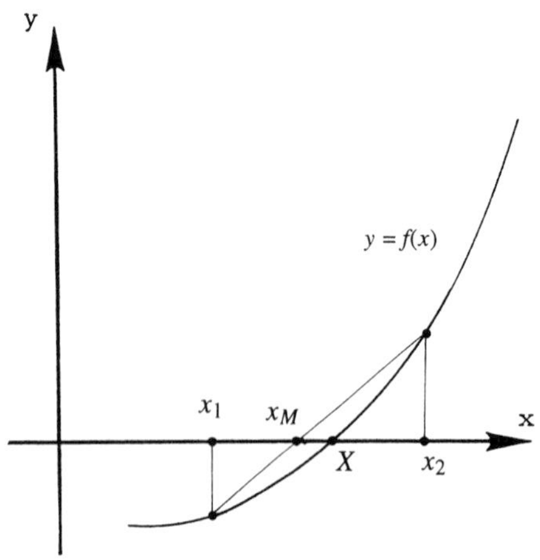

4.2.2 ***Bisection method for two equations.***

Devise a bisection algorithm for solving a system of two scalar equations for two scalar unknowns, and discuss its implementation.

4.3 One-point Iterations

In Chapter 3, we outlined the general principles of iterative methods for solving systems of *linear* equations, and developed specific procedures. We now proceed to extend these methods to the more general framework of nonlinear equations.

Fixed-point Iterations

Working in complete analogy with the linear case, we consider the vector equation $\boldsymbol{f}(\boldsymbol{x}) = \mathbf{0}$, recast it into the equivalent form

$$\boldsymbol{x} = \boldsymbol{g}(\boldsymbol{x}) \tag{4.3.1}$$

and then perform fixed-point iterations.

The simplest implementation of the method involves *one-point iterations*: We select an initial vector $\boldsymbol{x}^{(0)}$ and compute the sequence of vectors $\boldsymbol{x}^{(k)}$ using the *nonlinear* algebraic mapping,

$$\boldsymbol{x}^{(k+1)} = \boldsymbol{g}\left(\boldsymbol{x}^{(k)}\right) \tag{4.3.2}$$

also called an *iterated mapping*. It is possible that the sequence $\boldsymbol{x}^{(k)}$ will converge to a fixed point $\boldsymbol{X}$ of the *iteration function* $\boldsymbol{g}$, which, by definition, satisfies the equations

$$\boldsymbol{X} = \boldsymbol{g}(\boldsymbol{X}), \qquad \boldsymbol{f}(\boldsymbol{X}) = \mathbf{0} \tag{4.3.3}$$

The crux of the method is: *Instead of computing the solution of the equation* $\boldsymbol{f}(\boldsymbol{x}) = \mathbf{0}$, *compute the fixed point of the nonlinear mapping expressed by the iteration function* $\boldsymbol{g}(\boldsymbol{x})$, *and do so by the method of successive substitutions.*

Unfortunately, a general theorem concerning the existence or the number of fixed points of a general nonlinear mapping is not available, and we must carry our investigations on an individual basis for each given system of equations. Note, however, that if $g(x)$ is a continuous scalar function mapping a certain interval $[a, b]$ onto another interval that is contained in $[a, b]$, then the corresponding mapping is guaranteed to have at least one fixed point, as illustrated in the diagram (e.g., Traub 1964, p. 15).

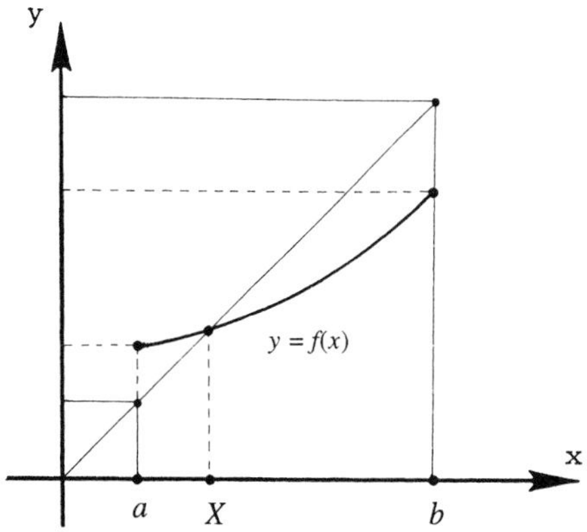

What Is the Optimal Iteration Function?

Given a function $\boldsymbol{f}(\boldsymbol{x})$, there are many ways of defining the mapping function $\boldsymbol{g}(\boldsymbol{x})$. The remainder of this and of the next section will be dedicated to devising a function that is best fitted for the job, within pragmatic constraints. We want to find an iteration function that makes the sequence of approximations always converge toward a fixed point for any reasonable initial guess, without requiring an excessive amount of work. Is this too much to ask? We shall see that the answer is *no*.

One Equation

To develop expertise and gain insights, we consider one equation with one unknown,

$$f(x) = 0 \tag{4.3.4}$$

and recast it into the equivalent form

$$x = g(x) \tag{4.3.5}$$

Following our stated policy, we select a certain initial guess $x^{(0)}$ and compute the sequence of numbers

$$x^{(k+1)} = g\left(x^{(k)}\right) \tag{4.3.6}$$

in the hope that it will converge toward a fixed point.

First case study: things look good

Consider the equation

$$f(x) = e^x - 3x^2 = 0 \tag{4.3.7}$$

and solve it for the x of the exponent to find

$$x = \ln 3 + \ln x^2 \tag{4.3.8}$$

The associated iteration function is $g(x) = \ln 3 + \ln x^2$. The one-point iterations are carried out using the formula

$$x^{(k+1)} = \ln 3 + \ln x^{(k)^2} \tag{4.3.9}$$

Beginning with the initial guess $x^{(0)} = 1$, we find

$$x^{(1)} = 1.099, \quad \ldots, \quad x^{(5)} = 2.527, \quad \ldots, \quad x^{(10)} = 3.651, \quad \ldots, \quad x^{(15)} = 3.730, \quad \ldots, \quad x^{(20)} = 3.733$$

The sequence clearly converges to a fixed point whose exact value is equal to 3.733, accurate to the third decimal place. Thus $X = 3.733$ is both the fixed point of the iterated mapping (4.3.9) and a solution of equation (4.3.7).

An alternative mapping function arises by solving equation (4.3.7) for the squared x. Keeping the plus sign in front of the square root, we find $g(x) = 3^{-\frac{1}{2}} \exp(\frac{1}{2}x)$. The associated iterated mapping is expressed by

$$x^{(k+1)} = \frac{1}{\sqrt{3}} \exp\left(\tfrac{1}{2}x^{(k)}\right) \tag{4.3.10}$$

Beginning with the initial guess $x^{(0)} = 3$, we find

$$x^{(1)} = 2.588, \quad \ldots, \quad x^{(5)} = 2.117, \quad \ldots, \quad x^{(10)} = 0.914, \quad \ldots, \quad x^{(15)} = 0.910$$

The sequence clearly converges to a fixed point whose exact value is equal to 0.910, accurate to the third decimal place. Thus $X = 0.910$ is a fixed point of the iterated map (4.3.10) and a second root of equation (4.3.7).

Second case study: things are not so rosy

Consider now the equation

$$f(x) = e^{-x} - 3x = 0 \tag{4.3.11}$$

and solve it for the x of the linear term to find

$$x = \tfrac{1}{3}e^{-x} \tag{4.3.12}$$

which suggests the mapping function $g(x) = \frac{1}{3}e^{-x}$. The one-point iterations are carried out using the formula

$$x^{(k+1)} = \tfrac{1}{3} \exp\left(-x^{(k)}\right) \tag{4.3.13}$$

Beginning with the initial guess $x^{(0)} = 1$, we find

$$x^{(1)} = 0.123, \quad \ldots, \quad x^{(5)} = 0.257, \quad \ldots, \quad x^{(10)} = 0.258$$

The sequence converges to a fixed point whose exact value is equal to 0.258, accurate to the third decimal place. Thus $X = 0.258$ is a solution of equation (4.3.11).

An alternative mapping function arises by splitting the x of the linear term in equation (4.3.11) into two terms, yielding $g(x) = e^{-x} - 2x$. The associated iterated mapping is computed from

$$x^{(k+1)} = \exp\left(-x^{(k)}\right) - 2x^{(k)} \tag{4.3.14}$$

Beginning with the initial guess $x^{(0)} = 0.258$, which is identical to the fixed point to the third significant figure, we find

$$x^{(1)} = 0.257, \quad \ldots, \quad x^{(5)} = 0.197, \quad \ldots, \quad x^{(10)} = 27.780$$

The sequence does not converge to a fixed point. This was a bad choice of an iteration function.

Graphical Interpretation of One-point Iterations for One Equation

To explain the failure of the method of one-point iterations in the second part of the second case study, we introduce the xy plane and draw the diagonal straight line $y = x$ and the graph of the function $y = g(x)$ as shown in Figure 4.3.1. The sequence of numbers $x^{(k)}$ can be traced by moving in a staircase-like manner: Vertically and upward until we hit the graph of $g(x)$, and then horizontally until we meet the diagonal.

Figure 4.3.1(*a*) illustrates a sequence of approximations that converges toward the fixed point X, whereas Figure 4.3.1(*b*) illustrates a sequence that diverges away from the fixed point X, even though the initial guess is very close to it. Cursory inspection reveals that in the first case $|(\partial g/\partial x)_{x=X}| > 1$, whereas in the second case $|(\partial g/\partial x)_{x=X}| < 1$. We shall show that these inequalities provide us with sufficient conditions for convergence.

Convergence of One-point Iterations for One Equation

The convergence of the method can be investigated in a more rigorous manner by subtracting from both sides of the equation $x^{(k+1)} = g(x^{(k)})$ the fixed point X. Introducing the error $e^{(k)} = x^{(k)} - X$, we obtain

$$e^{(k+1)} = g(x^{(k)}) - X \tag{4.3.15}$$

We proceed by assuming that the point $x^{(k)}$ is sufficiently close to X, expand $g(x^{(k)})$ in a Taylor series about the point X, and note that $g(X) = X$, to find

$$e^{(k+1)} = g'(X)e^{(k)} + \tfrac{1}{2}\, g''(X)e^{(k)^2} + \ldots \tag{4.3.16}$$

where a prime designates a derivative with respect to x. When the magnitude of $e^{(k)}$ is sufficiently small, the second and subsequent terms represented by the dots on the right-hand side may be neglected, yielding the linear behavior

$$e^{(k+1)} \approx g'(X)e^{(k)} \tag{4.3.17}$$

Taking the norm of both sides, we find that, every time we carry out one iteration, we effectively multiply the norm of the error by the factor $|g'(X)|$. Thus the iterations will converge provided that

$$|g'(X)| < 1 \tag{4.3.18}$$

which is another way of saying that the function $g(x)$ defines a *contraction* mapping. This inequality is the

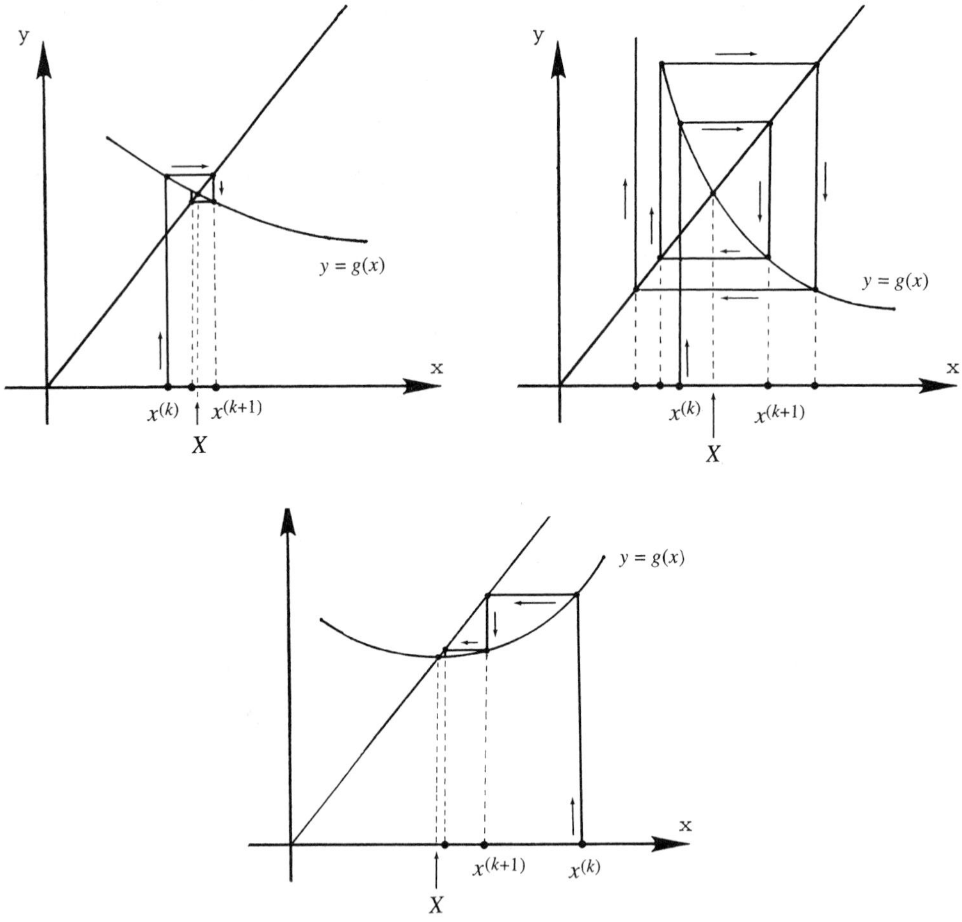

Figure 4.3.1 Graphical construction of the sequence of numbers $x^{(k)}$ computed with the method of one-point iterations. (*a*) A sequence converging towards the fixed point X with $|g'(X)| < 1$; (*b*) A sequence diverging away from the fixed point X with $|g'(X)| > 1$ (*c*) A converging sequence with $g'(X) = 0$; the iteration function is constructed using Newton's method.

sole condition for the convergence of the numerical method, provided that the initial guess is close enough to the root.

When criterion (4.3.17) is fulfilled, the rate of convergence will be at least linear, which means that, when the current approximation is sufficiently close to the root, carrying out one iteration will cause the error to be multiplied approximately by a constant factor. When the current approximation is *not* sufficiently close to the root, the preceding analysis ceases to be valid, and conditions for convergence cannot be established.

PROBLEM

4.3.1 *Explain a failure.*

Explain why the second sequence in the second case study of the preceding subsection failed to converge.

Achieving Quadratic Convergence

Equation (4.3.17) seems to indicate that, when $g'(X) = 0$, the error $e^{(1)}$ will vanish independently of the magnitude of $e^{(0)}$, that is, independently of the initial guess. In that case, it appears that we can find the solution in a single step. But equation (4.3.17) arose by neglecting the second- and higher-order terms on the right-hand side of equation (4.3.16). Keeping the leading-order contribution, we obtain

$$e^{(k+1)} \approx \tfrac{1}{2} g''(X) e^{(k)^2} \tag{4.3.19}$$

which shows that the rate of convergence is quadratic: When the current approximation is sufficiently close to the root, carrying out one iteration causes the error to be raised to the second power, and then multiplied by the constant factor $\frac{1}{2} g''(X)$.

The successive application of formula (4.3.19) leads to the expression

$$e^{(k+1)} \approx \frac{1}{\alpha} \left(\alpha \, e^{(0)} \right)^{2^{k+1}} \tag{4.3.20}$$

where

$$\alpha = \tfrac{1}{2} g''(X) \tag{4.3.21}$$

which shows that the magnitude of $e^{(k+1)}$ will tend to diminish as long as $|e^{(0)}| < 1/|\alpha|$. Thus the mapping will converge as long as the norm of $e^{(0)}$ is sufficiently small, that is, as long as the initial guess is sufficiently close to the solution. This restriction can be expressed in more precise terms, as discussed in monographs on nonlinear algebraic equations (e.g., Ostrowski 1966, Rabinowitz 1970).

For example, if $e^{(0)}$ is on the order of 10^{-1}, and if $\frac{1}{2} g''(X)$ is on the order of unity, then $e^{(1)}$ will be on the order of 10^{-2}, $e^{(2)}$ will be on the order of 10^{-4}, and $e^{(3)}$ will be on the order of 10^{-8}: Eighth decimal place accuracy is achieved in only three iterations! Figure 4.3.1(c) illustrates the way in which the corresponding sequence of approximations converges toward the fixed point X.

Design of the Iteration Function

We found that, in order for the sequence to converge, the iteration function $g(x)$ must be such that inequality (4.3.18) is fulfilled. But since the value of X is *a priori* unknown, testing against this criterion presents us with a vicious circle.

In practice, the testing can be done with an approximate value of the root that can be obtained using the methods discussed in Section 4.8. Furthermore, if the range of variation of the function $g'(x)$ is known to be confined within the interval $(-1, 1)$, then the iterations are guaranteed to converge provided that the initial guess is sufficiently close to the root.

Computing different roots may require using different iteration functions $g(x)$, as seen in the first case study discussed earlier in this section. Ideally, we would like to have an iteration function such that $g'(X) = 0$ for *all* roots, so that the sequences not only converge, but they do so in a quadratic fashion. Is it possible that an iteration function with such a remarkable property can be found? The answer is affirmative, thanks to Newton, as will be discussed in Section 4.4.

Aitken Extrapolation

Let us return to the general case of linear convergence. When the iterations converge, we can take advantage of our knowledge that the error behaves in a linear manner, as shown in equation (4.3.17), in order to expedite the approach toward the fixed point. Specifically, using the method of Aitken extrapolation discussed in Section 1.7, we obtain the improved approximate value

$$\hat{x}^{(k+1)} = x^{(k+1)} - \frac{\left(x^{(k+1)} - x^{(k)}\right)^2}{x^{(k+1)} - 2x^{(k)} + x^{(k-1)}} \tag{4.3.22}$$

whose computation requires values of a triplet of successive approximations. It is important to note that this formula does not, effectively, produce a second-order method: When the iterated map diverges, the extrapolation will make them diverge possibly at an even faster rate.

Finally, a word of caution: When using formula (4.3.22), hats and no-hats should not be mixed on the right-hand side.

PROBLEM

4.3.2 ***Successive substitutions for computing three roots of an equation.***

Find all three roots of the function $f(x) = e^x - 3x^2$ using the method of one-point iterations enhanced with the Aitken extrapolation according to Steffensen discussed in Section 1.7.

Systems of Equations

The handling of two or more equations is straightforward. Consider, for example, the system of two equations (4.1.4). A straightforward rearrangement of the various terms yields the equivalent forms

$$x_1 = 2\,\exp(-x_1 - x_2), \qquad x_2 = 0.10\,\exp(-x_2)\,/\,x_1 \tag{4.3.23}$$

which suggest the iteration functions

$$g_1(x_1, x_2) = 2\,\exp(-x_1 - x_2), \qquad g_2(x_1, x_2) = 0.10\,\exp(-x_2)\,/\,x_1 \tag{4.3.24}$$

The successive approximations are computed from the formulas

$$x_1^{(k+1)} = 2\,\exp(-x_1^{(k)} - x_2^{(k)}), \qquad x_2^{(k+1)} = 0.10\,\exp(-x_2^{(k)})\,/\,x_1^{(k)} \tag{4.3.25}$$

For example, beginning with $\boldsymbol{x}^{(0)} = (1, 0)^T$, we compute

$$\boldsymbol{x}^{(0)} = (0.736, 0.100)^T, \quad \ldots, \quad \boldsymbol{x}^{(5)} = (0.750, 0.103)^T, \quad \boldsymbol{x}^{(10)} = (0.841, 0.118)^T, \quad \ldots,$$
$$\boldsymbol{x}^{(15)} = (0.775, 0.107)^T, \quad \ldots, \quad \boldsymbol{x}^{(100)} = (0.802, 0.112)^T$$

The sequence converges to the fixed point $\boldsymbol{X} = (0.802, 0.112)^T$; this value is accurate to the third decimal place.

But we will not always be so lucky. Many other iteration functions $g_1(x_1, x_2)$ and $g_2(x_1, x_2)$ lead to sequences that do not converge to a fixed point.

PROBLEM

4.3.3 *Graphical interpretation.*

Discuss the graphical interpretation of the method of one-point iterations for a system of two equations, and present schematic illustrations analogous to those shown in Figure 4.3.1.

Convergence of One-point Iterations for More than One Equation

To derive a rigorous criterion for the convergence of a sequence constructed by the method of one-point iterations, we subtract from both sides of the recursion formula $\mathbf{x}^{(k+1)} = \mathbf{g}(\mathbf{x}^{(k)})$ the fixed point $\mathbf{X}$, and define the error vector $\mathbf{e}^{(k)} = \mathbf{x}^{(k)} - \mathbf{X}$ to obtain

$$\mathbf{e}^{(k+1)} = \mathbf{g}\left(\mathbf{x}^{(k)}\right) - \mathbf{X} \tag{4.3.26}$$

Next, we assume that the current approximation $\mathbf{x}^{(k)}$ is sufficiently close to $\mathbf{X}$, expand $\mathbf{g}(\mathbf{x}^{(k)})$ in a Taylor series about the point $\mathbf{X}$, and note that $\mathbf{g}(\mathbf{X}) = \mathbf{X}$ to find

$$e_i^{(k+1)} = J_{i,j}(\mathbf{X}) e_j^{(k)} + \tfrac{1}{2}\left(\frac{\partial J_{i,j}}{\partial x_l}\right)_{\mathbf{X}} e_j^{(k)} e_l^{(k)} + \cdots \tag{4.3.27}$$

where $\mathbf{J}$ is the *Jacobian matrix* of the vector function $\mathbf{g}$, defined as

$$J_{i,j} = \frac{\partial g_i}{\partial x_j} \tag{4.3.28}$$

In vector notation,

$$\mathbf{J} = (\nabla \mathbf{g})^T \tag{4.3.29}$$

where ∇ is the gradient operator, and $\nabla \mathbf{g}$ is the matrix of partial derivatives.

For example, for the iteration functions displayed in equations (4.3.24),

$$\begin{aligned} J_{1,1}(x_1, x_2) &= -2\,\exp(-x_2 - x_1), \quad & J_{1,2}(x_1, x_2) &= -2\,\exp(-x_2 - x_1), \\ J_{2,1}(x_1, x_2) &= -0.10\frac{\exp(-x_2)}{x_1^2}, \quad & J_{2,2}(x_1, x_2) &= -0.10\frac{\exp(-x_2)}{x_1} \end{aligned} \tag{4.3.30}$$

When the magnitude of $\mathbf{e}^{(k)}$ is sufficiently small, the second and subsequent terms represented by the dots on the right-hand side of equation (4.3.27) can be neglected. The resulting simplified formula shows that after an iteration has been carried out, the error vector has roughly been premultiplied by the constant matrix $\mathbf{J}(\mathbf{X})$. Using the results of Section 2.11 we find that the length of the error vector will tend to decrease, and thus the sequence will converge toward the fixed point, provided that the spectral radius of $\mathbf{J}(\mathbf{X})$ is less than unity. This observation provides us with a *necessary and sufficient condition* for convergence, always provided that the initial guess is sufficiently close to the root. In the case of one equation, we recover the scalar criterion (4.3.18).

Evaluating the spectral radius of $\mathbf{J}(\mathbf{X})$ requires computing all of its eigenvalues, which is a demanding task. In practice, the spectral radius of $\mathbf{J}(\mathbf{X})$ can be estimated on the basis of Gershgorin's first theorem, which provides us with a *sufficient* condition for convergence expressed by the inequality

$$\sum_{j=1}^{N} |J_{i,j}(X)| < 1 \qquad \text{or} \qquad \sum_{j=1}^{N} |J_{j,i}(X)| < 1 \tag{4.3.31}$$

for all values of i. An estimate of the root $\boldsymbol{X}$ for the evaluation of the right-hand side can be obtained using the methods described in Section 4.8. Because of the lurking uncertainties, however, the method of fixed-point iterations is hardly ever used for solving systems of equations.

Further analyses of the theory and discussion of the performance of iterative methods for systems of nonlinear equations can be found in dedicated monographs (e.g., Ortega and Rheinboldt 1970).

PROBLEMS

4.3.4 ***Convergence criteria for a linear system.***

Use the sufficient criterion for convergence expressed by inequalities (4.3.31) for the system of linear equations $\boldsymbol{A}\boldsymbol{x} = \boldsymbol{b}$, to show that a sufficient condition for Jacobi's method to succeed is that the matrix $\boldsymbol{A}$ be diagonally dominant.

4.3.5 ***Testing the convergence criteria.***

Evaluate the spectral radius of the matrix $\boldsymbol{J}(\boldsymbol{X})$ shown in equations (4.3.30) at the fixed point $\boldsymbol{X} = (0.802, 0.112)$, and discuss the significance of your results.

4.3.6 ***Reversible chemical reactions.***

Solve the system of nonlinear equations (4.1.21) using the method of fixed-point iterations. Report your failed and victorious efforts.

Gauss–Seidel Modification

The basic implementation of the method of one-point iterations can be modified in the spirit of Gauss–Seidel in order to expedite the rate of convergence. The added feature is that the new value of each scalar unknown replaces the old one as soon as it is available, and it is then used for updating the rest of the unknowns. When the method converges, this modification accelerates the rate of convergence toward the fixed point.

What Happens When the Iterations Do Not Converge?

In the present context of solving one or a system of nonlinear equations, a divergent sequence computed by the method of fixed-point iterations is, frankly, useless. But monitoring its behavior is interesting from a different, somewhat esoteric viewpoint: Sequences generated from a certain class of seemingly innocuous iteration functions are known to yield unexpected patterns that serve as precursors to the theory of dynamical systems, to be discussed in Chapter 9.

Consider, for example, the scalar iteration function

$$g(x) = \lambda\, x(1 - x) \tag{4.3.32}$$

where λ is a positive constant. The fixed points of $g(x)$ are also roots of the function $f(x) = x - g(x)$, which can readily be found to be $X = 0$ and $(\lambda - 1)/\lambda$. The one-point iteration sequence is computed from the *logistic mapping* equation

$$x^{(k+1)} = \lambda\, x^{(k)} \left(1 - x^{(k)}\right) \tag{4.3.33}$$

Graphs of $x^{(k)}$ versus k, starting from $x^{(0)} = 0.01$, for $\lambda = 2.0, 3.0, 3.5, 3.9, 4.0$, are shown in Figure 4.3.2(a). An apparently random pattern is established for the higher values of λ.

To understand the behavior of the sequences, it is helpful to perform a numerical experiment involving the following steps:

- Introduce the λx plane.
- Select a value of $x^{(0)}$ that lies between, but is not exactly equal to, 0 or 1.
- Choose a value for λ.
- Compute a few hundred terms of the sequence using equation (4.3.33).
- Skip the first hundred terms and graph the rest of them in the λx plane with dots.

The result is the graph shown in Figure 4.3.2(b). A cascade of bifurcations and, so to speak, chaotic behavior are apparent.

PROBLEMS

4.3.7 *Stability of the fixed points of the logistic mapping iteration function.*

(a) Establish the range of λ where the one-point iterations of the logistic mapping iteration function converge to the two fixed points. (b) Reproduce the graphs shown in Figure 4.3.2.

4.3.8 *A numerical experiment.*

If you enter a number into a calculator and keep pressing the cosine button on the radian mode, what number will eventually appear and why?

4.4 Newton's and Higher-order Methods for One Equation

Consider one nonlinear equation for a single unknown, $f(x) = 0$. In the preceding section, we posed the question: Does an iteration function that satisfies the property $g'(X) = 0$ exist? And if it does, is it easy to devise? The answer to both questions is affirmative, thanks to Newton.

Graphical Derivation

To derive the desired iteration function, we introduce the xy plane, draw the curve $y = f(x)$, and then approximate it with the line that is tangential to it at the point $x^{(k)}$. Elementary calculus says that the tangential line is described by the equation

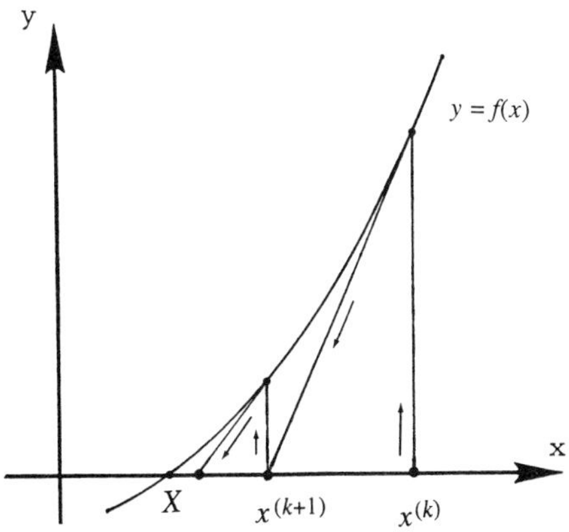

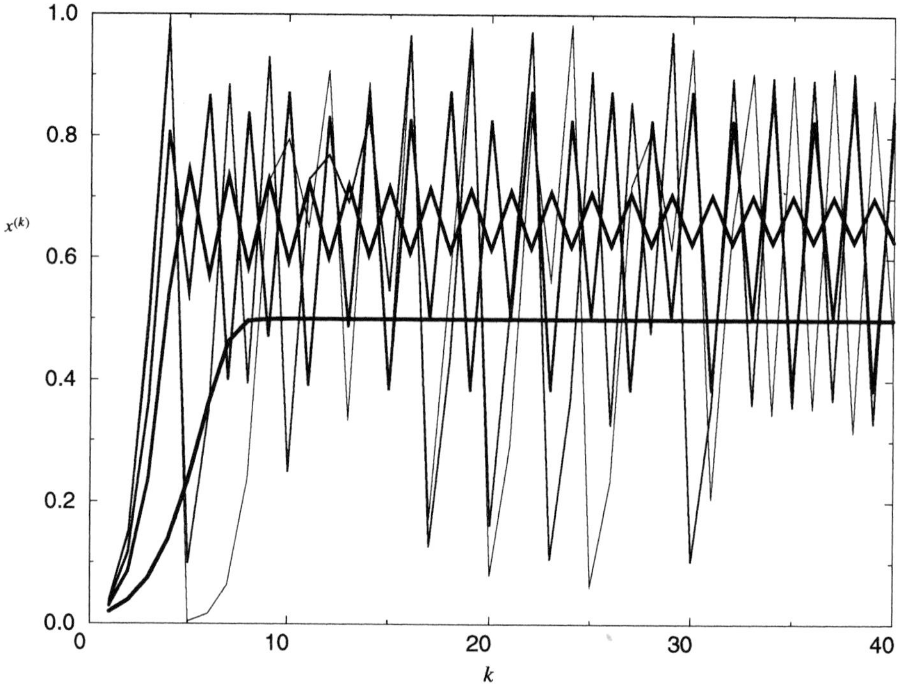

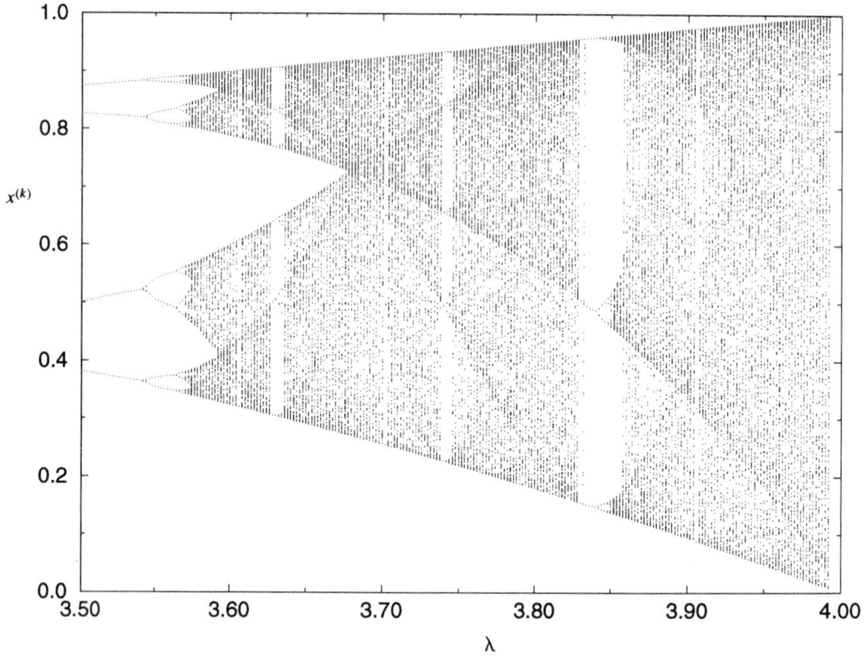

Figure 4.3.2 Behavior of sequences computed from the logistic mapping equation (4.3.33). (*a*) Thickness of line connecting $x^{(k)}$ decreases as λ is increased from 2.0 to 3.0, to 3.5, to 3.9, to 4.0. (*b*) Cascade of bifurcations in the λx-plane.

$$y = f\left(x^{(k)}\right) + f'\left(x^{(k)}\right)\left(x - x^{(k)}\right) \tag{4.4.1}$$

Finally, we identify the point $x^{(k+1)}$ with the intersection of the x axis and the tangential line, finding

$$x^{(k+1)} = x^{(k)} - \frac{f\left(x^{(k)}\right)}{f'\left(x^{(k)}\right)} \tag{4.4.2}$$

When the function $f(x)$ is linear, this formula provides us with the exact solution after one iteration.

Newton Iteration Function

Equation (4.4.2) produces a sequence of approximations based on the iteration function

$$g(x) = x - \frac{f(x)}{f'(x)} \tag{4.4.3}$$

For this iteration function to be acceptable, the equation $x = g(x)$ must imply the equation $f(x) = 0$, which can be shown to be true with half a line of algebra.

It remains to confirm that the rate of convergence is quadratic. Taking the first derivative of both sides of equation (4.4.3), we find

$$g'(x) = f(x)\frac{f''(x)}{f'^2(x)} \tag{4.4.4}$$

We note that, by definition, $f(X) = 0$, assume that $f'(X) \neq 0$, and find $g'(X) = 0$, which confirms the desired quadratic behavior. The problematic case $f'(X) = 0$ will be discussed later in this section.

Convergence

We proceed to compute the coefficient $g''(X)$ on the right-hand side of equation (4.3.19) in terms of the function f. Taking the first derivative of both sides of equation (4.4.4), we find

$$g''(x) = \frac{f''(x)}{f'(x)} + f(x)\frac{f'''(x)}{f'^2(x)} - 2f(x)\frac{f''^2(x)}{f'^3(x)} \tag{4.4.5}$$

Noting that $f(X) = 0$, and assuming that $f'(X) \neq 0$, we find $g''(X) = f''(X)\,/\,f'(X)$. Substituting this expression into the the right-hand side of equation (4.3.19), we find that, when $f'(X) = 0$, the error evolves according to the rule

$$e^{(k+1)} \approx \tfrac{1}{2}\frac{f''(X)}{f'(X)}e^{(k)2} \tag{4.4.6}$$

When $f''(X) = 0$ but $f'(X) \neq 0$, the fraction on the right-hand side of equation (4.4.6) vanishes, and the rate of convergence is cubic: After an iteration has been carried out, the error has been roughly raised to the third power, and then multiplied by a constant. This occurs, for example, for the function $f(x) = \sin\, x$ at each root (Problem 4.4.1).

PROBLEM

4.4.1 ***Cubic rate of convergence.***

(*a*) Verify, by numerical computation, that the rate of convergence of Newton's method for the function $f(x) = \sin x$ is cubic. (*b*) What is the rate of convergence toward the roots of the function $f(x) = \tan x$?

Implementation of Newton's Method

In the practical implementation of Newton's method, the derivative f' is often computed by numerical differentiation setting, for example, $f' = [f(x+\varepsilon) - f(x)]/\varepsilon$, where ε is a sufficiently small number. This approximation requires two function evaluations, instead of the evaluation of the function and its derivative at each iteration, and is thus preferable when the evaluation of the derivative is costly, sensitive, or otherwise undesirable.

Steffensen's method

Identifying, in particular, ε with $f(x)$, whose magnitude is small as long as the current approximation is close enough to the root, yields a variation of Newton's method proposed buy Steffensen. When $f'(X) \neq 0$, the rate of convergence is quadratic.

PROBLEMS

4.4.2 ***Computing the nth root of a number.***

Calculators love to compute functions using iterative methods. For instance, the square root of a real and positive number a is computed as the limit of the sequence

$$x^{(k+1)} = \frac{1}{2}\left(x^{(k)} + \frac{a}{x^{(k)}}\right) \tag{4.4.7}$$

which arises by applying Newton's method to the equation $f(x) = x^2 - a = 0$. Devise a similar algorithm, with quadratic convergence, for computing the *n*th root of a real and positive number.

4.4.3 ***Division of numbers or inverse of a matrix in terms of multiplication.***

Certain computers perform division in terms of multiplication. To compute the ratio $r = a\,/\,b$, where b is a positive number, we rewrite it in the form $r = ax$, where $x = 1/b$, and then identify x with the solution of the algebraic equation $f(x) = b - 1/x = 0$.
(*a*) Apply Newton's method to show that x arises as the limit of the sequence $x^{(k+1)} = x^{(k)}(2 - bx^{(k)})$. Show that a necessary and sufficient condition for convergence is that the initial guess $x^{(0)}$ lies within the interval $(0, 2/b)$.
(*b*) Let $\boldsymbol{B}$ be a square matrix. By extending Newton's method, show that $\boldsymbol{B}^{-1}$ can be computed as the limit of the sequence of matrices $\boldsymbol{X}^{(k)}$, where $\boldsymbol{X}^{(k+1)} = \boldsymbol{X}^{(k)}(2\boldsymbol{I} - \boldsymbol{B}\boldsymbol{X}^{(k)})$, for a reasonable initial guess $\boldsymbol{X}^{(0)}$.

Convergence to a Multiple Root

We return to examining the performance of Newton's method in the problematic case of a multiple root, where $f(X) = 0$ and $f'(X) = 0$. Under these circumstances, the evaluation of the right-hand sides of equations (4.4.4) and (4.4.5) requires careful consideration.

For a broad class of functions $f(x)$, when x is close to the root X, $f(x)$ behaves like $(x - X)^m$, where m is a real constant greater than unity (but see also Problem 4.4.4). Substituting this asymptotic form into equation (4.4.4), we find

$$g'(X) = \frac{m-1}{m} \tag{4.4.8}$$

which can be substituted into equation (4.3.17) to give

$$e^{(k+1)} = \frac{m-1}{m} e^{(k)} \tag{4.4.9}$$

Since $\alpha \equiv (m-1)/m < 1$, the magnitude of the error will keep decreasing during the iterations, but the rate of convergence will be linear instead of quadratic. For example, for a double root corresponding to $m = 2$, we find $\alpha = \frac{1}{2}$.

To understand why the rate of convergence is linear, we note that, when $f'(X) = 0$, as x tends to X, the graph of the function $f(x)$ tends to become parallel to the x axis, and this decelerates the approach toward their crossing.

Conversely, if the numerical results show a linear rate of approach toward the fixed point with a roughly constant convergence rate equal to $e^{(k+1)}/e^{(k)} = \alpha$, then the exponent m can be computed as $m = 1/(1-\alpha)$. Note, however, that this computation must be done *a posteriori*, after the root has been found with satisfactory accuracy.

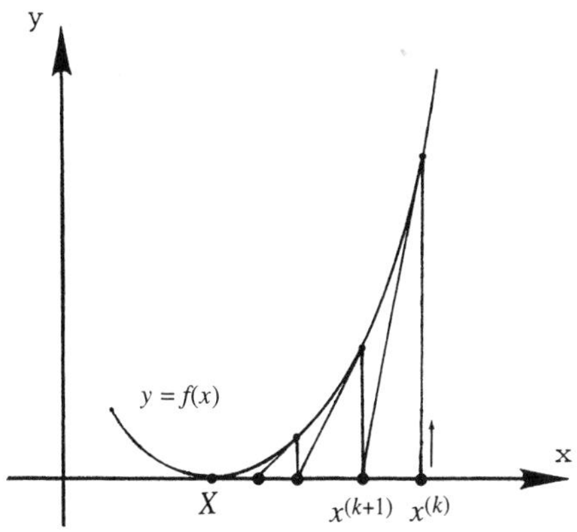

PROBLEM

4.4.4 *Convergence of Newton's method.*

The function $f(x) = x \ln x$ has a root at $x = 0$. What is the rate of convergence of Newton's method, and will the Newton sequence converge?

Summary of Results on Convergence

We found that Newton's method converges quadratically as long as the initial guess is sufficiently close to a single root X, where $f'(X) \neq 0$. Each time one iteration is carried out, the error is roughly squared and then multiplied by a constant factor.

If the root happens to be multiple, that is, $f'(X) = 0$, then the rate of convergence becomes linear: Each time one iteration is carried out, the error is roughly multiplied by a constant factor that is less than unity; for a double root the factor is equal to $\frac{1}{2}$. Thus the Newton sequence converges as long the initial guess is sufficiently close to the root.

Rectification of Quadratic Convergence for a Multiple Root

Newton's algorithm can be modified in a simple way to achieve quadratic convergence toward a multiple root with $m > 1$. According to Schröder, this is done simply by inserting the scalar coefficient m in front of the fraction on the right-hand side of equation (4.4.2), obtaining

$$x^{(k+1)} = x^{(k)} - m \frac{f(x^{(k)})}{f'(x^{(k)})} \tag{4.4.10}$$

A practical difficulty is that the value of m must somehow be known at the outset, or else estimated during the course of the computation. One way of circumventing this complication is to apply the standard Newton's method to the function $q = f/f'$, which is guaranteed to have a single root, but this has the disadvantage that the evaluation of f'', which is necessary for carrying out the iterations, may be precarious or undesirable (Traub 1964, Section 7.6). A compromise must be reached by exercising one's judgment.

Further Roots with Deflation

Once a root X has been found, it can be put aside by applying Newton's method to the modified equation $q(x) = 0$, where

$$q(x) = \frac{f(x)}{(x - X)^m} \tag{4.4.11}$$

and m is the multiplicity of X. The value of m may be assessed by monitoring the behavior of the error as discussed in a preceding subsection.

If $f(x)$ is a polynomial, then $q(x)$ will be another polynomial whose coefficients may be obtained using the synthetic division algorithm to be discussed in Section 4.7.

PROBLEMS

4.4.5 *Newton's method for one equation.*

(*a*) Write a computer program that produces one real root of a nonlinear equation $f(x) = 0$, using Newton's method. The program should be able to handle a variety of functions $f(x)$ offered in a menu. The derivative df/dx should be computed by numerical differentiation, setting $df/dx = [f(x+\varepsilon) - f(x)]/\varepsilon$, where ε is a sufficiently small number. The program should prompt you for the desired relative error of the solution, defined as $(x^{(k+1)} - x^{(k)})/x^{(k+1)}$, as well as for the initial guess.
(*b*) Confirm the validity of the program by running at least two test cases for which an analytical solution is available. Some choices are provided by the quadratic function $f(x) = ax^2 + bx + c$, the exponential function $f(x) = e^x - a$, and the logarithmic function $f(x) = \ln|x| - a$. In all cases, examine whether your results exhibit quadratic convergence. Investigate whether the choice of ε has a strong effect on the accuracy or rate of convergence of the solution, and discuss your observations.
(*c*) Compute all roots of the equation $\ln|x| + 3 - 3.10\ x^2 = 0$ accurate to the sixth decimal place. Explain your choice of initial guesses.

4.4.6 *Studies of convergence.*

(*a*) Apply Newton's method to compute one root of the function $f(x) = 1 - (1/x)\tan\ x$ with initial guess $x^{(0)} = 0.50$. Examine and discuss the rate of convergence.
(*b*) Repeat part (*a*) for the function $f(x) = (e^{x-1} - 1)\ln|x|$, using as initial guess $x^{(0)} = 2$.
(*c*) Repeat part (*a*) for the function $f(x) = (x-1)^{1/2}\ln|x|$, using as initial guess $x^{(0)} = 2$.

4.4.7 *Redlich–Kwong equation of state.*

Write a computer program that produces and prints a table showing the molecular volume of hydrogen for 15 combinations corresponding to pressure values $p = 1, 2, 3, 4, 5$ atm and temperature values $T = 200, 300, 400$ K, based on the the Redlich–Kwong equation of state (4.1.14). For initial guess, you should use the predictions of the ideal gas law. Discuss the physical significance of your results.
Note: *Perry's Chemical Engineer's Handbook* (McGraw-Hill, 5th edition pp. 3–41, 3–104) gives the following information about hydrogen: chemical formula; H_2; boiling point at 1 atm is equal to $-$ 252.7 °C; critical conditions; $T_c = -239.9$ °C, P_c =12.8 atm.

4.4.8 ***Viscous flow in a corner.***

The equation

$$\sin[2(x-1)\alpha] = (1-x)\sin(2\alpha) \tag{4.4.12}$$

describes viscous flow in a corner with aperture angle 2α. The real variable x is a measure of the strength of the flow (e.g., Pozrikidis 1997, Chapter 6). A trivial solution for any value of α is $X = 1$. Find and plot another solution branch $X(\alpha)$ in the range $0 < \alpha < \pi$.

4.4.9 ***A singular matrix.***

Find all real and complex values of x for which the following matrix is singular:

$$\mathbf{A} = \begin{bmatrix} 3 & 2 & x \\ -1 & x & -1 \\ x & -2 & -1 \end{bmatrix} \tag{4.4.13}$$

Methods with Cubic Convergence

Our next goal is to design an iteration function that yields a cubic rate of convergence. Specifically, we want to find an iteration function $g(x)$ that satisfies both conditions

$$g'(X) = 0, \qquad g''(X) = 0 \tag{4.4.14}$$

so that equation (4.3.16) yields

$$e^{(k+1)} = \tfrac{1}{6}\, g'''(X) e^{(k)^3} + \cdots \tag{4.4.15}$$

There are many avenues to follow, as described in considerable detail by Traub (1964, p. 76), and one of them leads to the iteration function

$$g(x) = x - \frac{f(x)}{f'(x)}\left(1 + \frac{1}{2}\,\frac{f''(x)f(x)}{f'^2(x)}\right) \tag{4.4.16}$$

This emerges by approximating the graph of the function $f(x)$ with a parabola whose slope and curvature at the point $x^{(k)}$ are equal to those of $f(x)$, and then identifying $x^{(k+1)}$ with an approximation to the closest root.

Another iteration function with cubic convergence, which has the advantage that it circumvents the computation of the second derivative $f''(x)$, is

$$g(x) = x - \frac{1}{f'(x)}\left[f(x) + f\left(x - \frac{f(x)}{f'(x)}\right)\right] \tag{4.4.17}$$

(Traub 1964, p. 174). The iterations are carried out in the predictor–corrector sense in three steps:

1. Compute $x^{(k+1)}$ using Newton's method.
2. Compute $f(x^{(k+1)})$.
3. Replace $x^{(k+1)}$ with $x^{(k+1)} - f(x^{(k+1)})/f'(x^{(k)})$.

Note that this amounts to carrying out two Newton iterations, where the value of the derivative $f'(x)$ is held constant at the second step.

Methods with High-order Convergence

Working in the same vein, we can devise algorithms with fourth or even higher-order convergence. A wealth of formulas and methods of deriving them have been compiled and discussed by Traub (1964). These include the *one-point iteration* family, the *one-point iteration with memory* family, and the *many-point iteration* family. Gerlach (1994) describes a modified Newton method where the function whose root is to be computed is redefined during the iterations to yield a faster rate of convergence. In practice, the complexity of the algebraic expressions discourages the application of higher-order methods. Newton's method and its variations, discussed in Section 4.6, are the standard choice.

PROBLEM

4.4.10 *Cubic convergence.*

Verify theoretically, and confirm by numerical experimentation, that the iteration function (4.4.17) yields cubic convergence.

4.5 *Newton's Method for a System of Equations*

Addressing now the more general case of a system of equations, we target a vector iteration function $\boldsymbol{g}(\boldsymbol{x})$ whose Jacobian matrix vanishes at the fixed point, that is,

$$\boldsymbol{J}(\boldsymbol{X}) = \boldsymbol{0} \tag{4.5.1}$$

When this occurs, the leading term on the right-hand side of equation (4.3.27) disappears, the rate of convergence becomes quadratic, and the error behaves as

$$e_i^{(k+1)} = \frac{1}{2}\left(\frac{\partial J_{i,j}}{\partial x_l}\right)_{\boldsymbol{X}} e_j^{(k)} e_l^{(k)} + \cdots \tag{4.5.2}$$

Repeating the arguments of Section 4.4, we find that, in general, the Newton sequence will converge provided the initial guess is sufficiently close to a root. If the root is multiple, the sequence will still converge, but the rate of convergence will be linear (see Problem 4.5.1).

Newton Iteration Function

The Newton iteration function arises by replacing the vector function $\boldsymbol{f}(\boldsymbol{x})$ with its linearized Taylor series expansion about the current estimate $\boldsymbol{x}^{(k)}$. This simplification transforms the nonlinear equation $\boldsymbol{f}(\boldsymbol{x}) = \boldsymbol{0}$ to the linear equation

$$\boldsymbol{f}\left(\boldsymbol{x}^{(k)}\right) + \boldsymbol{F}\left(\boldsymbol{x}^{(k)}\right)\left(\boldsymbol{x} - \boldsymbol{x}^{(k)}\right) = \boldsymbol{0} \tag{4.5.3}$$

where $\boldsymbol{F}$ is the *Jacobian matrix* of the vector function $\boldsymbol{f}$, defined as

$$F_{i,j} \equiv \frac{\partial f_i}{\partial x_j} \tag{4.5.4}$$

In vector notation, $\boldsymbol{F} \equiv (\nabla \boldsymbol{f})^T$. Solving equation (4.5.3) for $\boldsymbol{x}$, and identifying the solution with $\boldsymbol{x}^{(k+1)}$, we obtain

$$\boldsymbol{x}^{(k+1)} = \boldsymbol{g}(\boldsymbol{x}^{(k)}) \tag{4.5.5}$$

where

$$\boldsymbol{g}(\boldsymbol{x}) = \boldsymbol{x} - \boldsymbol{F}^{-1}(\boldsymbol{x})\boldsymbol{f}(\boldsymbol{x}) \tag{4.5.6}$$

is the Newton iteration function, and $\boldsymbol{F}^{-1}$ is the inverse of the matrix $\boldsymbol{F}$. It is clear that a solution of the equation $\boldsymbol{x} = \boldsymbol{g}(\boldsymbol{x})$ is also a root of the equation $\boldsymbol{f}(\boldsymbol{x}) = \boldsymbol{0}$.

To validate the claim that $\boldsymbol{J}(\boldsymbol{X}) = \boldsymbol{0}$, we take the partial derivatives of both sides of the ith component of equation (4.5.6) with respect to x_j and find

$$\begin{aligned} J_{i,j} \equiv \frac{\partial g_i}{\partial x_j} &= \frac{\partial x_i}{\partial x_j} - \frac{\partial F_{i,k}^{-1}}{\partial x_j} f_k - F_{i,k}^{-1} \frac{\partial f_k}{\partial x_j} = \delta_{i,j} - \frac{\partial F_{i,k}^{-1}}{\partial x_j} f_k - F_{i,k}^{-1} F_{k,j} \\ &= \delta_{i,j} - \frac{\partial F_{i,k}^{-1}}{\partial x_j} f_k - \delta_{i,j} \end{aligned} \tag{4.5.7}$$

Because $\boldsymbol{f}(\boldsymbol{X}) = \boldsymbol{0}$, the right-hand side of equation (4.5.7), vanishes when $\boldsymbol{x} = \boldsymbol{X}$, yielding the quadratic behavior shown in equation (4.5.2).

An exception occurs when $\boldsymbol{F}(\boldsymbol{X})$ is singular: $\boldsymbol{F}^{-1}(\boldsymbol{X})$ does not exist, and the rate of convergence is linear.

Algorithm

The algorithm involves making an initial guess $\boldsymbol{x}^{(0)}$, and then computing the sequence of vectors $\boldsymbol{x}^{(k)}$. At the $k+1$ stage, we solve the linear system of equations

$$\boldsymbol{F}\left(\boldsymbol{x}^{(k)}\right) \Delta\boldsymbol{x} = -\boldsymbol{f}\left(\boldsymbol{x}^{(k)}\right) \tag{4.5.8}$$

for the correction vector $\Delta\boldsymbol{x}$, using one of the methods described in Chapter 3, and then set

$$\boldsymbol{x}^{(k+1)} = \boldsymbol{x}^{(k)} + \Delta\boldsymbol{x} \tag{4.5.9}$$

Omission of the minus sign on the right-hand side of equation (4.5.8) is a frequent source of frustration.

PROBLEMS

4.5.1 ***A system with a multiple root.***

The system of two equations

$$\begin{aligned} f_1(x_1, x_2) &= (x_1 - 1)^2 + (x_2 - 1)^2 - 2 = 0 \\ f_2(x_1, x_2) &= x_1 x_2 - 4 = 0 \end{aligned} \tag{4.5.10}$$

has the solution $X_1 = 1, X_2 = 2$. Recover this solution numerically with a reasonable initial guess; assess and discuss the rate of convergence of Newton's method.

4.5.2 ***Linear equations.***

Show that Newton's method for a system of linear equations yields the exact solution after only one iteration.

Graphical Interpretation

Newton's method for one equation was developed from a graph, whereas its extension to many equations was developed from a formal mathematical approximation. To reconcile these two approaches and explain why we have chosen one or the other, we consider a system of two equations $f_1(x_1, x_2) = 0$, $f_2(x_1, x_2) = 0$, introduce the x_1, x_2, y Cartesian axes, and draw the two surfaces that are described by the functions

$$y = f_1(x_1, x_2), \qquad y = f_2(x_1, x_2) \tag{4.5.11}$$

as shown in Figure 4.5.1.

Next, we consider the current values $x_1^{(k)}$ and $x_2^{(k)}$ and draw the two planes that are tangential to the two surfaces at the point $x_1 = x_1^{(k)}$, $x_2 = x_2^{(k)}$. In general, these planes will intersect at a straight line; the intersection of this line and the x_1x_2 plane provides us with the new values $x_1^{(k+1)}$ and $x_2^{(k+1)}$.

Thus Newton's method arises by approximating the graphs of the functions with planes that are tangential to the graphs at the points $x_1 = x_1^{(k)}$ and $x_2 = x_2^{(k)}$. The extension to larger systems of equations is evident although hard to visualize.

PROBLEM

4.5.3 ***Pathologies for two equations.***

Discuss possible pathologies of Newton's method with reference to the intersection of the tangential planes in the x_1x_2y space.

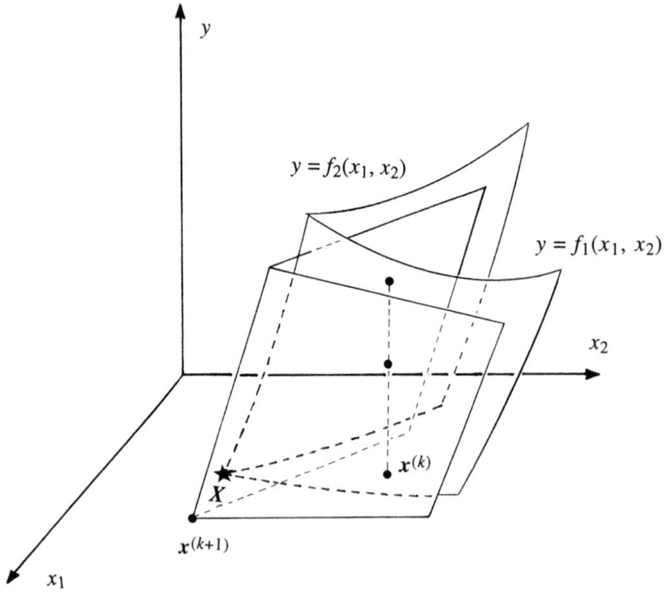

Figure 4.5.1 Graphical interpretation of Newton's method for a system of two equations.

Steffensen's Choice

In the practical implementation of Newton's method, the partial derivatives $\partial f_i/\partial x_j$ are often computed by numerical differentiation, setting, for example,

$$\partial f_i/\partial x_j = [f_i(x_1, x_2, \ldots, x_j + \varepsilon_j, \ldots, x_N) - f_i(x_1, x_2, \ldots, x_j, \ldots, x_N)]/\varepsilon_j \qquad (4.5.12)$$

where ε_j are sufficiently small numbers. With this modification, each Newton iteration requires $N(N+1)$ scalar function evaluations, instead of N function evaluations and N^2 evaluations of partial derivatives at each step. Thus this modification is preferred when the analytical evaluation of the partial derivative is expensive, sensitive, or undesirable.

In *Steffensen's method*, we set

$$\varepsilon_j = f_j(x_1, x_2, \ldots, x_N) \qquad (4.5.13)$$

whose magnitude is small as long as $\boldsymbol{x}$ is sufficiently close to the root. When the Jacobian $\boldsymbol{F}(\boldsymbol{X})$ is nonsingular, the rate of convergence remains quadratic.

PROBLEM

4.5.4 *Reversible chemical reactions.*

Solve the system of nonlinear equations (4.1.21) using (*a*) the Newton method and (*b*) the Newton method with Steffensen's modification. Discuss and compare the rates of convergence.

4.6 *Modified Newton Methods*

In practice, the evaluation of the derivative f' for one equation or of the Jacobian $\boldsymbol{F}$ for two or more equations, necessary for carrying out the Newton iterations, can be prohibitively expensive or inaccurate and thus undesirable. To overcome this difficulty we craftily alter the basic algorithm so as to bypass these evaluations, thereby deriving *modified*, *quasi*, or *inexact* Newton methods. The rate of convergence of these methods is no longer quadratic for a simple root, but in many cases this is a fair compromise.

Jacobian Is Held Constant During the Iterations

In the simplest variation of Newton's method, we evaluate the Jacobian only once, at the beginning of the iterations, and then keep it constant. If the initial guess is reasonably accurate, the iterations will converge at a less-than-quadratic rate.

PROBLEMS

4.6.1 *Graphical interpretation.*

Discuss the graphical interpretation of the simplified Newton method discussed in the texts for one equation.

4.6.2 *A system of two equations.*

Compute one solution of the system

$$(x-2)^2 + (y-3)^2 + (x-2.1)(y-3.1) = 2.81$$
$$10e^{-x} + 5e^{1-y} = 0.7468 \tag{4.6.1}$$

using (*a*) Newton's method and (*b*) Newton's method with the Jacobian evaluated only at the beginning and then held constant, and compare the rates of convergence.

Secant Method for One Equation

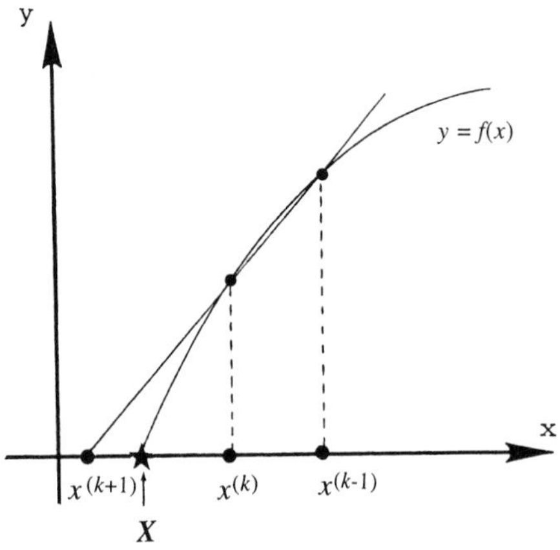

In a second variation of Newton's method, we approximate the derivative of the function $f(x)$ whose zeros we seek to compute with a finite difference that emerges by replacing the graph of $f(x)$ with a straight line passing through two consecutive points $(x^{(k)}, f^{(k)})$ and $(x^{(k-1)}, f^{(k-1)})$, as shown in the diagram; this amounts to setting $f'(x^{(k)}) = (f^{(k)} - f^{(k-1)})/(x^{(k)} - x^{(k-1)})$. We then perform *two-point iterations* based on the formula

$$x^{(k+1)} = x^{(k)} - f^{(k)} \frac{x^{(k)} - x^{(k-1)}}{f^{(k)} - f^{(k-1)}} \tag{4.6.2}$$

Two guesses must be made to initiate the iterations, but only *one function evaluation* is required at each subsequent step. It is then understandable why the secant method is preferred when the cost of function evaluations is high.

To investigate the convergence of the method, we work as in Section 4.4 for Newton's method, and find that, for a single root,

$$e^{(k+1)} = \frac{1}{2} \frac{f''(X)}{f'(X)} e^{(k)} e^{(k-1)} + \cdots \tag{4.6.3}$$

(Problem 4.6.3). This result guarantees that the method will converge provided that the two initial guesses are sufficiently close to the root.

To quantify the rate of convergence, we assume that, sufficiently close to the root, the magnitudes of two successive errors are related according to the power law

$$|e^{(k+1)}| = \alpha |e^{(k)}|^p \tag{4.6.4}$$

where α is a positive constant, and p is a positive exponent expressing the order of the method. For example, $p = 2$ yields a quadratic method. Writing

$$|e^{(k-1)}| = |\alpha|^{-1/p} |e^{(k)}|^{1/p} \tag{4.6.5}$$

and substituting the right-hand sides of the last two equations into equation (4.6.3), we find that p satisfies the quadratic equation $p = 1 + 1/p$. The root with the largest absolute value is $p = 1.62$.

We thus find that each time we carry out one iteration, the magnitude of the error is raised approximately to the 1.62 power, and then multiplied by a nearly constant factor. Clearly, the secant method will converge for any reasonable initial guess.

PROBLEM

4.6.3 ***Convergence of the secant method.***

(*a*) Derive equation (4.6.3). (*b*) Study the convergence of the secant method for one equation with a multiple root.

Muller's Method for One Equation

This is a straightforward extension of the secant method. The graph of the function $f(x)$ is approximated with a parabola that passes through the three consecutive points $(x^{(k)}, f^{(k)})$, $(x^{(k-1)}, f^{(k-1)})$ and $(x^{(k-2)}, f^{(k-2)})$, and one of the roots of the approximating binomial is identified with the next iterate $x^{(k+1)}$. The method is best implemented by computing (Press et al. 1992, p. 364):

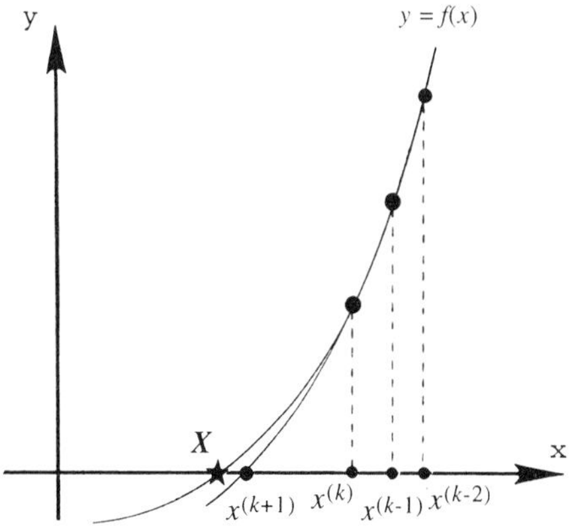

$$
\begin{aligned}
q &= \frac{x^{(k)} - x^{(k-1)}}{x^{(k-1)} - x^{(k-2)}} \\
A &= q f^{(k)} - q(q+1) f^{(k-1)} + q^2 f^{(k-2)} \\
B &= (1+2q) f^{(k)} - (1+q)^2 f^{(k-1)} + q^2 f^{(k-2)} \\
C &= (1+q) f^{(k)} \\
x^{(k+1)} &= x^{(k)} - \left(x^{(k)} - x^{(k-1)}\right) \frac{2C}{B \pm (B^2 - 4AC)^{1/2}}
\end{aligned}
\tag{4.6.6}
$$

The sign in the denominator of the last equation is chosen so as to make the magnitude of the denominator as large as possible. Provision should be made for the event that the quantity under the square root is negative, and the next iterate may be complex.

One significant advantage of Muller's method is its ability to capture a generally complex root, even with a real initial guess. The error behaves as shown in equation (4.6.4) with $p = 1.84$, which is a slight improvement over the value $p = 1.62$ of the secant method.

PROBLEM

4.6.4 ***Implementation of Muller's method.***

Repeat Problem 4.4.5 with Muller's method.

Regula-falsi Method for One Equation

Falsi in Latin means *error*; *falso* is a related work that describes the action of misstepping on a musical note; *regula falsi* describes the practice of making systematic errors, a practice that is *not* advocated in this book. The regula-falsi method involves the following steps:

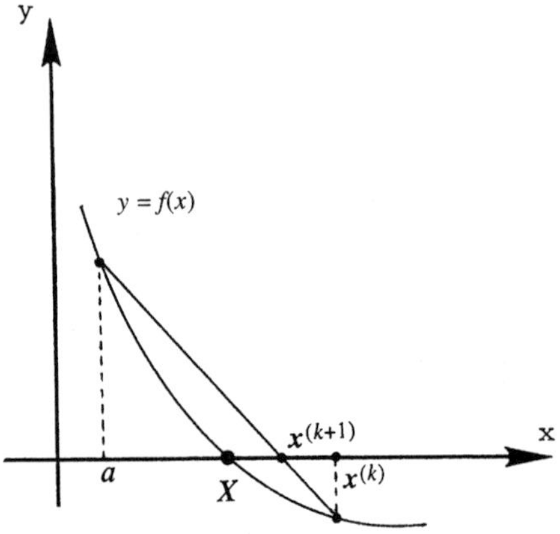

- Select a certain anchor point a.
- Make an initial selection $x^{(0)}$, so that the signs of $f(x^{(0)})$ and $f(a)$ are opposite.
- Approximate the derivative $f'(x^{(k)})$ with the finite-difference formula

$$f'(x^{(k)}) = \frac{f(a) - f\left(x^{(k)}\right)}{a - x^{(k)}} \tag{4.6.7}$$

- Compute $x^{(k+1)}$ from Newton's equation (4.4.2).
- Reset the anchor point a to the most recent value of the sequence $x^{(k)}$ for which the sign of $f(x^{(k)})$ is opposite to that of $f(a)$. When the function $f(x)$ is locally convex, that is, $f''(x) > 0$ and x is close to the root, the anchor point remains fixed during the iterations.

The main advantage of the regula-falsi method is that it requires only *one function evaluation* at each step, but this comes at the expense of a *linear rate of convergence.* The iterations are nevertheless guaranteed to converge as long as the initial guess is close enough to a root.

Secant Method for a System of Equations

The secant method for a system of equations is somewhat trickier to implement. The basic algorithm involves the following steps:

1. Make guesses for the first $N + 1$ members of Newton's sequence $\boldsymbol{x}^{(0)}, \boldsymbol{x}^{(1)}, \ldots, \boldsymbol{x}^{(N)}$.
2. Evaluate the $N + 1$ vector functions $\boldsymbol{f}(\boldsymbol{x}^{(0)}), \boldsymbol{f}(\boldsymbol{x}^{(1)}), \ldots, \boldsymbol{f}(\boldsymbol{x}^{(N)})$.
3. Approximate the Jacobian $\boldsymbol{F}$ by solving N linear system of equations for the individual rows of $\boldsymbol{F}$,

$$f_i\left(\boldsymbol{x}^{(j)}\right) - f_i\left(\boldsymbol{x}^{(0)}\right) = F_{i,k}\left(x_k^{(j)} - x_k^{(0)}\right) \tag{4.6.8}$$

where $j = 1, \ldots, N$, and the value of i is held constant in each solution.

4. Compute the new vector $\boldsymbol{x}^{(N+1)}$ using equations (4.5.8) and (4.5.9) with $k = N$.
5. Repeat the procedure with the new set $\boldsymbol{x}^{(1)}, \ldots, \boldsymbol{x}^{(N+1)}$.

Unfortunately, numerical instabilities render the method inferior to its alternatives discussed in the remainder of this section (e.g., Ortega and Rheinboldt, 1970).

PROBLEM

4.6.5 ***Secant method for a system of linear equations.***

Show that for a system of linear equations, the secant method yields the exact solution after only one iteration.

Other quasi-Newton Methods for System of Equations

Approximating or constructing the Jacobian in a systematic manner during the iterations produces a new class of intelligent quasi-Newton methods.

Violent diagonalization

In certain applications, mathematical or physical considerations suggest that the Jacobian matrix $\boldsymbol{F}$ is diagonally dominant. It is then reasonable to expect that setting its off-diagonal elements equal to zero will not introduce a serious error. When this approximation is made, the updating of the individual unknowns is decoupled, and the iterations proceed as if we were solving a system of N loosely connected but not entirely decoupled equations.

Consider, for example, the system of two equations

$$\begin{aligned} f_1(x_1, x_2) &= x_1^2 + x_2^2 \exp(-5x_1) - 2 = 0 \\ f_2(x_1, x_2) &= x_2^2 + x_1^2 \exp(-7x_2) - 1 = 0 \end{aligned} \tag{4.6.9}$$

whose Jacobian is given by

$$\boldsymbol{F} = \begin{bmatrix} 2x_1 - 5x_2^2 \exp(-5x_1) & 2x_2 \exp(-5x_1) \\ 2x_1 \exp(-7x_2) & 2x_2 - 7x_1^2 \exp(-7x_2) \end{bmatrix} \tag{4.6.10}$$

Under the pretense that both x_1 and x_2 are on the order of unity, we set the off-diagonal terms equal to zero and carry out the iterations on the basis of the *algorithm*

$$x_1^{(k+1)} = x_1^{(k)} - \frac{f_1\left(x_1^{(k)}, x_2^{(k)}\right)}{F_{1,1}\left(x_1^{(k)}, x_2^{(k)}\right)}$$
$$x_2^{(k+1)} = x_2^{(k)} - \frac{f_2\left(x_1^{(k+1)}, x_2^{(k)}\right)}{F_{2,2}\left(x_1^{(k+1)}, x_2^{(k)}\right)} \tag{4.6.11}$$

Note that we have chosen to update x_2 using the updated value of x_1 in the spirit of Gauss–Seidel.

More sophisticated versions of this method incorporate relaxation factors in the spirit of the successive over-relaxation method discussed in Section 3.7 (e.g., Ecker and Gross 1986).

PROBLEM

4.6.6 ***Nonlinear Gauss–Seidel method.***

Solve the system of equations (4.6.9) using the full Newton method, and then using the nonlinear Gauss–Seidel algorithm (4.6.11), and compare the two approaches.

Broyden's method

Broyden (1965, 1967) devised a clever method of constructing the Jacobian in the course of the iterations, by adding a rank-one matrix to a previous approximation. For one equation, the method reduces to the secant method discussed in a preceding subsection. More generally, the method proceeds according to the following steps:

1. Make an arbitrary guess for the inverse of the Jacobian $(\boldsymbol{F}^{(0)})^{-1}$.
2. Compute $\boldsymbol{x}^{(1)}$ using equations (4.5.8) and (4.5.9).
3. Replace $(\boldsymbol{F}^{(0)})^{-1}$ with $(\boldsymbol{F}^{(1)})^{-1}$ by working as described in the remainder of this section.
4. Repeat the computation.

To compute the kth approximation to the Jacobian, $\boldsymbol{F}^{(k)}$, we expand $\boldsymbol{f}(\boldsymbol{x}^{(k)})$ in a Taylor series about the point $\boldsymbol{x}^{(k-1)}$ and require that

$$\boldsymbol{f}\left(\boldsymbol{x}^{(k)}\right) - \boldsymbol{f}\left(\boldsymbol{x}^{(k-1)}\right) = \boldsymbol{F}^{(k)}\left(\boldsymbol{x}^{(k)} - \boldsymbol{x}^{(k-1)}\right) \tag{4.6.12}$$

It is instructive to note the partial similarity between equations (4.6.12) and (4.6.8). For one equation, combining the approximation (4.6.12) with the updating algorithm expressed by equations (4.5.8) and (4.5.9) yields the secant method.

But the preceding condition alone is not sufficient for determining $\boldsymbol{F}^{(k)}$ for more than one equation; additional constraints must be imposed. These constraints arise by stipulating that the null space of the difference matrix

$$\Delta\boldsymbol{F} \equiv \boldsymbol{F}^{(k)} - \boldsymbol{F}^{(k-1)} \tag{4.6.13}$$

be orthogonal to the correction vector

$$\Delta \boldsymbol{x} \equiv \boldsymbol{x}^{(k)} - \boldsymbol{x}^{(k-1)} \tag{4.6.14}$$

which is equivalent to requiring that for every vector $\boldsymbol{p}$ that satisfies $\boldsymbol{p}^T \Delta \boldsymbol{x} = 0$, where the superscript T designates the transpose,

$$\Delta \boldsymbol{F} \ \boldsymbol{p} = \boldsymbol{0} \tag{4.6.15}$$

The solution of the problem just described may be expressed in the form

$$\Delta F_{i,j} = -\frac{1}{\Delta \boldsymbol{x}^T \Delta \boldsymbol{x}} \left(F_{i,k}^{(k-1)} \Delta x_k - \Delta f_i \right) \Delta x_j \tag{4.6.16}$$

where we have defined

$$\Delta \boldsymbol{f} \equiv \boldsymbol{f}\left(\boldsymbol{x}^{(k)}\right) - \boldsymbol{f}\left(\boldsymbol{x}^{(k-1)}\right) \tag{4.6.17}$$

We can actually do better than that, and compute $\boldsymbol{F}^{(k)^{-1}}$ directly, thereby eliminating the need for solving the system of linear equations (4.5.8). Using the Sherman–Morrison formula (2.3.3), we find

$$\left(F_{i,j}^{(k)}\right)^{-1} = \left(F_{i,j}^{(k-1)}\right)^{-1} - \frac{1}{\Delta x_l \left(F_{l,m}^{(k-1)}\right)^{-1} \Delta f_m} \left(\left(F_{i,l}^{(k-1)}\right)^{-1} \Delta f_l - \Delta x_i \right) \Delta x_p \left(F_{p,j}^{(k-1)}\right)^{-1} \tag{4.6.18}$$

(Problem 4.6.7). The updating of the solution is done on the basis of the relaxation formula

$$\boldsymbol{x}^{(k+1)} = \boldsymbol{x}^{(k)} - \lambda^{(k)} (\boldsymbol{F}^{(k)})^{-1} \boldsymbol{f}\left(\boldsymbol{x}^{(k)}\right) \tag{4.6.19}$$

where $\lambda^{(k)}$ is a positive relaxation parameter. In practice, $\lambda^{(k)}$ is initially set equal to unity, and its value is reduced repeatedly and sequentially, as necessary, in order to ensure that the magnitude of $\boldsymbol{f}(\boldsymbol{x}^{(k+1)})$ decreases monotonically during the iterations.

PROBLEM

4.6.7 *Broyden's method.*

Derive equation (4.6.18) on the basis of the Sherman–Morrison formula (2.3.3).

Further methods

A plethora of further inexact Newton methods have been developed in the spirit of Broyden. Reviews and algorithms are discussed by Dennis and Schnabel (1983) and Eisenstat and Walker (1994).

4.7 *Zeros of Polynomials*

The zeros of an Nth degree polynomial $P_N(x)$ can be computed by any of the methods described in the previous sections for a single nonlinear equation $f(x) = 0$. In this case, we simply set

$$f(x) \equiv P_N(x) = a_1 x^N + a_2 x^{N-1} + a_3 x^{N-2} + \cdots + a_N x + a_{N+1} \tag{4.7.1}$$

where $a_i, i = 1, \ldots, N+1$, is a collection of $N+1$ real or complex coefficients; we have assumed that $a_1 \neq 0$. The fundamental theorem of algebra guarantees that $P_N(x)$ has exactly N zeros residing in the complex plane, some of which may appear up to N times. The multiplicity of a root m takes an integral value, that is, the mth derivative of $P_N(x)$ evaluated at the root has a nonzero value. If the polynomial coefficients are real, the roots appear in pairs of complex conjugates.

The simplicity of the polynomial expression (4.7.1), coupled with the extensive body of theory on polynomial behaviors (e.g., Householder 1970), allows us not only to expedite the root-finding process but also to devise powerful special-purpose methods.

Evaluation of a Polynomial and Its Derivatives by Synthetic Division

An efficient method of computing the value of a polynomial for a particular value of x is by *nested multiplication*, implemented according to *Horner's algorithm* discussed in Sections 1.3 and 1.4. This algorithm is, in fact, an integral part of the method of *synthetic division*, which is used to divide a polynomial by a monomial producing the constant remainder.

To demonstrate the connection between nested multiplication and synthetic division, we express the polynomial $P_N(w)$ in the form

$$P_N(w) = (w - x)(b_1 w^{N-1} + b_2 w^{N-2} + b_3 w^{N-2} + \cdots + b_{N-1} w + b_N) + b_{N+1} \tag{4.7.2}$$

where $b_i, i = 1, \ldots, N+1$ is a collection of real or complex coefficients. Equating the right-hand sides of equations (4.7.1) and (4.7.2), where the former equation is written with w in place of x, we obtain the formulas

$$\begin{aligned} b_1 &= a_1 \\ b_2 &= a_2 + x b_1 \\ &\cdots \\ b_{N+1} &= a_{N+1} + x b_N \end{aligned} \tag{4.7.3}$$

It is clear that

$$P_N(x) = b_{N+1} \tag{4.7.4}$$

The sequence (4.7.3) is identical to that computed using Horner's algorithm. If x happens to be a zero of the polynomial, then $b_{N+1} = 0$.

Differentiating equation (4.7.2) with respect to w, and evaluating the resulting expression at $w = x$, we find

$$P_N'(x) = b_1 x^{N-1} + b_2 x^{N-2} + \cdots + b_{N-1} x + b_N \tag{4.7.5}$$

Computing the right-hand side by the method of synthetic division yields the new sequence

$$
\begin{aligned}
c_1 &= b_1 \\
c_2 &= b_2 + x c_1 \\
&\cdots \\
c_N &= b_N + x c_{N-1}
\end{aligned}
\tag{4.7.6}
$$

where

$$P_N'(x) = c_N \tag{4.7.7}$$

Working in a similar manner, we produce the sequence

$$
\begin{aligned}
d_1 &= c_1 \\
d_2 &= c_2 + x d_1 \\
&\cdots \\
d_{N-1} &= c_{N-1} + x d_{N-2}
\end{aligned}
\tag{4.7.8}
$$

and compute

$$P_N''(x) = d_{N-1} \tag{4.7.9}$$

If x happens to be a double zero of the polynomial, then $c_N = 0$, and if it is a triple root, $c_N = 0$ and $d_{N-1} = 0$.

PROBLEM

4.7.1 ***Synthetic division for high-order derivatives.***

Develop an algorithm for computing the lth derivative $P_N^{(l)}(x)$, based on synthetic division.

Sequential Computation of All Roots by Deflation

After a root of multiplicity m has been found, it can be put aside by replacing the original polynomial $P_N(x)$ with the $(N - m)$ degree polynomial

$$Q_{N-m}(x) = d_1 x^{N-m} + d_2 x^{N-m-1} + \cdots + d_{N-m} \tag{4.7.10}$$

whose coefficients may be found by the method of synthetic-division. The process may then be repeated a number of times to obtain all roots.

To minimize the effect of the round-off error, the roots should be computed in the order of ascending magnitude. Furthermore, after a root has been found, it should be refined by applying, for example, Newton's method to the original polynomial, using the alleged root of the descendant polynomial as the initial guess. Other pieces of practical advice are offered by Wilkinson (1984).

Implementation of Newton's Method and Its Variations

Newton's method and its variations are implemented in a straightforward manner as discussed in the preceding sections. A complex initial guess is required to capture complex zeros. When a good initial guess is available, the secant method competes with, or even supersedes the performance of, Newton's method.

PROBLEMS

4.7.2 *Computing* π.

There are many ways of computing an approximation to the number π (Beckmann 1970). One ingenious method due to the great Greek thinker Αρχιμήδης, is to consider the limit of the perimeter P_m of an m-sided regular polygon inscribed within a circle whose radius is equal to $\frac{1}{2}$, which is clearly equal to π.
(a) Derive the expression $P_m = m\ \sin(\pi/m)$, demonstrate that when m is very large

$$P_m = \pi + \frac{\alpha}{m^2} + \frac{\beta}{m^4} + \frac{\delta}{m^6} + \cdots \tag{4.7.11}$$

and express the coefficients α, β, and γ in terms of π.
(b) Compute π keeping only the four terms shown on the right-hand side of equation (4.7.11), and the knowledge that $p_2 = 2$ or $p_6 = 3$, and compare the two answers.

4.7.3 *Eigenvalues of a tridiagonal matrix using a finite-difference method.*

Consider an $N \times N$ tridiagonal matrix with diagonal elements equal to $1 - 2\alpha$, and superdiagonal and subdiagonal elements equal to α, where α is a constant. This matrix arises in the numerical solution of the one-dimensional unsteady diffusion equation using the Crank–Nicolson finite-difference method, as will be discussed in Section 11.3. A stability analysis shows that the finite-difference method is stable when the norm of all eigenvalues is less than unity.

Find all eigenvalues for $\alpha = 0.10$ and $N = 5$ by computing the roots of the characteristic polynomial using Newton's method combined with deflation. Arrange the computed eigenvalues in order of decreasing magnitude, and plot λ_i versus i, where λ_1 is the eigenvalue with the maximum norm, and λ_5 is the eigenvalue with the minimum norm. Do you observe a pattern?

Bairstow's Method

This powerful method allows us to compute a pair of roots at a time. The method proceeds by expressing the polynomial shown in equation (4.7.1) in the seemingly cumbersome form

$$\begin{aligned} P_N(x) = (x^2 - rx - s)(b_1 x^{N-2} + b_2 x^{N-3} + b_3 x^{N-4} + \cdots + b_{N-2}x + b_{N-1}) \\ + \ b_N(x - r) + b_{N+1} \end{aligned} \tag{4.7.12}$$

where $b_i, i = 1, \ldots, N + 1$, is a collection of real or complex coefficients. The idea is to compute the constants r and s so that the coefficients b_N and b_{N+1} vanish. When this has been accomplished, the roots of the binomial $x^2 - rx - s$ will also be roots of the polynomial $P_N(x)$ and may thus be computed using formula (1.1.2).

Algorithm

The problem has been reduced to solving a system of two nonlinear equations

$$b_N(r, s) = 0, \qquad b_{N+1}(r, s) = 0 \tag{4.7.13}$$

where the coefficients b_N and b_{N+1} are polynomial functions of r and s; the solution is found using Newton's method. To compute the values of b_N and b_{N+1}, as well as their partial derivatives with respect to r and s, which are necessary for the evaluation of the 2×2 Jacobian matrix $\boldsymbol{F}$, we first construct the sequence

$$\begin{aligned} b_1 &= a_1 \\ b_2 &= a_2 + r b_1 \\ b_3 &= a_3 + r b_2 + s b_1 \\ &\dots \\ b_{N+1} &= a_{N+1} + r b_N + s b_{N-1} \end{aligned} \tag{4.7.14}$$

and then the sequence

$$\begin{aligned} c_1 &= 0 \\ c_2 &= b_1 \\ c_3 &= b_2 + r c_2 \\ c_4 &= b_3 + r c_3 + s c_2 \\ &\dots \\ c_{N+1} &= b_N + r c_N + s c_{N-1} \end{aligned} \tag{4.7.15}$$

It can be shown by straightforward substitution that

$$\frac{\partial b_k}{\partial r} = c_k, \qquad \frac{\partial b_k}{\partial s} = c_{k-1} \tag{4.7.16}$$

for $k = 1, \dots, N + 1$.

The method proceeds by making a guess for r and s, computing the preceding two sequences, solving the linear system

$$\begin{bmatrix} c_N & c_{N-1} \\ c_{N+1} & c_N \end{bmatrix} \begin{bmatrix} \Delta r \\ \Delta s \end{bmatrix} = - \begin{bmatrix} b_N \\ b_{N+1} \end{bmatrix} \tag{4.7.17}$$

for Δr and Δs using, for example, Cramer's rule, and then replacing r and s, respectively, with $r + \omega \Delta r$ and $s + \omega \Delta s$, where ω is a relaxation parameter.

After two roots have been found, the original polynomial $P_N(x)$ is replaced by the lower-degree polynomial

$$P_{N-2}(x) = b_1 x^{N-2} + b_2 x^{N-3} + \dots + b_{N-1} \tag{4.7.18}$$

whose coefficients are computed from the sequence (4.7.14) using the converged values of r and s. The procedure is repeated to obtain a further pair of roots.

PROBLEM

4.7.4 ***Zeros of orthogonal polynomials with Bairstow's method.***

(*a*) Compute all 12 zeros of the 12th degree Legendre polynomial presented in Table B.2 of Appendix B.
(*b*) Repeat part (*a*) for the 12th degree Chebyshev polynomial listed in Table B.3.

Laguerre's Method

This exotic method is guaranteed to converge to a real or complex root at a *cubic* rate, provided that the initial guess is close enough to the root. Every time we carry out one iteration, the error is raised to the third power and then multiplied by a constant factor. This remarkable behavior is in contrasted with the heuristic development of the algorithm shown below.

To develop the method, we express the polynomial $P_N(x)$ in a factorized form in terms of its zeros $x_1, x_2, \ldots, x_N$, as

$$P_N(x) = a_1(x - x_1)(x - x_2)\cdots(x - x_N) \tag{4.7.19}$$

where a_1 is the leading-power coefficient. Straightforward algebraic manipulations yield the two functions

$$G(x) \equiv \frac{P_N'}{P_N} = \frac{1}{x - x_1} + \frac{1}{x - x_2} + \cdots + \frac{1}{x - x_N} \tag{4.7.20}$$

and

$$H(x) \equiv -\left(\frac{P_N'}{P_N}\right)' = -\frac{P_N''}{P_N} + \left(\frac{P_N'}{P_N}\right)^2 = \frac{1}{(x - x_1)^2} + \frac{1}{(x - x_2)^2} + \cdots + \frac{1}{(x - x_N)^2} \tag{4.7.21}$$

where a prime signifies a derivative with respect to x.

Assume now that the current estimate $x^{(k)}$ is close enough to the ith zero x_i. The corresponding fractions on the right-hand sides of the preceding two equations will dominate the sums, and one may argue that a precise account of the contributions from the rest of the terms is not necessary. Accordingly, we replace these terms, respectively, with the quantities $1/b$ and $1/b^2$, where b is a certain fudging factor, and obtain a system of two nonlinear algebraic equations for x_i and b,

kern-1pt

$$\begin{aligned} G\left(x^{(k)}\right) &= \frac{1}{x^{(k)} - x_i} + \frac{N-1}{b} \\ H\left(x^{(k)}\right) &= \frac{1}{\left(x^{(k)} - x_i\right)^2} + \frac{N-1}{b^2} \end{aligned} \tag{4.7.22}$$

Solving for x_i, and identifying the solution with the new estimate $x^{(k+1)}$, yields Laguerre's algorithm expressed by the formula

$$x^{(k+1)} = x^{(k)} - \frac{N}{G\left(x^{(k)}\right) \pm \left\{(N-1)\left[NH\left(x^{(k)}\right) - G^2\left(x^{(k)}\right)\right]^{1/2}\right\}} \tag{4.7.23}$$

The sign at the denominator is selected so as to maximize its norm. An interesting feature of the method is that the sequence can move into the complex plane to capture complex roots, even when the initial guess is real.

To compute the functions G and H, we must evaluate the polynomial and its first two derivatives as required by equations (4.7.20) and (4.7.21). This can be done by the method of synthetic division discussed at the beginning of this section.

PROBLEMS

4.7.5 ***Zeros of orthogonal polynomials with Laguerre's method.***

Repeat Problem 4.7.4 using Laguerre's method.

4.7.6 ***Laguerre's method for a nonpolynomial equation?***

Discuss whether and how Laguerre's method can be extended to finding the roots of a nonpolynomial equation $f(x) = 0$.

Other Methods

A host of other methods for finding roots of polynomials are discussed in specialized monographs and texts on numerical analysis including those by Householder (1970), Henrici (1974, Vol. I), and Stoer and Bulirsch (1980, pp. 270–90).

4.8 Estimating the Location of a Root

In solving one nonlinear equation or a system of nonlinear equations using iterative methods, it is helpful, and sometimes critical, to have a good estimate for the location of a root. In certain cases, this can be done by examining whether any terms of the individual functions f_i assume *small* values for *small or large* values of the scalar components of $\mathbf{x}$. If they do, we ignore them, solve the resulting simplified system, and check *a posteriori* to see whether or not the approximations that lead us to the solution are justified. If they are, the solution of the simplified system provides us with a good estimate for the exact solution. Correspondingly, if some terms of the individual functions f_i assume *large* values for *small or large values* of the scalar components of $\mathbf{x}$, we discard the rest of the terms and work as just described.

Consider, for example, the equation

$$e^{-x} + 0.100\ x - 1.01 = 0 \tag{4.8.1}$$

and assume that one root has a large positive value. If this turns out to be true, the exponential term will be much smaller than the other terms and may be neglected yielding the approximate solution $X = 10.1$. For this value of X, the exponential term on the left-hand side of equation (4.8.1) is equal to 0.00004108, which is indeed small compared to unity. Since the original assumption is justified, we have a legitimate approximation to a root.

Next, we assume that a root is close to zero, expand the exponential term in its Maclaurin series, and approximate it with $1 - x$. Equation (4.8.1) then assumes the linear form $-0.90x - 0.010 = 0$, whose solution, $X = -0.0111$, is small compared to unity. Since the original assumption that X is close to zero is justified, we have a legitimate approximation to another root.

As a further example, we consider the system of two equations

$$x_2^3 + 0.123x_1 - 2.001 = 0, \qquad \ln x_2 + 0.010x_1^2 + 3.060 = 0 \tag{4.8.2}$$

To estimate the solution, we inspect the second equation. Since the second and third terms on the left-hand side are positive, the first term must be negative, that is, $x_2 < 1$. But then the first term on the left-hand side of the first equation must be small and could be neglected, yielding a linear equation whose solution is $x_1 = 16.268$. Substituting this value into the second equation, we find $X_2 = 0.003324$. All assumptions are justified and we have a reasonable approximation to a root.

PROBLEM

4.8.1 ***Fixed-point iterations.***

(*a*) Compute two roots of equation (4.8.1) using the method of fixed-point iterations, and discuss the selection of the iteration functions.
(*b*) Compute all roots of the equation $e^{-2x} - 3e^{-x} + 2 = 0$ accurate to the sixth decimal place. Explain the choice of your initial guesses, and discuss the selection of the iteration functions.

Estimating the Roots of Polynomials

The fundamental and practical importance of polynomial equations has motivated a substantial body of literature on the approximate location of their roots (e.g., Householder 1970, Hildebrand 1974, Stoer and Bulirsch 1980, Conte and de Boor 1980). Four important theorems are summarized in the remainder of this section.

Consider the Nth degree polynomial shown in equation (4.7.1). In what follows, a root of multiplicity m is counted m times.

Descartes' rule

This rule ensures that if the coefficients of the polynomial a_i are all real, then the number of positive real roots is equal to the number of changes in the sign of the sequence a_i, or is less than that by an even integer. The number of negative real roots is related to the coefficients of the polynomial $P_N(-x)$ in the same way.

Budan's rule

Assume that all coefficients of the polynomial are real. The number q of real roots lying within a certain interval (a, b) of the real axis can be deduced as follows. Let k and l be the number of changes of sign of the two sequences

$$\begin{aligned} &P_N(a),\, P_N'(a),\, \ldots,\, P_N^{(N)}(a) \\ &P_N(b),\, P_N'(b),\, \ldots,\, P_N^{(N)}(b) \end{aligned} \tag{4.8.3}$$

Then $q = k - l - 2s$, where s is zero or a positive integer.

Newton's product–sum identities

Carrying out the multiplications of the factors on the right-hand side of equation (4.7.19), and setting the result equal to the right-hand side of equation (4.7.1), provides us with $N + 1$ Newton product–sum identities relating the coefficients to the roots of the polynomial,

$$\sum_{i=1}^{N} x_i = -\frac{a_2}{a_1}$$

$$\sum_{i=1}^{N}\sum_{j=i+1}^{N} x_i x_j = \frac{a_3}{a_1} \tag{4.8.4}$$

$$\cdots$$

$$\prod_{i=1}^{N} x_i = (-1)^N \frac{a_{N+1}}{a_1}$$

When all roots are real, we begin with the inequality

$$\mathrm{Max}(x_i^2) \le \sum_{i=1}^{N} x_i^2 = \left(\sum_{i=1}^{N} x_i\right)^2 - 2\sum_{i=1}^{N}\sum_{j=1+1}^{N} x_i x_j \tag{4.8.5}$$

and use the first two relations (4.8.4) to find

$$\mathrm{Max}(x_i^2) \le \left(\frac{a_2}{a_1}\right)^2 - 2\frac{a_3}{a_1} \tag{4.8.6}$$

Circular disk theorem

All zeros of the polynomial (4.7.1), where the polynomial coefficients are allowed to be complex, lie within a circular disk that is centered at the origin of the complex plane. The radius of the disk is equal to

$$R = 1 + \frac{\mathrm{Max}_{i=2,\ldots,N}\{|a_i|\}}{|a_1|} \tag{4.8.7}$$

4.9 Difficult Problems, Continuation, and Embedding

There is a class of difficult problems whose solution requires a very accurate initial guess. In the case of one equation, this can be obtained by plotting and inspecting, or by using bracketing methods, as discussed in Section 4.2. For a higher number of equations, these methods are difficult to implement, but a good initial guess can still be generated using the programmable method of *parameter continuation*, otherwise known as *problem embedding*.

The idea is to perturb the equation $\boldsymbol{f}(\boldsymbol{x}) = \boldsymbol{0}$ whose roots are required, forming the modified equation

$$\boldsymbol{q}(\boldsymbol{x}, \varepsilon) = \boldsymbol{0} \tag{4.9.1}$$

The vector function $\boldsymbol{q}(\boldsymbol{x}, \varepsilon)$ is designed so that

$$\boldsymbol{q}(\boldsymbol{x}, \varepsilon = 1) = \boldsymbol{f}(\boldsymbol{x}) \tag{4.9.2}$$

and the roots of the equation $\boldsymbol{q}(\boldsymbol{x}, \varepsilon = 0) = \boldsymbol{0}$ are easy to find. We then solve a sequence of equations, for instance,

$$\begin{aligned} &\boldsymbol{q}(\boldsymbol{x}, \varepsilon = 0) = \boldsymbol{0} \\ &\boldsymbol{q}(\boldsymbol{x}, \varepsilon = 0.10) = \boldsymbol{0} \\ &\ldots \\ &\boldsymbol{q}(\boldsymbol{x}, \varepsilon = 1.0) = \boldsymbol{0} \end{aligned} \tag{4.9.3}$$

using one of the methods discussed in the preceding sections, where the initial guess for each equation is taken to be the converged solution of the previous equation. In a more advanced implementation of the method, the initial guess for a problem at a particular value of ε is found by extrapolating the converged results for two or more previous values, using the methods described in Chapter 6. This, of course, assumes that solution of $\boldsymbol{q}(\boldsymbol{x}, \varepsilon)$ does not exhibit a singular behavior or splits into two solution branches with respect to ε; that is, it follows a smooth path.

For example, the equation

$$f(x) = 3x^2 + 5 \ln|\pi - x| + 1 = 0 \tag{4.9.4}$$

has a root very close to $x = \pi$, but its computation requires a very good initial guess. The perturbed equation

$$q(x, \varepsilon) = 3x^2 + \varepsilon \;\, 5 \ln|\pi - x| + 1 + 3(\varepsilon - 1)x^3 = 0 \tag{4.9.5}$$

is easy to solve for $\varepsilon = 0$ and can be used to obtain the desired root by continuation.

Since we are not really interested in the solutions of the intermediate problems, we can save computational time by performing only a small number of iterations, until we reach the final equation corresponding to $\varepsilon = 1$, where we perform the number of iterations that are necessary to achieve the desired accuracy. The procedure can be automated by increasing the value of ε dynamically during the iterations, according to a certain protocol. For example, we may set

$$\varepsilon = 1 - \exp(-\alpha\, k) \tag{4.9.6}$$

where k is the iteration count, and α is an appropriate positive constant.

Continuation methods find important applications in continuum mechanics. In fluid mechanics, for example, the parameter ε can be identified with the Reynolds number, which is proportional to the characteristic velocity of a fluid (e.g., Pozrikidis, 1997). The difficulty of a computational problem typically increases as the value of the Reynolds number becomes higher and culminates when the flow becomes turbulent. Increasing the value of ε is the counterpart of physical experimentation, where the velocity of the fluid is increased slowly but gradually up to the desired value. For flow in a cavity driven by a moving lid, this can be done by increasing the speed of the lid.

PROBLEM

4.9.1 ***Ramped one-point iterations***

Find a root of equation (4.9.4) by dynamic ramping of equation (4.9.5), using the protocol (4.9.6) with a suitable value for α.

References

Beckmann, P., 1970, *The History of* π. Golem Press.

Broyden, C. G., 1965, A class of methods for solving nonlinear simultaneous equations. *Math. Comp.* **19**, 577–93.

Broyden, C. G., 1967, Quasi-Newton methods and their application to function minimisation. *Math. Comp.* **21**, 368–81.

Dennis, J. E., and Schnabel, R. B., 1983, *Numerical Methods for Unconstrained Optimization and Nonlinear Equations.* Prentice-Hall.

Conte, S. D., and de Boor, C., 1980, *Elementary Numerical Analysis.* McGraw-Hill.

Ecker, A., and Gross, D. H. E., 1986, A system of simultaneous non-linear equations in three-thousand variables. *J. Comput. Phys.* **64**, 246–52.

Eisenstat, A. C., and Walker, H., 1994, Globally convergent inexact Newton methods. *SIAM J. Optimization* **4**, 393–422.

Gerlach, J., 1994, Accelerated convergence in Newton's method. *SIAM Rev.* **36**, 272–76.

Gill, P. W., Murray, W., and Wright, M., 1981, *Practical Optimization.* Academic Press.

Henrici, P., 1974, *Applied and Computational Complex Analysis.* Wiley.

Hildebrand, F. B., 1974, *Introduction to Numerical Analysis.* Reprinted by Dover, 1987.

Householder, A. S., 1970, *The Numerical Treatment of a Single Nonlinear Equation.* McGraw-Hill.

Lanczos, C., 1988, *Applied Analysis.* Dover.

Moore, W. J., 1972, *Physical Chemistry.* Longman.

Ostrowski, A. M., 1966, *Solutions of Equations and Systems of Equations.* Academic Press.

Ortega, J., and Rheinboldt, W., 1970, *Iterative Solution of Nonlinear Equations in Several Variables.* Academic Press.

Ostrowski, A., 1966, *Solution of Equations and Systems of Equations.* Academic Press.

Pozrikidis, C., 1997, *Introduction to Theoretical and Computational Fluid Dynamics.* Oxford University Press.

Prausnitz, J. M., 1969, *Molecular Thermodynamics of Fluid-Phase Equilibria.* Prentice-Hall.

Press, W. H., Flannery, B. P., Teukolsky, S. A., and Vetterling, W. T., 1992, *Numerical Recipes.* 2nd ed. Cambridge University Press.

Rabinowitz, P., 1970, *Numerical Methods for Non-linear Algebraic Equations.* Gordon & Breach.

Smith, J. M., 1981, *Chemical Engineering Kinetics.* McGraw-Hill.

Stoer, J., and Bulirsch, R., 1980, *Introduction to Numerical Analysis.* Springer-Verlag.

Traub, J. F., 1964, *Iterative Methods for the Solution of Equations.* Prentice-Hall.

Treybal, R. E., 1968, *Mass Transfer Operations.* McGraw-Hill.

Wilkinson, J. H., 1984, The perfidious polynomial. In *Studies in Numerical Analysis,* G. H. Golub (ed.), pp. 1–28. Washington, D.C.: MAA.

Chapter 5

Eigenvalues of Matrices

5.1 Mathematical and Physical Context

Eigenvalues and eigenvectors of matrices are encountered under different guises in diverse branches of basic and applied science and engineering, and their computation defines an important subfield of numerical linear algebra. For example, in the preceding chapters, we saw that computing the spectral radius of a matrix, defined as the maximum of the norm of its eigenvalues, is necessary for assessing the convergence of mappings that arise in solving systems of linear and nonlinear equations. More applied examples can be drawn from the areas of quantum mechanics, system stability, vibration dynamics, and the theory of oscillations. To illustrate the way in which these algebraic eigenvalue problems are formulated in practice, we devote this section to discussing several characteristic examples.

Inclination and Aspect Ratio of an Elongated Body

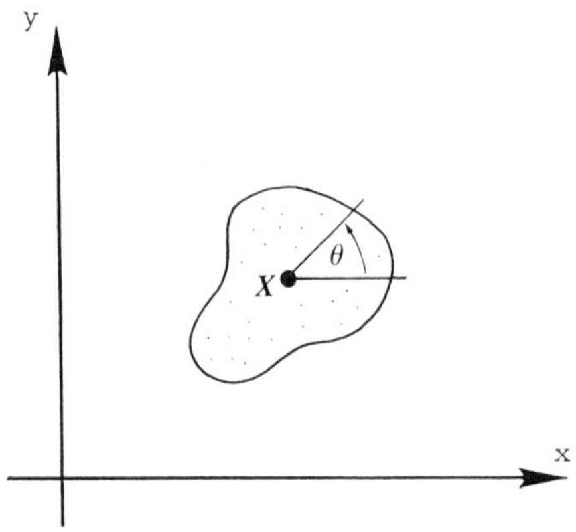

First, we want to quantify the orientation and aspect ratio of a two-dimensional elongated body, shown in the diagram. This can be done by several methods with varying degrees of generality and sophistication.

In one method, we introduce the symmetric matrix of the second moments of inertia $\boldsymbol{M}$, defined as

$$\boldsymbol{M} = \int_{Body} \begin{bmatrix} \hat{x}^2 & \hat{x}\hat{y} \\ \hat{x}\hat{y} & \hat{y}^2 \end{bmatrix} dA \tag{5.1.1}$$

where dA is a differential area in the xy plane, $\hat{\boldsymbol{x}} = \boldsymbol{x} - \boldsymbol{X}$, $\boldsymbol{X}$ is the center of the mass of the body given by

$$X = \frac{1}{A_B} \int_{Body} x \, dA \tag{5.1.2}$$

and A_B is the area of the body. In general, the areal integrals in equations (5.1.1) and (5.1.2) must be computed using the numerical methods discussed in Chapter 7. We then proceed in three stages:

- Compute the real eigenvalues and eigenvectors of $\boldsymbol{M}$.
- Identify the effective inclination angle θ with the angle that is subtended between the x axis and the eigenvector corresponding to the eigenvalue with the maximum norm. Since $\boldsymbol{M}$ is symmetric, the corresponding angle of the second eigenvector will be equal to $\theta + \pi/2$.
- Set the effective aspect ratio of the body equal to the ratio of the maximum to the minimum eigenvalue.

In a modified version of this method, the areal integrals in equations (5.1.1) and (5.1.2) are replaced by line integals around the perimeter of the body in the xy plane.

PROBLEMS

5.1.1 ***Inclination of a rectangle or an ellipse.***

Compute the effective inclination and aspect ratio of (a) a rectangle and (b) an ellipse, using the method described in the text, and compare the predictions with the exact values.

5.1.2 ***Orientation of a three-dimensional body.***

Develop a method for computing the inclination and two aspect ratios of an elongated three-dimensional body. Validate the method by showing that it reproduces reasonable results for a rectangular parallelepiped.

Vibration of Bodies Connected by Springs

Consider now the horizontal vibration of two bodies with masses m_1 and m_2, connected between themselves and two fixed walls with linear springs of stiffness k_1, k_2, and k_3, as shown in the diagram.

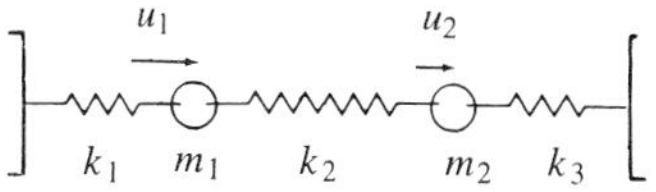

To describe the motion of the bodies, we introduce the displacements from the equilibrium positions, u_1 and u_2. Using Newton's second law to balance the rate of change of the linear momentum of the bodies and the sum of the forces exerted on them, we obtain the evolution equations

$$\begin{aligned} m_1 \frac{d^2 u_1}{dt^2} + (k_1 + k_2)\, u_1 - k_2\, u_2 = 0 \\ m_2 \frac{d^2 u_2}{dt^2} - k_2\, u_1 + (k_2 + k_3)\, u_2 = 0 \end{aligned} \tag{5.1.3}$$

where t stands for time. Assuming that the displacements vary harmonically in time with the same angular frequency ω, we set

$$u_1 = a_1 \sin \omega t, \qquad u_2 = a_2 \sin \omega t \tag{5.1.4}$$

where a_1 and a_2 are the amplitudes of the motion. Substituting expressions (5.1.4) into equations (5.1.3), we find that ω^2 is an eigenvalue of the 2×2 matrix

$$\mathbf{A} = \begin{bmatrix} \dfrac{k_1 + k_2}{m_1} & -\dfrac{k_2}{m_1} \\ -\dfrac{k_2}{m_2} & \dfrac{k_2 + k_3}{m_2} \end{bmatrix} \tag{5.1.5}$$

Note that if $m_1 = m_2$, $\mathbf{A}$ is symmetric.

Systems of bodies connected by springs are used in the modeling of the motion of macromolecules consisting of polymeric chains. The springs emulate the action of intermolecular force fields.

PROBLEM

5.1.3 ***Horizontal vibration of N bodies connected by springs.***

Derive the extended form of the matrix $\mathbf{A}$ shown in equation (5.1.5) for the more general case of N vibrating bodies connected in series.

An Eigenvalue Problem from an Ordinary Differential Equation

Algebraic eigenvalue problems are close relatives of eigenvalue problems involving ordinary and partial differential equations to be discussed in Section 10.6. To exemplify this connection, we consider the *homogeneous* linear ordinary differential equation

$$\frac{d^2 f}{dx^2} + \kappa^2 \pi^2 f = 0 \tag{5.1.6}$$

to be solved subject to the *homogeneous* boundary conditions $f(0) = 0$ and $f(1) = 0$, where κ is an unspecified parameter. An elementary computation reveals the existence of a trivial solution $f = 0$, for any value of κ, and of a nontrivial solution $f = a\,\sin(\kappa\pi x)$, when κ is an integer; a is an arbitrary constant. Accordingly, $\kappa = n$, where n is an integer, is one eigenvalue of the differential equation (5.1.6) with the stated boundary conditions.

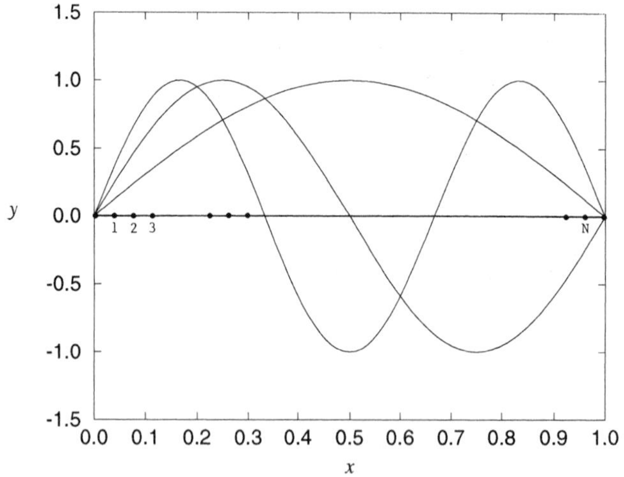

Pretending that we are unable to perform the analytical computation, we set out to obtain the eigenvalues numerically using a *finite-difference method.* We begin by discretizing the solution domain (0, 1) into $N + 1$ evenly spaced intervals that are separated by the nodal points $x_i = i\,\Delta x$, where $\Delta x = 1/(N + 1)$ and $i = 0, \ldots, N + 1$, applying the differential equation (5.1.6) at the ith nodal point, and approximating the second derivative using the second-order central-difference formula, which amounts to setting $(d^2 f/dx^2)_{x_i} = (f_{i+1} - 2f_i + f_{i-1})/\Delta x^2$, where $f_i \equiv f(x_i)$ (see Table 6.11.1). This approximation converts the differential equation to the finite-difference equation

$$-f_{i+1} + 2f_i - f_{i-1} = \lambda f_i \tag{5.1.7}$$

where we have defined

$$\lambda = \kappa^2 \pi^2 \, \Delta x^2 \tag{5.1.8}$$

Writing equation (5.1.7) for $i = 1, \ldots, N$, and requiring the boundary conditions $f_1 = 0$ and $f_{N+1} = 0$, we obtain a homogeneous system of linear algebraic equations for the N unknown values f_i, $i = 1, \ldots, N$, of the form

$$\boldsymbol{Af} = \lambda \boldsymbol{f} \tag{5.1.9}$$

The vector $\boldsymbol{f}$ contains the sequentially numbered values of f_i, and $\boldsymbol{A}$ is the $N \times N$ tridiagonal matrix

$$\boldsymbol{A} = \begin{bmatrix} 2 & -1 & 0 & \cdots & 0 & 0 & 0 \\ -1 & 2 & -1 & 0 & \cdots & 0 & 0 \\ 0 & -1 & 2 & -1 & 0 & \cdots & 0 \\ \cdots & \cdots & \cdots & \cdots & \cdots & \cdots & \cdots \\ 0 & \cdots & 0 & -1 & 2 & -1 & 0 \\ 0 & 0 & \cdots & 0 & -1 & 2 & -1 \\ 0 & 0 & 0 & \cdots & 0 & -1 & 2 \end{bmatrix} \tag{5.1.10}$$

Clearly, λ is an eigenvalue of $\boldsymbol{A}$.

The results of Section 5.3 will show that the eigenvalues of $\boldsymbol{A}$ can be found explicitly in closed form and are given by

$$\lambda_m = 2\left[1 + \cos\left(\frac{m\pi}{N+1}\right)\right] = 4\cos^2\left(\frac{m\pi}{2(N+1)}\right) \tag{5.1.11}$$

where $m = 1, \ldots, N$. For example, when $N = 1$, we find the eigenvalue $\lambda_1 = 2$.

Using the definition of λ from equation (5.1.8), we obtain numerical approximations to a finite set of N eigenvalues of the differential equation

$$\kappa_m^2 = \frac{4(N+1)^2}{\pi^2} \cos^2\left(\frac{m\pi}{2(N+1)}\right) \tag{5.1.12}$$

where $m = 1, \ldots, N$.

Setting, in particular, $m = N$, we derive an approximation to the *smallest* eigenvalue whose exact value is known from the analytical solution of the differential equation to be equal to unity. The values $N = 1, 2, 3$ yield, respectively, the numerical approximations $\kappa_N^2 = 0.811, 0.912, 0.950$. As N becomes increasingly large, we derive the asymptotic expansion

$$\kappa_N^2 = 1 - \frac{\pi^2}{12} \frac{1}{(N+1)^2} + \cdots \tag{5.1.13}$$

which confirms that the numerical results converge to the exact value of unity, and the error scales with N^{-2}.

PROBLEMS

5.1.4 *Numerical versus exact solution.*

(*a*) Plot the eigenvalues κ_m given in equation (5.1.12) for $N = 2, 4, 8, 16, 32$, and discuss the results with reference to the exact solution.
(*b*) Derive the asymptotic expression (5.1.13).

5.1.5 *A linear differential equation with nonconstant coefficients.*

Formulate the algebraic eigenvalue problem for the differential equation

$$\frac{d^2 f}{dx^2} + \kappa^2 \pi^2 [1 + 0.10 \cos(\pi x)] f = 0 \tag{5.1.14}$$

with boundary conditions $f(0) = 0$ and $f(1) = 0$, using the finite-difference method discussed in the text. Discuss the structure of the matrix $\mathbf{A}$ with reference to the structure of the matrix shown in equation (5.1.10).

Periodic Solutions of Mathieu's Equation

As a final example, we consider periodic solutions of Mathieu's second-order ordinary differential equation for the function $x(t)$. In an appropriate dimensionless form, Mathieu's equation has the standard form

$$\frac{d^2 x}{dt^2} + (p - 2q \cos 2t)x = 0 \tag{5.1.15}$$

where p and q are two constants (e.g., Abramowitz and Stegun 1972, Chapter 20). The equation is linear, but one of the coefficients is a harmonic function of the independent variable t, with period equal to π.

Mathieu's equation describes a host of physical phenomena characterized by oscillatory behavior, including the vibration of a pendulum whose pivot undergoes vertical oscillations, the propagation of acoustic waves through an elastic cylinder with elliptical cross section, and the small-amplitude oscillations of the interface of a liquid that is subjected to external vibrations. The latter can be realized by dragging a cup of coffee over a rough surface, thereby causing a vertical oscillatory motion. In the following discussion, we shall identify the independent variable t with time and the dependent variable $x(t)$ with the amplitude of the oscillatory motion.

Our objective is to compute the fundamental mode of response with period equal to π, and the first subharmonic mode of response with period equal to 2π. To make the mathematical formulation compact, we introduce the parameter l, where $l = 0$ or 1 corresponds, respectively, to solutions with period equal to π or 2π. Both types of solution can be represented in terms of a Fourier series, as

$$x(t) = \sum_{m=-\infty}^{\infty} c_{2m+l} \exp[-(2m + l)\ It] \tag{5.1.16}$$

where c_n are complex Fourier coefficients, and I is the imaginary unit. Stipulating that

$$c_{-n} = c_n^* \tag{5.1.17}$$

where an asterisk designates the complex conjugate, ensures that $x(t)$ is real.

Substituting the expansion (5.1.16) into equation (5.1.15), and collecting similar Fourier coefficients, we obtain an *infinite* system of linear *homogeneous* equations for c_i. Nontrivial solutions exist only for particular combinations of the constants p and q that form two families of functions:

$$p = a_r(q), \qquad p = b_r(q) \tag{5.1.18}$$

The integer subscript r labels solution branches, as will be discussed in the following paragraphs.

When the coefficients c_i are real, we obtain solutions $x(t)$ that are *even* with respect to the origin of time. When c_i are imaginary, we obtain solutions that are *odd* with respect to the origin of time. In either case, the mathematical formulation results in the homogeneous system of linear equations

$$(\boldsymbol{A} - p\,\boldsymbol{I})\,\boldsymbol{c} = \boldsymbol{0} \tag{5.1.19}$$

where $\boldsymbol{A}$ is a complex tridiagonal matrix, p is an eigenvalue of $\boldsymbol{A}$, and $\boldsymbol{c}$ is a Fourier-coefficient eigenvector to be defined in the following subsections.

Solutions with period π

For solutions with period equal to π, corresponding to $l = 0$, the Fourier-coefficient eigenvector is given by

$$\boldsymbol{c} = (c_0, c_2, \ldots)^T \tag{5.1.20}$$

Even solutions correspond to

$$\begin{gathered} A_{1,1} = 0, \qquad A_{1,2} = 2q \\ A_{i,i-1} = q, \qquad A_{i,i} = 4(i-1)^2, \qquad A_{i,i+1} = q, \qquad \text{for } i \geq 2 \end{gathered} \tag{5.1.21}$$

When $q = 0$, the matrix $\boldsymbol{A}$ is diagonal, and its eigenvalues are

$$a_0(q=0) = 0, \quad a_2(q=0) = 4, \quad a_4(q=0) = 16, \quad \ldots, \quad a_{2i}(q=0) = 4i^2, \quad \ldots \tag{5.1.22}$$

Solving the eigenvalue problem for nonzero values of q provides us with an infinite family of even solutions emanating from the points (5.1.22).

Odd solutions correspond to

$$\begin{gathered} A_{1,1} = 4, \qquad A_{1,2} = q, \\ A_{i,i-1} = q, \qquad A_{i,i} = 4i^2, \qquad A_{i,i+1} = q, \qquad \text{for } i \geq 2 \end{gathered} \tag{5.1.23}$$

When $q = 0$, the matrix $\boldsymbol{A}$ is diagonal, and its eigenvalues are

$$b_2(q=0) = 4, \quad b_4(q=0) = 16, \quad \ldots, \quad b_{2i}(q=0) = 4i^2, \quad \ldots \tag{5.1.24}$$

Solving the eigenvalue problem for nonzero values of q provides us with an infinite family of odd solutions emanating from the points (5.1.24).

Solutions with period 2π

For solutions with period equal to 2π, corresponding to $l = 1$, the Fourier-coefficient eigenvector is

$$\mathbf{c} = (c_1, c_3, \ldots)^T \tag{5.1.25}$$

Even solutions correspond to

$$\begin{aligned} &A_{1,1} = 1 + q, \qquad A_{1,2} = q \\ A_{i,i-1} = q, \qquad &A_{i,i} = (2i-1)^2, \qquad A_{i,i+1} = q, \qquad \text{for } i \geq 2 \end{aligned} \tag{5.1.26}$$

When $q = 0$, the matrix $\mathbf{A}$ is diagonal, and its eigenvalues are

$$a_1(q=0) = 1, \quad a_3(q=0) = 9, \quad \ldots, \quad a_{2i-1}(q=0) = (2i-1)^2, \quad \ldots \tag{5.1.27}$$

Solving the eigenvalue problem for nonzero values of q provides us with an infinite family of even solutions emanating from the points (5.1.27).

Odd solutions correspond to

$$A_{1,1} = 1 - q, \qquad A_{1,2} = q \tag{5.1.28}$$

Subsequent values are identical to those for even solutions given in the second set of equations (5.1.26).

When $q = 0$, the matrix $\mathbf{A}$ is diagonal, and its eigenvalues are

$$b_1(q=0) = 1, \quad b_3(q=0) = 9, \quad \ldots, \quad b_{2i-1}(q=0) = (2i-1)^2, \quad \ldots \tag{5.1.29}$$

Solving the eigenvalue problem for nonzero values of q provides us with an infinite family of odd solutions emanating from the points (5.1.29).

The aforementioned solution branches can be computed by truncating the infinite matrices at a sufficiently large level, and then using one of the numerical methods to be discussed in subsequent sections. Fortunately, the results converge rapidly with respect to the level of trunction. Figure 5.1.1 shows a compilation of solution branches after Abramowitz and Stegun (1972, p. 724).

PROBLEM

5.1.6 ***Algebraic eigenvalue problem from Mathieu's equation.***

Formulate the algebraic eigenvalue problem for the four cases discussed in the text, and rederive the coefficients (5.1.21), (5.1.23), (5.1.26), (5.1.28).

5.2 *Complex Matrices, Generalized, and Nonlinear Eigenvalue Problems*

The majority of the numerical methods for computing eigenvalues, to be discussed in subsequent sections, work for real as well as complex matrices. But since programming in complex variables may cause a severe case of cyberphobia, it might be preferable to decompose all complex quantities into their real and imaginary parts, and compute the eigenvalues of real matrices with twice as large sizes.

Figure 5.1.1 Solution branches for periodic solutions of Mathieu's equation.

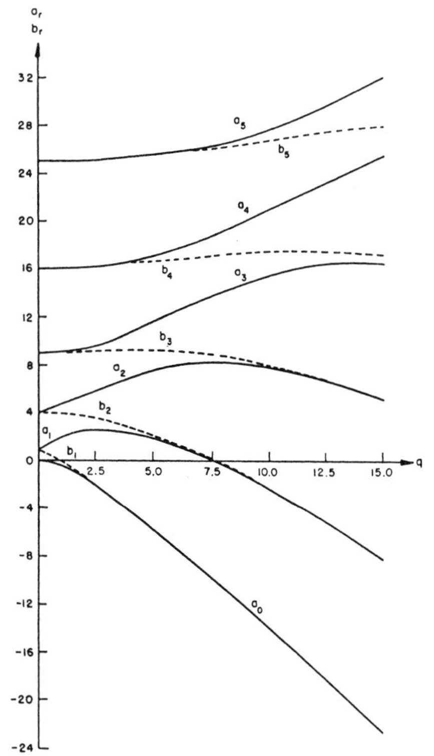

To begin, let us assume that a complex matrix $\boldsymbol{A}$ has a *real* eigenvalue λ, which is guaranteed when the matrix $\boldsymbol{A}$ is real and symmetric, or more generally Hermitian. Writing

$$\boldsymbol{A} = \boldsymbol{A}_R + I\,\boldsymbol{A}_I, \qquad \boldsymbol{u} = \boldsymbol{u}_R + I\,\mathbf{u}_I \tag{5.2.1}$$

where $\boldsymbol{u}$ is the associated eigenvector, I is the imaginary unit, and the subscripts R and I have the obvious meanings, and using the definition $\boldsymbol{A}\boldsymbol{u} = \lambda\boldsymbol{u}$, we obtain

$$\boldsymbol{M}\boldsymbol{w} = \lambda\,\boldsymbol{w} \tag{5.2.2}$$

where

$$\boldsymbol{M} \equiv \begin{bmatrix} \boldsymbol{A}_R & -\boldsymbol{A}_I \\ \boldsymbol{A}_I & \boldsymbol{A}_R \end{bmatrix}, \qquad \boldsymbol{w} \equiv \begin{bmatrix} \boldsymbol{u}_R \\ \boldsymbol{u}_I \end{bmatrix} \tag{5.2.3}$$

which shows that λ is an eigenvalue of the real $2N \times 2N$ matrix $\boldsymbol{M}$. If the matrix $\boldsymbol{A}$ is Hermitian, $\boldsymbol{A}_R$ will be symmetric, $\boldsymbol{A}_I$ will be skew-symmetric, and $\boldsymbol{M}$ will be symmetric. Thus the eigenvalues of $\boldsymbol{M}$ and therefore of $\boldsymbol{A}$ will be real.

In the more general case where λ is *complex*, the matrix $\boldsymbol{M}$ defined in the first of equations (5.2.3) has the following important properties discovered by Muller (1947) and discussed by Bodewig (1959, p. 267); an s-fold eigenvalue is an eigenvalue of multiplicity s:

- If A has the s-fold *real* eigenvalue λ, then M has the $2s$-fold real eigenvalue λ. If $u^{(j)} = u_R^{(j)} + I u_I^{(j)}$ are the eigenvectors of A, then $[u_R^{(j)}, u_I^{(j)}]$ and $[u_I^{(j)}, -u_R^{(j)}]$ are the eigenvectors of M. These properties are apparent when A is real.
- If A has the s-fold *complex* eigenvalue λ, then M has an s-fold complex eigenvalue λ, and the s-fold complex eigenvalue λ^*; an asterisk designates the complex conjugate.
- If A has the *single* complex eigenvalue λ with associated eigenvector u, then the eigenvector of M corresponding to this eigenvalue is $[u, -Iu]^T$, whereas the eigenvector corresponding to the complex conjugate eigenvalue λ^* is $[u^*, Iu^*]^T$.

  For example, the 1×1 matrix $[1 + 2I]$ has the complex eigenvalue $\lambda = 1 + 2I$; the corresponding eigenvector is any arbitrary scalar. The associated 2×2 real matrix

$$M = \begin{bmatrix} 1 & -2 \\ 2 & 1 \end{bmatrix} \tag{5.2.4}$$

  has the eigenvalue $\lambda = 1 + 2I$ with associated eigenvector $[1, -I]^T$, and the complex conjugate eigenvalue $\lambda = 1 - 2I$ with associated eigenvector $[1, I]^T$.

- If A has the s-fold *complex* eigenvalue λ with corresponding eigenvectors $u^{(j)}$, and if λ^* is *not* an eigenvalue of A, then $[u^{(j)}, -Iu^{(j)}]^T$ are eigenvectors of M.
- If A has the s-fold *complex* eigenvalue λ with corresponding eigenvectors $u^{(j)}$, and if λ^* is a t-fold complex eigenvalue of A with corresponding eigenvectors $v^{*(l)}$, then $[u^{(j)}, -Iu^{(j)}]^T$ and $[v^{(l)}, Iv^{(l)}]^T$ are eigenvectors of M corresponding to the eigenvalue λ.

PROBLEM

5.2.1 ***Muller properties.***

Verify the truthfulness of the Muller properties by means of numerical examples of your choice.

Generalized Eigenvalue Problems

The numerical solution of the differential equations of mathematical physics often results in generalized eigenvalue problems expressed by

$$Au = \lambda Bu \tag{5.2.5}$$

where A and B are two $N \times N$ matrices. An introductory example will be discussed in Problem 5.2.2, and a more advanced example concerning hydrodynamic stability will be discussed in Section 10.6.

Premultiplying both sides of equation (5.2.5) by B^{-1} yields the standard eigenvalue problem expressed by the familiar form

$$Cu = \lambda u \tag{5.2.6}$$

where

$$C = B^{-1}A \tag{5.2.7}$$

It is clear that the generalized problem will have N eigenvalues if and only if Rank($\boldsymbol{B}$) $= N$. If the rank of $\boldsymbol{B}$ is deficient, then the number of eigenvalues may be finite, zero, or infinite. For example, the problem expressed by

$$\begin{bmatrix} 1 & 2 \\ 0 & 3 \end{bmatrix} \boldsymbol{u} = \lambda \begin{bmatrix} 0 & 1 \\ 0 & 0 \end{bmatrix} \boldsymbol{u} \tag{5.2.8}$$

does not have an eigenvalue.

In Sections 5.6 and 5.7, we shall see that the eigenvalues of real and symmetric, and more generally Hermitian, matrices are much easier to compute than those of nonsymmetric matrices. Unfortunately, since the product of two Hermitian matrices is not necessarily Hermitian, the matrices $\boldsymbol{A}$ and $\boldsymbol{B}$ be may be Hermitian, but the matrix $\boldsymbol{C}$ will not necessarily be so.

Fortunately, when $\boldsymbol{A}$ *is Hermitian* and $\boldsymbol{B}$ *is Hermitian and positive-definite*, we can play a little trick. First, we compute the Cholesky decomposition of the matrix $\boldsymbol{B}$ discussed in Section 2.6; that is, we set $\boldsymbol{B} = \boldsymbol{L}\boldsymbol{L}^A$, where the matrix $\boldsymbol{L}$ is lower triangular. Second, we express equation (5.2.5) in the form

$$(\boldsymbol{L}^{-1} \boldsymbol{A}\, \boldsymbol{L}^{-1^A}) \; (\boldsymbol{L}^A \boldsymbol{u}) = \lambda \, (\boldsymbol{L}^A \boldsymbol{u}) \tag{5.2.9}$$

which shows that λ is an eigenvalue of the Hermitian matrix $\boldsymbol{L}^{-1}\boldsymbol{A}\boldsymbol{L}^{-1^A}$ corresponding to the eigenvector $\boldsymbol{L}^A\boldsymbol{u}$.

QZ decomposition

The possibly problematic computation of the inverses $\boldsymbol{B}^{-1}$ or $\boldsymbol{L}^{-1}$ has motivated a distinctive class of methods for solving the generalized eigenvalue problem expressed by equation (5.2.5), based on the *generalized Schur decomposition*, also called the $\boldsymbol{QZ}$ decomposition (e.g., Kerner 1989, Golub and van Loan 1989).

If $\boldsymbol{A}$ and $\boldsymbol{B}$ are two complex matrices with matching dimensions $N \times N$, then there exist two *unitary* matrices $\boldsymbol{Q}$ and $\boldsymbol{Z}$ so that the matrices

$$\boldsymbol{T} = \boldsymbol{Q}^A\boldsymbol{A}\boldsymbol{Z}, \qquad \boldsymbol{S} = \boldsymbol{Q}^A\boldsymbol{B}\boldsymbol{Z} \tag{5.2.10}$$

are both upper triangular. Writing

$$\boldsymbol{A} = \boldsymbol{Q}\boldsymbol{T}\boldsymbol{Z}^A, \qquad \boldsymbol{B} = \boldsymbol{Q}\boldsymbol{S}\boldsymbol{Z}^A \tag{5.2.11}$$

we obtain

$$\text{Det}(\boldsymbol{A} - \lambda\, \boldsymbol{B}) = \text{Det}(\boldsymbol{Q})\, \text{Det}(\boldsymbol{Z}^A)\, \text{Det}(\boldsymbol{T} - \lambda\, \boldsymbol{S}) \tag{5.2.12}$$

which shows that

$$\lambda = T_{i,i}/S_{i,i} \tag{5.2.13}$$

where $S_{i,i} \neq 0$ and summation is not implied over i. The problem is thus reduced to computing the matrices $\boldsymbol{T}$ and $\boldsymbol{S}$, which can be done using an algorithm due to Moler and Stewart (1973) and its descendants reviewed by Kerner (1989) and Golub and van Loan (1989, pp. 394–406).

PROBLEM

5.2.2 *A generalized eigenvalue problem from an ordinary differential equation.*

Formulate the generalized algebraic eigenvalue problem arising from the differential equation

$$\frac{d^4 f}{dx^4} + \kappa^2 \pi^2 \frac{d^2 f}{dx^2} + f = 0 \tag{5.2.14}$$

with boundary conditions $f(0) = 0$, $f'(0) = 0$, $f'(1) = 0$, and $f(1) = 0$, using the finite-difference method discussed in Section 5.1. For the finite-difference approximation of the second and fourth derivatives, consult Table 6.11.1.

Nonlinear Eigenvalue Problems

There is a class of eigenvalue problems where the matrix $\boldsymbol{A}$ is itself a function of λ, denoted as $\boldsymbol{A}(\lambda)$. These nonlinear problems are expressed by the equation

$$\boldsymbol{A}(\lambda)\boldsymbol{u} = \lambda\, \boldsymbol{u} \tag{5.2.15}$$

In general, the computation of λ requires combining algorithms for solving linear eigenvalue problems, to be discussed in subsequent sections, and algorithms for solving nonlinear algebraic equations discussed in Chapter 4. Fortunately, when $\boldsymbol{A}(\lambda)$ is or can be approximated with a polynomial, this twofold approach can be bypassed with the introduction of intermediate vectors.

Consider, for example, the linear second-order ordinary differential equation

$$\boldsymbol{B}\frac{d^2\boldsymbol{x}}{dt^2} + \boldsymbol{C}\frac{d\boldsymbol{x}}{dt} + \boldsymbol{D}\boldsymbol{x} = \frac{d\boldsymbol{x}}{dt} \tag{5.2.16}$$

for the N-dimensional vector function $\boldsymbol{x}(t)$, where $\boldsymbol{B}$, $\boldsymbol{C}$, and $\boldsymbol{D}$ are constant $N \times N$ matrices. To search for exponentially varying solutions, we set $\boldsymbol{x}(t) = \boldsymbol{u}\, \exp(\lambda t)$, where $\boldsymbol{u}$ is a constant vector, and obtain the nonlinear eigenvalue problem expressed by

$$(\lambda^2 \boldsymbol{B} + \lambda\, \boldsymbol{C} + \boldsymbol{D})\, \boldsymbol{u} = \lambda\, \boldsymbol{u} \tag{5.2.17}$$

To produce an equivalent linear eigenvalue problem, we define $\boldsymbol{v} \equiv \lambda\, \boldsymbol{u}$ and obtain

$$\boldsymbol{D}\boldsymbol{u} + \boldsymbol{C}\boldsymbol{v} = \lambda\, (\boldsymbol{u} - \boldsymbol{B}\boldsymbol{v}) \tag{5.2.18}$$

Combining this equation with the preceding definition of $\boldsymbol{v}$, we find

$$\begin{bmatrix} \boldsymbol{D} & \boldsymbol{C} \\ \boldsymbol{0} & \boldsymbol{I} \end{bmatrix} \begin{bmatrix} \boldsymbol{u} \\ \boldsymbol{v} \end{bmatrix} = \lambda \begin{bmatrix} \boldsymbol{I} & -\boldsymbol{B} \\ \boldsymbol{I} & \boldsymbol{0} \end{bmatrix} \begin{bmatrix} \boldsymbol{u} \\ \boldsymbol{v} \end{bmatrix} \tag{5.2.19}$$

which presents us with a generalized linear eigenvalue problem involving matrices with twice as large dimensions.

In certain applications, the quadratic term on the left-hand side of equation (5.2.17) involves some but not all of the scalar components of $\boldsymbol{u}$. Inspection of the right-hand side of equation (5.2.19) shows that, in these cases, all elements along one or more columns of the matrix $\boldsymbol{B}$ are equal to zero, and the corresponding block matrix is singular. To prevent this occurence, we define $\boldsymbol{v} = \lambda \boldsymbol{u}'$, where the vector $\boldsymbol{u}'$ contains only those components of $\boldsymbol{u}$ that are involved in the quadratic expression, and work with the extended vector $[\boldsymbol{u}, \boldsymbol{v}]$.

Consider, for example, a 2×2 system with

$$\boldsymbol{B} = \begin{bmatrix} B_{1,1} & 0 \\ B_{2,1} & 0 \end{bmatrix} \tag{5.2.20}$$

where both $B_{1,1}$ and $B_{2,1}$ have nonzero values. The quadratic term in equation (5.2.17) involves only the first component of $\boldsymbol{u}$. In this case, we introduce the new variable $u_3 = \lambda u_1$ and derive the associated linear generalized eigenvalue problem:

$$\begin{bmatrix} D_{1,1} & D_{1,2} & C_{1,1} \\ D_{2,1} & D_{2,2} & C_{2,1} \\ 0 & 0 & 1 \end{bmatrix} \begin{bmatrix} u_1 \\ u_2 \\ u_3 \end{bmatrix} = \lambda \begin{bmatrix} 1 & -C_{1,2} & -B_{1,1} \\ 0 & 1 - C_{2,2} & -B_{2,1} \\ 1 & 0 & 0 \end{bmatrix} \begin{bmatrix} u_1 \\ u_2 \\ u_3 \end{bmatrix} \tag{5.2.21}$$

The block matrix on the right-hand side is generally nonsingular.

PROBLEMS

5.2.3 ***Quadratic eigenvalue problem with sparse quadratic dependence.***

Consider a 3×3 system where the elements in the third column of the matrix $\boldsymbol{B}$ are equal to zero, and formulate the associated nonsingular, linear, generalized eigenvalue problem.

5.2.4 ***Cubic eigenvalue problems.***

Formulate the generalized eigenvalue problem corresponding to a nonlinear eigenvalue problem where $\boldsymbol{A}(\lambda)$ is a third degree polynomial.

5.3 Analytical Results for Diagonal, Triangular, and Tridiagonal Matrices; Circulants

We begin discussing the computation of eigenvalues by presenting analytical results for four classes of matrices with simple structures. The eigenvalues of matrices that do not fall in these categories must generally be computed using numerical methods.

Diagonal and Triangular Matrices

Expressing the determinant of the diagonal or triangular matrix $\boldsymbol{A} - \lambda \boldsymbol{I}$ in terms of the cofactors, we obtain the characteristic polynomial of $\boldsymbol{A}$ in the form of expression (2.9.5). Setting the right-hand side equal to zero, we find that *the eigenvalues of a diagonal or triangular matrix are equal to the diagonal elements.* This category of matrices includes, as a special case, *bidiagonal* matrices. Multiple eigenvalues appear twice or more frequently along the diagonal. If one of the diagonal elements is equal to zero, the matrix is singular.

PROBLEM

5.3.1 ***Eigenvectors of diagonal and triangular matrices.***

Discuss the directions of the eigenvectors of diagonal and triangular matrices.

Tridiagonal Matrices with Constant Diagonal Elements

Consider next an $N \times N$ tridiagonal matrix $\boldsymbol{T}$ with *constant* diagonal, superdiagonal, and subdiagonal elements equal to a, b, and c, respectively,

$$\boldsymbol{T} = \begin{bmatrix} a & b & 0 & 0 & 0 & 0 & \dots & 0 \\ c & a & b & 0 & 0 & 0 & \dots & 0 \\ 0 & c & a & b & 0 & 0 & \dots & 0 \\ \dots & \dots & \dots & \dots & \dots & \dots & \dots & \dots \\ 0 & \dots & 0 & 0 & c & a & b & 0 \\ 0 & \dots & 0 & 0 & 0 & c & a & b \\ 0 & \dots & 0 & 0 & 0 & 0 & c & a \end{bmatrix} \tag{5.3.1}$$

This is a special case of *Toeplitz* matrix; the elements along any diagonal of a general Toeplitz matrix are constant. For reasons that will soon become apparent, we introduce the matrix $\boldsymbol{T}'$ that is equal to $\boldsymbol{T}$, except that the diagonal elements have been replaced by zeros, and the super- and subdiagonal elements have been divided by $(bc)^{1/2}$ so that their product is equal to unity. Thus

$$\boldsymbol{T}' = \begin{bmatrix} 0 & (b/c)^{1/2} & 0 & 0 & 0 & 0 & \dots & 0 \\ (c/b)^{1/2} & 0 & (b/c)^{1/2} & 0 & 0 & 0 & \dots & 0 \\ 0 & (c/b)^{1/2} & 0 & (b/c)^{1/2} & 0 & 0 & \dots & 0 \\ \dots & \dots & \dots & \dots & \dots & \dots & \dots & \dots \\ 0 & \dots & 0 & 0 & (c/b)^{1/2} & 0 & (b/c)^{1/2} & 0 \\ 0 & \dots & 0 & 0 & 0 & (c/b)^{1/2} & 0 & (b/c)^{1/2} \\ 0 & \dots & 0 & 0 & 0 & 0 & (c/b)^{1/2} & 0 \end{bmatrix} \tag{5.3.2}$$

It is clear that

$$\boldsymbol{T} = a\,\boldsymbol{I} + (bc)^{1/2}\boldsymbol{T}' \tag{5.3.3}$$

where $\boldsymbol{I}$ is the $N \times N$ identity matrix. Using the results of Section 2.9, we find that the eigenvalues of $\boldsymbol{T}'$ and $\boldsymbol{T}$ are related by

$$\lambda_T = a + (bc)^{1/2}\,\lambda_{T'} \tag{5.3.4}$$

Next, we introduce the characteristic polynomial of $\boldsymbol{T}'$, denoted as $P_{T'}(\lambda)$, set $\lambda_{T'} = -2\mu$, and find

$$P_{T'}(\lambda_B) = P_{T'}(-2\mu) = \mathrm{Det}(\boldsymbol{T}'') \tag{5.3.5}$$

where $\boldsymbol{T}'' = \boldsymbol{T}' + 2\mu\,\boldsymbol{I}$.

To compute the determinant of the tridiagonal matrix $\boldsymbol{T}''$, we use algorithm (2.4.3) and obtain

$$\boxed{\begin{array}{l} P_0 = 1 \\ P_1 = 2\mu \\ \text{Do } i = 2, N \\ P_i = 2\mu P_{i-1} - P_{i-2} \\ \text{END DO} \\ \text{Det}(\boldsymbol{T}'') = P_N \end{array}} \tag{5.3.6}$$

An independent inspection of the recursion relation for the computation of the *Chebyshev polynomials of the second kind*, shown in Table B.4 of Appendix B, shows that μ is a root of the Nth degree Chebyshev polynomial of the second kind U_N. Closed-form expressions for the roots of these polynomials are listed in Table B.4. Recalling that $\lambda_{T'} = -2\mu$, and substituting the results back into equation (5.3.4), we find that the eigenvalues of $\boldsymbol{T}$ are given by

$$\lambda_{T_m} = a + 2\sqrt{bc}\cos\left(\frac{m\pi}{N+1}\right) \tag{5.3.7}$$

where $m = 1, \ldots, N$. Recall that the application of this formula with $a = 2$, $b = -1$, and $c = -1$ led us to the expression shown in equation (5.1.11). One may readily verify that the associated eigenvectors are given by

$$u_{m,l} = \left(\frac{c}{b}\right)^{1/2} \sin\left(\frac{lm\pi}{N+1}\right) \tag{5.3.8}$$

where $l = 1, \ldots, N$ (Problem 5.3.3).

Unfortunately, an analytical method of computing the eigenvalues of a general tridiagonal matrix with nonconstant elements is not available. The best we can do is identify classes of tridiagonal matrices whose eigenvalues are tabulated zeros of orthogonal polynomials, as described in the following subsection.

PROBLEMS

5.3.2 *Can a tridiagonal matrix with constant elements be singular?*

Discuss the conditions under which a tridiagonal matrix with constant but different elements along the three diagonals can be singular.

5.3.3 *Eigenvectors.*

Verify that the vectors given in equation (5.3.8) are the eigenvectors corresponding to the eigenvalues shown in equation (5.3.7).

Tridiagonal Matrices Whose Eigenvalues Are Roots of Orthogonal Polynomials

Before reading this subsection, please pause and take a look at Appendix B. Equation (B.9), in particular, produces the value of an orthogonal polynomial in terms of the determinant of the tridiagonal matrix $\boldsymbol{T}(t)$ shown in equation (B.11). Expressions for the constant A_0 and for the Greek coefficients corresponding to different families of polynomials arise by comparing the general formulas (B.7) with the recursive formulas listed in Tables B.2–B.9.

Having referred to Appendix B, we introduce the diagonal matrix

$$\boldsymbol{D} = \begin{bmatrix} -\alpha_0 & 0 & 0 & 0 & \dots & 0 \\ 0 & -\alpha_1 & 0 & 0 & \dots & 0 \\ \dots & \dots & \dots & \dots & \dots & \dots \\ 0 & \dots & 0 & 0 & -\alpha_{n-1} & 0 \\ 0 & \dots & 0 & 0 & 0 & -\alpha_n \end{bmatrix} \tag{5.3.9}$$

and the tridiagonal matrix

$$\boldsymbol{T}' = \begin{bmatrix} -\alpha_0\beta_0 & 1 & 0 & 0 & 0 & \dots & 0 \\ \gamma_1 & -\alpha_1\beta_1 & 1 & 0 & 0 & \dots & 0 \\ 0 & \gamma_2 & -a_2\beta_2 & 1 & 0 & \dots & 0 \\ \dots & \dots & \dots & \dots & \dots & \dots & \dots \\ 0 & \dots & 0 & \gamma_{n-2} & -\alpha_{n-2}\beta_{n-2} & 1 & 0 \\ 0 & \dots & 0 & 0 & \gamma_{n-1} & -\alpha_{n-1}\beta_{n-1} & 1 \\ 0 & \dots & 0 & 0 & 0 & \gamma_n & -\alpha_n\beta_n \end{bmatrix} \tag{5.3.10}$$

and recast the matrix $\boldsymbol{T}(t)$ shown in equation (B.11) into the form

$$\boldsymbol{T}(t) = \boldsymbol{T}' - \boldsymbol{D}t \tag{5.3.11}$$

Setting

$$\text{Det}[\boldsymbol{T}(t)] = \text{Det}(\boldsymbol{T}' - \boldsymbol{D}t) = \text{Det}(\boldsymbol{T}'\boldsymbol{D}^{-1} - \boldsymbol{I}t)\,\text{Det}(\boldsymbol{D}) \tag{5.3.12}$$

shows that the eigenvalues of the tridiagonal matrix

$$\boldsymbol{T}'' = \boldsymbol{T}'\boldsymbol{D}^{-1} \tag{5.3.13}$$

are also roots of the $(n+1)$ degree polynomial $p_{n+1}(t)$ corresponding to the matrix $\boldsymbol{T}(t)$. Explicitly, the matrix $\boldsymbol{T}''$ is given by

$$\boldsymbol{T}'' = \begin{bmatrix} \beta_0 & -1/\alpha_1 & 0 & 0 & 0 & \dots & 0 \\ -\gamma_1/\alpha_0 & \beta_1 & -1/\alpha_2 & 0 & 0 & \dots & 0 \\ 0 & -\gamma_2/\alpha_1 & \beta_2 & -1/\alpha_3 & 0 & \dots & 0 \\ \dots & \dots & \dots & \dots & \dots & \dots & \dots \\ 0 & \dots & 0 & -\gamma_{n-2}/\alpha_{n-3} & \beta_{n-2} & -1/\alpha_{n-1} & 0 \\ 0 & \dots & 0 & 0 & -\gamma_{n-1}/\alpha_{n-2} & \beta_{n-1} & -1/\alpha_n \\ 0 & \dots & 0 & 0 & 0 & -\gamma_n/\alpha_{n-1} & \beta_n \end{bmatrix} \tag{5.3.14}$$

The roots of orthogonal polynomials have been tabulated in a number of sources, as discussed in Appendix B.

The discussion in the last subsection of Section 2.9 indicates that the eigevalues of the matrix $\boldsymbol{T}''$ are the same as those of the simpler tridiagonal matrix

$$T''' = \begin{bmatrix} \beta_0 & 1 & 0 & 0 & 0 & \dots & 0 \\ \gamma_1/(\alpha_0\alpha_1) & \beta_1 & 1 & 0 & 0 & \dots & 0 \\ 0 & \gamma_2/(\alpha_1\alpha_2) & \beta_2 & 1 & 0 & \dots & 0 \\ \dots & \dots & \dots & \dots & \dots & \dots & \dots \\ 0 & \dots & 0 & \gamma_{n-2}/(\alpha_{n-3}\alpha_{n-2}) & \beta_{n-2} & 1 & 0 \\ 0 & \dots & 0 & 0 & \gamma_{n-1}/(\alpha_{n-2}\alpha_{n-1}) & \beta_{n-1} & 1 \\ 0 & \dots & 0 & 0 & 0 & \gamma_n/(\alpha_{n-1}\alpha_n) & \beta_n \end{bmatrix} \quad (5.3.15)$$

Note that the products of pairs of off-diagonal elements of T'' and T''' across the diagonal are the same.

Chebyshev polynomials

For example, using the relations given in Table B.3, we find that the eigenvalues of the $(n+1) \times (n+1)$ matrix

$$T''' = \begin{bmatrix} 0 & 1 & 0 & 0 & 0 & \dots & 0 \\ \frac{1}{2} & 0 & 1 & 0 & 0 & \dots & 0 \\ 0 & \frac{1}{4} & 0 & 1 & 0 & \dots & 0 \\ \dots & \dots & \dots & \dots & \dots & \dots & \dots \\ 0 & \dots & 0 & \frac{1}{4} & 0 & 1 & 0 \\ 0 & \dots & 0 & 0 & \frac{1}{4} & 0 & 1 \\ 0 & \dots & 0 & 0 & 0 & \frac{1}{4} & 0 \end{bmatrix} \quad (5.3.16)$$

are the roots of the $(n+1)$ degree Chebyshev polynomial $T_{n+1}(t)$.

Similarly, the eigenvalues of the Toeplitz tridiagonal matrix

$$T''' = \begin{bmatrix} 0 & 1 & 0 & 0 & 0 & \dots & 0 \\ \frac{1}{4} & 0 & 1 & 0 & 0 & \dots & 0 \\ 0 & \frac{1}{4} & 0 & 1 & 0 & \dots & 0 \\ \dots & \dots & \dots & \dots & \dots & \dots & \dots \\ 0 & \dots & 0 & \frac{1}{4} & 0 & 1 & 0 \\ 0 & \dots & 0 & 0 & \frac{1}{4} & 0 & 1 \\ 0 & \dots & 0 & 0 & 0 & \frac{1}{4} & 0 \end{bmatrix} \quad (5.3.17)$$

are the roots of the $(n+1)$ degree Chebyshev polynomial of the second kind $U_{n+1}(t)$.

PROBLEM

5.3.4 ***Tridiagonal matrices with known eigenvalues.***

Construct families of tridiagonal matrices whose eigenvalues are the roots of the Legendre and Laguerre polynomials discussed in Appendix B.

Circulant matrices

A right circulant matrix is a square matrix with the property that each row derives from the previous one by shifting all elements to the right by one place, and bringing the last element to the first place; the elements of the first row are arbitrary. Thus, the elements along any super- or sub-diagonal line are constant. For example, a 3×3 right circulant matrix has the form

$$\boldsymbol{A} = \begin{bmatrix} a & b & c \\ c & a & b \\ b & c & a \end{bmatrix} \tag{5.3.18}$$

Left circulant matrices are defined in a similar fashion. It is not suprising that circulant matrices arise from the mathematical modeling of problems involving some kind of temporal or spatial periodic behavior.

Remarkably, all eigenvalues and matrix eigenvectors of an arbitrary right circulant matrix may be found explicitly in closed form. Let $\boldsymbol{A}$ be a $N \times N$ right circulant, and let θ_k be the kth Nth-root of unity satisfying $\theta_k^N = 1$; that is, $\theta_k = \exp(2\pi i\, k\, /\, N)$ where i is the imaginary unit and $k = 1, \ldots, N$. Direct substitution shows that the eigenvalues of $\boldsymbol{A}$ are given by

$$\lambda_k = A_{1,1} + A_{1,2}\,\theta_k + A_{1,3}\,\theta_k^2 + \ldots + A_{1,N}\theta_k^{N-1} \tag{5.3.19}$$

and the corresponding eigenvectors are

$$\boldsymbol{u}^{(k)} = (1, \theta_k, \theta_k^2, \ldots, \theta_k^{N-1})^T \tag{5.3.20}$$

(Problem 5.3.5). For example, λ_N is equal to the sum of the elements in a row; the corresponding eigenvector is $\boldsymbol{u}^{(N)} = (1, 1, \ldots, 1)^T$.

PROBLEM

5.3.5 ***Properties of circulant matrices.***

(*a*) Can a diagonal or a tridiagonal matrix be a circulant?
(*b*) Explain why if the sum the elements in a row is equal to zero, a circulant matrix is singular.
(*c*) Verify the eigenvalue-eigenvector pairs given in equations (5.3.19), (5.3.20).
(*d*) Compare the eigenvalues and eigenvectors of left circulant matrices.

5.4 *Computing the Roots of the Characteristic Polynomial*

The most straightforward method of finding the eigenvalues of an arbitrary matrix $\boldsymbol{A}$ is by computing the roots of its characteristic polynomial

$$P(\lambda) \equiv \mathrm{Det}(\boldsymbol{A} - \lambda \boldsymbol{I}) = 0 \tag{5.4.1}$$

In principle, this can be done using a general-purpose numerical method for computing the roots of a polynomial, as discussed in Chapter 4. In practice, however, unless the degree of the polynomial is low, and the roots are well separated, this method suffers from the amplification of the round-off error and is not recommended.

In any case, before the root-finding procedure can be implemented, a programmable method of evaluating the characteristic polynomial must be available.

Evaluation of the Coefficients of the Characteristic Polynomial

Householder (1964, Chapter 6) discusses methods for computing the coefficients of the characteristic polynomial of an $N \times N$ matrix $\boldsymbol{A}$.

Leverrier method

A method originally due to Leverrier, expresses $P(\lambda)$ in the form

$$P(\lambda) = (-1)^N(\lambda^N + c_1\lambda^{N-1} + \ldots + c_{N-1}\lambda + c_N) \tag{5.4.2}$$

where $c_N = (-1)^N \mathrm{Det}(\boldsymbol{A})$, and then produces the polynomial coefficients c_i according to the algorithm

$$\boxed{\begin{array}{l} \boldsymbol{B} = \boldsymbol{I} \\ \text{Do } i = 1, N \\ \quad \boldsymbol{D} = \boldsymbol{AB} \\ \quad c_i = -(1/i)\,\mathrm{Tr}(\boldsymbol{D}) \\ \quad \boldsymbol{B} = \boldsymbol{D} + c_i\boldsymbol{I} \\ \text{END DO} \end{array}} \tag{5.4.3}$$

where $\boldsymbol{I}$ is the $N \times N$ identity matrix. For a 2×2 matrix, the algorithm produces the formula for the determinant shown in equation (2.3.7).

Unfortunately, for matrices of large size, the demand on the CPU time is substantial. The computation of the coefficients requires a number of multiplications on the order of N^4.

Krylov-sequence method

Another method of computing the coefficients of the characteristic polynomial, somewhat related to the power method to be discussed in Section 5.5, proceeds by forming a *Krylov sequence*.

Let us select a generally complex nonnull vector $\boldsymbol{x}^{(0)}$ and compute the sequence: $\boldsymbol{x}^{(k)}, k = 1, \ldots, N$, using $\boldsymbol{x}^{(k+1)} = \boldsymbol{A}\boldsymbol{x}^{(k)}$. Assuming that the first N vectors form a complete basis, we express $\boldsymbol{x}^{(N)}$ as a linear combination of them, writing

$$\boldsymbol{x}^{(N)} = c_0\boldsymbol{x}^{(0)} + c_1\boldsymbol{x}^{(1)} + \cdots + c_{N-1}\boldsymbol{x}^{(N-1)} \tag{5.4.4}$$

Next, we note that $\boldsymbol{x}^{(k)} = \boldsymbol{A}^k\boldsymbol{x}^{(0)}$, and obtain

$$(\boldsymbol{A}^N - c_{N-1}\boldsymbol{A}^{N-1} - \cdots - c_1\boldsymbol{A} - c_0)\boldsymbol{x}^{(0)} = \boldsymbol{0} \tag{5.4.5}$$

Since $\boldsymbol{x}^{(0)}$ is arbitrary, the matrix enclosed by the parentheses on the left-hand side must be equal to the null matrix $\mathbf{0}$. Invoking the Hamilton–Cayley theorem, we then find that the characteristic polynomial is given by

$$P(\lambda) = (-1)^N(\lambda^N - c_{N-1}\lambda^{N-1} - \cdots - c_1\lambda - c_0) \tag{5.4.6}$$

The emerging algorithm is as follows:

(5.4.7)

1. Select a generally complex initial vector $\boldsymbol{x}^{(0)}$.
2. Compute the Krylov sequence: $\boldsymbol{x}^{(k)}$, $k = 1, \ldots, N$, using $\boldsymbol{x}^{(k+1)} = \boldsymbol{A}\boldsymbol{x}^{(k)}$.
3. Put the vectors $\boldsymbol{x}^{(l)}$, $l = 0, \ldots, N-1$ at successive columns of the matrix $\boldsymbol{B}$.
4. Solve the $N \times N$ linear system of equations $\boldsymbol{B}\boldsymbol{d} = \boldsymbol{x}^{(N)}$ for the vector $\boldsymbol{d}$.

The characteristic polynomial is given by

$$P(\lambda) = (-1)^N(\lambda^N - d_N\lambda^{N-1} - \cdots - d_2\lambda - d_1) \tag{5.4.8}$$

The operation count is substantial. The computation of all coefficients requires a number of multiplications on the order of N^4. Furthermore, in Section 5.5 we shall see that when N is large, the columns of $\boldsymbol{B}$ are nearly linearly dependent, and this makes $\boldsymbol{B}$ nearly singular.

It is then not surprising that the explicit computation of the coefficients of the characteristic polynomial is designated only for matrices of small or moderate size.

Evaluation of the Characteristic Polynomial in Terms of the Determinant

One way of evaluating $P(\lambda)$ without explicitly referring to its coefficients is by performing the $\boldsymbol{LU}$ decomposition of the matrix $\boldsymbol{B} = \boldsymbol{A} - \lambda\boldsymbol{I}$ as discussed in Sections 2.4, 2.6, and 3.5, using, for instance, the method of Gauss elimination. A serious drawback is that the numerical computations may become prohibitively expensive, and the accuracy may degrade because of the amplification of the round-off error.

In practice, this method is used for *tridiagonal* or nearly triangular *Hessenberg* matrices, for which efficient algorithms for computing $P(\lambda)$ are available.

Tridiagonal matrices

When the matrix $\boldsymbol{A}$, and thus $\boldsymbol{A} - \lambda\boldsymbol{I}$, is *tridiagonal*, Det($\boldsymbol{A} - \lambda\boldsymbol{I}$) may be computed efficiently using algorithm (2.4.3).

When, in addition, $\boldsymbol{A}$ is symmetric, the location of the zeros of Det($\boldsymbol{A} - \lambda\boldsymbol{I}$) can be estimated based on the following theorem. If m is the number of times that the sign of the quantity P_i defined in algorithm (2.4.3) is the same as that of the quantity P_{i+1}, for $i = 0, \ldots, N-1$, then the number of eigenvalues that are greater than λ is equal to m.

Hessenberg matrices

When $\boldsymbol{A}$ and thus $\boldsymbol{A} - \lambda \boldsymbol{I}$ is a *Hessenberg* matrix, $\text{Det}(\boldsymbol{A} - \lambda \boldsymbol{I})$ may be evaluated efficiently using the iterative algorithm (2.4.11).

PROBLEMS

5.4.1 *Eigenvalues of a matrix.*

(*a*) Compute the characteristic polynomial and all three eigenvalues of the following matrix, accurate to the third significant figure, using a method of your choice:

$$\boldsymbol{A} = \begin{bmatrix} 1 & 2 & 3 \\ 3 & 1 & 2 \\ 2 & 3 & 1 \end{bmatrix} \tag{5.4.6}$$

Then compute the corresponding eigenvectors.
(*b*) Compute the matrix $\boldsymbol{B} = \cosh(\boldsymbol{A})$.

5.4.2 *Mathieu's equation.*

Consider Mathieu's equation discussed in Section 5.1. Compute the solution branches $p = a_r(q)$ and $p = b_r(q)$ for $r = 0, 1, 2$ over the range $0 < q < 10$, and compare your numerical results with the values displayed in Figure 5.1.1. The infinite system should be truncated at an appropriate level that permits the computation of the eigenvalues accurate to the third significant figure.

5.5 The Power Method

In its basic no-frills implementation, the power method produces *one* eigenvalue of a real or complex matrix $\boldsymbol{A}$, with some exceptions. But with various enhancements, the method can produce further or all eigenvalues, one eigenvalue at a time. Perhaps more importantly, the method can be used to capture an eigenvalue that is known to reside within a certain targeted area of the complex plane.

The key idea is to successively map a certain initial vector $\boldsymbol{x}^{(0)}$ using the linear mapping associated with the matrix $\boldsymbol{A}$ or a properly modified version of it, until it becomes an eigenvector corresponding to the eigenvalue with the *maximum norm* λ_1, as discussed in Section 2.11. By definition, the spectral radius of the matrix $\boldsymbol{A}$ is $\rho = |\lambda_1|$.

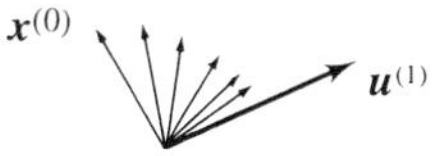

Nonsymmetric Matrices

We begin by assuming that the matrix $\boldsymbol{A}$ is nonsymmetric. The algorithm involves selecting an arbitrary initial vector $\boldsymbol{x}^{(0)}$ and computing the *Krylov sequence* of vectors

$$\boldsymbol{x}^{(k+1)} = \boldsymbol{A}\boldsymbol{x}^{(k)} \tag{5.5.1}$$

Using the asymptotic formula (2.11.8), bearing in mind its underlying assumptions, we find

$$\boldsymbol{x}^{(k+1)} \approx c_1 \lambda_1^{k+1} \boldsymbol{u}^{(1)}, \qquad \boldsymbol{x}^{(k)} \approx c_1 \lambda_1^{k} \boldsymbol{u}^{(1)} \tag{5.5.2}$$

Next, we form the inner product of both sides of these equations with an appropriate vector $\boldsymbol{w}$, and combine the resulting expression to derive the approximation

$$\lambda_1 \cong \frac{\boldsymbol{w}^T \boldsymbol{x}^{(k+1)}}{\boldsymbol{w}^T \boldsymbol{x}^{(k)}} \tag{5.5.3}$$

It is standard practice to identify $\boldsymbol{w}$ with $\boldsymbol{x}^{(k)}$, in which case the right-hand side of equation (5.5.3) reduces to the (1, 1) *Rayleigh–Schwartz quotient*, denoted by $\mu_{0,1}$. It can be shown that this choice yields the best results in the sense of the least-squares method discussed in Section 8.2.

Furthermore, in order to suppress the continuous expansion or growth of $\boldsymbol{x}^{(k)}$, we normalize it *after* λ_1 has been computed according to equation (5.5.3), by dividing its components with its Euclidean norm. The emerging algorithm is as follows:

$$\begin{array}{l} \text{Choose } \boldsymbol{x}^{(0)} \\ \text{Do } k = 0, Max \\ \quad s = (\boldsymbol{x}^{(k)^T} \boldsymbol{x}^{(k)})^{1/2} \\ \quad \boldsymbol{x}^{(k)} \leftarrow \frac{1}{s} \boldsymbol{x}^{(k)} \\ \quad \boldsymbol{x}^{(k+1)} = \boldsymbol{A} \boldsymbol{x}^{(k)} \\ \quad \lambda^{(k+1)} = \boldsymbol{x}^{(k+1)^T} \boldsymbol{x}^{(k)} \\ \text{END DO} \end{array} \tag{5.5.4}$$

A few important remarks regarding the performance and implementation of the method are in order:

1. If the matrix $\boldsymbol{A}$ is real and the dominant eigenvector is complex, we must supply a complex starting vector $\boldsymbol{x}^{(0)}$.

2. The rate of convergence of the sequence $\lambda^{(k)}$ is generally linear, and the convergence speed is determined by the norm of the ratio λ_2/λ_1, where λ_2 is the eigenvalue with the second largest norm. Applying the Aitken extrapolation method described in Section 1.7 expedites the approach to the limit.

3. A problematic situation may occur when $\boldsymbol{A}$ has two dominant complex eigenvalues λ_1 and λ_2 that have the *same norm but possibly different arguments*, corresponding to two *different* eigenvectors $\boldsymbol{u}^{(1)}$ and $\boldsymbol{u}^{(2)}$. For example, λ_1 may be a double eigenvalue, or λ_1 and λ_2 may be complex conjugate. In general, the ratio $\lambda_2/\lambda_1 = \exp(i\theta)$ is a complex number with unit norm and argument θ; $(\lambda_2/\lambda_1)^k = \exp(ki\theta)$, and the counterpart of the second of equations (5.5.2) becomes

$$\boldsymbol{x}^{(k)} \approx c_1 \lambda_1^k (\boldsymbol{u}^{(1)} + (c_2/c_1) \exp(ki\theta)\, \boldsymbol{u}^{(2)}) \tag{5.5.5}$$

   As the iterations continue, $\exp(ki\theta)$ wanders about the unit circle in the complex plane in a commensurate or incommensurate fashion.

   (*a*) When $\theta = 0$, corresponding to a double eigenvalue with two distinct eigenvectors, the vectors $\boldsymbol{x}^{(k)}$ tend to become parallel to the vector $\boldsymbol{u}^{(1)} + (c_2/c_1)\boldsymbol{u}^{(2)}$ whose direction depends on the choice of the starting vector $\boldsymbol{x}^{(0)}$.

   (*b*) When $\theta = \pi$, the iterated vectors flip-flop between the two eigenvectors, yielding

$$
\begin{aligned}
x^{(k)} &\approx c_1\lambda_1^k\left(u^{(1)} + \alpha u^{(2)}\right)\\
x^{(k+1)} &\approx c_1\lambda_1^{k+1}\left(u^{(1)} - \alpha u^{(2)}\right)\\
x^{(k+2)} &\approx c_1\lambda_1^{k+2}\left(u^{(1)} + \alpha u^{(2)}\right)
\end{aligned}
\tag{5.5.6}
$$

where $\alpha = (c_2/c_1)\,\exp(ki\pi)$. Combining these expressions, we find

$$
\lambda_1^2 \cong \frac{x^{(k+2)^T} x^{(k+1)}}{x^{(k)^T} x^{(k+1)}}
\tag{5.5.7}
$$

The computation of the three successive vectors $x^{(k)}$, $x^{(k+1)}$, and $x^{(k+2)}$ is followed by normalization of $x^{(k+2)}$, after the extraction of λ_1. With λ_1 being a known, the eigenvectors $v^{(1)} = c_1 u^{(1)}$ and $v^{(2)} = c_1\alpha u^{(2)}$ emerge by rewriting the first two equations of (5.5.6) as

$$
x^{(k)} \approx \lambda_1^k\left(v^{(1)} + v^{(2)}\right), \qquad x^{(k+1)} \approx \lambda_1^{k+1}\left(v^{(1)} - v^{(2)}\right)
\tag{5.5.8}
$$

and then solving for $v^{(1)}$ and $v^{(2)}$. Thus, in this case, the power method produces two eigenvalues and two eigenvectors.

(*c*) For other values of θ, the Rayleigh–Schwartz quotient does not converge, and the power method fails.

4. When the geometric multiplicity of the dominant eigenvalue is less than its algebraic multiplicity, that is, the set of eigenvectors is deficient, the sequence $x^{(k)}$ converges, but the rate of convergence is slow.

Real-symmetric Matrices

If the matrix A is real and symmetric, the eigenvalues are real, the set of eigenvectors are orthogonal, and the magnitude of each eigenvector can be adjusted to unity yielding an orthonormal set. Using expansion (2.11.7), we find

$$
x^{(m)^T} x^{(l)} = \sum_{i=1}^{N} c_i^2 \lambda_i^{m+e} = \lambda_1^{m+l} \sum_{i=1}^{N} c_i^2 \left(\frac{\lambda_i}{\lambda_1}\right)^{m+l}
\tag{5.5.9}
$$

for any value of m and l. Without loss of generality, we have assumed that the eigenvalues are ordered in a sequence of decreasing magnitude, with λ_1 being the eigenvalue with the maximum norm.

Applying expression (5.5.9) for several values of m and l, and truncating the sum on the right-hand side at an appropriate level, provides us with a system of algebraic equations that can be used to obtain not only the dominant, but also subsequent, eigenvalues.

Truncating, in particular, the sum after the second term, and keeping only the terms that are constant or depend linearly on the parameter

$$
\alpha = (c_2/c_1)^2(\lambda_2/\lambda_1)^{2k}
\tag{5.5.10}
$$

which is small as long as k is sufficiently large, we obtain the *Rayleigh–Schwartz quotients*

$$\mu_{0,1} \equiv \frac{\boldsymbol{x}^{(k)^T}\boldsymbol{x}^{(k+1)}}{\boldsymbol{x}^{(k)^T}\boldsymbol{x}^{(k)}} = \lambda_1\left(1-\left(1-\frac{\lambda_2}{\lambda_1}\right)\alpha\right) \tag{5.5.11}$$

$$\mu_{1,1} \equiv \frac{\boldsymbol{x}^{(k+1)^T}\boldsymbol{x}^{(k+1)}}{\boldsymbol{x}^{(k+1)^T}\boldsymbol{x}^{(k)}} = \lambda_1\left(1-\left(1-\frac{\lambda_2}{\lambda_1}\right)\frac{\lambda_2}{\lambda_1}\alpha\right) \tag{5.5.12}$$

$$\mu_{1,2} \equiv \frac{\boldsymbol{x}^{(k+1)^T}\boldsymbol{x}^{(k+2)}}{\boldsymbol{x}^{(k+1)^T}\boldsymbol{x}^{(k+1)}} = \lambda_1\left(1-\left(1-\frac{\lambda_2}{\lambda_1}\right)\frac{\lambda_2^2}{\lambda_1^2}\alpha\right) \tag{5.5.13}$$

Eliminating α among these equations, and solving for λ_1 and λ_2, we find

$$\lambda_1 = \mu_{1,2} - \frac{(\mu_{1,2}-\mu_{1,1})^2}{\mu_{1,2}-2\mu_{1,1}+\mu_{0,1}} \tag{5.5.14}$$

and

$$\lambda_2 = \lambda_1\frac{\mu_{1,2}-\mu_{1,1}}{\mu_{1,1}-\mu_{0,1}} \tag{5.5.15}$$

The first of these equations simply expresses the Aitken extrapolation; the second equation offers an additional piece of information.

The algorithm involves performing two successive projections, computing λ_1 and λ_2 from equations (5.5.14) and (5.5.15), normalizing $\boldsymbol{x}^{(k+1)}$ by the Euclidean norm, and repeating the computation. In practice, we find that the right-hand side of equation (5.5.15) is notably sensitive to the round-off error. For this reason, this equation is used only to obtain a rough estimate.

Computing Further Eigenvalues

Assume that we have computed the eigenvalue of $\boldsymbol{A}$ with the maximum norm λ_1, and that we want to compute the second largest eigenvalue. In theory, this can be done by using an initial vector $\boldsymbol{x}^{(0)}$ such that $c_1 = 0$; but, in practice, the round-off error will cause c_1 to obtain a nonzero value after only a few iterations.

If the matrix $\boldsymbol{A}$ is *real and symmetric*, the condition $c_1 = 0$ implies that $\boldsymbol{x}^{(0)}$ is and remains, orthogonal to the corresponding eigenvector $\boldsymbol{u}^{(1)}$. The round-off error, however, will destroy the orthogonality after a few iterations. A remedy is to remove the non-orthogonal component by replacing $\boldsymbol{x}^{(k)}$ with

$$\boldsymbol{x}^{(k)} - \frac{\boldsymbol{x}^{(k)^T}\boldsymbol{u}^{(1)}}{\boldsymbol{u}^{(1)^T}\boldsymbol{u}^{(1)}}\boldsymbol{u}^{(1)} \tag{5.5.16}$$

after each or a few iterations.

The procedure just described may be generalized to make further coefficients c_i vanish, thereby allowing up to compute a third and subsequent eigenvalues. Less cumbersome and more attractive methods for achieving this goal, however, are available, as will be described in the remainder of this section.

Eigenvalue Shifting

The simplest way to compute a second eigenvalue is by applying the power iteration to the singular matrix

$$\boldsymbol{B} = \boldsymbol{A} - \lambda_1 \boldsymbol{I} \tag{5.5.17}$$

Note that at least one eigenvalue of $\boldsymbol{B}$ is equal to zero. Since the eigenvalues of $\boldsymbol{B}$ are shifted with respect to those of $\boldsymbol{A}$ by the amount λ_1, the dominant eigenvalue λ_1 has been moved to the origin and does not play a role in the iterations. The power method produces the eigenvalue of $\boldsymbol{B}$ with the maximum norm, λ_B. The corresponding eigenvalue of $\boldsymbol{A}$ is $\lambda_A = \lambda_B + \lambda_1$.

To compute the eigenvalue of $\boldsymbol{A}$ with the minimum norm λ_A, we apply the power method to the inverse matrix $\boldsymbol{B} = \boldsymbol{A}^{-1}$ and thus obtain the corresponding eigenvalue with the maximum norm, λ_B. The eigenvalue of $\boldsymbol{A}$ with the minimum norm is $\lambda_A = 1/\lambda_B$. The success of this method depends on our ability to compute the inverse $\boldsymbol{A}^{-1}$ with adequate precision.

Inverse Iteration

More generally, in order to compute the eigenvalue of $\boldsymbol{A}$ that is closest to the complex number c, we apply the power method to the matrix

$$\boldsymbol{B} = (\boldsymbol{A} - c\boldsymbol{I})^{-1} \tag{5.5.18}$$

and thus obtain the eigenvalue of $\boldsymbol{B}$ with the maximum norm λ_B. The theory of eigenvalue shifting ensures that $\lambda_A = c + 1/\lambda_B$ is the eigenvalue of $\boldsymbol{A}$ that is closest to c. Setting $c = 0$ produces the eigenvalue of $\boldsymbol{A}$ with the minimum norm.

To avoid the demanding computation of the inverse $(\boldsymbol{A} - c\boldsymbol{I})^{-1}$, which may also erode the numerical accuracy, it is preferable to compute the vector $\boldsymbol{x}^{(k+1)}$ by the method of *inverse iteration*, which involves selecting a starting vector $\boldsymbol{x}^{(0)}$ and then solving the linear system of equations

$$(\boldsymbol{A} - c\boldsymbol{I})\boldsymbol{x}^{(k+1)} = \boldsymbol{x}^{(k)} \tag{5.5.19}$$

An expedient way of carrying out these iterations is to perform the $\boldsymbol{LU}$ decomposition of the coefficient matrix $\boldsymbol{A} - c\boldsymbol{I}$ at the outset, and then compute $\boldsymbol{x}^{(k+1)}$ by forward and backward substitution as described in Section 3.3. Practice has shown that the solution of the linear system proceeds without difficulties even when c is close to an eigenvalue, in which case the matrix $\boldsymbol{A} - c\boldsymbol{I}$ is nearly singular. But pathologies do arise when the eigenvalues are tightly spaced. Practical advice and alternative procedures are offered by Press et al. (1992, pp. 487–489).

Size Reduction with a Householder Matrix

We continue to assume that the eigenvalue of $\boldsymbol{A}$ with the maximum norm, λ_1, and the corresponding eigenvector, $\boldsymbol{u}^{(1)}$, are available, and that we have normalized the eigenvector so that $\boldsymbol{u}^{(1)^T}\boldsymbol{u}^{(1)} = 1$. All variables are assumed to be real.

Next, we introduce the real Householder matrix defined in equation (2.3.16),

$$H_{i,j} = \delta_{i,j} - 2w_i\, w_j \tag{5.5.20}$$

where w is a certain real vector of unit length, that is, $w^T w = 1$. Note that $H^{-1} = H$. We want to compute the vector w so that the elements in the first column of the matrix

$$B = HAH \qquad (5.5.21)$$

$$\begin{bmatrix} \lambda_1 & \bigcirc & \bigcirc & & \bigcirc & \bigcirc \\ 0 & \bigcirc & \bigcirc & & \bigcirc & \bigcirc \\ 0 & \bigcirc & \bigcirc & & \bigcirc & \bigcirc \\ & & & & & \\ 0 & \bigcirc & \bigcirc & & \bigcirc & \bigcirc \\ 0 & \bigcirc & \bigcirc & & \bigcirc & \bigcirc \end{bmatrix}$$

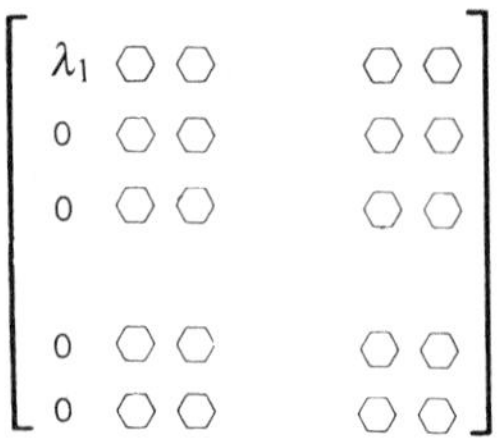

which is similar to A, are equal to zero, except for the very first element $B_{1,1}$ that is equal to λ_1, as shown in the diagram. Recall that this type of transformation is described as a *plane reflection*.

The key observation is that the eigenvalues of the bottom $(N-1) \times (N-1)$ diagonal block of B are also eigenvalues of A. In effect, the problem of computing further eigenvalues of A is reduced to the problem of computing the eigenvalues of a matrix whose dimensions are smaller than those of A by one square unit.

Considering the computation of w, we make two preliminary observations:

- The unit vector

$$e = (1, 0, \ldots, 0)^T \qquad (5.5.22)$$

is the eigenvector of B corresponding to the eigenvalue λ_1.

- The theory of similarity transformations requires that the two corresponding eigenvectors $u^{(1)}$ and e are related by $\alpha e = H u^{(1)}$, where α is a scalar. But since the length of $H u^{(1)}$ is equal to the length of $u^{(1)}$, which is equal to unity, $\alpha = \pm 1$.

Combining these observations, and using the definition of H stated in equation (5.5.20), we obtain

$$\pm e = u^{(1)} - 2w \left(w^T u^{(1)} \right) \qquad (5.5.23)$$

Taking the inner product of both sides of the last equation with w, we find $\pm w_1 = -w^T u^{(1)}$. Finally, we substitute the negative of the left-hand side in place of the term in the parentheses in equation (5.5.23) and obtain

$$w_1 = \left(\frac{1 \pm u_1^{(1)}}{2} \right)^{1/2}$$

$$w_i = \pm \frac{u_i^{(1)}}{2 w_1} \qquad \text{for } i = 2, \ldots, N$$

$$(5.5.24)$$

which can be implemented in a few lines of a computer code. The plus or minus sign is selected so as to send w_1 as far away from zero as possible, and thus reduce the round-off error. If it happens that $u^{(1)} = e$, then $w = 0$, in which case there is no need to carry out the deflation.

For example, the eigenvalue with the maximum norm of the matrix A defined in equation (2.10.3) is $\lambda_1 = \frac{5}{2}$, and the corresponding normalized eigenvector is $u^{(1)} = 37^{-1/2}(6, 1)^T$. Using equations (5.5.24), we find $w_1 = [(1 + 6 \times 37^{-1/2})/2]^{1/2} = 0.997$, and $w_2 = 37^{-1/2}/(2w_1) = 0.082$. The corresponding Householder matrix is

$$H = \begin{bmatrix} 1 - 2w_1^2 & -2w_1 w_2 \\ -2w_1 w_2 & 1 - 2w_2^2 \end{bmatrix} = \begin{bmatrix} -0.986 & -0.164 \\ -0.164 & 0.986 \end{bmatrix} \tag{5.5.25}$$

Finally, the matrix $\boldsymbol{B}$ computed according to equation (5.5.21) is

$$\boldsymbol{B} = \begin{bmatrix} 2.5 & -8.75 \\ 0 & -0.50 \end{bmatrix} \tag{5.5.26}$$

Cleary, $\lambda_2 = -\frac{1}{2}$ is the second eigenvalue of $\boldsymbol{A}$.

PROBLEM

5.5.1 ***Reduction of the Hilbert matrix.***

(*a*) Use the power method to compute one eigenvalue of the Hilbert matrix $\boldsymbol{H}^{(3)}$ defined in equation (2.6.5). Compute the corresponding eigenvector and normalize it to make its length equal to one.
(*b*) Perform the Householder deflation to derive a 2×2 matrix $\boldsymbol{B}$ whose eigenvalues are also eigenvalues of $\boldsymbol{H}^{(3)}$.
(*c*) Compute the eigenvalues of $\boldsymbol{B}$ analytically, and verify that they are also eigenvalues of $\boldsymbol{H}^{(3)}$.

Size Reduction with an Eigenvector

We describe now another method of deflation by size reduction that may be applied to a generally complex matrix. Suppose that we have available one eigenvalue λ_1 of the matrix $\boldsymbol{A}$, and the associated eigenvector $\boldsymbol{u}^{(1)}$. We proceed in three stages:

1. We normalize the eigenvector $\boldsymbol{u}^{(1)}$ so that its maximum element, located at the mth position, is equal to unity. This particular way of selecting m is not obligatory but helps reduce the numerical error.
2. We introduce the matrix

$$B_{i,j} = A_{i,j} - u_i^{(1)} A_{m,j} \tag{5.5.27}$$

   It can be shown that (*a*) the mth row of $\boldsymbol{B}$ is filled up with zeros, (*b*) $\boldsymbol{B}$ shares the eigenvalues of $\boldsymbol{A}$ with the exception of the eigenvalue λ_1 that has been replaced by zero (Problem 5.5.2), and (*c*) the eigenvectors of $\boldsymbol{B}$ corresponding to the unaltered eigenvalues have zeros in their mth entries.
3. Based on the third observation, we find that the eigenvalues of the $(N-1) \times (N-1)$ matrix that arises by rejecting the mth row and the mth column of $\boldsymbol{B}$, and then compressing the matrix, are also eigenvalues of $\boldsymbol{A}$.

For example, the eigenvalue with the maximum norm of the matrix $\boldsymbol{A}$ defined in equation (2.10.3) is $\lambda_1 = \frac{5}{2}$, and the corresponding eigenvector, normalized as described in the text, is $\boldsymbol{u}^{(1)} = (1, \frac{1}{6})^T$, corresponding to $m = 1$. Equation (5.5.27) yields the matrix

$$\boldsymbol{B} = \begin{bmatrix} 1 & 9 \\ \frac{1}{4} & 1 \end{bmatrix} - \begin{bmatrix} 1 \\ \frac{1}{6} \end{bmatrix} [1 \quad 9] = \begin{bmatrix} 0 & 0 \\ \frac{1}{12} & -\frac{1}{2} \end{bmatrix} \tag{5.5.28}$$

Discarding the first row and the first column produces the 1×1 matrix $[-\frac{1}{2}]$, which contains the second eigenvalue of $\boldsymbol{A}$.

PROBLEM

5.5.2 ***Size reduction with an eigenvector.***

Normalize all eigenvectors of a matrix $\boldsymbol{A}$ so that elements in the mth position are equal to unity. Then explain why $A_{m,j}\, u_j^{(p)} = \lambda_p$, and show that (*a*) $\boldsymbol{u}^{(1)}$ falls into the null space of $\boldsymbol{B}$, and (*b*) $\boldsymbol{v}^{(p)} = \boldsymbol{u}^{(p)} - \boldsymbol{u}^{(1)}$ are eigenvectors of $\boldsymbol{B}$ corresponding to the eigenvalues λ_p, where $p \neq 1$. Explain why all $\boldsymbol{v}^{(p)}$ have zeros in their mth entry.

Wielandt Deflation

A different type of deflation that leaves the size of the matrix unchanged can be implemented based on the Wielandt deflation theorem discussed in Section 2.10. The sequential application of this method shifts successively computed eigenvalues to the origin and thus reduces the spectral radius of the matrix $\boldsymbol{A}$ in a progressive fashion. The method can be applied to real as well as complex eigenvalues or matrices. The application to real and symmetric or, more generally, Hermitian matrices, is known as the *Hotelling deflation.*

5.6 *Methods for Computing all Eigenvalues by Similarity Transformations*

The relative ease of computing eigenvalues of diagonal, triangular, tridiagonal, and nearly triangular Hessenberg matrices has motivated the development of a sophisticated class of methods, to be discussed in the remainder of this chapter, whose objective is to transform an arbitrary matrix into a similar matrix that has one of the aforementioned forms. In the second stage, we obtain the eigenvalues of the simpler matrices using one of the methods described earlier in this chapter.

Consecutive similarity transformations

In Section 2.9, we showed that if $\boldsymbol{P}^{(1)}$ is a nonsingular matrix, then the matrix $\boldsymbol{A}^{(1)}$ computed by the *similarity transformation*

$$\boldsymbol{A}^{(1)} = \boldsymbol{P}^{(1)^{-1}}\, \boldsymbol{A}\, \boldsymbol{P}^{(1)} \tag{5.6.1}$$

shares the eigenvalues of $\boldsymbol{A}$. Furthermore, if $\boldsymbol{u}^{(1)}$ is an eigenvector of $\boldsymbol{A}^{(1)}$, then $\boldsymbol{u} = \boldsymbol{P}^{(1)}\, \boldsymbol{u}^{(1)}$ is the corresponding eigenvector of $\boldsymbol{A}$.

We can transform the matrix $\boldsymbol{A}$ in a similar manner once more using the new transformation matrix $\boldsymbol{P}^{(2)}$, to obtain the new matrix

$$\boldsymbol{A}^{(2)} = \boldsymbol{P}^{(2)^{-1}}\, \boldsymbol{A}^{(1)}\, \boldsymbol{P}^{(2)} = \boldsymbol{P}^{(2)^{-1}}\, \boldsymbol{P}^{(1)^{-1}}\, \boldsymbol{A}\, \boldsymbol{P}^{(1)}\boldsymbol{P}^{(2)} \tag{5.6.2}$$

and the eigenvalues of $\boldsymbol{A}$ will still be preserved. If $\boldsymbol{u}^{(2)}$ is an eigenvector of $\boldsymbol{A}^{(2)}$, then $\boldsymbol{u} = \boldsymbol{P}^{(1)}\boldsymbol{P}^{(2)}\, \boldsymbol{u}^{(2)}$ is the corresponding eigenvector of $\boldsymbol{A}$.

Repeating the similarity transformations k times using the transformation matrices $\boldsymbol{P}^{(1)}, \boldsymbol{P}^{(2)}, \ldots, \boldsymbol{P}^{(k)}$, we find that the transformed matrix

$$\boldsymbol{A}^{(k)} = \boldsymbol{P}^{(k)^{-1}} \cdots \boldsymbol{P}^{(2)^{-1}}\boldsymbol{P}^{(1)^{-1}}\boldsymbol{A}\, \boldsymbol{P}^{(1)}\boldsymbol{P}^{(2)} \cdots \boldsymbol{P}^{(k)} \tag{5.6.3}$$

shares the eigenvalues of $\boldsymbol{A}$. If $\boldsymbol{u}^{(k)}$ is an eigenvector of $\boldsymbol{A}^{(k)}$, then $\boldsymbol{u} = \boldsymbol{P}^{(1)}\boldsymbol{P}^{(2)} \ldots \boldsymbol{P}^{(k)}\, \boldsymbol{u}^{(k)}$ is the corresponding eigenvector of $\boldsymbol{A}$.

The problem of computing the eigenvalues of $\boldsymbol{A}$ is thus reduced to the problem of finding the transformation matrices $\boldsymbol{P}^{(i)}$ that render $\boldsymbol{A}^{(k)}$ diagonal, triangular, tridiagonal, or nearly triangular. Clearly, the efficiency of this approach hinges on our ability to compute the inverse matrices $\boldsymbol{P}^{(i)^{-1}}$. It is then not surprising that orthogonal transformation matrices are highly desirable; their inverses are equal to their adjoints.

Targeted forms

Unfortunately, putting an arbitrary matrix into the diagonal form, or more generally into the Jordan canonical form, by similarity transformations, cannot be achieved in a *finite* number of steps. As a compromise, we target either the *tridiagonal* or the nearly triangular *Hessenberg* form, both of which can be achieved with a *finite* number of transformations. For symmetric matrices, the Hessenberg form is equivalent to the symmetric tridiagonal form.

There are two classes of methods for carrying out the transformations. In the first class, the projection matrices are chosen to be either orthogonal *plane rotation* matrices or Householder *plane reflection* matrices, discussed in Section 2.3. The implementation of these methods will be discussed in Sections 5.7 and 5.8. In the second class, the projection matrices arise from matrix factorization. The leading performer in this class is the method of $\boldsymbol{QR}$ decomposition.

$\boldsymbol{QR}$ Decomposition

Let us assume that we have factored a matrix $\boldsymbol{A}$ into the product of two other matrices $\boldsymbol{B}$ and $\boldsymbol{C}$, so that

$$\boldsymbol{A} = \boldsymbol{BC} \tag{5.6.4}$$

We note that the new matrix

$$\boldsymbol{A}^{(1)} = \boldsymbol{CB} = \boldsymbol{B}^{-1}\boldsymbol{AB} \tag{5.6.5}$$

is similar to $\boldsymbol{A}$, with a pertinent transformation matrix equal to $\boldsymbol{B}$.

In the method of $\boldsymbol{QR}$ decomposition, we stipulate that $\boldsymbol{B}$ be orthogonal, denoted as $\boldsymbol{Q}$, and $\boldsymbol{C}$ be upper (right) triangular, denoted as $\boldsymbol{R}$. Three ways of carrying out this decomposition were discussed in Section 2.7. We then proceed according to the following repetitive steps:

- Decompose $\boldsymbol{A} = \boldsymbol{QR}$ and compute the matrix $\boldsymbol{A}^{(1)} = \boldsymbol{RQ}$ that is similar to $\boldsymbol{A}$.
- Decompose $\boldsymbol{A}^{(1)} = \boldsymbol{Q}^{(2)}\boldsymbol{R}^{(2)}$, and compute the matrix $\boldsymbol{A}^{(2)} = \boldsymbol{R}^{(2)}\boldsymbol{Q}^{(2)}$ that is similar to $\boldsymbol{A}$.
- Decompose $\boldsymbol{A}^{(2)} = \boldsymbol{Q}^{(3)}\boldsymbol{R}^{(3)}$, and compute the matrix $\boldsymbol{A}^{(3)} = \boldsymbol{R}^{(3)}\boldsymbol{Q}^{(3)}$ that is similar to $\boldsymbol{A}$.
- Continue in this manner to form the sequence of similar matrices $\boldsymbol{A}^{(k)}$, $k = 1, 2, \ldots$.

It can be shown that (e.g., Wilkinson 1965, Stoer and Bulirsch 1980):

- If the original matrix $\boldsymbol{A}$ is symmetric, tridiagonal, or nearly triangular, the descendant matrices $\boldsymbol{A}^{(k)}$ will also be symmetric, tridiagonal, or nearly triangular.
- When the eigenvalues of $\boldsymbol{A}$ have different magnitudes, as the transformations continue, the transformed matrices $\boldsymbol{A}^{(k)}$ tend to become *upper triangular* with the eigenvalues displayed along the diagonal in ascending order of magnitude. If the matrix $\boldsymbol{A}$ is symmetric, the transformed matrices tend to become diagonal.

- An eigenvalue of algebraic multiplicity m causes the occurrence of a diagonal block with dimension m.

The convergence of the method can be accelerated by eigenvalue shifting, as discussed in Section 5.5 with reference to the power method.

In practice, the method is applied after a matrix has been reduced to the tridiagonal or Hessenberg form, using the methods to be described in the next two sections. The reason is that the $\boldsymbol{QR}$ decomposition of an $N \times N$ tridiagonal or Hessenberg matrix can be effected with only $N - 1$ plane rotations in the planes $(1, 2), (2, 3), \ldots, (N - 1, N)$, which requires a reduced number of operations on the order of $2N^2$ (Press et al. 1992, pp. 480–486). The implementation of the method for real, symmetric, and tridiagonal matrices, and for real Hessenberg matrices, is described in detail by Press et al. (1992, pp. 470, 480).

We also note that selecting the second factor to be an upper triangular matrix $\boldsymbol{R}$ is not obligatory. We could have chosen a lower triangular matrix $\boldsymbol{L}$ instead, obtaining the decomposition $\boldsymbol{A} = \boldsymbol{QL}$ and the similar matrix $\boldsymbol{B} = \boldsymbol{Q}^T\boldsymbol{AQ} = \boldsymbol{LQ}$.

Algorithms and Further References

The challenge of computing all eigenvalues of a generally nonsymmetric matrix has spawned a large body of literature. A wealth of algorithms have been collected by Wilkinson and Reinsch (1971), and useful information on practical implementation is given in the volume *Matrix Eigensystems Routines* of the EISPACK Guide (Vol. 6 of Lecture Notes in Computer Science; Springer-Verlag). In the remainder of this chapter, we discuss what we consider to be a minimal exposition.

5.7 *Transforming a Symmetric Matrix to a Simpler One*

In light of our discussion in the preceding section, we consider methods for transforming a matrix to a similar diagonal, tridiagonal, or Hessenberg matrix, by means of a sequence of transformations expressing *plane rotations* or *plane reflections.* In this section, we consider real and symmetric matrices, and in the next section we consider nonsymmetric matrices. All methods for real and symmetric matrices discussed in this section can be modified to handle Hermitian matrices.

Jacobi's Method of Diagonalizing a Matrix by Sequential Rotations

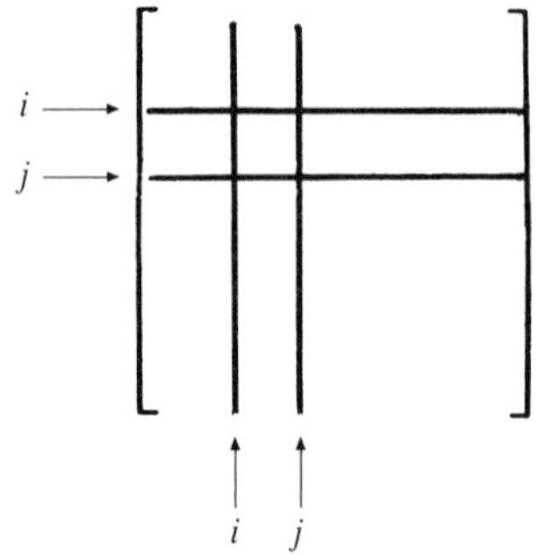

Jacobi's method reduces the magnitude of the off-diagonal elements of a symmetric matrix $\boldsymbol{A}$ in a systematic manner, by performing consecutive *plane rotations.* Although the perfect diagonal form cannot be achieved in a finite number of steps, the matrix obtains a virtually diagonal form after only a moderate number of transformations. The nearly diagonal form allows us to obtain good approximations to the eigenvalues, and then refine them by the method of inverse iteration discussion in Section 5.5.

Considering the $A_{i,j}$ off-diagonal element, where $i \neq j$, we identify the transformation matrix $\boldsymbol{P}$ with the orthogonal plane rotation matrix $\boldsymbol{P}^{(i,j)}$ defined in equation (2.3.15). The elements of the symmetric transformed matrix $\boldsymbol{B} = \boldsymbol{P}^{-1}\boldsymbol{AP}$ are identical to corresponding elements of $\boldsymbol{A}$, except for the elements in the ith and jth columns, and the elements in the the ith and jth rows.

We find, in particular,

$$B_{i,j} = B_{j,i} = -\tfrac{1}{2}(A_{i,i} - A_{j,j}) \sin 2\theta + A_{i,j} \cos 2\theta \tag{5.7.1}$$

which reveals that $B_{i,j}$ and $B_{j,i}$ will vanish provided that θ satisfies the equation

$$\cot 2\theta = \frac{A_{i,i} - A_{j,j}}{2A_{i,j}} \equiv w \tag{5.7.2}$$

We shall take θ to lie in the range $-\pi/4 < \theta < \pi/4$. With the value of θ computed in this manner, the rest of the elements in the ith and jth columns and rows are given by

$$\begin{aligned}
B_{i,i} &= A_{i,i} + A_{i,j} \tan\theta \\
B_{j,j} &= A_{j,j} - A_{i,j} \tan\theta \\
B_{i,k} = B_{k,i} &= A_{i,k} + (A_{j,k} - rA_{i,k}) \sin\theta \qquad \text{for } k = 1, \ldots, N, \text{ for } k \neq i, j \\
B_{k,j} = B_{j,k} &= A_{k,j} - (A_{i,k} + rA_{j,k}) \sin\theta \qquad \text{for } k = 1, \ldots, N, \text{ for } k \neq i, j
\end{aligned} \tag{5.7.3}$$

where we have defined

$$r = \sin\theta/(1 + \cos\theta) \tag{5.7.4}$$

Algorithm

In the implementation of the method, there is no need to solve for θ explicitly. Instead, we carry out the following steps:

1. Compute w from its definition in equation (5.7.2).
2. Recover $\tan\theta$ from the equation

$$\tan\theta = \frac{\mathrm{Sign}(w)}{|w| + \sqrt{w^2 + 1}} \tag{5.7.5}$$

3. Set

$$\cos\theta = \frac{1}{\sqrt{\tan^2\theta + 1}}, \qquad \sin\theta = \cos\theta \tan\theta \tag{5.7.6}$$

4. Compute r from equation (5.7.4).
5. Compute the elements of $\boldsymbol{B}$ corresponding to the amended elements of $\boldsymbol{A}$ from equations (5.7.3); set the rest of the elements of $\boldsymbol{B}$ equal to those of $\boldsymbol{A}$.

The algorithm involves sweeping the off-diagonal elements of the upper or lower triangular block of the evolving matrix according to a certain protocol. As the similarity transformations continue, the transformed matrix tends to become a diagonal matrix at a *quadratic* rate: Each time a full sweep has been completed, the sum of the squares of the norms of the off-diagonal elements has roughly been squared (Problem 5.7.1). In practice, only a few sweeps are necessary to reduce the magnitude of the off-diagonal elements below a negligible level.

Eigenvectors

If the eigenvectors of $\boldsymbol{A}$ are desired, they can be found at the columns of a matrix $\boldsymbol{U}$ that arises by multiplying the transformation matrices with each other, where the evolving product is updated after each iteration:

- Begin by setting $\boldsymbol{U} = \boldsymbol{I}$.
- Update the elements in the ith and jth column of $\boldsymbol{U}$ by setting

$$
\begin{aligned}
U_{k,i}^{New} &= U_{k,i}\ \cos\ \theta + U_{k,j}\ \sin\ \theta \\
U_{k,j}^{New} &= -U_{k,i}\ \sin\ \theta + U_{k,j}\ \cos\ \theta
\end{aligned}
\tag{5.7.7}
$$

for $k = 1, \ldots, N$.

Performance

In practice, Jacobi's method is designated for matrices of moderate size, for example, on the order of 10×10. For larger matrices, the alternative methods discussed later in this section are more efficient.

PROBLEMS

5.7.1 ***Convergence of Jacobi's method.***

Following the notation of the text, show that the sum of the squares of all off-diagonal elements of the matrices $\boldsymbol{A}$ and $\boldsymbol{B}$, respectively denoted as $s(\boldsymbol{A})$ and $s(\boldsymbol{B})$, are related by

$$
s(\boldsymbol{B}) = s(\boldsymbol{A}) - 2|A_{i,j}|^2 \tag{5.7.8}
$$

whereas the sum of the squares of the diagonal elements, respectively denoted as $d(\boldsymbol{A})$ and $d(\boldsymbol{B})$, are related by

$$
d(\boldsymbol{B}) = d(\boldsymbol{A}) + 2|A_{i,j}|^2 \tag{5.7.9}
$$

Since $s(\boldsymbol{B}) \le s(\boldsymbol{A})$ and $d(\boldsymbol{B}) \ge d(\boldsymbol{A})$, the method is guaranteed to converge. Show that, at the final stages of iteration, the rate of convergence is quadratic, as stated in the text.

5.7.2 ***Eigenvalues of a matrix***

(a) Use Jacobi's method to compute the eigenvalues and eigenvectors of the symmetric 3×3 matrix

$$
\boldsymbol{A} = \begin{bmatrix} 2 & 3 & 4 \\ 3 & 5 & 4 \\ 4 & 4 & 2 \end{bmatrix} \tag{5.7.10}
$$

Is this matrix positive-definite?
(b) Compute the matrix $\boldsymbol{B} = \sin(\boldsymbol{A})$.

Given's Method for Making a Symmetric Matrix Tridiagonal

Given's method puts the matrix into the tridiagonal instead of the diagonal form, by executing a sequence of plane rotations. The reduction to the tridiagonal form can be achieved *in a single sweep.*

More specifically, we use the plane rotation matrices $\boldsymbol{P}^{(i,j)}$ with appropriate values for the angle θ to annihilate the $(i-1, j)$ element, proceeding along successive rows. For example, in the first pass, we use the rotation matrices

$$\boldsymbol{P}^{(2,3)}, \boldsymbol{P}^{(2,4)}, \boldsymbol{P}^{(2,5)}, \ldots, \boldsymbol{P}^{(2,N)}$$

to annihilate the elements

$$(1,3), (1,4), (1,5), \ldots, (1,N)$$

The crux of the method is that once an element has been zeroed, it is not resurrected.

In general, however, Given's method requires a higher computational effort than its rivals discussed in the remainder of this section. Improvements of, and pragmatic difficulties with, the method are discussed by Golub and van Loan (1989).

Lanczos's Method for Recasting a Real Symmetric Matrix into the Symmetric Tridiagonal Form

Let us select an initial vector $\boldsymbol{p}^{(1)}$ and transform it $N-1$ times in a certain manner that involves the $N \times N$ matrix $\boldsymbol{A}$, to obtain a set of N linearly independent vectors $\boldsymbol{p}^{(i)}$, $i = 1, \ldots, N$. The changing orientations of the continuously transformed vectors allow us to explore all directions in the N-dimensional space and ought to permit the computation of the eigenvalues of $\boldsymbol{A}$. This idea was put into practice in Section 5.4 to compute the coefficients of the characteristic polynomial from the Krylov sequence $\boldsymbol{p}^{(i+1)} = \boldsymbol{A}\boldsymbol{p}^{(i)}$.

The Krylov sequence, however, biases the vectors toward the eigenvector corresponding to the dominant eigenvalue and does not allow us to explore the whole N-dimensional space with adequate resolution. A much better sequence is constructed by stipulating that the vectors $\boldsymbol{p}^{(i)}$ form an orthonormal set, which is produced modifying the Krylov sequence using a variant of the Gram–Schmidt orthogonalization method discussed at the end of Section 2.1, according to Algorithm 5.7.1.

A key observation is that since the vectors $\boldsymbol{p}^{(i)}$, $i = 1, \ldots, N$, provide us with a complete basis, the vector $\boldsymbol{A}\boldsymbol{p}^{(N)}$ is expressible as a linear combination of them, and $\boldsymbol{p}^{(N+1)}$ must necessarily vanish. This observation suggests a test for the debugging of the code that programs Algorithm 5.7.1.

Inspection of the equations involved in Algorithm 5.7.1. shows that

$$\boldsymbol{A}\boldsymbol{P} = \boldsymbol{P}\boldsymbol{T} \tag{5.7.11}$$

where $\boldsymbol{P}$ is an orthogonal matrix whose ith column is the vector $\boldsymbol{p}^{(i)}$, and $\boldsymbol{T}$ is a symmetric tridiagonal matrix similar to $\boldsymbol{A}$ given by:

$$\boldsymbol{T} = \begin{bmatrix} \alpha_1 & \beta_1 & 0 & 0 & 0 & \ldots & 0 \\ \beta_1 & \alpha_2 & \beta_2 & 0 & 0 & \ldots & 0 \\ 0 & \beta_2 & \alpha_3 & \beta_3 & 0 & \ldots & 0 \\ \ldots & \ldots & \ldots & \ldots & \ldots & \ldots & \ldots \\ 0 & \ldots & 0 & \beta_{N-3} & \alpha_{N-2} & \beta_{N-2} & 0 \\ 0 & \ldots & 0 & 0 & \beta_{N-2} & \alpha_{N-1} & \beta_{N-1} \\ 0 & \ldots & 0 & 0 & 0 & \beta_{N-1} & \alpha_N \end{bmatrix} \tag{5.7.12}$$

Thus the eigenvalues, the trace, and the determinant of $\boldsymbol{T}$ are equal to those of $\boldsymbol{A}$.

ALGORITHM 5.7.1 Lanczos's method for transforming a symmetric matrix $\boldsymbol{A}$ into the symmetric tradiagonal matrix $\boldsymbol{T}$ given in equation (5.7.12).

Specify $\boldsymbol{p}^{(1)}$

$$\beta_0 = \left(\boldsymbol{p}^{(1)^T}\boldsymbol{p}^{(1)}\right)^{1/2}$$

$$\boldsymbol{p}^{(1)} \leftarrow \frac{1}{\beta_0}\boldsymbol{p}^{(1)} \qquad \text{Normalize } \boldsymbol{p}^{(1)}$$

$$\alpha_1 = \boldsymbol{p}^{(1)^T}\boldsymbol{A}\boldsymbol{p}^{(1)}$$

$$\boldsymbol{p}^{(2)} = \boldsymbol{A}\boldsymbol{p}^{(1)} - \alpha_1\boldsymbol{p}^{(1)} \qquad \text{Second Krylov vector}$$

$$\beta_1 = \left(\boldsymbol{p}^{(2)^T}\boldsymbol{p}^{(2)}\right)^{1/2}$$

$$\boldsymbol{p}^{(2)} \leftarrow \frac{1}{\beta_1}\boldsymbol{p}^{(2)} \qquad \text{Normalize } \boldsymbol{p}^{(2)}$$

Do $i = 2, N$

$$\alpha_i = \boldsymbol{p}^{(i)^T}\boldsymbol{A}\boldsymbol{p}^{(i)}$$

$$\boldsymbol{p}^{(i+1)} = \boldsymbol{A}\boldsymbol{p}^{(i)} - \alpha_i\boldsymbol{p}^{(i)} - \beta_{i-1}\boldsymbol{p}^{(i-1)} \qquad i+1 \text{ Krylov vector}$$

$$\beta_i = \left(\boldsymbol{p}^{(i+1)^T}\boldsymbol{p}^{(i+1)}\right)^{1/2}$$

$$\boldsymbol{p}^{(i+1)} \leftarrow \frac{1}{\beta_i}\boldsymbol{p}^{(i+1)}$$

END DO

There are two circumstances under which $\boldsymbol{p}^{(i+1)}$ may vanish for $i < N$. The first one occurs when one or more coefficients in the linear expansion of $\boldsymbol{p}^{(1)}$ in terms of the eigenvectors vanish, but this can be avoided by restarting the computation with a different initial vector. The second exception occurs when $\boldsymbol{A}$ has repeated eigenvalues; in this case, we simply continue the computation with a different $\boldsymbol{p}^{(i+1)}$ that is orthogonal to the incomplete set $\boldsymbol{p}^{(j)}$, $j = 1, \ldots, i-1$.

Householder's Method for Recasting a Real Symmetric Matrix into the Symmetric Tridiagonal Form

The Householder method described in this subsection transforms a symmetric matrix into a similar symmetric tridiagonal matrix by executing $N-2$ sequential reflections using the familiar Householder matrices $\boldsymbol{H} = \boldsymbol{I} - 2\boldsymbol{w}\boldsymbol{w}^T$, where $\boldsymbol{w}^T\boldsymbol{w} = 1$. Counting the operations shows that the Householder method is more efficient than the Lanczos method described in the preceding subsection. The advantage of the Lanczos method is that it can readily be extended to nonsymmetric matrices, as will be discussed in Section 5.8, whereas the Householder method is restricted to symmetric and, more generally, Hermitian matrices. A modified version of the Householder method, however, can reduce an arbitrary nonsymmetric matrix to a nearly triangular Hessenberg matrix, as will be described in Section 5.8.

We begin developing the Householder method by stipulating that the first element w_1 of the vector $\boldsymbol{w}$ defining the Householder matrix be equal to zero. As a consequence, all elements of the first row and first column of $\boldsymbol{H}$ are equal to zero, except for the very first diagonal element $H_{1,1}$ that is equal to unity. Since the length of $\boldsymbol{w}$ must be equal to unity, we are left with $N-2$ degrees of freedom, which we can use to make all but the first and second elements in the first row and first column of the similar matrix $\boldsymbol{B} = \boldsymbol{HAH}$ vanish.

Examining the structure of the matrix $\boldsymbol{B}$, we find that our goal will be fulfilled, provided that: (*a*) we set

$$\boldsymbol{w} = (0, \alpha A_{1,2},\ \beta A_{1,3},\ \beta A_{1,4},\ \ldots,\ \beta A_{1,N})^T \tag{5.7.13}$$

and (*b*) we require that the two scalar constants α and β satisfy the nonlinear algebraic equation

$$\alpha^2 A_{1,2}^2 + \beta^2 c^2 = 1 \tag{5.7.14}$$

which expresses the normalization condition $\boldsymbol{w}^T \boldsymbol{w} = 1$, and the equation

$$\alpha A_{1,2}^2 + \beta c^2 = \frac{1}{2\beta} \tag{5.7.15}$$

which arises from the requirement that $B_{1,N} = 0$. We have defined

$$c^2 \equiv \sum_{j=3}^{N} A_{1,j}^2 \tag{5.7.16}$$

Fortunately, the solution of the preceding algebraic system can be found in analytical form, and the result is

$$\alpha = \beta \frac{A_{1,2} + s}{A_{1,2}}, \qquad \beta^2 = \frac{1}{2s(A_{1,2} + s)} \tag{5.7.17}$$

where we have defined

$$s^2 = c^2 + A_{1,2}^2 \tag{5.7.18}$$

To reduce the round-off error, it is preferable to choose the sign of s to be the same as that of $A_{1,2}$. The other choice may make the denominator of the second fraction of equations (5.7.17) dangerously close to zero.

After some straightforward algebraic manipulations, we find that the transformed matrix $\boldsymbol{B} = \boldsymbol{HAH}$ is given by

$$B_{i,j} = A_{i,j} - u_i\, v_j - v_i\, u_j$$

where

$$\boldsymbol{u} = (0,\ A_{1,2} + s,\ A_{1,3}, \ldots,\ A_{1,N})^T$$

$$\boldsymbol{v} = \boldsymbol{z} - \gamma \boldsymbol{u} \tag{5.7.19}$$

$$\boldsymbol{z} = \frac{1}{s(A_{1,2} + s)} \boldsymbol{A}\boldsymbol{u}, \qquad \gamma = \frac{1}{2s(A_{1,2} + s)} \boldsymbol{z}^T \boldsymbol{u}$$

Having completed this similarity transformation, we repeat the computation with the reduced $(N-1)\times(N-1)$ lower diagonal part of $\boldsymbol{B}$ that arises by neglecting its first row and first column, and we continue the process until the matrix reduces to the desired tridiagonal form.

At the kth pass, the first k elements of the vector $\boldsymbol{w}$ are set equal to zero. The elements in the first k columns and rows of the associated Householder matrix are equal to zero, except for the diagonal elements that are equal to unity. The kth transformation leaves previously reduced columns and rows of the transformed matrix unaffected. The whole process requires a total of $N-2$ reductions.

PROBLEM

5.7.3 ***Lanczos and Householder methods.***

(*a*) Use the Lanczos method to reduce the matrix $\boldsymbol{A}$ given in equation (5.7.10) to a similar tridiagonal matrix. Compute the trace, the eigenvalues, and the determinant of the tridiagonal matrix using a method of your choice, and confirm that they are equal to the corresponding quantities of the matrix $\boldsymbol{A}$.
(*b*) Repeat part (*a*) using the Householder method.

5.8 *Transforming an Arbitrary Matrix to a Tridiagonal or Hessenberg Matrix*

Culminating our efforts, we finally discuss methods for computing *all* eigenvalues of an *arbitrary*, not necessarily symmetric, matrix. Our strategy is to first transform the matrix to a similar tridiagonal or a nearly triangular Hessenberg matrix, and then apply one of the methods described in the preceding sections. In practice, the spectrum of eigenvalues of the tridiagonal or Hessenberg matrix is usually found by the method of $\boldsymbol{QR}$ decomposition discussed in Section 5.6, and its advanced implementations discussed in the cited monographs.

Lanczos's Method for Making a Matrix Tridiagonal

The general idea underlying the Lanczos method was described in Section 5.7 in our discussion of symmetric matrices; a quick review is recommended to help set up the grounds for the present developments.

Considering a nonsymmetric $N\times N$ matrix $\boldsymbol{A}$, we select two initial vectors $\boldsymbol{p}^{(1)}$ and $\boldsymbol{s}^{(1)}$ that are *not* orthogonal, and simultaneously construct the two sequences of vectors $\boldsymbol{p}^{(i)}, i=1,\ldots,N$, and $\boldsymbol{s}^{(i)}, i=1,\ldots,N$, so that

$$\boldsymbol{s}^{(i)T}\boldsymbol{p}^{(j)}=\delta_{i,j} \tag{5.8.1}$$

where $\delta_{i,j}$ is the Kronecker delta, working according to Algorithm 5.8.1. Inspection of the equations involved in this algorithm shows that

$$\boldsymbol{SAP}=\boldsymbol{T} \tag{5.8.2}$$

where the ith *column* of the matrix $\boldsymbol{P}$ is the vector $\boldsymbol{p}^{(i)}$, the ith *row* of the matrix $\boldsymbol{S}$ is the vector $\boldsymbol{s}^{(i)}$, and $\boldsymbol{T}$ is the tridiagonal matrix

ALGORITHM 5.8.1 Lanczos method for transforming an arbitrary matrix $\boldsymbol{A}$ to a similar tridiagonal matrix $\boldsymbol{T}$ defined in equation (5.8.3).

Specify $\boldsymbol{p}^{(1)}$ and $\boldsymbol{s}^{(1)}$ — They should not be orthogonal

$$\boldsymbol{s}^{(1)} \leftarrow \frac{1}{\boldsymbol{s}^{(1)^T}\boldsymbol{p}^{(1)}}\,\boldsymbol{s}^{(1)}$$

$$\alpha_1 = \boldsymbol{s}^{(1)^T}\boldsymbol{A}\boldsymbol{p}^{(1)}$$

$$\boldsymbol{p}^{(2)} = \boldsymbol{A}\boldsymbol{p}^{(1)} - \alpha_1\,\boldsymbol{p}^{(1)}$$

$$\beta_1 = \left(\boldsymbol{p}^{(2)^T}\boldsymbol{p}^{(2)}\right)^{1/2}$$

$$\boldsymbol{s}^{(2)} = \boldsymbol{A}^T\boldsymbol{s}^{(1)} - \alpha_1\,\boldsymbol{s}^{(1)}$$

$$\gamma_1 = \frac{\boldsymbol{s}^{(2)^T}\boldsymbol{p}^{(2)}}{\beta_1}$$

$$\boldsymbol{p}^{(2)} \leftarrow \frac{1}{\beta_1}\,\boldsymbol{p}^{(2)}$$

$$\boldsymbol{s}^{(2)} \leftarrow \frac{1}{\gamma_1}\,\boldsymbol{s}^{(2)}$$

Do $i = 2, N$

$$\alpha_i = \boldsymbol{s}^{(i)^T}\boldsymbol{A}\boldsymbol{p}^{(i)}$$

$$\boldsymbol{p}^{(i+1)} = \boldsymbol{A}\boldsymbol{p}^{(i)} - \alpha_i\boldsymbol{p}^{(i)} - \beta_{i-1}\,\boldsymbol{p}^{(i-1)}$$

$$\beta_i = \left(\boldsymbol{p}^{(i+1)^T}\boldsymbol{p}^{(i+1)}\right)^{1/2}$$

$$\boldsymbol{s}^{(i+1)} = \boldsymbol{A}^T\boldsymbol{s}^{(i)} - \alpha_i\boldsymbol{s}^{(i)} - \beta_{i-1}\boldsymbol{s}^{(i-1)}$$

$$\gamma_i = \frac{\boldsymbol{s}^{(i+1)^T}\boldsymbol{p}^{(i+1)}}{\beta_i}$$

$$\boldsymbol{p}^{(i+1)} \leftarrow \frac{1}{\beta_i}\boldsymbol{p}^{(i+1)}$$

$$\boldsymbol{s}^{(i+1)} \leftarrow \frac{1}{\gamma_i}\,\boldsymbol{s}^{(i+1)}$$

End Do

$$
T = \begin{bmatrix}
\alpha_1 & \gamma_1 & 0 & 0 & 0 & \cdots & 0 \\
\beta_1 & \alpha_2 & \gamma_2 & 0 & 0 & \cdots & 0 \\
0 & \beta_2 & \alpha_3 & \gamma_3 & 0 & \cdots & 0 \\
\cdots & \cdots & \cdots & \cdots & \cdots & \cdots & \cdots \\
0 & \cdots & 0 & \beta_{N-3} & \alpha_{N-2} & \gamma_{N-2} & 0 \\
0 & \cdots & 0 & 0 & \beta_{N-2} & \alpha_{N-1} & \gamma_{N-1} \\
0 & \cdots & 0 & 0 & 0 & \beta_{N-1} & \alpha_N
\end{bmatrix}
\tag{5.8.3}
$$

The stipulation (5.8.1) ensures that $\boldsymbol{S}$ is the inverse of $\boldsymbol{P}$, and thereby guarantees that $\boldsymbol{T}$ is similar to $\boldsymbol{A}$. Thus the eigenvalues, the trace, and the determinant of $\boldsymbol{T}$ are equal to the corresponding quantities of $\boldsymbol{A}$.

The theory behind the method and possible pathologies of its implementation are discussed by Wilkinson (1965, pp. 388–405) and Golub and van Loan (1989, pp. 502–503).

PROBLEM

5.8.1 ***Lanczos method.***

Use the Lanczos method to reduce the matrix $\boldsymbol{A}$ given in equation (2.1.23) to a similar tridiagonal matrix. Compute the trace, the eigenvalues, and the determinant of the tridiagonal matrix using a method of your choice, and confirm that they are equal to the corresponding quantities of the matrix $\boldsymbol{A}$.

Reduction to the Hessenberg Form

Upper Hessenberg matrix:

$$
A = \begin{bmatrix}
A_{1,1} & A_{1,2} & A_{1,3} & \cdots & \cdots & A_{1,N} \\
A_{2,1} & A_{2,2} & A_{2,3} & \cdots & \cdots & A_{2,N} \\
0 & A_{3,2} & A_{3,3} & \cdots & \cdots & \cdots \\
0 & 0 & \ddots & \cdots & \cdots & \cdots \\
\cdots & 0 & 0 & A_{N-1,N-2} & A_{N-1,N-1} & A_{N-1,N} \\
0 & \cdots & 0 & 0 & A_{N,N-1} & A_{N,N}
\end{bmatrix}
$$

Reducing a matrix of large size to a similar nearly triangular Hessenberg matrix can be done much more efficiently than reducing it to a tridiagonal matrix using, for example, the Lanczos method. Recall that the Hessenberg form is the triangular form, but the elements along the sub- or super-diagonal line are not necessarily equal to zero. The upper Hessenberg form is shown in the diagram. A symmetric Hessenberg matrix is a symmetric tridiagonal matrix.

There are several methods of carrying out the reduction to the Hessenberg form, with varying degrees of efficiency and sophistication. One way is to execute $N-2$ sequential plane rotations using a straightforward variation of Given's method discussed in Section 5.7. But this method is not the best performer.

Another way is to carry out $N - 2$ successive plane reflections using the Householder matrices $\boldsymbol{H} = \boldsymbol{I} - 2\boldsymbol{w}\boldsymbol{w}^T$. The implementation of the method for symmetric matrices was described in Section 5.7; a related method was described in Section 5.5 in the context of eigenvalue deflation. At the kth step, we choose the first k elements of $\boldsymbol{w}$ to be equal to zero. We then note that premultiplying or postmultiplying a matrix by $\boldsymbol{H}$ leaves, respectively, the first $k - 1$ columns or rows unchanged, and compute the last $N - k$ components of $\boldsymbol{w}$ so as to make the last $N - k - 1$ components of the kth column of $\boldsymbol{HAH}$ vanish, while satisfing the required condition $\boldsymbol{w}^T\,\boldsymbol{w} = 1$.

The most efficient method of reducing a matrix to the Hessenberg matrix is by a variant of Gauss elimination. To motivate the method, consider reducing the arbitrary $N \times N$ matrix $\boldsymbol{A}$ to the upper Hessenberg matrix

$$
\boldsymbol{B} = \boldsymbol{U} - \boldsymbol{S} \tag{5.8.4}
$$

where $\boldsymbol{U}$ is an upper triangular matrix and $\boldsymbol{S}$ is a subdiagonal matrix. We introduce the upper triangular matrix $\boldsymbol{L}$, and require that

$$\boldsymbol{B} = \boldsymbol{L}^{-1}\boldsymbol{A}\boldsymbol{L} \tag{5.8.5}$$

or

$$\boldsymbol{A}\boldsymbol{L} = \boldsymbol{L}\boldsymbol{U} - \boldsymbol{L}\boldsymbol{S} \tag{5.8.6}$$

This shows that the elements of $\boldsymbol{L}$ may be computed one step ahead of those of $\boldsymbol{U}$, column by column; the first column of $\boldsymbol{L}$ may be chosen arbitrarily.

In practice, the method is implemented as a variant of the Gauss elimination method with row pivoting according to the following $N - 2$ steps executed for $m = 1, \ldots, N - 2$ (Press et al. 1992, p. 478):

1. At the mth stage, find the element in the mth column below the diagonal, with the maximum norm. If it is zero, increase the value of m by one and repeat this step.
2. Suppose the element with the maximum norm in the mth colum is in the kth row. Interchange rows k and $m + 1$, and columns k and $m + 1$.
3. For $i = m + 2, m + 3, \ldots, N$, execute three substeps:
   - Compute the multiplier $c_{i,m+1} = A_{i,m}/A_{m+1,m}$.
   - Subtract from the ith row the $m + 1$ row multiplied by $c_{i,m+1}$.
   - Add to the $m + 1$ column the ith column multiplied by $c_{i,m+1}$.

PROBLEM

5.8.2 *Householder method.*

Develop an algorithm for reducing a matrix $\boldsymbol{A}$ to a similar upper Hessenberg matrix using the Householder method, and verify its reliability by means of a numerical example. *Hint*: Should you get desperate, refer to Cullen (1994, p. 139).

References

Abramowitz, M., and Stegun, I. A., 1972, *Handbook of Mathematical Functions.* Dover.
Bodewig, E., 1959, *Matrix Calculus.* North-Holland.
Cullen, C. G., 1994, *An Introduction to Numerical Linear Algebra.* PWS-Kent.
Householder, A. S., 1964, *The Theory of Matrices in Numerical Analysis.* Reprinted by Dover in 1975.
Golub, G. H., and van Loan, C. F., 1989, *Matrix Computations.* The Johns Hopkins University Press.
Kerner, W., 1989, Large-scale complex eigenvalue problems. *J. Comput. Phys.* **85**, 1–84.
Moler, C. B., and Stewart, G. W., 1973, An algorithm for generalized matrix eigenvalue problems. *SIAM J. Numer. Anal.* **10**, 241–56.
Muller, W. H., 1947, Handelingen van het XXX Nederlandse Natuurk. en Geneeskunding Congres. Delft, 109.
Press, W. H., Flannery, B. P., Teukolsky, S. A., and Vetterling, W. T., 1992, *Numerical Recipes.* 2nd ed. Cambridge University Press.
Stoer, J., and Bulirsch, R., 1980, *Introduction to Numerical Analysis.* Springer-Verlag.
Wilkinson, J. H., and Reinsch, C., 1971, *Linear Algebra, Vol. II of Handbook for Automatic Computation.* Springer-Verlag.
Wilkinson, J. H., 1965, *The Algebraic Eigenvalue Problem.* Oxford University Press.

Chapter 6

Function Interpolation and Differentiation

6.1 Interpolation and Extrapolation

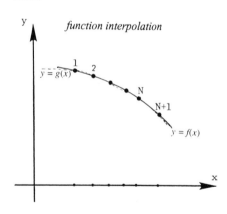

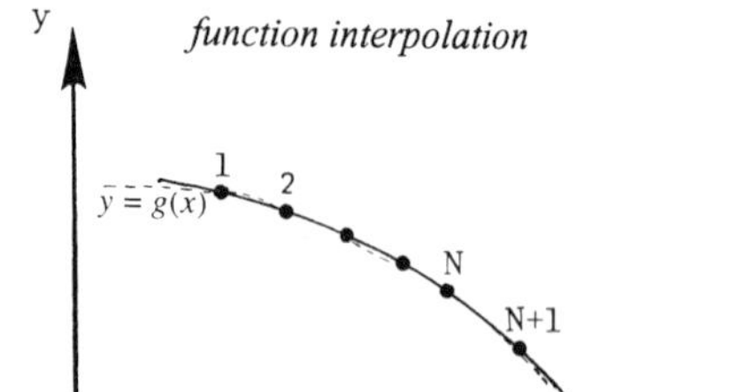

Given the values of a function $f(x)$ at the $N+1$ data points x_i, $i = 1, \ldots, N+1$, as shown in the diagram, we want to compute its value at an intermediate point, at a point that lies on the left of the first point x_1, or on the right of the last point x_{N+1}. In the first case, we are faced with the problem of *function interpolation*, and in the second and third cases, we are faced with the problem of *function extrapolation.*

Problems of this nature, and their generalizations to higher dimensions involving functions of more than one variable, arise routinely in various mathematical, computational, scientific, and engineering applications. For example, the data points may represent the entries of a mathematical or thermodynamics table published in a handbook. Thus the independent variable x may represent the temperature, and the dependent variable $y = f(x)$ may represent the specific entropy of a particular chemical compound at a certain pressure.

Furthermore, function interpolation is an essential component of a broad class of numerical methods for solving ordinary and partial differential equations, including *finite-difference, finite-element,* and *boundary-element* methods. The objective of these methods is to compute the value of an unknown function at a number of discrete points; interpolation is then invoked to produce or represent the solution in the rest of its domain of definition.

General Policy

Function interpolation and extrapolation are carried out by replacing the unknown function $f(x)$ with another smooth function $g(x)$ that passes through the data points, represented by the dashed line in the last drawing. This means that the interpolating function $g(x)$ is required to satisfy the matching conditions

$$f(x_i) = g(x_i) \tag{6.1.1}$$

for $i = 1, \ldots, N+1$, with *no additional stipulations or constraints.*

Polynomial and trigonometric interpolation

Two popular choices for the interpolating function $g(x)$ is an Nth degree polynomial $P_N(x)$, or a Fourier series containing N trigonometric functions. In the first case we perform *polynomial interpolation* to be discussed in the subsequent sections of this chapter, and in the second case we perform *trigonometric interpolation* to be discussed in Sections 8.6–8.8.

A explanation is in order regarding our seemingly bizarre notation. We use $N+1$ data points $x_i, i = 1, \ldots, N+1$, because we want to obtain an Nth degree interpolating polynomial. Many texts of numerical analysis number the points $x_i, i = 0, \ldots, N$; but why should the first point be penalized by indexing it with an unnatural number? Some authors use N points $x_i, i = 1, \ldots, N$, to obtain an $N-1$ degree interpolating polynomial; unfortunately, this attractive notation has not found many subscribers, and we are not about to take any chances.

Interpolation with rational functions

Although we shall not discuss it, interpolation and, more importantly, extrapolation with rational functions, that is, the ratio between two polynomials, has significant advantages for a certain class of functions. Theoretical and practical issues of rational function interpolation are discussed by Rivlin (1969, Chapter 5) and Stoer and Bulrisch (1980, pp. 58–72).

Derivatives and Integrals

Once the interpolating function $g(x)$ is available, it can be differentiated or integrated to yield approximations to the derivatives or to a definite integral of the interpolated function $f(x)$ over a certain interval. The results are expressed in terms of the coordinates of the data points. Differentiation produces *finite-difference approximations,* to be discussed later in this section, and integration produces *numerical rules* and *quadratures,* to be discussed in Chapter 7.

Interpolation Versus Approximation

In Chapter 8, we shall discuss the somewhat related but distinct topic of *function approximation.* The important difference between function interpolation and function approximation is that the graph of the approximating function $q(x)$ is not required to pass through the data points, which means that we forfeit the matching conditions (6.1.1). Instead, $q(x)$ is required to represent the unknown function $f(x)$ in a certain optimal sense over a particular domain of interest. It is then not surprising that function interpolation is sometimes called an *exact fit,* whereas function approximation is called an *approximate fit.*

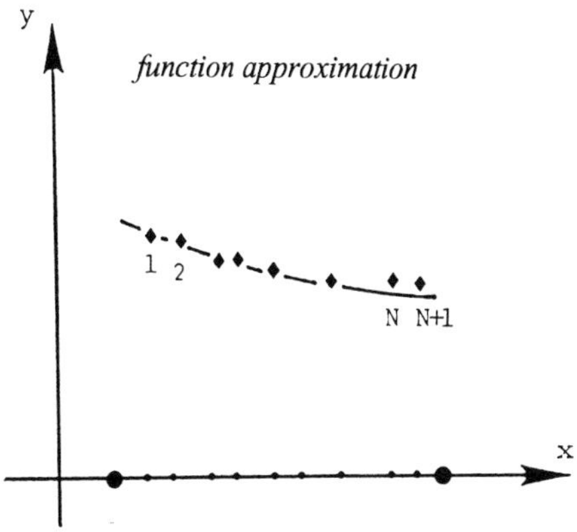

The approximate fit is preferred when the data points are suspected to be in appreciable error, or the abscissas x_i are distributed in a manner that forces the interpolating function to exhibit artificial oscillations.

In Chapter 8, we shall discuss the theory and practice of *polynomial* and *trigonometric approximations*, which are generalizations of polynomial and trigonometric interpolations. It is more natural to develop trigonometric interpolation within the framework of trigonometric approximation, and we defer its discussion to Sections 8.6–8.8.

6.2 The Interpolating Polynomial and its Computation

A standard method of function interpolation proceeds by replacing the unknown function $f(x)$ with the Nth degree *interpolating polynomial*

$$P_N(x) = a_1 x^N + a_2 x^{N-1} + a_3 x^{N-2} + \ldots + a_N x + a_{N+1} \tag{6.2.1}$$

that passes through the $N+1$ data points, so that

$$f(x_i) = P_N(x_i) \tag{6.2.2}$$

for $i = 1, \ldots, N+1$. The problem is reduced to computing the $N+1$ coefficients $a_1, a_2, \ldots, a_{N+1}$ so that $P_N(x)$ satisfies the constraints (6.2.2). Having completed this computation, we can evaluate $P_N(x)$ for any value of x, preferably by the method of nested multiplication or synthetic division discussed in Section 4.7, and thus obtain an approximation to the interpolated function $f(x)$.

It may happen that $a_1 = 0$, in which case the order of the interpolating polynomial is reduced from N to $N-1$ or even further. This will occur, for example, when the function $f(x)$ is a polynomial of degree less than N. If all $f(x_i)$ are equal, then all coefficients of the polynomial but the last one will vanish; $a_{N+1} = f(x_i)$, and the polynomial will express a constant function (see also Problem 6.2.1).

Vandermonde-system Method

The simplest way to compute the polynomial coefficients is by enforcing the constraints (6.2.2) at the $N+1$ data points, thereby deriving the system of $N+1$ linear algebraic quations

$$\mathbf{V}^T \boldsymbol{a} = \boldsymbol{f} \tag{6.2.3}$$

where

$$\boldsymbol{a} = (a_{N+1}, a_N, \ldots, a_2, a_1)^T, \qquad \boldsymbol{f} = (f(x_1), f(x_2), \ldots, f(x_{N+1}))^T \tag{6.2.4}$$

V is the $(N+1) \times (N+1)$ *Vandermonde matrix* with elements $V_{i,j} = x_j^{i-1}$, where $i, j = 1, \ldots, N+1$. Explicitly,

$$\mathbf{V}^T = \begin{bmatrix} 1 & x_1 & x_1^2 & \cdots & x_1^{N-1} & x_1^N \\ 1 & x_2 & x_2^2 & \cdots & x_2^{N-1} & x_2^N \\ \cdots & \cdots & \cdots & \cdots & \cdots & \cdots \\ 1 & x_N & x_N^2 & \cdots & x_N^{N-1} & x_N^N \\ 1 & x_{N+1} & x_{N+1}^2 & \cdots & x_{N+1}^{N-1} & x_{N+1}^N \end{bmatrix} \tag{6.2.5}$$

Note that the abscissas x_i do *not* have to be arranged in any particular order, for example, in the order of ascending or descending magnitude.

It can be shown that

$$\text{Det}(\boldsymbol{V}) = \prod_{j=1}^{N} \prod_{i>j} (x_i - x_j) \tag{6.2.6}$$

which guarantees that, as long as the abscissas of the data points are distinct, that is, $x_i \neq x_j$, the Vandermonde matrix is nonsingular, and the solution of the linear system (6.2.3) is unique.

When $N = 1$ or 2, the solution of the linear system (6.2.3) can be derived readily in closed form. The corresponding formulas will be given in Section 6.6. For a higher number of points, the solution should be computed using numerical methods.

Suppose now that we hold the abscissas of the first and last points x_1 and x_{N+1} fixed, and we start increasing the value of N from its minimum value of unity corresponding to two data points. Since the distances between consecutive data points are being reduced, the factors on the left-hand side of equation (6.2.6) tend to become smaller. In practice, we find that the determinant of the Vandermonde matrix becomes exceedingly small, and thus the linear system (6.2.3) becomes nearly singular, at only moderate values of N. This behavior places serious limits on the usefulness of this approach.

The aforementioned difficulties of the Vandermonde system do not imply that the problem of polynomial interpolation is ill-posed, in the sense that the results are hopelessly sensitive to the round-off error. We must simply find another way of computing the interpolating polynomial and, fortunately, there are a variety of alternatives. *When a numerical method does not work, one must not immediately blame the problem.*

PROBLEMS

6.2.1 ***Vandermonde system for special arrangements of the data points.***

Discuss the solution of the linear system (6.2.3) when the data points fall on a straight line or a parabola.

6.2.2 ***Vandermonde matrix with evenly spaced points.***

Derive the simplified form of the Vandermonde matrix when the data points are spaced evenly along the x axis with separation h, and $x_1 = 0$.

Lagrange Interpolation

Lagrange's method produces the interpolating polynomial in a manner that circumvents the explicit computation of the polynomial coefficients.

Lagrange polynomials

To develop the method, we consider a set of $N + 1$ data points and introduce the Nth degree ith *Lagrange polynomials* $l_{N,i}(x)$ that are defined in terms of the abscissas of the data points. (Some authors use the notation $L_{N,i}$ or even L_i, but we reserve the symbol L for the Legendre polynomials. Can you name another mathematician whose name begins with an L?) By construction, $l_{N,i}(x)$ is equal to zero at all data points, except at the ith data point where it is equal to unity, that is,

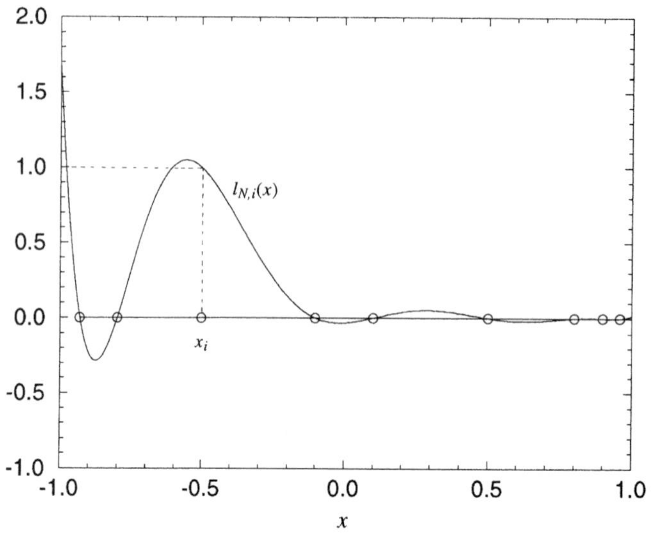

$$l_{N,i}(x_j) = \delta_{i,j} \tag{6.2.7}$$

where $\delta_{i,j}$ is Kronecker's delta. One may readily verify that

$$l_{N,i}(x) \equiv \frac{(x - x_1)(x - x_2)\cdots(x - x_{i-1})(x - x_{i+1})\cdots(x - x_N)(x - x_{N+1})}{(x_i - x_1)(x_i - x_2)\cdots(x_i - x_{i-1})(x_i - x_{i+1})\cdots(x_i - x_N)(x_i - x_{N+1})} \tag{6.2.8}$$

As a pneumonic aid, we note that the factor $x - x_i$ is missing from the numerator, and the corresponding factor $x_i - x_i$ is missing from the denominator. Furthermore, *the denominator is a constant*, whereas *the numerator is an Nth degree polynomials in* x. The denominator arises from the numerator by replacing x with x_i.

The interpolating polynomial

The interpolating polynomial is given by the linear expansion

$$P_N(x) = \sum_{i=1}^{N+1} f(x_i)\, l_{N,i}(x) \tag{6.2.9}$$

Equation (6.2.7) ensures that the right-hand side of expression (6.2.9) satisfies the matching conditions (6.2.2) and is thus the desired interpolating polynomial. Remarkably, we have circumvented solving a system of equations.

For example, the interpolating polynomial corresponding to the three data points (0, 1), (1, 2), (2,4) is

$$P_2(x) = 1\frac{(x-1)(x-2)}{(0-1)(0-2)} + 2\frac{(x-0)(x-2)}{(1-0)(1-2)} + 4\frac{(x-0)(x-1)}{(2-0)(2-1)} = \tfrac{1}{2}(x^2 + x + 2) \tag{6.2.10}$$

Lagrange's method has significant advantages: It is easy to program, it is numerically well-behaved with respect to the round-off error, and it does not require solving a system of equations. Furthermore, it can be used to find a root of the interpolated function by the method of *inverse Lagrange interpolation* to be discussed later in this subsection.

One disadvantage of Lagrange's method is that it does not yield the polynomial coefficients in an explicit manner, apart from the last coefficient, which may be computed as $a_{N+1} = P_N(0)$, although clever algorithms for extracting the rest of the coefficients can be devised (e.g., Press et al. 1992, pp. 113–116). Another disadvantage is that, if one more datum point is introduced in order to improve the accuracy, the whole computation must be repeated; this redundancy can be avoided by using the Neville, Aitken, or Newton interpolation methods discussed later in this section. But unless a large amount of data points are involved, these are not serious flaws, and Lagrange's method is recommended for most applications.

Lagrange generating polynomial

An alternative expression for the Lagrange polynomials, which is useful in developing the theory of interpolation, emerges by introducing the $(N+1)$ degree generating polynomial

$$\Phi_{N+1}(x) = (x - x_1)(x - x_2)\cdots(x - x_N)(x - x_{N+1}) \tag{6.2.11}$$

Note that $\Phi_{N+1}(x)$ vanishes at *all* data points. Straightforward algebra shows that

$$l_{N,i}(x) = \frac{\Phi_{N+1}(x)}{(x - x_1)\Phi'_{N+1}(x_i)} \tag{6.2.12}$$

where a prime denotes a derivative with respect to x.

Cardinal functions

The members of a class of functions $\phi_{N,i}(x)$ associated with a set of data points x_i, $i = 1, \ldots, N+1$, with the interpolating property

$$\phi_{N,i}(x_j) = \delta_{i,j} \tag{6.2.13}$$

are called cardinal functions. Equation (6.2.7) shows that the Legrange polynomials qualify to be included in this prestigious class. Examples of nonpolynomial cardinal functions are given in Problem 6.2.4.

Inverse Lagrange interpolation

To find the point X where $f(x) = 0$, we introduce the function $y = f(x)$ and consider its inverse function $x = w(y)$. We note that $X = w(y = 0)$, and this reduces the problem of finding the zero of $f(x)$ to the problem of evaluating the inverse function at the origin of its independent variable y. The second problem can be solved by approximating $w(y)$ with the interpolating polynomial, writing

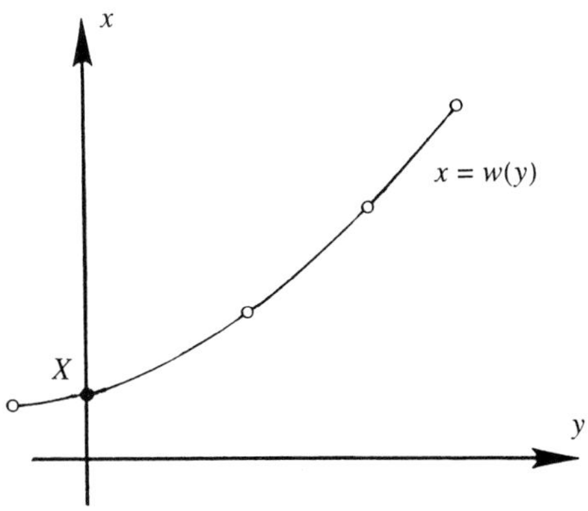

$$w(y) \approx P_N(y) = \sum_{i=1}^{N+1} x_i \, l_{N,i}\,(y) \tag{6.2.14}$$

where $l_{N,i}(x)$ is given by the right-hand side of equation (6.2.8), with x replaced throughout by y. The evaluation of $w(y = 0)$ is straightforward.

A word of caution: The method can be applied only if $w(y)$ is a single-valued function of its argument between the first and the last data points. If not, discarding some of the data points can bring the problem to the desired form.

As an example, we compute the root of a function that is represented by the three data points $(0, -1)$, $(1, 0.264)$, $(2, 0.729)$. Applying equation (6.2.14), we find

$$w(y) \approx 0\frac{(y-0.264)(y-0.729)}{(-1-0.264)(-1-0.729)} + 1\frac{(y+1)(y-0.729)}{(0.264+1)(0.264-0.729)} + 2\frac{(y+1)(y-0.264)}{(0.729+1)(0.729-0.264)} \tag{6.2.15}$$

An approximation to the desired root is $X \approx w(0) = 0.584$. The data points were designed to correspond to the function $f(x) = 1 - 2\,e^{-x}$, which has a zero at $X = \ln 2 = 0.693$. The numerical value 0.584 is a fair estimate.

PROBLEMS

6.2.3 *Lagrange interpolation with evenly spaced points.*

Assume that $N + 1$ data points x_i are distributed evenly along the x axis with separation $\Delta x = h$, where the range of i is as follows:

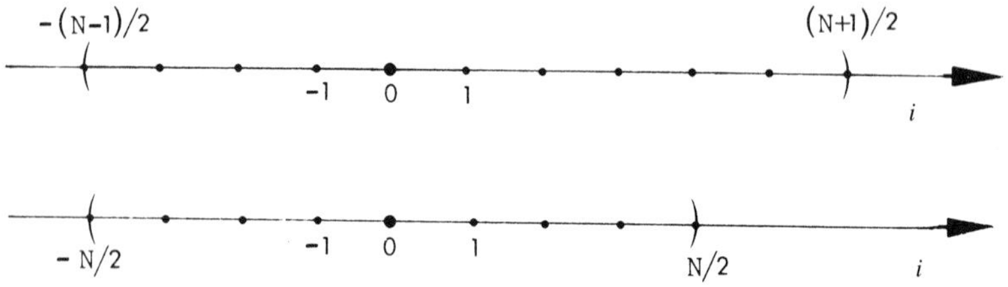

- When N is odd, $i = -(N-1)/2, \ldots, (N+1)/2$.
- When N is even, $i = -N/2, \ldots, N/2$.

Applying the Lagrange interpolation method, and carrying out a little algebra, we arrive at the formula

$$f(x) = \sum_i A_{N,i}(p) f(x_i) \tag{6.2.16}$$

where $p = (x - x_0)/h$, and the sum over i spans the aforementioned ranges.

When N is odd,

$$A_{N,i}(p) = \frac{(-1)^{(N+1)/2+i}}{\left(\frac{N-1}{2}+i\right)!\left(\frac{N+1}{2}-i\right)!(p-i)} \prod_{q=1}^{N+1}\left(p - q + \frac{N+1}{2}\right) \tag{6.2.17}$$

When N is even,

$$A_{N,i}(p) = \frac{(-1)^{N/2+i}}{\left(\frac{N}{2}+i\right)!\left(\frac{N}{2}-i\right)!(p-i)} \prod_{q=0}^{N} \left(p - q + \frac{N}{2}\right) \tag{6.2.18}$$

Derive the explicit form of the right-hand side of equation (6.2.16) for $N = 1, 2, 3$, (e.g., Abramowitz and Stegun 1972, p. 878).

6.2.4 ***Cardinal functions.***

(*a*) The nonpolynomial generating function

$$\text{sinc}_N(x) \equiv \frac{1}{N} \frac{\sin(N\pi x/L)}{\sin(\pi x/L)} \tag{6.2.19}$$

where N is *odd*, can be used to produce a family of periodic cardinal functions with period equal to L, associated with a set of $N + 1$ evenly spaced data points. The cardinal functions are

$$\phi_{N,i}(x) = \text{sinc}_N(x - x_i) \tag{6.2.20}$$

where $L = x_{N+1} - x_1$. Plot and discuss the shape of these cardinal functions for $N = 1, 3, 5, 7, 9$.
(*b*) The nonpolynomial generating function

$$\text{sonc}_N(x) \equiv \frac{\sin(N\pi x/L)}{N\pi x/L} \tag{6.2.21}$$

where N can be even or odd, can be used to produce a family of cardinal functions associated with a set of $N + 1$ evenly spaced data points. The cardinal functions are

$$\phi_{N,i}(x) = \text{sonc}_N(x - x_i) \tag{6.2.22}$$

where $L = x_{N+1} - x_1$. Plot these cardinal functions for $N = 1, 2, 3$, and compare them with the corresponding Lagrange polynomials.

Neville Interpolation

Neville's algorithm evaluates the interpolating polynomial in a manner that may readily account for the effect of an added datum point. The basic idea is to build a sequence of polynomials of increasingly higher order that fit an increasingly higher number of data points. To illustrate the underlying logic, we make a preliminary observation.

Let $Q_{M-L}(x)$ be the $(M-L)$ degree interpolating polynomial whose graph passes through the $M-L+1$ data points,

$$x_L, \quad x_{L+1}, \quad \ldots \quad \ldots \quad x_{M-1}, \quad x_M,$$

which are *not* arranged in any particular order, and $S_{M-L}(x)$ be the $(M-L)$ degree interpolating polynomial whose graph passes through the $M - L + 1$ data points,

$$x_{L+1}, \quad \ldots \quad \ldots \quad x_{M-1}, \quad x_M, \quad x_{M+1}$$

One may readily verify that the $(M - L + 1)$ degree interpolating polynomial $R_{M-L+1}(x)$ whose graph passes through all $M - L + 2$ data points,

$$x_L, \quad x_{L+1}, \quad \dots \quad \dots \quad x_{M-1}, \quad x_M, \quad x_{M+1}$$

may be expressed as a weighted average of the two lower-order polynomials $Q_{M-L}(x)$ and $S_{M-L}(x)$:

$$R_{M-L+1}(x) = \frac{(x_{M+1} - x)Q_{M-L}(x) + (x - x_L)S_{M-L}(x)}{x_{M+1} - x_L} \tag{6.2.23}$$

The right-hand side may be recast into the computationally preferred form

$$R_{M-L+1}(x) = Q_{M-L}(x) + \frac{S_{M-L}(x) - Q_{M-L}(x)}{1 + \dfrac{x_{M+1} - x}{x - x_L}} \tag{6.2.24}$$

The cornerstone of Neville's algorithm is that $Q_{M-L}(x)$ and $S_{M-L}(x)$ may be constructed from two polynomials of degree $M - L - 1$, and the process may be repeated for the descendant polynomials leading us to polynomials of zero degree, which are simply constants. The steps are:

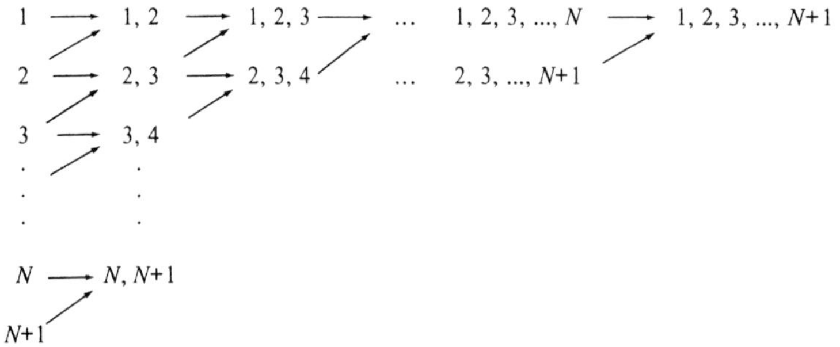

- Begin by considering the $N+1$ *zeroth degree* polynomials whose graphs pass through the individual $N + 1$ data points.
- Build the *first degree* polynomials whose graphs pass through successive doublets of data points.
- Build the *second degree* polynomials whose graphs pass through successive triplets of data points.
- Continue in this manner, to obtain the interpolating polynomial $P_N(x)$.

The intermediate polynomials may be evaluated systematically by updating N times the components of a $(N + 1)$ dimensional vector $\boldsymbol{u}$ according to Algorithm 6.2.1, where, initially, $u_i = f(x_i)$. The upper triangular matrix $A(i, m)$, where the index m counts the number of updates, forms the *Neville table.* The required value is

$$P_N(x) = A(1, N + 1) \tag{6.2.25}$$

The first row of the Neville table displays successively better approximations to the value $f(x)$, computed by taking into account an increasingly larger number of data points. For example, $A(1, 2)$ produces the value $f(x)$ by approximating the graph of the function $f(x)$ with a straight line that passes through the first two data points. If a new datum point is introduced, computing the new interpolated value requires only the values along the main back-diagonal line, and the rest of the table need not be stored.

ALGORITHM 6.2.1 Evaluation of the interpolating polynomial using *Neville's method.*

Do $i = 1, N+1$
  $u_i = f(x_i)$
  $A(i,1) = u_i$ — Optional, used for constructing the Neville table
END DO

Do $m = 2, N+1$
  Do $i = 1, N-m+2$

$$u_i \leftarrow u_i + \frac{u_{i+1} - u_i}{1 + \dfrac{x_{m+i-1} - x}{x - x_i}} \quad \text{or} \quad u_i \leftarrow \frac{(x - x_i)\,u_{i+1} + (x_{m+i-1} - x)\,u_i}{x_{m+i-1} - x_i}$$

  $A(i,m) = u_i$ — Optional, used for constructing the Neville table
  END DO
END DO
$P_N(x) = u_1$

Aitken Interpolation

Aitken's method differs from Neville's method in that it uses different partial sets of data points. Consider the $(N-1)$ degree interpolating polynomial $Q_{N-1}(x)$ whose graph passes through the first N data points

$$x_1, \quad x_2, \quad \dots \quad \dots \quad x_{N-1}, \quad x_N,$$

and the $(N-1)$ degree interpolating polynomial $S_{N-1}(x)$ whose graph passes through the N data points

$$x_1, \quad x_2, \quad \dots \quad \dots \quad x_{N-1}, \quad x_{N+1}$$

Note that the point x_N is missing from the second set. One may readily verify that the interpolating polynomial $P_N(x)$ may be constructed from $Q_{N-1}(x)$ and $S_{N-1}(x)$ as a weighted average, in the form

$$P_N(x) = \frac{(x_{N+1} - x)Q_{N-1}(x) - (x_N - x)S_{M-1}(x)}{x_{N+1} - x_N} \tag{6.2.26}$$

In turn, each one of $Q_{N-1}(x)$ and $S_{N-1}(x)$ may be constructed from two polynomials of degree $N-2$, and the procedure is repeated N times leading us to polynomials of zero degree that are simply constants. The steps are:

- Begin by considering the $N+1$ *zeroth degree* polynomials whose graphs pass through the individual $N+1$ data points.
- Build the *first degree* polynomials whose graphs pass through the *first,* and each one of the rest of the data points.
- Build the *second degree* polynomials whose graphs pass through the *first two,* and each one of the rest of the data points.

- Continue in this manner, to obtain the interpolating polynomial $P_N(x)$.

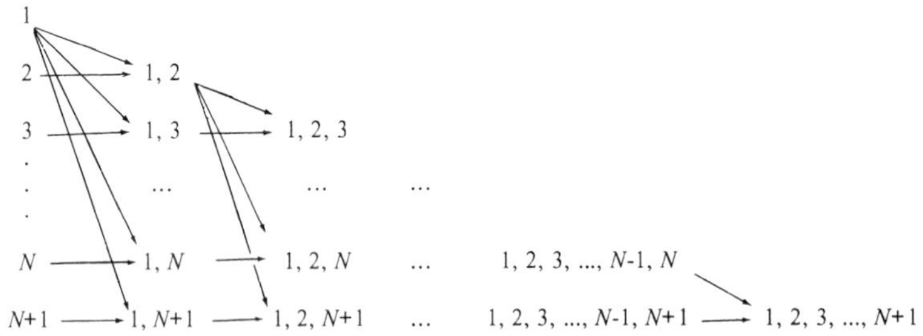

The intermediate polynomials are evaluated in a recursive manner, by updating N times the components of a $(N+1)$-dimensional vector $\boldsymbol{u}$, according to Algorithm 6.2.2, where, initially, $u_i = f(x_i)$. The *lower triangular* matrix $A(i, m)$ forms the *Aitken table*. The required value is

$$P_N(x) = A(N+1, N+1) \tag{6.2.27}$$

The main diagonal line of the Neville table displays successively better approximations to the value $f(x)$, computed taking into account an increasingly larger number of data points. For example, $A(N+1, 2)$ is the value of $f(x)$ that emerges by approximating the graph of the function $f(x)$ with a straight line passing through the first and the last data points. If a new datum point is introduced, the computations will require only values along the main diagonal, and the rest of the table need not be stored.

ALGORITHM 6.2.2 Evaluation of the interpolating polynomial using *Aitken's method*.

Do $i = 1, N+1$
  $u_i = f_i$
  $A(i, 1) = u_i$ — Optional, used for constructing the Aitken table
END DO
Do $m = 2, N+1$
    Do $i = m, N+1$

$$u_i \leftarrow \frac{(x - x_i)\, u_{m-1} - (x - x_{m-1})\, u_i}{x_{m-1} - x_i}$$

    $A(i, m) = u_i$ — Optional, used for constructing the Aitken table
    END DO
END DO
$P_N(x) = u_{N+1}$

Newton Interpolation with Divided Differences

Newton interpolation is similar in spirit to the Neville and Aitken interpolations, in the sense that the effect of an additional datum point may be included with little extra effort. Newton expressed the interpolating polynomial in terms of a linear expansion of a triangular family of elementary polynomials in the form

$$P_N(x) = \sum_{i=0}^{N} c_i \phi_i(x) \tag{6.2.28}$$

where c_i are constant coefficients, and $\phi_i(x)$ are ith degree polynomials defined in terms of the data points as

$$
\begin{aligned}
\phi_0(x) &= 1 \\
\phi_1(x) &= x - x_1 \\
\phi_2(x) &= (x - x_1)(x - x_2) \\
&\dots \\
\phi_i(x) &= (x - x_1)(x - x_2)\dots(x - x_i) \\
&\dots \\
\phi_N(x) &= (x - x_1)(x - x_2)\dots(x - x_i)\dots(x - x_N)
\end{aligned}
\tag{6.2.29}
$$

Note that the abscissa of the last datum point x_{N+1} does not appear in these expressions.

In practice, $P_N(x)$ is computed most efficiently by the method of nested multiplication, rewriting the polynomial as

$$P_N(x) = \Big[\big(\big(\dots\{c_N(x - x_N) + c_{N-1}\}\cdots\big)c_3(x - x_3) + c_2\big)(x - x_2) + c_1\Big](x - x_1) + c_0 \tag{6.2.30}$$

and using either the algorithm

$$
\boxed{
\begin{aligned}
&P = c_N \\
&\text{Do } i = N, 1, -1 \\
&P \leftarrow P(x - x_i) + c_{i-1} \\
&\text{END DO} \\
&P_N(x) = P
\end{aligned}
}
\tag{6.2.31}
$$

or the algorithm

$$
\boxed{
\begin{aligned}
&P = c_0 \\
&b = 1 \\
&\text{Do } i = 1, N \\
&b \leftarrow b\,(x - x_i) \\
&P \leftarrow P + c_i b \\
&\text{END DO} \\
&P_N(x) = P
\end{aligned}
}
\tag{6.2.32}
$$

The coefficients c_i are computed by constructing the divided-difference table:

$$
\begin{array}{llll}
f_1 & & & \\
 & f_{1,2} = (f_2 - f_1)/(x_2 - x_1) & & \\
f_2 & & f_{1,2,3} = (f_{2,3} - f_{1,2})/(x_3 - x_1) & \\
 & f_{2,3} = (f_3 - f_2)/(x_3 - x_2) & & f_{1,2,3,4} = (f_{2,3,4} - f_{1,2,3})/(x_4 - x_1) \\
f_3 & & f_{2,3,4} = (f_{3,4} - f_{2,3})/(x_4 - x_2) & \dots \\
 & f_{3,4} = (f_4 - f_3)/(x_4 - x_3) & & f_{2,3,4,5} = (f_{3,4,5} - f_{2,3,4})/(x_5 - x_2) \\
f_4 & & f_{3,4,5} = (f_{4,5} - f_{3,4})/(x_5 - x_3) & \\
\dots & \dots & \dots & \dots
\end{array}
\tag{6.2.33}
$$

It can be shown by straightforward algebraic manipulations that the top diagonal line provides us with the values of c_i, that is,

$$
\begin{aligned}
c_0 &= f_1 \\
c_1 &= f_{1,2} \\
c_2 &= f_{1,2,3} \\
&\dots \\
c_N &= f_{1,2,3,4,\dots,N+1}
\end{aligned}
\tag{6.2.34}
$$

Adding one datum point introduces a diagonal line at the bottom of the table but does not have an effect on the rest of the table.

The values of the coefficients c_i can be built in a recursive manner, by updating N times the components of a $(N + 1)$-dimensional vector $\boldsymbol{u}$, according to Algorithm 6.2.3, where, initially, $u_i = f(x_i)$. The upper triangular matrix $A(i, m)$ forms the *Newton divided-difference table.*

ALGORITHM 6.2.3 Evaluation of the coefficients c_i of the Newton expansion (6.2.28), by the method of divided differences.

Do $i = 1, N + 1$
  $u_i = f(x_i)$
  $A(i, 1) = u_i$ — Optional, used for constructing the Newton divided-difference table
END DO

$c_0 = u_1$
Do $m = 1, N$
  Do $i = 1, N - m + 1$
    $u_i \leftarrow \dfrac{u_{i+1} - u_i}{x_{i+m} - x_i}$
    $A(i, m + 1) = u_i$ — Optional, used for constructing the Newton divided-difference table
  END DO
  $c_m = u_1$
END DO

PROBLEMS

6.2.5 ***All polynomial interpolations produce the same answer.***

Use (*a*) the Vandermonde-system method, (*b*) the Lagrange interpolation method, (*c*) the Neville algorithm, (*d*) the Aitken algorithm, and (*e*) Newton's algorithm to compute the value of the function $f(x)$ at $x = 0.25$ from the four data points (0, 1.0), (0.5, 2.0), (1.0, 1.5), (1.5, 1.0). Verify that all methods produce the same answer.

6.2.6 ***Interpolation in thermodynamic tables.***

Evaluation of thermodynamic functions such as molar volume, temperature, pressure, and entropy is required for the design of various chemical and mechanical engineering processes. For instance, a knowledge of the temperature of a certain amount of gas that is adiabatically compressed from the atmospheric to a specified pressure is requisite for the design of internal combustion engines.

Thermodynamic functions have been tabulated in various manuals and handbooks. *Perry's Chemical Engineer's Handbook* gives the vapor pressure of saturated steam as a function of temperature, in the range 32.01 to 705.47 °F. Note that the vapor pressure at 100 °C is equal to 1 atm, which means that the normal boiling point of water is equal to 100 °C.

Write a computer program that computes the vapor pressure of saturated steam at a specified temperature between 100 and 150 °C, using four-point Lagrange interpolation. The program should receive the temperature in °C and produce the vapor pressure in atm. Run the program to compute the vapor pressure at 112, 114, 120, 136, 148 °C, and make sure that the results make good sense.

Fornberg's Method

In Section 6.11, we shall describe a powerful algorithm due to Fornberg (1996) that interpolates as well as produces the derivatives of the interpolating polynomial at the data points or at any intermediate point.

6.3 *Error and Convergence of Polynomial Interpolation*

Unless the interpolated function $f(x)$ happens to be a polynomial of degree N or less, the interpolating polynomial $P_N(x)$ and $f(x)$ will not agree between the data points. The difference between the approximate and the exact value,

$$e(x) \equiv P_N(x) - f(x) \tag{6.3.1}$$

is the *interpolation error*. The satisfaction of the matching conditions expressed by equation (6.2.2) ensures that

$$e(x_i) = 0 \tag{6.3.2}$$

for $i = 1, \ldots, N+1$. In general, however, $e(x) \neq 0$ if $x \neq x_i$.

Estimating the magnitude of the interpolation error is useful for three reasons: (*a*) it serves to accredit or discredit the results of the interpolation; (*b*) it can be used to get an idea of how much the error will be reduced by including more data points; and (*c*) it is the starting point for finding the optimal distribution of the abscissas of the data points that yields the best results, as will be discussed in Section 6.4.

We shall show that, when the function $f(x)$ is sufficiently smooth, the error associated with the polynomial interpolation is given by

$$e(x) = -\frac{f^{(N+1)}(\xi)}{(N+1)!}\Phi_{N+1}(x) \tag{6.3.3}$$

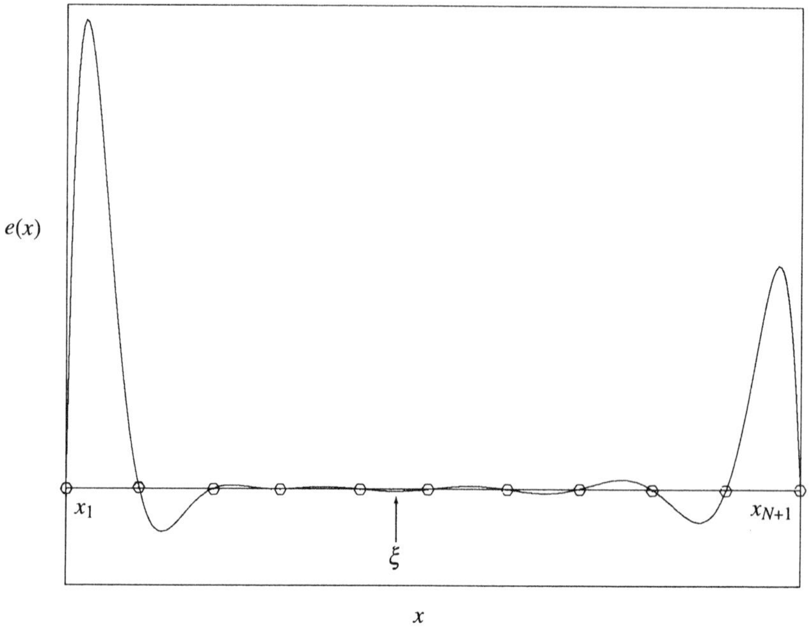

where the $(N+1)$ degree Lagrange generating polynomial $\Phi_{N+1}(x)$ was defined in equation (6.2.11), and $f^{(N+1)}(\xi)$ signifies the value of the $N+1$ derivative of the function f evaluated at a point ξ that lies somewhere between x_1 and x_{N+1}. *The precise location of* ξ *depends on the value of* x; that is, the quantity $f^{(N+1)}(\xi)$ is an implicit function of x.

Having an estimate for, and establishing bounds for the range of variation of the derivative $f^{(N+1)}(x)$, between x_1 and x_{N+1}, is thus useful for assessing the magnitude of the interpolation error.

To prove formula (6.3.3), we fix the value of x and introduce a new function of a new independent variable z, defined as

$$Q(z) = \text{Det}\left(\begin{bmatrix} \Phi_{N+1}(z) & \Phi_{N+1}(x) \\ e(z) & e(x) \end{bmatrix}\right) = \Phi_{N+1}(z)\, e(x) - \Phi_{N+1}(x)\, e(z) \tag{6.3.4}$$

Note that, having fixed the value of x, $\Phi_{N+1}(x)$ has become a constant. We observe that

$$Q(x_i) = 0 \qquad \text{for } i = 1, \ldots, N+1 \tag{6.3.5}$$

and

$$Q(x) = 0 \tag{6.3.6}$$

and this shows that the function $Q(z)$ has at least $N+2$ zeros and must therefore attain at least $N+1$ local maxima and mimima over the interval between x_1 and x_{N+1}. The first derivative $Q'(z)$ must have at least $N+1$ zeros, and the second derivative $Q''(z)$ must have at least N zeros between x_1 and x_{N+1}. Continuing this reduction, we find that the $N + 1$ derivative $Q^{(N+1)}(z)$ must have at least one zero between x_1 and x_{N+1}; one of these zeros is assumed to occur at $z = \xi$. By definition,

$$Q^{(N+1)}(\xi) = \Phi_{N+1}^{(N+1)}(\xi)\, e(x) - \Phi_{N+1}^{(N+1)}(x)\, e^{(N+1)}(\xi) = 0 \tag{6.3.7}$$

Differentiating $N + 1$ times the right-hand side of equation (6.3.1) with respect to x, noting that $P_N^{(N+1)}(x) = 0$, and evaluating the resulting expression at $x = \xi$, we find

$$e^{(N+1)}(\xi) = -f^{(N+1)}(\xi) \tag{6.3.8}$$

Differentiating $N + 1$ times the right-hand side of equation (6.2.11), we find

$$\Phi_{N+1}^{(N+1)}(\xi) = (N + 1)! \tag{6.3.9}$$

Substituting the last two expressions into equation (6.3.7), and solving for $e(x)$, produces the error estimate (6.3.3).

Interpolation of a Polynomial with a Lower-degree Polynomial

What happens when the interpolated function $f(x)$ is a Kth degree polynomial $W_K(x)$, where $K \geq N$? In that case, the error $e(x)$ defined in equation (6.3.1) is also a Kth degree polynomial, which we denote as $E_K(x)$. But since $E_K(x)$ must vanish at the $N + 1$ data points between x_1 and x_{N+1}, we can express it in the form

$$e(x) \equiv E_K(x) = \Phi_{N+1}(x)\Xi_{K-N-1}(x) \tag{6.3.10}$$

where $\Xi_{K-N-1}(x)$ is a $(K - N - 1)$ degree polynomial. Using expression (6.3.3), we find

$$\Xi_{K-N-1}(x) = -\frac{f^{(N+1)}(\xi)}{(N+1)!} \tag{6.3.11}$$

We note again that ξ is an implicit function of x.

Convergence

Next, we examine the behavior of the interpolation error with respect to the number of data points. Specifically, we keep the abscissas of the first and last points x_1 and x_{N+1} fixed and consider the behavior of the interpolation error with increasing N. Superficially, one might expect that, as N is increased, the interpolation error will be reduced at every point x that lies between x_1 and x_{N+1}. Perhaps surprisingly, we find that *the interpolation error is not necessarily reduced as N tends to infinity.* This is particularly true when the data points are evenly spaced.

One example where the interpolation error does not tend to vanish at every point within a fixed interpolating domain as the number of data points is increased, is provided by the seemingly innocuous Runge function

$$f(x) = \frac{1}{1 + 25\, x^2} \tag{6.3.12}$$

Distributing the data points *evenly* between $x_1 = -1$ and $x_{N+1} = 1$, we find that, as N is increased, the error $e(x)$ between the data points amplifies when $|x| > 0.73$, as shown in Figure 6.3.1.

To explain this behavior, we note that, according to the error estimate (6.3.3), in order for the interpolation error to decrease as N is raised, the product $f^{(N+1)}(\xi)\Phi_{N+1}(x)$ must tend to zero faster than $(N+1)!$ tends to infinity; this, however, is not guaranteed. The behavior of the product will depend on the particular functional form of $f(x)$, as well as on the functional form of the polynomial $\Phi_{N+1}(x)$; *the latter depends exclusively on the distribution of the data points.*

In Figure 6.3.2, we present graphs of the polynomials $\Phi_{N+1}(x)$ for the particular case where $x_1 = -1$, $x_{N+1} = 1$, with evenly spaced data points. As N is increased, strong oscillations occur near the end of the interpolation domain, and the function $\Phi_{N+1}(x)$ takes large values. This oscillatory behavior, known as the *Runge effect*, is responsible for slowing down the rate of convergence of interpolation may even lead to a divergent behavior.

Extrapolation

We saw that interpolation near the end of an interpolating domain can be dangerous. It is then not surprising that extrapolation beyond the boundaries of the domain can be even more unreliable. Unless the data points are distributed in a special way, to be discussed in the next section, *polynomial extrapolation is not recommended.* Extrapolation with rational functions discussed by Stoer and Bulirsch (1980, pp. 58–72) is safer.

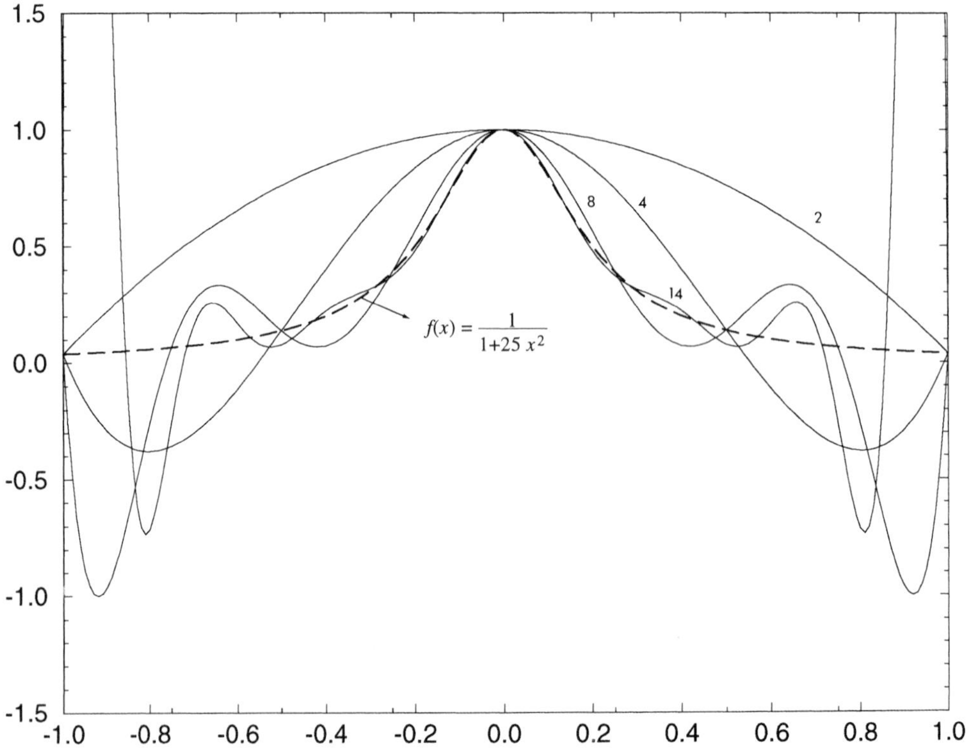

Figure 6.3.1 Divergence of polynomial interpolation with evenly spaced points for the Runge function given in equation (6.3.12), drawn with the dashed line. The solid lines correspond to an Nth degree interpolating polynomial with $N = 2, 4, 8, 14$.

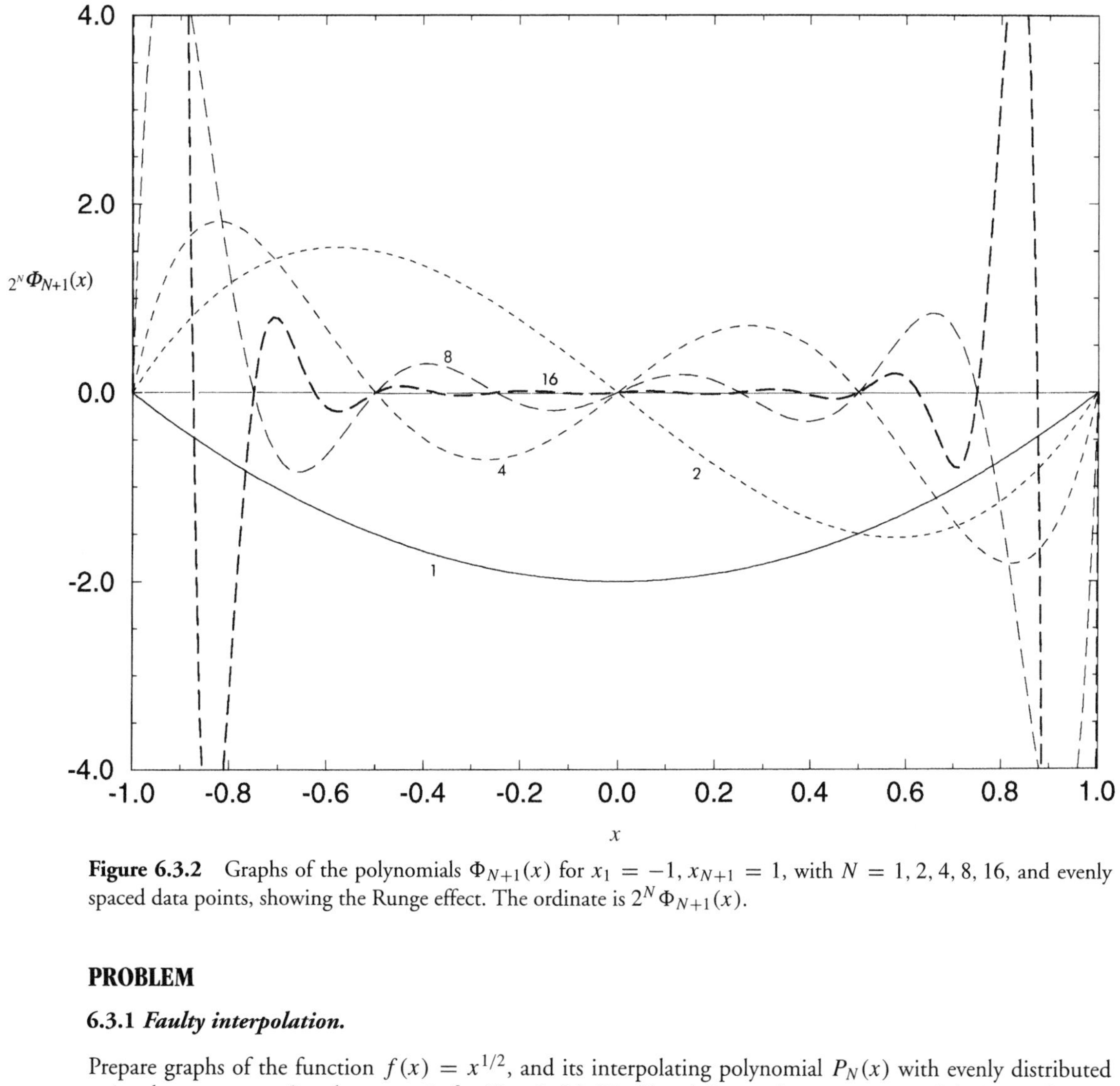

Figure 6.3.2 Graphs of the polynomials $\Phi_{N+1}(x)$ for $x_1 = -1, x_{N+1} = 1$, with $N = 1, 2, 4, 8, 16$, and evenly spaced data points, showing the Runge effect. The ordinate is $2^N \Phi_{N+1}(x)$.

PROBLEM

6.3.1 *Faulty interpolation.*

Prepare graphs of the function $f(x) = x^{1/2}$, and its interpolating polynomial $P_N(x)$ with evenly distributed points between $x_1 = 0$ and $x_{N+1} = 1$, for $N = 8, 16, 32, 64$ and discuss the convergence of the interpolation.

6.4 Optimal Positioning of Data Points

Suppose that we have the luxury of distributing the $N + 1$ data points in any way we desire over a certain closed interpolation interval $[a, b]$. Is there an optimal distribution that minimizes the interpolation error in some sense? And is there a systematic way of distributing the data points so that, as N is increased, the error decreases uniformly and at the fastest possible rate?

Before we can address these questions, we must introduce triangular families of orthogonal polynomials.

Orthogonal Polynomials

Consider a *triangular* family of orthogonal polynomials $p_i(t)$, $i = 0, 1, 2, \ldots$, where $p_i(t)$ is an ith degree polynomial of the independent variable t defined over a certain interval $[c, d]$, with associated *weighting function* $w(t)$. By definition, the polynomials of a particular family satisfy the orthogonality condition

$$(p_i, p_j) \equiv \int_c^d p_i(t)\, p_j(t)\, w(t)\, dt = D_{i,j} \tag{6.4.1}$$

where the matrix $\boldsymbol{D}$ is diagonal. One way of constructing the orthogonal polynomials is by using the counterpart of the Gram–Schmidt orthogonalization method discussed in Problem 6.4.1.

The salient properties of the orthogonal polynomials are discussed in Appendix B. We note, in particular, that any mth degree polynomial can be expressed as a linear combination of the $m+1$ orthogonal polynomials $p_0(t), p_1(t), \ldots, p_m(t)$.

Legendre and Chebyshev polynomials

Two important families of orthogonal polynomials, both defined over the domain $[-1, 1]$ corresponding to $c = -1$ and $d = 1$, distinguished by their weighting functions, are the *Legendre* polynomials $L_i(t)$ and the *Chebyshev* polynomials $T_i(t)$, described in Tables B.2 and B.3 of Appendix B. (The notation $T_i(t)$ was due to the French transiteration of Chebyshev's Cyrillic name чебкшев as *Tschebycheff*; the author of this book has seen his name spelled in many ways, but Chebyshev is a close competitor).

An important property of the Chebyshev polynomials is that their magnitude is less than unity for any value of t within $[-1, 1]$, even though the coefficient of the highest-power term of the nth degree polynomial is equal to 2^{n-1} for $n \geq 1$.

PROBLEMS

6.4.1 ***Orthogonal polynomials via Gram–Schmidt orthogonalization.***

Compute the coefficients of the first 20 members of a triangular family of orthogonal polynomials, defined over the interval $[-1, 1]$, with weighting function $w(x) = 1$, using the Gram–Schmidt orthogonalization process. Specifically, begin by considering the family of monomials

$$1, x, x^2, x^3, x^4, \ldots, x^{20} \tag{6.4.2}$$

and then set

$$\begin{aligned} p_0 &= 1 \\ p_1 &= x - a_{1,0}p_0 \\ p_2 &= x^2 - a_{2,1}p_1 - a_{2,0}p_0 \\ &\cdots \end{aligned} \tag{6.4.3}$$

where the coefficients $a_{n,m}$ are such that the orthogonality condition (6.4.1) is fulfilled. Discuss the relationship between these polynomials and the Legendre polynomials $L_i(t)$ shown in Table B.2 of Appendix B.

6.4.2 *Projection of orthogonal polynomials on polynomials of lower order.*

Let $Q_m(x)$ be an mth degree polynomial. Show that

$$\int_c^d Q_m(t)\; p_i(t)\; w(t)\; dt = 0 \tag{6.4.4}$$

for any $i > m$, where $p_i(t)$ belongs to a family of orthogonal polynomials defined over $[c, d]$ and corresponding to the weighting function $w(t)$. Thus an ith degree orthogonal polynomial is orthogonal to the space of all lower-order polynomials.

Chebyshev Interpolation

We are in a position now to tackle the issue of the optimal distribution of the data points. As a preliminary, we introduce the new independent variable t that is related to x by the linear transformation

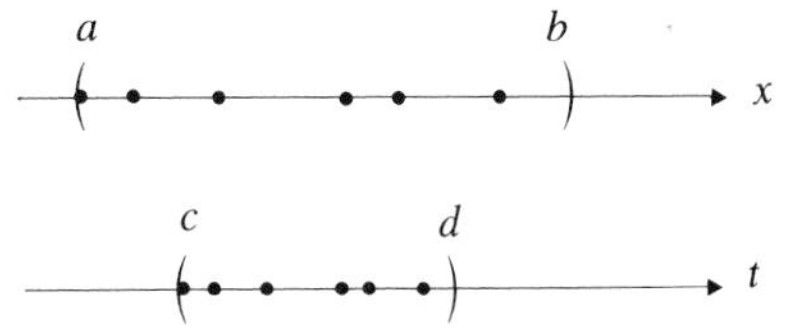

$$x = q(t) = \frac{b+a}{2} + \frac{b-a}{2}\frac{2t-c-d}{d-c} \tag{6.4.5}$$

As t increases from c to d, x increases from a to b. Having made this transformation, we regard f as a function of t, writing

$$f(x) = f(q(t)) \equiv h(t) \tag{6.4.6}$$

Using equation (6.3.3), we then find that the error introduced by the polynomial interpolation of $h(t)$, with $N+1$ data points t_i corresponding to x_i, is given by

$$e(t) \equiv P_N(t) - h(t) = -\frac{h^{(N+1)}(\xi)}{(N+1)!}(t-t_1)(t-t_2)\cdots(t-t_N)(t-t_{N+1}) \tag{6.4.7}$$

where ξ lies somewhere within the interval $[c, d]$.

Selecting $c = -1$, $d = 1$, stipulating that the data points coincide with the $N+1$ roots of the $(N+1)$ degree Chebyshev polynomial $T_{N+1}(t)$, which are given by

$$t_j = \cos\left(\frac{(j-\frac{1}{2})\pi}{N+1}\right) \tag{6.4.8}$$

where $j = 1, \ldots, N+1$, and noting that

$$T_{N+1}(t) = 2^N(t-t_1)(t-t_2)\cdots(t-t_N)(t-t_{N+1}) \tag{6.4.9}$$

for $N \geq 0$, allows us to express equation (6.4.7) in the form

$$e(t) = -\frac{h^{(N+1)}(\xi)}{2^N(N+1)!}T_{N+1}(t) \tag{6.4.10}$$

But since the magnitude of the Chebyshev polynomials is less than unity for any value of t throughout the interval $[-1, 1]$, the magnitude of the error is, at most, on the order of the fraction on the right-hand side of equation (6.4.10). Unless the function $h(t)$ is ill-mannered, $e(t)$ is a rapidly decreasing function of N.

More precisely, it can be shown that the error associated with the Chebyshev interpolation is not much worse than the error corresponding to the *mimimax approximation.* Specifically,

$$\text{Max}|e(t)| \leq [(2/\pi)\ln(N+1) + 2]\,\rho_N[h(t)] \tag{6.4.11}$$

where $\rho_N[h(t)]$ is the *minimax error*, defined as the minimum of the maximum value of $|e(t)|$ introduced by approximating the function $h(t)$ with a polynomial of degree less than or equal to N (e.g., Rivlin 1974, p. 13).

Evaluation of the Chebyshev expansion

Assume now that the data points t_i of the canonical function $h(t)$, corresponding to the data points x_i of the primary function $f(x)$ according to the transformation (6.4.5), coincide with the zeros of the $(N+1)$ degree Chebyshev polynomial $T_{N+1}(t)$, given in equation (6.4.8).

The most efficient way of evaluating the interpolating polynomial is to express it in the form

$$P_N(t) = \sum_{i=0}^{N} c_i T_i(t) \tag{6.4.12}$$

where c_i are a collection of $N+1$ constants, and then use the discrete orthogonality condition listed in Table B.3, to find

$$c_i = \frac{\alpha}{N+1} \sum_{j=1}^{N+1} h(t_j)\, T_i(t_j) \tag{6.4.13}$$

where

$$\alpha = \begin{cases} 1 & \text{when } i = 0 \\ 2 & \text{when } i \neq 0 \end{cases} \tag{6.4.14}$$

The computation of the sum on the right-hand side of equation (6.4.12) can be expedited by using a recurrence formula for the Chebyshev polynomials (see equation (B.5) and Table B.3 of Appendix B). The result is *Clenshaw's algorithm:*

$$\begin{array}{|l|}\hline d_N = c_N \\ d_{N-1} = 2\,tc_N + c_{N-1} \\ \text{Do } i = N-2, 0, -1 \\ d_i = 2\,td_{i+1} - d_{i+2} + c_i \\ \text{END DO} \\ P_N(t) = d_0 - td_1 \\ \hline \end{array} \tag{6.4.15}$$

Note the similarity with Horner's algorithm for evaluating polynomials.

Chebyshev Approximation

If we truncate the sum in equation (6.4.12) at a certain level $M < N$, but still compute the coefficients using equation (6.4.13), we shall obtain a *noninterpolating* polynomial that does not necessarily pass through the data points; this is an *approximating* polynomial.

A remarkable feature of the truncated expansion is that it provides us with an excellent representation of the function $f(x)$ even for small and moderate values of M. An example with $N = 5$ and $M = 1, 2, 3, 4, 5$ is illustrated in Figure 6.4.1. The origin of this behavior can be traced back to our earlier observation that the Chebyshev polynomials vary within the range $[-1, 1]$, with the consequence that successive terms in the expansion (6.4.12) make rapidly decreasing contributions. Applications of the approximating Chebyshev expansion will be discussed in Section 8.2.

PROBLEM

6.4.3 ***Chebyshev expansion.***

Compute the Chebyshev series expansion of the function $f(x) = \exp(x^2)$ over the interval $(0, 2)$ using the discrete projection method expressed by equation (6.4.13) for $N = 1, 2, 4, 8, 16$. Plot the truncated expansions for $M = 0, 1, \ldots, N$ and discuss the quality of the approximation.

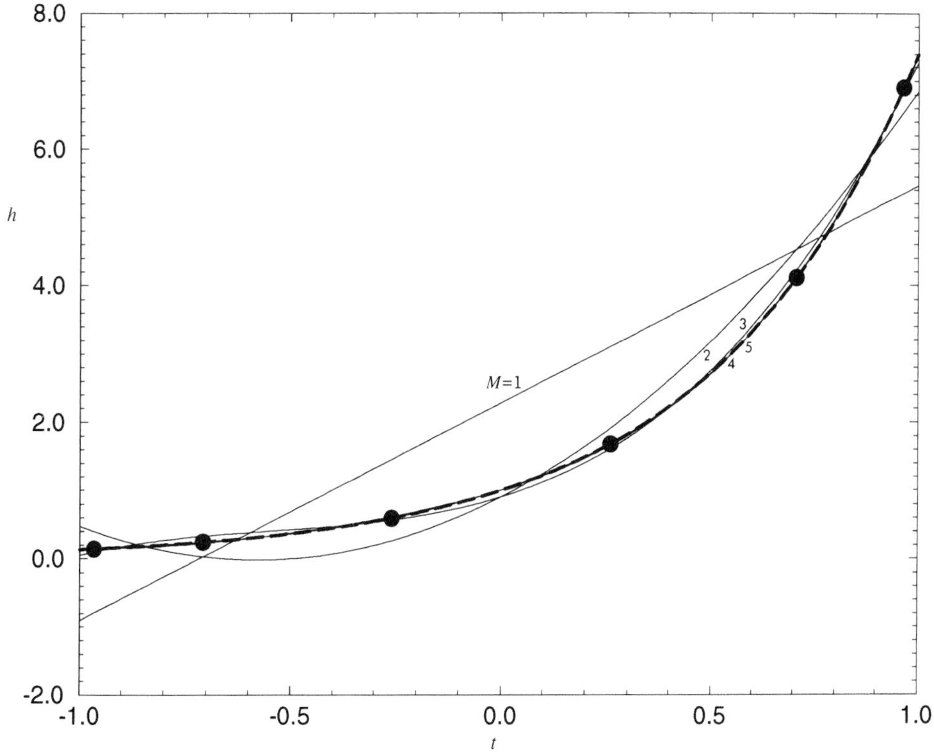

Figure 6.4.1 The interpolating Chebyshev expansion of the function $h(t) = \exp(2t)$ over the interval $[-1, 1]$, plotted with the dashed line, with $N = 5$ corresponding to the six data points shown with filled circles, and then truncated at the levels $M = 1, 2, 3, 4, 5$.

The Lebesque Constant

The error estimate (6.4.11) can be generalized and expressed in more formal terms. Interpolation through $N+1$ nodes introduces an error $e(t)$, where

$$\text{Max}|e(t)| \le (1+\Lambda_N)\rho_N[h(t)] \tag{6.4.16}$$

$\rho_N[h(t)]$ is the minimax error, and Λ_N is the Lebesque constant. The value of Λ_N is sensitive to the distribution of the abscissas of the data points. It can be shown that:

- In general, Λ_N grows at least as fast as $\ln N$.
- If the abscissas of the nodes coincide with the zeros of the *Chebyshev polynomials*, then Λ_N is on the order of $\ln N$ (e.g., Rivlin 1969, p. 90). Note that this statement is consistent with equation (6.4.11).
- If the abscissas of the nodes coincide with the zeros of the *Legendre polynomials*, then Λ_N is on the order of $N^{1/2}$ (e.g., Szegö 1959, p. 336).
- If the abscissas of the nodes are *evenly spaced*, then Λ_N is on the order of $2^N/(N \ln N)$. More specifically,

$$1+\Lambda_N = \text{Max} \sum_{i=1}^{N+1} |l_{N,i}(t)| \tag{6.4.17}$$

where $l_{N,i}(t)$ is the Lagrange interpolating polynomial defined in equation (6.2.8) (Fornberg 1996, p. 171).

The dependence of the interpolation error on the functional form of $h(t)$ is discussed by Powell (1981).

6.5 *Selection of the Interpolating Variable*

Let us consider an interpolated function $f(x)$, introduce a new independent variable t, and regard x as a function of t, $x = q(t)$, in which case f becomes a function of t, $h(t)$, as shown in equation (6.4.6). This change of variables can be effected, for example, by the linear transformation between x and t shown in equation (6.4.5), or by a nonlinear transformation expressed by the function $x = q(t) = \exp(-t)$.

x_1	f_1	t_1	h_1
x_2	f_2	t_2	h_2
...	...	...	...
x_N	f_N	t_N	h_N
x_{N+1}	f_{N+1}	t_{N+1}	h_{N+1}

Given a table that contains the values of the function $f(x)$ at the $N+1$ points x_i, we can readily construct another table that contains the values of the function $h(t)$ at the corresponding $N+1$ points t_i. Interpolation may then be carried out either with respect to x for the function $f(x)$, or with respect to t for the function $h(t)$. The results will *not* generally be identical.

To demonstrate the possibly dramatic effect that the interpolating variable might have on the accuracy of the interpolation, x versus t, consider an extreme case with $N = 1$, involving two data points that fall on the graph of the parabola $y = f(x) = x^2$. Polynomial interpolation for $f(x)$ amounts to replacing the section of the parabola between these points with a straight line, which introduces a nonzero interpolation error. Let us now regard x as a function of the new independent variable t defined as $x = q(t) = t^{1/2}$, which was designed so that the graph of the function $y = f(q(t)) = h(t) = t$ is a straight line. Polynomial interpolation for $h(t)$

amounts to replacing the section between the data points with a straight line, which does *not* introduce an interpolation error.

Thus, although the order of the error remains unchanged, the choice of the interpolation variable may have a significant effect on the accuracy of the interpolation. Formula (6.3.3) suggests that the best interpolating variable is the one that minimizes the magnitude of the $N+1$ derivative of the function $h(t)$: This is the variable that makes the graph of $h(t)$ appear more like a straight line.

There are circumstances where making a choice for the independent variable is not optional but imperative. Examples will be discussed in Sections 6.8 and 6.10 in the context of line and surface representation.

PROBLEM

6.5.1 ***Linear rescaling.***

Discuss whether the linear transformation $x = q(t) = c_1\, x + c_2$, where c_1 and c_2 are two constants, has an effect on the accuracy of the interpolation.

6.6 *Piecewise Polynomial Interpolation and Splines*

When the number of data points is large, or an appreciable error is suspected in the ordinates $f(x_i)$, polynomial interpolation may introduce a substantial error, especially near the ends of the domain of interpolation. A remedy is to replace the interpolating polynomial with a collection of low-order *local* interpolating polynomials, each involving a small group of data points. In contrast, an interpolating polynomial whose graph passes through all data points is a *global* interpolating polynomial.

The individual local interpolating polynomials are defined over intervals that span successive m-tuples of data points. An efficient method of finding the local polynomial that hosts a particular value of x is provided by the algorithm of *binary search* discussed in Section 1.4 (Problem 6.6.2).

Linear Interpolation

Linear interpolation uses two consecutive data points and produces the first degree polynomial

$$P_1^{(i)}(x) = f(x_i) + (x - x_i)\frac{f(x_{i+1}) - f(x_i)}{x_{i+1} - x_i}$$

$$= f(x_i)\frac{x - x_{i+1}}{x_i - x_{i+1}} + f(x_{i+1})\frac{x - x_i}{x_{i+1} - x_i} \tag{6.6.1}$$

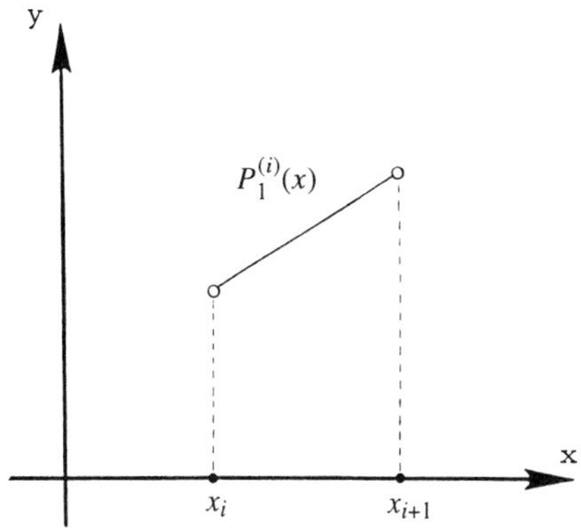

for $x_i < x < x_{i+1}$. The first expression is the familiar representation of a straight line in terms of the intercept and the slope. The second expression follows from Lagrange's interpolation method.

Error

The interpolation error is given by equation (6.3.3) with $N = 1$, in terms of the unknown value $\xi(x)$. An alternative, and somewhat more useful, expression arises by expanding $f(x)$ and $f'(x)$ in Taylor series about the left-end point x_i, and noting that the difference $x - x_i$ ranges between 0 and h_i, where $h_i \equiv x_{i+1} - x_i$, to obtain

$$f(x) = f(x_i) + f'(x_i)(x - x_i) + \tfrac{1}{2} f''(x_i)\,(x - x_i)^2 + O(h_i^3) \tag{6.6.2}$$

and

$$f'(x) = f'(x_i) + f''(x_i)\,(x - x_i) + O(h_i^2) \tag{6.6.3}$$

Next, we evaluate equation (6.6.3) at $x = x_{i+1}$, solve for $f'(x_i)$, substitute the result into the second term on the right-hand side of equation (6.6.2), and we find that the error is given by

$$P_1^{(i)}(x) - f(x) = -\tfrac{1}{2} f''(x_i)\,(x - x_i)\,(x - x_{i+1}) + O(h_i^3) \tag{6.6.4}$$

We note again that the difference $x - x_i$ ranges between 0 and h_i, whereas the difference $x - x_{i+1}$ ranges between $-h_i$ and 0, and this shows that the magnitude of the first term on the right-hand side is on the order of h_i^2.

The leading expression for the error shown in equation (6.6.4) could have been deduced from equation (6.3.3) by setting $N = 1$, expanding $f''(\xi_i)$ in a Taylor series about the point x_i, and retaining only the constant contribution. The present approach is more appealing, in the sense that it allows us to express higher-order corrections in terms of the derivatives of $f(x)$ evaluated at the data points, in a systematic fashion (Problem 6.6.1).

PROBLEMS

6.6.1 *Second contribution to the error.*

Deduce the second contribution to the error on the right-hand side of equation (6.6.4). *Hint*: Retain one more term on the right-hand sides of equations (6.6.2) and (6.6.3), and introduce the corresponding expansion for the second derivative.

6.6.2 *Binary search.*

Implement the method of binary search in a computer subroutine that finds the first degree polynomial hosting a specified value of x.

Triangular tent functions

Let us extend the domain of definition of the local interpolation polynomial $P_1^{(i)}(x)$ from its native terrain $x_i < x < x_{i+1}$ to the whole of the interpolation domain $x_1 < x < x_{N+1}$, so that $P_1^{(i)}(x) = 0$ when x is outside the terrain, that is, when $x < x_i$ or $x > x_{i+1}$. Having made this extension, we express the global interpolating polynomial in the form

$$P_1(x) = \sum_{i=1}^{N} P_1^{(i)}(x) = \sum_{i=1}^{N+1} f(x_i)\, \phi_i(x) \tag{6.6.5}$$

where $\phi_i(x)$ are a collection of *triangular global interpolating tent functions*, named after the shape of their graph:

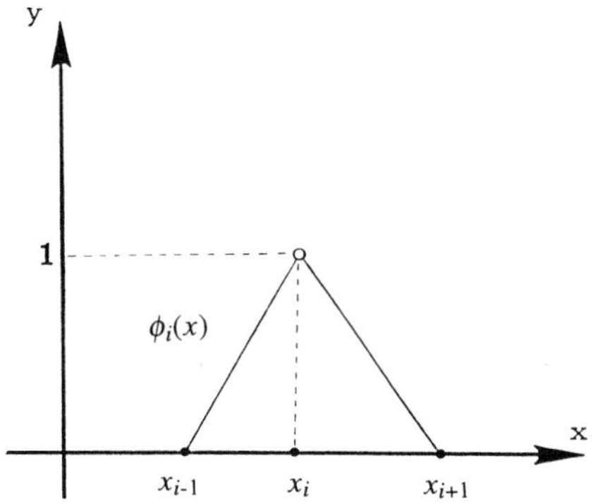

- $\phi_i(x)$ is equal to 0 when $x < x_{i-1}$ or $x > x_{i+1}$.
- It increases linearly from 0 to 1 between the points x_{i-1} and x_i.
- It decreases linearly from 1 to 0 between the points x_i and x_{i+1}.

It is instructive to note that the polygonal-like global interpolating functions $\phi_i(x)$ are crude representations of the smooth Lagrange interpolating polynomials $l_{N,i}(x)$ defined in equation (6.2.8).

Expansions of a function in terms of global interpolating functions are used in the numerical solution of ordinary and partial differential equations using *finite-element methods*, to be discussed in Section 10.4.

Quadratic Interpolation

Quadratic interpolation fits a parabola through a trio of consecutive data points and yields a second degree interpolating polynomial that can be expressed in the form

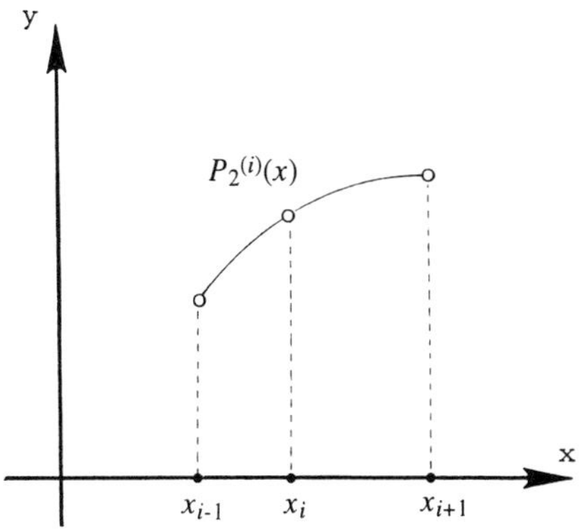

$$P_2^{(i)}(x) = f_i + \big(a_i(x - x_i) + b_i\big)(x - x_i) \tag{6.6.6}$$

for $x_{i-1} < x < x_{i+1}$. The coefficients a_i and b_i are best computed in a sequential manner as

$$\boxed{\begin{aligned} a_i &= \frac{1}{x_{i+1} - x_{i-1}} \left(\frac{f(x_{i+1}) - f(x_i)}{x_{i+1} - x_i} - \frac{f(x_{i-1}) - f(x_i)}{x_{i-1} - x_i} \right) \\ b_i &= \frac{f(x_{i+1}) - f(x_i)}{x_{i+1} - x_i} - a_i(x_{i+1} - x_i) \end{aligned}} \tag{6.6.7}$$

When the three points are spaced evenly along the x axis, that is, $x_{i+1} - x_i = x_i - x_{i-1} = h$, formulas (6.6.7) simplify to

$$a_i = \frac{f(x_{i+1}) - 2f(x_i) + f(x_{i-1})}{2h^2}$$
$$b_i = \frac{f(x_{i+1}) - f(x_{i-1})}{2h} \tag{6.6.8}$$

Error

Working as in the preceding subsection, we find that the interpolation error can be expressed in the asymptotic form

$$P_2^{(i)}(x) - f(x) = -\tfrac{1}{6} f''(x_i)\,(x - x_{i-1})\,(x - x_i)\,(x - x_{i+1}) + O(h_i^4) \tag{6.6.9}$$

where the magnitude of $h_{i+1} \equiv x_{i+2} - x_{i+1}$ is assumed to be comparable to that of $h_i \equiv x_{i+1} - x_i$.

Parabolic tent functions

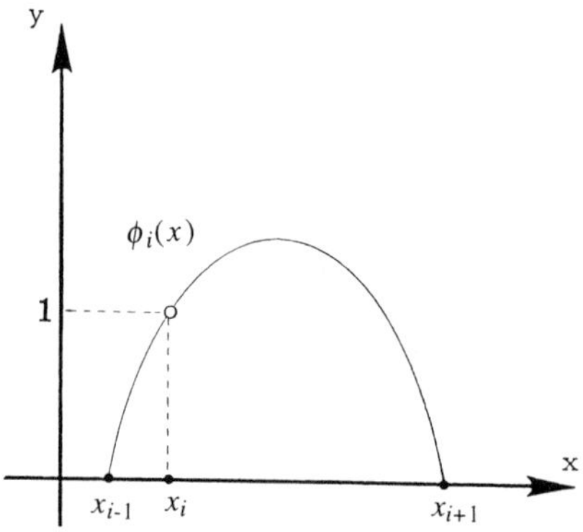

Global quadratic interpolating functions arise as in the case of linear interpolation; the global interpolating polynomial is expressed as the right-hand side of equation (6.6.5). A parabolic tent function $\phi_i(x)$ corresponding to the local quadratic interpolation is shown in the diagram. When the points are evenly spaced, the maximum value of unity occurs at the middle node. Otherwise, the location of the maximum value is shifted toward the larger adjacent interval.

PROBLEM

6.6.3 ***Tent functions and the finite-element method.***

Derive analytical expressions for the tent functions corresponding to local quadratic interpolation.

Cubic-spline Interpolation

Cubic-spline interpolation fits a third degree polynomial across each interval extending between two consecutive data points, and matches the first and second derivatives of adjacent polynomials at the seams. Thus the collection of the cubics yields a continuous function with continuous first and second derivatives, but discontinuous third derivative and vanishing higher-order derivatives.

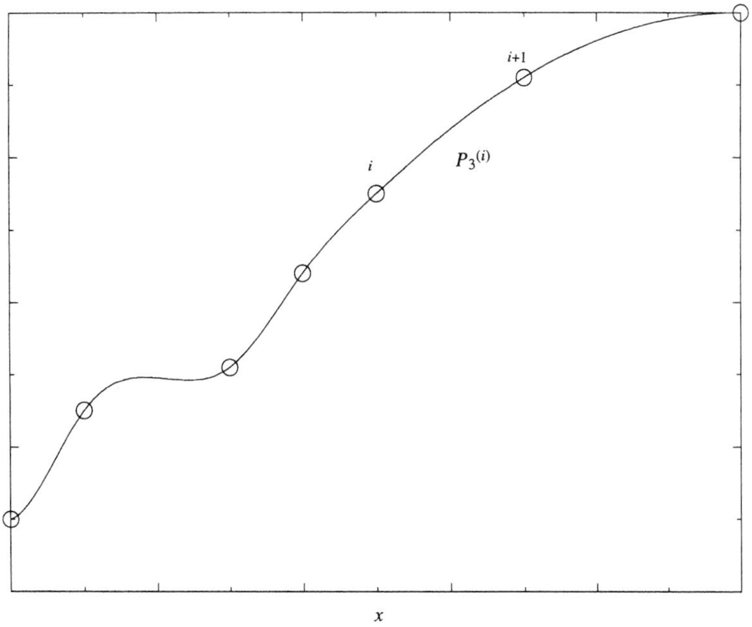

Literally, a spline is a strip of a flexible metal, wood, or a plastic sheet, used by draftpersons to draw smooth curves. This quaint device was particularly useful in the design of ships. Today, mathematical splines find extensive usage in computer graphics (e.g., Olfe 1995).

For simplicity of notation, we denote $f(x_i) \equiv f_i$. Proceeding with the computation of the polynomials, we express the ith cubic in the form

$$P_3^{(i)}(x) = \big[\big(a_i(x - x_i) + b_i\big)(x - x_i) + c_i\big](x - x_i) + f_i \tag{6.6.10}$$

for $x_i < x < x_{i+1}$, where $i = 1, \ldots, N$, and set out to compute the $3N$ unknown coefficients a_i, b_i, and c_i.

First, we require that $P_3^{(N)}(x_{N+1}) = f_{N+1}$

$$\begin{aligned} P_3^{(i)}(x_{i+1}) &= P_3^{(i+1)}(x_{i+1}) = f_{i+1} \\ P_3^{(i)'}(x_{i+1}) &= P_3^{(i+1)'}(x_{i+1}) \\ P_3^{(i)''}(x_{i+1}) &= P_3^{(i+1)''}(x_{i+1}) \end{aligned} \tag{6.6.11}$$

for $i = 1, \ldots, N - 1$, and we derive $2N$ expressions for the coefficients a_i and c_i in terms of b_i,

$$a_i = \frac{1}{3}\frac{b_{i+1} - b_i}{h_i}, \qquad c_i = \frac{f_{i+1} - f_i}{h_i} - \tfrac{1}{3}h_i(b_{i+1} + 2b_i) \tag{6.6.12}$$

where $h_i = x_{i+1} - x_i$ and $i = 1, \ldots, N$. In addition, we find that the $N + 1$ unknown values b_i satisfy the *tridiagonal* system of $N - 1$ equations

$$\frac{h_i}{3} b_i + \tfrac{2}{3}(h_i + h_{i+1})\, b_{i+1} + \frac{h_{i+1}}{3} b_{i+2} = \frac{f_{i+2} - f_{i+1}}{h_{i+1}} - \frac{f_{i+1} - f_i}{h_i} \tag{6.6.13}$$

for $i = 1, \ldots, N-1$, where we have defined

$$b_{N+1} \equiv \tfrac{1}{2} P_3^{(N)''}(x_{N+1}) \tag{6.6.14}$$

When the data points are spaced evenly along the x axis, separated by distance h, the system (6.6.13) takes the simpler form

$$b_i + 4b_{i+1} + b_{i+2} = 3\frac{f_{i+2} - 2f_{i+1} + f_i}{h^2} \tag{6.6.15}$$

for $i = 1, \ldots, N-1$.

Closure

Making two additional stipulations regarding the shape of the cubic splines completes the mathematical formulation of the problem and presents us with several options:

- In an uncoined method, we relate b_1 to b_2 and b_3, and b_{N+1} to b_N and b_{N-1}, by linear extrapolation, and then solve the completed tridiagonal system of equations (6.6.13) for b_i.
- To obtain a *clamped cubic spline*, we specify the slope of the first and last cubics at the first and last data points.
- If the interpolated function is periodic, we require that the first and second derivatives of the first cubic at the first point are equal to those of the last cubic at the last point.
- To obtain the *natural cubic spline*, we require that the curvature of the first and last cubics at the first and last data points vanishes.

Any of these choices leads to a nonsingular system of equations with a unique solution.

Further discussion

Applications of cubic-spline interpolation to line representation will be discussed in Section 6.8. Further discussion of cubic-spline interpolation, including theoretical and numerical studies of convergence and algorithms for numerical computation, can be found in monographs of computer graphics and computational geometry, and in advanced texts on numerical analysis (e.g., Ahlberg et al. 1967; de Boor 1978; Stoer and Bulrisch 1980, Lancaster and Šalkauskas 1986, Späth 1995a, Shikin and Plis 1995).

PROBLEMS

6.6.4 *Cubic splines.*

Derive the system of equations (6.6.12) and (6.6.13).

6.6.5 *Quadratic splines.*

Develop the mathematical formulation, and discuss the implementation of, quadratic-spline interpolation.

6.7 Hermite Interpolation

In certain applications, it is necessary to store and then accurately reproduce a function with as little information as possible, using only a small or moderate number of data points. One example is the function that describes the shape of a line or surface representing a physical boundary. The accurate representation of this function is required for the computation of boundary integrals encountered in solving differential or integral equations in domains with pronounced geometrical complexity. In these cases, Hermite interpolation is an excellent choice.

Interpolating Polynomial with Matching Slopes

First, we want to compute an interpolating polynomial whose graph not only passes through a set of data points but also preserves the specified slope of the interpolated function $f(x)$ at these points. Specifically, given the values of a function $f(x)$ and its first derivative $f'(x)$ at the $N+1$ data points x_i, $i = 1, \ldots, N+1$, we want to compute a *Hermite interpolating polynomial* of degree $2N+1$ or less, $P_{2N+1}(x)$, that satisfies the $2N+2$ matching constraints

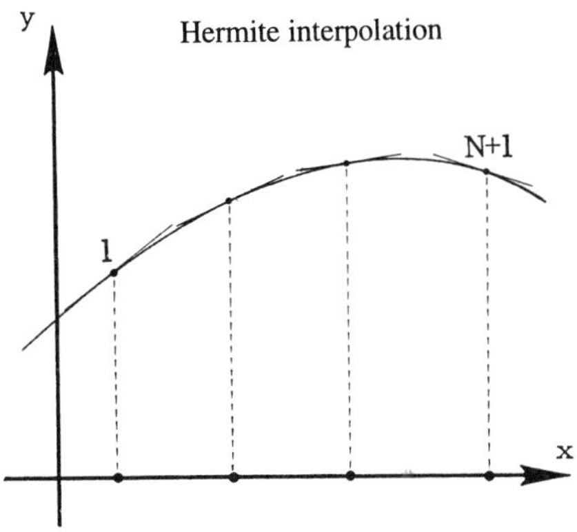

$$f(x_i) = P_{2N+1}(x_i), \qquad f'(x_i) = P'_{2N+1}(x_i) \tag{6.7.1}$$

for $i = 1, \ldots, N+1$, where a prime signifies a derivative with respect to x.

It can be shown that, as long as the abscissas x_i are distinct, an interpolating polynomial that satisfies conditions (6.7.1) exists and is unique (Problem 6.7.1). The polynomial coefficients can be computed with the Vandermonde-system method described in Section 6.2, but numerical instabilities arise even with a small number of data points. Fortunately, these difficulties are circumvented by using a variation of Lagrange interpolation.

Hermite–Lagrange interpolation

We begin by introducing the Nth degree Lagrange polynomials $l_{N,i}(x)$ defined in equation (6.2.8) and then define the $(2N+1)$ degree polynomials

$$\begin{aligned} h_{i,0}(x) &= \big(1 - 2\, l'_{N,i}(x_i)(x - x_i)\big)\, l^2_{N,i}(x) \\ h_{i,1}(x) &= (x - x_i)\, l^2_{N,i}(x) \end{aligned} \tag{6.7.2}$$

It can be shown by straightforward differentiation that

$$h_{i,0}(x_j) = \delta_{i,j}, \qquad h'_{i,0}(x_j) = 0, \qquad h_{i,1}(x_j) = 0, \qquad h'_{i,1}(x_j) = \delta_{i,j} \tag{6.7.3}$$

Thus the desired Hermite polynomial with matched first derivatives is given by

$$P_{2N+1}(x) = \sum_{i=1}^{N+1} f(x_i)\, h_{i,0}(x) + \sum_{i=1}^{N+1} f'(x_i)\, h_{i,1}(x) \tag{6.7.4}$$

The first derivative of $l_{N,i}(x)$, which is necessary for the evaluation of $h_{i,0}$, may be computed by expanding the derivative of the product of the N monomials, by numerical differentiation as discussed in Section 6.11, or by Fornberg's method discussed in Section 6.11.

When the function $f(x)$ is sufficiently smooth, the error associated with the Hermite interpolation is given by

$$P_{2N+1}(x) - f(x) = -\frac{f^{(2N+2)}(\xi)}{[2(N+1)]!}\,\Phi^2_{N+1}(x) \tag{6.7.5}$$

where $\Phi_{N+1}(x)$ is defined in equation (6.2.11) (Stoer and Bulirsch 1980, p. 56).

Cubic interpolation

For example, when $N = 1$, corresponding to two data points x_1 and x_2, we obtain a cubic interpolating polynomial. The functions $h_{i,j}(x)$ are given by

$$h_{i,0}(x) = H_{i,0}(\xi), \qquad h_{i,1}(x) = (x_2 - x_1) H_{i,1}(\xi) \tag{6.7.6}$$

where $\xi = (x - x_1)/(x_2 - x_1)$, and $H_{i,j}(\xi)$ are third degree polynomials given by

$$\begin{aligned} H_{1,0}(\xi) &= 1 - 3\xi^2 + 2\xi^3, \qquad & H_{2,0}(\xi) &= 3\xi^2 - 2\xi^3 \\ H_{1,1}(\xi) &= \xi - 2\xi^2 + \xi^3, \qquad & H_{2,1}(\xi) &= -\xi^2 + \xi^3 \end{aligned} \tag{6.7.7}$$

PROBLEMS

6.7.1 *Uniqueness of the Hermite polynomial.*

Assume that two Hermite polynomials exist, corresponding to a given set of data points, and show that their difference must necessarily be a polynomial of degree higher than $2N + 1$, which is a contradiction.

6.7.2 *Hermite interpolation.*

(*a*) Derive formulas (6.7.6) and (6.7.7). (*b*) Derive corresponding formulas for quintic interpolation with three evenly spaced points.

6.7.3 *Interpolation of a sine wave.*

Approximate the sinusoidal function $f(x) = \sin(x\pi/2)$ in the interval $0 \le x \le 1$ with a Hermite polynomial defined by a set of evenly spaced data points with $N = 1, 2, 4, 8, 16$, and discuss the convergence of the interpolation.

High-order Hermite Interpolation

In the most general case of Hermite interpolation, we are given a set of $N+1$ data points, and we want to find an Lth degree interpolating polynomial $P_L(x)$ that reproduces the values of the function $f(x)$ and its first $m_i - 1$ derivatives, $f^{(k)}(x)$ for $k = 0, 1, \ldots, m_i - 1$, at the ith datum point, where $i = 1, \ldots, N+1$. Note that the number of specified derivatives varies across the data points.

x_1	x_2	$\cdots$	x_i	$\cdots$	x_{N+1}
$f(x_1)$	$f(x_2)$	$\cdots$	$f(x_i)$	$\cdots$	$f(x_{N+1})$
$f'(x_1)$	$f'(x_2)$	$\cdots$	$f'(x_i)$	$\cdots$	$f'(x_{N+1})$
$\cdots$	$\cdots$	$\cdots$	$\cdots$	$\cdots$	$\cdots$
$\cdots$	$f^{(m_2-1)}(x_2)$	$\cdots$	$f^{(m_i-1)}(x_i)$	$\cdots$	$f^{(m_{N+1}-1)}(x_{N+1})$
$f^{(m_1-1)}(x_1)$		$\cdots$		—	
		—			

Matching the number of unknown polynomial coefficients with the number of constraints, we find that the maximum polynomial degree we can accommodate is

$$L = -1 + \sum_{i=1}^{N+1} m_i \tag{6.7.8}$$

For example, for the Hermite polynomial with matched first derivatives at all data points discussed earlier, $m_i = 2$ for all i, and $L = 2N + 1$. It can be shown that, as long as the abscissas x_i are distinct, such a Hermite polynomial exists and is unique (Stoer and Bulirsch 1980, p. 52).

The requisite polynomial may be expressed in the form

$$\begin{aligned} P_L(x) &= \sum_{k=0}^{m_1-1} f^{(k)}(x_1)\, h_{1,k}(x) + \cdots \\ &+ \sum_{k=0}^{m_i-1} f^{(k)}(x_i)\, h_{i,k}(x) + \cdots \\ &+ \sum_{k=0}^{m_{N+1}-1} f^{(k)}(x_{N+1})\, h_{N+1,k}(x) \\ &= \sum_{i=1}^{N+1} \sum_{k=0}^{m_i-1} f^{(k)}(x_i)\, h_{i,k}(x) \end{aligned} \tag{6.7.9}$$

where the superscript (k) signifies the kth derivative, and where $h_{i,k}(x)$ are Hermite cardinal polynomials of degree less than or equal to L, required to satisfy the cardinal properties

$$h_{i,k}^{(r)}(x_j) = \begin{cases} 1 & \text{if } i = j \text{ and } k = r \\ 0 & \text{otherwise} \end{cases} \tag{6.7.10}$$

When $m_i = 2$ for all i, expression (6.7.9) reduces to (6.7.4).

Recursive evaluation

The cardinal polynomials $h_{i,k}(x)$ can be evaluated in a recursive manner working as follows (Stoer and Bulirsch 1980, p. 53). First, we evaluate the ancillary polynomials $r_{i,k}(x)$ and their derivatives according to the algorithm

$$
\begin{aligned}
&\text{Do } i = 1, N+1 \\
&\quad \text{Do } k = 0, m_i - 1 \\
&\qquad r_{i,k}(x) = \frac{(x - x_i)^k}{k!} \prod_{\substack{j=1 \\ j \neq i}}^{N+1} \left(\frac{x - x_j}{x_i - x_j} \right)^{m_j} \\
&\qquad \text{Do } s = 1, m_i - 1 \\
&\qquad\quad \text{Evaluate } r_{i,k}^{(s)}(x_i) \\
&\qquad \text{End Do} \\
&\quad \text{End Do} \\
&\text{End Do}
\end{aligned}
\tag{6.7.11}
$$

The evaluation of the sth derivative $r_{i,k}^{(s)}(x)$ can be done by numerical differentiation, as discussed in Section 6.11. In the second stage, we compute

$$
\begin{aligned}
&\text{Do } i = 1, N+1 \\
&\quad \text{Do } k = m_i - 1, 0, -1 \\
&\qquad h_{i,k}(x) = r_{i,k}(x) - \sum_{s=k+1}^{m_i-1} r_{i,k}^{(s)}(x_i)\, h_{i,s}(x) \\
&\quad \text{End Do} \\
&\text{End Do}
\end{aligned}
\tag{6.7.12}
$$

PROBLEM

6.7.4 *High-order Hermite interpolation.*

Show that algorithms (6.7.11) and (6.7.12) reproduce formulas (6.7.2).

Explicit computation of the polynomial coefficients

Stoer and Bulirsch (1980, pp. 54–56) describe a method for computing the coefficients of the Hermite polynomial in an explicit manner, based on a modification of the Newton divided-difference table.

We begin by introducing the new interpolating variable t and the function $u(t)$, and we construct a two-column table with $L+1$ entries:

$$
\begin{array}{llll}
t_1 & = x_1 & u(t_1) & = f(x_1) \\
t_2 & = x_1 & u(t_2) & = f'(x_1) \\
\cdots & & & \\
t_{m_1} & = x_1 & u(t_{m_1}) & = f^{(m_1-1)}(x_1) \\
\hline
t_{m_1+1} & = x_2 & u(t_{m_1+1}) & = f(x_2) \\
t_{m_1+2} & = x_2 & u(t_{m_1+2}) & = f'(x_2) \\
\cdots & & & \\
t_{m_1+m_2} & = x_2 & u(t_{m_1+m_2}) & = f^{(m_2-1)}(x_2) \\
\hline
\cdots & & \cdots &
\end{array}
\tag{6.7.13}
$$

where L is defined in equation (6.7.8). Next, we express the Hermite polynomial in the familiar Newton form

$$
P_L(x) = P_L(t) = \sum_{i=0}^{L} c_i\, \phi_i(t) \tag{6.7.14}
$$

where the polynomials $\phi_i(t)$ are defined in equations (6.2.29) with t in place of x, t_i in place of x_i, and L in place of N. The coefficients c_i fall at the top line of a modified Newton difference table that is constructed using the formulas

$$
G_{t_i,t_{i+1},\ldots,t_{i+k}} = \begin{cases} \dfrac{u(t_r+k)}{k!} & \text{if } t_i = t_{i+k}, \text{ including } k=0 \\[2ex] \dfrac{G_{t_{i+1},\ldots,t_{i+k}} - G_{t_i,\ldots,t_{i+k-1}}}{t_{i+k}-t_i} & \text{otherwise} \end{cases} \tag{6.7.15}
$$

for $k = 0, \ldots, L$, where $0! = 1$, and r is the smallest index such that $t_r = t_{r+1} = \cdots = t_i$.

Stoer and Bulirsch (1980) discuss an example with $N = 1$ involving two data points, where

$$
\begin{array}{llll}
x_1 = 0 & f(x_1) = -1, & f'(x_1) = -2 & \\
x_2 = 1, & f(x_2) = 0, & f'(x_2) = 10, & f''(x_2) = 40
\end{array}
$$

In this case, $m_1 = 2$, $m_2 = 3$, $L = 4$. Table (6.7.13) takes the form

$$
\begin{array}{ll}
t_1 = 0 & u(t_1) = -1 \\
t_2 = 0 & u(t_2) = -2 \\
\hline
t_3 = 1 & u(t_3) = 0 \\
t_4 = 1 & u(t_4) = 10 \\
t_5 = 1 & u(t_5) = 40
\end{array}
$$

For $k = 0$, we find

$$G_{t_1} = G_{t_2} = u(t_1) = -1 \qquad \text{First equation in (6.7.15) with } r = 1$$

$$G_{t_3} = G_{t_4} = G_{t_5} = u(t_3) = 0 \qquad \text{First equation in (6.7.15) with } r = 3$$

The complete modified Newton table is:

i	t_i	$u(t_i)$	$k=0$	$k=1$	$k=2$	$k=3$	$k=4$
1	0	−1	$G_{t_1} = -1 = c_0$				
				$G_{t_1,t_2} = -2 = c_1$			
2	0	−2	$G_{t_2} = -1$		$G_{t_1,t_2,t_3} = 3 = c_2$		
				$G_{t_2,t_3} = 1$		$G_{t_1,t_2,t_3,t_4} = 6 = c_3$	
3	1	0	$G_{t_3} = 0$		$G_{t_2,t_3,t_4} = 9$		$G_{t_1,t_2,t_3,t_4,t_5} = 5 = c_4$
				$G_{t_3,t_4} = 10$		$G_{t_2,t_3,t_4,t_5} = 11$	
4	1	10	$G_{t_4} = 0$		$G_{t_3,t_4,t_5} = 20$		
				$G_{t_4,t_5} = 10$			
5	1	40	$G_{t_5} = 0$				

PROBLEM

6.7.5 ***Hermite polynomial by the modified Newton table.***

Write a program that produces the coefficients of the Hermite polynomial using the modified Newton difference table.

6.8 *Parametric Description of Lines*

Function interpolation finds important applications in the parametric description of planar and three-dimensional lines that are defined in terms of a collection of marker points, forming an open or closed loop. For example, the description of lines representing material interfaces is required in the solution of differential and integral equations governing two-fluid flow (e.g., Pozrikidis 1992). More applied examples can be drawn from the field of Computer-Aided Design (e.g., Olfe 1995).

Global Representation and Interpolation

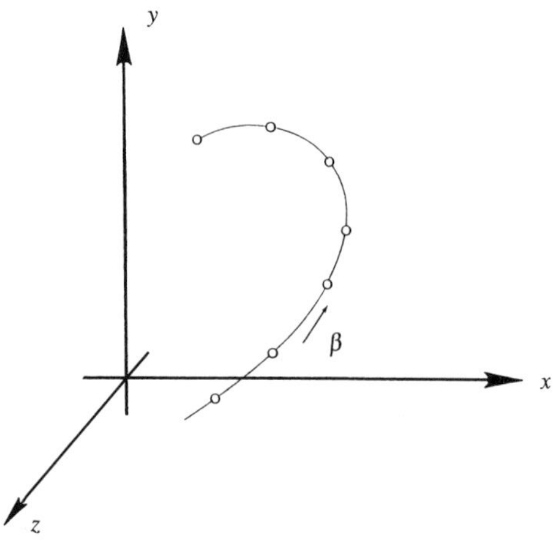

Let us assume that we are given the Cartesian coordinates $\mathbf{x} = (x, y, z)$ of a set of $N+1$ points that lie on a smooth three-dimensional line. To describe the line, we use the parametric representation

$$x = f_1(\beta), \quad y = f_2(\beta), \quad z = f_3(\beta) \tag{6.8.1}$$

where the parameter β increases along the line in a monotonic fashion. The functions $f_1(\beta)$, $f_2(\beta)$, and $f_3(\beta)$ can be evaluated by carrying out three independent interpolations for the three Cartesian coordinates.

Before we can perform these interpolations, we must assign values of the interpolating variable β to each one of the $N+1$ data points, thereby generating three independent sets of data points

$$(\beta_i, f_1(\beta_i)), \quad (\beta_i, f_2(\beta_i)), \quad (\beta_i, f_2(\beta_i)) \tag{6.8.2}$$

for $i = 1, \ldots, N+1$. To this end, we are faced with a number of options.

One simple choice is to set $\beta_i = i$. A better choice is to set β_i equal to the polygonal arc length, that is, the length of the polygonal line that connects successive points, also called a *polyline*. Other choices may be devised by exercising physical and geometrical intuition.

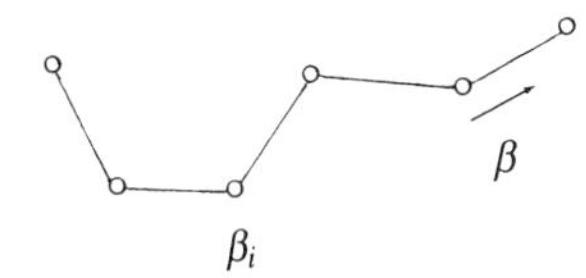

The quality of the representation generally improves as β becomes a better approximation to the arc length along the interpolated line. Although the latter is *a priori* unknown, a sequence of successively better approximations to it may be obtained by iteration:

- Begin with a crude choice for β_i.
- Compute the arc length as discussed in the following subsection, measured from the first point.
- Use the arc length as a new parameter, and return to the second step.

Computation of the arc length

An expression for the differential arc length dl along a line in terms of $d\beta$ is required for computing line integrals, including the line integral expressing the arc length. Using the Pythagorean theorem, we find

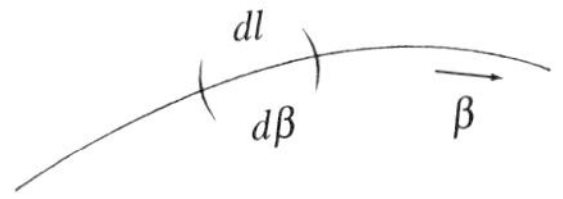

$$\frac{dl}{|d\beta|} = \left[\left(\frac{df_1}{d\beta} \right)^2 + \left(\frac{df_2}{d\beta} \right)^2 + \left(\frac{df_3}{d\beta} \right)^2 \right]^{1/2} \tag{6.8.3}$$

which allows us to express the line integral of a certain scalar, vector, or matrix function g between two points A and B as

$$I(A, B, g) = \int_A^B g(\mathbf{x})\, dl = \int_{\beta_A}^{\beta_B} g(\mathbf{x}(\beta)) \left[\left(\frac{df_1}{d\beta} \right)^2 + \left(\frac{df_2}{d\beta} \right)^2 + \left(\frac{df_3}{d\beta} \right)^2 \right]^{1/2} |d\beta| \tag{6.8.4}$$

The integral can be evaluated using the numerical methods to be discussed in Chapter 7.

Setting $g = 1$, we obtain an approximation to the arc length between the points A and B. An improved parametric representation arises by fixing the point A and setting $\beta(B) = I(A, B, g = 1)$.

PROBLEM

6.8.1 *Choice of an independent variable.*

Suggest a method of defining β_i that is different from those described in the text, and discuss its merits and possible pitfalls.

Local Interpolation

When a line has a convoluted shape, a local parametric representation is appropriate. The general idea is to describe the line in a piecewise manner using different representations for segments that are subtended across a small group of marker points.

Polyline

In the simplest approximation, we represent the section of the line between two successive marker points $\mathbf{x}_i$ and $\mathbf{x}_{i+1}$ with a straight segment that is described as

$$\mathbf{x} = \mathbf{x}_i + \beta \; (\mathbf{x}_{i+1} - \mathbf{x}_i) \tag{6.8.5}$$

where $0 \leq \beta \leq 1$. The differential arc length is given by

$$dl \equiv d|\mathbf{x}| = \beta|\mathbf{x}_{i+1} - \mathbf{x}_i| \tag{6.8.6}$$

The collection of the straight segments yields a *polyline*, also called a *linear B-spline* (see Section 8.4).

Cubic splines

The unit vector that is tangential to the line suffers a discontinuity across the vertices of the polyline. To obtain a parametric representation that produces continuous unit tangential, normal, and binormal vectors, as well as with continuous normal curvature, we resort to cubic splines.

When the line is closed, or is repeated in a periodic fashion, we use periodic boundary conditions for the first and second derivatives of the three individual functions $f_1(\beta)$, $f_2(\beta)$, $f_3(\beta)$ at the first and last marker points (Problem 6.8.2).

PROBLEM

6.8.2 ***Parametric representation of a curve using cubic splines.***

Write a program that uses cubic-spline interpolation to compute the Cartesian coordinates of a point lying on a closed line that is traced by a set of eight marker points, as a function of a suitably chosen parameter β that increases monotonically from 0 to 1 along the line, from start to finish. The Cartesian coordinates of the points are: (1.1, 1.0, 0.0), (2.2, 0.10, 0.05), (3.9, 1.1, 0.07), (4.1, 3.0, 0.07), (3.0, 3.9, 0.06), (2.5, 4.0, 0.05), (2.0, 4.1, 0.03), (1.3, 3.0, 1.0).

Specifically, your program should return the triplet (x, y, z) corresponding to a given value of β between 0 and 1. Note that $\beta = 0$ and 1 correspond to $x = 1.1$, $y = 1.0$, $z = 0.0$. Construct a table of 21 values of the quadruplets (β, x, y, z) at 20 evenly spaced intervals of β between 0 and 1, plot the curve in the three-dimensional space, and discuss the quality of the representation.

Circular arcs for a three-dimensional line

In this representation, we approximate the section of the line that is subtended across three consecutive marker points $\mathbf{x}_{i-1}, \mathbf{x}_i, \mathbf{x}_{i+1}$, with a circular arc of radius a centered at the point $\mathbf{x}_c = (x_c, y_c, z_c)$. An

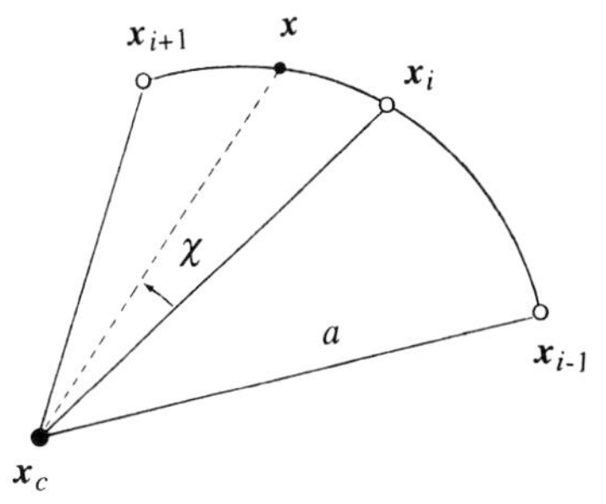

important feature of this representation is that the tangential, normal, and binormal vectors, as well as the curvature of the line, follow immediately and with no extra effort. In particular, the radius of curvature of the line is simply equal to $\pm a$. The torsion of the line, however, cannot be estimated.

An additional bonus of the arc representation is that certain important line integrals of mathematical physics may be evaluated analytically in closed form.

The center of the arc lies in the plane that passes through the three points. Assuming that the three points do not lie on a straight line, we compute the center of the arc by writing

$$(x - x_c)^2 + (y - y_c)^2 + (z - z_c)^2 = a^2 \tag{6.8.7}$$

for $\boldsymbol{x} = \boldsymbol{x}_{i-1}, \boldsymbol{x}_i, \boldsymbol{x}_{i+1}$. Subtracting the equation corresponding to $\boldsymbol{x}_i$ from the other two equations corresponding to $\boldsymbol{x}_{i-1}$ and $\boldsymbol{x}_{i+1}$, we obtain two linear equations for x_c, y_c, z_c:

$$\begin{aligned} 2(x_{i-1} - x_i)x_c + 2(y_{i-1} - y_i)y_c + 2(z_{i-1} - z_i)z_c &= x_{i-1}^2 + y_{i-1}^2 + z_{i-1}^2 - x_i^2 - y_i^2 - z_i^2 \\ 2(x_{i+1} - x_i)x_c + 2(y_{i+1} - y_i)y_c + 2(z_{i+1} - z_i)z_c &= x_{i+1}^2 + y_{i+1}^2 + z_{i+1}^2 - x_i^2 - y_i^2 - z_i^2 \end{aligned} \tag{6.8.8}$$

To derive a third equation, we note that the vectors $\boldsymbol{x}_c - \boldsymbol{x}_{i-1}, \boldsymbol{x}_{i+1} - \boldsymbol{x}_{i-1}$, and $\boldsymbol{x}_i - \boldsymbol{x}_{i-1}$ lie in the same plane, which requires that the vector $(\boldsymbol{x}_{i+1} - \boldsymbol{x}_{i-1}) \times (\boldsymbol{x}_i - \boldsymbol{x}_{i-1})$ be perpendicular to the vector $\boldsymbol{x}_c - \boldsymbol{x}_{i-1}$, where $\times$ signifies the outer vector product. Thus

$$(\boldsymbol{x}_c - \boldsymbol{x}_{i-1})^T[(\boldsymbol{x}_{i+1} - \boldsymbol{x}_{i-1}) \times (\boldsymbol{x}_i - \boldsymbol{x}_{i-1})] = 0 \tag{6.8.9}$$

where

$$(\boldsymbol{x}_{i+1} - \boldsymbol{x}_{i-1}) \times (\boldsymbol{x}_i - \boldsymbol{x}_{i-1}) = \begin{bmatrix} (y_{i+1} - y_{i-1})(z_i - z_{i-1}) - (z_{i+1} - z_{i-1})(y_i - y_{i-1}) \\ (z_{i+1} - z_{i-1})(x_i - x_{i-1}) - (x_{i+1} - z_{i-1})(z_i - z_{i-1}) \\ (x_{i+1} - x_{i-1})(y_i - y_{i-1}) - (y_{i+1} - y_{i-1})(x_i - x_{i-1}) \end{bmatrix} \tag{6.8.10}$$

Solving the system of three linear equations (6.8.8) and (6.8.9) for x_c, y_c, z_c yields the center of the arc. The radius a follows as the distance between the center and one of the three points.

To complete the parametric representation, we introduce the polar angle χ that is subtended between the vector $\boldsymbol{x} - \boldsymbol{x}_c$ and the vector $\boldsymbol{x}_i - \boldsymbol{x}_c$, as shown in the drawing above, where the point $\boldsymbol{x}$ lies on the arc, and regard $\boldsymbol{x}$ a function of χ. The values of χ at the end-points $\boldsymbol{x}_{i-1}$ and $\boldsymbol{x}_{i+1}$ are computed readily by invoking the definition and geometric interpretation of the inner product,

$$\begin{aligned} \chi_{i-1} &= -\arccos[(\boldsymbol{x}_{i-1} - \boldsymbol{x}_c)^T(\boldsymbol{x}_i - \boldsymbol{x}_c)/a^2] \\ \chi_{i+1} &= \arccos[(\boldsymbol{x}_{i+1} - \boldsymbol{x}_c)^T(\boldsymbol{x}_i - \boldsymbol{x}_c)/a^2] \end{aligned} \tag{6.8.11}$$

For a specified value of χ, the corresponding position vector $\boldsymbol{x}$ is found by solving the system of three linear equations

$$
\begin{aligned}
(\boldsymbol{x} - \boldsymbol{x}_c)^T(\boldsymbol{x}_{i-1} - \boldsymbol{x}_c) &= a^2 \cos(\chi - \chi_{i-1}) \\
(\boldsymbol{x} - \boldsymbol{x}_c)^T(\boldsymbol{x}_i - \boldsymbol{x}_c) &= a^2 \cos\chi \\
(\boldsymbol{x} - \boldsymbol{x}_c)^T[(\boldsymbol{x}_{i+1} - \boldsymbol{x}_{i-1}) \times (\boldsymbol{x}_i - \boldsymbol{x}_{i-1})] &= 0
\end{aligned}
\tag{6.8.12}
$$

for $\boldsymbol{x} - \boldsymbol{x}_c$, and then adding it to $\boldsymbol{x}_c$ to produce $\boldsymbol{x}$.

The differential arc length is simply given by $dl = a|d\chi|$.

PROBLEM

6.8.3 ***Parametric representation of a three-dimensional line with circular arcs.***

Write a program that uses the circular-arc approximation to represent a closed line that has been traced by an even number of marker points, working in two stages:

First, compute the centers of the *preliminary* arcs that pass through successive *triplets* of marker points.

The arc subtended between each *pair* of marker points should have the following properties: It should pass through these points; and its center should be the average of the centers of the two preliminary arcs the pass through the points—one point before, and one point after. This averaging technique is called *blending*.

Run the program to compute and display the curve that passes through the eight points given in Problem 6.8.2.

Circular arcs for a planar line

When the three marker points lie in the xy plane, equations (6.8.8) take the simplified forms

$$
\begin{aligned}
x_c + \alpha_{i-1} y_c &= \tfrac{1}{2}(x_{i-1} + x_i) + \tfrac{1}{2}\alpha_{i-1}(y_{i-1} + y_i) \\
x_c + \alpha_{i+1} y_c &= \tfrac{1}{2}(x_{i+1} + x_i) + \tfrac{1}{2}\alpha_{i+1}(y_{i+1} + y_i)
\end{aligned}
\tag{6.8.13}
$$

which provide us with a system of two equations for x_c and y_c, where

$$
\alpha_{i-1} = (y_{i-1} - y_i)/(x_{i-1} - x_i), \qquad \alpha_{i+1} = (y_{i+1} - y_i)/(x_{i+1} - x_i) \tag{6.8.14}
$$

When $\alpha_{i-1} = \alpha_{i+1}$ the system (6.8.13) becomes singular, but this simply means that the three points lie on a straight line.

To complete the parametric representation, we introduce the polar angle θ that is subtended between the vector $\boldsymbol{x} - \boldsymbol{x}_c$ and the x axis, and regard $\boldsymbol{x}$ a function of θ. We thus find

$$
x = x_c + a\cos\theta, \qquad y = y_c + a\sin\theta \tag{6.8.15}
$$

The differential arc length is simply given by $dl = a|d\theta|$.

Which Method?

The question of which method works best admits a multitude of answers. Factors that influence the performance of a method include complexity of shape and number of available marker points.

Four representations of a line passing through a set of six points are compared in Figure 6.8.1, after Olfe (1995). Readers are invited to draw their own conclusions regarding the best approach. Methods of line approximation will be discussed in Chapter 8, and readers should familiarize themselves with that material before making a selection.

6.9 *Interpolation of a Function of Two Variables*

Polynomial interpolation of a function of two or more variables is carried out by methods and procedures that are direct but not necessarily straightforward extensions of those for a function of one variable described in the previous sections. In the case of one variable, the abscissas of the data points lie at the nodes of a one-dimensional grid along the x axis, but in the case of two or more variables they may lie at the nodes of a *structured* or *unstructured* grid in the two-dimensional plane or a higher-dimensional space. Each case requires individual attention.

Bivariate Interpolation Through a Cartesian Grid

Interpolation in two variables is sometimes called *bivariate.* Consider a two-dimensional Cartesian grid, and assume that we are given the values of the function $f(x, y)$ at grid points located at the intersections of x and y grid lines, as shown in the diagram. Suppose now that a point $\mathbf{x} = (x, y)$ lies within the rectangle that is confined by the vertical grid lines $x = x_i, x_{i+1}$, and the horizontal grid lines $y = y_j, y_{j+1}$. The values of i and j can be found most efficiently using the method of binary search discussed in Section 1.4.

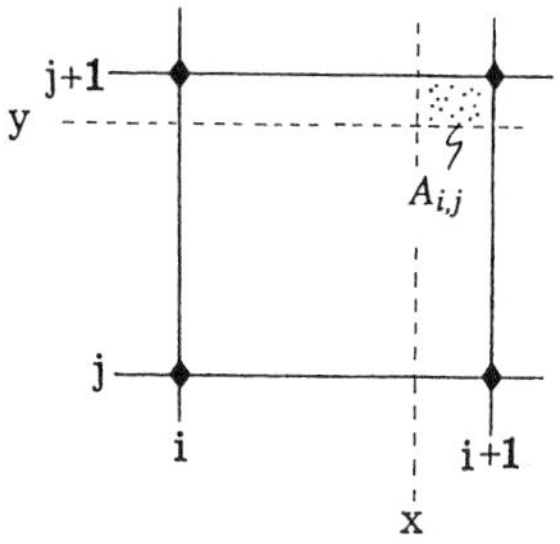

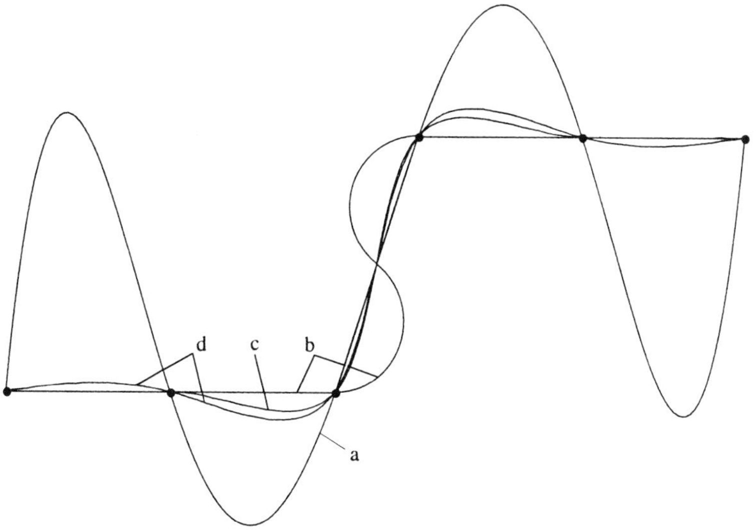

Figure 6.8.1 Interpolating lines passing through six marker points: (*a*) high-order polynomial interpolation, (*b*) circular-arc fit, (*c*) parabolic blend, and (*d*) natural cubic spline.

Bilinear interpolation

This is the two-dimensional counterpart of the one-dimensional linear local interpolation discussed in Section 6.6. For each value of x, the interpolating function is assumed to vary linearly with respect to y, and vice versa.

The value $f(x, y)$ is computed as a weighted average of the values of f at the four closest grid points. The weight corresponding to the grid point (x_i, y_j) is equal to $A_{i,j}/A$, where A is the area of the rectangular cell, and $A_{i,j}$ is the dotted area shown in the last drawing. The weights of the other three points are defined in a similar manner. The interpolating formula is

$$P_1^{(i,j)}(x, y) = \begin{bmatrix} (1-p)q & pq \\ (1-p)(1-q) & p(1-q) \end{bmatrix} : \begin{bmatrix} f_{i,j+1} & f_{i+1,j+1} \\ f_{i,j} & f_{i+1,j} \end{bmatrix} \tag{6.9.1}$$

where the semicolon indicates the double-dot matrix product defined in equation (2.1.24), and where

$$p = \frac{x - x_i}{x_{i+1} - x_i}, \qquad q = \frac{y - y_j}{y_{j+1} - y_j} \tag{6.9.2}$$

Bilinear interpolation yields a continuous function with discontinuous first derivatives across the grid lines.

It can be shown that when $f(\mathbf{x}) = a + bx + cy + dxy$, where a to d are four constants, bilinear interpolation produces the exact answer; that is, it interpolates exactly linear functions and a particular class of quadratic functions. More generally, the leading contribution to the error is on the order of Δx^2 or Δy^2.

Biquadratic interpolation

To improve the accuracy of the bilinear interpolation, we involve a higher number of grid points. A general quadratic function of x and y,

$$f(x, y) = a + bx + cy + dx^2 + exy + fy^2 \tag{6.9.3}$$

involves six constants a to f. To achieve second-order accuracy, we must involve an equal number of grid points. Depending on the location of these points, we obtain a variety of finite-difference formulas (e.g., Abramowitz and Stegun 1972, p. 882).

For example, the six-point formula corresponding to the stencil shown in the diagram, with uniformly spaced grid lines, yields

$$P_2(x, y) = \begin{bmatrix} 0 & \frac{1}{2}q(q-2p+1) & pq \\ \frac{1}{2}p(p-1) & 1+pq-p^2-q^2 & \frac{1}{2}p(p-2q+1) \\ 0 & \frac{1}{2}q(q-1) & 0 \end{bmatrix} : \begin{bmatrix} 0 & f_{i,j+1} & f_{i+1,j+1} \\ f_{i-1,j} & f_{i,j} & f_{i+1,j} \\ 0 & f_{i,j-1} & 0 \end{bmatrix} \tag{6.9.4}$$

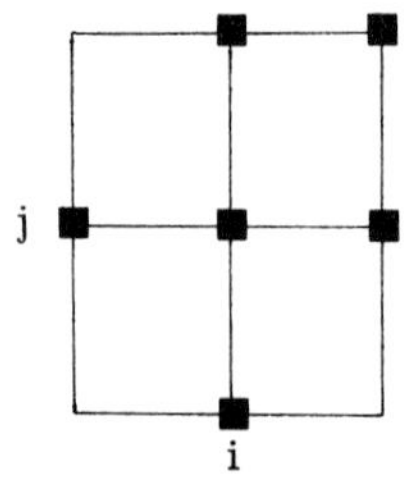

where p and q are defined in equations (6.9.2). The leading contribution to the error is on the order of Δx^3 or Δy^3.

PROBLEM

6.9.1 ***Biquadratic interpolation.***

Derive quadratic six-point formulas corresponding to the stencils shown in the diagram, for uniformly spaced grid lines.

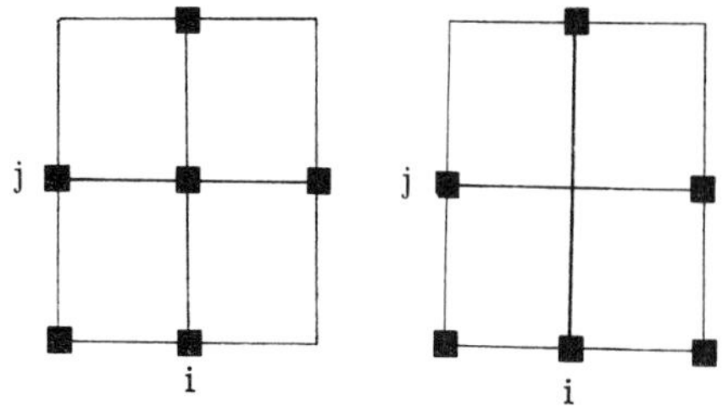

High-order interpolation

High-order bivariate interpolation can be implemented by several methods including the use of two-dimensional cardinal functions constructed in terms of Boolean sums, Hermite polynomials in two variables, and surface cubic splines (e.g., Lancaster and Šalkauskas 1986; Shikin and Plis 1995; Späth 1995b; Kincaid and Cheney 1996, pp. 449–467). The subtlety of high-order interpolation becomes apparent by observing that it is not generally possible to find a *unique* polynomial cardinal function that takes the value of unity at a particular grid point and vanishes at an arbitrary number of other grid points. Many commonly used interpolation methods have been developed for applications in computer graphics, but most of them are also appropriate for scientific and mainstream engineering computation.

Bivariate Interpolation Through a Quadrilateral

Consider a quadrilateral in the xy plane defined by its four vertices $\mathbf{x}_1, \mathbf{x}_2, \mathbf{x}_3, \mathbf{x}_4$, numbered in the counterclockwise sense, where $\mathbf{x} = (x, y)$. To interpolate a function $f(x, y)$ over the area of the quadrilateral given its values at vertices, we first map the quadrilateral to a square centered at the origin of the $\xi\eta$ plane, and then carry out the interpolation with respect to ξ and η using one of the methods described in the preceding subsection.

In particular, the transformation

$$\mathbf{x} = \sum_{i=1}^{4} \mathbf{x}_i \, \phi_i(\xi, \eta) \tag{6.9.5}$$

where

$$\begin{aligned} \phi_1(\xi, \eta) &= \tfrac{1}{4}(1 - \xi)(1 - \eta), \qquad & \phi_2(\xi, \eta) &= \tfrac{1}{4}(1 + \xi)(1 - \eta) \\ \phi_3(\xi, \eta) &= \tfrac{1}{4}(1 + \xi)(1 + \eta), \qquad & \phi_4(\xi, \eta) &= \tfrac{1}{4}(1 - \xi)(1 + \eta) \end{aligned} \tag{6.9.6}$$

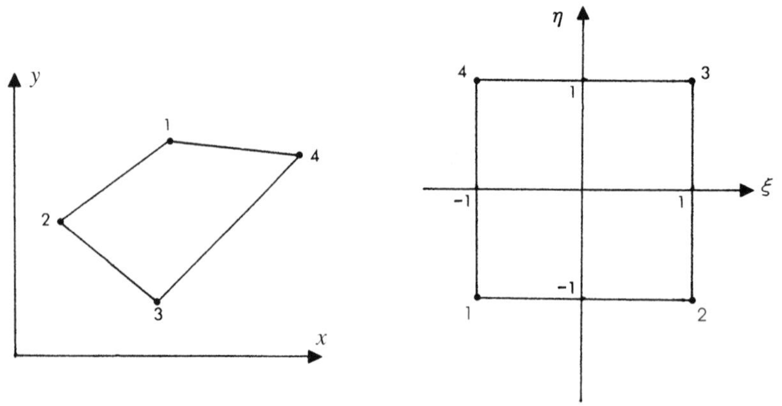

maps the interior of a quadrilateral to the square $-1 \le \xi \le 1$ and $-1 \le \eta \le 1$, as shown in the diagram. To find the values of ξ and η corresponding to a specified point $\boldsymbol{x}$, we solve a system of two quadratic equations for two unknowns, based on equation (6.9.5).

Bivariate Interpolation Through an Unstructured Triangular Grid

Several classes of numerical methods for solving partial differential equations over a planar surface in the xy plane tile the surface with triangles, thereby producing an unstructured triangular grid, and then generate the values of an unknown function at the vertices. Examples are finite-element, finite-volume, spectral-element, and boundary-element methods. The process of subdivision is called *triangulation.*

Linear interpolation within a triangle with straight sides

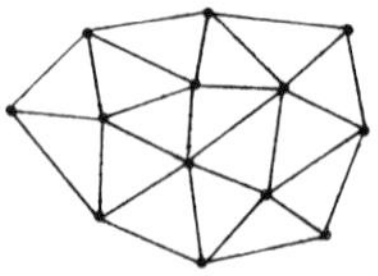

In the simplest triangulation method, we tessellate the interpolation surface in the xy plane into a collection of triangles with straight sides; the value of the interpolated function $f(x, y)$ is assumed to be known at the vertices. Our objective is to compute $f(x, y)$ at a specified point.

Before we begin the interpolation, we must identify the triangle that hosts a particular point of interest in the xy plane. This can be done by several methods; one method was discussed in Section 1.5, and another method will be described shortly; other methods are reviewed by Milgram (1989) and Löhner (1995).

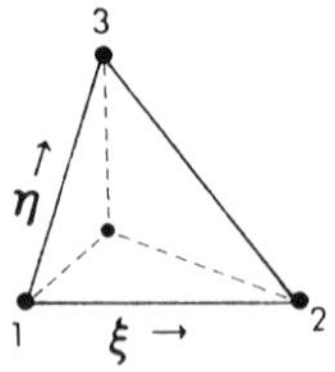

A triangle with straight sides is defined by the location of its three vertices in the xy plane, $\boldsymbol{x}_1, \boldsymbol{x}_2$, and $\boldsymbol{x}_3$, numbered in the counterclockwise sense. The known values of the interpolated function at the vertices are denoted, respectively, by f_1, f_2, and f_3. To obtain the value of f at an arbitrary point $\boldsymbol{x} = (x, y)$ that lies within the triangle, we first compute the areas of the original triangle and of two daughter triangles with vertices at the ordered triplets of points

$$(\boldsymbol{x}_1, \boldsymbol{x}_2, \boldsymbol{x}_3), \quad (\boldsymbol{x}, \boldsymbol{x}_3, \boldsymbol{x}_1), \quad (\boldsymbol{x}, \boldsymbol{x}_1, \boldsymbol{x}_2)$$

which are respectively equal to

$$\begin{aligned} A &= (x_2 - x_1)(y_3 - y_1) - (y_2 - y_1)(x_3 - x_1) \\ A_2 &= (x_3 - x)(y_1 - y) - (y_3 - y)(x_1 - x) \\ A_3 &= (x_1 - x)(y_2 - y) - (y_1 - y)(x_2 - x) \end{aligned} \tag{6.9.7}$$

Next, we compute the local triangle coordinates

$$\xi = A_2/A, \qquad \eta = A_3/A \tag{6.9.8}$$

Note that $\eta = 0$ along the straight segment that is subtended between the points $\boldsymbol{x}_1$ and $\boldsymbol{x}_2$, while ξ increases linearly from 0 to 1 with respect to arc length between these points. Similarly, $\xi = 0$ along the straight segment that is subtended between the points $\boldsymbol{x}_1$ and $\boldsymbol{x}_3$, while η increases linearly from 0 to 1 with respect to arc length between these points.

The two parameters ξ and η take values that fall within the flat isosceles orthogonal triangle shown in the diagram. A straight segment parallel to the ξ or η axis in the $\xi\eta$ plane corresponds to another straight segment parallel to the 1–2 or 1–3 side in the xy plane.

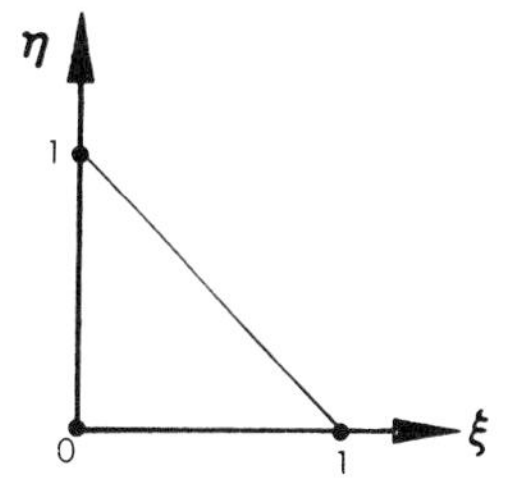

Linear interpolation is effected by setting

$$f(\xi, \eta) = f_1\zeta + f_2\xi + f_3\eta \tag{6.9.9}$$

where

$$\zeta = 1 - \xi - \eta \tag{6.9.10}$$

The variables ξ, η, ζ are sometimes called the *triangle barycentric coordinates.* When the point $\boldsymbol{x}$ lies at the centroid of the triangle, ξ, η, ζ are all equal to $\frac{1}{3}$. When $\boldsymbol{x}$ lies within the triangle, ξ, η, ζ are all positive, whereas if $\boldsymbol{x}$ lies outside the triangle, at least one of them is negative. This property provides us with a method of assessing whether a particular point lies inside or outside a triangle.

When $f(x, y)$ is constant or varies in a linear manner with respect to x and y, formula (6.9.9) produces the exact answer.

PROBLEMS

6.9.2 *Inside or outside a triangle?*

Write a program that, given three vertices, deduces whether a specified point is located inside or outside the triangle using the method discussed in the text.

6.9.3 *Inside or outside a polygon?*

Develop and implement a method for assessing whether a point lies inside a polygon. You will benefit from the wisdom of the papers by Milgram (1989) and Löhner (1995).

Triangular tepees

Let us compile the linear local interpolation functions of all M triangles that arise by triangulating a certain area, each one expressed by equation (6.9.9), and extend their domain of definition so that the right-hand side of equation (6.9.9) is equal to zero when the point $\boldsymbol{x}$ is located outside the host triangle. Having made this extension, we express the interpolated function in the form

$$f(x, y) = \sum_{i=1}^{M} f_i \, \phi_i(x, y) \tag{6.9.11}$$

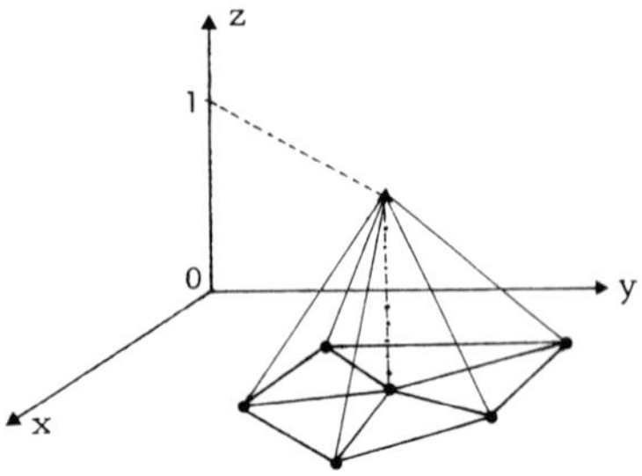

where $\phi_i(x, y)$ are prismoidal tent functions, formally called *shape factors.* Such expansions of a function of two variables in terms of tent functions are used in the numerical solution of partial differential equations using finite-element and boundary-element methods.

Quadratic interpolation through triangles with straight sides

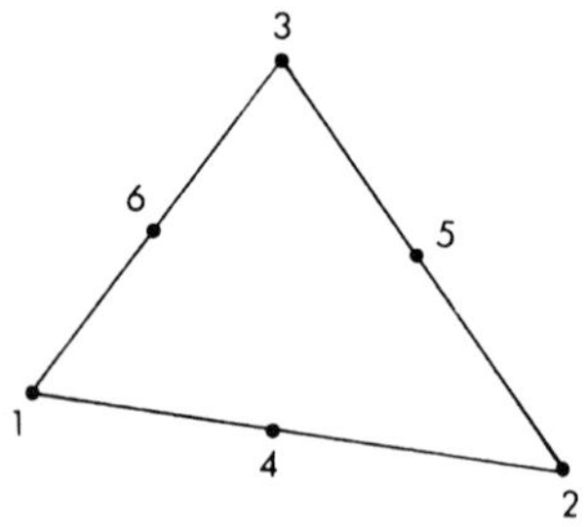

Next, we consider interpolating a function $f(x, y)$ through a triangle with straight sides, where the value of $f(x, y)$ is known at six points: The three vertices $\mathbf{x}_1, \mathbf{x}_2$, and $\mathbf{x}_3$, numbered in the counterclockwise sense, and three side-points $\mathbf{x}_4, \mathbf{x}_5$, and $\mathbf{x}_6$. The corresponding values of the interpolated function are denoted by $f_i, i = 1, 2, \ldots, 6$.

Quadratic interpolation is implemented by computing the values of the parameters ξ and η corresponding to a certain point $\mathbf{x}$ that lies within the triangle using equations (6.9.8), and then setting

$$f(\xi, \eta) = \sum_{i=1}^{6} f_i \, \phi_i(\xi, \eta) \tag{6.9.12}$$

where

$$\begin{aligned}
\phi_2 &= \frac{1}{1-\alpha}\xi\left(\xi - \alpha + \frac{\alpha-\gamma}{1-\gamma}\eta\right) \\
\phi_3 &= \frac{1}{1-\beta}\eta\left(\eta - \beta + \frac{\beta+\gamma-1}{\gamma}\xi\right) \\
\phi_4 &= \frac{1}{\alpha(1-\alpha)}\xi(1-\xi-\eta) \\
\phi_5 &= \frac{1}{\gamma(1-\gamma)}\xi\eta \\
\phi_6 &= \frac{1}{\beta(1-\beta)}\eta(1-\xi-\eta) \\
\phi_1 &= 1 - \phi_2 - \phi_3 - \phi_4 - \phi_5 - \phi_6
\end{aligned} \tag{6.9.13}$$

and

$$\alpha = \frac{1}{1 + \dfrac{|\boldsymbol{x}_4 - \boldsymbol{x}_2|}{|\boldsymbol{x}_4 - \boldsymbol{x}_1|}}, \qquad \beta = \frac{1}{1 + \dfrac{|\boldsymbol{x}_6 - \boldsymbol{x}_3|}{|\boldsymbol{x}_6 - \boldsymbol{x}_1|}}, \qquad \gamma = \frac{1}{1 + \dfrac{|\boldsymbol{x}_5 - \boldsymbol{x}_2|}{|\boldsymbol{x}_5 - \boldsymbol{x}_3|}} \tag{6.9.14}$$

The (ξ, η) coordinates of the side-points $\boldsymbol{x}_4, \boldsymbol{x}_5$, and $\boldsymbol{x}_6$ are given, respectively, by

$$(\alpha, 0), \quad (\gamma, 1-\gamma), \quad (0, \beta) \tag{6.9.13}$$

When the mid-points are located in the middle of the corresponding sides, in which case $\alpha = \beta = \gamma = 0.5$, we obtain the simpler forms

$$\begin{aligned} \phi_1 &= \xi(2\xi - 1) \\ \phi_2 &= \zeta(2\zeta - 1) \\ \phi_3 &= \eta(2\eta - 1) \\ \phi_4 &= 4\xi\zeta \\ \phi_5 &= 4\xi\eta \\ \phi_6 &= 4\eta\zeta \end{aligned} \tag{6.9.16}$$

where $\zeta = 1 - \xi - \eta$.

Quadratic interpolation through a triangle with curved sides

To accommodate interpolation domains with curved boundaries, we use triangles with curved sides. In their simplest representation, these are defined in terms of the three vertices and three side-points, as shown in the diagram.

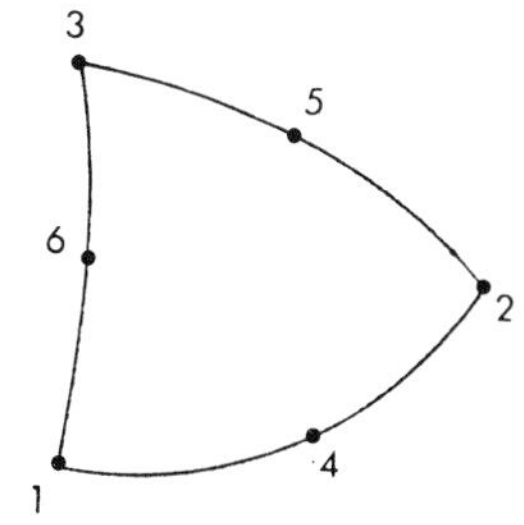

Equations (6.9.12) to (6.9.16) hold, but we can no longer compute the values of the parameters ξ and η corresponding to a certain point $\boldsymbol{x}$ that lies within the triangle from equations (6.9.8). Instead, we must apply equation (6.9.12) with $f = x$ or $f = y$, and solve the resulting system of quadratic equations for ξ and η using one of the methods discussed in Chapter 4.

Higher-order Interpolation, Nontriangular Elements, and Orthogonal Expansions

Formulas and procedures for performing high-order interpolation over elements with triangular, rectangular, and other shapes, and methods of interpolating through three-dimensional elements are discussed extensively in the literature of finite-element methods (e.g., Chung 1978).

Orthogonal polynomial bases over rectangles arise as pairwise products of orthogonal polynomials of one variable. Orthogonal expansions over disks and triangles are discussed by Dubiner (1991) and Sherwin and Karniadakis (1995).

Interpolation from Scattered Data

In certain applications, the data points are scattered in an irregular manner over a certain region of a plane or a volume of the three-dimensional or higher-dimensional space. One systematic way of carrying out the interpolation is to divide the plane into triangles whose vertices coincide with the data points using the *Delaunay triangulation* method, and then to use the interpolation methods discussed earlier in this section; similarly for the space.

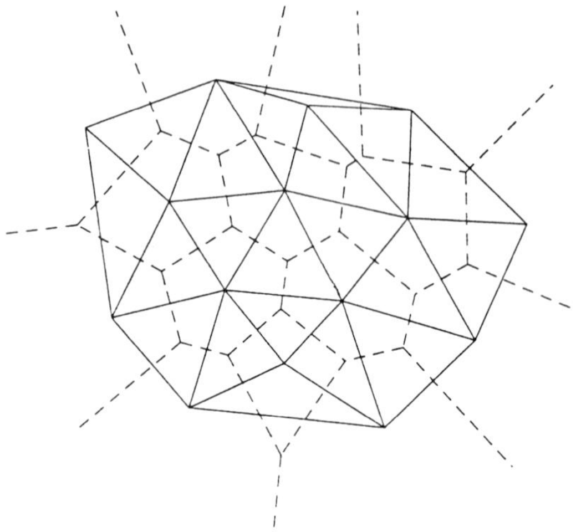

One method of carrying out the Delaunay triangulation emanates from the *Dirichlet–Voronoi–Thiessen tessellation*, DVT, which divides the plane into polygons, each polygon containing one datum point. The distance of a point in a polygon from the host datum point is smaller than the distance from any other datum point. It can be shown that the sides of the polygons are perpendicular bisectors to the straight segments connecting pairs of data points, as shown in the diagram. Algorithms for the DVT tessellation are described by Green and Sibson (1978).

Once the DVT tessellation has been completed, the Delaunay triangulation emerges by connecting each datum point with all other data points that share with it a polygon side. Direct Delaunay triangulation methods are discussed by Rebay (1993).

6.10 Parametric Description of Surfaces

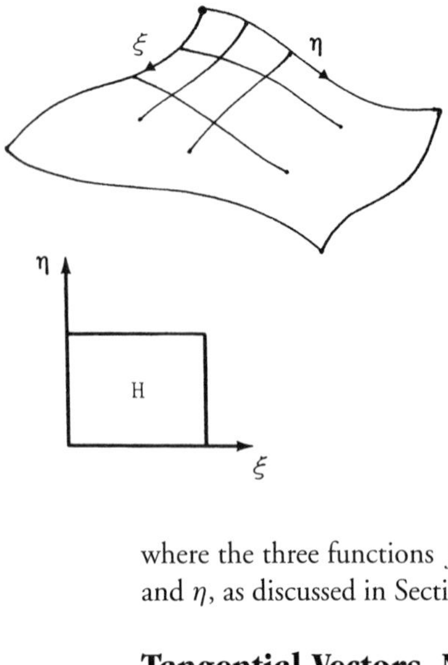

Consider an open or closed, infinite or finite surface whose position has been traced with a collection of marker points. One way to describe the surface is to introduce two variables ξ and η forming a right-handed coordinate system, called *surface curvilinear coordinates*, and regard the position vector $\boldsymbol{x}$ as a function of ξ and η, writing $\boldsymbol{x}(\xi, \eta)$. This parametric representation establishes a generally nonlinear mapping between the curved material surface in physical space and a planar area H in the parametric (ξ, η) plane, where ξ and η take values over a specified range.

More specifically, the parametric representation is effected by setting

$$x = f_1(\xi, \eta), \quad y = f_2(\xi, \eta), \quad z = f_3(\xi, \eta) \tag{6.10.1}$$

where the three functions f_1, f_2, f_3 are evaluated by methods of bivariate interpolation with respect to ξ and η, as discussed in Section 6.9.

Tangential Vectors, Normal Vector, and Surface Integrals

In order to establish relations between the geometrical properties of the surface and the position of the marker points in the surface, $\boldsymbol{x}(\xi, \eta)$, we introduce the tangential unit vectors

$$t_\xi = \frac{1}{h_\xi}\frac{\partial \boldsymbol{x}}{\partial \xi}, \qquad t_\eta = \frac{1}{h_\eta}\frac{\partial \boldsymbol{x}}{\partial \eta} \tag{6.10.2}$$

illustrated in Figure 6.10.1, where

$$h_\xi = \left|\frac{\partial \boldsymbol{x}}{\partial \xi}\right|, \qquad h_\eta = \left|\frac{\partial \boldsymbol{x}}{\partial \eta}\right| \tag{6.10.3}$$

are the *metric coefficients* associated with the curvilinear coordinates. The arc length of an infinitesimal section of a ξ or η line is, respectively, equal to

$$dl_\xi = h_\xi \, d\xi \ \text{ or } \ dl_\eta = h_\eta \, d\eta \tag{6.10.4}$$

Any linear combination of the tangential vectors defined in equations (6.10.2) is also a tangential vector.

The unit vector normal to the surface is given by

$$\boldsymbol{n} = \frac{1}{h_S}\frac{\partial \boldsymbol{x}}{\partial \xi} \times \frac{\partial \boldsymbol{x}}{\partial \eta} \tag{6.10.5}$$

where

$$h_s = \left|\frac{\partial \boldsymbol{x}}{\partial \xi} \times \frac{\partial \boldsymbol{x}}{\partial \eta}\right| \tag{6.10.6}$$

is the *surface metric* coefficient. It should be noted that the surface normal $\boldsymbol{n}$ is different from the normal vectors to the lines of constant ξ or η, as illustrated in Figure 6.10.1.

Combining equations (6.10.5) and (6.10.2), we obtain

$$\boldsymbol{n} = \frac{h_\xi h_\eta}{h_S} \boldsymbol{t}_\xi \times \boldsymbol{t}_\eta \tag{6.10.7}$$

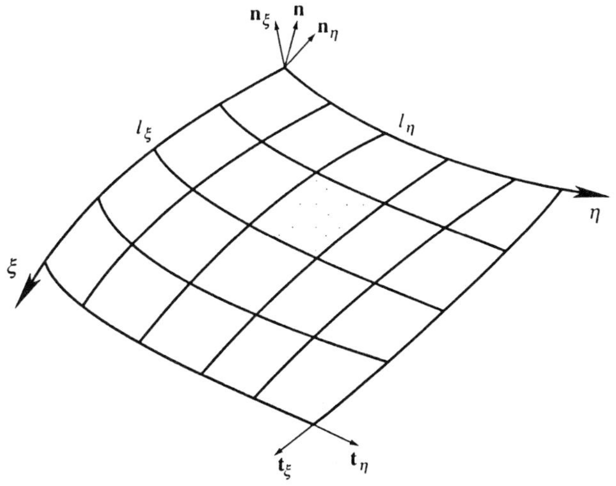

Figure 6.10.1 A system of two curvilinear axes (ξ, η) on a three-dimensional surface; $\boldsymbol{n}$ is the unit vector normal to the surface, and $\boldsymbol{n}_\xi$, $\boldsymbol{n}_\eta$ are the principal unit vectors normal to the ξ and to the η axis. From *Introduction to Theoretical and Computational Fluid Dynamics*, by C. Pozrikidis. Copyright © 1997 by Oxford University Press. Used by permission of Oxford University Press, Inc.

When $\boldsymbol{t}_\eta$ is perpendicular to $\boldsymbol{t}_\xi$, in which case $\boldsymbol{t}_\xi \cdot \boldsymbol{t}_\eta = 0$, the system of the surface coordinates (ξ, η) is orthogonal, $\boldsymbol{n} = \boldsymbol{t}_\xi \times \boldsymbol{t}_\eta$ and $h_S = h_\xi h_\eta$. The $\cdot$ and the $\times$ indicate, respectively, the inner and the outer product of three-dimensional vectors.

The area of a differential element of the surface is given by $dS = h_S \, d\xi \, d\eta$, and the surface integral of a function $f(\boldsymbol{x})$ is given by

$$\int f(\boldsymbol{x}) \, dS = \int_H f\big(\boldsymbol{x}(\xi, \eta)\big) h_S \, d\xi \, d\eta \tag{6.10.8}$$

where H is the domain of variation of ξ and η over the surface.

PROBLEM

6.10.1 ***Local representation with a small section of a sphere.***

It might appear reasonable to represent the portion of a surface in the vicinity of four neighboring points with a section of a sphere that passes through these points. Explain why this approximation may lead to gross error even when the points are located very close to one another.

Surface Discretization

In many applications, it is convenient to represent a surface with a collection of *surface elements* of various shapes, including triangles and quadrilaterals with straight or curved sides. The process of generating the surface elements is called *surface discretization*; the result of the discretization is a *structured*, a *semistructured*, or *unstructured surface grid*. The appropriate qualifier depends on the regularity of the grid as perceived by the naked eye. The selection of the element shape and size, and the way in which the discretization is carried out depend on the particular application that motivated the discretization.

To keep track of the relative location of the elements, we label them using sequential integer values and introduce a *connectivity table* that contains the labels of the elements and the corresponding labels of their vertices, arranged in the counterclockwise sense.

For example, one way of discretizing the surface of the sphere into rectangular elements whose union provides us with a structured surface grid is to inscribe it within a cube, discretize the sides of the cube working with four individual Cartesian grids, and then project the grid lines onto the surface of the sphere. The result is shown in the illustration.

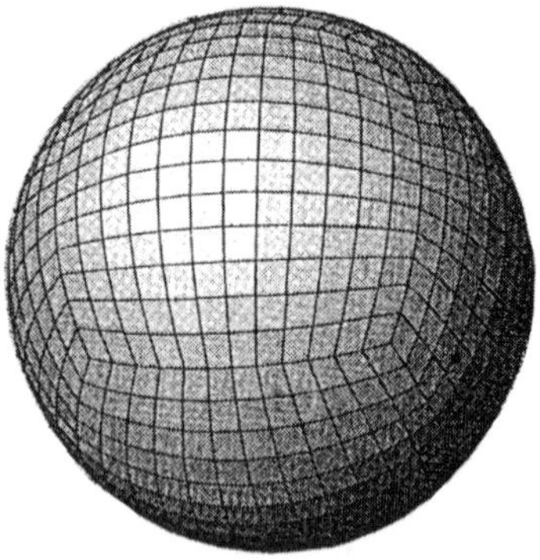

Representation with Triangles

An important theorem of differential geometry guarantees that any surface can be discretized into a collection of triangles whose union provides us with a structured or unstructured surface grid. First, we discuss the parametric representation of the triangles, and then we discuss the process of *triangulation*.

Flat triangles

In the simplest approximation, we use flat triangles defined by their vertices $\mathbf{x}_1, \mathbf{x}_2$, and $\mathbf{x}_3$, numbered in the counterclockwise sense as we look at the surface from a specified side. The parametric representation is effected by stipulating that $\eta = 0$ while ξ increases from 0 to 1 along the straight segment that is subtended between the points $\mathbf{x}_1$ and $\mathbf{x}_2$; and $\xi = 0$ while η increases from 0 to 1 along the straight segment that is subtended between the points $\mathbf{x}_1$ and $\mathbf{x}_2$.

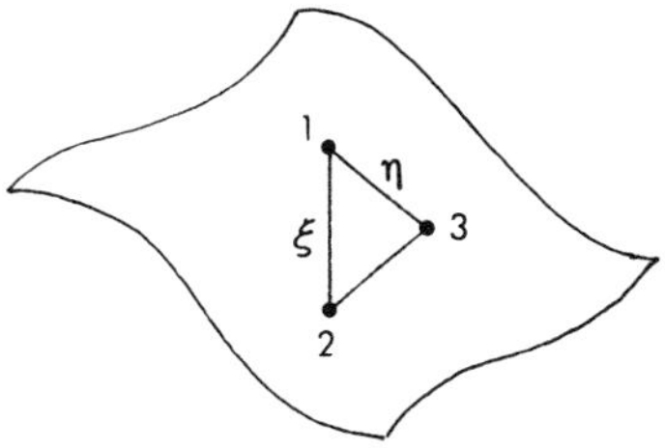

The parametric representation is expressed by equation (6.9.9) with $\mathbf{x}$ in place of f, where ξ and η vary over the planar isosceles orthogonal triangle depicted in the last drawing before equation (6.9.9). A straight line in the interior of the $\xi\eta$ triangle is mapped to a straight segment in the interior of the physical triangle. In general, the mapping is not conformal, which means that the angles of the triangle in physical space are not the same as those in the parametric space. Since, however, all partial derivatives of the right-hand side of equation (6.9.9) with respect to ξ or η are constant, the mapping is nonsingular even when the triangle is notably skewed. The surface metric coefficient h_S defined in equation (6.10.6) is constant and equal to $2A$, where A is the area of the triangle.

Curved triangles

A better description is effected by using triangles with curved sides defined by six marker points, as shown in the diagram. The parametric representation is expressed by equations (6.9.12)–(6.9.16) with $\mathbf{x}$ in place of f; ξ and η vary over the planar isosceles orthogonal triangle depicted in the last drawing before equation (6.9.9). The surface metric coefficient h_S defined in equation (6.10.6) varies over the triangle; that is, it is a function of ξ and η.

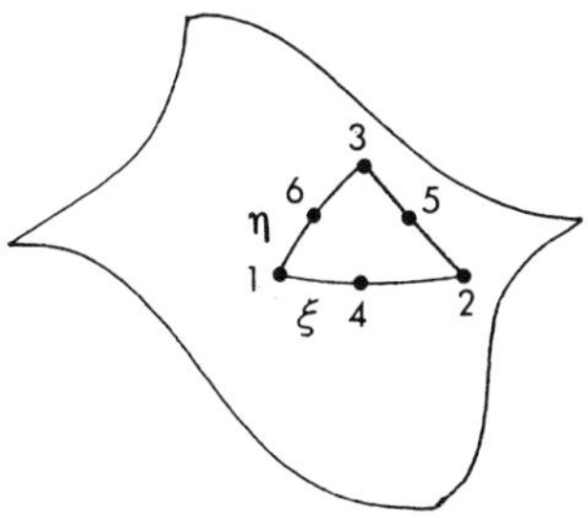

Triangulation

Triangulating a planar or curved surface can be done by several methods, with varying degrees of accuracy and sophistication.

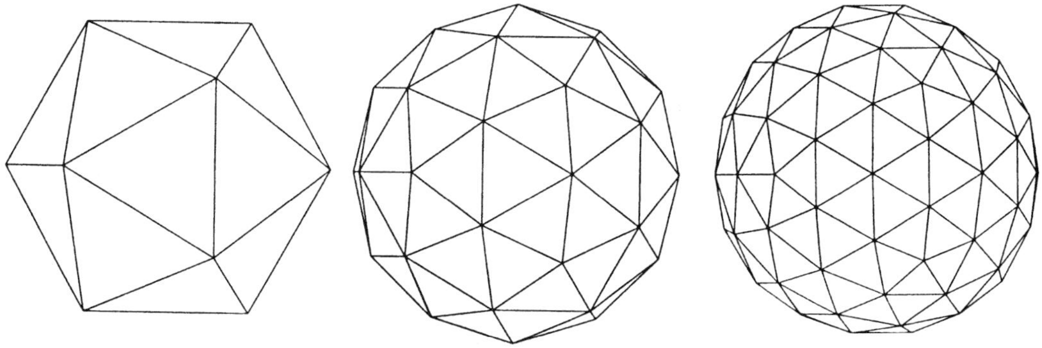

A simple way of discretizing the surface of a sphere into a collection of triangles with comparable dimensions is to begin from a crude discretization and then refine it. For example, we may discretize the surface of the whole sphere into eight flat or curved triangles with vertices at the polar and equatorial planes forming a *regular*

octahedron; subdivide each parental triangle into four descendant subtriangles; project the new vertices onto the sphere radially or parallel to a specified plane; and repeat the process until a certain level of refinement has been achieved. The surface grid may then be deformed to yield a nonspherical shape, such as a spheroid or an ellipsoid.

An alternative triangulation method departs from a *regular icosahedron* and proceeds by carrying out the refinement discussed in the previous paragraph, as shown in the last illustration after Olfe (1995); εἴκοσι in Greek means *twenty*, and ἕδρα translates into *base*.

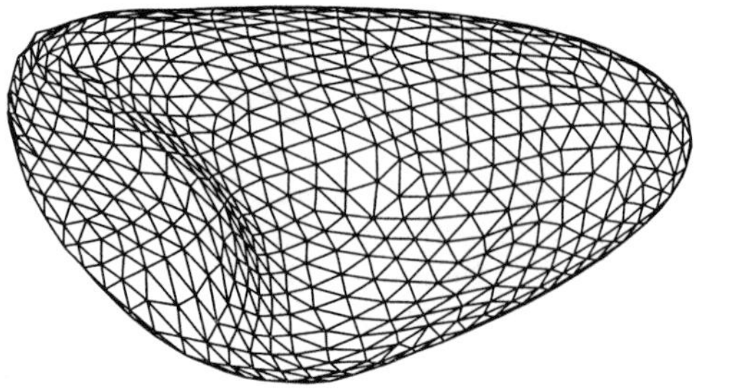

More sophisticated methods of adaptive plane and surface triangulation have been developed to allow for refined representation at regions of high curvature. Examples are the *Delaunay triangulation* method, the *advancing-front* method, and their variations (e.g., Hinton et al. 1991; Okabe et al. 1992; Rebay 1993; Nakahashi and Sharov 1995). The diagram shows the triangulation of a deformed prolate spheroid with a dimple using the advancing-front method with curvature adaptation (Kwak and Pozrikidis 1998).

6.11 Numerical Differentiation of a Function of One Variable

Having computed a polynomial or another interpolating function $g(x)$, we may differentiate it or integrate it to obtain approximations to the derivatives or to definite integrals of the interpolated function $f(x)$. In fact, function interpolation discussed previously in this chapter and function approximation to be discussed in Chapter 8 are often motivated by the necessity to perform function differentiation and integration. Function differentiation will be the subject of the present and of the following section, and function integration will be the exclusive subject of Chapter 7.

The Problem of Numerical Differentiation

The one-dimensional version of the problem is generally stated as follows: Given the values of a function $f(x)$ at the $N+1$ arbitrarily distributed data points x_i, $i = 1, \ldots, N+1$, compute its first or higher-order derivatives at the data points or at intermediate points. This is typically done by approximating the function with an Nth degree interpolating polynomial using the numerical methods discussed in the preceding sections, and then differentiating the interpolating polynomial to obtain *finite-difference* approximations; that is, expressions for the derivatives in terms of the abscissas and ordinates of the data points.

In previous sections, we saw that the accuracy of the polynomial interpolation can be notably sensitive to the number and distribution of the data points. We saw, in particular, that unless the data points are distributed in a special manner that is dictated by the zeros of orthogonal polynomials, the use of a large number of data points can introduce significant numerical error. It is understandable then that, in practice, function differentiation with arbitrarily spaced data points is carried out with a small or moderate value of N; that is, by employing a low-order interpolating polynomial.

Backward, Centered, and Forward Differences

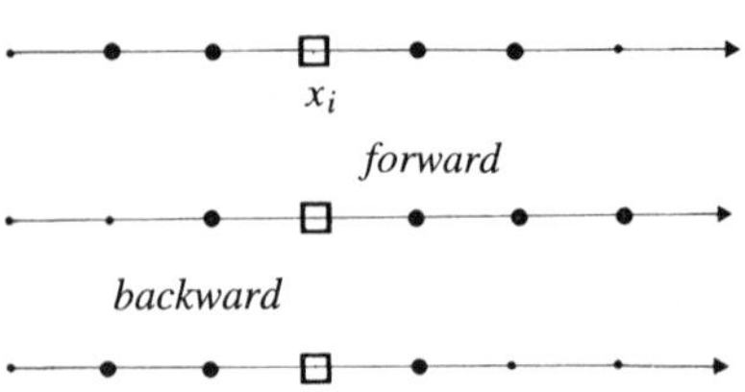

Consider the computation of the derivatives of a function at the ith data point. If we use an equal number of data points on either side of x_i in the process of constructing the interpolating polynomial, we obtain *central* or *centered differences*. Otherwise, we obtain spatially biased *forward* or *backward differences*. The former employ a higher number of points on the right, and the latter employ a higher number of points on the left of the point x_i.

The choice of backward, central, and forward differences is dictated by the physics of the problem whose mathematical modeling introduced the derivative, as well as by the availability of data points on either side of x_i. These vague statements will become more precise in Chapters 10 and 11, where we shall discuss the numerical solution of differential equations using finite-difference methods.

There are several ways of deriving finite-difference approximations but, if done correctly, they all lead to the same formulas. Simplicity of derivation and ease in estimating the numerical error are two important factors in making a selection. We begin by addressing analytical derivations and then discuss the use of tables and programmable computation.

Finite-difference Formulas from Linear Interpolation

Linear interpolation provides us with a rudimentary approximation to the first derivative. The second- and higher-order derivatives are estimated to be equal to zero—that is, they are computed with an error on the order of unity; this is clearly unacceptable by any standards.

Differentiating the right-hand side of equation (6.6.1), evaluating the resulting expression at the point x_i, and using the error formula (6.6.4), we derive the *forward-difference* formula

$$f'(x_i) = \frac{f(x_{i+1}) - f(x_i)}{x_{i+1} - x_i} + O(h_i) \tag{6.11.1}$$

Working in a similar manner with the counterpart of equation (6.6.1) for the $i - 1$ interval, we obtain the *backward-difference* formula

$$f'(x_i) = \frac{f(x_i) - f(x_{i-1})}{x_i - x_{i-1}} + O(h_{i-1}) \tag{6.11.2}$$

The *centered-difference* formula will be discussed in the next subsection.

Finite-difference Formulas from Quadratic Interpolation

Quadratic interpolation provides us with approximations to both the first and second derivatives; higher-order derivatives are computed with an error on the order of unity. Using equation (6.6.6) and the error estimate given in equation (6.6.9), we derive the *centered-difference* approximations

$$f'(x_i) = b_i + O(h_i^2), \qquad f''(x_i) = 2a_i + O(h_i) \tag{6.11.3}$$

where a_i and b_i are computed from equations (6.6.7). If the data points are spaced evenly along the x axis, a_i and b_i are given by the simplified expressions (6.6.8), and the error in the computation of the second derivative is elevated to $O(h_i^2)$.

Finite-difference Formulas from High-order Interpolation

To exemplify the derivation of finite-difference formulas by high-order interpolation, we consider the computation of $f''(x_0)$ in terms values of $f(x)$ at the four *evenly spaced* points located at

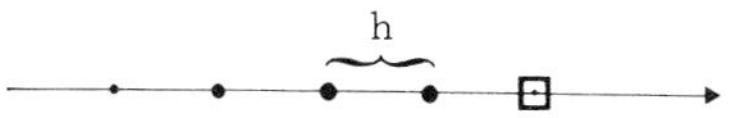

$$x_0 - 3h, \quad x_0 - 2h, \quad x_0 - h, \quad x_0$$

respectively denoted as $f_{-3}, f_{-2}, f_{-1}, f_0$. Using Lagrange's method, we find that the interpolating polynomial is given by

$$
\begin{aligned}
P_4(x) = & f_{-3}\frac{(x-x_0+2h)(x-x_0+h)(x-x_0)}{(-h)(-2h)(-3h)} + f_{-2}\frac{(x-x_0+3h)(x-x_0+h)(x-x_0)}{(h)(-h)(-2h)} \\
& + f_{-1}\frac{(x-x_0+3h)(x-x_0+2h)(x-x_0)}{(2h)(h)(-h)} \\
& + f_0\frac{(x-x_0+3h)(x-x_0+2h)(x-x_0+h)}{(3h)(2h)(h)}
\end{aligned}
\tag{6.11.4}
$$

Straightforward differentiation gives

$$
\begin{aligned}
P_4''(x) = & -f_{-3}\frac{(x-x_0+2h)+(x-x_0+h)+(x-x_0)}{3h^3} + f_{-2}\frac{(x-x_0+3h)+(x-x_0+h)+(x-x_0)}{h^3} \\
& - f_{-1}\frac{(x-x_0+3h)+(x-x_0+2h)+(x-x_0)}{h^3} \\
& + f_0\frac{(x-x_0+3h)+(x-x_0+2h)+(x-x_0+h)}{3h^3}
\end{aligned}
\tag{6.11.5}
$$

where a prime designates a derivative with respect to x. Evaluating the right-hand side of equation (6.11.5) at $x = x_0$ produces the desired result

$$
f''(x_0) \cong P_4''(x_0) = \frac{-f_{-3} + 4f_{-2} - 5f_{-1} + 2f_0}{h^2}
\tag{6.11.6}
$$

The numerical error may be estimated by several methods. One method involves differentiating the asymptotic expression for the interpolation error given in equation (6.3.3) with respect to x and retaining the terms with the lowest exponent of h; this method was already practiced previously in this section.

An a *posteriori* error estimate can be derived by expressing f_{-3}, f_{-2}, f_{-1}, on the right-hand side of equation (6.11.6) in Taylor series about x_0. After some straightforward simplifications, we obtain

$$
P_4''(x_0) = f''(x_0) - \tfrac{11}{12} f''''(x_0)\, h^2 + \cdots
\tag{6.11.7}
$$

which shows that the error is on the order of h^2: Reducing h by a factor of 2 causes the error to be reduced by a factor of 4.

PROBLEMS

6.11.1 *Consistency of interpolation.*

Show that formula (6.11.6) is consistent; that is, show that as h tends to zero, the right-hand side reduces to $f''(x_0)$ and the leading-order contribution to the error behaves as shown in equation (6.11.7).

6.11.2 *Backward-difference methods for the second derivative.*

(*a*) Given the values of a function $f(x)$ at the four evenly spaced points $x_0 - 4h, x_0 - 3h, x_0 - 2h, x_0 - h$, respectively denoted as f_{-4}, f_{-3}, f_{-2}, f_{-1}, show that it is consistent to compute $f''(x_0)$ using the finite-difference formula

$$
f''(x_0) \cong P_4''(x_0) = \frac{-2f_{-4} + 7f_{-3} - 8f_{-2} + 3f_{-1}}{h^2}
\tag{6.11.8}
$$

and the associated numerical error is on the order of h^2.

(*b*) Apply formulas (6.11.6) and (6.11.8) to compute $f''(3)$ for $f(x) = \ln x$, where h is a small interval of your choice, and compare the numerical results with the exact value. By way of this example, demonstrate that the error associated with these formulas is indeed on the order of h^2.

Method of Undetermined Coefficients

Another way to derive finite-difference formulas proceeds by expressing the derivatives of $f(x)$ as linear combinations of the values of $f(x)$ at a number of neighboring data points, and then computing the coefficients of the linear expansions by requiring (*a*) consistency and (*b*) a desired level of accuracy. Consistency demands that, in the limit as the spacing between the data points tends to zero, the finite-difference formulas produces the exact values of the corresponding derivatives.

To exemplify the method, let us approximate the first and second derivatives of the function $f(x)$ at the point x_0 with the centered differences:

$$f'(x_0) = af_{-1} + bf_0 + cf_1, \qquad f''(x_0) = Af_{-1} + Bf_0 + Cf_1 \tag{6.11.9}$$

where f_{-1}, f_0, f_1 designate, respectively, the values of $f(x)$ at the points $x_0 - h, x_0, x_0 + h$, and where a, b, c, A, B, C are six undetermined coefficients. Expressing f_{-1}, f_0, f_1, on the right-hand sides in terms of $f_0, f_0', f_0'', \ldots$ using Taylor series expansions, we obtain

$$f'(x_0) = (a + b + c)f_0 + h\,(c - a)f_0' + \tfrac{1}{2}h^2(c + a)f_0'' + \ldots \tag{6.11.10}$$

and

$$f''(x_0) = (A + B + C)f_0 + h\,(C - A)f_0' + \tfrac{1}{2}h^2(C + A)f_0'' + \ldots \tag{6.11.11}$$

Consistency requires two relations involving a, b, c,

$$a + b + c = 0, \qquad h(c - a) = 1 \tag{6.11.12}$$

and three relations involving A, B, C,

$$A + B + C = 0, \qquad h(C - A) = 0, \qquad \tfrac{1}{2}h^2(C + A) = 1 \tag{6.11.13}$$

We can afford to impose one more constraint involving a, b, c, and it is natural to demand that the coefficient of the second-order term on the right-hand side of equation (6.11.10) vanish, that is, $c + a = 0$, so that the error is elevated to a higher order. The solutions of the resulting linear systems are readily found to be

$$a = -\frac{1}{2h}, \qquad b = 0, \qquad c = \frac{1}{2h} \tag{6.11.14}$$

and

$$A = \frac{1}{h^2}, \qquad B = -\frac{2}{h^2}, \qquad C = \frac{1}{h^2} \tag{6.11.15}$$

Substituting these expressions into equations (6.11.9) provides us with consistent centered-difference approximations.

PROBLEMS

6.11.3 ***Consistency and accuracy.***

(a) Compute the coefficients a, b, c so that the errors of the following finite-difference formulas are of second order in Δx:

$$f'(x) = af(x - 2\Delta x) + bf(x - \Delta x) + cf(x), \qquad f'(x) = af(x) + bf(x + \Delta x) + cf(x + 2\Delta x)$$

(b) Compute the coefficients a, b, c, d, e so that the error of the following finite-difference approximation is of fourth order in Δx:

$$f'(x) = af(x - 2\Delta x) + bf(x - \Delta x) + cf(x) + df(x + \Delta x) + ef(x + 2\Delta x)$$

6.11.4 ***Heat capacity and concentration of a binary solution.***

The specific heat at constant pressure c_p of an aqueous solution of methyl alcohol is a function of the alcohol mole percentage ϕ. *Perry's Chemical Engineer's Handbook* (5th ed., p. 3–136, Table 3–196) gives the following data for 40 °C and atmospheric pressure:

ϕ (%)	c_p(cal/g. °C)
5.88	0.995
12.3	0.98
27.3	0.92
45.8	0.83
69.6	0.726
100	0.617

The magnitude of the partial derivative $\partial c_p/\partial\phi$ is a measure of the difference in the inensity of the intermolecular forces between the water and the methyl alcohol molecules: For an ideal solution, $\partial c_p/\partial\phi$ is constant no matter what the value of ϕ.
Compute and plot the function $\partial c_p/\partial\phi$ at $\phi = 0, 10, 20, 30, \ldots, 90, 100\%$ using a method of your choice. Based on your results would you say that the methyl alcohol solution is an ideal solution?

Tables of Finite-difference Formulas

Table 6.11.1 displays finite-difference approximations to the derivatives of the function $f(x)$ at the data point x_i, in terms of the values of the function at a collection of evenly spaced data points separated by the distance $\Delta x = h$. These formulas may be used to derive partially-biased, forward or backward difference approximations.

Consider, for example, computing $f'''(x_0)$ from knowledge of the values of $f(x)$ at the four evenly spaced points $x_0 - 2h, x_0 - h, x_0, x_0 + h$, respectively denoted as f_{-2}, f_{-1}, f_0, f_1. Using Table 6.11.1, we write

$$\begin{aligned} f'''(x_0) &\cong \frac{1}{2h}[f''(x_0 + h) - f''(x_0 - h)] \\ &\cong \frac{1}{2h}\left(\frac{-f_{-2} + 4f_{-1} - 5f_0 + 2f_1}{h^2} - \frac{f_{-2} - 2f_{-1} + f_0}{h^2}\right) \\ &= \frac{-2f_{-2} + 6f_{-1} - 6f_0 + 2f_1}{2h^3} \end{aligned} \tag{6.11.16}$$

which is a partially-biased backward difference approximation.

Table 6.11.1 Finite-difference formulas for computing the derivatives of a function $f(x)$ at the point x_i, in terms of the values of the function at a set of evenly spaced points separated by distance $\Delta x = h$.

Backward differences with accuracy $O(h)$

$$\begin{bmatrix} hf_i' \\ h^2 f_i'' \\ h^3 f_i''' \\ h^4 f_i'''' \end{bmatrix} = \begin{bmatrix} 0 & 0 & 0 & -1 & 1 \\ 0 & 0 & 1 & -2 & 1 \\ 0 & -1 & 3 & -3 & 1 \\ 1 & -4 & 6 & -4 & 1 \end{bmatrix} \begin{bmatrix} f_{i-4} \\ f_{i-3} \\ f_{i-2} \\ f_{i-1} \\ f_i \end{bmatrix}$$

Backward differences with accuracy $O(h^2)$

$$\begin{bmatrix} 2hf_i' \\ h^2 f_i'' \\ 2h^3 f_i''' \\ h^4 f_i'''' \end{bmatrix} = \begin{bmatrix} 0 & 0 & 0 & 1 & -4 & 3 \\ 0 & 0 & -1 & 4 & -5 & 2 \\ 0 & 3 & -14 & 24 & -18 & 5 \\ -2 & 11 & -24 & 26 & -14 & 3 \end{bmatrix} \begin{bmatrix} f_{i-5} \\ f_{i-4} \\ f_{i-3} \\ f_{i-2} \\ f_{i-1} \\ f_i \end{bmatrix}$$

Centered differences with accuracy $O(h^2)$

$$\begin{bmatrix} 2hf_i' \\ h^2 f_i'' \\ 2h^3 f_i''' \\ h^4 f_i'''' \end{bmatrix} = \begin{bmatrix} 0 & -1 & 0 & 1 & 0 \\ 0 & 1 & -2 & 1 & 0 \\ -1 & 2 & 0 & -2 & 1 \\ 1 & -4 & 6 & -4 & 1 \end{bmatrix} \begin{bmatrix} f_{i-2} \\ f_{i-1} \\ f_i \\ f_{i+1} \\ f_{i+2} \end{bmatrix}$$

Centered differences with accuracy $O(h^4)$

$$\begin{bmatrix} 12hf_i' \\ 12h^2 f_i'' \\ 8h^3 f_i''' \\ 6h^4 f_i'''' \end{bmatrix} = \begin{bmatrix} 0 & 1 & -8 & 0 & 8 & -1 & 0 \\ 0 & -1 & 16 & -30 & 16 & -1 & 0 \\ 1 & -8 & 13 & 0 & -13 & 8 & -1 \\ -1 & 12 & -39 & 56 & -39 & 12 & -1 \end{bmatrix} \begin{bmatrix} f_{i-3} \\ f_{i-2} \\ f_{i-1} \\ f_i \\ f_{i+1} \\ f_{i+2} \\ f_{i+3} \end{bmatrix}$$

Forward differences with accuracy $O(h)$

$$\begin{bmatrix} hf_i' \\ h^2 f_i'' \\ h^3 f_i''' \\ h^4 f_i'''' \end{bmatrix} = \begin{bmatrix} -1 & 1 & 0 & 0 & 0 \\ 1 & -2 & 1 & 0 & 0 \\ -1 & 3 & -3 & 1 & 0 \\ 1 & -4 & 6 & -4 & 1 \end{bmatrix} \begin{bmatrix} f_i \\ f_{i+1} \\ f_{i+2} \\ f_{i+3} \\ f_{i+4} \end{bmatrix}$$

Forward differences with accuracy $O(h^2)$

$$\begin{bmatrix} 2hf_i' \\ h^2 f_i'' \\ 2h^3 f_i''' \\ h^4 f_i'''' \end{bmatrix} = \begin{bmatrix} -3 & 4 & -1 & 0 & 0 & 0 \\ 2 & -5 & 4 & -1 & 0 & 0 \\ -5 & 18 & -24 & 14 & -3 & 0 \\ 3 & -14 & 26 & -24 & 11 & 11 \end{bmatrix} \begin{bmatrix} f_i \\ f_{i+1} \\ f_{i+2} \\ f_{i+3} \\ f_{i+4} \\ f_{i+5} \end{bmatrix}$$

A different kind of tabulation arises by expressing the kth derivative of a function $f(x)$ at the data point x_i in terms of its values at a number of evenly spaced data points separated by the distance $\Delta x = h$, in the form

$$\left(\frac{d^k f}{dx^k}\right)_{x=x_i} = \frac{1}{h^k} \sum_{l=-l_i}^{l_2} c_l f_{i+1} + O(h^p) \tag{6.11.17}$$

where p is the order of the approximation. The values of the constants c_l are given in Tables 6.11.2 and 6.11.3 after Fornberg (1996), for centered and forward differences corresponding, respectively, to $l_1 = l_2$ and $l_1 = 0$; backward differences emerge by a straightforward change in notation.

PROBLEM

6.11.5 *Consistency of interpolation.*

Show that formula (6.11.16) is consistent and derive the associated error.

Fornberg's Algorithm for Interpolation and Differentiation

Fornberg (1996, pp. 15–23) developed a compact algorithm for polynomial interpolation or extrapolation, and for computing the first m derivatives of a function $f(x)$ at an *arbitrary* point x, in terms of the values of the function at the $N+1$ points x_i, $i = 1, 2, \ldots, N+1$. The finite-difference approximations are expressed in the form

Table 6.11.2 The coefficients c_l associated with the finite-difference formula (6.11.17) for centered differences.

$p \backslash l$	-4	-3	-2	-1	0	1	2	3	4
Function									
∞					1				
First derivative									
2				$-\frac{1}{2}$	0	$\frac{1}{2}$			
4			$\frac{1}{12}$	$-\frac{2}{3}$	0	$\frac{2}{3}$	$-\frac{1}{12}$		
6		$-\frac{1}{60}$	$\frac{3}{20}$	$-\frac{3}{4}$	0	$\frac{3}{4}$	$-\frac{3}{20}$	$\frac{1}{60}$	
8	$\frac{1}{280}$	$-\frac{4}{105}$	$\frac{1}{5}$	$-\frac{4}{5}$	0	$\frac{4}{5}$	$-\frac{1}{5}$	$\frac{4}{105}$	$-\frac{1}{280}$
Second derivative									
2				1	-2	1			
4			$-\frac{1}{12}$	$\frac{4}{3}$	$-\frac{5}{2}$	$\frac{4}{3}$	$-\frac{1}{12}$		
6		$\frac{1}{90}$	$-\frac{3}{20}$	$\frac{3}{2}$	$-\frac{49}{18}$	$\frac{3}{2}$	$-\frac{3}{20}$	$\frac{1}{90}$	
8	$-\frac{1}{560}$	$\frac{8}{315}$	$-\frac{1}{5}$	$\frac{8}{5}$	$-\frac{205}{72}$	$\frac{8}{5}$	$-\frac{1}{5}$	$\frac{8}{315}$	$-\frac{1}{560}$
Third derivative									
2			$-\frac{1}{2}$	1	0	-1	$\frac{1}{2}$		
4		$\frac{1}{8}$	-1	$\frac{13}{8}$	0	$-\frac{13}{8}$	1	$-\frac{1}{8}$	
6	$-\frac{7}{240}$	$\frac{3}{10}$	$-\frac{169}{120}$	$\frac{61}{30}$	0	$-\frac{61}{30}$	$\frac{169}{120}$	$-\frac{3}{10}$	$\frac{7}{240}$
Fourth derivative									
2			1	-4	6	-4	1		
4		$-\frac{1}{6}$	2	$-\frac{13}{2}$	$\frac{28}{3}$	$-\frac{13}{2}$	2	$-\frac{1}{6}$	
6	$\frac{7}{240}$	$-\frac{2}{5}$	$\frac{169}{60}$	$-\frac{122}{15}$	$\frac{91}{8}$	$-\frac{122}{15}$	$\frac{169}{60}$	$-\frac{2}{5}$	$\frac{7}{240}$

$$\frac{d^k f}{dx^k} = \sum_{i=1}^{N+1} c_{k,N+1,i} f_i \tag{6.11.18}$$

Table 6.11.3 The coefficients c_l associated with the finite-difference formula (6.11.17) for forward differences. The corresponding coefficients for backward differences arise by a straightforward change in notation.

$p \backslash l$	0	1	2	3	4	5	6	7	8
Function									
∞	1								
First derivative									
1	-1	1							
2	$-\frac{3}{2}$	2	$-\frac{1}{2}$						
3	$-\frac{11}{6}$	3	$-\frac{3}{2}$	$\frac{1}{3}$					
4	$-\frac{25}{12}$	4	-3	$\frac{4}{3}$	$-\frac{1}{4}$				
5	$-\frac{137}{60}$	5	-5	$\frac{10}{3}$	$-\frac{5}{4}$	$\frac{1}{5}$			
6	$-\frac{49}{20}$	6	$-\frac{15}{2}$	$\frac{20}{3}$	$-\frac{15}{4}$	$\frac{6}{5}$	$-\frac{1}{6}$		
7	$-\frac{363}{140}$	7	$-\frac{21}{2}$	$\frac{35}{3}$	$-\frac{35}{4}$	$\frac{21}{5}$	$-\frac{7}{6}$	$\frac{1}{7}$	
8	$-\frac{761}{280}$	8	-14	$\frac{56}{3}$	$-\frac{35}{2}$	$\frac{56}{5}$	$-\frac{14}{3}$	$\frac{8}{7}$	$-\frac{1}{8}$
Second derivative									
1	1	-2	1						
2	2	-5	4	-1					
3	$\frac{35}{12}$	$-\frac{26}{3}$	$\frac{19}{2}$	$-\frac{14}{3}$	$\frac{11}{12}$				
4	$\frac{15}{4}$	$-\frac{77}{6}$	$\frac{107}{6}$	-13	$\frac{61}{12}$	$-\frac{5}{6}$			
5	$\frac{203}{45}$	$-\frac{87}{5}$	$\frac{117}{4}$	$-\frac{254}{9}$	$\frac{33}{2}$	$-\frac{27}{5}$	$\frac{137}{180}$		
6	$\frac{469}{90}$	$-\frac{223}{10}$	$\frac{879}{20}$	$-\frac{949}{18}$	41	$-\frac{201}{10}$	$\frac{1019}{180}$	$-\frac{7}{10}$	
7	$\frac{29531}{5040}$	$-\frac{962}{35}$	$\frac{621}{10}$	$-\frac{4006}{45}$	$\frac{691}{8}$	$-\frac{282}{5}$	$\frac{2143}{90}$	$-\frac{206}{35}$	$\frac{363}{560}$
Third derivative									
1	-1	3	-3	1					
2	$-\frac{5}{2}$	9	-12	7	$-\frac{3}{2}$				
3	$-\frac{17}{4}$	$\frac{71}{4}$	$-\frac{59}{2}$	$\frac{49}{2}$	$-\frac{41}{4}$	$\frac{7}{4}$			
4	$-\frac{49}{8}$	29	$-\frac{461}{8}$	62	$-\frac{307}{8}$	13	$-\frac{15}{8}$		
5	$-\frac{967}{120}$	$\frac{638}{15}$	$-\frac{3929}{40}$	$\frac{389}{3}$	$-\frac{2545}{24}$	$\frac{268}{5}$	$-\frac{1849}{120}$	$\frac{29}{15}$	
6	$-\frac{801}{80}$	$\frac{349}{6}$	$-\frac{18353}{120}$	$\frac{2391}{10}$	$-\frac{1457}{6}$	$\frac{4891}{30}$	$-\frac{561}{8}$	$\frac{527}{30}$	$-\frac{469}{240}$
Fourth derivative									
1	1	-4	6	-4	1				
2	3	-14	26	-24	11	-2			
3	$\frac{35}{6}$	-31	$\frac{137}{2}$	$-\frac{242}{3}$	$\frac{107}{2}$	-19	$\frac{17}{6}$		
4	$\frac{28}{3}$	$-\frac{111}{2}$	142	$-\frac{1219}{6}$	176	$-\frac{185}{2}$	$\frac{82}{3}$	$-\frac{7}{2}$	
5	$\frac{1069}{80}$	$-\frac{1316}{15}$	$\frac{15289}{60}$	$-\frac{2144}{5}$	$\frac{10993}{24}$	$-\frac{4772}{15}$	$\frac{2803}{20}$	$-\frac{536}{15}$	$\frac{967}{240}$

where the coefficients $c_{k,N+1,i}$ are functions of the abscissa of the interpolated point x. Having specified the value of the highest-order derivative desired m, the coefficients $c_{k,j,i}$ are computed for $k = 0, 1, \dots, m$, and $j = k+1, \dots, N+1$ according to the algorithm

Initialize all variables to zero

$c_{0,1,1} = 1$

$\alpha = 1$

Do $j = 2, N+1$

  $\beta = 1$

  Do $i = 1, j-1$

    $\beta \leftarrow \beta(x_j - x_i)$

    Do $k = 0, \text{Min}(j-1, m)$

$$c_{k,j,i} = \frac{(x_j - x)\, c_{k,j-1,i} - k c_{k-1,j-1,i}}{x_j - x_i} \tag{6.11.19}$$

    END DO

  END DO

  Do $k = 0, \text{Min}(j-1, m)$

$$c_{k,j,i} = \frac{\alpha}{\beta}\left(k c_{k-1,j-1,i-1} - (x_{j-1} - x)\, c_{k,j-1,i-1}\right)$$

  END DO

  $\alpha = \beta$

END DO

Fornberg finds that for $m = 0$, corresponding to interpolation, this algorithm is significantly faster than the Neville or the Aitken algorithm discussed in Section 6.2. More generally, Fornberg shows that the formula

$$\frac{d^k f}{dx^k} = \sum_{i=1}^{j} c_{k,j,i} f_i \tag{6.11.20}$$

where $k = 0, 1, \dots, m$ and $j = k+1, \dots, N+1$, produces the exact answer for polynomials of as high degree as possible for the number of data points employed.

For example, when $k = 0$ and $j = 1$ or 2, we obtain, respectively,

$$f(x) = c_{0,1,1} f_1 \qquad \text{or} \qquad f(x) = c_{0,2,1} f_1 + c_{0,2,2} f_2 \tag{6.11.21}$$

corresponding to a constant or linear interpolating polynomial whose graph passes through the first or first and second data points. When $k = 1$ and $j = 2$, we find

$$\frac{df}{dx}(x) = c_{1,2,1} f_1 + c_{1,2,2} f_2 \tag{6.11.22}$$

representing differentiation by linear interpolation.

6.12 Numerical Differentiation of a Function of Two Variables

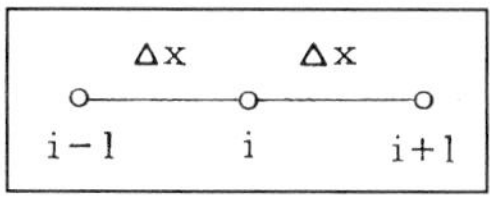

Partial derivatives of a function of two variables $f(x, y)$, with respect to one of the independent variables x or y may be computed using either the formulas of Table 6.11.1 or the equations accompanying Tables 6.11.2 and 6.11.3, subject to a straightforward change in notation.

For example, the centered-difference approximation with evenly spaced grid lines along the x axis, separated by the distance Δx, yields

$$\frac{\partial^2 f_{i,j}}{\partial x^2} = \frac{f_{i+1,j} - 2f_{i,j} + f_{i-1,j}}{\Delta x^2} \tag{6.12.1}$$

The associated error is on the order of Δx^2.

Laplacian

Finite-difference approximations to the Laplacian

$$\nabla^2 f = \frac{\partial^2 f}{\partial x^2} + \frac{\partial^2 f}{\partial y^2} \tag{6.12.2}$$

in terms of values of $f(x, y)$ at a number of neighboring grid points are required in the numerical solution of the Laplace, Poisson, Helmholtz, and similar equations using finite-difference methods, to be discussed in Chapter 11.

Five-point formula for a Cartesian grid

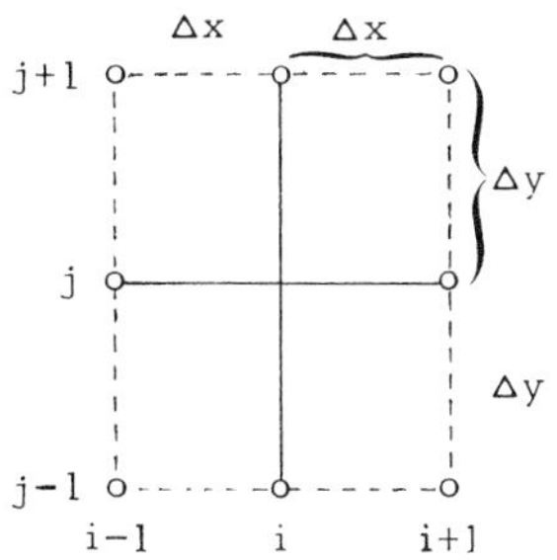

The standard five-point formula arises by applying the second-order centered-difference finite-difference operator to compute the two partial derivatives on the right-hand side of equation (6.12.2). The result is an expression for the Laplacian at the point (x_i, y_i) in terms of the values of $f(x, y)$ at the five closest neighboring grid points, as indicated at the *computational molecule* shown in the diagram, also called a *finite-difference stencil.* When the grid spacings are uniform, we find

$$(\nabla^2 f)_{i,j} = \frac{1}{\Delta x^2} \begin{bmatrix} 0 & \beta & 0 \\ 1 & -2(1+\beta) & 1 \\ 0 & \beta & 0 \end{bmatrix} : \begin{bmatrix} f_{i-1,j+1} & f_{i,j+1} & f_{i+1,j+1} \\ f_{i-1,j} & f_{i,j} & f_{i+1,j} \\ f_{i-1,j-1} & f_{i,j-1} & f_{i+1,j-1} \end{bmatrix} \tag{6.12.3}$$

where

$$\beta = (\Delta x/\Delta y)^2 \tag{6.12.4}$$

Note that the sum of the elements of the first matrix on the right-hand side of equation (6.12.3) is equal to zero; this is a consequence of the mean-value theorem of harmonic functions.

The leading contribution to the error is

$$\tfrac{1}{12}\left(\frac{\partial^4 f}{\partial x^4} + \frac{1}{\beta^2}\frac{\partial^4 f}{\partial y^4}\right)_{i,j} \Delta x^2 \tag{6.12.5}$$

The five-point formula produces the exact value of the Laplacian of third or lower degree polynomials in x or y.

Diagonal form of the five-point formula for a Cartesian grid

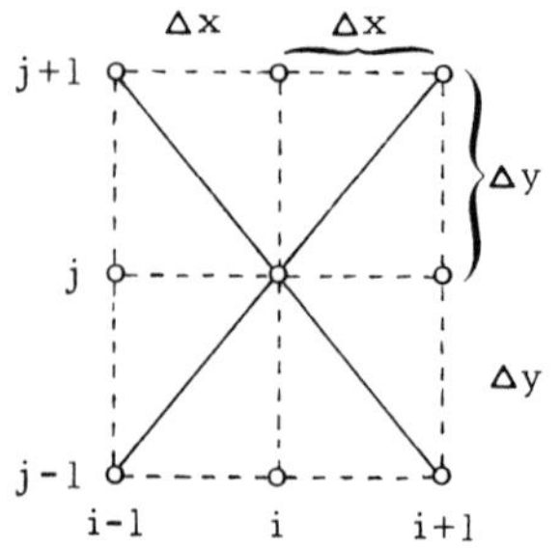

The diagonal form of the five-point formula involves the values of $f(x, y)$ along the diagonal lines passing through the point (x_i, y_i), where the Laplacian is to be evaluated, as shown in the diagram. Assuming that the grid spacings are uniform and $\Delta x = \Delta y$ or $\beta = 1$, we find

$$(\nabla^2 f)_{i,j} = \frac{1}{2\Delta x^2}\begin{bmatrix} 1 & 0 & 1 \\ 0 & -4 & 0 \\ 1 & 0 & 1 \end{bmatrix} : \begin{bmatrix} f_{i-1,j+1} & f_{i,j+1} & f_{i+1,j+1} \\ f_{i-1,j} & f_{i,j} & f_{i+1,j} \\ f_{i-1,j-1} & f_{i,j-1} & f_{i+1,j-1} \end{bmatrix} \tag{6.12.6}$$

(Problem 6.12.1). The leading contribution to the error is

$$\tfrac{1}{2}\left[\frac{\partial^4 f}{\partial x^2\,\partial y^2} + \tfrac{1}{6}\left(\frac{\partial^4 f}{\partial x^4} + \frac{\partial^4 f}{\partial y^4}\right)\right]_{i,j} \Delta x^2 \tag{6.12.7}$$

Nine-point formula

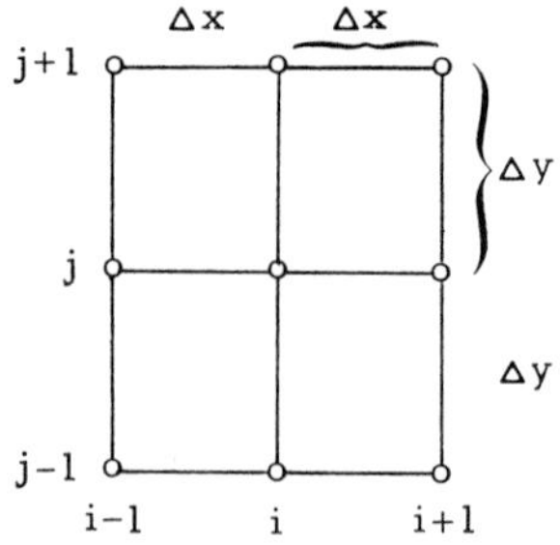

The *nine-point formula* produces the value of the Laplacian at the point (x_i, y_i) in terms of the values of $f(x, y)$ at the nine closest neighboring points of a Cartesian grid. When $\Delta x = \Delta y$ or $\beta = 1$, we find

$$(\nabla^2 f)_{i,j} = \frac{1}{6\Delta x^2}\begin{bmatrix} 1 & 4 & 1 \\ 4 & -20 & 4 \\ 1 & 4 & 1 \end{bmatrix} : \begin{bmatrix} f_{i-1,j+1} & f_{i,j+1} & f_{i+1,j+1} \\ f_{i-1,j} & f_{i,j} & f_{i+1,j} \\ f_{i-1,j-1} & f_{i,j-1} & f_{i+1,j-1} \end{bmatrix} \tag{6.12.8}$$

The right-hand side of formula (6.12.8) is a weighted average of the right-hand sides of formula (6.12.3) with

$\beta = 1$, and formula (6.12.6), with weights equal to $\frac{2}{3}$ and $\frac{1}{3}$. Weighting the leading-order contributions to the error in a similar manner, we find

$$\frac{1}{12}\left(\left\langle \frac{\partial^2}{\partial x^2} + \frac{\partial^2}{\partial y^2} \right\rangle \left\langle \frac{\partial^2}{\partial x^2} + \frac{\partial^2}{\partial y^2} \right\rangle f \right)_{i,j} \Delta x^2 \qquad (6.12.9)$$

where the angular brackets designate that the enclosed quantity is an operator.

The nine-point formula appears to carry no particular advantage over the five-point formulas regarding the order of the error. We note, however, that the leading contribution to the error vanishes when $f(x, y)$ is a *harmonic* or a *biharmonic* function; that is, when $f(x, y)$ satisfies the Laplace equation $\nabla^2 f = 0$ or the biharmonic equation $\nabla^4 f = 0$. In these cases, the nine-point formula introduces an error on the order of Δx^6 (Problem 6.12.2).

Non-Cartesian grids

In certain applications, it is advantageous to use a non-Cartesian grid involving polygonal cells. An isotropic uniform triangular grid and an isotropic uniform hexagonal grid are shown in Tables 6.12.1 and 6.12.2 along with a family of finite-difference approximations to the Laplacian evaluated at the point numbered 0, after Beyer (1987).

PROBLEMS

6.12.1 *Method of undetermined coefficients.*

Derive the diagonal form of the five-point formula using the method of undetermined coefficients.

6.12.2 *Nine-point formula.*

Show that the function $f(x, y) = \exp(ax)\cos(ay)$, where a is an arbitrary constant, is harmonic. Evaluate its Laplacian at the origin using the nine-point formula with $\Delta x = \Delta y$, and verify by numerical experimentation that the error behaves like Δx^6.

Table 6.12.1 Finite-difference approximations to the Laplacian evaluated at the point numbered 0, with a uniform isotropic triangular grid.

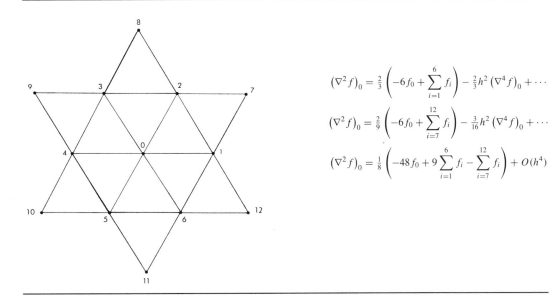

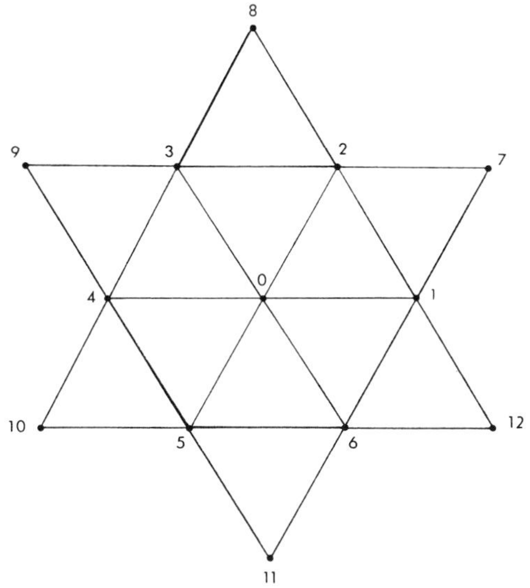

$$(\nabla^2 f)_0 = \frac{2}{3}\left(-6 f_0 + \sum_{i=1}^{6} f_i\right) - \frac{2}{3} h^2 (\nabla^4 f)_0 + \cdots$$

$$(\nabla^2 f)_0 = \frac{2}{9}\left(-6 f_0 + \sum_{i=7}^{12} f_i\right) - \frac{3}{16} h^2 (\nabla^4 f)_0 + \cdots$$

$$(\nabla^2 f)_0 = \frac{1}{8}\left(-48 f_0 + 9\sum_{i=1}^{6} f_i - \sum_{i=7}^{12} f_i\right) + O(h^4)$$

Tables of Finite-difference Formulas

Several finite-difference approximations to partial derivatives and differential operators for evenly spaced grid points, with $\Delta x = \Delta y$, are shown in Table 6.12.3 after Abramowitz and Stegun (1972, pp. 884–885). A more extended compilation is presented by Beyer (1987, pp. 497–508).

Table 6.12.2 Finite-difference approximations to the Laplacian evaluated at the point numbered 0, with a uniform isotropic hexagonal grid.

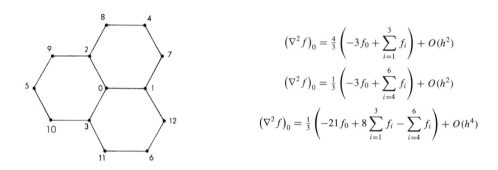

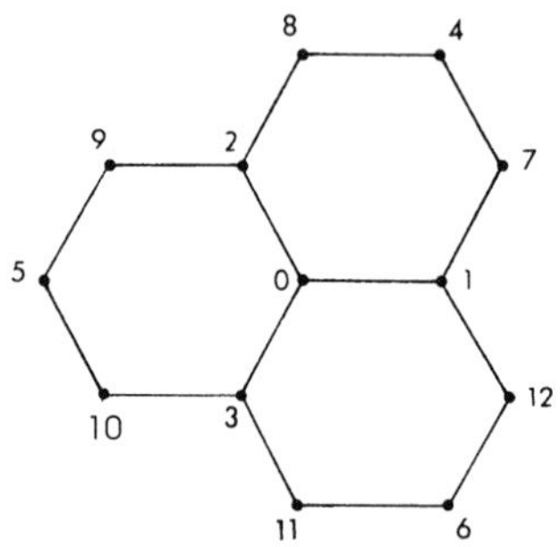

$$(\nabla^2 f)_0 = \tfrac{4}{3}\left(-3 f_0 + \sum_{i=1}^{3} f_i\right) + O(h^2)$$

$$(\nabla^2 f)_0 = \tfrac{1}{3}\left(-3 f_0 + \sum_{i=4}^{6} f_i\right) + O(h^2)$$

$$(\nabla^2 f)_0 = \tfrac{1}{3}\left(-21 f_0 + 8\sum_{i=1}^{3} f_i - \sum_{i=4}^{6} f_i\right) + O(h^4)$$

Table 6.12.3 Finite-difference approximations to partial derivatives and differential operators of a function $f(x, y)$ at the grid point (x_0, y_0), in terms of values of the function at a set of evenly spaced grid points separated by the distance $\Delta x = \Delta y = h$, after Abramowitz and Stegun (1972, pp. 884–885).

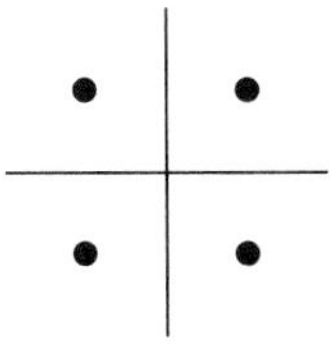

$$\left(\frac{\partial f}{\partial x}\right)_{0,0} = \frac{1}{4h}(f_{1,1} - f_{-1,1} + f_{-1,-1}) + O(h^2)$$

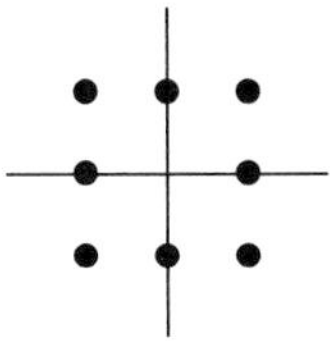

$$\left(\frac{\partial^2 f}{\partial x}\right)_{0,0} = \frac{1}{3h^2}(f_{1,1} - 2f_{0,1} + f_{-1,-1} + f_{1,0} - 2f_{0,0} + f_{-1,0} + f_{1,-1} - 2f_{2,-1} + f_{-1,-1}) + O(h^2)$$

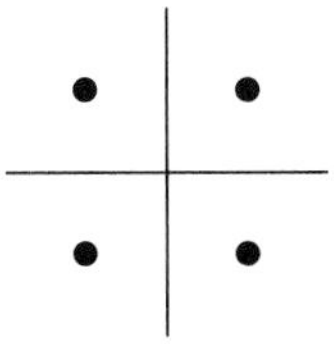

$$\left(\frac{\partial^2 f}{\partial x \partial y}\right)_{0,0} = \frac{1}{4h^2}(f_{1,1} - f_{1,-1} + f_{-1,1} + f_{-1,-1}) + O(h^2)$$

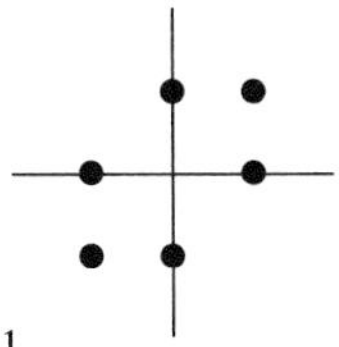

$$\left(\frac{\partial^2 f}{\partial x \partial y}\right)_{0,0} = -\frac{1}{2h^2}(f_{1,0} + f_{-1,0} + f_{0,1} + f_{0,-1} - 2f_{0,0} - 2f_{1,1} - f_{-1,-1}) + O(h^2)$$

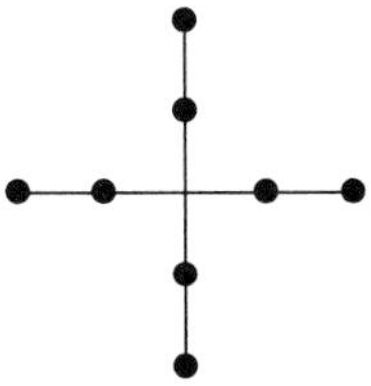

$$\nabla^2 f_{0,0} = -\frac{1}{12h^2}[-60f_{0,0} + 16(f_{1,0} + f_{0,1} + f_{-1,0} + f_{0,-1}) - (f_{2,0} + f_{0,2} + f_{-2,0} + f_{o,-2})] + O(h^2)$$

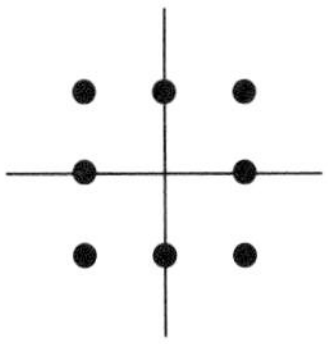

$$\left(\frac{\partial^4 f}{\partial x^2 \partial y^2}\right)_{0,0} = \frac{1}{h^4}(f_{1,1} + f_{-1,1} + f_{1,-1} + f_{-1,-1} - 2f_{1,0} - 2f_{-1,0} - 2f_{0,1} - 2f_{0,-1} + 4f_{0,0}) + O(h^2)$$

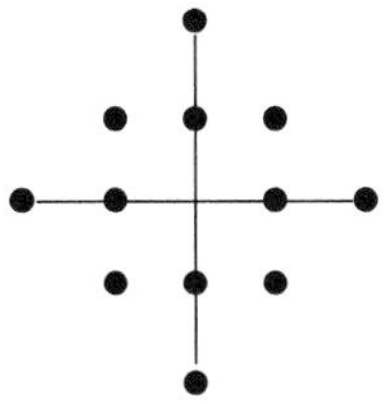

$$\nabla^4 f_{0,0} = \left(\frac{\partial^4 f}{\partial x^4} + 2\frac{\partial^4 f}{\partial x^2 y^2} + \frac{\partial^4 f}{\partial y^4}\right)_{0,0} = \frac{1}{h^4}[20f_{0,0} - 8(f_{1,0} + f_{0,1} + f_{-1,0} + f_{0,-1}) + 2(f_{1,1} + f_{1,-1} + f_{-1,1} + f_{-1,-1}) + (f_{0,2} + f_{2,0} + f_{-2,0} + f_{0,-2})] + O(h^2)$$

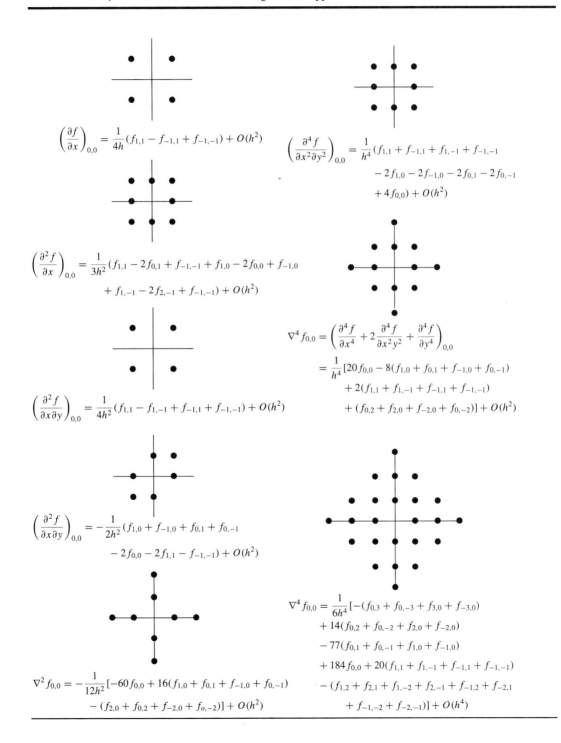

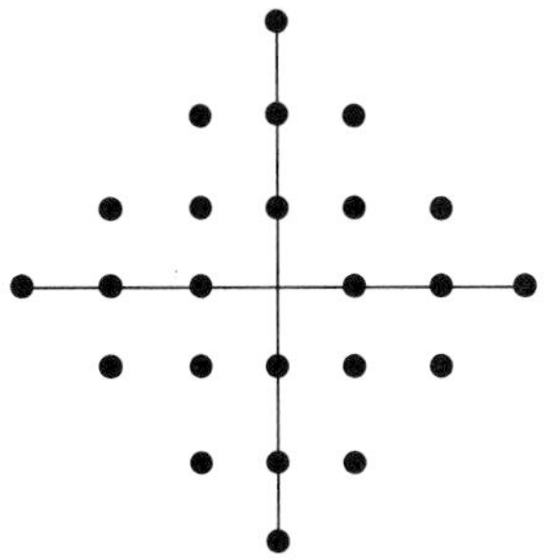

$$\nabla^4 f_{0,0} = \frac{1}{6h^4}[-(f_{0,3} + f_{0,-3} + f_{3,0} + f_{-3,0}) + 14(f_{0,2} + f_{0,-2} + f_{2,0} + f_{-2,0}) - 77(f_{0,1} + f_{0,-1} + f_{1,0} + f_{-1,0}) + 184f_{0,0} + 20(f_{1,1} + f_{1,-1} + f_{-1,1} + f_{-1,-1}) - (f_{1,2} + f_{2,1} + f_{1,-2} + f_{2,-1} + f_{-1,2} + f_{-2,1} + f_{-1,-2} + f_{-2,-1})] + O(h^4)$$

References

ABRAMOWITZ, M., and STEGUN, I. A., 1972, *Handbook of Mathematical Functions.* Dover.

AHLBERG, J. H., NILSON, E. N., and WALSH, J. L., 1967, *The Theory of Splines and Their Applications.* Academic Press.

BEYER, W. H., 1987, *CRC Standard Mathematical Tables.* CRC Press.

CHUNG, T. J., 1978, *Finite Element Analysis in Fluid Dynamics.* McGraw-Hill.

DAVIS, P. J., 1963, *Interpolation and Approximation.* Reprinted by Dover, 1975.

DUBINER, M., 1991, Spectral methods on triangles and other domains. *J. Sci. Comput.* **6**, 345–90.

DE BOOR, C., 1978, *A Practical Guide to Splines.* Springer-Verlag.

FORNBERG, B., 1996, *A Practical Guide to Pseudospectral Methods.* Cambridge University Press.

GREEN, P. J., and SIBSON, R., 1978, Computing Dirichlet tessellations in the plane. *Comput. J.* **21**, 168–73.

HINTON, E., RAO, N. V. R., and ÖZAKÇA, M., 1991, Mesh generation with adaptive finite element analysis. *Adv. Eng. Software* **13**, 238–62.

KINCAID, D., and CHENEY, W., 1996, *Numerical Analysis.* Brooks/Cole.

KWAK, S., and POZRIKIDIS, C., 1998, Triangulation of an evolving open or closed surface by the advancing-front method. *Research Report.*

LANCASTER, P., and ŠALKAUSKAS, K., 1986, *Curve and Surface Fitting.* Academic Press.

LÖHNER, R., 1995, Robust, vectorized search algorithms for interpolation on unstructured grids. *J. Comput. Phys.* **118**, 380–387.

MILGRAM, M. S., 1989, Does a point lie inside a polygon? *J. Comput. Phys.* **84**, 134–44.

NAKAHASHI, K., and SHAROV, D., 1995, Direct surface triangulation using the advancing front method. *AIAA* **95–1686**, 442–51.

OKABE, A., BOOTS, B., and SUGIHARA, K., 1992, *Spatial Tessellations: Concepts and Applications of Voronoi Diagrams.* Wiley.

OLFE, D. B., 1995, *Computer Graphics for Design.* Prentice-Hall.

POWELL, M. J. D., 1981, *Approximation Theory and Methods.* Cambridge University Press.

POZRIKIDIS, C., 1992, *Boundary Integral and Singularity Methods for Linearized Viscous Flow.* Cambridge University Press.

PRESS, W. H., FLANNERY, B. P., TEUKOLSKY, S. A., and VETTERLING, W. T., 1992, *Numerical Recipes.* 2nd ed. Cambridge University Press.

REBAY, S., 1993, Efficient unstructured mesh generation by means of Delaunay triangulation and Bower–Watson algorithm. *J. Comput. Phys.* **106**, 125–38.

RIVLIN, T. J., 1969, *An Introduction to the Approximation of Functions.* Reprinted by Dover, 1981.

RIVLIN, T. J., 1974, *The Chebyshev Polynomials.* Wiley.

SHERWIN, S. J., and KARNIADAKIS, G. E., 1995, A triangular spectral element method; applications to incompressible Navier–Stokes equations. *Comput. Meth. Appl. Mech. Eng.* **123**, 189–229.

SHIKIN, E. V., and PLIS, A. I., 1995, *Handbook of Splines for the User.* CRC Press.

SPÄTH, H., 1995a, *One-Dimensional Spline Interpolation Algorithms.* A. K. Peters.

SPÄTH, H., 1995b, *Two-Dimensional Spline Interpolation Algorithms.* A. K. Peters.

STOER, J., and BULIRSCH, R., 1980, *Introduction to Numerical Analysis.* Springer-Verlag.

SZEGÖ, G., 1959, *Orthogonal Polynomials.* Amer. Math. Soc. Coll. Publ., New York.

Chapter 7

Numerical Integration

7.1 Computation of One-dimensional Integrals

Function interpolation finds extensive applications in the computation of definite integrals of functions of one or more variables, over one- or multidimensional domains. The diametrically opposite problem of function differentiation was addressed in the last two sections of Chapter 6.

In the one-dimensional version of the problem, we want to compute a number expressing the definite integral of a function $f(x)$ between two specified limits a and b, denoted as

$$I = \int_a^b f(x)\ dx \tag{7.1.1}$$

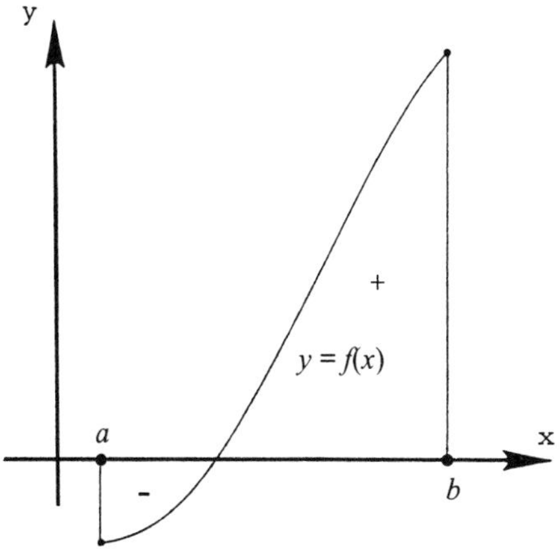

Rudimentary calculus shows that I expresses the signed area between the graph of the function $y = f(x)$, the x axis, and the vertical lines $x = a$ and $x = b$. When $b > a$, and thus $dx > 0$, the area above the x axis is counted as positive, and the area below the x axis is counted as negative. When $b < a$, the signs are switched. The integrated function $f(x)$ is called the *integrand.*

At the outset, we allow for the unpleasant possibility that $f(x)$ may exhibit a singular behavior at one or more points within the integration domain $[a, b]$. Furthermore, we allow a to be shifted to $-\infty$, or b to be shifted to ∞, or both. In each case, we are faced with *improper integrals* of various sorts.

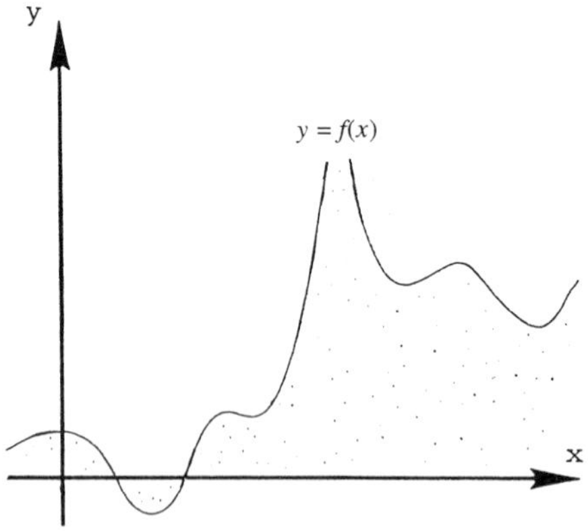

Analytical Integration with the Antiderivative

The *indefinite integral* of the function $f(x)$, also called the *antiderivative*, is another function $F(x)$ satisfying the differential equation

$$\frac{dF}{dx} = f(x) \tag{7.1.2}$$

Thus, by definition, the slope of the graph of the function $F(x)$ at the point x is equal to $f(x)$. If the indefinite integral is known, then the definite integral can be computed simply as

$$I = F(b) - F(a) \tag{7.1.3}$$

The indefinite integral of a function is defined only up to an arbitrary constant, which, however, is subtracted out in computing the right-hand side of equation (7.1.3) and does not make a net contribution.

Unfortunately, the indefinite integrals of a large number of functions are not known in closed form, and even browsing through the comprehensive tables of Gradshteyn and Ryzhik (1980), and the even more comprehensive tables of Prudmikov et al. (1986), or using algebraic manipulation programs, may not produce an answer.

Analytical Integration by the Theory of Residues

A powerful method of computing definite integrals of a large number of complicated functions relies on the theory of residues of a function of a complex variable (e.g., Dettman 1965). Extensive tables of definite integrals computed in this manner are compiled in the references cited in the last paragraph.

Standard Integrals of Mathematical Physics

The frequent occurrence of certain types of integrals in various applications of mathematical physics has motivated their tabulation or approximate representation in terms of polynomial and other functions. The terminology *special functions* is often a euphemism for definite integrals that cannot be found in closed form. The handbook of mathematical functions by Abramowitz and Stegun (1972) contains a wealth of information on such functions, and two examples follow.

Error function

The indefinite integral of the Gaussian function $f(t) = \exp(-t^2)$ is not available in closed form. The definite integral can be expressed in terms of the error function, defined as

$$\mathrm{Erf}(x) \equiv \frac{2}{\sqrt{\pi}} \int_0^x e^{-t^2}\, dt \tag{7.1.4}$$

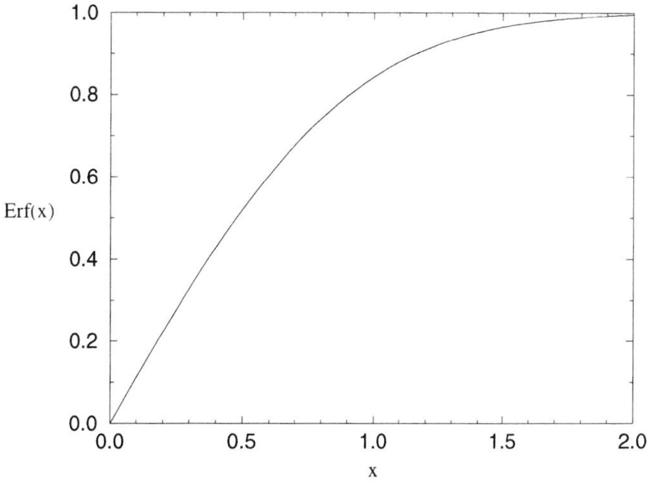

for $x > 0$, and $\mathrm{Erf}(x) = \mathrm{Erf}(-x)$ for $x < 0$. The normalization constant before the integral has been chosen so that

$$\mathrm{Erf}(0) = 0, \qquad \mathrm{Erf}(\infty) = 1 \tag{7.1.5}$$

(Problem 7.1.1). The complementary error function is defined as

$$\mathrm{Erfc}(x) \equiv 1 - \mathrm{Erf}(x) \tag{7.1.6}$$

for $x > 0$, and $\mathrm{Erfc}(x) = \mathrm{Erfc}(-x)$ when $x < 0$. Equations (7.1.5) show that

$$\mathrm{Erfc}(0) = 1, \qquad \mathrm{Erf}(\infty) = 0 \tag{7.1.7}$$

The complementary error function can be computed accurately and efficiently using polynomial approximations (e.g., Abramowitz and Stegun 1972, p. 299; Vedder 1987). For example, the formula

$$\begin{aligned}\mathrm{Erfc}(x) = t(0.254829592 + t\{-0.284496736 + t[1.421413741+ \\ t(-1.453152027 + 1.061405429t)]\})\; e^{-x^2}\end{aligned} \tag{7.1.8}$$

where

$$t = 1/(1 + 0.3275911\;\; x) \tag{7.1.9}$$

produces values that are accurate at least up to the seventh decimal place. The error function follows from equation (7.1.6).

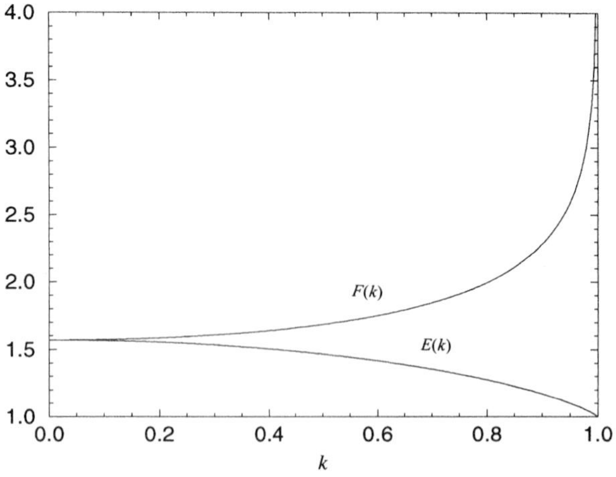

Another important pair of special functions, defined in terms of definite integrals, are the *complete elliptic integrals of the first and second kind*, respectively denoted by F and E, defined as

$$F(k) = \int_0^{\pi/2} \frac{d\varphi}{\sqrt{1-k^2\sin^2\varphi}},$$

$$E(k) = \int_0^{\pi/2} \sqrt{1-k^2\sin^2\varphi}\; d\varphi \tag{7.1.10}$$

for $0 < k < 1$. An efficient method of computing F and E proceeds by generating the sequence

$$K_0 = k$$

$$K_p = \frac{1-(1-K_{p-1}^2)^{1/2}}{1+(1-K_{p-1}^2)^{1/2}} \tag{7.1.11}$$

for $p = 1, 2, \ldots,$ and then setting

$$F = \frac{\pi}{2}(1+K_1)(1+K_2)(1+K_3)\ldots$$

$$E = F\left(1-\frac{k^2}{2}P\right) \tag{7.1.12}$$

where

$$P = 1 + \frac{K_1}{2}\left(1+\frac{K_2}{2}\left(+\frac{K_3}{2}(\ldots)\ldots\right)\right) \tag{7.1.13}$$

Abramowitz and Stegun (1972, Chapter 17) provide alternative polynomial approximations.

PROBLEMS

7.1.1 *Error function.*

Show that $\text{Erf}(\infty) = 1$. *Hint*: Consider the xy plane, introduce plane polar coordinates with radial distance r and polar angle θ, and write

$$\int_{All\ plane} e^{-r^2} dA = 2\pi \int_0^\infty e^{-r^2} r\, dr = \pi$$

$$= \int_{-\infty}^{\infty}\int_{-\infty}^{\infty} e^{-x^2-y^2}\, dx\, dy = \left(\int_{-\infty}^{\infty} e^{-t^2}\, dt\right)^2 \tag{7.1.14}$$

7.1.2 *Flow due to an impulsively started plate.*

Consider a water tank that is closed at the top with a flat lid. Suddenly, the lid starts moving with constant velocity V, generating a fluid flow. The velocity of the fluid u at a point in the tank is a function of distance from the lid y and time t. Solving the equations of fluid flow (e.g., Pozrikidis 1997), we find

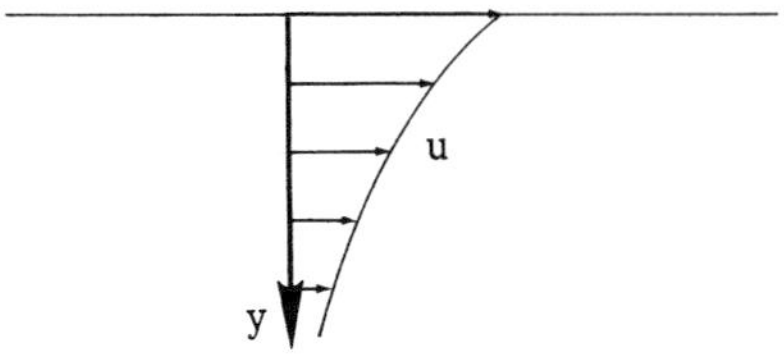

$$u = V\left(1 - \frac{2}{\sqrt{\pi}}\int_o^\eta e^{-z^2}\, dz\right) \tag{7.1.15}$$

where $\eta = y/(4\nu t)^{1/2}$ is a similarity variable, $\nu = \mu/\rho$ is a physical constant called the kinematic viscosity, μ is the viscosity, and ρ is the density of the fluid.

Calculate the velocity u at $t = 30$ s accurate to the third decimal place, for $V = 4$ cm / s, at distances $y = 0.25, 0.50, 0.75, 1.0, 1.25, 1.50, 1.75, 2.0$ cm, when the temperature is 20 °C. The kinematic viscosity of water at 20 °C is 0.010037 cm^2/s.

7.1.3 *Electrostatic potential due to a charged ring.*

The electrostatic potential V at the point $\boldsymbol{x}$ due to a uniformly charged circular ring of radius a, with charge per unit length equal to q, is given by

$$V(\boldsymbol{x}) = -\frac{q}{4\pi}\int_{Ring}\frac{1}{|\boldsymbol{x} - \boldsymbol{x}'|}\, dl(\boldsymbol{x}') \tag{7.1.16}$$

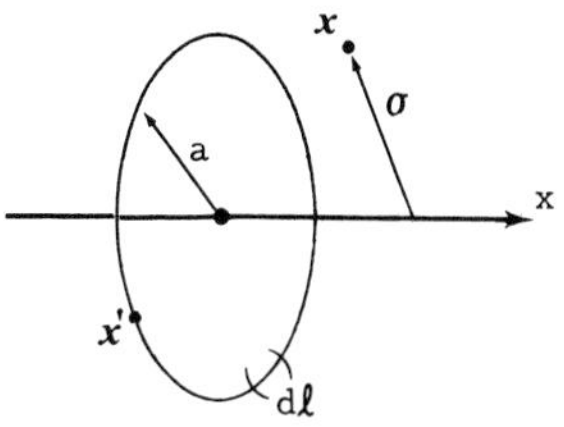

where $dl(\boldsymbol{x}')$ is the differential arc length along the ring around the point $\boldsymbol{x}'$ that lies on the ring. Express $V(\boldsymbol{x})$ in terms of complete elliptic integrals whose arguments are defined in terms of the polar cylindrical coordinates (x, σ) of the point $\boldsymbol{x}$, as shown in the diagram.

When Do We Need Numerical Integration?

Even with ready access to a comprehensive table of integrals, analytical integration may not be feasible. Numerical integration becomes imperative under the following circumstances:

- The integrand $f(x)$ is complicated enough, so that its indefinite or even its definite integral over a specified range cannot be found. Recall that the indefinate integral is a function, whereas the definite integral is a number.

- The indefinite or the definite integral can be found, but they have unwieldy forms that discourage their evaluation.

- The integrand is given in terms of a table. That is, we are provided with discrete values of $f(x)$ at $N+1$ *data* or *base* points x_i, $i = 1, \ldots, N+1$, that lie within the domain of integration.

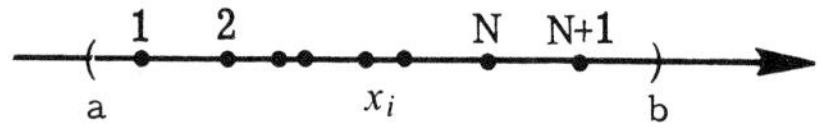

- Numerical integration is an indispensable component of methods for solving differential and integral equations, as will be discussed at the last section of this chapter, and then in Sections 10.4 and 10.5.

General Strategy of Numerical Integration

A typical method of numerical integration involves two steps:

1. Approximate the integrand $f(x)$ with either a global interpolating polynomial defined over the whole of the domain of integration, or a collection of local interpolating polynomials defined over intervals that are subtended by a small group of data points, as discussed in Chapter 6.
2. Integrate the interpolating polynomials over their individual domains of definition.

A modification of this basic procedure is required when the integrand is singular at one or more points over the domain of integration, or when one or both limits of integration are infinite, as will be discussed in the following sections.

The use of a global interpolating polynomial is discouraged by the possible lack of convergence with increasing number of data points. In Section 6.3, we saw that when the data points are distributed in an arbitrary manner, as N is increased, the interpolating polynomial does not necessarily converge to the interpolated function, and neither does its integral. Thus integration in terms of local interpolating polynomials is preferred in developing general-purpose integration methods.

Furthermore, the representation of the integrand in terms of a global or local interpolating polynomial, instead of an approximating polynomial, a rational function, or an interpolating Fourier series to be discussed in Chapter 8, allows us to optimize the efficiency of the numerical method and improve accuracy of the results in a systematic manner, taking advantage of the distinctive properties of orthogonal polynomials. A prerequisite is that the integrand can be evaluated with a high degree of accuracy, or the error in the data points be sufficiently small. When these conditions are not met, the aforementioned alternatives may produce more reliable answers.

PROBLEM

7.1.4 ***Integration via interpolation and approximation.***

Consider the integral

$$I = \int_0^{0.5} e^x \, dx \tag{7.1.17}$$

Replace e^x with (*a*) its quadratic Maclaurin series expansion, and (*b*) a quadratic interpolating polynomial passing through three evenly spaced points, and spanning the whole of the domain of integration. Perform the integration in each case; compare and discuss the results with reference to the exact value.

7.2 Integration by Local Polynomial Interpolation: Newton–Cotes Rules

We begin by assuming that neither the integrand $f(x)$ nor any of its derivatives become infinite at any point over the domain of integration $[a, b]$, and that the limits of integration a and b are finite. Furthermore, we assume that $f(x)$ can be computed, or its values are known, at $N+1$ *data* or *base* points x_i, $i = 1, \ldots, N+1$ that are distributed in some manner over the $[a, b]$, with $x_1 = a$ and $x_{N+1} = b$.

Trapezoidal Rule

Approximating the graph of $f(x)$ with a polygonal line passing through pairs of successive data points leads us to the trapezoidal rule. The area confined between the polygonal line, the x axis, and the two vertical lines passing through two successive data points resembles a *trapezoid*, named after the Greek word τραπέζιον, which means *table.*

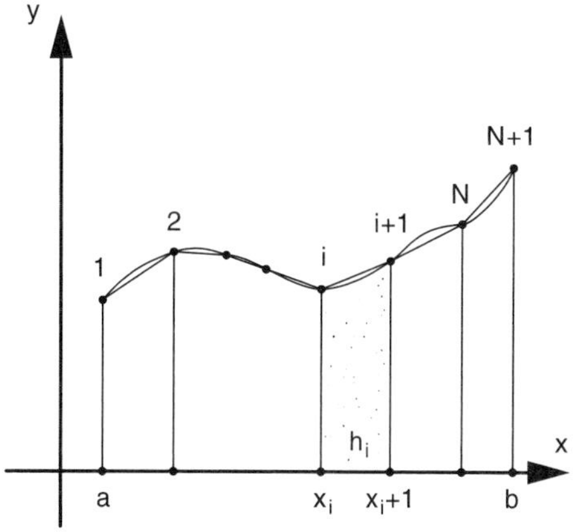

Considering, in particular, the section of the function $f(x)$ between the points x_i and x_{i+1}, we use the expression for the first degree polynomial shown in equations (6.6.1), and the error formula (6.6.4), and compute

$$\int_{x_i}^{x_{i+1}} f(x)\,dx \cong \int_{x_i}^{x_{i+1}} P_1^{(i)}(x)\,dx + \tfrac{1}{2} f''(x_i) \int_{x_i}^{x_{i+1}} (x - x_i)(x - x_{i+1})\,dx \qquad (7.2.1)$$

Evaluating the integrals on the right-hand side, we find

$$\int_{x_i}^{x_{i+1}} f(x)\,dx \cong \tfrac{1}{2} h_i (f_i + f_{i+1}) - \tfrac{1}{12} f''(x_i) h_i^3 \qquad (7.2.2)$$

where $h_i \equiv x_{i+1} - x_i$.

Composite trapezoidal rule

Adding up the contributions from all intervals, we obtain the *composite, closed trapezoidal rule* expressed by

$$I \cong \tfrac{1}{2}\left[h_1 f_1 + (h_1 + h_2) f_2 + \cdots + (h_{N-1} + h_N) f_N + h_N f_{N+1}\right] - \tfrac{1}{12} \sum_{i=1}^{N} f''(x_i) h_i^3 \qquad (7.2.3)$$

The qualifier *closed* signifies that the integration rule involves the values of the integrand at both end-points. *Semiclosed* and *open* composite rules will be discussed in later sections.

Evenly spaced points

When the base points are distributed evenly with uniform spacings $h_i = h = (b - a)/N$, the composite trapezoidal rule reduces to the trapezoidal version of the *closed Newton–Cotes rule* expressed by

$$I \cong h\left(\tfrac{1}{2}f_1 + f_2 + \cdots + f_N + \tfrac{1}{2}f_{N+1}\right) - \tfrac{1}{12}h^3 \sum_{i=1}^{N} f''(x_i) \tag{7.2.4}$$

A general Newton–Cotes rule produces a numerical approximation to an integral in terms of values of the integrand at a string of evenly spaced data points.

Integration error

One way to estimate the magnitude of the error represented by the last term on the right-hand side of equation (7.2.4), is to use the *mean-value theorem* discussed in Appendix A, finding

$$\sum_{i=1}^{N} f''(x_i) = Nf''(\xi) = \frac{b-a}{h} f''(\xi) \tag{7.2.5}$$

where the point ξ lies somewhere between x_1 and x_{N+1}. Thus if I is the exact value of the integral, then

$$I^{TR}(h) = I + E^{TR}(h) \tag{7.2.6}$$

where

$$I^{TR}(h) = h\left(\tfrac{1}{2}f_1 + f_2 + f_3 + \cdots + f_N + \tfrac{1}{2}f_{N+1}\right) \tag{7.2.7}$$

and

$$E^{TR}(h) \cong \tfrac{1}{12}(b-a)\, f''(\xi)\, h^2 \tag{7.2.8}$$

is the leading-order contribution to the error.

As an example, consider the integral

$$I = \int_0^{0.5} \ln\left(1 - 0.99 \cos^2 \frac{x\pi}{2}\right) dx \tag{7.2.9}$$

The trapezoidal rule with $N = 5$, corresponding to $h = 0.10$, yields

$$\begin{aligned} I^{TR}(0.2) = 0.10 \Bigg\{ & \tfrac{1}{2}\ln[1 - 0.99\cos^2 0] + \ln\left[1 - 0.99\cos^2\left(\frac{0.10\pi}{2}\right)\right] \\ & + \ln\left[1 - 0.99\cos^2\left(\frac{0.20\pi}{2}\right)\right] + \ln\left[1 - 0.99\cos^2\left(\frac{0.30\pi}{2}\right)\right] \\ & + \ln\left[1 - 0.99\cos^2\left(\frac{0.40\pi}{2}\right)\right] + \tfrac{1}{2}\ln\left[1 - 0.99\cos^2\left(\frac{0.50\pi}{2}\right)\right]\Bigg\} \\ = -1.0863 & \end{aligned}$$

The exact value is equal to -1.0870. The magnitude of the error, $|-1.0863 + 1.0870| = 0.0007$, is on the order of $\frac{1}{12}(b-a)h^2 = 0.0004$.

To further illustrate the significance of the error formula (7.2.8), in Figure 7.2.1 we plot, with a solid line, the error introduced by applying the trapezoidal rule to integrate the exponential function $f(x) = e^x$, between $a = 0$ and $b = 2$. Plotting the absolute value of the error against h on a logarithmic scale yields a straight line whose slope is equal to 2, which is consistent with the predictions of equation (7.2.8).

A similar but more extended error analysis reveals that the integration error can be expressed as an asymptotic series in h^2, in the form

$$E^{TR}(h) = c_2\, h^2 + c_4\, h^4 + \cdots \tag{7.2.10}$$

Equation (7.2.8), in particular, shows that $c_2 = \frac{1}{12}(b-a)f''(\xi)$. More generally, the coefficient c_i is proportional to the ith derivative of $f(x)$ evaluated at some point within the domain of integration. The absence of odd powers of h, with coefficients that would have been proportional to odd-order derivatives of $f(x)$, is attributed to the smoothing action of integration: A small ripple in the graph of the function $f(x)$ has an equal probability of producing a positive or negative error, with a vanishing net contribution.

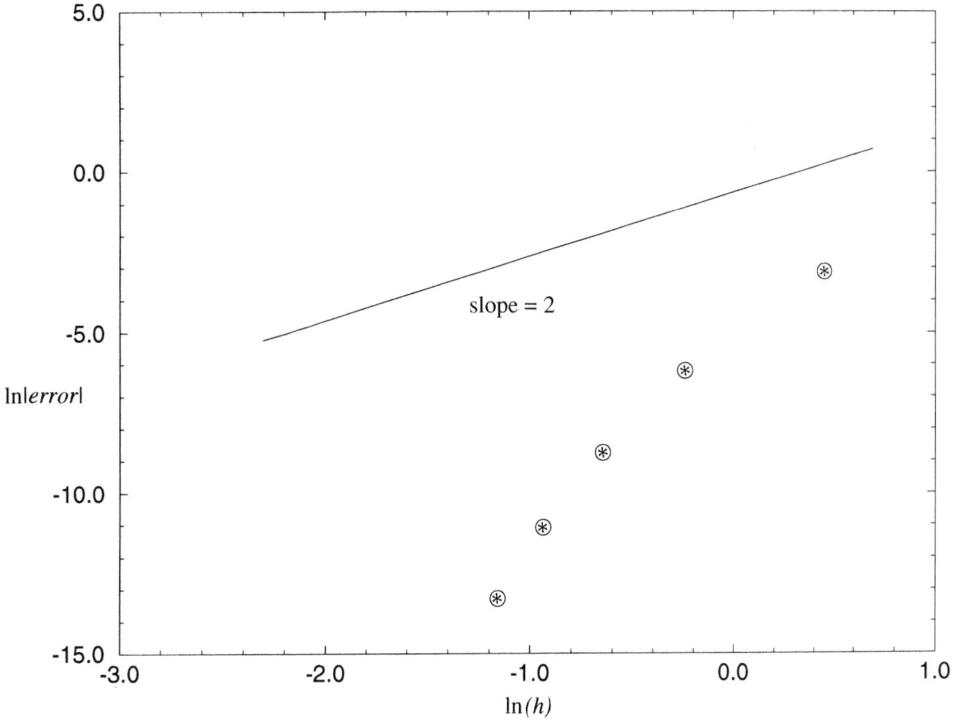

Figure 7.2.1 The error introduced by applying the trapezoidal rule, as a function of step-size h, on a logarithmic scale. The solid line corresponds to the integral of the exponential function $f(x) = e^x$, between $a = 0$ and $b = 2$; the error is quadratic in h. The circles correspond to the complete elliptic integral of the first kind defined in the second of equations (7.1.10), for $k = 0.90$; because the integrand is periodic, the error decreases at a rate that is faster than any power of h.

Romberg integration

Knowledge of the asymptotic behavior of the error displayed in equation (7.2.10) allows us to use the method of Richardson extrapolation, discussed in Section 1.6, in order to improve the accuracy of the numerical computation.

The Romberg integration method uses a series of results with different intervals h, typically forming a geometric sequence, to improve the accuracy to fourth or higher order in h. The method is implemented as discussed in Section 1.6 in the more general context of Richardson extrapolation.

For example, the first stage of Richardson extrapolation produces the improved Romberg value

$$I^{ROM}\left(h = \frac{\varepsilon}{2}\right) = \tfrac{1}{3}\left(4\, I^{TR}\left(h = \frac{\varepsilon}{2}\right) - I^{TR}(h = \varepsilon)\right) \tag{7.2.11}$$

where ε is sufficiently small. Note that the values of $f(x)$ required for the computation of the integral with $h = \varepsilon$ are already available from the more accurate computation of the integral with $h = \frac{1}{2}\varepsilon$. Thus applying formula (7.2.11) does not require any more function evaluations than those required for the computation of the integral with the finer step. The effective integration formula produced by the right-hand of equation (7.2.11) is identical to that of Simpson's 1/3 rule to be discussed in the next subsection, which will be derived on the basis of quadratic local polynomial approximation (Problem 7.2.4).

As a numerical example, we consider the integral (7.2.9). Setting $\varepsilon = 0.05$, or 0.025 corresponding to $N = 10$, or 20, we obtain $I^{TR}(h = 0.05) = -1.0864$ and $I^{TR}(h = 0.025) = -1.0869$. Formula (7.2.11) gives $I^{ROM}(h = 0.025) = \frac{1}{3}(4(-1.0869) - (-1.0864)) = -1.0870$, which is equal to the exact value to the shown accuracy.

Euler–Maclaurin formula

We can derive a more useful expression for the leading-order contribution to the error in terms of derivatives of the integrand evaluated at the beginning and at the end of the domain of integration.

We begin by applying formula (7.2.4) to approximate the definite integral of the function $f''(x)$ between a and b, whose exact value is, of course, equal to $f'(b) - f'(a)$. The result is

$$f'_{N+1} - f'_1 \cong h\left(\tfrac{1}{2} f''_1 + f''_2 + \cdots + f''_N + \tfrac{1}{2} f''_{N+1}\right) - \tfrac{1}{12} h^3 \sum_{i=1}^{N} f''''_i \tag{7.2.12}$$

Rearranging, we obtain

$$\sum_{i=1}^{N} f''_i \cong \frac{f'_{N+1} - f'_1}{h} - \tfrac{1}{2}(f''_{N+1} - f''_1) + \tfrac{1}{12} h^2 \sum_{i=1}^{N} f''''_i \tag{7.2.13}$$

Substituting the right-hand side of this equation in place of the sum on the right-hand side of equation (7.2.4), and retaining only the term with the lowest power in h, we obtain the error estimate

$$E^{TR}(h) \cong \tfrac{1}{12}(f'_{N+1} - f'_1)\, h^2 \tag{7.2.14}$$

which is useful when the slope of the integrand at the ends of the integration domain is somehow known.

The right-hand side of equation (7.2.14) is the first term of a series named after *Euler* and *Maclaurin* (e.g., Abramowitz and Stegun 1972, Chapter 23). Assume that $f(x)$ and its first $2m+2$ derivatives are continuously differentiable over the closed interval $[a, b]$, for some $m \geq 0$. One version of the Euler–Maclaurin formula expresses the error of the trapezoidal rule in the form

$$E^{TR}(h) = b_2 \tfrac{1}{2}(f'_{N+1} - f'_1)h^2 + \cdots + b_{2m}\frac{1}{(2m)!}\left(f_{N+1}^{(2m-1)} - f_1^{(2m-1)}\right)h^{2m}$$

$$+ b_{2m+2}\frac{1}{(2m+2)!}f^{(2m+2)}(\xi)(b-a)h^{2m+2} \tag{7.2.15}$$

where $a < \xi < b$, a superscipt (l) designates the lth derivative, and b_n are the *Bernoulli numbers* defined in terms of the Maclaurin series

$$\frac{t}{e^t - 1} = \sum_{n=0}^{\infty} b_n \frac{t^n}{n!} \tag{7.2.16}$$

The first few Bernoulli numbers are:

$$b_0 = 1, \quad b_2 = \tfrac{1}{6}, \quad b_4 = -\tfrac{1}{30}, \quad b_6 = \tfrac{1}{42}, \quad b_8 = -\tfrac{1}{30}, \quad \ldots$$
$$b_1 = -\tfrac{1}{2}, \quad b_3 = 0, \quad b_5 = 0, \quad b_7 = 0, \quad \ldots$$

The Bernoulli numbers also arise from the *Bernoulli polynomials* $B_n(x)$, which are defined in terms of the Maclaurin series

$$\frac{te^{xt}}{e^t - 1} = \sum_{n=0}^{\infty} B_n(x)\frac{t^n}{n!} \tag{7.2.17}$$

The first few Bernoulli polynomials are

$$\begin{aligned} B_0(x) &= 1 \\ B_1(x) &= x - \tfrac{1}{2} \\ B_2(x) &= x^2 - x + \tfrac{1}{6} \\ B_3(x) &= x^3 - \tfrac{3}{2}x^2 + \tfrac{1}{2}x \end{aligned} \tag{7.2.18}$$

The Bernoulli numbers are $b_n = B_n(0)$.

Trapezoidal rule for periodic functions

When the integrand $f(x)$ is repeated periodically with period equal to the length of the integration domain $b - a$, formula (7.2.7) takes the simpler form

$$I^{TR}(h) = h(f_1 + f_2 + f_3 + \cdots + f_N) \tag{7.2.19}$$

Considering the magnitude of the error, we note that the first two terms on the right-hand side of equation (7.2.13) vanish, and this shows that the error depends on h in a manner that is higher than quadratic. More importantly, using the Euler–Maclaurin formula, we find that *if the integrand $f(x)$ is infinitely differentiable, then the integration error associated with the trapezoidal rule decays at a superalgebraic rate; that is, it decreases faster than any power of h.*

It is then not surprising that *when integrating periodic functions, the trapezoidal rule works much better than any other general-purpose method, even better than the otherwise powerful quadratures described in Section 7.3.*

To illustrate the significance of these results, in Figure 7.2.1 we plot with circles the error introduced by using the trapezoidal rule to compute the complete elliptic integral of the first kind given in the second of equations (7.1.10), for $k = 0.90$. Plotting the absolute value of the error against h on a logarithmic scale, we obtain a curved line that tends to become vertical at small values of h, which is consistent with the statement in the preceding paragraph.

There is a somewhat more striking behavior: Using the discrete orthogonality condition for trigonometric functions listed in Table 8.6.2, we find that *the $(N + 1)$-point trapezoidal rule with evenly spaced points integrates exactly periodic functions that can be represented by a trigonometric Fourier series that terminates at the Nth term.*

PROBLEMS

7.2.1 *Properties of the Bernoulli polynomials.*

Prove the following properties of the Bernoulli polynomials

$$B_l'(x) = l B_{l-1}(x)$$

$$B_p(x) = \sum_{l=0}^{p} \binom{p}{l} b_p x^{p-l} \tag{7.2.20}$$

$$B_p(x+1) - B_p(x) = p x^{p-1} \quad \text{for } p \geq 2$$

$$1^l + 2^l + \cdots + l^p = \frac{B_{p+1}(l+1) - B_{p+1}(0)}{p+1}$$

The combinational on the right-hand side of the second equation is defined in equation (1.4.6).

7.2.2 *Use of the Euler–Maclaurin summation formula.*

(*a*) With formula (7.2.15) given, derive the following version of the Euler–Maclaurin summation formula:

$$\sum_{i=1}^{N+1} f(x_i) = \frac{1}{h}\int_{x_1}^{x_{N+1}} f(x)\,dx + \tfrac{1}{2}(f_{N+1} + f_1) + b_2\tfrac{1}{2}(f'_{N+1} - f'_1)h$$
$$+ \cdots + b_{2m}\frac{1}{(2m)!}\left(f_{N+1}^{(2m-1)} - f_1^{(2m-1)}\right)h^{2m-1}$$
$$+ b_{2m+2}\frac{1}{(2m+2)!} f^{(2m-2)}(\xi)(b-a)\,h^{2m+1} \tag{7.2.21}$$

(*b*) Based on formula (7.2.21), show that

$$\sum_{n=1}^{\infty} \frac{1}{n^{3/2}} = 2.61238 \tag{7.2.22}$$

accurate to the fifth decimal place (Atkinson 1989, p. 289).

7.2.3 *Behavior of the error.*

Consider the integral shown in equation (7.2.9), but with the upper limit of integration set equal to 2. Compute the integral using the trapezoidal rule for several values of N, plot the magnitude of the error against N on a logarithmic scale, and discuss its behavior.

Simpson's 1/3 Rule

Approximating the graph of the integrand $f(x)$ with a collection of parabolas that are subtended across successive triplets of consecutive data points, corresponding to a second degree local interpolating polynomial, leads us to Simpson's 1/3 rule.

Using Lagrange's method to construct the interpolating polynomial, we find that the counterpart of equation (7.2.2) for evenly spaced points separated by distance h is

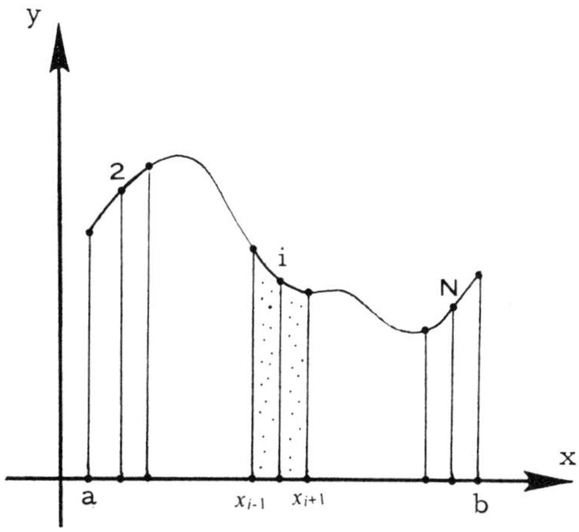

$$\int_{x_{i-1}}^{x_{i+1}} f(x)\,dx \cong \tfrac{1}{3}h(f_{i-1} + 4f_i + f_{i+1}) - \tfrac{1}{90} f^{(4)}(x_i)\,h_i^5 \tag{7.2.23}$$

Composite rule

Assuming that the number of intervals N is even, we add the contributions from the $\frac{1}{2}N$ nonoverlapping parabolas that are subtended across the successive triplets of points $(1, 2, 3)$, $(3, 4, 5)$, $\ldots$, $(N - 1, N, N + 1)$ and obtain the closed composite Simpson rule

$$I^{SIMP}(h) = I^{SIMP} + E^{SIMP}(h) \tag{7.2.24}$$

where

$$I^{SIMP}(h) = \tfrac{1}{3}h(f_1 + 4f_2 + 2f_3 + 4f_4 + \cdots + 2f_{N-1} + 4f_N + f_{N+1}) \tag{7.2.25}$$

and $E^{SIMP}(h)$ is the associated error.

Consider, for example, the integral (7.2.9). Simpson's 1/3 rule with $N = 10$, corresponding to $h = 0.05$, yields

$$
\begin{aligned}
I^{SIMP}(0.1) = {} & \tfrac{1}{3}0.05 \left\{ \ln[1 - 0.99 \cos^2 0] + 4 \ln \left[1 - 0.99 \cos^2 \left(\frac{0.05\pi}{2} \right) \right] \right. \\
& + 2 \ln \left[1 - 0.99 \cos^2 \left(\frac{0.10\pi}{2} \right) \right] + \cdots + 4 \ln \left[1 - 0.99 \cos^2 \left(\frac{0.40\pi}{2} \right) \right] \\
& \left. +2 \ln \left[1 - 0.99 \cos^2 \left(\frac{0.45\pi}{2} \right) \right] + \ln \left[1 + 4 \ln \left[1 - 0.99 \cos^2 \left(\frac{0.50\pi}{2} \right) \right] \right] \right\} \\
= {} & -1.0864
\end{aligned}
$$

This value is identical to the one obtained previously by applying the Richardson extrapolation method to improve the results of the trapezoidal rule with $N = 5$ and 10. The exact value is equal to -1.0870.

Error

The leading-order contribution to the error can be expressed as

$$
E^{SIMP} \cong \tfrac{1}{180}(b - a) f''''(\xi) h^4 \cong \tfrac{1}{180}(f'''_{N+1} - f'''_1) h^4 \tag{7.2.26}
$$

where the point ξ lies somewhere in the domain of integration. A more extended analysis shows that the error is expressible as a series containing even powers of $h$as,

$$
E^{SIMP}(h) = c_4 h^4 + c_6 h^6 + \cdots \tag{7.2.27}
$$

The coefficient c_i is proportional to the ith derivative of $f(x)$ evaluated at some point over the domain of integration.

Simpson's 1/3 rule is a popular method for integrals whose integrands can be evaluated or are known at a sequence of evenly spaced points. The amount of required work is comparable to that of the trapezoidal rule: Comparing formulas (7.2.7) and (7.2.25), we observe differences only in the weights.

Romberg integration

Romberg integration uses the results of two computations conducted with different step sizes, to improve the accuracy to sixth order in h setting, for example,

$$
I^{ROM}(h = \tfrac{\varepsilon}{2}) = \tfrac{1}{15} \left(16 I^{SIMP} \left(h = \tfrac{\varepsilon}{2} \right) - I^{SIMP}(h = \varepsilon) \right) \tag{7.2.28}
$$

where ε is a sufficiently small interval. Further improvements can be made as discussed in Section 1.6 in the more general context of Richardson extrapolation.

Modified Simpson rules

The preferential treatment of the odd- and even-numbered data points in constructing the local interpolating polynomials can be prevented by first computing the areas below the parabolas that pass through *all* triplets of successive data points, and then averaging the contributions from the overlapping portions. In practice, however, the straight Simpson's rule is adequate in most applications.

PROBLEMS

7.2.4 Romberg–trapezoidal method and Simpson's rule.

Show that the effective integration formula produced by the right-hand side of equation (7.2.11) is identical to that of Simpson's 1/3 rule applied with $h = \frac{1}{2}\varepsilon$.

7.2.5 Comparison of integration methods.

Compute the integral

$$\int_0^{\pi/4} \frac{d\theta}{\cos\theta} \tag{7.2.29}$$

accurate to the fourth significant figure, using the trapezoidal rule. Repeat using Simpson's 1/3 rule, and compare the number of intervals that are necessary in each case. Confirm the accuracy of your computations by comparing the numerical with the exact value obtained by elementary analytical methods.

Simpson's 3/8, Boole's, and Higher-order Rules

Approximating the integrand $f(x)$ with a *cubic* polynomial, whose graph passes through quadruplets of consecutive data points, yields *Simpson's 3/8 rule.* The counterpart of equation (7.2.2) for evenly spaced points is given in the first entry of Table 7.2.1.

Approximating $f(x)$ with a *fourth degree* polynomial, whose graph passes through successive quintets of consecutive data points, yields *Boole's rule.* The counterpart of equation (7.2.2) for evenly spaced points is given in the second entry of Table 7.2.1.

Approximating $f(x)$ with a high-order local interpolating polynomial, whose graph passes through a higher number of evenly spaced points, yields high-order *closed Newton–Cotes formulas*; several of them are shown in Table 7.2.1. The qualifier *closed* signifies that both end-points of the integration interval are involved in the computation of the integral. Later in this section we shall discuss *open* rules.

The corresponding composite rules arise by adding the contributions from the local polynomials, possibly averaging the overlapping portions. Because of the lurking oscillations associated with the Gibbs effect, however, high-order formulas are not used for practical numerical integration, although they are useful for the numerical solution of ordinary differential equations (Chapter 9).

Table 7.2.1 Closed Newton–Cotes formulas

$$\int_{x_i}^{x_{i+3}} f(x)\,dx \cong \tfrac{3}{8}h(f_i + 3f_{i+1} + 3f_{i+2} + f_{i+3}) - \tfrac{3}{80}f^{(4)}(x_i)\,h^5$$

$$\int_{x_i}^{x_{i+4}} f(x)\,dx \cong \tfrac{2}{45}h(7f_i + 32f_{i+1} + 12f_{i+2} + 32f_{i+3} + 7f_{i+4}) - \tfrac{8}{945}f^{(6)}(x_i)\,h^7$$

$$\int_{x_i}^{x_{i+5}} f(x)\,dx \cong \tfrac{5}{288}h(19f_i + 75f_{i+1} + 50f_{i+2} + 50f_{i+3} + 75f_{i+4} + 19f_{i+5}) - \tfrac{275}{12096}f^{(6)}(x_i)\,h^7$$

$$\int_{x_i}^{x_{i+6}} f(x)\,dx \cong \tfrac{1}{140}h(41f_i + 216f_{i+1} + 27f_{i+2} + 272f_{i+3} + 27f_{i+4} + 216f_{i+5} + 41f_{i+6}) - \tfrac{9}{1400}f^{(8)}(x_i)\,h^9$$

PROBLEMS

7.2.6 *Line integrals.*

(a) Consider the following line integral along a circle of unit radius centered at the origin of the xyplane

$$I(x_0, y_0) = \int_{Circle} \ln\,[(x(l) - x_0)^2 + (y(l) - y_0)^2]\,dl \tag{7.2.30}$$

where l is the arc length along the circle, and (x_0, y_0) is a specified point. Evaluate the integral at two different points (x_0, y_0) located inside the circle, and two other points located outside the circle, and discuss your results. Note that along a circle of unity radius $dl = d\theta$, where θ is the polar angle measured in the counterclockwise sense.

(b) Compute the *line integral*

$$I = \int x\,dl = \int x\sqrt{1 + \left(\frac{dy}{dx}\right)^2}\,dx \tag{7.2.31}$$

where l is the arc length along the segment of the line $y = \sin\,x$, confined in the interval $0 \le x \le \frac{1}{2}\pi$, using a method of your choice. Your result should be accurate to the second significant figure.

7.2.7 *Potential flow via a vortex sheet.*

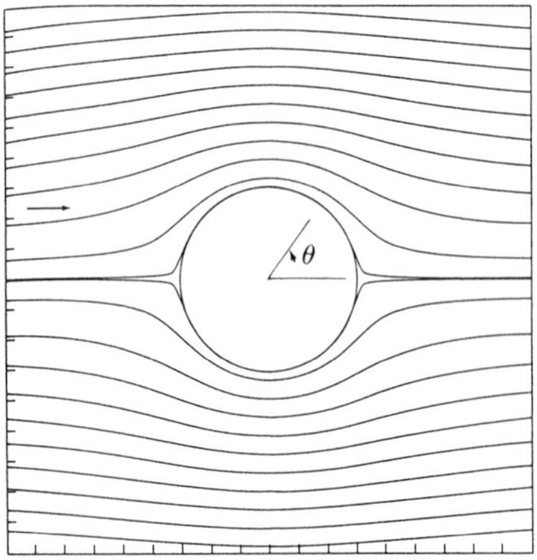

Two-dimensional potential flow due to a moving body can be expressed in terms of a vortex sheet situated over the body. Specifically, the velocity $\boldsymbol{u}$ may be express as the sum of the unperturbed incident velocity $\boldsymbol{u}^\infty$ that prevails at large distances from the body, and the velocity induced by the vortex sheet $\boldsymbol{u}^V$. The strength of the vortex sheet γ is equal to the tangential component of boundary velocity, viewed in a frame of reference under which the body appears to be stationary (e.g., Pozrikidis 1997).

If γ is known, $\boldsymbol{u}^V$ may be computed at any point in the flow by straightforward numerical integration. In complex variables, the velocity $\boldsymbol{u}^V$ at the complex position $z_0 = x_0 + I y_0$, where I is the imaginary unit, is given by

$$u^V - I v^V = \frac{1}{2\pi I}\int_C \frac{1}{z - z_0}\gamma\,dl \tag{7.2.32}$$

where C is the contour of the body, l is the arc length along the contour of the body, and $z = x + I y$.

Consider flow along the x axis with velocity V past a stationary circular cylinder of radius a with center at the origin. The strength of the vortex sheet is known to be $\gamma = 2V\,\sin\,\theta$, where θ is the polar angle.

(a) Develop a numerical method for computing the integral on the right-hand side of equation (7.2.32), and write a computer program.

(b) Compute and plot profiles of the x and y velocity components at several points along the segment $x = 2, -2 < y < 2$.

(c) Compute the velocity at several points inside the cylinder and discuss your findings.

Seemingly but not Truly Singular Integrands

Consider the integral

$$I = \int_0^{\pi} \frac{\sin x}{x} \exp(-x^2)\, dx \tag{7.2.33}$$

The integrand appears to have a singularity at $x = 0$, but by using the L' Hôpital rule, or expanding sin x in a Maclaurin series and then dividing by x, we find that, as x tends to zero, the integrand tends to the value of unity.

It is then permissible to compute the integral using, for example, the trapezoidal rule, but this will require the evaluation of the integrand at $x = 0$. The computer will be unable to essentially divide zero by zero and will protest with a message. To prevent this dispiriting action, we can work in three alternative ways:

- Start the integration from a small positive number ε instead of zero. This amounts to having the computer effectively perform the Maclaurin series expansion and introduces an error on the order of ε.
- Supply the value $f(0) = 1$ directly into the definition of the integrand.
- Replace the trapezoidal rule with the midpoint rule discussed in the next subsection.

PROBLEM

7.2.8 ***Seemingly singular integrands.***

Show that the following integrands are seemingly, but not truly, singular:

$$\int_0^1 \frac{1 - x - (1+x)\,\mathrm{e}^{-x}}{x}\, dx, \qquad \int_0^1 \frac{1 - \mathrm{e}^{-x}(1+x)}{x^2}\, dx \tag{7.2.34}$$

and then evaluate the integrals accurate to the third significant figure using a closed Newton–Cotes rule of your choice.

Midpoint Rule and Open Formulas

This method proceeds by replacing equation (7.2.2) with its cousin

$$\int_{x_i}^{x_{i+1}} f(x)\, dx \cong \tfrac{1}{2} h_i f_{i+1/2} - \tfrac{1}{24} f''(x_i)\, h_i^3 \tag{7.2.35}$$

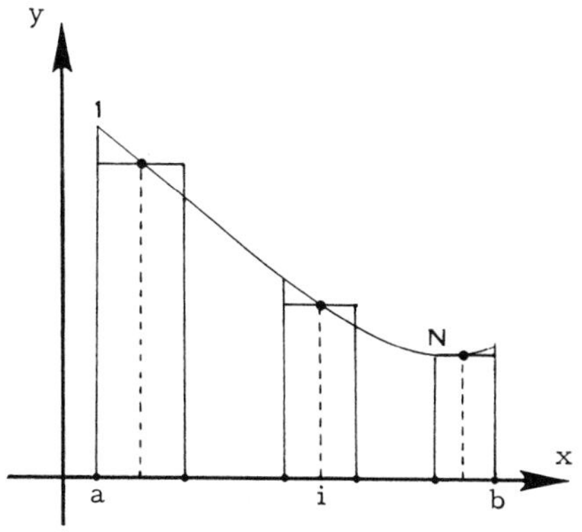

where

$$f_{i+1/2} = f\left(\tfrac{1}{2}(x_i + x_{i+1})\right) \tag{7.2.36}$$

Summing up all contributions produces an *open composite rule* whose error is about half that of the trapezoidal rule. The qualifier *open* signifies that the end-points are not involved in the computation.

The midpoint rule is the first member of a family of open Newton–Cotes rules corresponding to quadratic, cubic, and higher-order local interpolating polynomials with evenly spaced base points. Several other members of the family are shown in Table 7.2.2. These formulas find applications in the numerical solution of ordinary differential equations, to be discussed in Chapter 9.

Integrands with Singular Derivatives

If we apply a standard Newton–Cotes rule to evaluate the integral

$$\int_0^2 x \ln x \, dx \tag{7.2.37}$$

we shall observe unexpectedly low accuracy and slow rate of convergence. The reason becomes evident by using the L' Hôpital rule to find that, as x tends to zero, the product $x \ln x$ tends to vanish, but its first and higher derivatives tend to become infinite. The implicit presence of these singularities erodes the accuracy of the polynomial approximation that underlies the derivation of the Newton–Cotes rules; the penalty is slow convergence (e.g., Lyness and Ninham 1967, Problem 7.2.9).

The treatment of integrals with regular integrands whose derivatives become singular at some point will be discussed in Section 7.4.

PROBLEM

7.2.9 *Integrands with singular derivatives.*

(*a*) Compute the integral (7.2.37) analytically and then numerically using the trapezoidal rule and the Simpson 1/3 rule. Study and discuss the convergence of the methods.
(*b*) By means of numerical experimentation using the Simpson 1/3 rule, show that the error in the computation of the integral

Table 7.2.2 Open Newton–Cotes formulas.

$$\int_{x_i}^{x_{i+2}} f(x)\,dx \cong 2h\, f_{i+1} + \tfrac{1}{3} f''(x_i)\, h^3$$

$$\int_{x_i}^{x_{i+3}} f(x)\,dx \cong \frac{3}{2}h\,(f_{i+1} + f_{i+2}) + \tfrac{3}{4} f''(x_i)\, h^3$$

$$\int_{x_i}^{x_{i+4}} f(x)\,dx \cong \frac{4}{3}h\,(2f_{i+1} - f_{i+2} + 2f_{i+3}) + \tfrac{14}{45} f^{(4)}(x_i)\, h^5$$

$$\int_{x_i}^{x_{i+5}} f(x)\,dx \cong \frac{5}{24}h\,(11f_{i+1} + f_{i+2} + f_{i+3} + 11f_{i+4}) + \tfrac{95}{144} f^{(4)}(x_i)\, h^5$$

$$\int_{x_i}^{x_{i+6}} f(x)\,dx \cong \frac{3}{10}h\,(11f_{i+1} - 14f_{i+2} + 26f_{i+3} - 14f_{i+4} + 11f_{i+5}) + \tfrac{41}{140} f^{(6)}(x_i)\, h^7$$

$$\int_0^1 x^{3/2}\, dx \tag{7.2.38}$$

behaves like $h^{2.5}$. In this case, because the derivatives of the integrand are singular, the order of the method is reduced from 4 to 2.5.

7.3 Optimal Distribution of Base Points, and the Gauss–Legendre Quadrature

We continue to assume that the integrand and its derivatives are free of singularities over the whole of the domain of integration $[a, b]$, and the limits of integration are finite, and approximate the integrand with a global interpolating polynomial whose graph passes through $N + 1$ data points. Moreover, we assume that we have the luxury of distributing the $N + 1$ data points in any way we desire, and ask: Is there an optimal way of distributing them so as to minimize the integration error in some appropriate sense, and thereby ensure that, as N is increased, the results of the numerical computation converge to the exact value at the fastest possible rate?

A similar question was posed in Section 6.4 in the context of polynomial interpolation. The analysis showed that, in order to minimize the interpolation error, the data points over the scaled interpolation domain $[-1, 1]$ should correspond to the roots of the $(N + 1)$ degree Chebyshev polynomial.

Canonical Form

Before tackling the issue of optimal point distribution, we bring the problem to its canonical form. This is done by introducing the new independent variable t that is related to the integration variable x by the linear transformation

$$x = q(t) = \frac{b+a}{2} + \frac{b-a}{2}\, t \tag{7.3.1}$$

As t increases from -1 to 1, x increases from a to b. Having made this transformation, we regard f as a function of t, writing

$$f(x) = f(q(t)) = h(t) \tag{7.3.2}$$

and

$$I \equiv \int_a^b f(t)\, dx = \frac{(b-a)}{2} \int_{-1}^{1} h(t)\, dt \tag{7.3.3}$$

The problem has been reduced to evaluating the integral on the right-hand side of equation (7.3.3) over the canonical domain $[-1, 1]$.

For example, to recast the integral (7.2.9) into its canonical form, we note that $a = 0$ and $b = 0.50$, write $x = 0.25 + 0.25\, t$ and $dx = 0.25\, dt$, and obtain

$$I = 0.25 \int_{-1}^{1} \ln\left(1 - 0.99 \cos^2 \frac{(1+t)\pi}{8}\right) dt \tag{7.3.4}$$

The function $h(t)$ is the integrand of the integral (7.3.4).

Integration by Lagrange Interpolation

Using Lagrange's method to approximate the function $h(t)$ with an Nth degree interpolating polynomial whose graph passes through the $N+1$ data points t_i, $i = 1, \ldots, N+1$, where $-1 \le t_i \le 1$, and then integrating the interpolating polynomial between -1 and 1 as dictated by the right-hand side of equation (7.3.3), we obtain the *Gauss quadrature*

$$\int_{-1}^{1} h(t)\, dt \cong \sum_{i=1}^{N+1} h(t_i)\, w_i \tag{7.3.5}$$

The weights w_i are given by

$$w_i = \int_{-1}^{1} l_{N,i}(t)\, dt \tag{7.3.6}$$

The function $l_{N,i}(t)$ is the ith, Nth degree Lagrange polynomial defined in terms of the $N+1$ data points as

$$l_{N,i}(t) = \frac{(t-t_1)\ldots(t-t_{i-1})(t-t_{i+1})\ldots(t-t_{N+1})}{(t_i-t_1)\ldots(t_i-t_{i-1})(t_i-t_{i+1})\ldots(t_i-t_{N+1})} = \frac{1}{t-t_i}\frac{\Phi_{N+1}(t)}{\Phi'_{N+1}(t_i)} \tag{7.3.7}$$

where

$$\Phi_{N+1}(t) \equiv (t-t_1)(t-t_2)\ldots(t-t_N)(t-t_{N+1}) \tag{7.3.8}$$

Gauss–Legendre Quadrature

To this end, we regard the locations of the base points as free parameters of the numerical method. No matter how we distribute the $N+1$ data points, the quadrature (7.3.5) will integrate *exactly* polynomials of degree N or less. But we want to do better than that: We have at our disposal $N+1$ degrees of freedom, the abscissas of the base points, and we want to adjust them so that the quadrature integrates *exactly* polynomials of degree $2N+1$ or less.

One point

To get our feet wet, we consider the simplest case of one base point corresponding to $N = 0$ and assume that $h(t)$ is the first degree polynomial

$$h(t) = a_1 t + a_0 \tag{7.3.9}$$

Formula (7.3.6) with $l_{0,1}(t) = 1$ yields $w_1 = 2$. Requiring that formula (7.3.5) produces the exact answer, we find

$$2\, a_0 = 2\, (a_1 t_1 + a_0) \tag{7.3.10}$$

which is true for any value of a_1 and a_0, provided that $t_1 = 0$.

Thus the optimal location of the base point is at the middle of the domain of integration. This could have been argued at the outset on the basis of spatial nondiscrimination.

Two points

Considering next the case $N = 1$, we set out to compute the location of the two base points so that the quadrature (7.3.5) integrates exactly constant, first, second, and third degree polynomials expressed by

$$h(t) = a_3 t^3 + a_2 t^2 + a_1 t + a_0 \tag{7.3.11}$$

Formula (7.3.6) applied for

$$l_{1,1}(t) = \frac{t - t_2}{t_1 - t_2}, \qquad l_{1,2}(t) = \frac{t - t_1}{t_2 - t_1} \tag{7.3.12}$$

yields

$$w_1 = 2\frac{t_2}{t_2 - t_1}, \qquad w_2 = -2\frac{t_1}{t_2 - t_1} \tag{7.3.13}$$

Requiring that the quadrature (7.3.5) produces the exact answer, we find

$$\tfrac{2}{3}a_2 + 2\,a_0 = (a_3\,t_1^3 + a_2\,t_1^2 + a_1\,t_1 + a_0)w_1 + (a_3\,t_2^3 + a_2\,t_2^2 + a_1\,t_2 + a_0)\,w_2 \tag{7.3.14}$$

which will be true for any values of the polynomial coefficients provided that

$$\begin{aligned} t_1^3 w_1 + t_2^3 w_2 &= 0 \\ t_1^2 w_1 + t_2^2 w_2 &= \tfrac{2}{3} \\ t_1 w_1 + t_2 w_2 &= 0 \\ w_1 + w_2 &= 2 \end{aligned} \tag{7.3.15}$$

Eliminating w_1 and w_2 in favor of t_1 and t_2, and using equations (7.3.13), we find that all four equations will be satisfied provided that

$$t_1 = -\frac{1}{\sqrt{3}}, \qquad t_2 = \frac{1}{\sqrt{3}} \tag{7.3.16}$$

for which

$$w_1 = 1, \qquad w_2 = 1 \tag{7.3.17}$$

For example, applying the two-point quadrature (7.3.5) to compute the integral (7.3.4), we find

$$\begin{aligned} I^{GL}\,(N = 1) =& 0.25\{\ln[1 - 0.99\cos^2 \tfrac{1}{8}\left(1 - \frac{1}{\sqrt{3}}\right)\pi] \times 1 \\ &+ \ln[1 - 0.99\cos^2 \tfrac{1}{8}\left(1 + \frac{1}{\sqrt{3}}\right)\pi] \times 1\} \\ =& -1.0911 \end{aligned}$$

which is close to the exact value -1.0870.

Many points

It must have become clear by now that the proceeding method of computing the base points and corresponding weights is cumbersome, to say the least. A more efficient approach is required. Fortunately, this is much easier to develop than one might think.

We begin by assuming that $h(t)$ is an $(2N + 1)$ degree polynomial,

$$h(t) = Q_{2N+1}(t) \tag{7.3.18}$$

and combine equations (6.3.1) and (6.3.10) with $h(t)$ in place of $f(x)$, t in place of x, and $K = 2N + 1$, to obtain the interpolation error

$$e(t) = E_{2N+1}(t) = \Phi_{N+1}(t)\ \Xi_N(t) \tag{7.3.19}$$

where $\Xi_n(t)$ is an Nth degree polynomial, and $\Phi_{N+1}(t)$ was defined in equation (7.3.8). We want to distribute the base points so that the integral of the error vanishes, that is,

$$\int_{-1}^{1} \Phi_{N+1}(t)\ \Xi_N(t)\, dt = 0 \tag{7.3.20}$$

This relation requires that the $(N + 1)$ degree polynomial $\Phi_{N+1}(t)$ is orthogonal to all lower-degree polynomials. Expressing $\Xi_N(x)$ as a linear combination of the first $N + 1$ Legendre polynomials $L_0(x)$, $L_1(x), \ldots, L_N(x)$, and then using the orthogonality property of the Legendre polynomials listed in Table B.2, we find that the orthogonality requirement will be satisfied provided that $\Phi_{N+1}(t)$ is proportional to the $(N + 1)$ degree Legendre polynomial $L_{N+1}(t)$.

How do we ensure that $\Phi_{N+1}(t)$ is proportional to $L_{N+1}(t)$? This follows immediately from the definition of $\Phi_{N+1}(t)$ in equation (7.3.8). All we need to do is *identify the base points with the* $N+1$ *zeros of* $L_{N+1}(t)$. Note that these results are consistent with those of the preceding two subsections corresponding to $N = 0$ and 1: The zeros of the second-degree Legendre polynomial are given in equations (7.3.16).

Gauss–Legendre base points

Table B.2 of Appendix B contains analytical expressions for the locations of the base points t_i. Explicit values are given in Table 7.3.1. Alternatively, these values may be computed using one of the polynomial root-finder methods discussed in Section 4.7, including Bairstow's method.

Gauss–Legendre weights

Taking advantage of the distinctive properties of the Legendre polynomials to evaluate the intergals (7.3.6), we obtain the relations for the weights w_i shown at the end of Table B.2, Appendix B.

Table 7.3.1 displays the weights for several values of N. Note that for each value of N, the w_i add up to two, which is necessary if the right-hand side of equation (7.3.5) is to produce the exact value when the function $h(t)$ is a constant.

A more general method of obtaining the weights, which is applicable in the more general context of Gaussian quadratures to be discussed in the following sections, is to apply the quadrature (7.3.5) for a suitable group of $N + 1$ test polynomials of degree less than or equal to $2N + 1$, compute the left-hand side exactly by analytical means, and then solve the resulting system of $N + 1$ linear equations for w_i. A suitable

Table 7.3.1 Abscissas and weights for the *Gauss–Legendre* quadrature with $N + 1$ points,

$$\int_{-1}^{1} h(t)\, dt \cong \sum_{i=1}^{N+1} h(t_i)\, w_i$$

The function $h(t)$ and its derivatives are assumed to be nonsingular over the whole of the canonical integration domain $[-1, 1]$.

One-point formula $N = 0$	
$t_1 = 0$	$w_1 = 2$
Two-point formula, $N = 1$	
$t_1 = -0.57735\ 02691$	$w_1 = 1$
$t_2 = -t_1$	$w_2 = w_1$
Three-point formula, $N = 2$	
$t_1 = -0.77459\ 66692$	$w_1 = 0.55555\ 55555$
$t_2 = 0$	$w_2 = 0.88888\ 88888$
$t_3 = -t_1$	$w_3 = w_1$
Four-point formula, $N = 3$	
$t_1 = -0.86113\ 63115$	$w_1 = 0.34785\ 48451$
$t_2 = -0.33998\ 10435$	$w_2 = 0.65214\ 51548$
$t_3 = -t_2$	$w_3 = w_2$
$t_4 = -t_1$	$w_4 = w_1$
Five-point formula, $N = 4$	
$t_1 = -0.90617\ 98459$	$w_1 = 0.23692\ 68850$
$t_2 = -0.53846\ 93191$	$w_2 = 0.47862\ 86704$
$t_3 = 0$	$w_3 = 0.56888\ 88888$
$t_4 = -t_2$	$w_4 = w_2$
$t_5 = -t_1$	$w_5 = w_1$
Six-point formula, $N = 5$	
$t_1 = -0.93246\ 95142$	$w_1 = 0.17132\ 44923$
$t_2 = -0.66120\ 93864$	$w_2 = 0.36076\ 15730$
$t_3 = -0.23861\ 91860$	$w_3 = 0.46791\ 39345$
$t_4 = -t_3$	$w_4 = w_3$
$t_5 = -t_2$	$w_5 = w_2$
$t_6 = -t_1$	$w_6 = w_1$

choice of test polynomials are the first $N + 1$ Legendre polynomials $L_0(t), \ldots, L_N(t)$. The integral of all of these polynomials between -1 and 1 is equal to zero, except for the integral of $L_0(t)$ that is equal to 2. The aforementioned linear system takes the form

$$L_i(t_1)\, w_1 + L_i(t_2)\, w_2 + \ldots + L_i(t_{N+1})\, w_{N+1} = 2\delta_{i,0} \tag{7.3.21}$$

where $i = 0, \ldots, N$. Note that the first equation, corresponding to $i = 0$, simply states that the sum of the weights must be equal to 2. A word of caution: For large values of N, the solution may be susceptible to the amplification of the round-off error.

Error of the Gauss–Legendre quadrature

It can be shown that the numerical error associated with the Gauss–Legendre quadrature (7.3.5) is given by

$$E^{GL} = -\frac{[\,(N+1)!]^4}{(2N+3)\,[(2N+2)!]^4}\, f^{(2N+2)}(\xi) \tag{7.3.22}$$

where the point ξ lies somewhere within the domain of integration. The fast decay with respect to N is apparent.

PROBLEMS

7.3.1 ***Tedious derivation of the Gauss–Legendre abscissas and weights.***

Derive the generalized form of the system (7.3.15) for an arbitrary value of N.

7.3.2 ***Integral of Legendre polynomials.***

Based on the quadrature (7.3.5), show that the integral of $L_{N+1}(t)$ over $[-1, 1]$ is equal to zero. *Hint*: Identify $h(t)$ with $L_{N+1}(t)$.

7.3.3 ***Gauss–Legendre quadrature.***

(*a*) Compute the integral (7.2.9) using the three-point Gauss–Legendre quadrature, and comment on the accuracy of the method.
(*b*) Use the eight-point Gauss–Legendre integration formula to compute Erf(2), and compare your result to the approximation produced by formula (7.1.8).

7.3.4 ***A nonlinear algebraic equation.***

Compute one solution of the following equation for x, accurate to the fourth decimal place, using the Gauss–Legendre quadrature:

$$\int_0^x \frac{\sin\, t}{t} \exp(-t^2)\, dt = 0.09 \tag{7.3.23}$$

Gauss–Chebyshev Quadrature of the Second Kind

The problem of finding the optimal location of the data points may be tackled from a slightly different perspective that leads us to a different type of quadrature. Taking the absolute value of both sides of the error formula (6.3.3) written for the function $h(t)$, and integrating the result between -1 and 1, we find

$$\int_{-1}^{1} |P_N(T) - h(t)\,|\, dt \cong \frac{1}{(N+1)!} \int_{-1}^{1} |\, h^{(N+1)}(\xi)\,(t-t_1)\,(t-t_2)\ldots(t-t_N)\,(t-t_{N+1})\,|\, dt \tag{7.3.24}$$

Next, we denote the maximum value of the function $|h^{(N+1)}(t)|$ over the interval $[-1, 1]$ by c and derive the inequality

$$\int_{-1}^{1} |P_N(t) - h(t)\,|\, dt \le \frac{c}{(N+1)!} \int_{-1}^{1} |\,(t-t_1)\,(t-t_2)\ldots(t-t_N)\,(t-t_{N+1})\,|\, dt \tag{7.3.25}$$

The problem has been reduced to computing the set of the $N + 1$ data points that minimizes the integral on the right-hand side.

It is known that of all $(N + 1)$ degree polynomials whose leading-order coefficient is equal to unity, the polynomial $2^N U_{N+1}(t)$, where $U_{N+1}(t)$ is the $(N+1)$ degree second-kind Chebyshev polynomial discussed in Table B.4 of Appendix B, minimizes the aforementioned integral (e.g., Todd 1962, p. 149). Accordingly, the data points should be identified with the zeros of $U_{N+1}(t)$. In general, however, the Gauss–Legendre quadrature produces more accurate results. The second-kind Gauss–Chebyshev quadrature is appropriate for integrals whose integrand behaves like the weighting function of the second-kind Chebyshev polynomials shown in Table B.4, as will be discussed in Section 7.4.

Gauss–Radau and Gauss–Lobatto Quadratures

A related family of quadratures arises by specifying the position of some but not all of the $N+1$ data points.

- The *Gauss–Radau* quadrature arises by specifying $t_1 = -1$, and minimizing the error with respect to the abscissas of the remaining N points, in the spirit of Gauss–Legendre.
- The *Gauss–Lobatto* quadrature arises by specifying $t_1 = -1$, $t_{N+1} = 1$, and minimizing the error with respect to the abscissas of the remaining $N-1$ points, in the spirit of Gauss–Legendre.

The abscissas of the unknown data points are the roots of the Radau or Lobatto orthogonal polynomials whose weighting functions are, respectively, equal to $w(t) = 1 + t$, and $w(t) = 1 - t^2$ (see Tables B.6 and B.7, Appendix B) (Hildebrand 1974, pp. 406- 411). The locations of the base points and the corresponding weights are displayed in Tables 7.3.2 and 7.3.3, and formulas are given in Tables B.6 and B.7 (see also Abramowitz and Stegun 1972, pp. 888, 920).

Extensions

The Clenshaw–Curtis (1960) quadrature discussed by Evans (1993, pp. 51–55) arises by approximating the integrand with a truncated series of Chebyshev polynomials, and then integrating the approximation. The $(N+1)$-point quadrature integrates exactly polynomials of degree up to and including N; in this respect, the quadrature is similar to the Newton–Cotes rule. But the magnitude of the error is by far lower than might be expected; in practice, it is comparable to that of the Gauss quadrature.

Kronrod (1965) showed that an M-point quadrature can be supplemented with a further set of $M+1$ points to yield a rule that integrates exactly polynomials of degree up to and including $3M+1$ when M is even, or $3M+2$ when M is odd (Evans 1993, pp. 58–60). Patterson (1968) developed an extension of Kronrod's method.

Table 7.3.2 Abscissas and weights for the *Gauss–Radau* quadrature with $N+1$ points,

$$\int_{-1}^{1} h(t)\, dt \cong \frac{2}{(N+1)^2} h(-1) + \sum_{i=1}^{N} h(t_i)\, w_i$$

The function $h(t)$ and its derivatives are assumed to be nonsingular within the whole of the canonical integration domain $[-1, 1]$.

Two-point formula, $N = 1$	
$t_1 = \frac{1}{3}$	$w_1 = \frac{3}{2}$
Three-point formula, $N = 2$	
$t_1 = -0.289898\ldots$	$w_1 = 0.75280\,6$
$t_2 = 0.689898\ldots$	$w_2 = 1.02497\,2$
Four-point formula, $N = 3$	
$t_1 = -0.57531\,9$	$w_1 = 0.65768\,9$
$t_2 = 0.18106\,6$	$w_2 = 0.77638\,7$
$t_3 = 0.82282\,4$	$w_3 = 0.44092\,5$
Five-point formula, $N = 4$	
$t_1 = -0.72048\,0$	$w_1 = 0.44620\,7$
$t_2 = -0.16718\,1$	$w_2 = 0.62365\,3$
$t_3 = 0.44631\,4$	$w_3 = 0.56271\,2$
$t_4 = 0.88579\,2$	$w_4 = 0.28742\,7$

Table 7.3.3 Abscissas and weights for the *Gauss–Lobatto* quadrature with $N+1$ points,

$$\int_{-1}^{1} h(t)\, dt \cong \frac{2}{N(N+1)} h(-1) + \sum_{i=1}^{N-1} h(t_i)\, w_i + \frac{2}{N(N+1)} h(1)$$

The function $h(t)$ and its derivatives are assumed to be nonsingular within the whole of the canonical integration domain $[-1, 1]$.

Three-point formula, $N = 2$	
$t_1 = 0.0$	$w_1 = \frac{4}{3}$
Four-point formula, $N = 3$	
$t_1 = -0.44721\,360 = -1/\sqrt{5}$	$w_1 = \frac{5}{6}$
$t_2 = -t_1$	$w_2 = w_1$
Five-point formula, $N = 4$	
$t_1 = -0.65465\,367$	$w_1 = \frac{49}{90}$
$t_2 = 0.0$	$w_2 = \frac{32}{45}$
$t_3 = -t_1$	$w_3 = w_1$
Six-point formula, $N = 5$	
$t_1 = -0.76505\,532$	$w_1 = 0.37847\,496$
$t_2 = -0.28523\,152$	$w_2 = 0.55485\,838$
$t_3 = -t_2$	$w_3 = w_2$
$t_4 = -t_1$	$w_4 = w_1$
Seven-point formula, $N = 6$	
$t_1 = -0.83022\,390$	$w_1 = 0.27682\,604$
$t_2 = -0.46884\,879$	$w_2 = 0.43174\,538$
$t_3 = 0.0$	$w_3 = 0.48761\,904$
$t_4 = -t_2$	$w_4 = w_2$
$t_5 = -t_1$	$w_5 = w_1$
Eight-point formula, $N = 7$	
$t_1 = -0.87174\,015$	$w_1 = 0.21070\,422$
$t_2 = -0.59170\,018$	$w_2 = 0.34122\,270$
$t_3 = -0.20929\,922$	$w_3 = 0.41245\,880$
$t_4 = -t_3$	$w_4 = w_3$
$t_5 = -t_2$	$w_5 = w_2$
$t_6 = -t_1$	$w_6 = w_1$

7.4 *Singular Integrands*

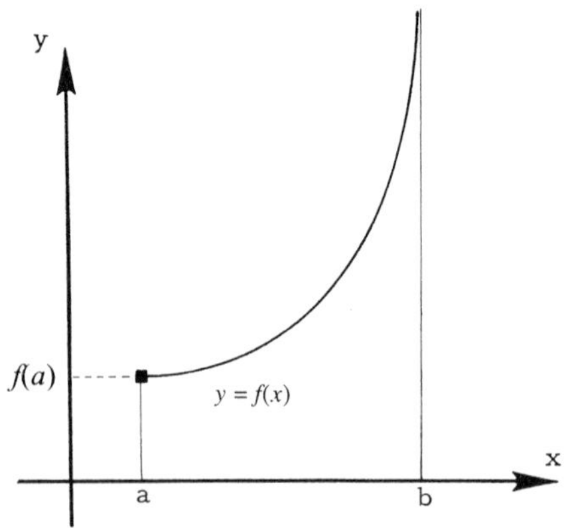

What happens when the integrand $f(x)$ takes an infinite value at some point at the beginning, in the interior, or at the end of the domain of integration? The definite integral may either have an infinite value, in which case we say that it diverges or does not exist and the singularity of the integrand is *nonintegrable*; or it may have a finite value, in which case the singularity is *integrable.*

When the singularity is integrable, we obtain the seemingly paradoxical behavior that the graph of the function $y = f(x)$ shoots off to infinity, yet the area confined between its graph, the x axis, and the two vertical lines $x = a$ and $x = b$ is finite. The paradox is resolved by noting that the integral of a singular integrand can be restated as the integral of a nonsingular integrand over an infinite domain of integration.

For example, using the geometrical interpretation of a definite integral, and assuming that $f(b) = \infty$, we write

$$\int_a^b f(x)\,dx = \int_0^\infty g(t)\,dt \tag{7.4.1}$$

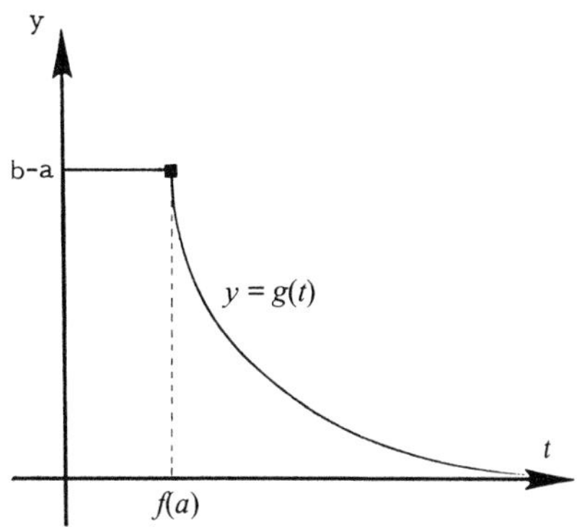

where

$$g(t) = \begin{cases} b - a & \text{when } 0 < t < f(a) \\ b - f^{Inv}(t) & \text{when } t > f(a) \end{cases} \tag{7.4.2}$$

and $f^{Inv}(t)$ is the inverse of the function $f(x)$, that is, $f^{Inv}(f(t)) = t$. As long as the function $g(t)$ decays sufficiently fast, the integral on the right-hand side of equation (7.4.1) is finite.

Is the Singularity Integrable?

To assess whether an integrand is integrable, we investigate its functional form close to the singular point with reference to prototypical classes of elementary singular integrals that serve as benchmarks.

One fundamental class of improper integrals whose integrands exhibit algebraic singularities are

$$I_m = \int_a^b \frac{1}{(x-a)^m}\,dx \tag{7.4.3}$$

where the exponent m is positive. Graphs of the integrand $f(x) = 1/(x-a)^m$ for several values of m and $a = 0$ are shown in the diagram.

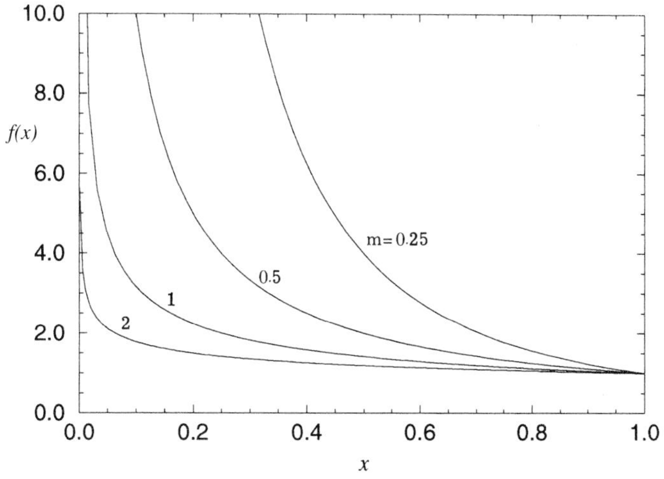

Elementary calculus shows that, when $m < 1$,

$$I_m = \frac{1}{1-m}(b-a)^{1-m} \tag{7.4.4}$$

and thus the singularity is integrable. When $m \geq 1$, the integral diverges and the singularity is nonintegrable.

Another common class of improper integrals are the integrals

$$I_m = \int_a^b \frac{\ln(x-a)}{(x-a)^m}\, dx \tag{7.4.5}$$

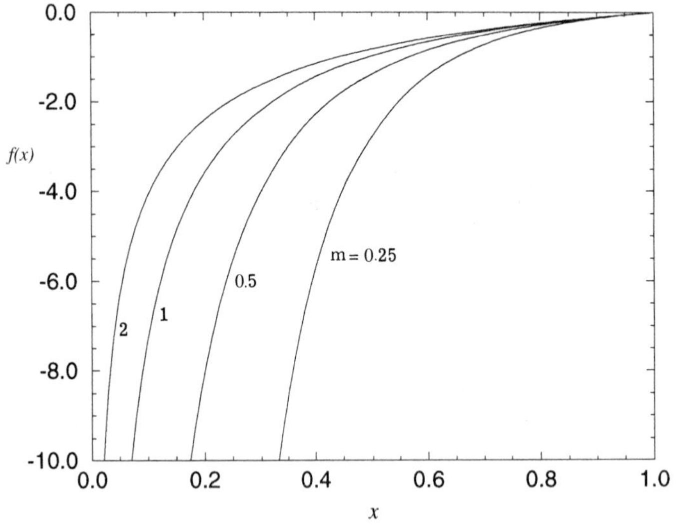

Graphs of the integrand $f(x) = \ln(x-a)/(x-a)^m$ for several values of m and $a = 0$ are shown in the diagram. These integrals can be shown to exist for any value of $m < -1$, but not for $m \geq 1$.

For example, when $m = 0$, we obtain

$$I_0 = \int_a^b \ln(x-a)\, dx = (b-a)\,(\ln(b-a)\; - 1) \tag{7.4.6}$$

Assessing the nature of the singularity

To assess whether the singularity of an integrand is integrable, we expand the nonsingular functions that define the integrand in Taylor series about the singular point, simplify the resulting expressions, and examine the nature of the singularity with reference to prototypical integrals, such as the integrals shown in equations (7.4.3) and (7.4.5).

Consider, for example, the integral

$$\int_0^{\pi/4} \ln(e^x + \cos\, x - 2.52)^2\, dx \tag{7.4.7}$$

whose integrand is plotted with the solid line in the following diagram. The argument of the logarithmic term has a single root at $x = 0.4946$; at that point, the integrand becomes infinite. Expanding the argument in a Taylor series about the root, and retaining the linear term, we find that the integrand behaves like

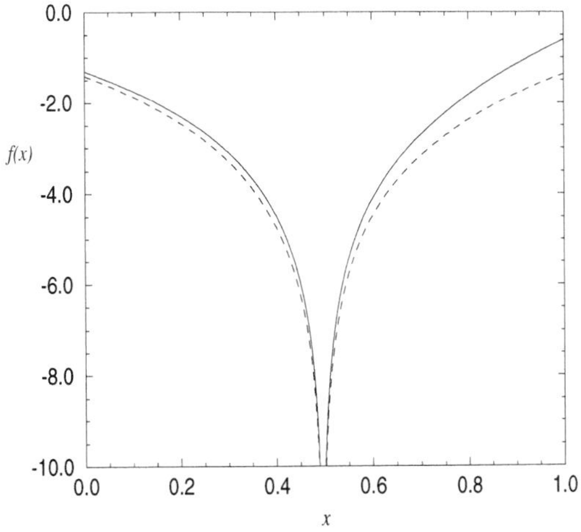

$$\ln(x - 0.4946)^2 = 2\ \ln|x - 0.4946| \tag{7.4.8}$$

The right-hand side is plotted with the dashed line in the last illustration. Equation (7.4.6) indicates that the singularity is integrable and the integral has a finite value.

Dealing with an Integrable Singularity

There are several general-purpose methods of computing integrals whose integrands have integrable singularities. Two simple methods that are *not* recommended are:

- *Play dumb*, that is, *ignore* the singularity; this may lead to gross numerical error.
- *Use a large number of base points* in the vicinity of the singularity. Apart from not being elegant, this method often requires an unaffordable computational cost.

Three recommended methods, to be discussed in the following subsections, are:

- *Desingularize* the integral so that the singularity is either ameliorated or disappears.
- Use a *product–integration rule.*
- Use a *specialized quadrature.*

These methods are also applicable to integrands that are nonsingular but have singular derivatives; examples are the integrals shown in equations (7.2.37) and (7.2.38).

Desingularizing a Singular Integrand

There are several ways of desingularizing a singular integrand, and we illustrate them by examples.

Subtract off the singularity

Considering the integral (7.4.7), we subtract off and then add back the logarithmic singularity to obtain

$$\int_0^{\pi/4} \ln \frac{(e^x + \cos x - 2.52)^2}{(x - 0.4946)^2} \, dx + \int_0^{\pi/4} \ln(x - 0.4946)^2 \, dx \tag{7.4.9}$$

The first integral on the left-hand side is nonsingular and may be computed using a standard numerical method, such as the trapezoidal rule. The second integral may be found exactly by elementary methods.

It is important to note that, even after the singularity has been removed, the derivatives of the regularized integrand might still be singular. The performance of integration rules for such integrals was discussed at the end of Section 7.2. For example, subtracting off the singularity, we obtain

$$\int_0^1 e^x \ln x \, dx = \int_0^1 (e^x - 1) \ln x \, dx + \int_0^1 \ln x \, dx \tag{7.4.10}$$

Close to the origin, the integrand of the first integral on the right-hand side behaves like $x \ln x$; as x tends to zero, its derivatives tend to become singular. The second integral on the right-hand side can be computed analytically using elementary methods.

Integration by parts

Alternatively, we consider the integral (7.4.7), rewrite it as

$$I = \int_0^{\pi/4} \ln\,(e^x + \cos x - 2.52)^2 \, d(x - 0.4946) \tag{7.4.11}$$

and then integrate by parts to obtain

$$\begin{aligned} I = {} & \left[(x - 0.4946)\ \ln(e^x + \cos x - 2.52)^2\right]_0^{\pi/4} \\ & - \int_0^{\pi/4} (x - 0.4946)\, d\ \ln(e^x + \cos x - 2.52)^2 \end{aligned} \tag{7.4.12}$$

The last integral may be recast into the form

$$2 \int_0^{\pi/4} \frac{(x - 0.4946)\,(e^x - \sin x)}{e^x + \cos x - 2.52} \, dx \tag{7.4.13}$$

which has a nonsingular integrand.

Change the variable of integration

A singularity can be made to disappear by stretching the variable of integration by an appropriate factor. For example,

$$\int_0^4 \frac{e^x + \cos x}{\sqrt{x}} \, dx = 2 \int_0^2 (e^x + \cos x) \, d\sqrt{x} = 2 \int_0^2 (e^{y^2} + \cos y^2) \, dy \tag{7.4.14}$$

where we have defined $y = x^{1/2}$. The last integral is nonsingular and may be computed using, for example, a composite Newton–Cotes rule.

The singularities in the derivatives of an integrand may also be eliminated by this method. For example,

$$\int_0^4 e^{-x^6}\sqrt{x}\,dx = \frac{2}{3}\int_0^8 e^{-x^6}\,dx^{3/2} = \frac{2}{3}\int_0^8 e^{-y^4}\,dy \tag{7.4.15}$$

where we have defined $y = x^{3/2}$. The first derivative of the integrand on the left-hand side becomes infinite at the origin, whereas that of the integrand on the right-hand side is equal to zero.

A large class of singularities occurring, for example, at the beginning of the domain of integration, $x = a$, can be eliminated by introducing a new variable of integration t that is related to the primary variable x by the differential equation

$$\frac{dx}{dt} = \exp\left(-\frac{\delta}{t}\right) \tag{7.4.16}$$

where δ is a positive constant, and $x = a$ when $t = 0$. Integrating, we obtain the nonlinear transformation

$$x - a = \int_0^t \exp\left(-\frac{\delta}{u}\right)\,du \tag{7.4.17}$$

Changing the variable of integration from x to t eliminates the singular behavior. The method is developed and discussed in a more general context by Iri et al. (1987).

Isolate the singularity and treat it using asymptotic methods

Considering the integral on the left-hand side of equation (7.4.14), we break up the domain of integration into two parts, $[0, \varepsilon]$ and $[\varepsilon, 4]$, where ε is a small positive number. The integral over the second subdomain is nonsingular and may be computed using a standard method. The integral over the first subdomain is computed using an asymptotic method as

$$\int_0^\varepsilon \frac{e^x + \cos\,x}{\sqrt{x}}\,dx = \int_0^\varepsilon \frac{1}{\sqrt{x}}(2 + x + \ldots)\,dx = 4\sqrt{\varepsilon} + \ldots \tag{7.4.18}$$

PROBLEM

7.4.1 ***Handling a logarithmic singularity.***

Consider the integral

$$\int_0^{\pi/2} \ln\,[\sin\,(2x)\,]\,dx \tag{7.4.19}$$

First, ignore the logarithmic singularity of the integrand, and compute the integral using the eight-point Gauss–Legendre quadrature. Then, subtract off the singularity and integrate it analytically prior to performing numerical integration. Compare and discuss the results of the two methods.

Product–Integration Rules

The three distinguishing steps involved in deriving these rules are:

1. Identify the nature of the singularity $w(x)$; in the neighborhood of the singular point, the integrand $f(x)$ behaves like the elementary singular function $w(x)$.
2. Set $f(x) = g(x)w(x)$, where $g(x)$ is a nonsingular function, and approximate $g(x)$ either with a global interpolating polynomial or with a collection of local interpolating polynomials. For example, for the integral shown on the left-hand side of equation (7.4.10), $w(x) = \ln x$ and $g(x) = e^x$.
3. Integrate the approximated product $g(x)w(x)$ by analytical means.

Composite product–trapezoidal rule

With reference to step 3, if we approximate the function $g(x)$ with a collection of N linear local interpolating polynomials defined by the $N+1$ base points x_i, $i = 1, \ldots, N+1$, we shall obtain the *composite product–trapezoidal rule*

$$I^{PR-TR} = \sum_{i=1}^{N} \int_{x_i}^{x_{i+1}} \left(g(x_i) \frac{x - x_{i+1}}{x_i - x_{i+1}} + g(x_{i+1}) \frac{x - x_i}{x_{i+1} - x_i} \right) w(x)\, dx = \sum_{i+1}^{N+1} g(x_i)\, w_i \qquad (7.4.20)$$

The weights w_i are given by

$$w_1 = -\frac{1}{h_1} \int_{x_1}^{x_2} (x - x_2)\, w(x)\, dx$$

$$w_j = \frac{1}{h_{j-1}} \int_{x_{j-1}}^{x_j} (x - x_{j-1})\, w(x)\, dx - \frac{1}{h_j} \int_{x_j}^{x_{j+1}} (x - x_j)\, w(x)\, dx \qquad \text{for } j = 2, \ldots, N \qquad (7.4.21)$$

$$w_{N+1} = \frac{1}{h_N} \int_{x_N}^{x_{N+1}} (x - x_N)\, w(x)\, dx$$

where $h_i = x_{i+1} - x_i$.

For example, when $w(x) = \ln |x - c|$, where $a \le c \le b$, we find

$$w_1 = \tfrac{1}{2} h_1 \left(J(A) + \ln\, h_1 \right) \qquad (7.4.22)$$

where $A = (x_1 - c)/h_1$, and where

$$J(A) \equiv \int_0^1 (1 - u)\, \ln(A + u)^2\, du \qquad (7.4.23)$$

The integral on the right-hand side may be computed exactly by elementary methods.

PROBLEM

7.4.2 *Product–integration rules.*

Compute the integral shown on the left-hand side of equation (7.4.10), accurate to the third decimal place, using the trapezoidal Newton–Cotes product–integration method. How many intervals do you require?

Gauss–Chebyshev Quadrature for Integrands with a $(x-b)^{-1/2}$ Singularity

Consider an integrand $f(x)$ that has an integrable singularity at the right end-point $x = b$, behaving like $(x-b)^{-1/2}$, but is otherwise nonsingular.

As a preliminary, we recast the integral into a canonical form in terms of the new independent variable t as shown in equation (7.3.3), with the singularity occurring at $t = 1$. Without compromising generality, we write

$$\int_a^b f(x)\,dx = \frac{b-a}{2}\int_{-1}^{1} h(t)\,dt = \frac{b-a}{2}\int_{-1}^{1}\frac{g(t)}{\sqrt{1-t}}\,dt = \frac{b-a}{2}\int_{-1}^{1}\frac{q(t)}{\sqrt{1-t^2}}\,dt \tag{7.4.24}$$

where we have defined

$$g(t) = \sqrt{1-t}\,h(t), \qquad q(t) \equiv \sqrt{1+t}\,g(t) \tag{7.4.25}$$

The functions $g(t)$ and $q(t)$ are nonsingular over the whole of the domain of integration $[-1, 1]$.

For example,

$$\int_2^7 \frac{e^x}{\sqrt{7-x}}\,dx = \tfrac{5}{2}\int_{-1}^{1}\frac{\exp\left(\frac{9+5t}{2}\right)}{\left[\frac{5}{2}(1-t)\right]^{1/2}}\,dt = \tfrac{5}{2}\int_{-1}^{1}\frac{\sqrt{1+t}\,\exp\left(\frac{9+5t}{2}\right)}{\left[\frac{5}{2}(1-t^2)\right]^{1/2}}\,dt \tag{7.4.26}$$

In this case

$$h(t) = \left(\frac{2}{5}\right)^{1/2}\frac{\exp\left(\frac{9+5t}{2}\right)}{\sqrt{1-t}}, \qquad g(t) = \left(\frac{2}{5}\right)^{1/2}\exp\left(\frac{9+5t}{2}\right),$$
$$q(t) = \left(\frac{2}{5}\right)^{1/2}\sqrt{1+t}\,\exp\left(\frac{9+5t}{2}\right) \tag{7.4.27}$$

If the singularity occurs at the beginning of the integration domain at $x = a$, with the integrand behaving like $(x-a)^{-1/2}$, we work in a similar manner to derive the nonsingular function $q(t)$. If the singularity occurs in the middle of the interval, we break up the integral into two parts and consider each individual part.

Similar arrangements can be made to accommodate integrals that are singular at both ends $x = a$ and b, behaving like $(x-a)^{-1/2}\,(x-b)^{-1/2}$. For example,

$$\int_2^7 \frac{e^x}{\sqrt{(x-2)\,(7-x)}}\,dx = \tfrac{5}{2}\int_{-1}^{1}\frac{\exp\left(\frac{9+5t}{2}\right)}{\frac{5}{2}\sqrt{1-t^2}}\,dt \tag{7.4.28}$$

Returning to the right-hand side of equation (7.4.24), we approximate the function $q(t)$ with an interpolating polynomial that passes through the $N+1$ data points and compute the last integral analytically to obtain a Gaussian quadrature.

But we want to do better than that: We want to find the optimal distribution of the $N+1$ base points, so that the quadrature produces that exact answer when $q(t)$ is a polynomial of degree $2N+1$ or less. Working as in Section 7.3 in deriving the Gauss–Legendre quadrature, which is applicable to nonsingular integrands, we find that the optimal abscissas t_i are the zeros of the $(N+1)$ degree Chebyshev polynomial $T_{N+1}(t)$, given by

$$t_i = \cos\left(\frac{\left(i - \frac{1}{2}\right)\pi}{N+1}\right) \tag{7.4.29}$$

where $i = 1, \ldots, N+1$ (Table B.3, Appendix B). The corresponding weights turn out to be constant, equal to

$$w_i = \frac{\pi}{N+1} \tag{7.4.30}$$

The Gauss–Chebyshev quadrature is then

$$\int_{-1}^{1} \frac{q(t)}{\sqrt{1-t^2}} dt \cong \frac{\pi}{N+1} \sum_{i=1}^{N+1} q(t_i) \tag{7.4.31}$$

The simple change of variables $t = \cos\,\theta$ transforms the integral on the left-hand side of equation (7.4.31) into the integral

$$\int_0^{\pi} q(\cos\,\theta)\, d\theta \tag{7.4.32}$$

In this light, the quadrature (7.4.31) is equivalent to the composite midpoint rule applied to the last integral on the right-hand side of equation (7.4.32). This is another case where an even distribution of base points is the best choice. A previously discussed case concerned the computation of regular integrals with periodic integrands.

Error of the Gauss–Chebyshev quadrature

It can be shown that the numerical error associated with the Gauss–Chebyshev quadrature (7.4.31) is given by

$$E^{GC} = -\frac{2\,\pi}{2^{N+2}(2N+2)!}\, q^{(2N+2)}(\xi) \tag{7.4.33}$$

where the point ξ lies somewhere within the domain of integration.

Quadratures for Integrands with Singularities of Various Sorts

Integrands with other types of integrable singularities and nonsingular integrands with singular derivatives can be handled in a similar fashion. We begin by introducing the new independent variable t and associated family of orthogonal polynomials $p_i(t)$, defined over the interval $[c, d]$, whose weighting function $w(t)$ corresponds to the functional form of the singularity (Appendix B). Under the linear transformation (6.4.5), the integrand $f(x)$ becomes a function of t, $f(x) = f(q(t)) = h(t)$, and we write

$$I = \int_a^b f(x)\, dx = \frac{b-a}{d-c} \int_c^d h(t)\, dt \tag{7.4.34}$$

The problem is reduced to evaluating the integral on the right-hand side, which is done with the quadrature

$$\int_c^d h(t)\, dt = \int_c^d q(t)\, w(t)\, dt \cong \sum_{i=1}^{N+1} q(t_i)\, w_i \tag{7.4.35}$$

The modulating function

$$q(t) = \frac{h(t)}{w(t)} \tag{7.4.36}$$

has no singularities in the interval $[c, d]$. The abscissas t_i are the zeros of the $(N+1)$ degree polynomial $p_{N+1}(t)$. When $q(t)$ is a polynomial of degree $(2N+1)$ or less, the right-hand side of equation (7.4.35) produces that exact answer.

One way of computing the weights w_i is to identify $q(t)$ with the first $N+1$ orthogonal polynomials $p_i(t)$, and then use the quadrature (7.4.35) to obtain the linear system

$$p_i(t_1)\, w_1 + p_i(t_2)\, w_2 + \ldots + p_i(t_{N+1})\, w_{N+1} = \alpha\, \delta_{i,0} \tag{7.4.37}$$

for $i = 0, \ldots, N$, where

$$\alpha = p_0 \int_c^d w(t)\, dt \tag{7.4.38}$$

In practice, the weights are read off mathematical tables. Comprehensive analyses of numerical integration by Gaussian quadratures, and tables of base point abscissas and corresponding weights can be found in the monographs of Krylov (1962), Stroud and Secrest (1966), Engels (1980), and Davis and Rabinowitz (1984). An important special case follows.

Integrand with a logarithmic singularity

Integrands with logarithmic singularities arise in the solution of the Laplace, Poisson, Helmholtz, and related partial differential equations in two-dimensional domains using boundary-integral methods. To derive the associated quadrature, we introduce a family of orthogonal polynomials defined over the interval [0, 1], with weighting function

$$w(t) = -\ln\, t \tag{7.4.39}$$

The associated Gaussian quadrature is

$$-\int_0^1 q(t)\, \ln t\, dt \cong \sum_{i=1}^{N+1} q(t_i)\, w_i \tag{7.4.40}$$

Values of the abscissas and weights for several values of N are given in Table 7.4.1, after Stroud and Secrest (1966). Note that, for each value of N, the sum of the weights is equal to unity, which is necessary if the quadrature is to produce the exact answer when $q(t)$ is constant.

To exemplify the application of the quadrature, consider the integral

Table 7.4.1 Abscissas and weights of the Gaussian quadrature for an integral with a logarithmic singularity, with $N+1$ points,

$$-\int_0^1 q(t)\,\ln t\,dt \cong \sum_{i=1}^{N+1} q(t_i) w_i$$

The function $h(t)$ and its derivatives are assumed to be nonsingular over the whole of the canonical integration domain [0, 1].

One-point formula, $N = 0$	
$t_1 = 0.25$	$w_1 = 1.0$
Two-point formula, $N = 1$	
$t_1 = 0.11200\ 88062$	$w_1 = 0.71853\ 93190$
$t_2 = 0.60227\ 69081$	$w_2 = 0.28146\ 06810$
Three-point formula, $N = 2$	
$t_1 = 0.06389\ 07930$	$w_1 = 0.51340\ 45522$
$t_2 = 0.36899\ 70637$	$w_2 = 0.39198\ 00412$
$t_3 = 0.76688\ 03039$	$w_3 = 0.09461\ 54066$
Four-point formula, $N = 3$	
$t_1 = 0.04144\ 84802$	$w_1 = 0.38346\ 40681$
$t_2 = 0.24527\ 49143$	$w_2 = 0.38687\ 53178$
$t_3 = 0.55616\ 54536$	$w_3 = 0.19043\ 51270$
$t_4 = 0.84898\ 23945$	$w_4 = 0.03922\ 54871$
Five-point formula, $N = 4$	
$t_1 = 0.02913\ 44722$	$w_1 = 0.29789\ 34718$
$t_2 = 0.17397\ 72133$	$w_2 = 0.34977\ 62265$
$t_3 = 0.41170\ 25203$	$w_3 = 0.23448\ 82901$
$t_4 = 0.67731\ 41746$	$w_4 = 0.09893\ 04595$
$t_5 = 0.89477\ 13610$	$w_5 = 0.01891\ 15521$

$$I = \int_0^{\pi/2} \mathrm{e}^x \ln x\,dx \tag{7.4.41}$$

introduce the canonical variable t defined by $x = \pi t/2$, and recast the integral into the form

$$I = \frac{\pi}{2}\left(\int_0^1 \mathrm{e}^{\pi t/2}\ \ln\ t\ dt + \ln\frac{\pi}{2}\int_0^1 \mathrm{e}^{\pi t/2}dt\right) \tag{7.4.42}$$

The second integral on the right-hand side may be computed by elementary methods. The first integral has the standard form shown in equation (7.4.40), provided that we define

$$q(t) \equiv -\mathrm{e}^{\pi t/2} \tag{7.4.43}$$

Applying the two- and three-point quadrature we find the values -1.58167 and -1.58302; the second value is exact to shown accuracy.

PROBLEM

7.4.3 *Quadrature for an integrand with a logarithmic singularity.*

Compute the integral (7.4.19) using an appropriate Gaussian quadrature with two, three, and four points, and discuss the accuracy of your results in each case.

7.5 Integrals Over Infinite Domains

Modifications of the integration rules described in the preceding sections or recasting of the integral are necessary in order to compute integrals over semi-infinite domains corresponding to $a = -\infty$ or $b = \infty$, and integrals over an infinite domain corresponding to $a = -\infty$ and $b = \infty$.

The question of whether these improper integrals are finite can be resolved by preparing the graph of the integrand $f(x)$ in the xy plane, recalling that the integral is equal to the area confined between this graph, the x axis, and the two vertical lines corresponding to the limits of integration, and then rotating the xy plane by 90°, as shown in the first two illustrations of Section 7.4. Regarding y as the independent variable, we obtain a singular integral defined over a finite domain and revert to the discussion of Section 7.4. Working in this manner, we find, for example, that the tail-end of an integral will give a finite contribution provided that $f(x)$ decays at an algebraic rate that is faster than $1/x$.

An integral over a semi-infinite or infinite domain can be computed by three general methods:

- Sensible domain truncation.
- Change of the variable of integration.
- Use of a specialized Gaussian quadrature.

We describe the implementation of these methods in the following three subsections.

Domain Truncation

We can certainly truncate the domain of integration at a finite level, but if the integrand does not decay sufficiently fast, this simplification will introduce a substantial amount of error. The decay of the integrand may be accelerated by subtracting off one or more dominant modes of decay.

Consider, for example, the integral

$$I = \int_1^\infty \frac{\sqrt{x}}{1+x^2}\, dx = \int_1^\infty x^{-3/2} \frac{1}{1+1/x^2}\, dx \tag{7.5.1}$$

For large values of x, the integrand decays like $^{-3/2}$, which is slow. To accelerate the rate of decay, we expand the fraction on the right-hand side in a Maclaurin series with respect to $1/x^2$, and then subtract off and add back the two leading terms to obtain

$$I = \int_1^\infty \left(\frac{\sqrt{x}}{1+x^2} - x^{-3/2} + x^{-7/2} \right) dx + \int_1^\infty (x^{-3/2} - x^{-7/2})\, dx \tag{7.5.2}$$

Since the integrand of the first integral decays at the improved rate $x^{-9/2}$, the upper limit of integration may be truncated at a moderate level. The second integral can be computed analytically using elementary methods.

PROBLEM

7.5.1 *Domain truncation.*

Compute the two integrals on the right-hand side of equation (7.5.2) using a method of your choice, both accurate to the fourth significant figure, and discuss the tolerated level of of domain truncation.

Change of Variables

Another method of computing an integral over a semi-infinite or infinite domain is to compress the x axis using a nonlinear transformation. Consider, for example, the integral of a function $f(x)$ over the semi-infinite interval (a, ∞), and introduce the new independent variable t defined by the transformation

$$t = \frac{x - 2a - \lambda}{x + \lambda} \tag{7.5.3}$$

where λ is a constant and $a + \lambda > 0$, so that the denominator does not vanish when $x > a$. As x increases from a to ∞, t increases from -1 to 1. Solving for x, we find $x = [2a + \lambda(t+1)] \,/\, (1-t)$ and then write

$$I = \int_a^\infty f(x)\, dx = 2(\lambda + a) \int_{-1}^{1} f\left(\frac{2a + \lambda(t+1)}{1-t}\right) \frac{1}{(1-t)^2}\, dt \tag{7.5.4}$$

It appears that we have introduced a strong singularity at $t = 1$, but if the integral has a finite value, then as t approaches 1, the function $f(t)$ will tend to zero fast enough so that the singularity will either disappear or become integrable. The integral on the right-hand side of equation (7.5.4) may be computed using, for example, the Gauss–Legendre quadrature, leading to the so-called *Gauss–rational* rule.

Other transformation rules used in practice are $x = a - \ln t$ and $x = a + \tanh(t/2)$ (e.g., Murota and Iri 1982). Note that the second transformation does not make the interval of integration finite but causes instead the integrand to decay at a much faster rate.

PROBLEM

7.5.2 *Infinite domain of integration.*

Compute the following integral, accurate to the fourth significant figure, using a method of your choice:

$$I = \int_0^\infty \frac{\sqrt{x}}{1 + \frac{1}{12}x^2 + \left(1 + \frac{1}{12}x^2\right)^{1/2}}\, dx \tag{7.5.5}$$

This integral arises in the theory of heat or mass transport from a small particle suspended in a steady simple shear flow.

Gauss–Laguerre quadrature for an Exponential Decaying Integrand

Consider now an intregral with respect to the independent variable t over the semi-infinite domain $[0, \infty)$. As t tends to infinity, the integrand $h(t)$ is assumed to decay at an exponential rate, behaving like

$$h(t) \approx q(t)\, \exp(-t) \tag{7.5.6}$$

In this limit, the function $q(t)$ tends to a constant value, increase, or decrease at a rate that is slower than exponential, so that it may be approximated with a polynomial.

Working as in Section 7.4 in deriving the Gauss–Legendre quadrature for a nonsingular integrand over a finite domain, we derive the Gauss–Laguerre quadrature expressed by

$$\int_0^\infty h(t)\, dt \cong \int_0^\infty q(t)\, e^{-t}\, dt = \sum_{i=1}^{N+1} q(t_i)\, w_i \tag{7.5.7}$$

The base points t_i coincide with the zeros of the $(N+1)$ degree Laguerre polynomial $\mathcal{L}_{N+1}(t)$ discussed in Table B.8 of Appendix B. Expressions for the corresponding weights w_i are also included in Table B.8 (see also Problem 7.5.3). Several sets of abscissas and weights are listed in Table 7.5.1, after Stroud and Secrest (1966).

The error associated with the Gauss–Laguerre quadrature is given by

$$E^{GLA} = \frac{[(N+1)!]^2}{[2(N+1)]!}\, q^{(2N+2)}(\xi) \tag{7.5.8}$$

and ξ is a positive number (e.g., Hildebrand 1974, p. 392). Thus when $q(t)$ is a polynomial of degree $2N+1$ or less, the quadrature produces the exact answer.

For example, to compute the integral

$$I = \int_0^\infty (x^4+1)e^{-\sqrt{4x^2+1}}\, dx \tag{7.5.9}$$

we recast it into the canonical form shown on the left-hand side of equation (7.5.7), by defining $t = 2x$ and writing

$$I = \int_0^\infty \tfrac{1}{2}\left(\tfrac{1}{16}t^4+1\right)e^{-\sqrt{t^2+1}}\, dt = \int_0^\infty \left[\tfrac{1}{2}\left(\tfrac{1}{16}t^4+1\right)e^{t-\sqrt{t^2+1}}\right] e^{-t}\, dt \tag{7.5.10}$$

The function $q(t)$ is the expression enclosed by the square brackets on the right-hand side. Applying the two-point, the three-point, and the four-point quadrature we find, respectively, $I = 0.8446, 0.9660, 0.9660$.

PROBLEMS

7.5.3 *Computation of weights.*

One way of computing the Gauss–Laguerre weights is to apply the quadrature $N+1$ times, identifying $q(t)$ with the first $N+1$ Laguerre polynomials $\mathcal{L}_0(t), \ldots, \mathcal{L}_N(t)$. We note that the integral of all but the first of these polynomials is equal to zero, and the integral of $\mathcal{L}_0(t)$ is equal to 1, and thus obtain a system of linear equations for the weights

$$\mathcal{L}_i(t_1)\, w_1 + \mathcal{L}_i(t_2)\, w_2 + \ldots + \mathcal{L}_i(t_{N+1})\, w_{N+1} = \delta_{i,0} \tag{7.5.11}$$

where $i = 0, \ldots, N$. The first equation corresponding to $i = 0$ simply states that the sum of the weights must be equal to unity independent of N.

Compute the Gauss–Laguerre weights for $N = 2$ using the method just described, and make sure that they agree with the values shown in Table 7.5.1.

7.5.4 *A system of two nonlinear equations.*

Solve the following system of algebraic equations for x and y:

$$e^x + y^2 = 10, \qquad e^y + 2x^2 = 18 + \int_0^\infty \frac{1}{1+u^2}\exp(-xu)\, du \tag{7.5.12}$$

Your results should be accurate to the third significant figure.

Table 7.5.1 Abscissas and weights of the *Gauss–Laguerre* quadrature, for an exponentially decaying integrand, with $N+1$ points,

$$\int_0^\infty h(t)\,dt = \int_0^\infty q(t)\,\mathrm{e}^{-t}\,dt \cong \sum_{i=1}^{N+1} q(t_i)\,w_i$$

One-point formula, $N = 0$	
$t_1 = 1.0$	$w_1 = 1.0$
Two-point formula, $N = 1$	
$t_1 = 0.58578\ 64376$	$w_1 = 0.85355\ 33906$
$t_2 = 3.41421\ 3562$	$w_2 = 0.14644\ 66094$
Three-point formula, $N = 2$	
$t_1 = 0.41577\ 45568$	$w_1 = 0.71109\ 30099$
$t_2 = 2.29428\ 0360$	$w_2 = 0.27851\ 77336$
$t_3 = 6.28994\ 5082$	$w_3 = 0.01038\ 92565$
Four-point formula, $N = 3$	
$t_1 = 0.32254\ 76896$	$w_1 = 0.60315\ 41043$
$t_2 = 1.74576\ 1101$	$w_2 = 0.35741\ 86924$
$t_3 = 4.53662\ 0297$	$w_3 = 0.03888\ 79085\ 2$
$t_4 = 9.39507\ 0912$	$w_4 = 0.00053\ 92947\ 056$
Five-point formula, $N = 4$	
$t_1 = 0.26356\ 03197$	$w_1 = 0.52175\ 56106$
$t_2 = 1.41340\ 3059$	$w_2 = 0.39866\ 68111$
$t_3 = 3.59642\ 5771$	$w_3 = 0.07594\ 24496\ 8$
$t_4 = 7.08581\ 0006$	$w_4 = 0.00361\ 17586\ 80$
$t_5 = 12.64080\ 084$	$w_5 = 0.00002\ 33699\ 7239$

Generalized Gauss–Laguerre Quadrature

Consider an integral with respect to the independent variable t over the semi-infinite domain $[0, \infty)$. As t tends to infinity, the integrand $h(t)$ behaves like

$$h(t) \approx q(t)\, t^{\beta} \exp(-t) \tag{7.5.13}$$

where $\beta > -1$. In this limit, the function $q(t)$ is assumed to tend to a constant value, increase or decrease at a rate that is slower than exponential, so that it can be approximated with a polynomial.

Working as in the precrding sections, we derive the generalized Gauss–Laguerre quadrature expressed by the right-hand side of equation (7.5.7). The base points t_i coincide with the zeros of the $(N+1)$ degree generalized Laguerre polynomials $\mathcal{L}_{N+1}^{\beta}(t)$ defined over the interval $[0, \infty)$, When $\beta = 0$, we recover the Gauss–Laguerre quadrature discussed in the preceding subsection. When $q(t)$ is a $2N+1$ or lower-degree polynomial, the quadrature produces the exact answer.

PROBLEMS

7.5.5 *Computation of weights.*

The counterpart of equation (7.5.11) for the generalized Gauss–Laguerre quadrature is

$$\mathcal{L}_i^{\beta}(t_1)\, w_1 + \mathcal{L}_i^{\beta}(t_2)\, w_2 + \ldots + \mathcal{L}_i^{\beta}(t_{N+1})\, w_{N+1} = \Gamma(\beta+1)\delta_{i,0} \tag{7.5.14}$$

where $i = 0, \ldots, N$, and Γ is the Gamma function (e.g., Abramowitz and Stegun 1972).
Compute the generalized Gauss–Laguerre weights for $\beta = 1$ and $N = 2$ using the method described in Problem 7.5.3.

Gauss–Hermite Quadrature for Integrals Decaying at a Gaussian Rate

As a final case, we consider an integral with respect to the independent variable t over the infinite domain $(-\infty, \infty)$. As t tends to $\pm\infty$, the integrand $h(t)$ behaves in a Gaussian manner as

$$h(t) \approx q(t)\ \exp(-t^2) \tag{7.5.15}$$

It is assumed that, in this limit, the function $q(t)$ may be approximated with a polynomial.

Working in the familiar way, we derive the Gauss–Hermite quadrature

$$\int_{-\infty}^{\infty} h(t)\, dt = \int_{-\infty}^{\infty} q(t)\, e^{-t^2}\, dt \cong \sum_{i=1}^{N+1} q(t_i)\, w_i \tag{7.5.16}$$

The base points t_i coincide with the zeros of the $(N + 1)$ degree Hermite polynomial $H_{N+1}(t)$whose properties are listed in Table B.9 (see also Problem 7.5.6). Expressions for the corresponding weights are also in Table B.9 Several sets of abscissas and weights are listed in Table 7.5.2 after Stroud and Secrest (1966). When $q(t)$ is a $2N + 1$ or lower degree polynomial, the quadrature produces the exact answer.

Table 7.5.2 Abscissas and weights of the *Gauss–Hermite* quadrature, for an integrand that decays in a Gaussian-like manner, with $N + 1$ points,

$$\int_{-\infty}^{\infty} h(t)\, dt = \int_{-\infty}^{\infty} q(t)\, e^{-t^2}\, dt \cong \sum_{i=1}^{N+1} q(t_i)\, w_i$$

One-point formula, $N = 0$	
$t_1 = 0$	$w_1 = \sqrt{\pi}$
Two-point formula, $N = 1$	
$t_1 = -0.70710\ 67812$	$w_1 = 0.88622\ 69255$
$t_2 = -t_1$	$w_2 = w_1$
Three-point formula, $N = 2$	
$t_1 = -1.22474\ 4871$	$w_1 = 0.29540\ 89752$
$t_2 = 0$	$w_2 = 1.18163\ 5901$
$t_3 = -t_1$	$w_3 = w_1$
Four-point formula, $N = 3$	
$t_1 = -1.65068\ 0123$	$w_1 = 0.08131\ 28354\ 5$
$t_2 = -0.52464\ 76232$	$w_2 = 0.80491\ 40900$
$t_3 = -t_2$	$w_3 = w_2$
$t_4 = -t_1$	$w_4 = w_1$
Five-point formula, $N = 4$	
$t_1 = -2.02018\ 2870$	$w_1 = 0.01995324206$
$t_2 = -0.95857\ 24646$	$w_2 = 0.3936193232$
$t_3 = 0$	$w_3 = 0.9453087205$
$t_4 = -t_2$	$w_4 = w_2$
$t_5 = -t_1$	$w_5 = w_1$

PROBLEMS

7.5.6 ***Computation of weights.***

The counterpart of equation (7.5.11) for the Gauss–Hermite quadrature is

$$H_i(t_1)\, w_i + H_i(t_2)\, w_2 + \ldots + H_i(t_{N+1})\, w_{N+1} = \pi^{1/2}\, \delta_{i,0} \tag{7.5.17}$$

The first equation corresponding to $i = 0$ shows that the sum of the weights are equal to $\pi^{1/2}$ independently of the value of N. Compute the Gauss–Hermite weights for $N = 2$ using the method described in Problem 7.5.3.

7.5.7 ***Temperature field due to a steady point source of heat.***

Consider a two-dimensional point source of heat of constant strength equal to α, located at the position $\boldsymbol{x}_0$, generating a steady temperature field in the presence of a two-dimensional flow, as described in Problem 1.2.2. In appropriate dimensionless units, the temperature field is given by

$$T(\boldsymbol{x}, \boldsymbol{x}_0) = \alpha \int_0^\infty G(\boldsymbol{x}, \tau;\ \boldsymbol{x}_0)\, d\tau \tag{7.5.18}$$

where G is the Green's function given in equation (1.2.3).
(a) Write a program that computes the temperature at a specified point $\boldsymbol{x}$ using the Gauss–Hermite quadrature.
(b) Consider uniform convection corresponding to $k = 0$ but $U \neq 0$, and prepare contour plots of the Green's function. Discuss the physical significance of your results.
(c) Repeat part (b) for convection in a symmetric shear flow corresponding to $U = -ky_0$.

7.6 *Rapidly Varying and Oscillatory Integrands*

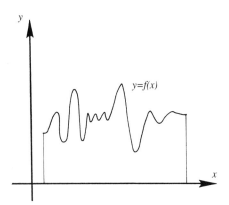

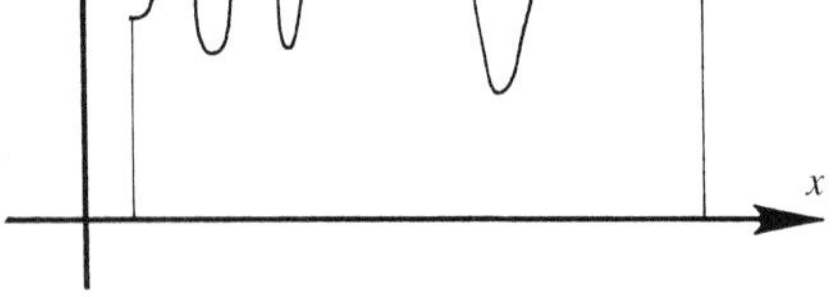

When the integrand $f(x)$ exhibits pronounced variations at certain regions between the two limits of integration a and b, common sense suggests breaking up the domain of integration into a number of subdomains and carrying out the individual integrations using the same or different methods. A large number of base points should be used at the subdomains that support the pronounced variations. The methodology of breaking up the domain of integration, and adjusting the number of base points according to some error estimate, is called *adaptive integration.*

Reformulation as an Ordinary Differential Equation

The problem of computing the definite integral (7.1.1) is intimately related to the problem of solving the first-order ordinary differential equation

$$\frac{dy}{dx} = f(x) \tag{7.6.1}$$

subject to the initial condition $y(x = a) = 0$. The required value is $I = y(x = b)$.

Computing I is thus reduced to integrating equation (7.6.1) from the departure point $x = a$ to the arrival point $x = b$, which can be done using the adaptive methods for solving initial-value problems involving ordinary differential equations discussed in Chapter 9. This problem recasting is designated for integrals whose integrands exhibit strong variations.

Oscillatory Integrands

The Fourier decomposition or transform of a function, their inverses, and the mathematical modeling of physical processes involving wave diffraction and particle motions in quantum mechanics, are examples of processes that produce oscillatory integrals of the type

$$I^c = \int_a^b q(x)\,\cos\,kx\,dx, \qquad I^s = \int_a^b q(x)\,\sin\,kx\,dx \tag{7.6.2}$$

where k is the wave number, and $q(x)$ is a smooth modulating function. There are other classes of integrals involving integrands in the form of the product of a smooth and an oscillatory function, such as a trigonometric or a Bessel function.

Product–integration rules for handling integrals with oscillatory integrands over finite or infinite domains are reviewed by Evans (1993, Chapters 3 and 4). The Filon integration method, described next, is an example.

Filon algorithm

The Filon algorithm is used to compute the integrals (7.6.2) by means of a Simpson-type product–integration rule (e.g., Hildebrand 1974, pp. 107–110). The method is particularly effective for large values of the dimensionless wave number $k(b - a)$.

We begin by discretizing the interval $[a, b]$ into an *even number* of N subintervals separated by $N + 1$ evenly spaced base points, and compute the cosine sums,

$$\begin{aligned} C_1 &= \tfrac{1}{2}q_1\cos\,kx_1 + q_3\,\cos\,kx_3 + q_5\,\cos\,kx_5 + \ldots + q_{N-1}\cos\,kx_{N-1} + \tfrac{1}{2}q_{N+1}\,\cos\,kx_{N+1} \\ C_2 &= q_2\,\cos\,kx_2 + q_4\,\cos\,kx_4 + \ldots + q_{N-2}\cos\,kx_{N-2} + q_N\,\cos\,kx_N \end{aligned} \tag{7.6.3}$$

as well as corresponding sine sums denoted by S_1 and S_2; we have defined $q(x_i) = q_i$. These sums express, respectively, the trapezoidal and midpoint approximations to I^C and I^S with point separation equal to $2h$, where $h = (b - a)/N$ is the spacing between the base points.

Next, we compute the quantities

$$\begin{aligned} w^s &= q_{N+1}\sin\,kx_{N+1} - q_1\sin\,kx_1 \\ w^C &= q_{N+1}\cos\,kx_{N+1} - q_1\cos\,kx_1 \end{aligned} \tag{7.6.4}$$

and then the quantities

$$\begin{aligned} \alpha(u) &= \frac{1}{u} + \tfrac{1}{2}\frac{\sin\,2u}{u^2} - 2\,\frac{\sin^2\,u}{u^3} \\ \beta(u) &= 2\,\frac{1+\cos^2\,u}{u^2} - 2\,\frac{\sin\,2u}{u^3} \\ \gamma(u) &= 4\,\frac{\sin\,u}{u^3} - 4\,\frac{\cos\,u}{u^2} \end{aligned} \tag{7.6.5}$$

where $u = kh$. The Filon integration rule is

$$I^C = h(\alpha w^S + \beta C_1 + \gamma C_2)$$
$$I^S = h(-\alpha w^C + \beta S_1 + \gamma S_2) \qquad (7.6.6)$$

When $q(x)$ is a constant, linear, or quadratic polynomial, the Filon rule produces the exact answer. More generally, the error is on the order of h^4.

The programmable extension of the Filon algorithm just described to an advanced rule with higher–order accuracy is discussed by Evans (1993, pp. 81–88).

7.7 *Area, Surface, and Multidimensional Integrals*

The design of numerical methods for computing the areal integral of a function of two variables $f(x, y)$ over a two-dimensional domain R in the xy plane,

$$I = \int_R f(x,\ y)\, dA \qquad (7.7.1)$$

is guided by two factors: The geometry of the domain of integration, and the smoothness of the integrand $f(x, y)$.

A domain with a complicated geometry is usually broken up into a number of subdomains with simpler geometry, also called surface elements, and the integral over each subdomain is evaluated using a custom-made numerical method.

Product–integration Rules

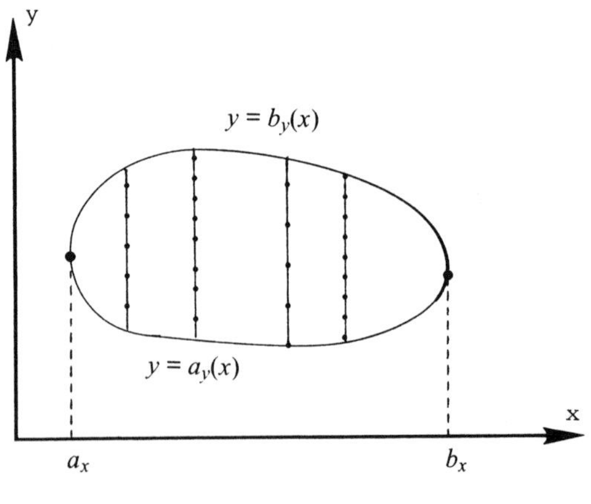

Let us assume that the boundary of the integration domain is confined between the graphs of the functions $y = a_y(x)$ and $y = b_y(x)$ for $a_x < x < b_x$. To compute the integral (7.7.1), we express the differential area as $dA = dx\, dy$, and recast the integral into the form

$$I = \int_{a_x}^{b_x} g(x)\, dx \qquad (7.7.2)$$

where

$$g(x) = \int_{a_y(x)}^{b_y(x)} f(x,\ y)\, dy \qquad (7.7.3)$$

Applying two generally independent numerical methods to compute the last two integrals provides us with product–integration rules of various sorts.

For example, when the integrand $f(x, y)$ is nonsingular, we can use the trapezoidal rule to integrate in the x direction, with N_x intervals of equal size $h_x = (b_x - a_x)/N_x$ spanning the range $a_x \le x \le b_x$, and thus obtain

$$I \cong \frac{b_x - a_x}{N_x} \sum_{i=1}^{N_x+1} u_i \, g(x_i) \tag{7.7.4}$$

The x base lines are described by $x_i = a_x + (i-1)h_x$, and all weights u_i are equal to 1, except that $u_1 = u_{N_x+1} = \frac{1}{2}$. The function $g(x)$ may be computed using one of the numerical methods discussed in the preceding sections. To reduce the number of function evaluations, we may set the number of base points in the y direction roughly proportional to the width of the integration domain $b_y(x) - a_y(x)$.

Integration Over a Rectangular Area

Consider a rectangular domain of integration confined between $a_x < x < b_x$ and $a_y < y < b_y$. Since the width of the y interval, $b_y(x) - a_y(x)$, is constant, it is reasonable to use a number of y base points that are independent of x. The result is a two-dimensional open or closed composite Newton–Cotes rule, or a Gaussian quadrature of a certain type.

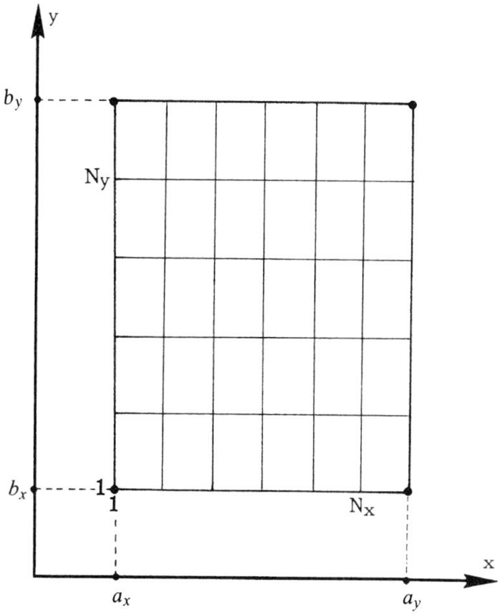

Trapezoidal product rule

When the integrand is nonsingular, we use the trapezoidal rule with evenly spaced points for integration in each direction, and thereby obtain the *two-dimensional composite trapezoidal product rule*

$$I = \frac{(b_x - a_x)\,(b_y - a_y)}{N_x \, N_y} \sum_{i=1}^{N_x+1} \sum_{j=1}^{N_y+1} w_{i,j} \; f\,(x_i, \; y_j) \tag{7.7.5}$$

where N_x and N_y are, respectively, the number of divisions in the x and y directions. The base points are located at

$$x_i = a_x + (i-1)h_x, \qquad y_j = a_y + (j-1)h_y$$

where

$$h_x = (b_x - a_x)/N_x, \qquad h_y = (b_y - a_y)/N_y$$

The weights are given by

$$w_{i,j} = u_i \, v_j \tag{7.7.6}$$

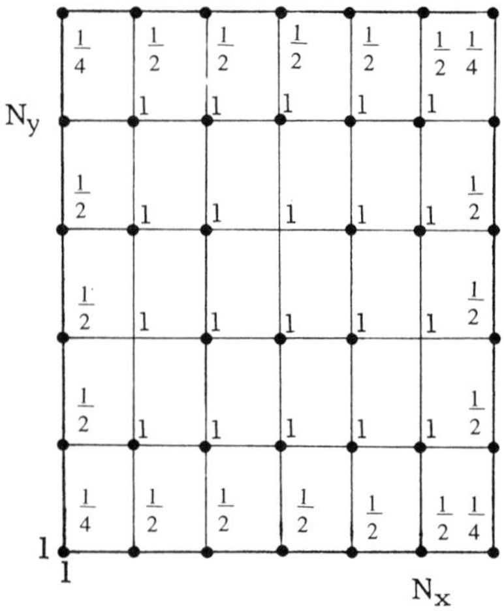

where all u_i and v_j equal 1, except for u_1, u_{N_x+1}, v_1, and v_{N_y+1} that are equal to $\frac{1}{2}$. Thus all elements of the $(N_x+1)\times(N_y+1)$ matrix $w_{i,j}$ are equal to 1, except for the bordering off-corner elements that are equal to $\frac{1}{2}$, and the four corner elements that are equal to $\frac{1}{4}$, as shown in the illustration.

Simpson's product rule

Using Simpson's 1/3 rule with evenly spaced points to integrate in both directions, we obtain a two-dimensional composite rule expressed by the right-hand side of equation (7.7.5); in this case, N_x and N_y are assumed to be even. The weights are given by equation (7.7.6) where

$$\boldsymbol{u} = \tfrac{1}{3}(1, 4, 2, 4, \ldots, 4, 1)^T, \qquad \boldsymbol{v} = \tfrac{1}{3}(1, 4, 2, 4, \ldots, 4, 1)^T \tag{7.7.7}$$

An example with $N_x = 2$ and $N_y = 2$ is depicted in the first column of Table 7.7.1.

Gauss–Legendre product–integration quadratures

We continue to assume that the function $f(x, y)$ is free of singularities in the domain of integration and apply the Gauss–Legendre quadrature to integrate in each direction. The result is the two-dimensional Gauss–Legendre quadrature

$$I \cong \tfrac{1}{4}(b_x - a_x)\,(b_y - a_y) \sum_{i=1}^{N_x+1} \sum_{j=1}^{N_y+1} f\,(x_i,\; y_j)\, u_i\, v_j \tag{7.7.8}$$

Table 7.7.1 Integration over a planar rectangle. The location of the base points and the corresponding weights for the Simpson product rule and for the quadrature (7.7.8).

$$\frac{1}{4h^2}\int_S\int f(x,y)\,dx\,dy = \sum_{i=1}^{N} w_i\, f(x_i, y_i) + R$$

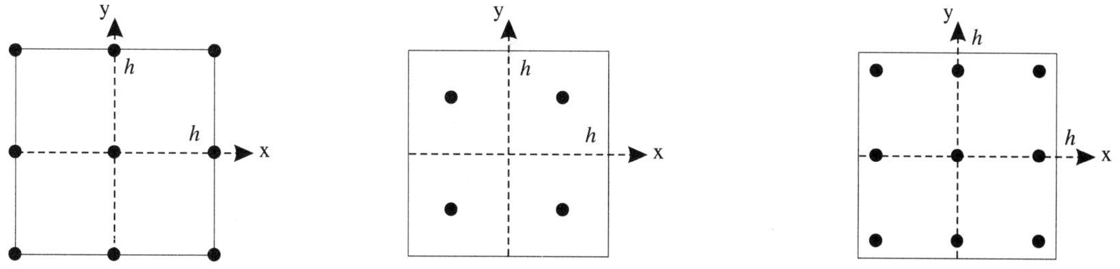

(x_i, y_i)	w_i	(x_i, y_i)	w_i	(x_i, y_i)	w_i
$(0, 0)$	4/9	$\left(\pm h\sqrt{\frac{1}{3}}, \pm h\sqrt{\frac{1}{3}}\right)$	1/4	$(0, 0)$	16/81
$(\pm h, \pm h)$	1/36			$\left(\pm h\sqrt{\frac{3}{5}}, \pm h\sqrt{\frac{3}{5}}\right)$	25/324
$(\pm h, 0)$	1/9			$\left(0, \pm h\sqrt{\frac{3}{5}}\right)$	10/81
$(0, \pm h)$	1/9			$\left(\pm h\sqrt{\frac{3}{5}}, 0\right)$	10/81
$R = O(h^4)$		$R = O(h^4)$		$R = O(h^6)$	

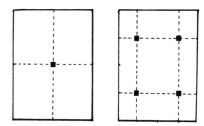

The base points x_i, y_j correspond to the zeros of the $(N_x + 1)$ and $(N_y + 1)$ degree Legendre polynomials, and $w_{i,j} \equiv u_i v_j$ are the corresponding weights. The location of the base points for $N_x = N_y = 1$ and $N_x = N_y = 2$ are shown in the diagram. The weights for $N_x = N_y = 2$ and $N_x = N_y = 3$ are listed in the second and third columns of Table 7.7.1.

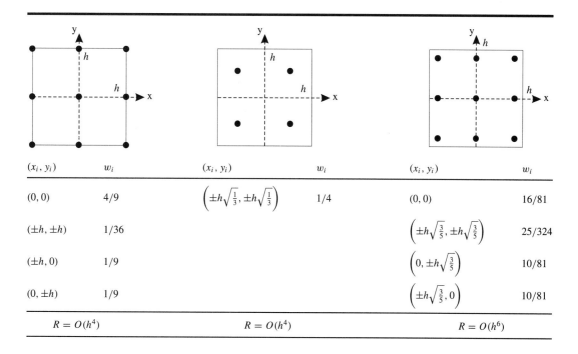

PROBLEMS

7.7.1 *Area integrals.*

Consider the integral

$$I = \int (x^2 + y^2)\, dA \tag{7.7.9}$$

over the rectangular area $0 < x < 1, 0 < y < 2$. Compute the integral (*a*) exactly, (*b*) using the trapezoidal rule for both the x and y integration with two intervals in each variable, (*c*) using the two-point Gauss–Legendre quadrature for both the x and y integration. Discuss the accuracy of the numerical results.

7.7.2 ***Gauss–Legendre–Chebyshev quadrature.***

Assume that the singular integrand $f(x, y)$ can be expressed in the form $(x - a_x)^{-1/2} g(x, y)$ where $g(x, y)$ is free of singularities over the rectangular domain of integration, and develop an appropriate Gauss–Legendre–Chebyshev quadrature.

Integral Over a Quadrilateral

To compute an integral over a quadrilateral, we first map it onto a rectangle using the transformation shown in equation (6.9.5), and then apply one of the methods discussed in the preceding section.

Genuine Two-dimensional Quadratures

The high-priced product–integration quadratures provide us with a higher level of accuracy than desired. We want to develop genuinely two-dimensional quadratures that allow us to integrate exactly all monomials $P_{n,m}(x, y) = x^n\, y^m$ of a specified overall order $l = n + m$, where n and m are two integers with $n + m \le l$. The product quadratures arise by combining two one-dimensional quadratures, each one integrating exactly all monomials x^n up to order l, and integrate exactly all monomials of overall order less than or equal to $2l$, which is more than required. For example, when $l = 2$, we want a two-dimensional quadrature that integrates exactly polynomials with (n, m) pairs: (0, 0), (0, 1), (1, 0), (0, 2), (1, 1), (2, 0), a total of six pairs. The product quadrature will integrate exactly all (n, m) polynomials with $n = 0, 1, 2$ and $m = 0, 1, 2$, a total of nine combinations.

The theory of genuine two-dimensional quadratures is discussed in considerable detail by Stroud (1971). More recent contributions are reviewed by Davis and Rabinowitz (1984), Cools and Rabinowitz (1993), and Capstick and Keister (1996). Other more esoteric methods of multiple integration are discussed by Sloan and Joe (1994).

PROBLEM

7.7.3 ***Multiple integration of a function of one variable.***

Prove the identity

$$\int_a^{x_N} \int_a^{x_{N-1}} \cdots \int_a^{x_2} \int_a^{x_1} f(x_0)\, dx_0\, dx_1 \ldots dx_{N-2}\, dx_{N-1} = \frac{(x_N - a)^N}{(N-1)!} \int_0^1 t^{N-1} f\left[a + (x_N - a)t\right] dt \tag{7.7.10}$$

Integration Over a Planar Triangle

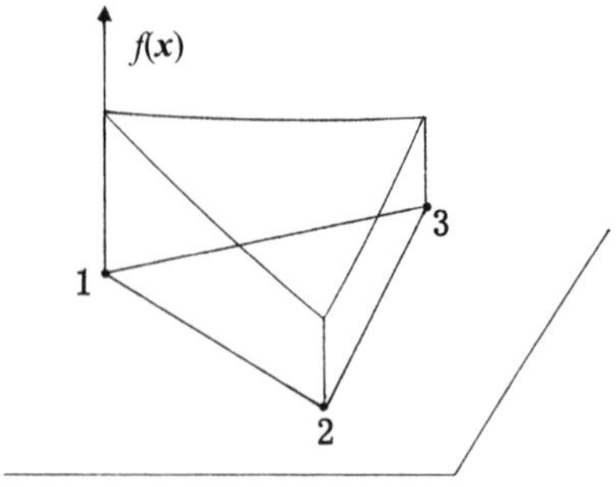

Integrals over triangular domains are encountered in the numerical solution of partial differential equations using finite-element, boundary-element, and related methods.

Consider the integral of a nonsingular function $f(\mathbf{x})$ over a planar triangle defined by its three vertices $\mathbf{x}_1, \mathbf{x}_2$, and $\mathbf{x}_3$. For generality, we allow the vertices and thus the triangle to be lifted off the xy plane into the three-dimensional space. The position vector $\mathbf{x}$ of a point in the triangle may be expressed in terms of the local triangle coordinates ξ and η defined in equations (6.9.8). The parametric representation is given in equation (6.9.9) with $\mathbf{x}$ in place of f.

It is useful to recast the areal integral into the canonical form shown in equation (6.10.8), where $h_S = 2A$, and $A = \frac{1}{2}|(x_2 - x_1)\,(y_3 - y_1) - (x_3 - x_1)\,(y_2 - y_1)|$ is the area of the triangle. We thus write

$$I = \int_{Triangle} f(\mathbf{x})\, dS = 2\, A \int_H f\,(\mathbf{x}(\xi, \eta))\; d\xi\; d\eta \tag{7.7.11}$$

where the parameter area H is the orthosonal isosceles triangle depicted immediately before equation (6.9.9).

The identity

$$\int_H \xi^m \eta^n \zeta^l d\xi\; d\eta = \frac{m!\,n!\,l!}{(m+n+l+2)!} \tag{7.7.12}$$

where m, n, and l are nonnegative integers, and $\zeta = 1 - \xi - \eta$ allows us to evaluate analytically integrals whose integrands vary in a polynomial-like fashion with respect to the triangle coordinates.

An integration rule approximates the integral with a weighted sum of the values of the integrand at N base points that lie within the triangle, in the form

$$I = \int_{Triangle} f(\mathbf{x})\, dS \cong A \sum_{i=1}^{N} f\,(\mathbf{z}_i)\; w_i \tag{7.7.13}$$

where

$$\mathbf{z}_i = \xi_i \mathbf{x}_1 + \eta_i \mathbf{x}_2 + \zeta_i \mathbf{x}_3 \tag{7.7.14}$$

describes the location of the base points, and w_i are the corresponding weights. For the point $\mathbf{z}_i$ to lie in the plane of the triangle, we must insist that $\xi_i + \eta_i + \zeta_i = 1$. The coefficients ξ_i, η_i, and ζ_i and the weights w_i are computed so that the integration rule generates the exact answer for as broad a class of functions as possible.

Centroid rule

$N = 1$ corresponds to one base point located at the centroid of the triangle, with triangle coordinates $\xi_1 = \eta_1 = \zeta_1 = \frac{1}{3}$. The associated weight is $w_1 = 1$, as shown in the first entry of Table 7.7.2.

The centroid rule is the counterpart of the midpoint rule for integration in one dimension. When the function $f(\mathbf{x})$ is constant or varies in a linear manner with respect to distance in the plane of the triangle, the centroid rule produces the exact answer.

Three-point rule

$N = 3$ corresponds to three base points located at the midpoints of the three vertices. The coefficients ξ_i, η_i, and ζ_i and associated weights w_i are given in the second entry of Table 7.7.2. When the function $f(\mathbf{x})$ is constant or varies in a linear or quadratic manner with respect to distance in the plane of the triangle, the three-point rule produces the exact answer.

Another version of the three-point rule with the data points located in the interior of the triangle was discovered by Cowper (1973).

Higher-point rules

The location of the base points for higher-order rules and the corresponding weights are shown in Table

Table 7.7.2 Integration over a planar triangle. The location of the base points and corresponding weights w_i of the quadrature (7.7.13); L is the typical length of a side of the triangle.

Base-point triangle coordinates ξ_i, η_i, ζ_i defined in equation (7.7.14)			Weights w_i	Location of base points
One-point formula with accuracy $O(L^2)$				
$\frac{1}{3}$	$\frac{1}{3}$	$\frac{1}{3}$	1.0	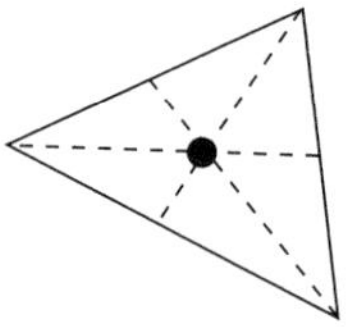
Three-point formula with accuracy $O(L^3)$				
$\frac{1}{2}$	$\frac{1}{2}$	0	$\frac{1}{3}$	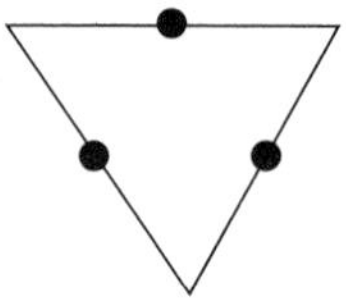
0	$\frac{1}{2}$	$\frac{1}{2}$	$\frac{1}{3}$	
$\frac{1}{2}$	0	$\frac{1}{2}$	$\frac{1}{3}$	
Four-point formula with accuracy $O(L^4)$				
$\frac{1}{3}$	$\frac{1}{3}$	$\frac{1}{3}$	$-\frac{27}{48}$	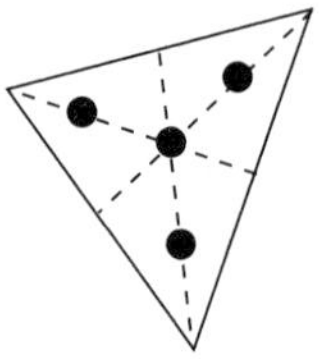
$\frac{1}{5}$	$\frac{1}{5}$	$\frac{3}{5}$	$\frac{25}{48}$	
$\frac{3}{5}$	$\frac{1}{5}$	$\frac{1}{5}$	$\frac{25}{48}$	
$\frac{1}{5}$	$\frac{3}{5}$	$\frac{1}{5}$	$\frac{25}{48}$	
Seven-point formula with accuracy $O(L^4)$				
$\frac{1}{3}$	$\frac{1}{3}$	$\frac{1}{3}$	$\frac{27}{60}$	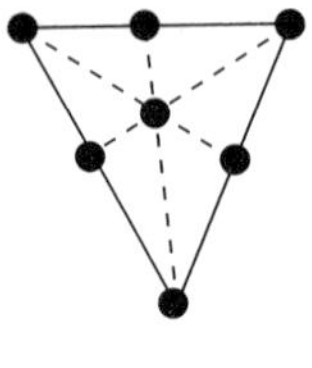
$\frac{1}{2}$	$\frac{1}{2}$	0	$\frac{8}{60}$	
0	$\frac{1}{2}$	$\frac{1}{2}$	$\frac{8}{60}$	
$\frac{1}{2}$	0	$\frac{1}{2}$	$\frac{8}{60}$	
1	0	0	$\frac{3}{60}$	
0	1	0	$\frac{3}{60}$	
0	0	1	$\frac{3}{60}$	
Seven-point formula with accuracy $O(L^6)$				
$\frac{1}{3}$	$\frac{1}{3}$	$\frac{1}{3}$	$\frac{9}{40}$	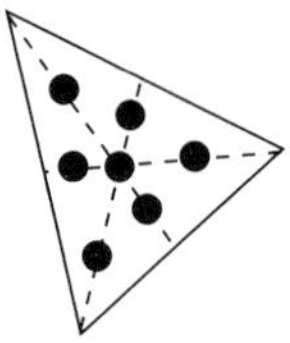
a	b	b	e	
b	a	b	e	
b	b	a	e	
c	d	d	f	
d	c	d	f	
d	d	c	f	

where: $a = 0.05971\,587$, $b = 0.47014\,206$, $c = 0.7974\,2699$, $d = 0.10128\,651$, $e = (155 + 15^{1/2})/1200$, $f = (155 - 15^{1/2})/1200$.

7.7.2 (e.g., Cowper, 1973). Numerous further rules have been developed. A comprehensive review of the literature is given by Cools and Rabinowitz (1993).

Conical product rules

In this method, the integral over the triangular parametric area H is converted into an integral over the unit square S by introducing the new variable of integration χ taking values over [0, 1], defined by

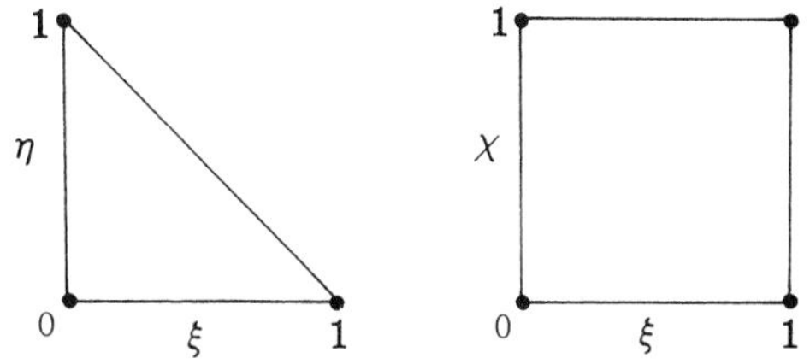

$$\eta = (1 - \xi)\chi \tag{7.7.15}$$

The integral (7.7.11) is expressed as

$$I = \int_{Triangle} f(\boldsymbol{x})\, dS = 2\,A \int_S f(\boldsymbol{x}[\xi, (1-\xi)\chi])\,(1-\xi)\,d\xi\,d\chi \tag{7.7.16}$$

and computed using the rules for integration over rectangular domains discussed in the preceding subsections. If a product rule is selected, then the integral with respect to ξ can be found using a one-dimensional quadrature with weighting function equal to $1 - \xi$.

Singular integrands

In many applications of mathematical physics, we encounter surface integrals with singular but integrable integrands $f(\boldsymbol{x})$ that behave like $1/r$. This means, for example, that as the integration point $\boldsymbol{x}$ approaches the vertex $\boldsymbol{x}_1$, the integrand behaves, to leading order, like $1/|\boldsymbol{x} - \boldsymbol{x}_1|$. The accurate evaluation of such integrals requires special attention.

To formalize the procedures, we introduce the familiar local triangle coordinates ξ and η and write

$$I = \int_{Triangle} f(\boldsymbol{x})\, dS = \int_{Triangle} \frac{g(\boldsymbol{x})}{|\boldsymbol{x} - \boldsymbol{x}_1|}\, dS = 2\,A \int_H \frac{g(\boldsymbol{x})}{|\boldsymbol{x} - \boldsymbol{x}_1|}\, d\xi\, d\eta \tag{7.7.17}$$

where the function $g(\boldsymbol{x})$ is nonsingular over the area of the triangle, including the sides. Using equation (6.9.9) with $\boldsymbol{x}$ in place of f, we write

$$\boldsymbol{x} - \boldsymbol{x}_1 = -(\xi + \eta)\,\boldsymbol{x}_1 + \xi \boldsymbol{x}_2 + \eta \boldsymbol{x}_3 = \xi\,(\boldsymbol{x}_2 - \boldsymbol{x}_1) + \eta\,(\boldsymbol{x}_3 - \boldsymbol{x}_1) \tag{7.7.18}$$

which can be used to express I in the form

$$I = 2\frac{A}{a} \int_H \frac{g(\xi, \eta)}{(\xi^2 + 2B\xi\eta + C\eta^2)^{1/2}}\, d\xi\, d\eta \tag{7.7.19}$$

where

$$B = b/a^2, \qquad C = c^2/a^2 \tag{7.7.20}$$

and

$$a = |\boldsymbol{x}_2 - \boldsymbol{x}_1|, \qquad b = (\boldsymbol{x}_2 - \boldsymbol{x}_1)^T\ (\boldsymbol{x}_3 - \boldsymbol{x}_1). \qquad c = |\boldsymbol{x}_3 - \boldsymbol{x}_1| \tag{7.7.21}$$

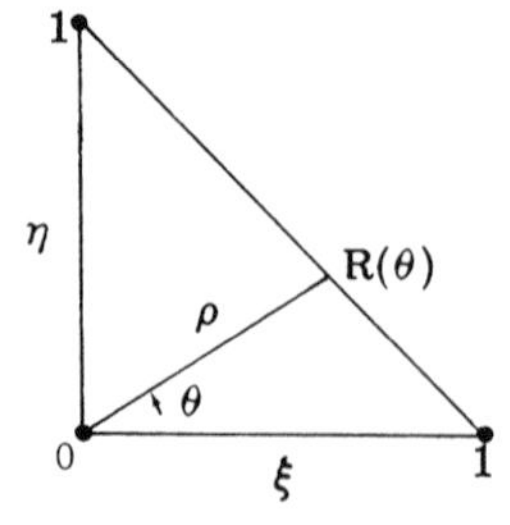

The idea behind the *polar integral rule* is to introduce plane polar coordinates (ρ, θ) corresponding to the ξ and η axes, so that

$$\xi = \rho\ \cos\ \theta, \qquad \eta = \rho\ \sin\ \theta, \qquad d\xi\ d\eta = \rho\ d\rho\ d\theta \tag{7.7.22}$$

We note that ρ varies between 0 and $R(\theta) = 1/(\cos\theta + \sin\theta)$, and we write

$$I = 2\frac{A}{a}\int_0^{\pi/2} \frac{1}{(\cos^2\ \theta + B\ \sin\ 2\,\theta + C\ \sin^2\ \theta)^{1/2}} \int_0^{R(\theta)} g(\rho,\ \theta)\ d\rho\ d\theta \tag{7.7.23}$$

The two-dimensional integral on the right-hand side is nonsingular and may be computed using one of the numerical methods discussed previously in this section.

Integration Over a Curved Triangle

The integral over a curved triangle whose sides are defined by six nodes can be expressed in a parametric manner with respect to the triangle coordinates ξ and η, as

$$I = \int_{Triangle} f(\boldsymbol{x})\ dS = \int_H f(\boldsymbol{x}(\xi, \eta))\ h_s\ d\xi\ d\eta \tag{7.7.24}$$

where $\boldsymbol{x}(\xi, \eta)$ is given by the quadratic interpolation formulas (6.9.12)–(6.9.14) with $\boldsymbol{x}$ in place of f; the quantity h_S is defined in equation (6.10.6). The parameters ξ and η vary over the planar isosceles orthogonal triangle shown immediately before equation (6.9.9).

An integrand with a $1/r$ singularity may be computed by the methods discussed in the preceding subsection.

Further Integration Rules

Quadratures for integrating functions over two-dimensional, three-dimensional, and higher-dimensional domains with various shapes were derived by Stroud (1971) and reviewed by Cools and Rabonowitz (1993). Some examples are collected in Tables 7.7.3–7.7.8 after Stroud (1971) and Abramowitz and Stegun (1972).

Table 7.7.3 Quadratures for integrating functions over domains with various shapes, after Stroud (1971) and Abramowitz and Stegun (1972). D: disk of radius h.

$$\frac{1}{\pi h^2} \iint_D f(x, y)\, dx\, dy = \sum_{i=1}^{N} w_i,\ f(x_i, y_i) + R$$

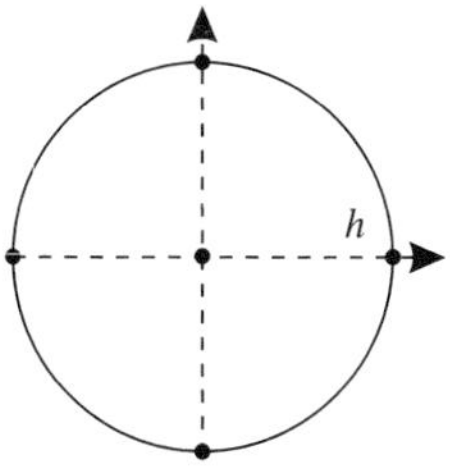

(x_i, y_i)	w_i
$(0, 0)$	$1/2$
$(\pm h, 0), (0, \pm h)$	$1/8$
$R = O(h^4)$	

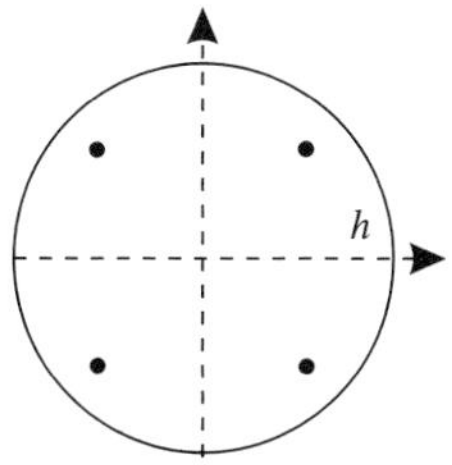

(x_i, y_i)	w_i
$\left(\pm\frac{h}{2}, \pm\frac{h}{2}\right)$	$1/4$
$R = O(h^4)$	

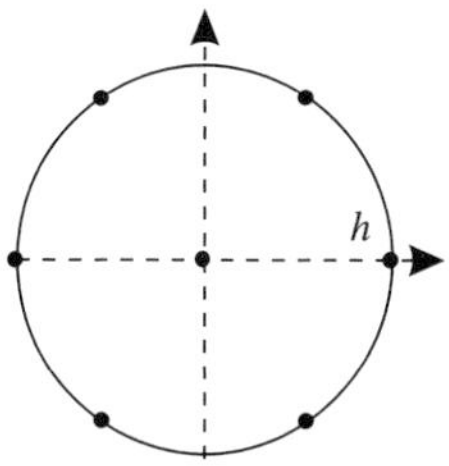

(x_i, y_i)	w_i
$(0, 0)$	$1/2$
$(\pm h, 0)$	$1/12$
$\left(\pm\frac{h}{2}, \pm\frac{h}{2}\sqrt{3}\right)$	$1/12$
$R = O(h^4)$	

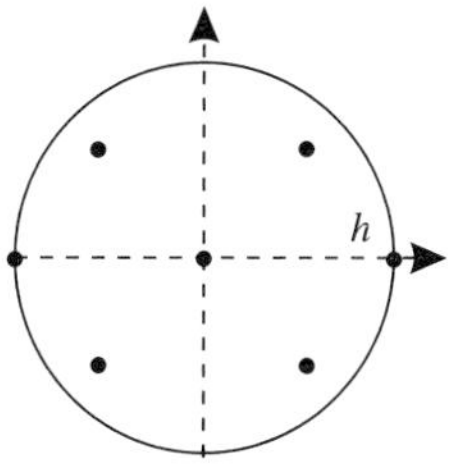

(x_i, y_i)	w_i
$(0, 0)$	$1/6$
$(\pm h, 0)$	$1/24$
$(0, \pm h)$	$1/24$
$\left(\pm\frac{h}{2}, \pm\frac{h}{2}\right)$	$1/6$
$R = O(h^{1}0)$	

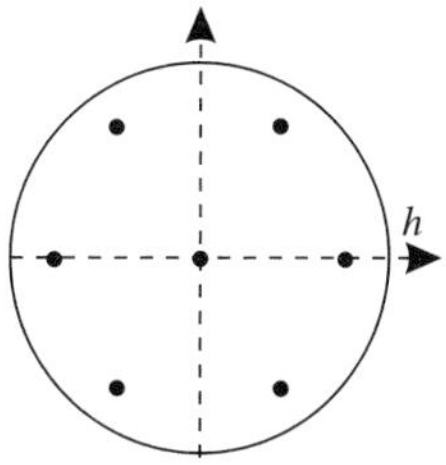

(x_i, y_i)	w_i
$(0, 0)$	$1/4$
$\left(\pm\sqrt{\frac{2}{3}}h, 0\right)$	$1/8$
$\left(\pm\sqrt{\frac{1}{6}}h, \pm\frac{h}{2}\sqrt{2}\right)$	$1/8$
$R = O(h^6)$	

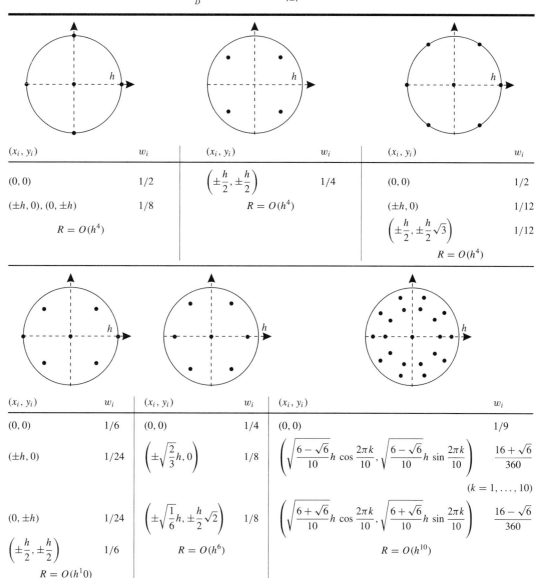

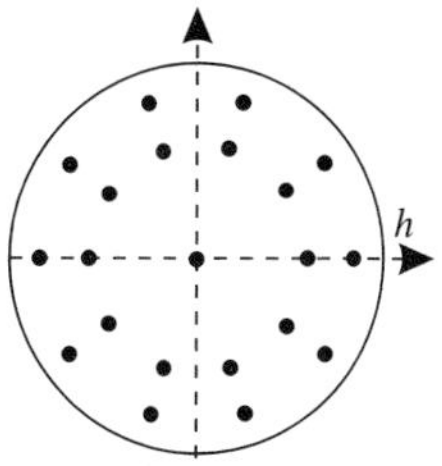

(x_i, y_i)	w_i
$(0, 0)$	$1/9$
$\left(\sqrt{\frac{6-\sqrt{6}}{10}}h \cos\frac{2\pi k}{10}, \sqrt{\frac{6-\sqrt{6}}{10}}h \sin\frac{2\pi k}{10}\right)$	$\frac{16+\sqrt{6}}{360}$
$(k = 1, \ldots, 10)$	
$\left(\sqrt{\frac{6+\sqrt{6}}{10}}h \cos\frac{2\pi k}{10}, \sqrt{\frac{6+\sqrt{6}}{10}}h \sin\frac{2\pi k}{10}\right)$	$\frac{16-\sqrt{6}}{360}$
$R = O(h^{10})$	

Table 7.7.4 Quadratures for integrating functions over domains with various shapes, after Stroud (1971) and Abramowitz and Stegun (1972). H: hexagon of side-length h.

$$\frac{1}{\frac{3}{2}\sqrt{3}h^2}\iint_H f(x,y)\,dx\,dy = \sum_{i=1}^{N} w_i,\, f(x_i, y_i) + R$$

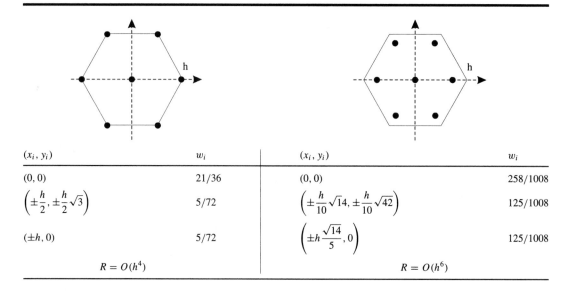

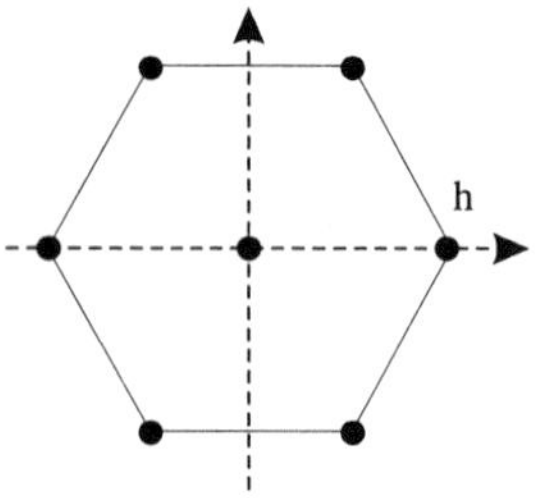

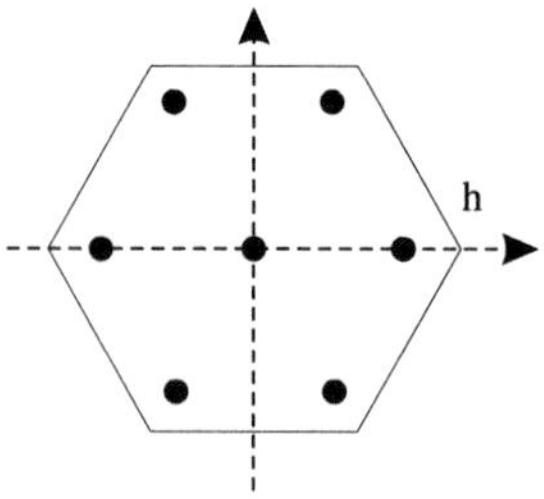

(x_i, y_i)	w_i	(x_i, y_i)	w_i
$(0, 0)$	21/36	$(0, 0)$	258/1008
$\left(\pm\frac{h}{2}, \pm\frac{h}{2}\sqrt{3}\right)$	5/72	$\left(\pm\frac{h}{10}\sqrt{14}, \pm\frac{h}{10}\sqrt{42}\right)$	125/1008
$(\pm h, 0)$	5/72	$\left(\pm h\frac{\sqrt{14}}{5}, 0\right)$	125/1008
$R = O(h^4)$		$R = O(h^6)$	

Table 7.7.5 Quadratures for integrating functions over domains with various shapes, after Stroud (1971) and Abramowitz and Stegun (1972). Σ: sphere of radius h.

$$\frac{1}{4\pi h^2}\iint_{\Sigma} f(x, y, z)\, dS = \sum_{i=1}^{N} w_i,\ f(x_i, y_i, z_i) + R$$

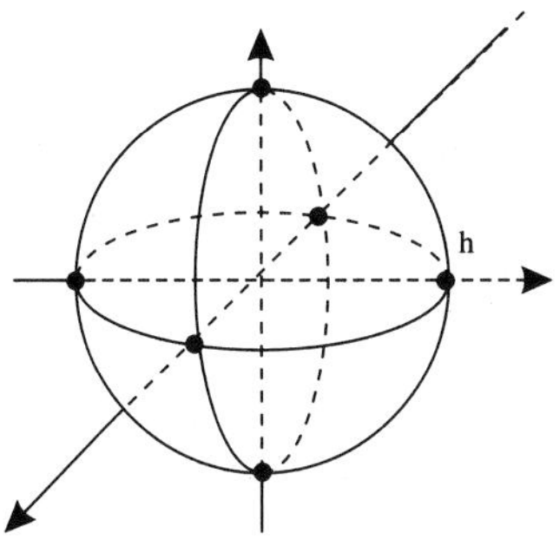

(x_i, y_i, z_i)	w_i
$(\pm h, 0, 0)$	1/6
$(0, \pm h, 0)$	1/6
$(0, 0, \pm h)$	1/6

$R = O(h^4)$

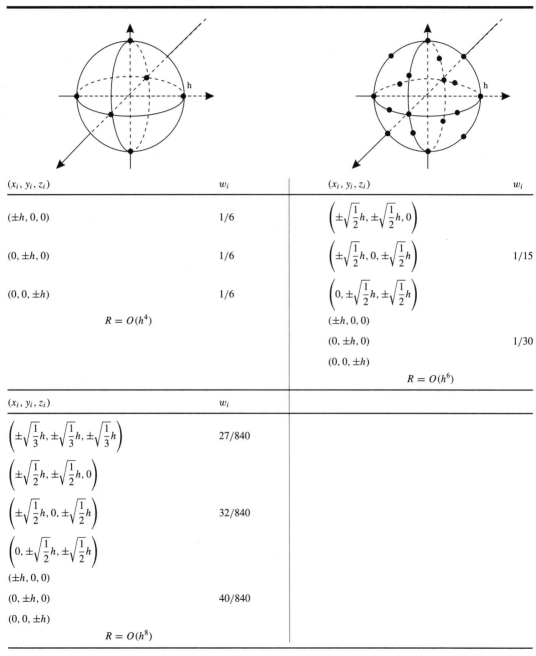

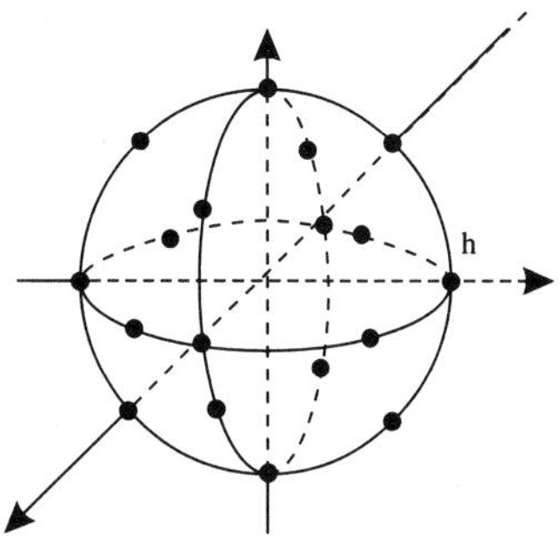

(x_i, y_i, z_i)	w_i
$\left(\pm\sqrt{\frac{1}{2}}h, \pm\sqrt{\frac{1}{2}}h, 0\right)$	
$\left(\pm\sqrt{\frac{1}{2}}h, 0, \pm\sqrt{\frac{1}{2}}h\right)$	1/15
$\left(0, \pm\sqrt{\frac{1}{2}}h, \pm\sqrt{\frac{1}{2}}h\right)$	
$(\pm h, 0, 0)$	
$(0, \pm h, 0)$	1/30
$(0, 0, \pm h)$	

$R = O(h^6)$

(x_i, y_i, z_i)	w_i
$\left(\pm\sqrt{\frac{1}{3}}h, \pm\sqrt{\frac{1}{3}}h, \pm\sqrt{\frac{1}{3}}h\right)$	27/840
$\left(\pm\sqrt{\frac{1}{2}}h, \pm\sqrt{\frac{1}{2}}h, 0\right)$	
$\left(\pm\sqrt{\frac{1}{2}}h, 0, \pm\sqrt{\frac{1}{2}}h\right)$	32/840
$\left(0, \pm\sqrt{\frac{1}{2}}h, \pm\sqrt{\frac{1}{2}}h\right)$	
$(\pm h, 0, 0)$	
$(0, \pm h, 0)$	40/840
$(0, 0, \pm h)$	

$R = O(h^8)$

Table 7.7.6 Quadratures for integrating functions over domains with various shapes, after Stroud (1971) and Abramowitz and Stegun (1972). C: cube with side-length h.

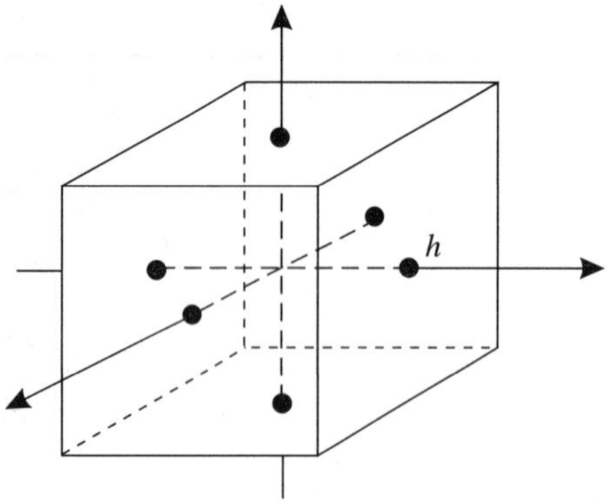

$$\frac{1}{8h^3}\int\int\int_C f(x,y,z)\,dx\,dy\,dz = \sum_{i=1}^{N} w_i\, f(x_i,y_i,z_i) + R$$

(x_i, y_i, z_i)	w_i
$(\pm h, 0, 0)$	1/6
$(0, \pm h, 0)$	1/6
$(0, 0, \pm h)$	1/6

$R = O(h^4)$

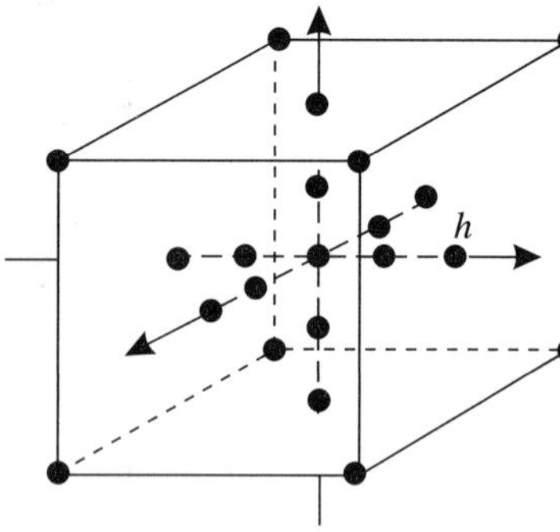

$$\frac{1}{8h^3}\int\int\int_C f(x,y,z)\,dx\,dy\,dz$$

$$= \tfrac{1}{360}[-496 f_m + 128\sum f_r + 8\sum f_f + 5\sum f_\sigma] + O(h^6)$$

$$= \tfrac{1}{450}[91\sum f_f - 40\sum f_\delta + 16\sum f_d] + O(h^6)$$

where $f_m = f(0,0,0)$.

$\sum f_r$ = sum of values of f at the 6 points midway from the center of C to the 6 faces.

$\sum f_f$ = sum of values of f at the 6 centers of the faces of C.

$\sum f_\sigma$ = sum of values of f at the 8 vertices of C.

$\sum f_\delta$ = sum of values of f at the 12 midpoints of edges of C.

$\sum f_d$ = sum of values of f at the 4 points on the diagonals of each face at a distance of $\frac{1}{2}\sqrt{5}h$ from the center of the face.

Table 7.7.7 Quadratures for integrating functions over domains with various shapes, after Stroud (1971) and Abramowitz and Stegun (1972). S: sphere of radius h.

$$\frac{1}{\frac{4}{3}\pi h^3}\int\int_S\int f(x, y, z)\,dx\,dy\,dz = \sum_{i=1}^{N} w_i,\ f(x_i, y_i, z_i) + R$$

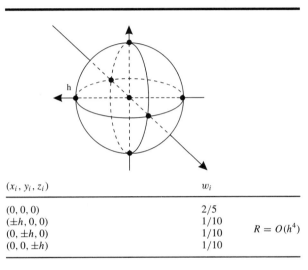

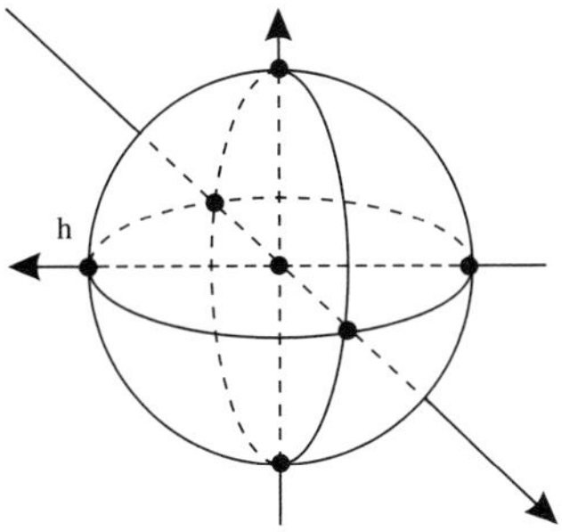

(x_i, y_i, z_i)	w_i	
$(0, 0, 0)$	2/5	
$(\pm h, 0, 0)$	1/10	$R = O(h^4)$
$(0, \pm h, 0)$	1/10	
$(0, 0, \pm h)$	1/10	

Table 7.7.8 Quadratures for integrating functions over domains with various shapes, after Stroud (1971) and Abramowitz and Stegun (1972).

Tetrahedron: $\mathcal{T}$

$$\frac{1}{V}\int\int_{\mathcal{T}}\int f(x, y, z)\,dx\,dy\,dz = \frac{1}{40}\sum f_\sigma + \frac{9}{40}\sum f_f + \text{ terms of fourth order}$$

$$= \frac{32}{60} f_m + \frac{1}{60}\sum f_\sigma + \frac{4}{60}\sum f_e + \text{ terms of fourth order}$$

where

V: Volume of $\mathcal{T}$

$\sum f_\sigma =$ sum of values of the function at the vertices of $\mathcal{T}$.

$\sum f_e =$ sum of values of the function at the midpoints of the edges of $\mathcal{T}$.

$\sum f_f =$ sum of values of the function at the center of gravity of the faces of $\mathcal{T}$.

$f_m =$ value of function at center of gravity of $\mathcal{T}$.

7.8 *Numerical Solution of Fredholm Integral Equations*

Numerical integration finds extensive applications in the solution of integral equations using numerical methods. A one-dimensional Fredholm integral equation us provides with an expression for the integral of an unknown function $f(x)$, over a fixed domain of integration, in terms of the function itself and other known functions. Such equations arise in a wide variety of theoretical and practical applications including wave scattering, elastostatics, and low-Reynolds-number hydrodynamics. A problem that has been formulated in terms of an integral equation is sometimes called an *inverse problem.*

Linear Equations of the First Kind

A linear Fredholm integral equation of the first kind has the standard form

$$\int_a^b K(x,\ x_0)\ f(x)\ dx + g(x_0) \tag{7.8.1}$$

where $K(x, x_0)$ is a given function of two variables called the *kernel*, and $g(x_0)$ is a specified forcing function. An example is the integral equation

$$\int_0^1 \exp(x x_0)\ f(x)\ dx = \frac{\exp(x_0 + 1) - 1}{x_0 + 1} \tag{7.8.2}$$

whose solution is $f(x) = e^x$ (Delves and Mohamed 1985, p. 11).

Linear Equations of the Second Kind

A linear Fredholm integral equation of the second kind has the standard form

$$f(x_0) = \alpha \int_a^b K(x,\ x_0)\ f(x)\ dx + g(x_0) \tag{7.8.3}$$

where α is a scalar constant. An example is the equation

$$f(x_0) = \alpha \int_{-1}^1 \cosh(x + x_0)\ f(x)\ dx - \cosh\ x_0 \tag{7.8.4}$$

whose solution is $f(x) = \cosh\ x/[\alpha(1 + \frac{1}{2}\sinh\ 2) - 1]$ (Delves and Mohamed 1985, p. 11).

Nÿstrom Methods

A powerful class of methods for solving integral equations of the first or second kind, named after their inventor Nÿstrom, proceed according to the following steps:

1. Approximate the integrals in equations (7.8.1) or (7.8.2) using a Newton–Cotes rule or a Gaussian quadrature.
2. Identify the point x_0 with each one of the $N + 1$ base points of the integration rule or quadrature.
3. Solve the resulting system of linear equations for the values of $f(x)$ at the base points using a direct or an iterative method.

Two important considerations in the implementation of the method are:

- The choice of the integration method is guided by the expected smoothness of the unknown function $f(x)$.
- The resulting system of linear equations may have a unique solution, an infinity of solutions, or no solution, reflecting corresponding behaviors of the exact solution of the integral equation.

For example, applying the two-point Gauss–Legendre quadrature to compute the integral on the right-hand side of equation (7.8.4), and evaluating the equation at the base points $x_0 = \pm\frac{1}{\sqrt{3}}$, we obtain the system

$$\begin{bmatrix} f\left(\frac{1}{\sqrt{3}}\right) \\ f\left(-\frac{1}{\sqrt{3}}\right) \end{bmatrix} = \alpha \begin{bmatrix} \cosh\left(-\frac{1}{\sqrt{3}}+\frac{1}{\sqrt{3}}\right) & \cosh\left(\frac{1}{\sqrt{3}}+\frac{1}{\sqrt{3}}\right) \\ \cosh\left(-\frac{1}{\sqrt{3}}-\frac{1}{\sqrt{3}}\right) & \cosh\left(\frac{1}{\sqrt{3}}-\frac{1}{\sqrt{3}}\right) \end{bmatrix} \begin{bmatrix} f\left(-\frac{1}{\sqrt{3}}\right) \\ f\left(\frac{1}{\sqrt{3}}\right) \end{bmatrix} - \begin{bmatrix} \cosh\left(\frac{1}{\sqrt{3}}\right) \\ \cosh\left(-\frac{1}{\sqrt{3}}\right) \end{bmatrix} \tag{7.8.5}$$

whose solution is

$$f\left(\frac{1}{\sqrt{3}}\right) = f\left(-\frac{1}{\sqrt{3}}\right) = \frac{\cosh\dfrac{1}{\sqrt{3}}}{\alpha\left(1+\cosh\dfrac{2}{\sqrt{3}}\right)-1} \tag{7.8.6}$$

We note that $\cosh(2/\sqrt{3}) = 1.744$ and $\frac{1}{2}\sinh 2 = 1.813$, and obtain fair agreement with the exact solution shown after equation (7.8.4).

Generalized Fredholm Method

A somewhat different class of methods founded by Fredholm originate from the product–integration rule and proceed according to the following steps:

1. Approximate the function $f(x)$ with an interpolating polynomial or another interpolating function defined in terms of the *a priori* unknown values of $f(x)$ at $N+1$ nodes x_i, $f(x_i)$. This amounts to expressing $f(x)$ as a linear combination of appropriate basis function $\phi_i(x)$, in the form

$$f(x) = \sum_{i=1}^{N+1} f(x_i)\,\phi_i(x) \tag{7.8.7}$$

   For example, the basis function may be identified with the triangular tent functions defined after equation (6.6.5).

2. Substitute the interpolating polynomial into the integral equation, and extract from the integral the unknown values $f(x_i)$. For example, for the integral equation (7.8.3), we obtain

$$f(x_0) = \alpha \sum_{i=1}^{N+1} f(x_i) \int_a^b K(x,\ x_0)\,\phi_i(x)\,dx + g(x_0) \tag{7.8.8}$$

3. Identify the point x_0 with each one of $N+1$ *collocation points* x_j^c that lie within the domain of integration, and use the expansion (7.8.7) once more to obtain a system of linear equations for the unknown values $f(x_i)$. For example, in the case of integral equation (7.8.3), we obtain

$$\sum_{i=1}^{N+1} f(x_i)\, \phi_i(x_j^c) = \alpha \sum_{i=1}^{N+1} f(x_i) \int_a^b K(x, x_j^c)\, \phi_i(x)\, dx + g(x_j^c) \tag{7.8.9}$$

which can be placed in the form

$$\sum_{i=1}^{N+1} \left(\phi_i(x_j^c) - \alpha \int_a^b K(x, x_j^c)\, \phi_i(x)\, dx \right) f(x_i) = g(x_j^c) \tag{7.8.10}$$

or more concisely,

$$\sum_{i=1}^{N} \left(\phi_i(x_j^c) - \alpha\, B_{j,i} \right) f(x_i) = g(x_j^c) \tag{7.8.11}$$

The *influence matrix* $\boldsymbol{B}$ is defined as

$$B_{j,i} \equiv \int_a^b K(x, x_j^c)\, \phi_i(x)\, dx \tag{7.8.12}$$

Staircase approximation

In the simplest implementation of the Fredholm method, we divide the solution interral $[a, b]$ into N evenly spaced subintervals separated by the nodes x_i, $i = 1, \ldots, N+1$, where $x_1 = a$ and $x_{N+1} = b$, and assume that the function $f(x)$ is constant over the ith interval, equal to f_i. Equation (7.8.8) becomes

$$f(x_0) = \alpha \sum_{i=1}^{N} f_i \int_{x_i}^{x_{i+1}} K(x, x_0)\, dx + g(x_0) \tag{7.8.13}$$

The integrals on the right-hand side are computed using an analytical or numerical method.

We then successively identify x_0 with the jth collocation point, which can be chosen to lie in the middle of the ith interval, $x_j^c = \frac{1}{2}(x_j + x_{j+1})$, $j = 1, \ldots, N$, note that $f(x_j^c) = f_j$, and thus obtain a system of linear equations for the unknown values f_j,

$$\sum_{i=1}^{N} \left(\delta_{j,i} - \alpha \int_{x_i}^{x_{i+1}} K(x, x_j^c)\, dx \right) f_i = g(x_j^c) \tag{7.8.14}$$

Computation of the influence matrix

Analytical complexities in deriving expressions for, and computing the influence matrix $\boldsymbol{B}$ for linear, quadratic, and higher-order global or local interpolating polynomials discourage the implementation of the Fredholm method. Fortunately, the procedure can be programmed in a manner that circumvents large amounts of analytical work as follows.

First, we note that the right-hand side of equation (7.8.8) is a linear function of the nodal values $f(x_i)$. Then, we write a subroutine that receives the position of the jth collocation point x_j^c and the nodal values of $f(x)$, computes the interpolating polynomial, and returns the value of the integral on the right-hand side of equation (7.8.3), $I(x_j^c;\, f_1, f_2, \ldots, f_{N+1})$. The influence coefficients are given by

$$B_{j,i} = \alpha I(x_j^c;\, f_i = 0,\, f_2 = 0, \ldots, f_{i-1} = 0,\, f_i = 1,\, f_{i+1} = 0, \ldots, f_{N+1} = 0) \qquad (7.8.15)$$

Further discussion of numerical methods for solving linear integral equations can be found in the monograph of Delves and Mohamed (1985). Applications in fluid mechanics are discussed by Pozrikidis (1997).

PROBLEMS

7.8.1 *Fredholm's method.*

Solve the integral equation (7.8.4) using the formulation of equation (7.8.14) for $N = 2, 4, 8, 16, 32$. Compare the numerical results with the exact solution, and discuss the rate of convergence.

7.8.2 *Heat conduction from an isothermal body.*

Consider a hot two-dimensional body whose surface is maintained at a constant temperature T_0, immersed in an infinite thermally conducting surrounding medium. Far from the body, the temperature has the reference value of zero. The temperature distribution in the ambient medium can be expressed as a boundary integral over the surface of the body, in the form

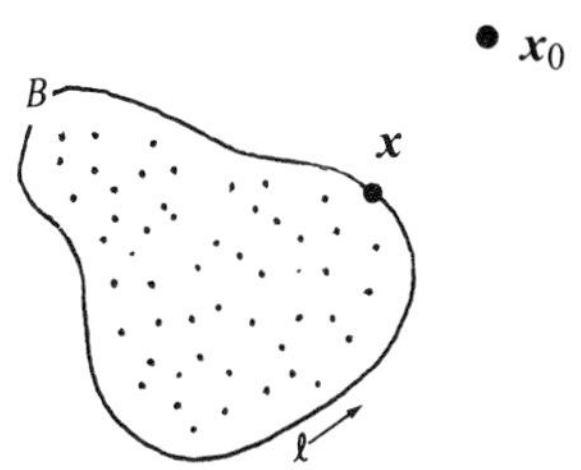

$$T(\boldsymbol{x}_0) = -\int_B G(\boldsymbol{x}_0, \boldsymbol{x})\, q(\boldsymbol{x})\, dl(\boldsymbol{x}) \qquad (7.8.16)$$

where $\boldsymbol{x}_0$ and $\boldsymbol{x}$ are position vectors, q is the heat flux, l is the arc length along the boundary of the body B, and G is the Green's function of Laplace's equation in two dimensions. All variables have been properly nondimensionalized. For two-dimensional conduction in an infinite ambient medium, the Green's function is given by

$$G(\boldsymbol{x}_0, \boldsymbol{x}) = -\frac{1}{2\pi} \ln |\boldsymbol{x}_0 - \boldsymbol{x}| \qquad (7.8.17)$$

Placing $\boldsymbol{x}_0$ at the surface of the body, and requiring that $T = T_0$, we obtain a Fredholm integral equation of the first kind for the flux q,

$$\int_B G(\boldsymbol{x}_0, \boldsymbol{x}) q(\boldsymbol{x})\, dl(\boldsymbol{x}) = -T_0 \qquad (7.8.18)$$

To solve this equation, we trace B with a set of marker points and represent it with N boundary elements, such as straight segments or circular arcs. Assuming that q is constant over each element, we reduce the integral representation (7.8.16) to the algebraic equation

$$T(\boldsymbol{x}_0) = \sum_{j=1}^{N} q_j \, A_j(\boldsymbol{x}_0) \tag{7.8.19}$$

where

$$A_j(\boldsymbol{x}_0) = -\int_{E_j} G(\boldsymbol{x}_0, \boldsymbol{x}) \, dl(\boldsymbol{x}) \tag{7.8.20}$$

are the *influence coefficients*, and E_j is the jth element.

(a) Derive the corresponding discrete form of the integral equation (7.8.18).

(b) Discretize B into a collection of straight segments that connect two successive marker points, and derive analytical expressions for the influence coefficients in terms of the position of the marker points.

Solve the corresponding discrete form of the integral equation derived in part (a) when B is an ellipse with axis ratios 1.0, 2.0, 5.0, 10.0 for several values of N, and discuss the convergence of your results with respect to N.

(c) Discretize B into a collection of circular arcs connecting three successive marker points. An arc is described by its center $\boldsymbol{x}_c = (x_c, y_c)$, radius R, and the polar angles of the first and third end-points (θ_1, θ_3).

Evaluate the influence coefficients analytically for a point $\boldsymbol{x}_0$ that lies *on* an arc. Investigate whether it is possible to compute the influence coefficients analytically over each arc, for a point $\boldsymbol{x}_0$ that lies *off* the arc. If not, write a subroutine that performs the integration numerically using the Gauss–Legendre quadrature.

Repeat the computations of part (b) and discuss the performance of the numerical method.

References

Abramowitz, M, and Stegun, I. A., 1972, *Handbook of Mathematical Functions.* Dover.

Atkinson, K. E., 1989, *Introduction to Numerical Analysis.* Wiley.

Capstick, S., and Keister, B. D., 1996, Multidimensional quadrature algorithms at higher degree and/or dimension. *J. Comput. Phys.* **123**, 267–73.

Clenshaw, C. W., and Curtis, A. R., 1960, A method for numerical integration in an automatic computer. *Numer. Math.* **12**, 197–205.

Cools, R., and Rabinowitz, P., 1993, Monomial cubature rules since Stroud: a compilation. *J. Comput. Appl. Math.* **43**, 309–26.

Cowper, G. R., 1973, Gaussian quadrature formulas for triangles. *Int. J. Numer. Meth. Eng.* 7, 405–08.

Davis, P. J., and Rabinowitz, P., 1984, *Methods of Numerical Integration.* Academic Press.

Delves, L. M., and Mohamed, J. L., 1985, *Computational Methods for Integral Equations.* Cambridge University Press.

Dettman, J. W. 1965, *Applied Complex Variables.* Dover.

Engels, H., 1980, *Numerical Quadrature and Cubature.* Academic Press.

Evans, G., 1993, *Practical Numerical Integration.* Wiley.

Gradshteyn, I. S., and Ryzhik, I. M., 1980, *Table of Integrals, Series, and Products.* Academic Press.

Hildebrand, F. B., 1974, *Introduction to Numerical Analysis.* Reprinted by Dover, 1987.

Iri, M., Moriguti, S., and Takasawa, Y., 1987, On a certain quadrature formula. *J. Comput. Appl. Math.* **17**, 3–20.

Kronrod, A. S., 1965, *Nodes and Weights of Quadrature Formulae.* Consultants Bureau, New York.

Krylov, V., 1962, *Approximation Calculation of Integrals.* MacMillan.

Lyness, J., and Ninham, G., 1967, Numerical quadrature and asymptotic expansions. *Math. Comput.* **21**, 162–78.

Murota, K., and Iri, M., 1982, Parameter tuning and repeated application of IMT type transformation in numerical quadrature. *Numer. Math.* **38**, 347–63.

Patterson, T. N. L., 1968, The optimum addition of points to quadrature formulae. *Math. Comput.* **22**, 847–56.
Pozrikidis, C., 1997, *Introduction to Theoretical and Computational Fluid Dynamics.* Oxford University Press.
Prudmikov, A. P., Brychov, Y. A., and Mariche, O. I., 1986, *Integrals and Series,* Vol I. Gordon & Breach.
Sloan, I. H., and Joe, S., 1994, *Lattice Methods for Multiple Integration.* Oxford University Press.
Stroud, A., 1971, *Approximate Calculation of Multiple Integrals.* Prentice-Hall.
Stroud, A., and Secrest, D., 1966, *Gaussian Quadrature Formulas.* Prentice-Hall.
Vedder, J. D., 1987, Simple approximations for the error function and its inverse. *Am. J. Phys.* **55**, 762–63.
Todd, J., 1962, *A Survey of Numerical Analysis.* McGraw-Hill.

Chapter 8

Approximation of Functions, Lines, and Surfaces

8.1 Problem Statement and Significance

We are given a certain amount of information about a function $f(x)$ within a closed interval $[a, b]$, including, for example, the values of $f(x)$ and its derivatives at a set of data points that are distributed in a certain manner over that interval. Our objective is to approximate or represent the function with another function $q(x)$ computed in some appropriate manner. The approximating function $q(x)$ is usually chosen to be a *polynomial*, but other choices, including ratios of polynomials, that is, *rational functions*, and expansions in *exponential* or *trigonometric* basis functions, may lead to satisfactory results and even produce better approximations.

Approximation versus Interpolation

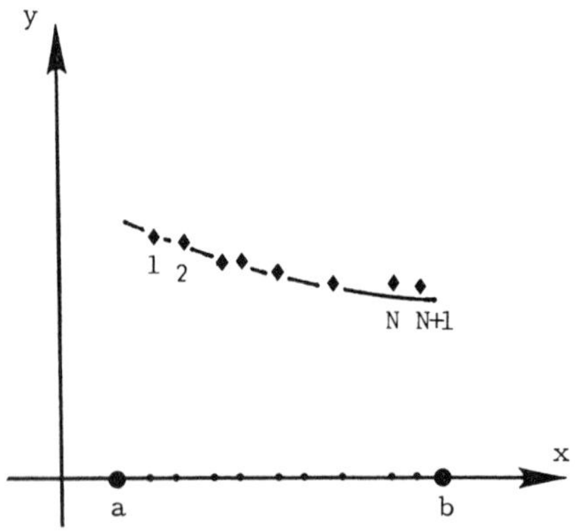

In one class of problems, we are provided with the values of the function $f(x)$ at $N + 1$ data points $x_i, i = 1, \ldots, N + 1$, that lie within the interval $[a, b]$, representing, for example, a time-series of a computed or measured variable. The distinguishing feature of function approximation compared to function interpolation is that the approximating function $q(x)$ does not necessarily have to pass through the data points, which means that *we do not use all of the available degrees of freedom.* The extra degrees of freedom are spent to ensure a smooth behavior. In contrast, in the case of function interpolation discussed in Chapter 6, the interpolating function $g(x)$ was required to pass through all data points, with the possible penalty of oscillatory behavior.

Approximation is preferred over interpolation when the data are suspected of carrying a significant amount of error, for example, they have been corrupted by noise. Furthermore, approximation can be used to represent a function in a compact fashion and thus expedite its computation. Finally, approximation of functions describing lines and surfaces can be used to generate smooth shapes and is therefore an important tool of computer-aided design and computer-aided manufacturing, CAD/CAM.

8.2 Polynomial Function Approximation

First, we consider approximating a function $f(x)$ with a polynomial. The polynomial choice is motivated by two reasons: Simplicity in the algebraic manipulations and, perhaps more importantly, availability of a substantial body of theory on polynomial behaviors. Nonpolynomial approximations will be discussed in subsequent sections.

Taylor Series Approximation

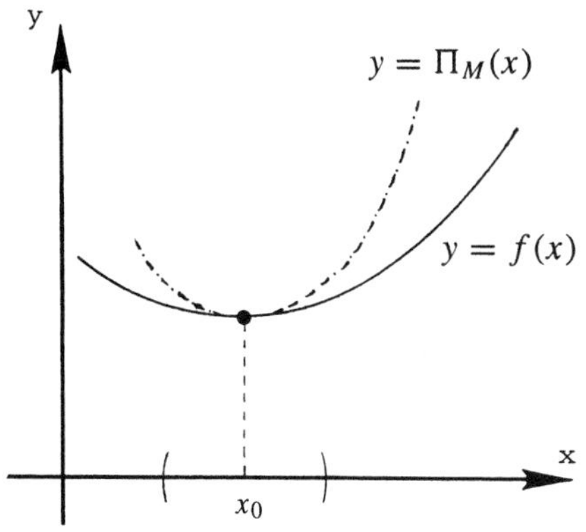

A simple approximating polynomial is provided by the Taylor series expansion of $f(x)$ about a certain point x_0, truncated at a certain level after $M + 1$ terms, yielding

$$\Pi_M(x; x_0) = \sum_{k=0}^{M} \frac{1}{k!} f^{(k)}(x_0)(x - x_0)^k \tag{8.2.1}$$

(Appendix A). The function $f(x)$ is assumed to be differentiable at least M times at the point x_0. It is clear that $\Pi_M(x)$ is an Mth degree polynomial in x.

The approximation is perfect at the point x_0 but, unless $f(x)$ happens to be a polynomial of degree less than or equal to M, there will be a certain amount of error

$$e(x) \equiv \Pi_M(x) - f(x) \tag{8.2.2}$$

at any other point. Provided the values of the derivatives $f^{(k)}(x_0)$ are available with high precision, as M is increased, the magnitude of the error $|e(x)|$ will diminish at every point x *in a certain neighborhood* of the point x_0, but not necessarily over the whole real axis.

To establish the radius of convergence of the Taylor series, we introduce the function of a complex variable $f(z)$, where $z = x + iy$ and i is the imaginary unit, and identify the location of the singularities of $f(z)$ in the complex plane. Let z_S be the singularity that is closest to x_0. As M is increased, the Taylor series $\Pi_M(x, x_0)$ will converge to $f(x)$ only if $|x - x_0| < |z_S - x_0|$. Accordingly, the distance $|z_S - x_0|$ is the *radius of convergence* of $\Pi_\infty(x, x_0)$.

Thus, as M is increased, the Taylor series will converge *uniformly* to the function $f(x)$ at every point within a specified interval $[a, b]$ that contains or is adjacent to the point x_0, only if the distances of the end-points of the interval from x_0 are less than the radius of convergence of $\Pi_\infty(x, x_0)$. This behavior is clearly unsatisfactory from the practical perspective of function approximation.

For example, as M is increased, the truncated Taylor series of the function $f(x) = 1/(1 + x^2)$ about the point $x_0 = 0$ does not converge at points that lie outside the interval $(-1, 1)$. The reason is that the complex function $f(z) = 1/(1 + z^2)$ has singularities at the complex conjugate points $z = \pm i$.

Weierstrass Theorem

Let $P_M(x)$ be an Mth degree polynomial approximating the function $f(x)$. We define the error

$$e(x) = P_M(x) - f(x) \tag{8.2.3}$$

and ask: Is there a sequence of polynomials $P_M(x)$ such that, as M is increased, the error $|e(x)|$ diminishes uniformly at every point within a particular interval of interest $[a, b]$? The answer is affirmative.

Weierstrass's theorem ensures that, if $f(x)$ is a continuous function, then there exists a polynomial of a sufficiently high degree M so that $|e(x)| < \varepsilon$ for any value of x between a and b, where ε is an arbitrary positive tolerance (e.g., Davis 1963, Rivlin 1969). The practical issue of how to compute this approximating polynomial, or class of approximating polynomials, must now be considered.

Bernstein Polynomials

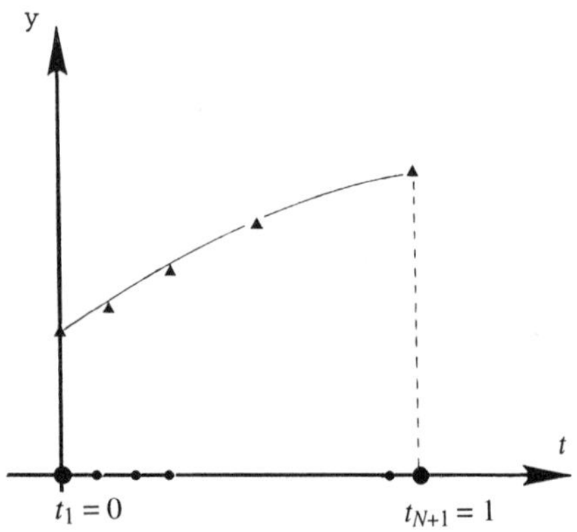

Suppose that we are provided with the values of the function $h(t)$ at $M + 1$ *evenly spaced* data points t_i, $i = 1, \ldots, M + 1$, that lie within the interval $[0, 1]$, where $t_1 = 0$ and $t_{M+1} = 1$; with reference to our previous notation, $a = 0$ and $b = 1$.

If we are given the values of the function $f(x)$ at a sequence of evenly spaced points over an interval that is different from $[0, 1]$, then linear scaling of the independent variable will bring the problem to its canonical form, producing the function $h(t)$.

It can be shown that, as M tends to infinity, the polynomial

$$P_M(t) = \sum_{i=1}^{M+1} h(t_i) B_{M,i}(t) \tag{8.2.4}$$

where

$$B_{M,i}(t) \equiv \binom{M}{i-1} t^{i-1} (1-t)^{M+1-i} \tag{8.2.5}$$

are the Mth degree *Bernstein polynomials*, approximates the function $h(t)$ uniformly and with increasing accuracy over the whole of the interval $[0, 1]$ (e.g., Davis 1963, pp. 108–118). The combinatorial, designated by the first set of parentheses and its enclosures on the right-hand side of equation (8.2.5), was defined in equation (1.4.6).

We note that

$$B_{M,i}(t_1) = \delta_{i,1}, \qquad B_{M,i}(t_{M+1}) = \delta_{i,M+1} \tag{8.2.6}$$

and this shows that the approximating polynomial $P_M(t)$ passes through the first and last data points, but does not necessarily pass through the intermediate points. Only when $M = 1$ is the graph of $P_1(x)$ a straight line passing through the two data points.

It is somewhat dispiriting to realize that the representation (8.2.4) is approximate even when $f(x)$ is a polynomial of degree M or less, except when $M = 1$, or if all $h(t_i)$ are equal.

PROBLEM

8.2.1 ***Properties of the Bernstein polynomials.***

Show that

$$\sum_{i=1}^{M+1} B_{M,i}(t) = 1 \tag{8.2.7}$$

for any value of M.

Recursive evaluation of the Bernstein polynomials

An efficient method for evaluating the Bernstein polynomials uses the recursion formulas

$$\begin{aligned} B_{0,0}(t) &= 0, \qquad B_{0,1}(t) = 1, \qquad B_{0,2}(t) = 0 \\ B_{l,i}(t) &= (1-t)B_{l-1,i}(t) + tB_{l-1,i-1}(t) \end{aligned} \tag{8.2.8}$$

for $l \geq 1$ and $i = 1, \ldots, l+1$, where

$$B_{l,0}(t) = 0, \qquad B_{l,l+2}(t) = 0$$

A sequence of Bernstein polynomials approximating the function $h(t) = \exp(2t)$ is shown in Figure 8.2.1.

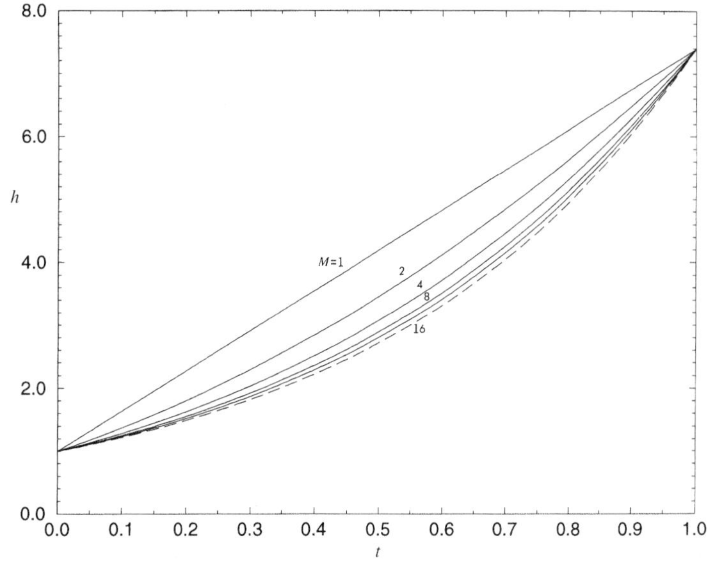

Figure 8.2.1 Approximation of the section of the function $h(t) = \exp(2t)$ in the interval [0, 1], represented by the dashed line, with the Bernstein polynomials of various degrees.

Limitations of the approximation

Unfortunately, the usefulness of the Bernstein approximation is abated by the slow convergence and the requirement that the approximated function $h(t)$ be known at $M+1$ *evenly spaced* points. Nevertheless, Bernstein polynomials of low order are useful in the approximation of lines and surfaces using the Bézier method to be discussed in Section 8.3.

Polynomials That Arise by Minimizing a Weighted Integral of the Approximation Error

We set out to develop a general-purpose method of computing the approximating polynomial $P_M(x)$; a method that is both efficient and free of pitfalls.

We begin by expressing $P_M(x)$ in the standard form

$$P_M(x) = a_1 x^M + a_2 x^{M-1} + a_3 x^{M-2} + \cdots + a_M x + a_{M+1} \tag{8.2.9}$$

and seek to compute the coefficients $a_i, i = 1, \ldots, M+1$, so as to minimize the approximation error defined in equation (8.2.3) in some appropriate global sense. This can be done, for example, by minimizing either one of the positive functionals

$$E(e(x)) \equiv \int_a^b e(x)w(x)\,dx, \qquad E(|e(x)|) \equiv \int_a^b |e(x)|w(x)\,dx \tag{8.2.10}$$

where $w(x)$ is a certain nonnegative *weighting function.* The role of $w(x)$ is to bias the computation toward a desired region where the data are either more valuable or more reliable. An unbiased computation corresponds to $w(x) = 1$.

The method is implemented by substituting the right-hand side of equation (8.2.9) into equation (8.2.3), and the result into one of the integrals (8.2.10), setting the partial derivatives of the integral with respect to a_i equal to zero, and then solving the resulting system of linear algebraic equations for the polynomial coefficients. Unfortunately, one can produce examples where this approach either fails to produce a unique solution or is extremely sensitive to the parameters of the problem.

Least-squares Polynomials

An infallible approach is based on the *least-squares method.* The polynomial coefficients are computed with the purpose of minimizing the nonnegative *least-squares functional*

$$E[e^2(x)] = \int_a^b e^2(x)w(x)\,dx = \int_a^b \left(\sum_{i=1}^{M+1} a_i x^{M+1-i} - f(x) \right)^2 w(x)\,dx \tag{8.2.11}$$

which is done by setting

$$\frac{\partial E}{\partial a_j} = \int_a^b \frac{\partial}{\partial a_j} \left(\sum_{i=1}^{M+1} a_i x^{M+1-i} - f(x) \right)^2 w(x)\,dx = 0 \tag{8.2.12}$$

for $j = 1, \ldots, M+1$. Carrying out the differentiations, we obtain a linear system of equations for $a_i, i = 1, \ldots, M+1$,

$$\sum_{i=1}^{M+1} a_i \int_a^b x^{2M+2-i-j} w(x)\, dx = \int_a^b x^{M+1-j} f(x)\, w(x)\, dx \tag{8.2.13}$$

where $j = 1, \ldots, M+1$. With the coefficients computed in this manner, it can be shown that, as M is increased, the approximating polynomial converges uniformly to $f(x)$ at every point within $[a, b]$ (e.g., Davis 1963).

Canonical form and the Hilbert matrix

One can shift and stretch or compress the independent variable x in a linear manner, transforming it into the canonical variable t that varies between 0 and 1, as discussed in Section 6.4. Setting $x = q(t) = a+(b-a)t$, writing $f(x) = f(q(t)) = h(t)$, and applying the least-squares method for $h(t)$ with $w(t) = 1$, we obtain the linear system

$$\boldsymbol{H}^{(M+1)} \boldsymbol{a} = \boldsymbol{F} \tag{8.2.14}$$

where

$$\boldsymbol{a} = (a_{M+1}, a_M, \ldots, a_1) \tag{8.2.15}$$

where the least-squares polynomial for $h(t)$ is given by the right-hand side of equation (8.2.9) with t in place of x. $\boldsymbol{H}^{(M+1)}$ is the $(M+1) \times (M+1)$ Hilbert matrix defined in equations (2.6.5) and (2.6.6). The components of the vector $\boldsymbol{F}$ are give by

$$F_i = \int_0^1 t^{i-1} h(t)\, dt \tag{8.2.16}$$

The solution of the linear system (8.2.14) proceeds uneventfully when M has a moderate value but, unfortunately, the Hilbert matrix is nearly singular when M is higher than about five (Problem 2.6.2). Clearly, a better approach is required when a higher-degree approximating polynomial is desired, and the omnipotent orthogonal polynomials come to our rescue. Before discussing the theory of approximation by orthogonal polynomials, we digress to outline methods of data regression; a reader who has no interest in it should skip the following two subsections.

Data Regression with Least Squares

In practice, the approximated function $f(x)$ is often defined in terms of a table of values, and the evaluation of the preceding integrals requires the use of numerical methods. In the least-squares data-regression method, the integrals in the preceding equations are replaced with sums over $N+1$ data points. In particular, the counterpart of equation (8.2.13) with $w(x) = 1$ is

$$\boldsymbol{L}^{(M+1)} \boldsymbol{a} = \boldsymbol{F} \tag{8.2.17}$$

where the vector $\boldsymbol{a}$ is defined in equation (8.2.15), $\boldsymbol{L}^{(M+1)}$ is an $(M+1) \times (M+1)$ symmetric matrix with elements

$$
L^{(M+1)} = \begin{bmatrix}
N & \sum_{k=1}^{N+1} x_k & \sum_{k=1}^{N+1} x_k^2 & \cdots & \sum_{k=1}^{N+1} x_k^M \\
\sum_{k=1}^{N+1} x_k & \sum_{k=1}^{N+1} x_k^2 & \cdots & \cdots & \sum_{k=1}^{N+1} x_k^{M+1} \\
\sum_{k=1}^{N+1} x_k^2 & \vdots & \cdots & \cdots & \vdots \\
\vdots & \vdots & \cdots & \cdots & \vdots \\
\sum_{k=1}^{N+1} x_k^M & \cdots & \cdots & \cdots & \sum_{k=1}^{N+1} x_k^{2M}
\end{bmatrix}
$$

$$
L_{i,j}^{(M+1)} = \sum_{k=1}^{N+1} x_k^{i+j-2} \tag{8.2.18}
$$

and

$$
F_i = \sum_{k=1}^{N+1} x_k^{i-1} f(x_k) \tag{8.2.19}
$$

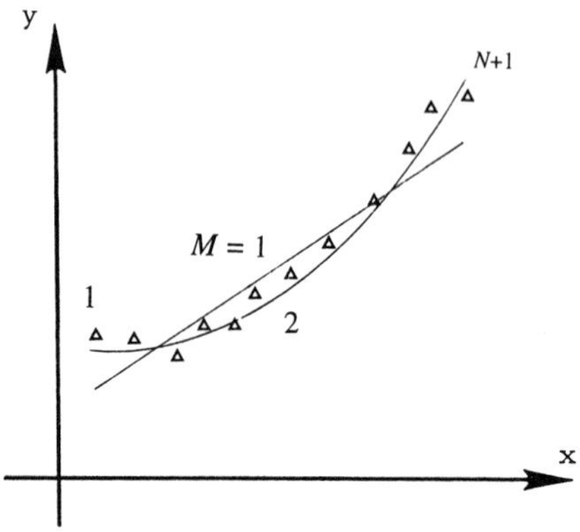

Linear and quadratic regression correspond, respectively, to $M = 1$ and 2. The method works well when M has a small or moderate value, but the linear system (8.2.17) typically becomes nearly singular when M is higher than about five. Thus, in order to avoid oscillations, the degree of the approximating polynomial M should to be significantly less than N.

When $M = N$, the approximating polynomial reduces to the interpolating polynomial, which means that the computation becomes extremely sensitive to the precise location of the data points (Problem 8.2.2).

PROBLEM

8.2.2 ***Regression and interpolation.***

Show that when $M = N$, the least-squares approximating polynomial reduces to the interpolating polynomial.

Nonlinear Data Regression

It may happen that the approximating polynomial is not appropriate for describing a particular set of data points, and another function will do a better job. One example is the exponential function

$$q(x) = a \exp(-bx) \tag{8.2.20}$$

where a and b are two adjustable parameters. An easy way of computing these two parameters is to take the logarithm of both sides of equation (8.2.20), finding

$$Q(x) \equiv \ln(q(x)) = A + Bx \tag{8.2.21}$$

where

$$A = \ln a, \qquad B = -b \tag{8.2.22}$$

The constants A and B can be computed by applying the linear regression method to the function $Q(x)$, based on the tabulation of $\ln(f(x))$ versus x.

As another example, approximating a function $f(x)$ with the function $q(x) = a + b/x$ can be done by performing linear regression on the function $q(1/t) \equiv a + bt$, where $t = 1/x$. In this case, we tabulate $f(x)$ versus $1/x$.

PROBLEM

8.2.3 *Transient operation of a heat exchanger.*

Measurements of the transient response of the temperature T at the outlet of a heat exchanger, as a function of time t, gave the following data:

t in min	T in °C
1.0	37.9
2.0	29.4
3.0	26.8
5.0	22.0
10.0	18.3

(*a*) Approximate the data using linear regression, setting $T = at + b$, and compute the constants a and b using the least-squares method. Plot the data points and the least-squares curve, and remark on the accuracy of the regression.
(*b*) Process engineer Justin Case suggested that a correlation of the form $T = (at^{1/2} + b)^{-1}$ might be more appropriate. What are the units of the constants a and b? Compute a and b using the least-squares method, and compare the measured values to the ones predicted by the regression. Plot the data points and the least-squares curve and discuss the accuracy of the regression. *Hint*: Introduce the new variables $y = 1/T$ and $x = t^{1/2}$.
(*c*) Repeat the linear regression problem, but this time compute the constants a and b so as to minimize the alternative least-quartics functional

$$E[e^4(x)] = \sum_{k=1}^{N+1} \left(a_1 x_k + a_2 - f(x_k)\right)^4 \tag{8.2.23}$$

Discuss the efficiency of this approach.

Approximation with Orthogonal Polynomials

The numerical difficulties encountered in solving the linear system (8.2.14) are not without a cure. As in the case of function interpolation with the Vandermonde-matrix approach discussed in Section 6.2, we must simply find a way of getting around the pronounced numerical sensitivity; the orthogonal polynomials described in Appendix B are *Deus exmachina.*

Let us consider a certain triangular family of orthogonal polynomials, $p_i(t)$, defined over the interval $[c, d]$ with a nonnegative weighting function $w(t)$. As a preliminary, we introduce the new independent variable t that is related to the primary variable x by means of the linear transformation (6.4.5). Under this transformation, $f(x)$ becomes a function of t, and we write $f(q(t)) = h(t)$.

Next, we approximate the canonical function $h(t)$ with an Mth degree polynomial $P_M(t)$ expressed as a weighted sum of orthogonal polynomials of a certain class, in the form

$$P_M(t) = \sum_{i=0}^{M} c_i p_i(t) \tag{8.2.24}$$

Finally, we compute the coefficients c_i using the orthogonality condition for $p_i(t)$ stated in equation (6.4.1), obtaining

$$c_i = \frac{1}{D_{i,i}} \int_c^d h(t) p_i(t) w(t)\, dt \tag{8.2.25}$$

where summation over i is *not* implied in the denominator. In practice, the integrals on the right-hand side of equation (8.2.25) are computed, for example, in terms of the values of $h(t)$ at a set of data points using the numerical methods discussed in Chapter 7.

Relation to the least-squares problem

Interestingly, one may show by straightforward algebraic manipulations that the approximating polynomial computed on the basis of the preceding two equations solves the least-squares minimization problem; that is, it minimizes the functional

$$E[e^2(t)] = \int_c^d \left(P_M(t) - h(t)\right)^2 w(t)\, dt \tag{8.2.26}$$

with respect to variations in the polynomial coefficients c_i. Thus, if the function $h(t)$ is a polynomial of degree less than or equal to M, and the integrals (8.2.25) are computed exactly, then the representation (8.2.24) is exact.

Chebyshev approximation

When $p_i(t)$ are chosen to be the Chebyshev polynomials $T_i(t)$ discussed in Table B.3 of Appendix B, the least-squares approximating polynomial is close to the *minimax* polynomial. By definition, of all Mth degree polynomials with the same leading-power coefficient, the Mth degree minimax polynomial yields the minimum of the maximum value of $|e(t)|$ (see also end of Section 6.4).

Furthermore, when the values of the function $h(t)$ are known or can be computed at the $N + 1$ data points t_j, $j = 1, \ldots, N + 1$, given in equation (6.4.8), corresponding to the zeros of $T_{N+1}(t)$, then the coefficients c_i can conveniently be computed using the relations (6.4.13) and (6.4.14). When $M = N$, the approximating polynomial passes through the data points; that is, it coincides with the interpolating polynomial, as illustrated in Figure 6.4.1.

Economization of Representation

The minimax property of the Chebyshev polynomials provides us with a boot-strapping method for representing an Nth degree polynomial

$$P_N(t) = a_1 t^N + a_2 t^{N-1} + \cdots + a_N t + a_{N+1} \tag{8.2.27}$$

in the range $[-1, 1]$, with a lower degree polynomial, while introducing the least amount of error.

The idea, discussed by Lanczos (1956), is to subtract from $P_N(t)$ the Nth degree polynomial $a_1 T_N(t)/2^{N-1}$, which takes up the monomial t^N and yields the new $(N-1)$ degree polynomial $Q_{N-1}(t)$. Since the difference between $P_N(t)$ and $Q_{N-1}(t)$ is a multiple of a Chebyshev polynomial, it fluctuates in the mildest possible way within the interval $[-1, 1]$. In that sense, $Q_{N-1}(t)$ is the best approximation to $P_N(t)$. We can continue this process, replacing $Q_{N-1}(t)$ with an $(N-2)$ degree polynomial $R_{N-2}(t)$, and this economizes the representation even further. But the fact that $R_{N-2}(t)$ is the best approximation to $Q_{N-1}(t)$ does not guarantee that it is also be the best approximation to $P_N(t)$.

In practice, the process of economization is carried out by expressing $P_N(x)$ in a Chebyshev expansion, as shown in equation (8.2.24), and then truncating the expansion. The first step is done by expressing the monomials in terms of Chebyshev polynomials of the same order, using the relations listed at the end of Table B.3.

Function representation

Polynomial economization is useful in a more general context: Considering an arbitrary function, we identify $P_N(t)$ with its truncated Taylor series retaining a large number of terms, and then we economize it by replacing it with a polynomial of lower order.

It is important to note that economization cannot improve the accuracy of a truncated Taylor series expansion, or convert a divergent Taylor series into a convergent one and vice versa. It simply provides us with a method of achieving a certain level of accuracy with fewer operations. Economization with rational functions, to be discussed in Section 8.5, can perform both of these tricks in a rather mystical way.

To exemplify the method, consider the truncated Maclaurin series expansion of the function $\ln(1+t)$ and retain five terms to obtain the polynomial approximation

$$\ln(1+t) \approx t - \tfrac{1}{2}t^2 + \tfrac{1}{3}t^3 - \tfrac{1}{4}t^4 + \tfrac{1}{5}t^5 \tag{8.2.28}$$

Expressing the five monomials in terms of Chebyshev polynomials using the relations listed at the end of Table B.3, we obtain the equivalent truncated expansion

$$\ln(1+t) \approx -\tfrac{11}{32}T_0(t) + \tfrac{11}{8}T_1(t) - \tfrac{3}{8}T_2(t) + \tfrac{7}{48}T_3(t) - \tfrac{1}{32}T_4(t) + \tfrac{1}{80}T_5(t) \tag{8.2.29}$$

Since the Chebyshev polynomials vary between -1 and $+1$, dropping the last term on the right-hand side of expansion (8.2.29) introduces less error than dropping the last term on the right-hand side of expansion (8.2.28) (Problem 8.2.4).

PROBLEM

8.2.4 ***Economization.***

(*a*) Drop the last term on the right-hand side of equation (8.2.29), recast the remaining expression into the standard form of a fourth degree polynomial, and compare it with the polynomial that arises by discarding the last term on the right-hand side of equation (8.2.28). Discuss the error incurred by replacing the function $f(t) = \ln(1+t)$ with either one of these polynomials in the range (0, 1). (*b*) Repeat part (*a*) dropping two terms.

Polynomial Approximation of a Function of More Than One Variable

The geometry of the domain of approximation and the distribution of data points are important factors in the design of polynomial approximation methods for functions of two or more variables. The problem is analogous to that of function interpolation discussed in the penultimate subsection of Section 6.9. The subject is treated in advanced monographs on function approximation.

8.3 *Bézier Representation of Lines and Surfaces*

Hard-core engineering applications related to computer-aided design and computer-aided manufacturing, CAD/CAM, require constructing lines and surfaces that follow the generally jagged trail of a small set of marker points (e.g., Olfe 1995).

One way to describe the position of these lines or surfaces is to demand that they pass through the marker points, and then represent them in a parametric manner using the *interpolation* methods discussed in Sections 6.8 and 6.10; this description, however, may lead to oscillatory behaviors. Methods of function *approximation* work much better but require large sets of data points. As a compromise, we adopt a graphics-oriented approach that might make the rigorous theoretician cringe but will please the practicing engineer.

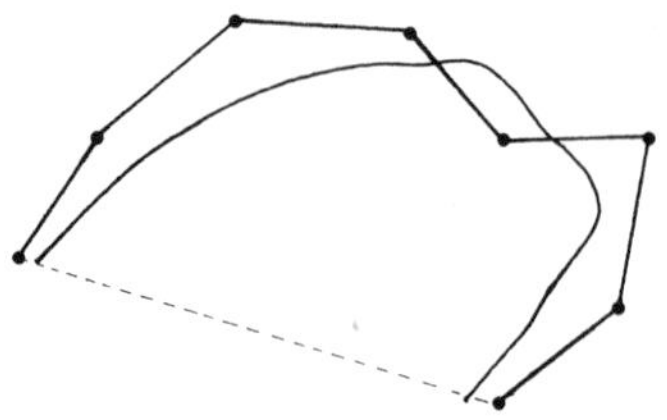

Planar *Bézier curves* discussed in this section, and *B-spline approximants* described in the next section, have the important property that they reside within the *convex hull* of a set of marker points that trace a planar line. The convex hull is the polygon that arises when a rubber band is stretched around the outermost marker points. This important property makes the Bézier curves and the *B*-spline approximants, as well as their corresponding two-dimensional counterparts, attractive in engineering practice.

Bézier Curves

A warning must be made at the outset: Bézier curves are primarily used by artists, designers, and stylists. Although the paths of art, science, and engineering often cross, and the author expects the readers to compute with style, the Bézier representation should be avoided when a high level of accuracy and mathematical rigor are required.

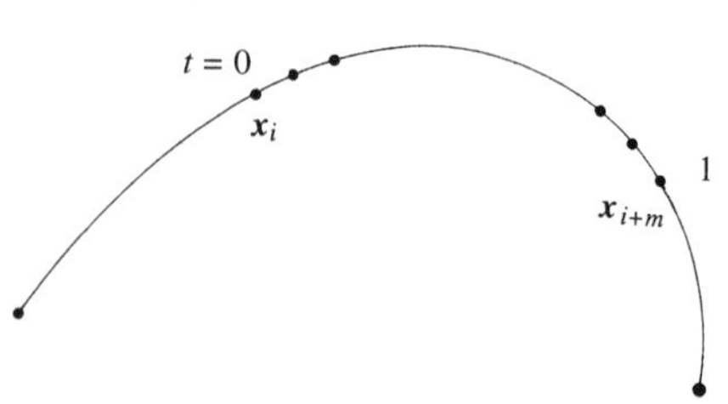

The Bézier curves were named after an engineer at the Renault company in France who invented them and used them in engineering design (Bézier 1971). Let us assume that we are given the Cartesian coordinates of a set of $N+1$ marker points $\boldsymbol{x}_i$, $i = 1, \ldots, N+1$, that trace the approximate location of a curve in the xy plane. Let us now select a group of $m+1$ successive marker points $\boldsymbol{x}_{i+j}$, $j = 0, \ldots, m$, and introduce the parameter t that takes the value of 0 at the first point $\boldsymbol{x}_i$, and the value of 1 at the last point $\boldsymbol{x}_{i+m}$. The corresponding Bézier curve is described by the parametric form

$$\boldsymbol{x}(t) = \sum_{l=1}^{m+1} \boldsymbol{x}_{i+l-1} B_{m,l}(t) \tag{8.3.1}$$

where $0 \le t \le 1$, and $B_{m,l}$ are the Bernstein polynomials defined in equation (8.2.5) and computed efficiently using the recursive formulas (8.2.8). Evaluating the right-hand side of equation (8.3.1) at $t = 0$ and 1 shows that the Bézier curve passes through the first and the last point, $\boldsymbol{x}_i$ and $\boldsymbol{x}_{i+m}$, but does not necessarily pass through the intermediate $m - 1$ points.

Since the Bézier curve follows the gross shape of the polygon with vertices at the m marker points, changing the position of one vertex alters the shape of the curve in an intuitive fashion. To raise the polynomial order of the Bézier curve, we simply introduce additional marker points. These features make the Bézier curves attractive alternatives to spline approximations.

For example, the cubic Bézier curve corresponding to $m = 3$ is described by the parametric form

$$\boldsymbol{x}(t) = \boldsymbol{x}_i(1-t)^3 + \boldsymbol{x}_{i+1}3t(1-t)^2 + \boldsymbol{x}_{i+2}3t^2(1-t) + \boldsymbol{x}_{i+3}t^3 \tag{8.3.2}$$

Several point distributions and the corresponding cubic Bézier segments are shown in the diagram after Rogers and Adams (1990).

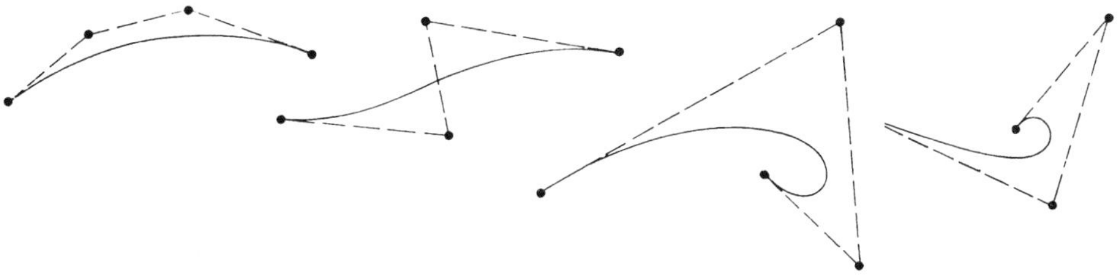

The Bézier curves have the interesting property that the tangential line at the first point $\boldsymbol{x}_i$ passes through the first doublet of points $\boldsymbol{x}_i$ and $\boldsymbol{x}_{i+1}$. Similarly, the tangential line at the last point $\boldsymbol{x}_{i+m}$ passes through the last doublet of points $\boldsymbol{x}_{i+m-1}$ and $\boldsymbol{x}_{i+m}$. More specifically, differentiating both sides of equation (8.3.1) with respect to t, and using the properties of the Bernstein polynomials, we find

$$\begin{aligned} \frac{d\boldsymbol{x}}{dt}(t) &= m \sum_{l=1}^{m} (\boldsymbol{x}_{i+l} - \boldsymbol{x}_{i+l-1}) B_{m-1,l}(t) \\ \frac{d^2\boldsymbol{x}}{dt^2}(t) &= m(m-1) \sum_{l=1}^{m} (\boldsymbol{x}_{i+l+1} - 2\boldsymbol{x}_{i+l} + \boldsymbol{x}_{i+l-1}) B_{m-2,l}(t) \end{aligned} \tag{8.3.3}$$

Bézier splines

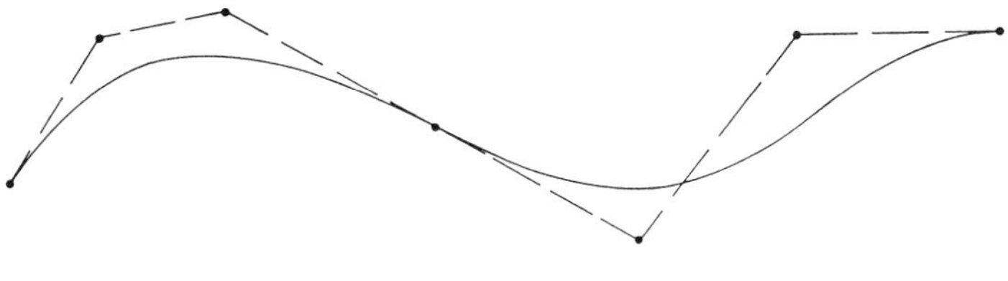

The collection of a sequence of Bézier segments, each defined by successive groups of $m + 1$ marker points, yields a continuous line with a generally discontinuous tangential vector at the first and last points of each group. To obtain a line with a continuous tangential vector, the three splicing points at the seams must be collinear, as shown in the last illustration after Rogers and Adams (1990). To obtain a line with continuous curvature, the five splicing points at the seams must be collinear.

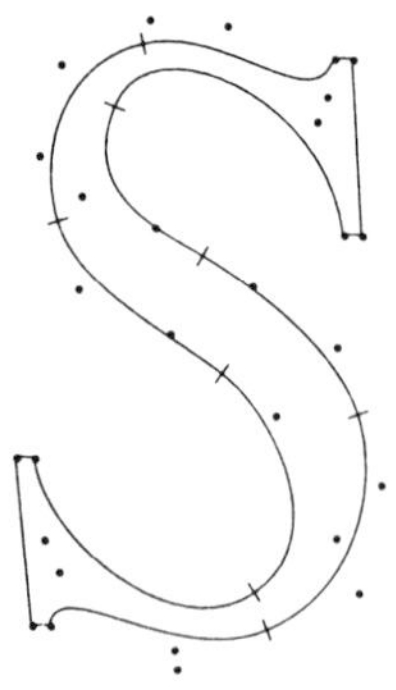

Bézier curves find a host of practical applications. For example, they are used to draw font shapes with the *PostScript*™ software. The contour of the letter S drawn with Bézier segments is shown in the diagram, after Olfe (1995).

In summary, the distinguishing features of the Bézier curves are: The order of the polynomial representation is determined by the number of marker points involved; and a change in the position of one marker point is felt through the entire span. These features may limit the maximum size of the data points that can be included in a collection. For large data sets, the B-splines described next in Section 8.4 are more appropriate.

PROBLEM

8.3.1 ***Bézier curves with continuous curvature.***

Prove the conditions on the distribution of the marker points stated in the text for two successive cubic Bézier curves to join smoothly with continuous tangential vectors and curvature.

Bézier Surfaces

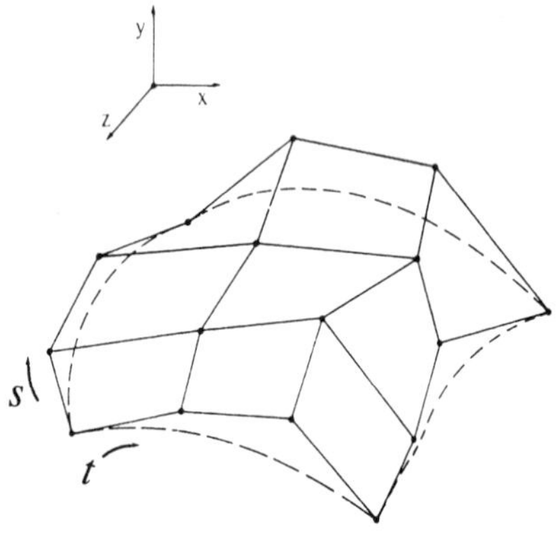

Bézier surface representations arise as straightforward extensions. Consider, for example, a $(m+1)\times(n+1)$ grid of points on a surface, forming a rectangular pattern. The Bézier surface element arises by introducing the curvilinear coordinates t and s, and describing the surface in the bivariate parametric form

$$\mathbf{x}(t) = \sum_{l_1=1}^{m+1}\sum_{l_2=1}^{n+1} \mathbf{x}_{i+l_1-1,j+l_2-1} B_{m,l_1}(t) B_{n,l_2}(s) \tag{8.3.4}$$

where t and s vary over the interval [0, 1] (e.g., Crow 1987). Further discussion of Bézier curves and surfaces can be found in texts of computational geometry and computer graphics (e.g., Farin 1990, Rogers and Adams 1990, Olfe 1995).

8.4 B-*spline Approximation and Representation of Lines and Surfaces*

The qualifier *B* is attributed to the word *Basis,* but some argue that it came from the word *Bell* because, as we shall see, *B*-splines look like bells. The word *Bogus* has been also suggested, but on a less serious note.

We emphasize again that the *B*-spline lines have the important property that they reside within the convex hull of a set of marker points that trace a line in a plane, as discussed in Section 8.3.

Influence Functions

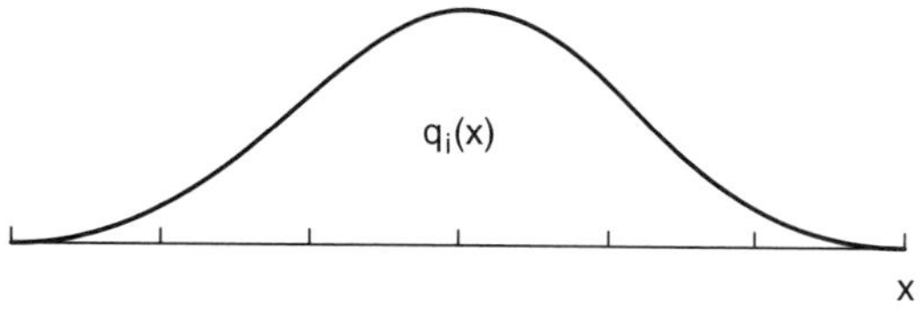

Consider the polynomial approximation of a function $f(x)$ that is described by $N + 1$ data points which, for reasons that will soon become apparent, we call *influence* points, $x_i^I, i = 1, 2, \ldots, N + 1$. To each one of these points, we assign a *basis function* $q_i(x)$ that takes nonnegative values and vanishes as x tends to $\pm\infty$; in the language of functional analysis, $q_i(x)$ are *distributions.* We then approximate the function $f(x)$ with the function

$$g(x) = \sum_{i=1}^{N+1} f(x_i^I) q_i(x) \tag{8.4.1}$$

To interpret this representation in physical terms, we identify the approximated function $f(x)$ with the temperature field induced by a collection of point sources of heat located at the positions x_i^I. In a qualitative sense, the functions $q_i(x)$ express the temperature fields associated with the individual point sources. For example, if the intervening medium is nonconductive, $q_i(x)$ will be delta functions, and the temperature field will consist of a sequence of spikes.

For the distributions $q_i(x)$ to be acceptable in the context of function approximation, they must conform to a number of constraints. The least we can ask is that $q_i(x)$ be such that the approximating function $g(x)$ is continuous. Moreover, if all $f(x_i)$ have the same value, we should demand that $g(x)$ also have the same value: If the function $f(x)$ is constant, $g(x)$ should also be a constant. In this light, the aforementioned choice of delta functions is not acceptable.

The *B*-spline approximation employs a family of basis functions $q_i(x)$ of different orders that exhibit the expected behaviors. The *B*-splines used in practice are similar to the interpolating cubic splines discussed in Section 6.6, but *do not always interpolate*; that is, they do *not* necessarily pass through the data points.

Linear B-splines

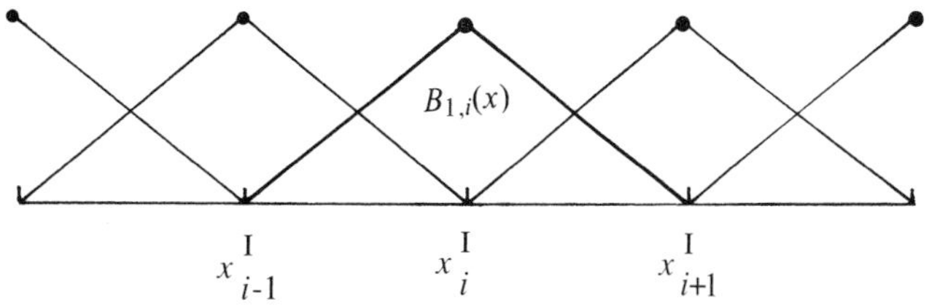

The first family of the *B*-spline basis functions are the linear tent-like functions shown in the diagram, familiar from Section 6.6. Denoting these functions as $q_i(x) = B_{1,i}(x)$, we restate equation (8.4.1) in the form

$$g(x) = \sum_{i=1}^{N+1} f(x_i^I) B_{1,i}(x) \tag{8.4.2}$$

which simply expresses the local linear interpolation discussed in Section 6.6. In this case, the B-spline approximation is actually an interpolation.

Evenly spaced points

When the data points are distributed evenly along the x axis, the right-hand side of equation (8.4.2) may be recast into an alternative form that expedites its numerical computation. This is done by introducing the *uniform linear spline function* $B_1(\xi)$ shown in the diagram, and describing the section of the interpolating function $g(x)$ between the points x_i^I and x_{i+1}^I in the form

$$g(\xi) = f(x_i^I) B_1(\xi) + f(x_{i+1}^I) B_1(\xi - 1) \tag{8.4.3}$$

where the parameter ξ varies between 0 and 1 in the span between the points x_i^I and x_{i+1}^I. For consistency with future notation, we recast equation (8.4.3) into the form

$$g(\xi) = f(x_i^I) b_{1,0}(\xi) + f(x_{i+1}^I) b_{1,1}(\xi) \tag{8.4.4}$$

where

$$b_{1,0}(\xi) = 1 - \xi, \qquad b_{1,1}(\xi) = \xi \tag{8.4.5}$$

are *linear basis functions*, also called *linear B-spline functions.*

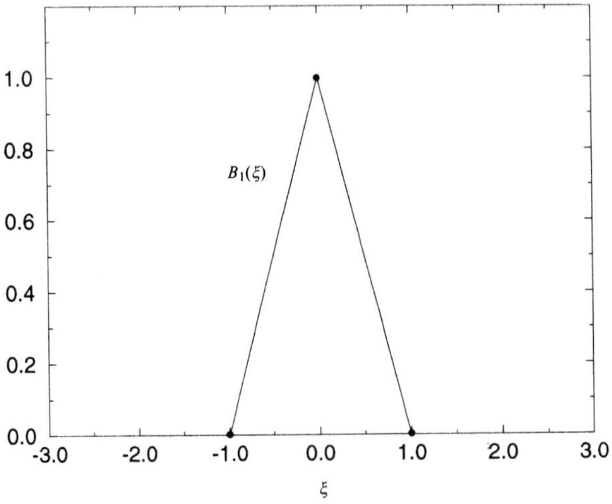

Quadratic B-splines

The second family of B-spline basis functions includes the *bell-shaped quadratic spline functions* $B_{2,i}(x)$ shown in the following diagram. Note that $B_{2,i}(x)$ is supported over the interval (x_{i-1}, x_{i+2}), where $x_i \le x_i^I \le x_{i+1}$; the absciscas x_i are called the *nodes* or *knots.* Thus, in the present case, we make a distinction between the *influence points,* x_i^I, and the *knots,* x_i. This distinction will be made for all even-degree B-splines to be discussed later in this section.

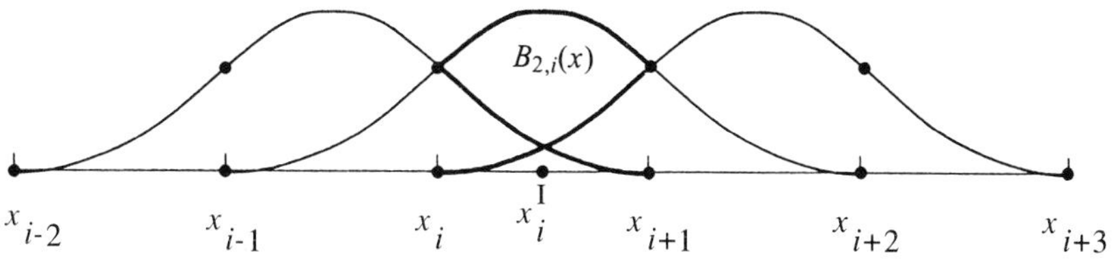

Each one of $B_{2,i}(x)$ consists of three quadratic segments; each segment is described by a distinct second-degree polynomial involving 3 coefficients, a total of 9 unknown coefficients for each spline. Equation (8.4.1) takes the form

$$g(x) = \sum_{i=1}^{N+1} f(x_i^I) B_{2,i}(x) \tag{8.4.6}$$

To compute the nine aforementioned coefficients, we require that $B_{2,i}(x)$ be a continuous function with continuous first derivatives, and we stipulate a normalization condition. The details of this computation will be illustrated shortly for the simple case of uniformly distributed points corresponding to *uniform* splines. A recursive algorithm for computing $B_{2,i}(x)$ in the more general case of unevenly spaced points will be discussed later in this section.

It is important to emphasize that the B-spline described by equation (8.4.6) does not necessarily pass through the marker points, although it does form a continuous line. This is a distinguishing feature of the B-spline approximation.

Evenly spaced points

Assuming that the data points are distributed evenly along the x axis, we introduce the independent variable ξ and the *uniform quadratic spline function* $B_2(\xi)$, which is supported over the interval $-1 \le \xi \le 2$. The graph of $B_2(\xi)$ consists of three quadratic segments; each segment is described by a second-degree polynomial involving 3 coefficients, a total of 9 unknowns.

To compute the 9 coefficients, we require the following:

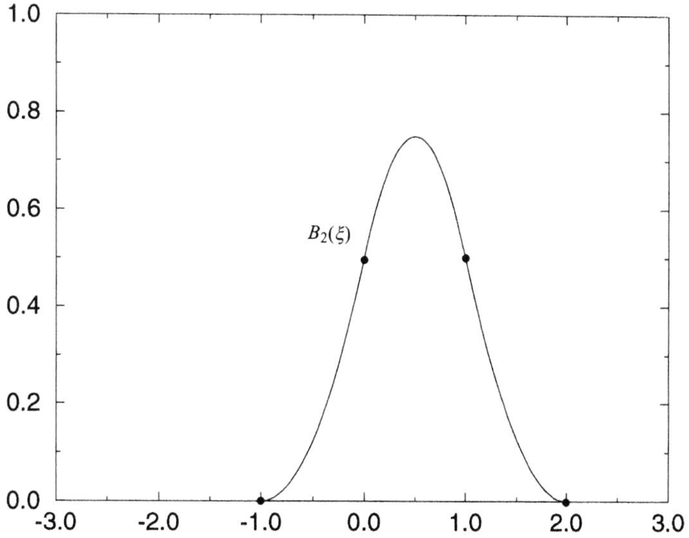

1. $B_2(\xi)$ and its first derivative be continuous at the interior knots $\xi = 0, 1$; these stipulations provide us with four equations.
2. $B_2(\xi) = 0$ and $B_2'(\xi) = 0$ at $\xi = -1, 2$; these stipulations provide us with four additional equations.
3. The normalization condition

$$B_2(\xi = 0) + B_2(\xi = 1) = 1.0 \tag{8.4.7}$$

whose origin will be revealed shortly.

Solving the resulting algebraic problem, we obtain

$$B_2(\xi) = \begin{cases} \beta_{2,1}(\xi) = \frac{1}{2}(1+\xi)^2 & \text{for } -1 \le \xi \le 0 \\ \beta_{2,0}(\xi) = \frac{1}{2}(1+2\xi-2\xi^2) & \text{for } 0 \le \xi \le 1 \\ \beta_{2,-1}(\xi) = \frac{1}{2}(2-\xi)^2 & \text{for } 1 \le \xi \le 2 \end{cases} \tag{8.4.8}$$

Returning to the representation (8.4.6), we describe the graph of the function between the points x_i and x_{i+1} in the parametric form

$$g(\xi) = f(x_{i-1}^I)B_2(\xi+1) + f(x_i^I)B_2(\xi) + f(x_{i+1}^I)B_2(\xi-1) \tag{8.4.9}$$

where the parameter ξ varies between 0 and 1 in the span between the points x_i and x_{i+1}.

The requirement (8.4.7) ensures that the right-hand side of equation (8.4.9) is a true weighted average; that is, the sum of the weights expressed by the β functions is equal to unity. Substituting equations (8.4.8) into the right-hand side of equation (8.4.9), and performing some algebra, we obtain the standard form

$$g(\xi) = f(x_{i-1}^I)b_{2,-1}(\xi) + f(x_i^I)b_{2,0}(\xi) + f(x_{i+1}^I)b_{2,1}(\xi) \tag{8.4.10}$$

where we have defined

$$\boxed{\begin{aligned} b_{2,-1}(\xi) &= \tfrac{1}{2}(1-\xi)^2 \\ b_{2,0}(\xi) &= \tfrac{1}{2}(1+2\xi-2\xi^2) \\ b_{2,1}(\xi) &= \tfrac{1}{2}\xi^2 \end{aligned}} \tag{8.4.11}$$

The reader may readily verify that $b_{2,-1}(\xi) + b_{2,0}(\xi) + b_{2,1}(\xi) = 1$.

Cubic *B*-splines

The third family of *B*-spline basis functions includes the *bell-shaped cubic-spline functions* $B_{3,i}(\xi)$ shown in the following diagram. Note that $B_{3,i}(x)$ is supported over the interval (x_{i-2}^I, x_{i+2}^I). In this case, the influence points coincide with the knots; this is a property of all odd-degree *B*-splines. Each $B_{3,i}(x)$ consists of four quadratic segments; each segment is described by a third degree polynomial involving four coefficients, a total of 12 unknown coefficients for each spline. Equation (8.4.1) takes the form

$$g(x) = \sum_{i=1}^{N+1} f(x_i^I) B_{3,i}(x) \tag{8.4.12}$$

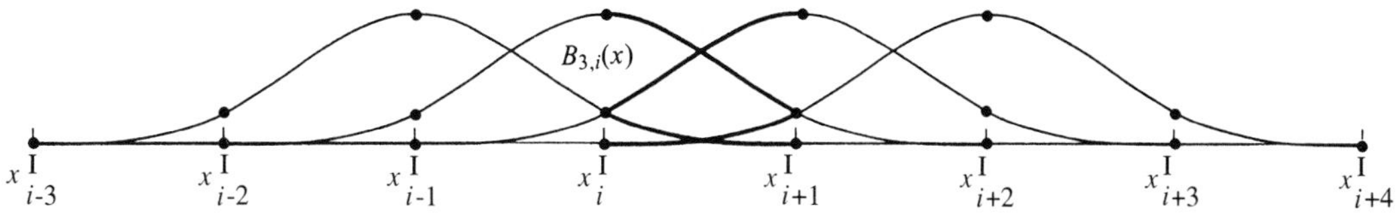

To compute the 12 coefficients, we require that each one of $B_{3,i}(x)$ be a continuous function with continuous first and second derivatives, and we stipulate a normalization condition, as will be explained shortly for uniform cubic splines. A recursive algorithm for computing $B_{3,i}(x)$ in the general case of unevenly spaced points will be discussed later in this section.

Evenly spaced points

Assuming that the abscissas of the data points are distributed evenly along the x axis, we introduce the *uniform cubic-spline function* $B_3(\xi)$ shown in the diagram. The graph of $B_3(\xi)$ consists of four cubic segments, each one expressible in terms of a third degree polynomial involving 4 coefficients, a total of 16 unknowns.

To compute the polynomial coefficients, we require the following:

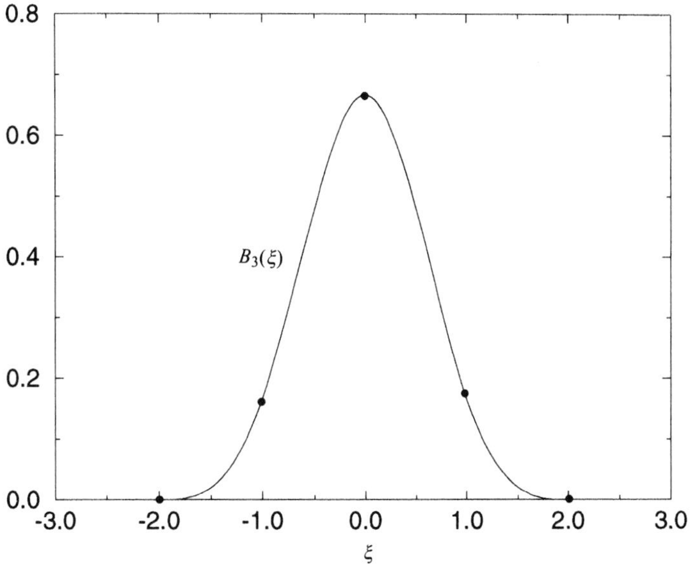

1. $B_3(\xi)$ and its first and second derivatives be continuous at the interior knots $\xi = -1, 0, 1$; these stipulations provide us with nine equations.
2. $B_3(\xi) = 0$, $B_3'(\xi) = 0$, and $B_3''(\xi) = 0$ at $\xi = \pm 2$; these stipulations provide us with six equations.
3. The normalization condition

$$B_3(\xi = -1) + B_3(\xi = 0) + B_3(\xi = 1) = 1.0 \tag{8.4.13}$$

whose origin will be revealed shortly.

Solving the algebraic system we obtain

$$B_3(\xi) = \begin{cases} \beta_{3,2}(\xi) = \frac{1}{6}(\xi+2)^2 & \text{for } -2 \le \xi \le -1 \\ \beta_{3,1}(\xi) = \frac{1}{6}(4 - 6\xi^2 - 3\xi^3) & \text{for } -1 \le \xi \le 0 \\ \beta_{3,0}(\xi) = \frac{1}{6}(4 - 6\xi^2 + 3\xi^3) & \text{for } 0 \le \xi \le 1 \\ \beta_{3,-1}(\xi) = \frac{1}{6}(2-\xi)^3 & \text{for } 1 \le \xi \le 2 \end{cases} \tag{8.4.14}$$

Applying the representation (8.4.12), we describe the section of the function between the data points x_i^I and x_{i+1}^I in the parametric form:

$$g(\xi) = f(x_{i-1}^I)B_3(\xi+1) + f(x_i^I)B_3(\xi) + f(x_{i+1}^I)B_3(\xi-1) + f(x_{i+2}^I)B_3(\xi-2) \tag{8.4.15}$$

where the parameter ξ varies between 0 and 1 between the points x_i^I and x_{i+1}^I. Substituting equations (8.4.14) into the right-hand side of equation (8.4.15), we obtain

$$g(\xi) = f(x_{i-1}^I)\beta_{3,-1}(\xi+1) + f(x_i^I)\beta_{3,0}(\xi) + f(x_{i+1}^I)\beta_{3,1}(\xi-1) + f(x_{i+2}^I)\beta_{3,2}(\xi-2) \tag{8.4.16}$$

The requirement (8.4.13) ensures that the right-hand side of equation (8.4.16) is a weighted average; the sum of the weights expressed by the β functions is equal to unity.

Finally, we recast equation (8.4.16) into the standard form

$$\boxed{g(\xi) = f(x_{i-1}^I)b_{3,-1}(\xi) + f(x_i^I)b_{3,0}(\xi) + f(x_{i+1}^I)b_{3,1}(\xi) + f(x_{i+2}^I)b_{3,2}(\xi)} \tag{8.4.17}$$

where

$$\boxed{\begin{aligned} b_{3,-1}(\xi) &= \tfrac{1}{6}(1-\xi)^3 \\ b_{3,0}(\xi) &= \tfrac{1}{6}(4 - 6\xi^2 + 3\xi^3) \\ b_{3,1}(\xi) &= \tfrac{1}{6}(1 + 3\xi + 3\xi^2 - 3\xi^3) \\ b_{3,2}(\xi) &= \tfrac{1}{6}\xi^3 \end{aligned}} \tag{8.4.18}$$

The reader may readily verify that $b_{3,-1}(\xi) + b_{3,0}(\xi) + b_{3,1}(\xi) + b_{3,2}(\xi) = 1$.

High-order *B*-splines

Generalizing the preceding discussion, we consider the kth degree B-splines $B_{k,i}(x)$ that take nonzero values over the interval (x_{i-l}, x_{i+m}) and are equal to zero outside this interval; the points x_j are the knots.

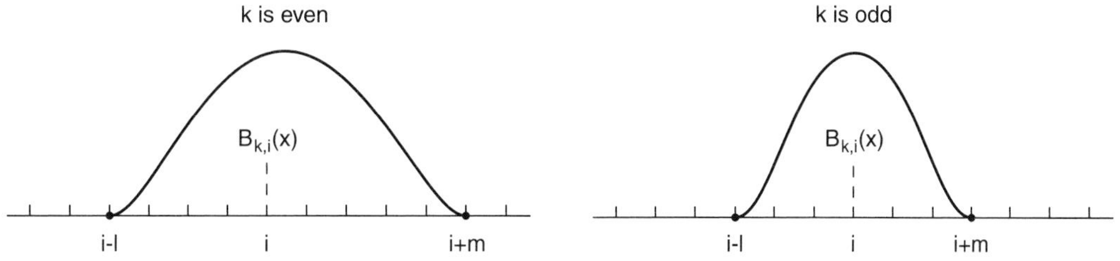

Odd-degree splines

When k is odd, we take $l = m = \frac{1}{2}(k+1)$, so that $l + m = k + 1$. For example, for the cubic splines discussed earlier, $k = 3, l = m = 2, l + m = 4$.

The influence points coincide with the knots. The B-spline approximation is effected by representing the function $f(x)$ between the points x_i^I and x_{i+1}^I with the function

$$g(x) = \sum_{j=i-(k-1)/2}^{i+(k+1)/2} f(x_j^I) A_{k,j-(k+1)/2}(x) \tag{8.4.19}$$

The spline functions $A_{k,q}$ can be computed recursively using the *Cox–de Boor algorithm*:

$$A_{0,q} = \begin{cases} 1 & \text{if } q = i \\ 0 & \text{otherwise} \end{cases}$$

Do $p = 1, k$

Do $q = i - p, i$

$$A_{p,q} = \frac{x - x_q}{x_{p+q} - x_q} A_{p-1,q} + \frac{x_{p+1+q} - x}{x_{p+1+q} - x_{q+1}} A_{p-1,q+1} \tag{8.4.20}$$

End Do

End Do

The B-spline approximation of a section of the function $f(x) = \cos(2\pi x)$ with nonuniform B-splines of degrees $k = 1, 3, 5$, is shown in Figure 8.4.1. The first degree line is the polyline.

Even-degree splines

When k is even, we take $l = \frac{1}{2}k$ and $m = \frac{1}{2}k + 1$, so that $l + m = k + 1$. For example, for the quadratic cubic splines discussed earlier, $k = 2, l = 1, m = 2, l + m = 3$.

The knots are placed between the influence points, so that $x_j \le x_j^I \le x_{j+1}$. The B-spline approximation is effected by approximating the function $f(x)$ between the knots x_i and x_{i+1} with the function

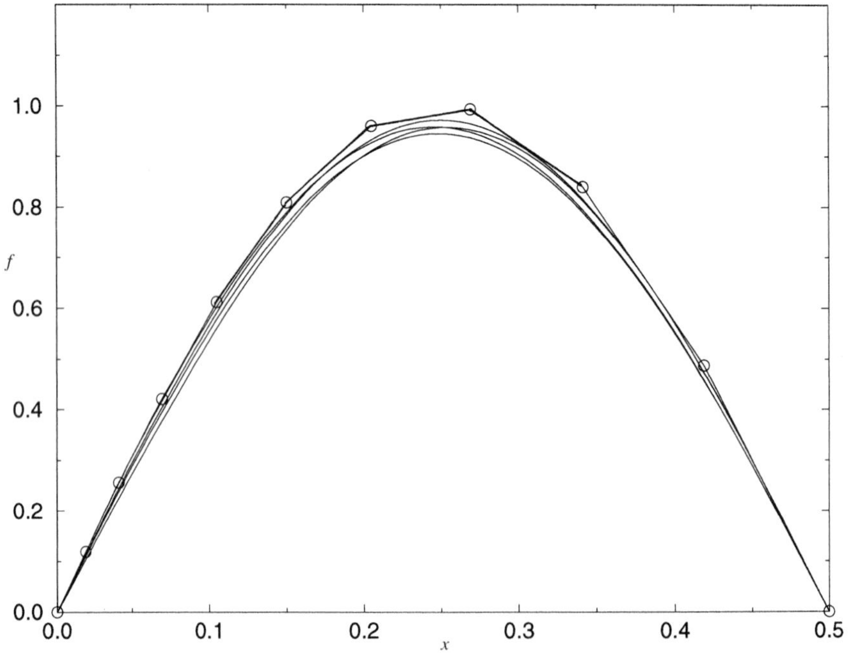

Figure 8.4.1 *B*-spline approximation of a section of the function $f(x) = \cos(2\pi x)$ with nonuniform *B*-splines of degrees $k = 1, 2, 3, 4, 5$. The knots are shown with circles. The first degree line is the polyline. For the even-degree splines, we have set $f(x_i^I) = \frac{1}{2}[f(x_i) + f(x_{i+1})]$.

$$g(x) = \sum_{j=i-k/2}^{i+k/2} f(x_j^I) A_{k,j-k/2}(x) \tag{8.4.21}$$

The spline functions $A_{k,q}$ can be computed recursively using the Cox–de Boor algorithm (8.4.20).

The *B*-spline approximation of a section of the function $f(x) = \cos(2\pi x)$ with nonuniform *B*-splines of degrees $k = 2, 4$, is shown in Figure 8.4.1.

B-spline Representation of Lines

To describe the shape of a line that has been traced with a trail of marker points, we express the coordinates $\boldsymbol{x} = (x, y, z)$ of a point along the line in a parametric manner using the parameter t, as $\boldsymbol{x}(t)$. We then assign values of t to the marker points, t_i, and carry out the *B*-spline approximation of the individual scalar functions $x(t)$, $y(t)$, $z(t)$. If the line is open, we count the end-points as multiples as many times as necessary.

The *B*-spline approximation replaces a curve with a collection of straight or curved segments yielding a *continuous* line. For example, using linear *B*-splines, we obtain a polygonal line that connects successive marker points.

Uniform splines

When the values of t_i are evenly spaced, we obtain uniform B-splines. Using the formulas developed previously in this section, we find the following:

- For linear B-splines:

$$\boxed{\boldsymbol{x}(t) = b_{1,0}(\xi)\,\boldsymbol{x}_i + b_{1,1}(\xi)\,\boldsymbol{x}_{i+1}} \tag{8.4.22}$$

  where $0 \le \xi \le 1$ corresponding to $t_i \le t \le t_{i+1}$. In this case, $\boldsymbol{x}_i$ are both influence points and knots.

- For quadratic B-splines:

$$\boxed{\boldsymbol{x}(t) = b_{2,-1}(\xi)\,\boldsymbol{x}_{i-1} + b_{2,0}(\xi)\,\boldsymbol{x}_i + b_{2,1}(\xi)\,\boldsymbol{x}_{i+1}} \tag{8.4.23}$$

  where $0 \le \xi \le 1$ corresponding to $t_{i-1/2} \le t \le t_{i+1/2}$. In this case, $\boldsymbol{x}_i$ are influence points, and the knots correspond to $t_{j+1/2}$.

- For cubic B-splines:

$$\boxed{\boldsymbol{x}(t) = b_{3,-1}(\xi)\,\boldsymbol{x}_{i-1} + b_{3,0}(\xi)\,\boldsymbol{x}_i + b_{3,1}(\xi)\,\boldsymbol{x}_{i+1} + b_{3,2}(\xi)\,\boldsymbol{x}_{i+2}} \tag{8.4.24}$$

  where $0 \le \xi \le 1$ corresponding to $t_i \le t \le t_{i+1}$. In this case, $\boldsymbol{x}_i$ are both influence points and knots.

Character outlines approximated with cubic B-splines from a rough frame are shown in the diagram after Olfe (1995); the left column shows the associated polylines. The smoothness of the representation is uncanny.

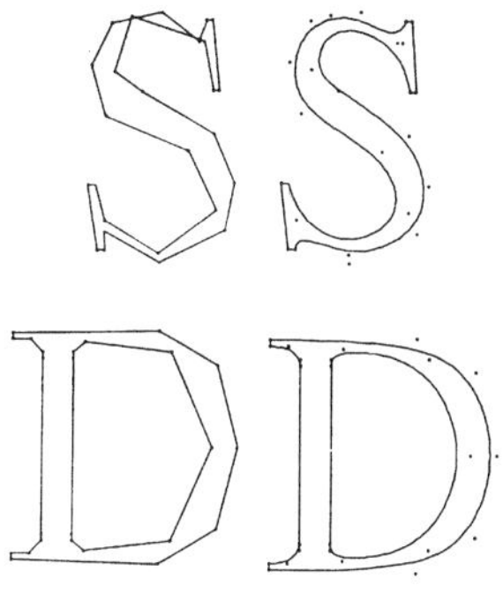

B-splines Representation of Surfaces

The theory and practice of bivariate B-spline approximation of surfaces arise as extensions of those for one-dimensional approximation. Details can be found in specialized monographs on computational geometry and computer graphics (e.g., Farin 1990, Rogers and Adams 1990, Olfe 1995, Shikin and Plis 1995).

8.5 Padé Approximation

The Padé approximantion represents a function $f(x)$ with the ratio of two polynomials. The approximation proceeds in two stages:

- First, $f(x)$ is approximated with an Nth degree polynomial

$$P_N(x) = c_0 + c_1 x + \cdots + c_N x^N \tag{8.5.1}$$

- Second, $P_N(x)$ is approximated with the *Padé approximant*

$$P_{(S/T)}(x) = \frac{W_S(x)}{Q_T(x)} \tag{8.5.2}$$

The numerator is an Sth degree polynomial,

$$W_S(x) = a_0 + a_1 x + \cdots + a_S x^S \tag{8.5.3}$$

and the denominator is the Tth degree polynomial,

$$Q_T(x) = 1 + b_1 x + \cdots + b_T x^T \tag{8.5.4}$$

with constant term equal to unity. The orders of the polynomials are such that

$$S + T = N \tag{8.5.5}$$

In practice, we usually select either $S = T$ or $T + 1$; in the first case we obtain a *diagonal approximant.*

The polynomial coefficients a_i and b_i are computed so that the Maclaurin series expansion of $R_{(S/T)}(x)$ truncated at the Nth power is identical to $P_N(x)$. Thus the error $e(x) = R_{(S/T)}(x) - P_N(x)$ can be expressed as an infinite power series in x whose first term involves the power x^{S+T+1}.

To exemplify the computation of the polynomial coefficients, we consider the function $f(x) = \arctan(x)$, and write out its Maclaurin series up to the ninth power (Gerald and Wheatly 1995),

$$P_9(x) = x - \tfrac{1}{3}x^3 + \tfrac{1}{5}x^5 - \tfrac{1}{7}x^7 + \tfrac{1}{9}x^9 \tag{8.5.6}$$

corresponding to $N = 9$. The $R_{(5/4)}(x)$ Padé approximant is

$$R_{(5/4)}(x) = (a_0 + a_1 x + a_2 x^2 + a_3 x^3 + a_4 x^4 + a_5 x^5)/(1 + b_1 x + b_2 x^2 + b_3 x^3 + b_4 x^4) \tag{8.5.7}$$

Setting the right-hand sides of equations (8.5.6) and (8.5.7) equal, we obtain

$$\begin{aligned}&\left(x - \tfrac{1}{3}x^3 + \tfrac{1}{5}x^5 - \tfrac{1}{7}x^7 + \tfrac{1}{9}x^9\right)(1 + b_1 x + b_2 x^2 + b_3 x^3 + b_4 x^4)\\ &\quad = a_0 + a_1 x + a_2 x^2 + a_3 x^3 + a_4 x^4 + a_5 x^5\end{aligned} \tag{8.5.8}$$

We have ten unknown coefficients and require an equal number of equations. Carrying out the multiplications on the left-hand side, and equating corresponding coefficients of the first ten monomials, we find

$$0 = a_0, \quad 1 = a_1, \quad b_1 = a_2, \quad b_2 - \tfrac{1}{3} = a_3, \quad b_3 - \tfrac{1}{3}b_1 = a_4, \quad b_4 - \tfrac{1}{3}b_2 + \tfrac{1}{5} = a_5$$

and

$$\tfrac{1}{3}b_3 - \tfrac{1}{5}b_1 = 0, \quad \tfrac{1}{3}b_4 - \tfrac{1}{5}b_2 + \tfrac{1}{7} = 0, \quad \tfrac{1}{5}b_3 - \tfrac{1}{7}b_1 = 0, \quad \tfrac{1}{5}b_4 - \tfrac{1}{7}b_2 + \tfrac{1}{9} = 0$$

The last four equations may readily be solved for the b_i, and the rest of them yield a_i. The Padé approximant is

$$R_{(5/4)}(x) = \left(x + \tfrac{7}{9}x^3 + \tfrac{64}{945}x^5\right) \Big/ \left(1 + \tfrac{10}{9}x^2 + \tfrac{5}{21}x^4\right) \tag{8.5.9}$$

To examine the accuracy of the approximation, we compare the exact value $f(1) = 0.78540$, with the value given by the nine-term Maclaurin series $P_9(1) = 0.83492$, and the value given by the Padé approximant $R_{(5/4)}(1) = 0.78558$. The Padé value is much more accurate than that computed from the nine-term Maclaurin series, even though the former arose as an approximation of the latter!

A programmable method of computing the coefficients will be discussed later in this section.

Improvement of series

In many applications, the explicit form of the approximated function $f(x)$ is unknown. Instead, the polynomial representation $P_N(x)$ is provided in the form of an asymptotic series. In these cases, the Padé approximation may be used to extend the range of accuracy of this series (e.g., van Dyke 1974).

Accuracy and Convergence

The ability of the Padé approximants to represent a certain class of functions with high accuracy and over an extended domain is intimately connected with the concept of analytic continuation. When the complex roots of the denominator $Q_T(z)$, where z is a complex variable, are located close to the singularities of the complex function $f(z)$, the Padé approximant $R_{(S/T)}(x)$ follows $f(x)$ more faithfully than the Maclaurin series. When this occurs, the Padé approximation produces the analytical continuation of the series outside its radius of convergence. When, on the other hand, the complex roots of $Q_T(z)$ are not close to the roots of $f(z)$, the Padé approximation does not have an advantage. Unfortunately, a criterion for distinguishing between these two possibilities is not available, and adopting the Padé approximation to improve the representation of a function requires a leap of faith.

The theoretical foundation of the Padé approximation is discussed in the comprehensive volumes of Wuytack (1979) and Baker and Graves-Morris (1996).

Computation of the Padé Approximants

The first two diagonal Padé approximants are readily found to be

$$R_{(1,1)}(x) = \frac{c_0c_1 + (c_1^2 - c_0c_2)x}{c_1 - c_2x} \tag{8.5.10}$$

and

$$R_{(2,2)}(x)$$
$$= \frac{c_0(c_2^2 - c_1c_3) + [c_1(c_2^2 - c_1c_3) + c_0(c_1c_4 - c_2c_3)]x + [c_0(c_3^2 - c_2c_4) + c_1(c_1c_4 - c_2c_3) + c_2(c_2^2 - c_1c_3)]x^2}{(c_2^2 - c_1c_3) + (c_1c_4 - c_2c_3)x + (c_3^2 - c_2c_4)x^2} \tag{8.5.11}$$

We can make the constant term in the denominator equal to 1 to conform with the standard form (8.5.4), simply by dividing both the numerator and the denominator by it, but this merely results in longer algebraic expressions.

To compute the coefficients of higher-order diagonal Padé approximants, with $S = T$, we set

$$a_0 = c_0, \qquad b_0 = 1 \tag{8.5.12}$$

solve the system of equations

$$\sum_{i=1}^{S} b_i c_{S-i+j} = -c_{S+j} \tag{8.5.13}$$

for the b_i, where $j = 1, \ldots, S$, and then compute the a_j from

$$a_j = \sum_{i=1}^{j} b_i c_{j-i} \tag{8.5.14}$$

where $j = 1, \ldots, S$.

Other algorithms for generating the polynomial coefficients of higher-order, and nondiagonal approximations, are described by Baker and Graves-Morris (1996).

Padé Approximants of the Exponential Function

Table 8.5.1 displays the first eight Padé approximants of the function $f(x) = \exp(x)$, along with the corresponding leading error, after Smith (1985). The approximation remains valid when x is replaced by a square matrix $\boldsymbol{A}$ to yield $\mathbf{exp}(\boldsymbol{A})$, in which case the reciprocal of a number is interpreted as the matrix inverse. For example,

$$\boldsymbol{R}_{(1/1)}(\boldsymbol{A}) = \left(\boldsymbol{I} + \tfrac{1}{2}\boldsymbol{A}\right)^{-1}\left(\boldsymbol{I} + \tfrac{1}{2}\boldsymbol{A}\right) \tag{8.5.15}$$

The formulas of Table 8.5.1 will be used in Chapter 9 when we discuss the solution of ordinary differential equations using numerical methods.

PROBLEM

8.5.1 *Deriving Padé approximants.*

(*a*) Write the Maclaurin series expansion of the function $f(x) = \ln(1+x)$, and retain terms up to and including the fifth power in x. Then, calculate $R_{(2/3)}$ and compare the exact value of $f(x)$, the value given by the truncated Taylor series expansion, and the value given by the Padé approximant at the points $x = -0.10, -0.50, -0.99$. Discuss the accuracy of the approximations.
(*b*) Derive the expression for $R_{(2/2)}(x)$ shown in Table 8.5.1.

Table 8.5.1 **The first eight Padé approximants of the function $f(x) = \exp(x)$ including the leading error terms, after Smith (1985).**

	Leading error term
$R_{(1/0)}(x) = 1 + x$	$\frac{1}{2}x^2$
$R_{(0/1)}(x) = 1/(1 - x)$	$-\frac{1}{2}x^2$
$R_{(2/0)}(x) = 1 + x + \frac{1}{2}x^2$	$\frac{1}{6}x^3$
$R_{(1/1)}(x) = \left(1 + \frac{1}{2}x\right) / \left(1 - \frac{1}{2}x\right)$	$-\frac{1}{12}x^3$
$R_{(0/2)}(x) = 1/\left(1 - x + \frac{1}{2}x^2\right)$	$\frac{1}{6}x^3$
$R_{(2/1)}(x) = \left(1 + \frac{2}{3}x + \frac{1}{6}x^2\right) / \left(1 - \frac{1}{3}x\right)$	$-\frac{1}{72}x^4$
$R_{(1/2)}(x) = \left(1 + \frac{1}{3}x\right) / \left(1 - \frac{2}{3}x + \frac{1}{6}x^2\right)$	$\frac{1}{72}x^4$
$R_{(2/2)}(x) = \left(1 + \frac{1}{2}x + \frac{1}{12}x^2\right) / \left(1 - \frac{1}{2}x + \frac{1}{12}x^2\right)$	$\frac{1}{72}x^4$

8.5.2 *Extending the accuracy of a series.*

After long calculations, a team of aerodynamicists came up with the asymptotic series $F = 1.0 + t + 0.5t^2$ for the lift force F on a new airfoil, as a function of the airfoil shape factor t. Improve the accuracy of this expansion by replacing it with a Padé approximant. Calculate F at $t = 1$ using the asympotic series and the Padé approximant and discuss the discrepancies.

Extensions

A more sophisticated version of the Padé approximation method replaces the Maclaurin expansions with Chebyshev series defined over the scaled interval $[-1, 1]$. The method is described by Ralston (1965) and Press et al. (1992, pp. 197–201).

Bivariate and multivariate Padé approximations are discussed by Prenter (1975) and Cuyt (1987).

8.6 *Trigonometric Approximation and Interpolation*

Consider the portion of a real function $f(x)$ over the interval $a < x < b$ of length $L = b - a$. Trigonometric or Fourier approximation and interpolation represents the function over that interval with the truncated *complete* Fourier series

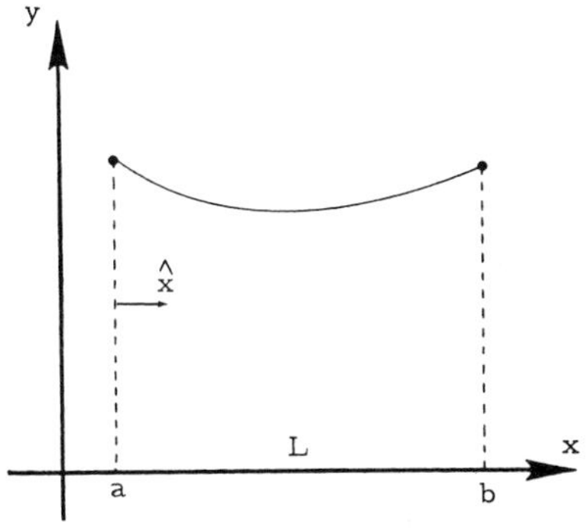

$$\begin{aligned} F_M(x) &= \tfrac{1}{2}a_0 + \sum_{p=1}^{M} a_p \cos(pk\hat{x}) + \sum_{p=1}^{M} b_p \sin(pk\hat{x}) \\ &= \sum_{p=-M}^{M} c_p \exp(-ipk\hat{x}) \end{aligned} \tag{8.6.1}$$

where $k = 2\pi/L$ is the wave number, M is a certain truncation level, a_p and b_p are real Fourier coefficients, c_p are complex Fourier coefficients, $\hat{x} = x - a$, and i is the imaginary unit. We can replace $\hat{x}$ with x without loss of generality, but this will only complicate the forthcoming algebraic manipulations.

To ensure that the right-hand side of the expansion (8.6.1) is real, we require that

$$c_{-p} = c_p^* \tag{8.6.2}$$

where an asterisk denotes the complex conjugate. Graphing $|c_p|^2$ against p produces the truncated *discrete power spectrum* of the function $f(x)$.

Using the Euler decomposition of the complex exponential, we find that the real and complex Fourier coefficients are related by

$$c_p = \tfrac{1}{2}(a_p + ib_p), \qquad a_p = 2\,\mathrm{Re}(c_p), \qquad b_p = 2\,\mathrm{Im}(c_p) \tag{8.6.3}$$

with the understanding that $b_0 = 0$, and therefore $c_0 = \frac{1}{2}a_0$ is real. The seemingly preferential treatment of the first term on the right-hand side of equation (8.6.1) is motivated by our desire to make the first of equations (8.6.3) hold true for all values of p, including zero.

Computation of the Fourier Coefficients

One way to compute the $2M+1$ Fourier coefficients a_p and b_p is to require that the graph of $F_M(x)$ passes through an equal number of data points, that is,

$$F_M(x_j) = f(x_j) \tag{8.6.4}$$

for $j = 1, \ldots, 2M+1$, thereby deriving a system of linear equations. This method, however, is not only computationally intensive but also suffers from serious difficulties similar to those encountered with the Vandermonde-matrix approach for polynomial approximation discussed in Section 6.2. A better method is required.

Fourier projection

The better method makes use of the Fourier orthogonality properties shown in Table 8.6.1. Multiplying all sides of the representation (8.6.1) with $\exp(iqk\hat{x})$, where q is an integer, using the first of the orthogonality properties listed in Table 8.6.1, and then relabeling q as p, we find

$$c_p = \frac{1}{L}\int_a^b f(x)\,\exp(ipk\hat{x})\,dx \tag{8.6.5}$$

Table 8.6.1 Orthogonality properties of trigonometric functions; $L = b - a$; $k = 2\pi/L$; p, q, s are integers; and $\hat{x} = x - a$.

$$\int_a^b \exp[i(p-q)k\hat{x}]\,dx = \int_0^L \exp[i(p-q)k\hat{x}]\,d\hat{x} = L\delta_{p,q}$$

$$\int_0^L \cos(pk\hat{x})\cos(qk\hat{x})\,d\hat{x} = \begin{cases} L & \text{if } p-q = sN \neq 0 \\ \frac{1}{2}L & \text{if } p = q = 0 \\ 0 & \text{otherwise} \end{cases}$$

$$\int_0^L \sin(pk\hat{x})\sin(qk\hat{x})\,d\hat{x} = \begin{cases} \frac{1}{2}L & \text{if } p = q \neq 0 \\ 0 & \text{otherwise} \end{cases}$$

$$\int_0^L \cos(pk\hat{x})\sin(qk\hat{x})\,d\hat{x} = 0$$

and thus

$$a_p = \frac{2}{L}\int_a^b f(x)\cos(pk\hat{x})\,dx, \qquad b_p = \frac{2}{L}\int_a^b f(x)\sin(pk\hat{x})\,dx \tag{8.6.6}$$

With the coefficients computed in this manner, it can be shown that, in the limit as M tends to infinity, the representation (8.6.1) becomes exact, even when the function $f(x)$ is discontinuous (e.g., Tolstov 1962).

Outside the interval (a, b), the Fourier series produces the periodic repetition of the section of $f(x)$ between (a, b), as depicted with the short dashed lines in Figure 8.6.1.

It is important to point out that the values of the Fourier coefficients computed from equations (8.6.5) and (8.6.6) are independent of how many terms in the Fourier expansion are retained; that is, they are independent of the value of M. Adding more terms does not affect the values of lower-order coefficients.

Consider, for example, a piecewise constant function defined as

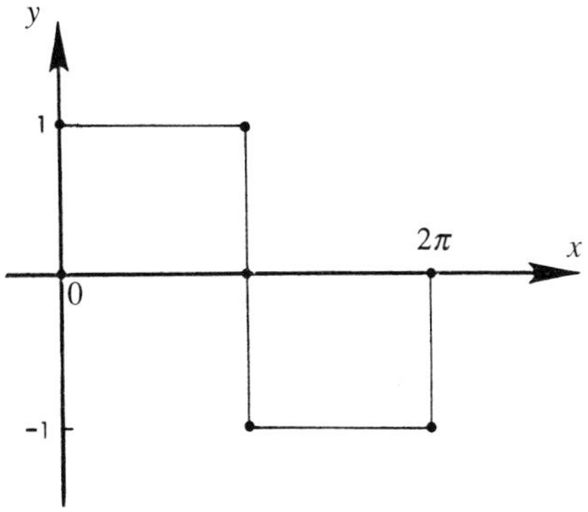

$$f(x) = \begin{cases} 0 & \text{for } x < 0 \\ 1 & \text{for } 0 < x < \pi \\ -1 & \text{for } \pi < x < 2\pi \\ 0 & \text{for } x > 2\pi \end{cases} \tag{8.6.7}$$

and concentrate on the section confined between $x = a = 0$ and $b = 2\pi$, corresponding to $L = 2\pi$ and $k = 1$. Straightforward evaluation of the integrals (8.6.6) yields $a_p = 0$ for any value of p, $b_p = 0$ when p is an even integer, and $b_p = 4/(p\pi)$ when p is an odd integer. The emerging sine Fourier series expansion is

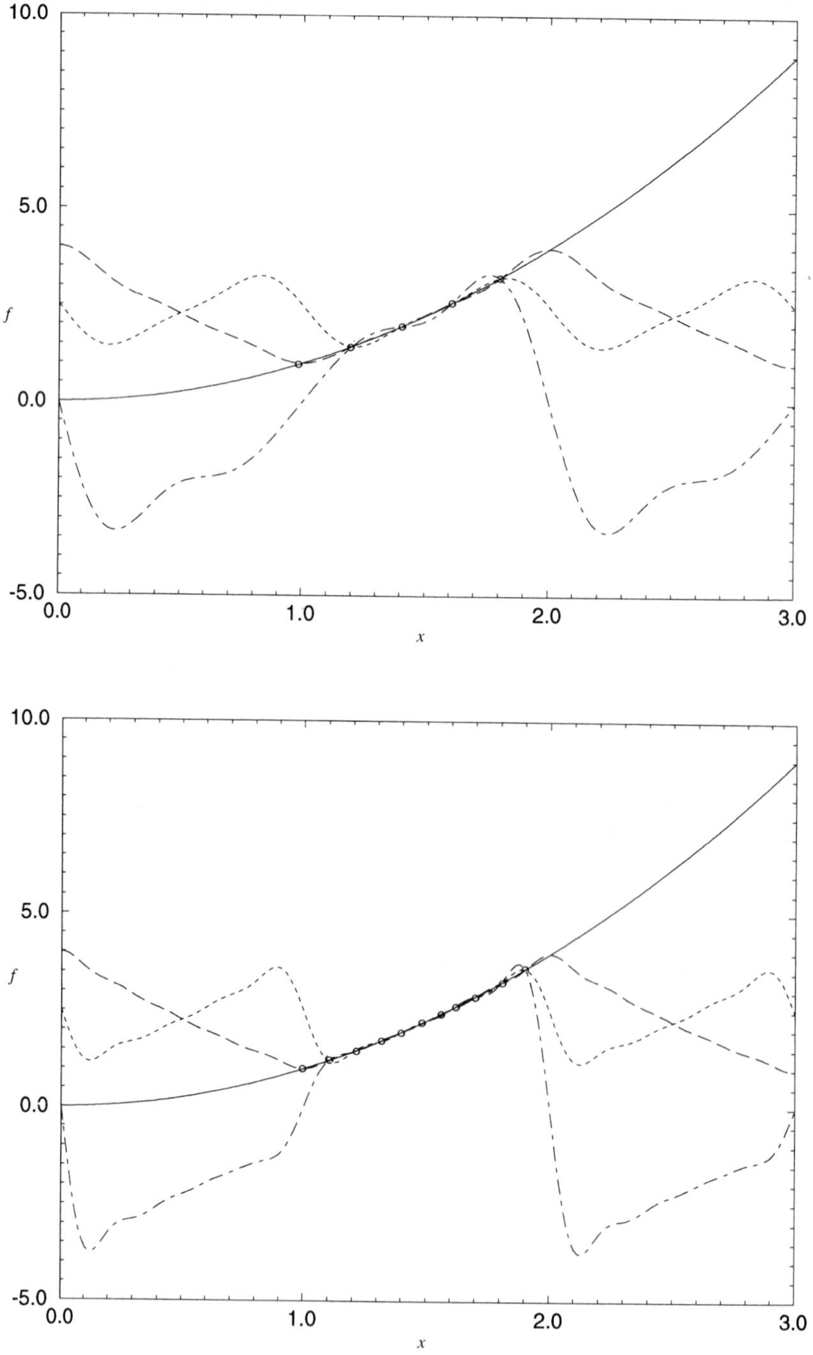

Figure 8.6.1 Representation of the section of the function $f(x) = x^2$ over the interval $1 \leq x \leq 2$, drawn with the solid line, with a truncated *complex Fourier series* drawn with the short dashed line, a truncated *cosine Fourier series* drawn with the long dashed line, and a truncated *sine Fourier series* drawn with the dot-dashed line. The Fourier coefficients are computed from values of the function at $M + 1$ evenly spaced data points; all series interpolate through the *interior* data points; (a) $M = 5$ and (b) $= 10$.

$$f(x) = \frac{4}{\pi}(\sin x + \tfrac{1}{3}\sin 3x + \tfrac{1}{5}\sin 5x + \cdots) \tag{8.6.8}$$

The periodic continuation of this function yields a square wave.

PROBLEMS

8.6.1 ***Square wave.***

Plot the truncated Fourier series $F_M(x)$ of the square wave function defined in equations (8.6.7) for $M = 1, 3, 5, 15, 30$, and discuss the oscillations near the points of discontinuity, described as the *Gibbs effect.*

8.6.2 ***Triangular wave.***

Derive the Fourier series of a function whose graph is a triangular wave of maximum amplitude equal to εL, where ε is a free parameter, as shown in the diagram.

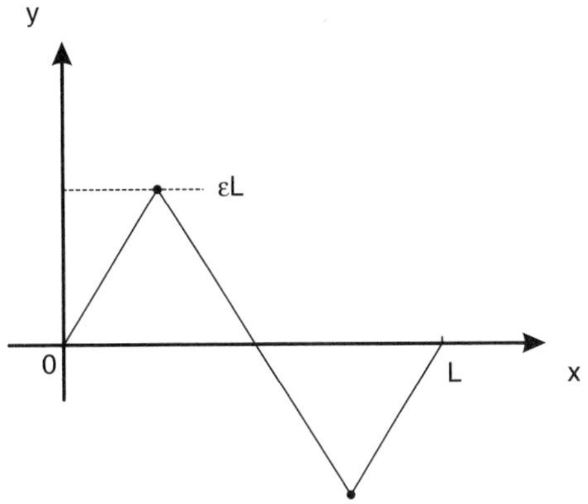

Least-squares approximation

It can be shown that the values of the coefficients displayed in equations (8.6.5) and (8.6.6) minimize the least-squares functional with a flat weighting function

$$E_M[e^2(x)] \equiv \int_a^b \left(F_M(x) - f(x)\right)^2 dx \tag{8.6.9}$$

Thus the Fourier projection method and the least-squares approximation method are *equivalent.*

Computation of the Fourier Coefficients

The integrals on the right-hand sides of expressions (8.6.5) and (8.6.6) may be computed using a standard method of numerical integration in terms of the values of the function $f(x)$ at a number of data points distributed over the interval $[a, b]$, as discussed in Chapter 7. We note that the values of the Fourier coefficients obtained in this manner will depend on the selected integration method, and this reveals that *the truncated Fourier series will not generally be an interpolating function,* that is, it will not pass through the data points.

Evenly Spaced Points

Curiously but remarkably, when the data points are evenly spaced, applying the trapezoidal rule to evaluate the integrals (8.6.5) and (8.6.6) guarantees that the truncated Fourier series is interpolating.

The requirement of evenly spaced points may appear to be strict, but it is not so; one can introduce a new independent variable t so that $x = q(t)$ as discussed in Section 6.4, so that the data points become evenly spaced with respect to t. The simplest way of effecting this transformation is to relabel the data points with successive integers. But a warning must be made: When the function $x = q(t)$ is not sufficiently smooth, the interpolating Fourier series may exhibit strong fluctuations.

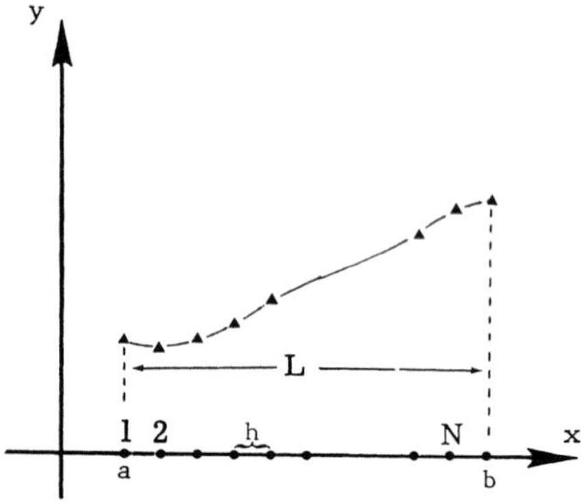

Let us then divide the interval L into N subintervals that are demarcated by the $N + 1$ data points $x_j = a + (j - 1)h$, $j = 1, \ldots, N + 1$, where $h = L/N$, and $x_1 = a$, $x_{N+1} = b$. In practice, x usually represents time, h is the *sampling time*, and $1/h$ is the *sampling rate.* Using the trapezoidal rule to evaluate the complex Fourier integral in equation (8.6.5), we obtain

$$c_p = \frac{1}{N}\left(\tfrac{1}{2}f(x_1) + \omega^p f(x_2) + \cdots + \omega^{p(N-1)} f(x_N) + \tfrac{1}{2}f(x_{N+1})\right) \tag{8.6.10}$$

where we have defined

$$\omega \equiv \exp(ikh) = \exp(2\pi i/N) \tag{8.6.11}$$

It is clear that

$$\omega^N = 1 \tag{8.6.12}$$

and therefore

$$c_{p+rN} = c_p \tag{8.6.13}$$

where r is an integer, which means that the Fourier coefficients are repeated periodically after N terms. The property (8.6.13) is called *aliasing.* Moreover, since the coefficients c_p and c_{N-p} are complex conjugate, we only have to compute about half of the complex coefficients involved in the truncated expansion, and the rest of them can be found by reflection with repect to the real axis.

Using the second and third of relations (8.6.3), we find

$$\begin{aligned} a_p &= \frac{2}{N}\left[\tfrac{1}{2}f(x_1) + \cos(pkh)f(x_2) + \cdots + \cos(pkh(N-1))f(x_N) + \tfrac{1}{2}f(x_{N+1})\right] \\ b_p &= \frac{2}{N}\left[\sin(pkh)f(x_2) + \cdots + \sin(pkh(N-1))f(x_N)\right] \end{aligned} \tag{8.6.14}$$

To this end, we distinguish two cases:

N is odd

We truncate the Fourier sum (8.6.1) at the value $M = \frac{1}{2}(N-1)$, and compute a_p and b_p for $p = 0, \ldots, M$, using equations (8.6.14).

N is even

We truncate the Fourier sum (8.6.1) at the value $M = \frac{1}{2}N$, compute a_p and b_p for $p = 0, \ldots, M-1$, using equations (8.6.14), and set

$$a_M = 2c_M = 2c_{-M} = \frac{1}{N}\left(\tfrac{1}{2}f(x_1) - f(x_2) + \cdots - f(x_N) + \tfrac{1}{2}f(x_{N+1})\right)$$
$$b_M = 0 \tag{8.6.15}$$

The alternating signs on the right-hand side of the penultimate equation are explained by observing that $\omega^{N/2} = \exp(\frac{1}{2}ikhN) = \exp(i\pi) = -1$.

With the Fourier coefficients computed in this manner, and using the identities shown in Table 8.6.2, it can be shown that

$$F_M(x_j) = f(x_j) \qquad \text{for } j = 2, \ldots, N \tag{8.6.16}$$

that is, the truncated Fourier series interpolates through the interior data points, and furthermore,

$$F_M(x_1) = F_M(x_{N+1}) = \tfrac{1}{2}[f(x_1) + f(x_{N+1})] \tag{8.6.17}$$

that is, the truncated Fourier series interpolates through the mean ordinate of the first and last points. An example of such an interpolating truncated Fourier series with $N = 5$ and 10 is shown with the short-dashed line in Figure 8.6.1.

Table 8.6.2 Discrete orthogonality properties of trigonometric functions; $x_j = a + (j-1)h$, $j = 1, \ldots, N+1$, is a sequence of evenly spaced points; $h = L/N$, $L = b - a$, $x_1 = a$, $x_{N+1} = b$, $k = 2\pi/L$, $\hat{x}_j = x_j - a$; i is the imaginary unit; p, q, s are integers.

$$\sum_{j=1}^{N} \exp(ipk\hat{x}_j) = \begin{cases} N & \text{if } l = sN \\ 0 & \text{otherwise} \end{cases}$$

$$\sum_{j=1}^{N} \exp[-ijk(x_p - x_q)] = \begin{cases} N & \text{if } p - q = sN \\ 0 & \text{otherwise} \end{cases}$$

$$\sum_{j=1}^{N} \cos(pk\hat{x}_j)\cos(qk\hat{x}_j) = \begin{cases} N & \text{if } p - q = sN \neq 0 \\ \dfrac{1}{2}N & \text{if } p = q \\ 0 & \text{otherwise} \end{cases}$$

$$\sum_{j=1}^{N} \sin(pk\hat{x}_j)\sin(qk\hat{x}_j) = \begin{cases} \dfrac{1}{2}N & \text{if } p - q = sN \\ 0 & \text{otherwise} \end{cases}$$

$$\sum_{j=1}^{N} \cos(pk\hat{x}_j)\sin(qk\hat{x}_j) = 0$$

PROBLEM

8.6.3 *Discrete Fourier orthogonality property.*

(*a*) Derive the first identity shown in Table 8.6.2. *Hint*: Note that the sum on the left-hand side is equal to $(1 - z^N)/(1 - z)$, where z is an appropriate complex number.
(*b*) Derive the second identity shown in Table 8.6.2 on the basis of the first identity.

Periodic functions

If the function $f(x)$ is repeated periodically with period L, $f(x_1) = f(x_{N+1})$, formula (8.6.10) takes the simpler form

$$c_p = \frac{1}{N} \sum_{j=1}^{N} \omega^{p(j-1)} f(x_j) \tag{8.6.18}$$

and formulas (8.6.14) undergo analogous simplifications. In this case, it can be shown that

$$F_M(x_j) = f(x_j) \tag{8.6.19}$$

for all $j = 1, \ldots, N + 1$, that is, the truncated Fourier series interpolates through all $N + 1$ data points.
In practice, when N is large, the sum on the right-hand side of equation (8.6.18) is computed most efficiently using the method of *fast Fourier transform* to be discussed in Section 8.7.

Cosine Fourier Series

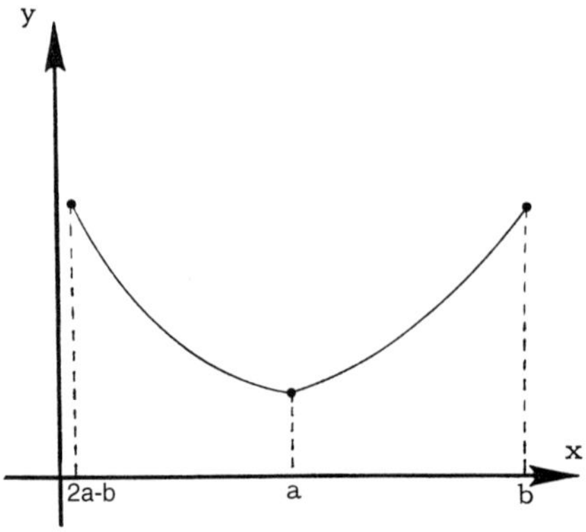

Consider now the portion of a function $f(x)$ over the interval $a < x < b$ of length $L = b - a$, and define its *even extension* over the adjacent interval $c < x < a$ of same length L, where $c = 2a - b$, so that

$$f(x) = f(2a - x) \tag{8.6.20}$$

for $c < x < b$. Introducing the Fourier representation (8.6.1) over the whole interval $c < x < b$ of length $b - c = 2L$, and requiring that the truncated Fourier series respects the reflection property (8.6.20), thereby discarding the sine terms, we obtain the cosine Fourier series expansion

$$F_M^c(x) = \tfrac{1}{2} A_0 + \sum_{p=1}^{M} A_p \cos\left(\tfrac{1}{2} p k \hat{x}\right) \tag{8.6.21}$$

where $k = 2\pi/L$, and $\hat{x} = x - a$. Outside the interval (c, b), the Fourier series yields the periodic repetition of the section of $f(x)$ between c and b, as illustrated in Figure 8.6.1.

One way to compute the $M+1$ coefficients A_p is to require that the graph of $F_M^c(x)$ pass through $M+1$ data points, that is, $F_M^c(x_j) = f(x_j)$, for $j = 1, \ldots, M+1$, thereby obtaining a system of linear equations. A much better method makes use of the orthogonality property of trigonometric functions displayed in Table 8.6.1.

Multiplying both sides of the representation (8.6.21) by $\cos(qk\hat{x}/2)$, where q is an integer, using the second of the orthogonality properties shown in Table 8.6.1, and then relabeling q as p, we find that the cosine Fourier coefficients are given by

$$A_p = \frac{2}{L} \int_a^b f(x) \cos\left(\tfrac{1}{2} pk\hat{x}\right) dx \tag{8.6.22}$$

As an example, we take up the seemingly absurd task of representing the section of the sine function $f(x) = \sin x$ that lies within the interval $0 < x < \pi$ with a cosine Fourier series; in this case, $a = 0, b = \pi, L = \pi$, and $k = 2$. Straightforward evaluation of the integrals in equation (8.6.22) yields $A_0 = 4/\pi$, and $A_p = 0$ when p is an odd integer. The expressions for $A_p = 0$ when p is an even integer may be derived with the help of standard mathematical tables (Problem 8.6.4).

PROBLEM

8.6.4 ***Cosine Fourier series of the sine.***

Plot the cosine Fourier series of $f(x) = \sin x$ over the interval $0 < x < \pi$, as well as its periodic extension outside the interval $0 < x < \pi$, for $M = 1, 3, 11, 17, 41$, and discuss the quality of the approximation (for the evaluation of the integrals see Gradshteyn and Ryzhik 1980, p. 373).

Evenly spaced points

When the data points are spaced evenly, computing the integrals on the right-hand side of equation (8.6.22) using the trapezoidal rule ensures that the truncated cosine Fourier series interpolates through the data points. The numerical procedure involves the following steps:

1. Divide the interval L into N subintervals that are demarcated by the $N+1$ data points $x_j = a + (j-1)h$, $j = 1, \ldots, N+1$, where $h = L/N$, $x_1 = a$, $x_{N+1} = b$.
2. Compute the Fourier coefficients

$$\begin{aligned} A_p = \frac{2}{N} \Big[& \tfrac{1}{2} f(x_1) + \cos\left(\tfrac{1}{2} pkh\right) f(x_2) + \cdots \\ & + \cos\left(\tfrac{1}{2} p(N-1)kh\right) f(x_N) + \tfrac{1}{2} \cos(p\pi) f(x_{N+1}) \Big] \end{aligned} \tag{8.6.23}$$

   for $p = 0, \ldots, N$. Replace A_N with $\frac{1}{2} A_N$.
3. Compute the series (8.6.21) truncated at $M = N$.

With the Fourier coefficients computed in this manner, and with the help of Table 8.6.2, it can be shown that the truncated cosine Fourier series interpolates through all data points. An example of such an interpolating series with $N = 5$ and 10 is shown with the long-dashed line in Figure 8.6.1.

Sine Fourier Series

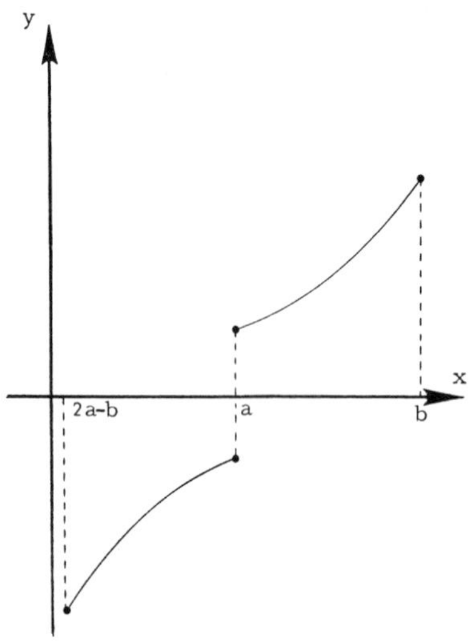

We reconsider the portion of the function $f(x)$ over the interval $a < x < b$ of length $L = b - a$; this time, we define its *odd extension* over the adjacent interval $c < x < a$, where $c = 2a - b$, so that

$$f(x) = -f(2a - x) \tag{8.6.24}$$

for $c < x < b$. Applying the Fourier representation (8.6.1) over the whole interval $c < x < b$ of length $b - c = 2L$, and insisting that the truncated Fourier series respects the property (8.6.24), thereby discarding the cosine terms, we obtain the sine Fourier series

$$F_M^s(x) = \sum_{p=1}^{M} B_p \sin\left(\tfrac{1}{2}pk\hat{x}\right) \tag{8.6.25}$$

where $k = 2\pi/L$ is the wave number, and $\hat{x} = x - a$. Outside the interval (c, b), the Fourier series yields the periodic repetition of the section of $f(x)$ between (c, b), as illustrated in Figure 8.6.1.

We can compute the M coefficients B_p by requiring that the graph of F_M^s pass through M data points, that is, $F_M^s(x_j) = f(x_j)$ for $j = 1, \ldots, M$, but problems with numerical accuracy will arise. A much better method uses the third of the orthogonality properties displayed in Table 8.6.1. Working in the familiar way, we find

$$B_p = \frac{2}{L}\int_a^b f(x) \sin\left(\tfrac{1}{2}pk\hat{x}\right) dx \tag{8.6.26}$$

As an example, consider the flat function $f(x) = 1$ in the interval between $a = 0$ and $b = \pi$, corresponding to $L = \pi$ and $k = 2$. The odd extension of this function yields a step function. Straightforward analytical evaluation of the integrals in equation (8.6.26) produces $B_p = 0$ when p is an even integer, and $B_p = 4/(p\pi)$ when p is an odd integer. These results are consistent with those shown in equation (8.6.8).

PROBLEM

8.6.5 ***Sine Fourier series yielding a square wave of period*** 2π.

Plot the sine Fourier series of the flat function $f(x) = 1$ in the interval $(0, \pi)$, as well as its odd periodic extension, for $N = 1, 3, 11, 17, 41$, and discuss the quality of the approximation.

Evenly spaced points

When the data points are spaced evenly, evaluating the integral on the right-hand side of equation (8.2.26) using the trapezoidal rule ensures that the truncated sine Fourier series passes through the interior data points. The numerical procedure involves the following steps:

1. Divide the interval L into N subintervals that are demarcated by the $N+1$ data points $x_j = a + (j-1)h$, $j = 1, \ldots, N+1$, where $h = L/N$, $x_1 = a$, $x_{N+1} = b$.
2. Compute the Fourier coefficients from

$$B_p = \frac{2}{N}\left[\sin\left(\tfrac{1}{2}pkh\right) f(x_2) + \cdots + \sin\left(\tfrac{1}{2}p(N-1)kh\right) f(x_N)\right] \tag{8.6.27}$$

   for $p = 1, \ldots, N-1$.
3. Compute the series (8.6.25) truncated at $M = N - 1$.

With the Fourier coefficients computed in this manner, and with the help of Table 8.6.2, it can be shown that the truncated sine Fourier series passes through the interior data points and interpolates through the end-points $f(x_1) = 0$ and $f(x_{N+1}) = 0$. Two such interpolating series with $N = 5$ and 10 are shown with the dot-dashed lines in Figure 8.6.1. The oscillations near the end of the interpolation interval are manifestations of the Gibbs effect.

Which Fourier Series?

Suppose that we are given the values of a function at $N+1$ data points evenly distributed over the interval $a < x < b$, $x_j = a + (j-1)h$, where $j = 1, \ldots, N+1$, $h = L/N$, $x_1 = a$, and $x_{N+1} = b$. We can use this information to compute the complete Fourier series (8.6.1) with an appropriate number of terms, the cosine Fourier series (8.6.21) with $M = N$, and the sine Fourier series (8.6.25) with $M = N - 1$. Which one of these series is more appropriate? Inspection of the graphs in Figure 8.6.1 reveals that this choice may have a significant effect on the accuracy of the interpolation.

The answer is: *Using the complete Fourier series one can never go wrong,* whereas using the sine or the cosine series requires hindsight. Whenever appropriate, the complete series reduces to the cosine or sine series in a natural manner. For example, we saw that it is possible to represent the function $f(x) = \cos x$ using either a one-term cosine series or a slowly convergent sine Fourier series; the complete Fourier series will automatically make the right choice. In general, unless $f(a) = 0$ and $f(b) = 0$, the sine series introduces significant Gibbs oscillations.

8.7 *Fast Fourier Transform*

When the number of data points N is large—in practice it can be on the order of several million—computing the sum on the right-hand side of equation (8.6.10) in the regular manner is prohibitively expensive. Fortunately, this stumbling block can be circumvented by the use of a powerful summation method known as the *fast Fourier transform*, FFT, attributed to Cooley and Tukey (1965) (for a historical account see Brigham 1974).

To simplify the notation, we define

$$g_{j-1} \equiv \frac{1}{N} f(x_j) \tag{8.7.1}$$

and rewrite formula (8.6.18), applicable to a periodic function, in the form

$$c_p = \sum_{\beta=0}^{N-1} \omega^{p\beta} g_\beta \tag{8.7.2}$$

where $\omega = \exp(2\pi i/N)$. The function $f(x)$ is not necessarily periodic, but we pretend that it is; as long as N is sufficiently large, this approximation introduces an insignificant error.

When N is even, we compute the coefficients c_p for $p = -\frac{1}{2}N + 1, \ldots, \frac{1}{2}N$; when N is odd, we compute the coefficients c_p for $p = -\frac{1}{2}(N-1), \ldots, \frac{1}{2}(N-1)$. But because of the property (8.6.13), we can shift these ranges to any other range of N consecutive integers. Selecting the range $0 \le p \le N-1$, and placing equations (8.7.2) into the vector–matrix form, we obtain

$$\begin{bmatrix} c_0 \\ c_1 \\ c_2 \\ \cdots \\ c_{N-1} \end{bmatrix} = \begin{bmatrix} 1 & 1 & 1 & 1 & \cdots & 1 \\ 1 & \omega & \omega^2 & \omega^3 & \cdots & \omega^{N-1} \\ 1 & \omega^2 & \omega^4 & \omega^6 & \cdots & \omega^{2N-2} \\ \cdots & \cdots & \cdots & \cdots & \cdots & \cdots \\ 1 & \omega^{N-1} & \omega^{2N-2} & \cdots & \cdots & \omega^{(N-1)^2} \end{bmatrix} \begin{bmatrix} g_0 \\ g_1 \\ g_2 \\ \cdots \\ g_{N-1} \end{bmatrix} \tag{8.7.3}$$

The matrix on the right-hand side is known as the *discrete Fourier transform* matrix, DFT.

Evaluating the polynomial-like sums on the left-hand side of equation (8.7.3) using Horner's rule (Section 1.4) requires a total of N^2 additions and an equal number of multiplications. When N is large, the cost of these operations is unaffordable. We shall see that the fast Fourier transform method requires a number of operations on the order of $N \log_2 N$, which allows for the efficient handling of systems with very large size.

The Idea Behind the Method

First, we illustrate the idea behind the method, and then we address its practical implementation. A key realization is that $\omega^N = 1$, which renders many powers of ω in the DFT matrix on the right-hand side of equation (8.7.3) identical.

For simplicity, we assume that N is equal to a power of 2, that is, $N = 2^s$, where s is a positive integer. We may then divide N by 2 a number of s times, each time obtaining an even integer as a divisor. Considering the summation index β introduced on the right-hand side of equation (8.7.2), we express it in the form

$$\beta = \begin{cases} 2\beta_H & \text{when } \beta \text{ is even} \\ 2\beta_H + 1 & \text{when } \beta \text{ is odd} \end{cases} \tag{8.7.4}$$

where $\beta_H = 0, 1, \ldots, N_H - 1$, and $N_H = \frac{1}{2}N$, and recast equation (8.7.2) into the form

$$c_p = \sum_{\beta_H=0}^{N_H-1} (\omega^2)^{p\beta_H} g_{2\beta_H} + \omega^p \sum_{\beta_H=0}^{N_H-1} (\omega^2)^{p\beta_H} g_{2\beta_H+1} \tag{8.7.5}$$

where $0 \le p \le N-1$. Note the similarity of the two sums on the right-hand side.

Next, we write $p = sN_H + p_H$, where $s = 0, 1$, and $p_H = 0, 1, \ldots, N_H - 1$, and observe that

$$(\omega^2)^{p\beta_H} = (\omega^2)^{sN\beta_H/2}(\omega^2)^{p_H\beta_H} = (\omega^N)^{s\beta_H}(\omega^2)^{p_H\beta_H} = (\omega^2)^{p_H\beta_H} \tag{8.7.6}$$

Substituting this expression into equation (8.7.5), we find

$$c_p = c_{p_H}^{(0)} + \omega^p c_{p_H}^{(1)} \tag{8.7.7}$$

where

$$c_{p_H}^{(0)} = \sum_{\beta_H=0}^{N_H-1} \omega_H^{p_H \beta_H} g_{2\beta_H}, \qquad c_{p_H}^{(1)} = \sum_{\beta_H=0}^{N_H-1} \omega_H^{p_H \beta_H} g_{2\beta_H+1} \tag{8.7.8}$$

and

$$\omega_H \equiv \exp(2\pi i/N_H) = \exp(4\pi i/N) \tag{8.7.9}$$

Comparing equations (8.7.9) and (8.7.2) reveals that $c_{pH}^{(1)}$ and $c_{pH}^{(2)}$ are the *Fourier coefficients corresponding to the even- and odd-numbered data points.*

Working in a similar manner, we find

$$c_{p_H}^{(0)} = c_{p_{HH}}^{(0,0)} + \omega_H^{p_1} c_{p_{HH}}^{(0,1)}, \qquad c_{p_H}^{(1)} = c_{p_{HH}}^{(1,0)} + \omega_H^{p_1} c_{p_{HH}}^{(1,1)} \tag{8.7.10}$$

where

$$\begin{aligned} c_{p_{HH}}^{(0,0)} &= \sum_{\beta_{HH}=0}^{N_{HH}-1} \omega_{HH}^{p_{HH}\beta_{HH}} g_{4\beta_{HH}}, \qquad & c_{p_2}^{(0,1)} &= \sum_{\beta_{HH}=0}^{N_{HH}-1} \omega_{HH}^{p_{HH}\beta_{HH}} g_{2(2\beta_{HH}+1)} \\ c_{p_{HH}}^{(1,0)} &= \sum_{\beta_{HH}=0}^{N_{HH}-1} \omega_{HH}^{p_{HH}\beta_{HH}} g_{4\beta_{HH}+1}, \qquad & c_{p_{HH}}^{(1,1)} &= \sum_{\beta_{HH}=0}^{N_{HH}-1} \omega_{HH}^{p_{HH}\beta_{HH}} g_{2(2\beta_{HH}+1)+1} \end{aligned} \tag{8.7.11}$$

where $N_{HH} = \frac{1}{4}N = \frac{1}{2}N_H$, $p_{HH} = sN_{HH} + p_{HH}$, $s = 0, 1$, $p_{HH} = 0, 1, \ldots, N_{HH} - 1$, and

$$\omega_{HH} \equiv \exp(2\pi i/N_{HH}) = \exp(8\pi i/N) \tag{8.7.12}$$

The cascade is now evident. Detailed inspection reveals that the algorithm requires a number of operations on the order of $N \log_2 N$, which is affordable even when N has a very large value (e.g., Brigham 1974, Henrici 1986).

Implementation

The method can be implemented in several different ways (van Loan 1992, Press et al. 1992, Chapter 12; Fornberg 1996, Appendix F), and we illustrate one of them by means of case studies.

The case $N = 4$

The sequence of computations is:

$$\omega = \exp(2\pi i/N) = i$$

$$
\begin{aligned}
g_0^{(1)} &= g_0 + g_2 \\
g_1^{(1)} &= g_1 + g_3 \\
g_2^{(1)} &= g_0 + \omega^2 g_2 \\
g_3^{(1)} &= g_1 + \omega^2 g_3
\end{aligned}
\qquad \text{First stage, } \textit{Stage} = 1
$$

$$
\begin{aligned}
g_0^{(2)} &= g_0^{(1)} + g_2^{(1)} \\
g_1^{(2)} &= g_1^{(1)} + \omega^2 g_3^{(1)} \\
g_2^{(2)} &= g_0^{(1)} + \omega g_2^{(1)} \\
g_3^{(2)} &= g_1^{(1)} + \omega^3 g_3^{(1)}
\end{aligned}
\qquad \text{Second stage, } \textit{Stage} = 2
$$

$$
\begin{aligned}
c_0 &= g_0^{(2)} \\
c_1 &= g_2^{(2)} \\
c_2 &= g_1^{(2)} \\
c_3 &= g_3^{(2)}
\end{aligned}
\qquad \text{Unscrambling}
$$

To understand the action of the method, we write out the DFT matrix for $N = 4$, and note that $\omega = \exp(\pi i/2) = i$, to obtain

$$
\begin{bmatrix} c_0 \\ c_1 \\ c_2 \\ c_3 \end{bmatrix} = \begin{bmatrix} 1 & 1 & 1 & 1 \\ 1 & \omega & \omega^2 & \omega^3 \\ 1 & \omega^2 & \omega^4 & \omega^6 \\ 1 & \omega^3 & \omega^6 & \omega^9 \end{bmatrix} \begin{bmatrix} g_0 \\ g_1 \\ g_2 \\ g_3 \end{bmatrix} \quad \text{or} \quad \begin{bmatrix} c_0 \\ c_1 \\ c_2 \\ c_3 \end{bmatrix} = \begin{bmatrix} 1 & 1 & 1 & 1 \\ 1 & i & -1 & -i \\ 1 & -1 & 1 & -1 \\ 1 & -i & -1 & i \end{bmatrix} \begin{bmatrix} g_0 \\ g_1 \\ g_2 \\ g_3 \end{bmatrix} \tag{8.7.13}
$$

Inspecting the preceding sequence of computations reveals that the method effectively produces the Fourier coefficents by factoring a permutated version of the DFT matrix, as

$$
\begin{bmatrix} c_0 \\ c_2 \\ c_1 \\ c_3 \end{bmatrix} = \begin{bmatrix} 1 & 1 & 0 & 0 \\ 1 & -1 & 0 & 0 \\ 0 & 0 & 1 & i \\ 0 & 0 & 1 & -i \end{bmatrix} \begin{bmatrix} 1 & 0 & 1 & 0 \\ 0 & 1 & 0 & 1 \\ 1 & 0 & -1 & 0 \\ 0 & 1 & 0 & -1 \end{bmatrix} \begin{bmatrix} g_0 \\ g_1 \\ g_2 \\ g_3 \end{bmatrix} \tag{8.7.14}
$$

Premultiplying the vector $\boldsymbol{g}$ by the second matrix on the right-hand side produces the vector $\boldsymbol{g}^{(1)}$. Premultiplying the vector $\boldsymbol{g}^{(1)}$ by the first matrix on the right-hand side produces the vector $\boldsymbol{g}^{(2)}$; this is then unscrambled to yield the Fourier coefficient vector $\boldsymbol{c}$. The computational savings, in this case by a factor of 2 compared to the direct multiplication method, are reflected in the sparse nature of the intermediate matrices.

Thus the FFT method relies on the shrewd factorization of the DFT matrix into the product of sparse matrices, followed by unscrambling. The factorization is particularly simple when N is a highly composite number, in particular, a power of 2.

The case $N = 8$

The sequence of computations is:

$$
\omega = \exp(2\pi i/N) = \exp(\pi i/4)
$$

First stage, $Stage = 1$

$$\begin{aligned}
g_0^{(1)} &= g_0 + g_4\\
g_1^{(1)} &= g_1 + g_5\\
g_2^{(1)} &= g_2 + g_6\\
g_3^{(1)} &= g_3 + g_7\\
g_4^{(1)} &= g_0 + \omega_1^4 g_4\\
g_5^{(1)} &= g_1 + \omega_1^4 g_5\\
g_6^{(1)} &= g_2 + \omega_1^4 g_6\\
g_7^{(1)} &= g_3 + \omega_1^4 g_7
\end{aligned}$$

Second stage, $Stage = 2$

$$\begin{aligned}
g_0^{(2)} &= g_0^{(1)} + g_2^{(1)}\\
g_1^{(2)} &= g_1^{(1)} + g_3^{(1)}\\
g_2^{(2)} &= g_4^{(1)} + \omega^4 g_6^{(1)}\\
g_3^{(2)} &= g_5^{(1)} + \omega^4 g_7^{(1)}\\
g_4^{(2)} &= g_0^{(1)} + \omega^2 g_2^{(1)}\\
g_5^{(2)} &= g_1^{(1)} + \omega^2 g_3^{(1)}\\
g_6^{(2)} &= g_4^{(1)} + \omega^6 g_6^{(1)}\\
g_7^{(2)} &= g_5^{(1)} + \omega^6 g_7^{(1)}
\end{aligned}$$

Third stage, $Stage = 3$

$$\begin{aligned}
g_0^{(3)} &= g_0^{(2)} + g_1^{(2)}\\
g_1^{(3)} &= g_2^{(2)} + \omega^4 g_3^{(2)}\\
g_2^{(3)} &= g_4^{(2)} + \omega^2 g_6^{(2)}\\
g_3^{(3)} &= g_6^{(2)} + \omega^6 g_7^{(2)}\\
g_4^{(3)} &= g_0^{(2)} + \omega g_1^{(2)}\\
g_5^{(3)} &= g_2^{(2)} + \omega^5 g_3^{(2)}\\
g_6^{(3)} &= g_4^{(2)} + \omega^3 g_6^{(2)}\\
g_7^{(3)} &= g_6^{(1)} + \omega^7 g_7^{(2)}
\end{aligned}$$

Unscrambling

$$\begin{aligned}
c_0 &= g_0^{(3)}\\
c_1 &= g_4^{(3)}\\
c_2 &= g_2^{(3)}\\
c_3 &= g_6^{(3)}\\
c_4 &= g_1^{(3)}\\
c_5 &= g_5^{(3)}\\
c_6 &= g_3^{(3)}\\
c_7 &= g_7^{(3)}
\end{aligned}$$

Unscrambling

What are the rules for the unscrambling? The recipe is as follows:

- Express the index j of $g_j^{(s)}$ in the binary system, where $\mathbf{g}^{(s)}$ corresponds to the last stage, $s = \ln_2 N$, adding, if necessary, zeros in front of it to make the total number of binary digits equal to s. For example, for $N = 8$ and $j = 1$, we find $1 = (001)_2$.
- Reverse the bits, and compute the associated decimal integer k. Then $g_j^{(s)}$ corresponds to c_k. For the aforementioned example, $k = (100)_2 = 4$.

Consider now the example with $N = 8$, write the column of indices of $g_j^{(3)}$ at the stage of unscrambling, pull out half of it, and repeat to obtain

0	0	0
4	4	4
2	2	
6	6	
1		
5		
3		
7		

The columns are simply the repeated exponents of the powers of ω at successive stages of the computation. These observations suggest a nifty method of unscrambling the indices, as well as of computing the exponents. An algorithm for generating the first column of the preceding table is:

Introduce the N-dimensional integer vector $\boldsymbol{p}$ and set all of its elements equal to 0.

$\ln_2 N = \ln N / \ln 2$

Do $i = 1, \ln_2 N$

  $\mu = 2^{i-1}$

  Do $j = 1, \mu$

    $p_j \leftarrow 2p_j$

  END DO

  Do $j = 0, \mu - 1$

    $p_{j+\mu} = p_j + 1$

  END DO

END DO

The desired column sits in the vector $\boldsymbol{p}$. (8.7.15)

PROBLEM

8.7.1 FFT ***for*** $N = 16$.

Lay out the sequence of computations for $N = 16$, including unscrambling.

Algorithm

The sequence of computations for the FFT method just described is given in Algorithm 8.7.1 after Gerald and Wheatley (1994). To make the algorithm more complete, we have allowed the data vector $\boldsymbol{g}$ to be complex. If $\boldsymbol{g}$ is real, we simply set the imaginary part equal to zero.

ALGORITHM 8.7.1 Fast Fourier transform algorithm of a complex input function f. The function g is computed from equation (8.7.1). The subscripts R and I designate, respectively, the real and imaginary parts. The function INT(a) extracts the integer part of a. The function MOD(a, b) produces the remainder of the division of a by b.

Read the complex N-dimensional complex data vector $\boldsymbol{f}$.
If $\boldsymbol{f}$ is real, set the imaginary part equal to zero.
Compute the complex data vector $\mathbf{g}$ from equation (8.7.1).
Execute algorithm (8.7.15) to produce the indexing vector $\boldsymbol{p}$.

Do $j = 0, N-1$
  $c_j = \cos(2\pi j/N)$
  $s_j = \sin(2\pi j/N)$
END DO

$Nsets = 1$
$Ndel = N/2$

Do $Stage = 1, \ln_2 N$
  $Ig = 0$ — Indexes g data points
  Do $Iset = 1, Nsets$
    Do $m = 0, (N/Nsets) - 1$
      $j = \text{MOD}(m, Ndel) + 2Ndel(Iset - 1)$
      $l = \text{INT}(Ig/Ndel)$
      $l = p_l$
      $gs_R(Ig) = g_{R_j} + c_l g_{R_{j+Ndel}} - s_l g_{I_{j+Ndel}}$ — s in g stands for *save*
      $gs_I(Ig) = f_{I_j} + c_l g_{I_{j+Ndel}} + s_l g_{R_{j+Ndel}}$
      $Ig \leftarrow Ig + 1$
    END DO
  END DO

  Do $i = 0, N-1$
    $g_{R_i} = gs_{R_i}$
    $g_{I_i} = gs_{I_i}$
  END DO

```
    Nsets = 2Nsets
    N = N/2
  END DO

    DO i = 0, N − 1
      a_i = 2.0 g_{Rp(i)}
      b_i = 2.0 g_{Ip(i)}
    END DO
```

8.8 Trigonometric Approximation of a Function of Two Variables

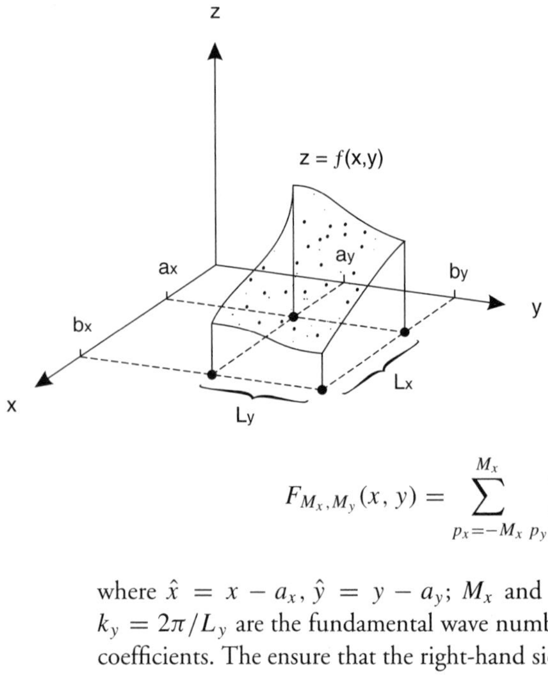

Generalizing the discussion of the preceding two sections, we consider a real function of two variables $f(x, y)$ defined over a rectangular domain in the xy plane. The domain is defined by the intersection of the x strip confined in $a_x < x < b_x$ with length $L_x = b_x - a_x$, and the y strip confined in $a_y < y < b_y$ with length $L_y = b_y - a_y$, as shown in the diagram.

Trigonometric approximation represents the function $f(x, y)$ with the truncated Fourier series:

$$F_{M_x,M_y}(x, y) = \sum_{p_x=-M_x}^{M_x} \sum_{p_y=-M_y}^{M_y} c_{p_x,p_y} \exp\big(-i(p_x k_x \hat{x} + p_y k_y \hat{y})\big) \tag{8.8.1}$$

where $\hat{x} = x - a_x$, $\hat{y} = y - a_y$; M_x and M_y are two specified truncation levels; $k_x = 2\pi/L_x$ and $k_y = 2\pi/L_y$ are the fundamental wave numbers; i is the imaginary unit; and $c_{px,py}$ are complex Fourier coefficients. The ensure that the right-hand side of equation (8.8.1) is real, we require that

$$c_{-p_x,-p_i} = c^*_{p_x,p_y} \tag{8.8.2}$$

where an asterisk signifies the complex conjugate. Thus the complex Fourier coefficients are conjugate antisymmetric with respect to the origin of the $p_x p_y$ plane.

Outside the rectangular domain, the Fourier series yields the periodic repetition of the function $F_{M_x,M_y}(x, y)$. It can be shown that, in the limit as M_x and M_y tend to infinity, the representation (8.8.1) becomes exact (e.g., Tolstov 1962).

Using the orthogonality properties of the trigonometric functions discussed in Section 8.6, we find that the complex Fourier coefficients are given by

$$c_{p_x,p_y} = \frac{1}{L_x L_y} \int_{a_x}^{b_x} \int_{a_y}^{b_y} f(x, y) \exp\big(i(p_x k_x \hat{x} + p_y k_y \hat{y})\big)\, dy\, dx \tag{8.8.3}$$

Note that this equation respects the property (8.8.2).

The integrals on the right-hand side of expression (8.8.3) may be computed by numerical integration in terms of the values of $f(x, y)$ at a grid of data points, as discussed in Chapter 7. We note that the values of the coefficients computed in this manner will depend on the selected integration rule or quadrature, and this reveals that the truncated Fourier series will not generally be an interpolating function, that is, it will not necessarily interpolate through the data points.

Fourier Coefficients for Evenly Spaced Points

When the data points are distributed uniformly on a grid consisting of evenly separated grid lines, evaluating the integrals in equation (8.8.3) by the trapezoidal rule provides us with a truncated Fourier series that interpolates through all interior grid points.

The restriction of evenly spaced points may appear to be demanding, but it is not necessarily so: One can introduce two new independent variables t and s, where $x = q_x(t, s)$ and $y = q_y(t, s)$, so that the data points are evenly spaced with respect to t and s, and then carry out the Fourier approximation with respect to t and s. For example, one may label the grid lines using sequential values of integers. Note, however, that when the functions $x = q_x(t, s)$ and $y = q_y(t, s)$ are not sufficiently smooth, the interpolating Fourier series may exhibit fluctuations between the data points.

Proceeding with the details of the computation, we divide the L_x interval into N_x subintervals separated by the $N_x + 1$ grid lines $x_j = a_x + (j - 1)h_x$, $j = 1, \ldots, N_x + 1$, where $h_x = L_x/N_x$, $x_1 = a_x$, $x_{Nx+1} = b_x$. Similarly, we divide the L_y interval into N_y subintervals separated by the $N_y + 1$ grid lines $y_k = a_y + (k-1)h_y$, $k = 1, \ldots, N_y + 1$, where $h_y = L_y/N_y$, $y_1 = a_y$, $y_{Ny+1} = b_y$. The intersections of the grid lines define a mesh of $(N_y + 1)(N_y + 1)$ grid points, as shown in the diagram.

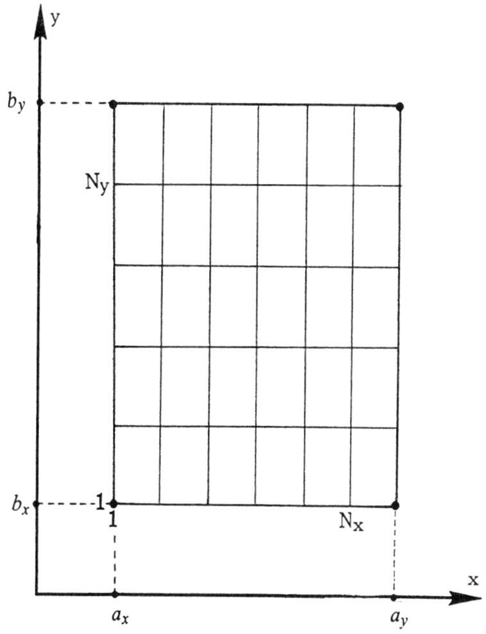

Using the trapezoidal rule to evaluate the complex Fourier integral in equation (8.8.3), we obtain

$$c_{p_x,p_y} = \frac{1}{N_x N_y} \left(\tfrac{1}{2} q_1 + \omega_x^{p_x} q_2 + \cdots + \omega_x^{p_x(N_x-1)} q_{N_x} + \tfrac{1}{2} q_{N_x+1}\right) \tag{8.8.4}$$

where

$$q_m(p_y) = \left(\tfrac{1}{2} f_{m,1} + \omega_y^{p_y} f_{m,2} + \cdots + \omega_y^{p_y(N_y-1)} f_{m,N_y} + \tfrac{1}{2} f_{m,N_y+1}\right) \tag{8.8.5}$$

we have defined $f_{n,m} \equiv f(x_n, y_m)$, and where

$$\omega_x = \exp(ik_x h_x), \qquad \omega_y = \exp(ik_y h_y) \tag{8.8.6}$$

Note that $\omega_x^{Nx} = 1$ and $\omega_y^{Ny} = 1$. The computations proceed according to the following rules:

- When the number of intervals N_x is odd, we truncate the Fourier sum (8.8.1) at the value $M_x = \frac{1}{2}(N_x - 1)$. When N_x is even, we truncate it at the value $M_x = \frac{1}{2}N_x$. Similar truncation levels pertain to M_y.
- When N_y is even, the right-hand side of equation (8.8.5) is multiplied by the factor $\frac{1}{2}$ for $p_y = -M_y$ and M_y, in each case for $m = 1, \ldots, N_x + 1$.
- When N_x is even, the right-hand side of equation (8.8.4) is multiplied by the factor $\frac{1}{2}$ for $p_x = -M_x$ and M_x, in each case for $p_y = -M_y, \ldots, M_y$.

With the Fourier coefficients computed in this manner, it can be shown that the truncated Fourier series interpolates through the interior data points and through the average value of the corresponding boundary points located on opposite sides of the rectangular strip.

Periodic Functions

If the function $f(x, y)$ is doubly periodic with periods equal to L_x and L_y, that is, $f_{1,m} = f_{Nx+1,m}$ and $f_{m,1} = f_{m,Ny+1}$, and if both N_x and N_y are odd, equations (8.8.4) and (8.8.5) combine to yield the compact form

$$c_{p_x,p_y} = \frac{1}{N_x N_y} \sum_{j_x=1}^{N_x} \sum_{j_y=1}^{N_y} \omega_x^{p_x(j_x-1)} \omega_y^{p_y(j_y-1)} f(x_{j_x}, y_{j_y}) \tag{8.8.7}$$

With the coefficients computed in this manner, the truncated Fourier series interpolates through all interior and boundary data points. When one or both of N_x and N_y are even, the factors of $\frac{1}{2}$ discussed in the preceding subsection should be incorporated into the right-hand side of equation (8.8.7) (Problem 8.8.2).

Fast Fourier transform methods for computing the Fourier coefficients are discussed by Nussbaumer (1982) and Press et al. (1992, pp. 515–519).

PROBLEMS

8.8.1 *Computation of Fourier representations.*

Consider the function $f(x, y) = x^2 + \frac{1}{2}y^2$ over the rectangular domain $1 < x < 2$ and $0 < y < 1$. Compute the interpolating Fourier series for $N_x = N_y = 2, 4, 8, 16$, plot it along with its doubly periodic extension, and discuss the quality of the approximation.

8.8.2 *Periodic functions.*

Derive the counterpart of expression (8.8.7) when one or both of N_x and N_y are even.

8.8.3 *Functions of three variables.*

Derive the counterpart of equations (8.8.4), (8.8.5), and (8.8.7) for a function of three variables.

References

Baker, G. A., and Graves-Morris, P., 1996, *Padé Approximants.* Cambridge University Press.

Bézier, P. E., 1971, Example of an existing system in the motion industry: the unisurf system. *Proc. R. Soc. London A* **321**, 207–18.

Brigham, A. O., 1974, *The Fast Fourier Transform.* Prentice-Hall.

Cooley, J., and Tukey, J. W., 1965, An algorithm for the machine calculation of complex Fourier series. *Math. Comput.* **19**, 297–301.

Crow, F., 1987, Origins of a teapot. *IEEE Comput. Graphics Appl.* **7**, 8–19.

Cuyt, A., 1987, A recursive computation scheme for multivariate rational interpolants. *SIAM J. Numer. Anal.* **24**, 228–39.

Davis, P. J., 1963, *Interpolation and Approximation.* Reprinted by Dover, 1975.

Farin, G., 1990, *Curves and Surfaces for Computer Aided Geometric Design: A Practical Guide.* Academic Press.

Fornberg, B., 1996, *A Practical Guide to Pseudospectral Methods.* Cambridge University Press.

Gerald, C. F., and Wheatley, P. P., 1994, *Applied Numerical Analysis.* Addison-Wesley.

Gradshteyn, I. S., and Ryzhik, I. M., 1980, *Tables of Integrals, Series, and Products.* Academic Press.

Henrici, P., 1986, *Applied and Computational Complex Analysis,* Vol. 3. Wiley.

Kincaid, D., and Cheney, W., 1996, *Numerical Analysis.* Brooks/Cole.

Lanczos, C., 1956, *Applied Analysis.* Reprinted by Dover, 1988.

Nussbaumer, H. J., 1982, *Fast Fourier Transform and Convolution Algorithms.* Springer-Verlag.

Olfe, D. B., 1995, *Computer Graphics for Design.* Prentice-Hall.

Prenter, P. M., 1975, *Splines and Variational Methods.* Wiley.

Press, W. H., Flannery, B. P., Teukolsky, S. A., and Vetterling, W. T., 1992, *Numerical Recipes.* 2nd ed. Cambridge University Press.

Ralston, A., 1965, *A First Course in Numerical Analysis.* McGraw-Hill.

Rivlin, T. J., 1969, *An Introduction to the Approximation of Functions.* Reprinted by Dover, 1981.

Rogers, D. F, and Adams, J. A., 1990, *Mathematical Elements for Computer Graphics.* McGraw-Hill.

Shikin, E. V., and Plis, A. Z., 1995, *Handbook of Splines for the User.* CRC Press.

Smith, G. D., 1985, *Numerical Solution of Partial Differential Equations: Finite-Difference Methods.* Oxford University Press.

Tolstov, G. P., 1962, *Fourier Series.* Dover.

Van Dyke, M., 1974, Analysis and improvement of perturbation series. *Q. J. Mech. Appl. Math.* **27**, 423–50.

Van Loan, C., 1992, *Computational Frameworks for the Fast Fourier Transform,* SIAM, Philadephia.

Wuytack, L. (ed.), 1979, *Padé Approximation and Its Applications.* Lecture Notes in Mathematics **765**. Springer-Verlag.

Chapter 9

Ordinary Differential Equations; Initial-Value Problems

9.1 Problem Formulation and Physical Context

Consider N unknown functions of one independent variable t, $x_i(t)$, $i = 1, \ldots, N$, and assume that we are given the slopes of their graphs, $\tan\theta_i = dx_i/dt$, as shown in Figure 9.1.1. This means that we are provided with N functions f_i that depend on t in two ways: Explicitly, and implicitly through their dependence on x_i, that is,

$$\frac{dx_i}{dt} = f_i(x_1, x_2, \ldots, x_N, t) \tag{9.1.1}$$

for $i = 1, \ldots, N$. The ith equation has the standard form of a first-order scalar ordinary differential equation for the variable x_i. An example with $N = 2$ is provided by equations (9.1.4).

The problem is to compute the functions $x_i(t)$ that satisfy the system (9.1.1) and, in addition, respect certain conditions or constraints.

Vector Formulation

To simplify the notation, we introduce the vector of unknown functions $x_i(t)$ and the vector of slope functions f_i,

$$\boldsymbol{x} = (x_1, \ldots, x_N)^T, \qquad \boldsymbol{f} = (f_1, \ldots, f_N)^T \tag{9.1.2}$$

and collect the N scalar equations (9.1.1) into the first-order vectorial ordinary differential equation

$$\frac{d\boldsymbol{x}}{dt} = \boldsymbol{f}(\boldsymbol{x}, t) \tag{9.1.3}$$

Given the vector slope-function $\boldsymbol{f}$, we want to compute the vector function $\boldsymbol{x}(t)$ that satisfies equation (9.1.3) subject to an appropriate number of conditions or constraints.

When the components of the vector slope-function $\boldsymbol{f}$ are available in analytical form, their evaluation should be done in a single subroutine. It is both unnecessary and wasteful to evaluate each one of them by calling separate subroutines.

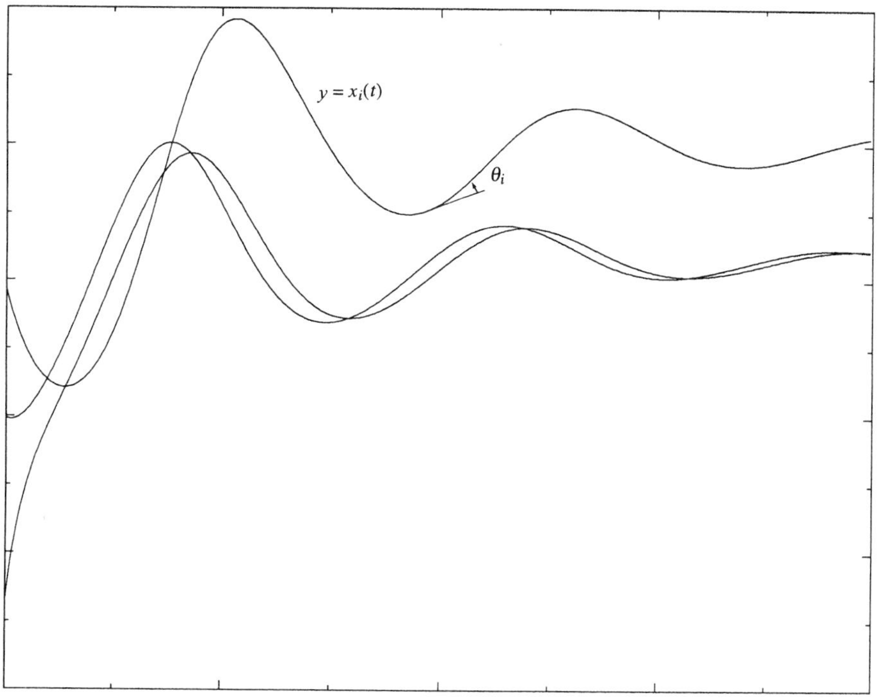

Figure 9.1.1 Graphs of N functions $x_i(t)$ that satisfy the first-order ordinary differential equations (9.1.1). The functions are not given explicitly; instead, we are provided with the slopes of their graphs in terms of $x_i(t)$ and t, as shown in equation (9.1.1)

The importance of function evaluations

In practice, the components of the vector slope-function $\mathbf{f}$ may not be given explicitly in the statement of the problem. Instead, their values may arise as the outcome of a numerical computation involving, for example, the solution of a system of linear or nonlinear algebraic equations. In some applications, the evaluation of $\mathbf{f}$ may even require physical observation or laboratory experimentation at specific conditions.

For example, $\mathbf{f}$ may represent the velocity of an aerosol particle settling in the atmosphere, or the velocity of an air bubble suspended in a bubbly liquid. In both cases, the evaluation of $\mathbf{f}$ requires solving a system of equations that describe the motion of the particle or bubble, and the motion of the ambient fluid.

When $\mathbf{f}$ is not available explicitly in a convenient form, the number of evaluations required for the numerical solution of the system (9.1.3) can be an important parameter of the numerical method.

Equations of Higher Order

The mathematical modeling of many physical phenomena and engineering processes leads to ordinary differential equations that are of second or higher order; that is, they involve second or higher-order derivatives of the vector function $\mathbf{f}$ with respect to the independent variable t. The second derivative is usually associated with acceleration; its origin can be traced back to Newton's second law for the motion of a finite or infinitesimal material parcel.

Mathieu's equation, for example, describing among other phenomena the propagation of waves through a tube with an elliptical cross section, has the form of the second-order differential equation (5.1.15). But by redefining $x_1 = x$, and introducing the intermediate variable $x_2 = dx_1/dt = d^2x/dt^2$, we can reduce the second-order equation (5.1.15) to two first-order equations

$$\frac{dx_1}{dt} = x_2, \qquad \frac{dx_2}{dt} = (-p + 2q \cos 2t)x_1 \tag{9.1.4}$$

We can work in a similar manner with any Nth order differential equation, reducing it to a system of N first-order differential equations, and this allows us to narrow down the scope of our study. We shall see, however, that it is sometimes more efficient to work with the unreduced second-order system, with the benefits of efficiency and reduced sensitivity to round-off error.

When an Nth order differential equation contains nonlinear functions of the derivatives d^kx/dt^k, where $k = 1, \ldots, N$, solving for the first derivatives of the intermediate functions in order to produce the N first-order differential equations may require inverting nonlinear algebraic equations, which may have to be done using numerical methods.

As an example, consider the second-order equation

$$\left(\frac{d^2x}{dt^2}\right)^2 + \exp\left(\frac{d^2x}{dt^2}\right) + t^4\left(\frac{dx}{dt}\right)^2 + x + \ln(t+2) + 5 = 0 \tag{9.1.5}$$

The equivalent system of two first-order equations is

$$\frac{dx_1}{dt} = x_2, \qquad \frac{dx_2}{dt} = f_2(x_1, x_2, t) \tag{9.1.6}$$

where we have set $x_1 = x$. The function f_2 must be evaluated by solving the nonlinear algebraic equation for f_2

$$f_2^2 + \exp(f_2) + t^4x_2^2 + x_1 + \ln(t+2) + 5 = 0 \tag{9.1.7}$$

for specified values of x_1, x_2, and t.

PROBLEM

9.1.1 ***Expansion into, and consolidation of, first-order*** ODE*s.*

(a) Consider the Nth-order linear equation: $x^{(N)} + a_1(t)x^{(N-1)} + \ldots + a_N(t)\,x = 0$, where the superscript (i) signifies the ith derivative. Expand this equation into a system of N first-order equations defined in terms of the Frobenius matrix (2.9.20).
(b) Show by a counterexample that it is not generally feasible to condense a system of N first-order differential equations to a single Nth order differential equation.

Evolution Equations, Autonomous Systems, Steady State, and Equilibrium Points

When the variable t represents time, equations (9.1.1) assume the identity of *evolution equations,* providing us with the *rate of change* of the dependent variables regarded as functions of their instantaneous values and time. The system (9.1.1) is then called a *dynamical system*, although this terminology is sometimes reserved for systems where $\boldsymbol{f}$ is a nonlinear function of $\boldsymbol{x}$.

If the slope-function $\boldsymbol{f}$ does not depend on t explicitly but only implicitly through its dependence on $\boldsymbol{x}$, the dynamical system is *autonomous*, otherwise it is *nonautonomous*. For example, the Mathieu system (9.1.4) is nonautonomous because of the term $\cos 2t$ on the right-hand side. If this term were absent, the system would have been autonomous.

A nonautonomous system may be transformed into an autonomous system simply by relabeling t on the right-hand side as x_{N+1}, t on the left-hand side as τ, and introducing the additional differential equation $dx_{N+1}/d\tau = f_{N+1} = 1$. This modification provides us with the associated *suspended autonomous system* of $N+1$ equations $d\boldsymbol{x}/d\tau = \boldsymbol{f}(\boldsymbol{x})$, where $\boldsymbol{x}$ and $\boldsymbol{f}$ are $(N+1)$-dimensional vectors. Converting a nonautonomous to an autonomous system, however, is not necessary in computing a numerical solution.

An autonomous system may have one or more *critical points* $\boldsymbol{x} = \boldsymbol{X}$. By definition, all components of the vector function $\boldsymbol{f}$ evaluated at $\boldsymbol{X}$ vanish, that is,

$$\boldsymbol{f}(\boldsymbol{X}) = \boldsymbol{0} \tag{9.1.8}$$

Depending on the physical context, a fixed point is also called an *equilibrium point*, or a *steady state*. In physical terms, the fixed point represents the set of conditions where a steady process or an equilibrium configuration can be established.

Start-up of a chemical reactor

To exemplify the way in which dynamical systems arise in practice, we discuss an important example from the field of chemical engineering. Consider a continuous stirred tank reactor, CSTR, used to carry out the reaction

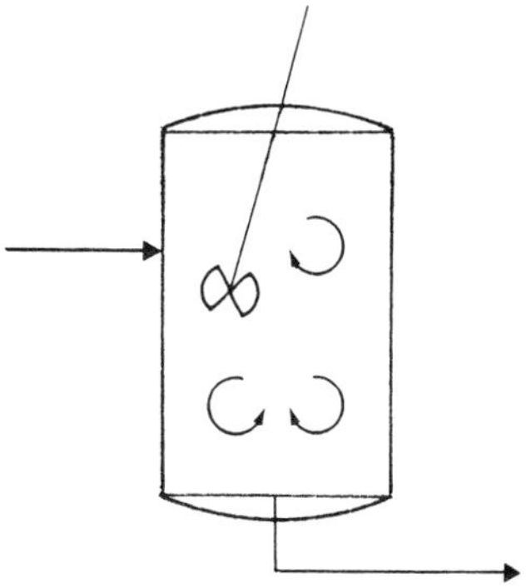

$$A + 2B \rightarrow P \tag{9.1.9}$$

where A, B, and P signify three chemical species. At the initial instant, the reactor is filled with an inert liquid that is free of A, B, and P; thus the initial concentrations of A, B, and P are equal to zero. As soon as the reactor is turned on, the flow rate into the reactor is kept equal to the flow rate out from the reactor, at the constant value F. As time progresses, A and B react producing P according to the reaction (9.1.9).

We want to develop evolution equations for the concentrations of the three species c_A, c_B, and c_P, measured in number of moles per volume, under the assumption that the concentration of each species within the reactor is uniform and equal to that at the exit due to the vigorous mixing. It is known that the reaction rates $r_i \equiv dc_i/dt$, $i = A, B, P$, are related to the concentrations by the kinetics law

$$r_A = \tfrac{1}{2} r_B = -r_P = k c_A c_B^2 \tag{9.1.10}$$

where k is a chemical reaction constant.

We begin the mathematical modeling by noting that the total number of moles of A within the reactor is equal to Vc_A, where V is the volume of the mixture that occupies the reactor. A molar balance for species A around the reactor gives

$$\text{(rate of change of } A) = \text{(entering } A) - \text{(exiting } A) - \text{(reacting } A) \tag{9.1.11}$$

or

$$\frac{d(Vc_A)}{dt} = Fc_{A,in} - Fc_A - kVc_Ac_B^2 \tag{9.1.12}$$

where $c_{A,in}$ is the inlet concentration of A. Assuming that V is constant, and dividing both sides of equation (9.1.12) by F, we obtain

$$\tau\frac{dc_A}{dt} = c_{A,in} - c_A - k\tau c_Ac_B^2 \tag{9.1.13}$$

where $\tau = V/F$ is the *residence time.*

It is furthermore useful to rewrite equation (9.1.13) in terms of the dimensionless variables

$$\hat{t} = \frac{t}{\tau}, \qquad \hat{c}_A = \frac{c_A}{c_{A,in}}, \qquad \hat{c}_B = \frac{c_B}{c_{A,in}} \tag{9.1.14}$$

and the dimensionless constant

$$\beta = k\tau c_{A,in}^2 \tag{9.1.15}$$

Dividing both sides of equation (9.1.13) by $c_{A,in}$, we find

$$\frac{d\hat{c}_A}{d\hat{t}} = 1 - \hat{c}_A - \beta\hat{c}_A\hat{c}_B^2 \tag{9.1.16}$$

Working in a similar manner, we derive analogous evolution equations for the dimensionless concentrations of B and P,

$$\frac{d\hat{c}_B}{d\hat{t}} = \alpha - \hat{c}_B - 2\beta\hat{c}_A\hat{c}_B^2 \tag{9.1.17}$$

and

$$\frac{d\hat{c}_P}{d\hat{t}} = -\hat{c}_P + \beta\hat{c}_A\hat{c}_B^2 \tag{9.1.18}$$

where we have defined

$$\hat{c}_P = \frac{c_P}{c_{A,in}}, \qquad \alpha = \frac{c_{B,in}}{c_{A,in}} \tag{9.1.19}$$

where $c_{B,in}$ is the inlet concentration of B.

Equations (9.1.16)–(9.1.18) comprise the desired autonomous system of evolution equations for the three dimensionless concentrations. The initial conditions are

$$\hat{c}_A = 0, \qquad \hat{c}_B = 0, \qquad \hat{c}_P = 0 \tag{9.1.20}$$

An analytical solution is not feasible, but a numerical solution may be computed readily using one of the numerical methods to be described in Section 9.3. At equilibrium or steady state, the right-hand sides of all three equations are equal to zero (Problem 9.1.4).

PROBLEMS

9.1.2 *Mass–damper–spring system.*

Consider a body of mass m attached with a spring and a damper to a fixed wall as shown in the diagram, and denote the x coordinate of a certain marker point on the body by $x(t)$. Newton's second law of motion requires

$$m\frac{d^2x}{dt^2} = F_S(t) + F_D(t) + F(t) \tag{9.1.21}$$

where $F_S(t)$, $F_D(t)$, and $F(t)$ are, respectively, the force due to the spring, the force due to the damper, and an externally applied force, all acting in the x direction. The spring is unstressed when $x = L$. Setting $F_S(t) = -k(x - L)$, where k is the spring constant, and $F_D(t) = -c\,dx/dt$, where c is the damping coefficient, we obtain a second-order equation governing the motion of the body,

$$m\frac{d^2x}{dt^2} + c\frac{dx}{dt} + k(x - L) = F(t) \tag{9.1.22}$$

Derive the corresponding first-order non-autonomous and suspended systems.

9.1.3 *Duffing's equation.*

Duffing's equation describes the motion of a vertical pendulum executing small-amplitude forced harmonic vibrations, in the absence of damping. In dimensionless form, Duffing's equation reads

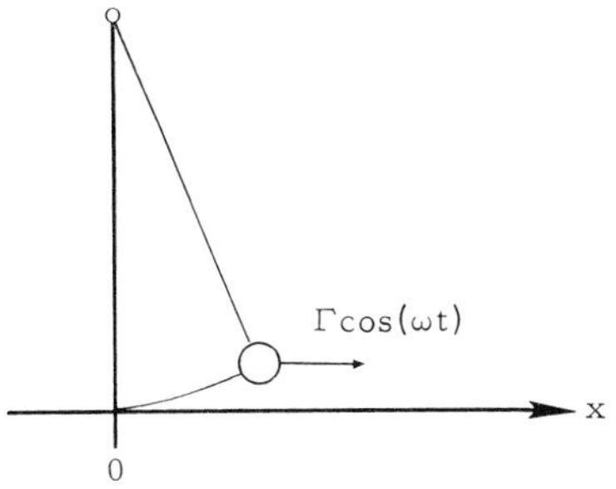

$$\frac{d^2x}{dt^2} + x - \tfrac{1}{3}x^3 = \Gamma \cos \omega t \tag{9.1.23}$$

where x is the reduced horizontal position of the pendulum, and Γ and ω are two physical parameters. Note the similarities and differences with equation (9.1.22). Derive the corresponding first-order nonautonomous and suspended systems.

9.1.4 ***Steady state of a chemical reactor.***

Compute the steady-state dimensionless concentrations within a CSTR of volume $V = 100$ L, for $k = 0.020\ \mathrm{L}^2/(\mathrm{mol}^2 \cdot \mathrm{min})$, $F = 1$ L/ min, $c_{A,in} = 1.5$ mol/L and $c_{B,in} = 3.0$ mol/L. These are the coordinates of the equilibrium points of the dynamical system described by equations (9.1.16)–(9.1.18), expressing the conditions at steady state.

9.1.5 ***Start-up of another chemical reactor.***

Consider the CSTR discussed in the text. Dr. George R. Frankenstein, the director of the laboratory, decided that the reaction (9.1.9) is not likely to produce life and should be replaced with the alternative reaction $A+3C \rightarrow E$. The reaction rate is given by $r_A = \frac{1}{3} r_C = -r_E = k c_A c_C^3$, where k is a reaction-rate constant.
(*a*) Derive evolution equations for properly defined dimensionless concentrations of the three chemical species.
(*b*) Compute the steady-state dimensionless concentrations of the three species in a reactor of volume $V = 100$ L, for $k = 0.020\ \mathrm{L}^3/(\mathrm{mol}^3 \cdot \mathrm{min})$, $F = 2$ L/ min, $c_{A,in} = 1.5$ mol/L and $c_{B,in} = 3.0$ mol/L.

Phase Space and Orbits

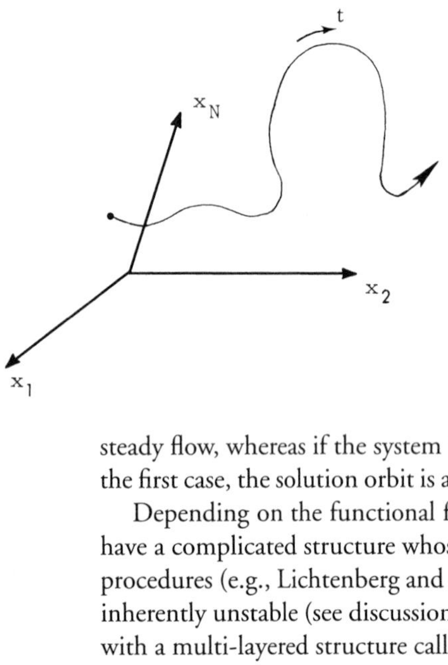

To analyze the solution of a system of ordinary differential equations, it is helpful to introduce the N-dimensional *phase space* with coordinates y_i. The equations $y_i = x_i(t)$ describe a curve called the *orbit* of the solution in the phase space. This method of displaying the solution is often preferred over plotting the functions $x_i(t)$ individually against t, as was done in Figure 9.1.1. The phase-space orbits of a dynamical system with $N = 2$ for several pairs of functions $x_1(t)$ and $x_2(t)$ are shown in Figure 9.1.2.

We may then regard t as the time elapsed in the phase space, and the slope-function f as the velocity of motion of a point particle in the phase space corresponding to a *flow*. This interpretation owes its origin to fluid mechanics, where the phase space is the physical space of an actual flow. If a system is autonomous, a point particle is advected in a steady flow, whereas if the system is nonautonomous, the point particle is advected in an unsteady flow. In the first case, the solution orbit is a *streamline*, and in the second case, it is a *path line* (e.g., Pozrikidis 1997).

Depending on the functional form of the phase-space velocity f, the orbit of a dynamical system may have a complicated structure whose analysis requires sophisticated theoretical concepts and computational procedures (e.g., Lichtenberg and Lieberman 1983). This is particularly true when the dynamical system is inherently unstable (see discussion at the end of this section). For example, an orbit may describe an object with a multi-layered structure called a *strange attractor.*

The Lorenz system

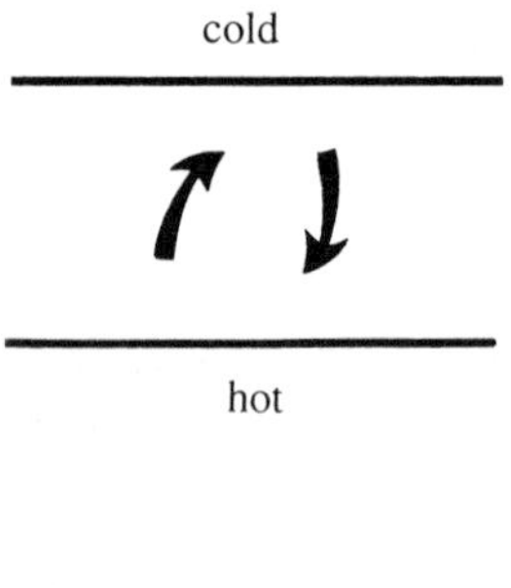

A dynamical system whose orbits exhibit a rich structure was discovered by Lorenz (1963). The system arises by considering a liquid layer resting between two horizontal plates, where the top plate is cold and the bottom plate is hot. Since, in general, a cold fluid parcel is denser than a hot fluid parcel, cold fluid elements at the top tend to move downward toward the bottom, and hot fluid elements at the bottom tend to move upward toward the top. These motions establish a cellular flow pattern known as the *Rayleigh–Benard* convection pattern. The amateur cook must have seen this pattern developing at the surface of a simmering pot.

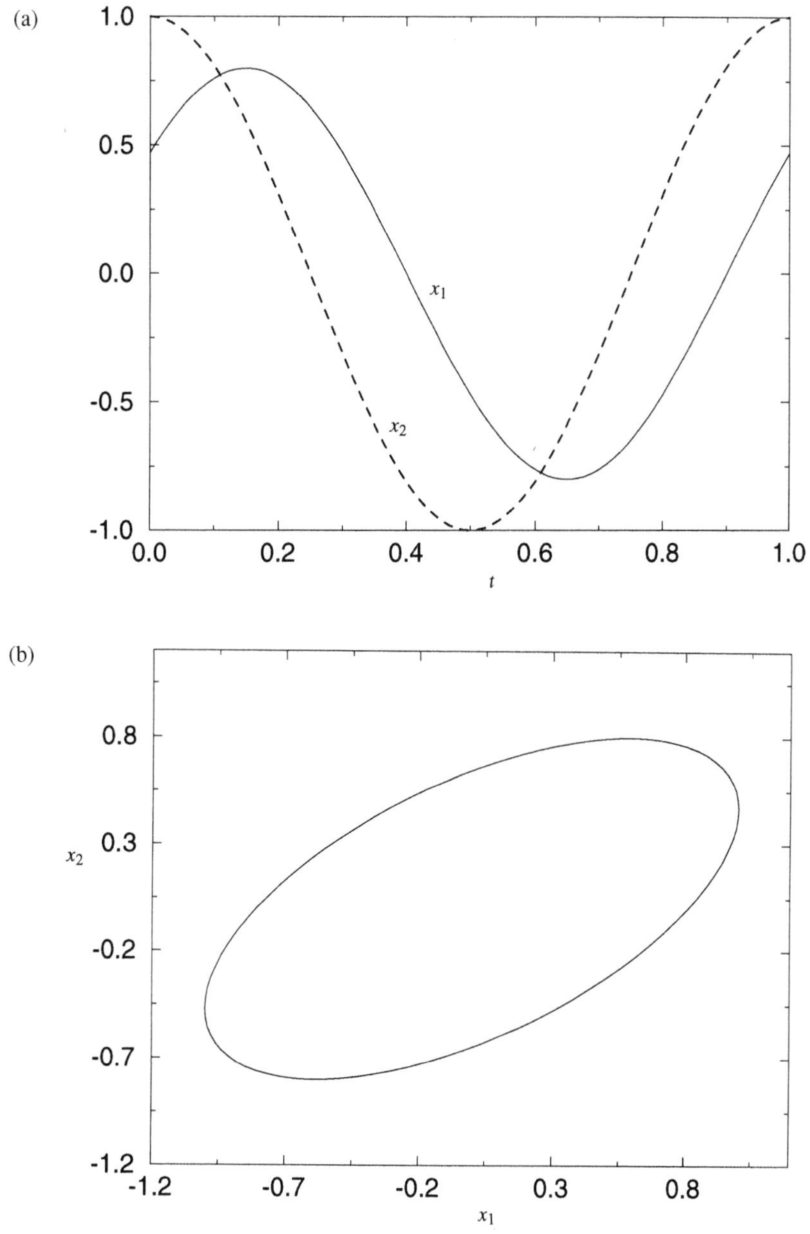

Figure 9.1.2 The behavior of two pairs of functions $x_1(t)$ and $x_2(t)$ corresponding to a dynamical system with $N = 2$, and the associated orbits in the phase plane.

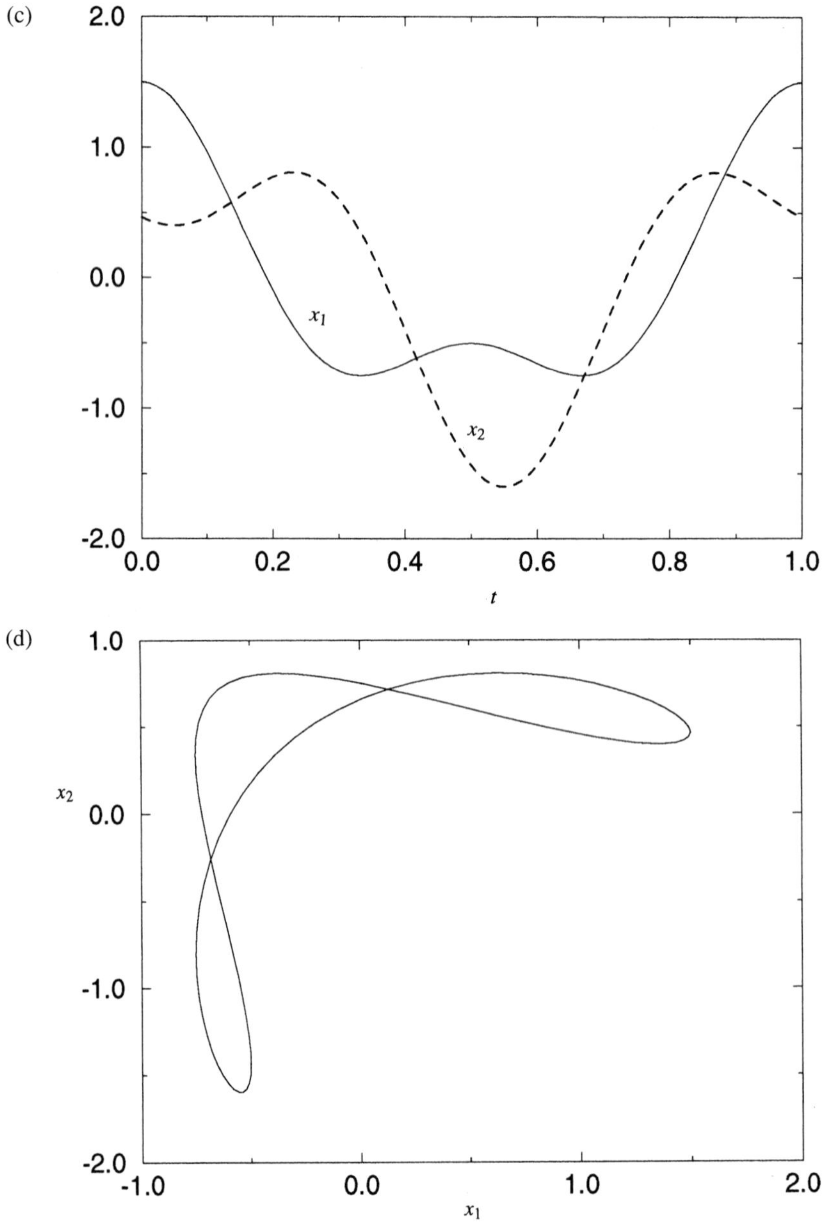

Figure 9.1.2 (continued)

The mathematical analysis of the Rayleigh–Benard convection pattern results in an intractable system of partial differential equations. Under certain drastic assumptions, one may describe the motion in terms of the following autonomous system of nonlinear ordinary differential equations, coined the *Lorenz system* after its inventor,

$$\frac{dA}{dt} = -k(A - T)$$
$$\frac{dT}{dt} = -AD + rA - T \qquad (9.1.24)$$
$$\frac{dD}{dt} = AT - bD$$

Roughly speaking, A is the amplitude of the convective motion; T is the temperature difference between the ascending and descending currents; D is the deviation of the vertical temperature profile from linearity; and k, r, and b are three physical and geometrical parameters with positive values.

It has been established that for combinations of values of k, r, and b that fall within certain ranges, the orbit of the solution in the phase space exhibits a complex structure that includes chaotic trajectories and strange attractors. The reader will have the opportunity to construct these orbits using the numerical methods described in later sections (see Problem 9.3.1).

Initial-value Problems

In a common class of problems, called *initial-value problems*, the phase-space orbit is required to emanate from a specified point in the phase space, $\boldsymbol{x}(t = 0)$. The start-up of the CSTR discussed previously in this section provides us with one example.

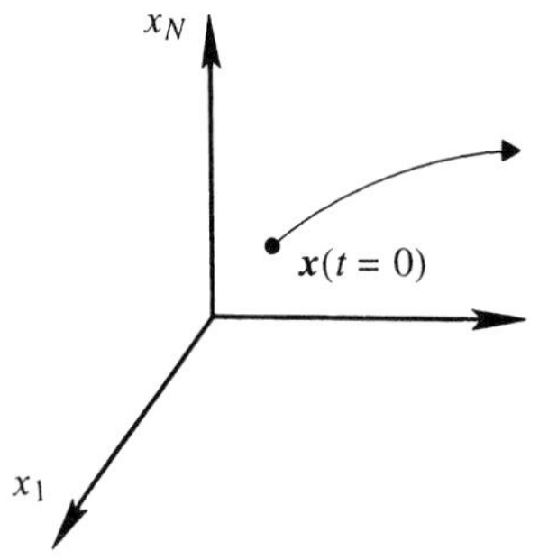

When the phase-space orbit is required to pass through a specified point in the phase space at some time $T, \boldsymbol{x}(t = T)$, we introduce the new time-like variable $\tau = T - t$ and obtain a standard initial-value problem with respect to τ, as defined in the previous paragraph. Consider, for example, the differential equation $dx_1/dt = \exp(at)$ with the condition $x_1(t = T) = b$. The corresponding standard initial-value problem is expressed by the equation $dx_1/d\tau = -\exp(a(T - \tau))$ with the initial condition $x_1(\tau = 0) = b$.

Solving initial-value problems

When the phase-space velocity $\boldsymbol{f}$ is independent of $\boldsymbol{x}$, and hence an explicit function of t, the solution of an initial-value problem can be found simply by integrating equation (9.1.3) from the initial instant $t = 0$ up to the present time t, obtaining

$$\boldsymbol{x}(t) = \boldsymbol{x}(t = 0) + \int_0^t \boldsymbol{f}(\tau)\, d\tau \qquad (9.1.25)$$

The integral on the right-hand side may be computed analytically in simple cases or numerically in more involved cases, using the methods discussed in Chapter 7. This integral form of the solution explains why the terms *solution* and *integration* of a differential equation are sometimes used interchangeably and without discrimination.

Conversely, the vector integral

$$\boldsymbol{I} = \int_0^T \boldsymbol{f}(\tau)\, d\tau \tag{9.1.26}$$

can be computed by solving the ordinary differential equation $d\boldsymbol{x}/dt = \boldsymbol{f}(t)$ with initial condition $\boldsymbol{x}(t=0) = 0$; then $\boldsymbol{I} = \boldsymbol{x}(t=T)$.

When the phase-space velocity $\boldsymbol{f}$ is not an explicit function of t, which means that the system is autonomous and, in addition, $\boldsymbol{f}$ is a linear function of $\boldsymbol{x}$, the solution of the initial-value problem may be computed analytically in closed form, as will be discussed in Section 9.2.

Under more general circumstances, the solution must be found using the numerical methods to be discussed in Sections 9.3–9.6.

The solution may become singular at a finite time

It might come as a surprise to realize that the solution of an initial-value problem may become *infinite* after a *finite* evolution time, even though the phase-space velocity is nonsingular. For example, the solution of the equation $dx/dt = -x^2$ with initial condition $x(t=0) = -1/c$, where c is a positive constant, is given by $x(t) = 1/(t-c)$, which becomes singular after a finite evolution time $t = c$. Conditions under which the solution of a system will remain nonsingular at all times have been established but are outside the scope of this book.

The solution may not be unique

It may also happen that the solution of an initial-value problem may not be unique, even though the phase-space velocity is a continuous function of its arguments. For example, the equation $dx/dt = x^{2/3}$ with initial condition $x(t=0) = 0$ has two solutions: $x(t) = 0$ and $\frac{1}{27}t^3$. To have a unique solution, it is necessary that the first partial derivatives of the phase-space velocity with respect to its argument are continuous functions of their arguments. This will be assumed to be the case in the following sections.

Behavior of the phase-space orbit near a point

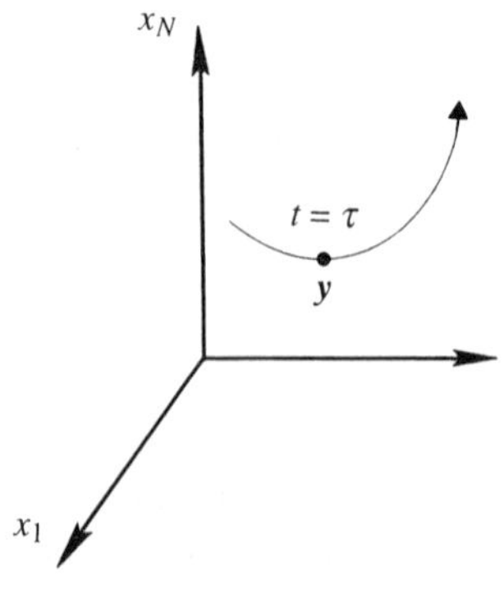

Although an analytical solution of an initial-value problem is not generally available, one can estimate the direction of the phase-space orbit in the neighborhood of a certain point $\boldsymbol{y}$ at the time instant τ, using analytical methods. This is done by expanding the phase-space velocity $\boldsymbol{f}$ in a Taylor series about the time $t = \tau$ and the position $\boldsymbol{x} = \boldsymbol{y}$, retaining only the constant and linear terms, and thereby obtaining the linearized system

$$\frac{d\hat{x}_i}{dt} = f_i(\boldsymbol{x} = \boldsymbol{y}, t = \tau) + \left(\frac{\partial f_i}{\partial x_j}\right)^{t=\tau}_{\boldsymbol{x}=\boldsymbol{y}} \hat{x}_j + \left(\frac{\partial f_i}{\partial t}\right)^{t=\tau}_{\boldsymbol{x}=\boldsymbol{y}} (t - \tau) \tag{9.1.27}$$

where $\hat{\boldsymbol{x}} = \boldsymbol{x} - \boldsymbol{y}$. The solution of this simplified system may be found analytically, as discussed in Section 9.2, but the results will be accurate only for a small period of time before or after the current time τ.

Stability of an equilibrium point

If the starting point of the phase-space orbit $\boldsymbol{x}(t = 0)$ coincides with an equilibrium point of an *autonomous* system, that is, $\boldsymbol{x}(t = 0) = \boldsymbol{X}$, then the initial rate of change of $\boldsymbol{x}$ with respect to t will vanish, and $\boldsymbol{x}$ will remain equal to $\boldsymbol{X}$ at all times.

Assume now that $\boldsymbol{x}(t = 0)$ is located very close to $\boldsymbol{X}$ but does not coincide with it. If the solution of the differential equations produces an orbit that takes us back to the equilibrium point for *any* initial condition that lies sufficiently close to $\boldsymbol{x}(t = 0)$, then the fixed point is *linearly stable*, otherwise it is *linearly unstable.*

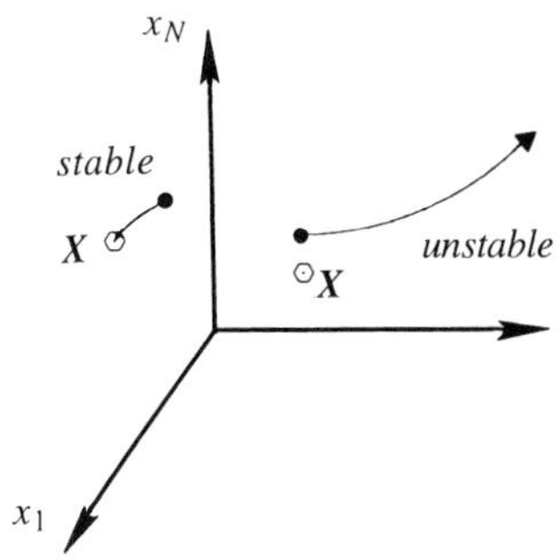

To assess the linear stability of a fixed point, we expand the phase-space velocity $\boldsymbol{f}$ in a Taylor series with respect to $\boldsymbol{x}$ about the equilibrium point $\boldsymbol{X}$. Retaining only the constant and linear terms, and noting that $\boldsymbol{f}(\boldsymbol{X}) = \boldsymbol{0}$, we obtain

$$\frac{d\hat{x}_i}{dt} = \left(\frac{\partial f_i}{\partial x_j}\right)_{\boldsymbol{x}=\boldsymbol{X}} \hat{x}_j \tag{9.1.28}$$

where $\hat{\boldsymbol{x}} = \boldsymbol{x} - \boldsymbol{X}$. The matrix of partial derivatives on the right-hand side is the *Jacobian matrix of the linearized system* at the equilibrium point and is denoted by $\boldsymbol{J}(\boldsymbol{X})$. By definition, $J_{i,j} = \partial f_i / \partial x_j$.

The results of Section 9.2 will reveal that the equilibrium point is linearly stable only when the real parts of all eigenvalues of $\boldsymbol{J}(\boldsymbol{X})$ are negative; otherwise it is linearly unstable. Finer classifications into elliptic and hyperbolic equilibrium points are based on the number of eigenvalues with a positive or negative real part.

For example, in the case of one equation, $N = 1$, the fixed point is stable when the value of the partial derivative $\partial f_1 / \partial x_1$ evaluated at the fixed point $x_1 = X_1$ is negative.

PROBLEM

9.1.6 *Stability of a chemical reactor.*

Compute the Jacobian $\boldsymbol{J}(\boldsymbol{X})$ of the system describing the operation of the CTSR discussed in Problem 9.1.4, and assess the stability of the operation. Will the reactor blow up?

Ill-conditioned or inherently unstable problems

There is a class of ill-conditioned or inherently unstable systems, where small changes in either the initial condition or phase-space velocity cause large variations in the *exact* solution at long times. This type of behavior occurs, in particular, when the general solution of the differential equations contains a dominant term that has been eliminated by a fortuitous or careful positioning of the point of departure in the phase space. Shifting the initial position by a small amount unleashes the dominant component and leads to a significantly different type of long-time behavior.

For example, the general solution of the equation $dx/dt = x - t$ is readily found to be $x(t) = ce^t + t + 1$, where c is a constant; stipulating that $x(t = 0) = 1$ requires that $c = 0$. Perturbing the initial condition, either willfully or inevitably due to the round-off error, forces c to take a nonzero value and allows for an exponential growth that dominates at long times.

It is then not surprising that computing a numerical solution of an inherently unstable system is extremely difficult. Meaningful results are obtained only when the numerical error is kept very low by using extended arithmetic precision or by employing a purifying mechanism. Even then, the results are bound to become inaccurate after a sufficiently long evolution time.

Boundary-value, Eigenvalue, and Free-boundary Problems

Apart from the initial-value problems discussed previously in this section, the mathematical modeling of a broad range of natural and engineering systems produces the three classes of problems stated in the above header. Solving these problems can be reduced to solving one or a family of initial-value problems, but this is not always the best approach: Specialized methods developed specifically for each class of problems work much better. The fundamental and practical importance of these problems justifies their separate discussion in Chapter 10.

9.2 Linear Autonomous Systems

When the components of the phase-space velocity $\boldsymbol{f}$ are *linear* functions of the components of the vector of unknown functions $\boldsymbol{x}$, and the system is autonomous, we can write

$$\boldsymbol{f}(\boldsymbol{x}) = \boldsymbol{A}\boldsymbol{x} - \boldsymbol{b} \tag{9.2.1}$$

where $\boldsymbol{A}$ is a constant $N \times N$ matrix and $\boldsymbol{b}$ is a constant N-dimensional column vector. The system (9.1.3) takes the simple form

$$\frac{d\boldsymbol{x}}{dt} = \boldsymbol{A}\boldsymbol{x} - \boldsymbol{b} \tag{9.2.2}$$

Under such circumstances, we can find the solution exactly in analytical form working as follows.

Diagonal systems

In the fortunate case where the matrix $\boldsymbol{A}$ is diagonal, denoted as $\boldsymbol{D}$, the individual scalar equations of the system (9.2.1) are decoupled, that is, each one contains only one unknown. The solution may then be found using elementary methods, and the result is

$$x_i(t) = X_i + \exp(t D_{i,i}) \left(x_i^{(0)} - X_i \right) \tag{9.2.3}$$

where $X_i = b_i / D_{i,i}$, and we have set $\boldsymbol{x}^{(0)} \equiv \boldsymbol{x}(t = 0)$; summation over the repeated index i is not implied. In vector notation, we obtain the equivalent statement

$$\boldsymbol{x}(t) = \boldsymbol{X} + \mathbf{exp}(t\boldsymbol{D}) \left(\boldsymbol{x}^{(0)} - \boldsymbol{X} \right) \tag{9.2.4}$$

where $\mathbf{exp}(t\boldsymbol{D})$ is a diagonal matrix function of t whose diagonal elements are the exponentials of the corresponding elements of $\boldsymbol{D}$, and

$$\boldsymbol{X} = \boldsymbol{D}^{-1}\boldsymbol{b} \tag{9.2.5}$$

is the equilibrium point of the linear system satisfying the equation $\boldsymbol{f}(\boldsymbol{X}) = \mathbf{0}$.

Triangular system

When the matrix $\boldsymbol{A}$ is lower triangular, the first equation contains only the first unknown, the second equation contains the first two unknowns, and the ith equation contains the first i unknowns. Accordingly, we solve the first equation for the first unknown, substitute the solution into all subsequent equations, solve the second equation for the second unknown, and proceed in this manner to recover the solution in a way that is reminiscent of forward substitution. For an upper triangular matrix, we use backward substitution.

The mathematical formalism and resulting expressions, however, can be simplified considerably working within a more general framework as follows.

Arbitrary systems

Let us consider the system (9.2.2) where the matrix $\boldsymbol{A}$ is arbitrary. We want to precondition the system, that is, redefine the Cartesian axes in the phase space, so that the problem takes a simpler form. This can be done by premultiplying both sides of equation (9.2.2) with a projection matrix $\boldsymbol{C}$, obtaining

$$\frac{d\boldsymbol{y}}{dt} = \boldsymbol{B}\boldsymbol{y} - \boldsymbol{C}\boldsymbol{b} \tag{9.2.6}$$

where we have defined

$$\boldsymbol{y} = \boldsymbol{C}\boldsymbol{x}, \qquad \boldsymbol{B} = \boldsymbol{C}\boldsymbol{A}\boldsymbol{C}^{-1} \tag{9.2.7}$$

Stipulating that the matrix $\boldsymbol{B}$ is diagonal takes us back to the problem of matrix diagonalization discussed in Section 2.9.

For simplicity, but with some loss of generality, we assume that $\boldsymbol{A}$ has N linearly independent eigenvectors and introduce a diagonal matrix $\boldsymbol{\Lambda}$ containing the possibly multiple eigenvalues of $\boldsymbol{A}$ along the diagonal, and the associated matrix of eigenvectors $\boldsymbol{U}$ defined in the paragraph before equation (2.9.15). Setting

$$\boldsymbol{C} = \boldsymbol{U}^{-1}, \qquad \boldsymbol{B} = \boldsymbol{\Lambda} \tag{9.2.8}$$

and using equations (2.9.21) and (2.9.22), we find that the system (9.2.6) has become diagonal. Using the solution (9.2.4), we then find

$$\boldsymbol{y}(t) = \boldsymbol{Y} + \mathbf{exp}(t\boldsymbol{\Lambda})\left(\boldsymbol{y}^{(0)} - \boldsymbol{Y}\right) \tag{9.2.9}$$

where

$$\boldsymbol{y} = \boldsymbol{U}^{-1}\boldsymbol{x}, \qquad \boldsymbol{y}^{(0)} = \boldsymbol{U}^{-1}\boldsymbol{x}^{(0)}, \qquad \boldsymbol{Y} = \boldsymbol{\Lambda}^{-1}\boldsymbol{U}^{-1}\boldsymbol{b} \tag{9.2.10}$$

To recover the desired solution $\boldsymbol{x}(t)$, we premultiply both sides of equation (9.2.9) by $\boldsymbol{U}$ and find

$$\boldsymbol{x}(t) = \boldsymbol{X} + \mathbf{exp}(t\,\boldsymbol{A})\left(\boldsymbol{x}^{(0)} - \boldsymbol{X}\right) \tag{9.2.11}$$

where

$$\mathbf{exp}(t\,\boldsymbol{A}) = \boldsymbol{U}\mathbf{exp}(t\boldsymbol{\Lambda})\boldsymbol{U}^{-1} \tag{9.2.12}$$

is the fixed point of the linear system. According to our discussion in Section 2.12, before we can evaluate the matrix $\mathbf{exp}(t\,\boldsymbol{A})$ we must have available all eigenvalues of the matrix $\boldsymbol{A}$.

The preceding two equations show that if the real part of at least one of the eigenvalues of $\boldsymbol{A}$ is positive, $\boldsymbol{x}(t)$ will amplify at an exponential rate. If the real part of all eigenvalues is negative, the difference $\boldsymbol{x}(t) - \boldsymbol{X}$ will shrink at an exponential rate, and $\boldsymbol{x}(t)$ will eventually tend to the fixed point $\boldsymbol{X}$.

PROBLEMS

9.2.1 ***System of linear*** ODE***s for a matrix function.***

(a) Let the elements of the $N \times N$ matrix $\boldsymbol{\psi}$ be functions of t. Show that the solution of the linear system of ordinary differential equations $d\boldsymbol{\psi}/dt = \boldsymbol{A}\boldsymbol{\psi} - \boldsymbol{B}$, where $\boldsymbol{A}$ and $\boldsymbol{B}$ are two constant $N \times N$ matrices, can be expressed in the form $\boldsymbol{\psi}(t) = \boldsymbol{\Psi} + \mathbf{exp}(t\,\boldsymbol{A})(\boldsymbol{\psi}^{(0)} - \boldsymbol{\Psi})$, where the matrix $\boldsymbol{\Psi} = \boldsymbol{A}^{-1}\boldsymbol{B}$ is the fixed point of the linear system.
(b) Show that, with the preceding definitions, the solution of the system $d\boldsymbol{\psi}/dt = \boldsymbol{\psi}\boldsymbol{A} - \boldsymbol{B}$ can be expressed in the form $\boldsymbol{\psi}(t) = \boldsymbol{\Psi} + (\boldsymbol{\psi}^{(0)} - \boldsymbol{\Psi})\,\mathbf{exp}(t\,\boldsymbol{A})$, where the matrix $\boldsymbol{\Psi} = \boldsymbol{B}\boldsymbol{A}^{-1}$ is the fixed point of the linear system.

9.2.2 ***Solving a system of two linear*** ODE***s.***

Compute the solution of the linear differential equation $d\boldsymbol{x}/dt = \boldsymbol{A}\boldsymbol{x} + (1, 2)^T$ at $t = 1$ when $\boldsymbol{A}$ is a 2×2 matrix with elements $A_{1,1} = 1$, $A_{1,2} = 2$, $A_{2,1} = 2$, $A_{2,2} = -1$, subject to the initial condition $\boldsymbol{x}(t = 0) = (1, 0)^T$.

Numerical Approximations

When the size of the matrix $\boldsymbol{A}$ is not small, computing all of its eigenvalues, which is necessary for the evaluation of the matrix function $\mathbf{exp}(t\,\boldsymbol{A})$, can be demanding. Additional complications arise when $\boldsymbol{A}$ does not have N linearly independent eigenvectors. In practice, it is more expedient to obtain an approximate solution using numerical methods.

We begin by noting that after a time interval Δt following time t has elapsed, then according to equation (9.2.11), the solution will be

$$\boldsymbol{x}(t + \Delta t) = \boldsymbol{X} + \mathbf{exp}(\Delta t\,\boldsymbol{A})\,\mathbf{exp}(t\,\boldsymbol{A})\left(\boldsymbol{x}^{(0)} - \boldsymbol{X}\right) \tag{9.2.13}$$

which may be recast into the form

$$\boldsymbol{x}(t + \Delta t) = \boldsymbol{X} + \mathbf{exp}(\Delta t\,\boldsymbol{A})(\boldsymbol{x}(t) - \boldsymbol{X}) \tag{9.2.14}$$

To simplify the notation, we denote

$$\boldsymbol{x}(t) = \boldsymbol{x}^n, \qquad \boldsymbol{x}(t + \Delta t) = \boldsymbol{x}^{n+1}, \qquad t = t^n, \qquad t + \Delta t = t^{n+1} \tag{9.2.15}$$

where the superscript n signifies the nth step of the numerical computation. Equation (9.2.14) assumes the simpler form

$$\mathbf{x}^{n+1} = \mathbf{X} + \mathbf{exp}(\Delta t\,\mathbf{A})(\mathbf{x}^n - \mathbf{X}) \tag{9.2.16}$$

The key idea is that, when Δt is sufficiently small, the matrix $\mathbf{exp}(\Delta t\,\mathbf{A})$ may be approximated with a polynomial or a rational function. Equation (9.2.16) then allows us to compute $\mathbf{x}(t + \Delta t)$ in terms of $\mathbf{x}(t)$ by carrying out matrix–vector multiplications or by solving systems of linear equations. In this manner, we can build the orbit of the solution in the phase space by marching in time from an initial point that is either specified or assumed at the outset.

Euler method

In the simplest approximation, we replace the matrix $\mathbf{exp}(\Delta t\,\mathbf{A})$ with its two-term Taylor expansion $\mathbf{I}+\Delta t\,\mathbf{A}$, which introduces a numerical error on the order of Δt^2, and obtain the Euler formula

$$\mathbf{x}^{n+1} = \mathbf{X} + (\mathbf{I} + \Delta t\,\mathbf{A})(\mathbf{x}^n - \mathbf{X}) = \mathbf{x}^n + \Delta t\,(\mathbf{A}\mathbf{x}^n - \mathbf{b}) \tag{9.2.17}$$

This approximation provides us with an *explicit* time-marching method involving the values of the solution at two time levels. Accordingly, it is classified as a *two-level explicit method.*

It is illuminating to recast equation (9.2.17) into the form

$$\frac{\mathbf{x}^{n+1} - \mathbf{x}^n}{\Delta t} = \mathbf{f}(\mathbf{x}^n) \equiv \mathbf{f}^n \tag{9.2.18}$$

and note that it represents the forward-time discretization of the left-hand side of the differential equation (9.2.2). Physically, according to equation (9.2.18), the orbit of the solution in the phase space is built by executing a sequence of steps, where the phase-space velocity during each step is kept constant, equal to its value at departure.

After M steps have been executed, the accumulated error E will be on the order of $M\Delta t^2$, and since $M = T/\Delta t$, where T is the total travel time, E will be on the order of $T\,\Delta t$, or simply Δt. A knowledge of the order of the error allows us to apply the method of Richardson extrapolation discussed in Section 1.6 to obtain an accurate solution from a crude and a refined solution.

Implicit Euler method

Selecting now a different type of approximation, we replace the matrix $\mathbf{exp}(\Delta t\,\mathbf{A})$ with its (0, 1) Padé approximant $(\mathbf{I} - \Delta t\,\mathbf{A})^{-1}$ shown in Table 8.5.1, which introduces a numerical error on the order of Δt^2, and obtain

$$\mathbf{x}^{n+1} = \mathbf{X} + (\mathbf{I} - \Delta t\,\mathbf{A})^{-1}(\mathbf{x}^n - \mathbf{X}) \tag{9.2.19}$$

Rearranging, we derive a system of linear equations for the unknown vector $\mathbf{x}^{n+1}$,

$$(\mathbf{I} - \Delta t\,\mathbf{A})\,\mathbf{x}^{n+1} = \mathbf{x}^n - \Delta t\,\mathbf{b} \tag{9.2.20}$$

Computing the solution at the end of a time step is thus done in an *implicit* manner, which means that it involves matrix inversion or solving a system of linear equations.

Equation (9.2.20) may be recast into the form

$$\frac{\boldsymbol{x}^{n+1} - \boldsymbol{x}^n}{\Delta t} = \boldsymbol{f}(\boldsymbol{x}^{n+1}) \equiv \boldsymbol{f}^{n+1} \tag{9.2.21}$$

which represents the backward-time discretization of the right-hand side of the differential equation (9.2.2). Physically, according to equation (9.2.21), the orbit of the solution in the phase space is built by traveling through a sequence of steps, where the phase-space velocity during each step is kept constant equal to the value at arrival. The accumulated error is on the order of Δt.

Crank–Nicolson method

Next, we approximate the matrix $\mathbf{exp}(\Delta t\, \boldsymbol{A})$ with the $(1, 1)$ Padé approximant shown in Table 8.5.1, setting it equal to $(\boldsymbol{I} - \frac{1}{2}\Delta t\, \boldsymbol{A})^{-1}(\boldsymbol{I} + \frac{1}{2}\Delta t\, \boldsymbol{A})$, which introduces an error on the order of Δt^3, and obtain

$$\boldsymbol{x}^{n+1} = \boldsymbol{X} + \left(\boldsymbol{I} - \tfrac{1}{2}\Delta t\, \boldsymbol{A}\right)^{-1} \left(\boldsymbol{I} + \tfrac{1}{2}\Delta t\, \boldsymbol{A}\right)(\boldsymbol{x}^n - \boldsymbol{X}) \tag{9.2.22}$$

Rearranging, we derive the Crank–Nicolson equation

$$\left(\boldsymbol{I} - \tfrac{1}{2}\Delta t\, \boldsymbol{A}\right)\boldsymbol{x}^{n+1} = \left(\boldsymbol{I} + \tfrac{1}{2}\Delta t\, \boldsymbol{A}\right)\boldsymbol{x}^n - \Delta t\, \boldsymbol{b} \tag{9.2.23}$$

which provides us with a system of linear equations for the unknown vector $\boldsymbol{x}^{n+1}$; the right-hand side is known.

Recasting equation (9.2.23) into the form

$$\frac{\boldsymbol{x}^{n+1} - \boldsymbol{x}^n}{\Delta t} = \tfrac{1}{2}(\boldsymbol{f}^n + \boldsymbol{f}^{n+1}) \tag{9.2.24}$$

shows that the orbit of the solution in the phase space is constructed by traveling through a sequence of steps, where the phase-space velocity during each step is kept constant and equal to the average value at arrival and departure. The accumulated error is on the order of Δt^2.

Modified Euler method

Another way of achieving third-order accuracy, while avoiding the implicit step, is to approximate the matrix $\mathbf{exp}(\Delta t\, \boldsymbol{A})$ with its three-term Taylor expansion $\boldsymbol{I} + \Delta t\, \boldsymbol{A} + \frac{1}{2}\Delta t^2 \boldsymbol{A}^2$, which introduces an error on the order of Δt^3. This approximation transforms equation (9.2.16) to

$$\boldsymbol{x}^{n+1} = \boldsymbol{X} + \left(\boldsymbol{I} + \Delta t\, \boldsymbol{A} + \tfrac{1}{2}\Delta t^2 \boldsymbol{A}^2\right)(\boldsymbol{x}^n - \boldsymbol{X}) \tag{9.2.25}$$

which can be rewritten as

$$\boldsymbol{x}^{n+1} = \boldsymbol{x}^n + \Delta t\, \left(\boldsymbol{I} + \tfrac{1}{2}\Delta t\, \boldsymbol{A}\right)(\boldsymbol{A}\boldsymbol{x}^n - \boldsymbol{b}) \tag{9.2.26}$$

It may readily be verified that the same result can be obtained by proceeding in a *predictor–corrector* sense, making one provisional step and one educated step, according to the algorithm:

$$
\boxed{\begin{array}{ll}
1.\ \text{Compute} & \boldsymbol{f}(\boldsymbol{x}^n) = \boldsymbol{A}\boldsymbol{x}^n - \boldsymbol{b} \\
2.\ \text{Set} & \boldsymbol{x}^{Temp} = \boldsymbol{x}^n + \Delta t\, \boldsymbol{f}(\boldsymbol{x}^n) \\
3.\ \text{Compute} & \boldsymbol{f}^{Temp} = \boldsymbol{f}(\boldsymbol{x}^{Temp}) = \boldsymbol{A}\boldsymbol{x}^{Temp} - \boldsymbol{b} \\
4.\ \text{Set} & \boldsymbol{f}^{Final} = \tfrac{1}{2}(\boldsymbol{f}(\boldsymbol{x}^n) + \boldsymbol{f}^{Temp}) \\
5.\ \text{Compute} & \boldsymbol{x}^{n+1} = \boldsymbol{x}^n + \Delta t\, \boldsymbol{f}^{Final}
\end{array}} \tag{9.2.27}
$$

The accumulated error is on the order of $\Delta t^{\,2}$. This is a special case of the second-order Runge–Kutta method to be discussed within the more general context of nonlinear and nonautonomous systems in Section 9.4.

Fourth-order Runge–Kutta method

The more terms we retain in the expansion of $\mathbf{exp}(\Delta t\, \boldsymbol{A})$, the higher the accuracy of the numerical method. For example, to raise the order of the error to $\Delta t^{\,5}$, we retain five terms and obtain

$$
\boldsymbol{x}^{n+1} = \boldsymbol{X} + \left(\boldsymbol{I} + \Delta t\, \boldsymbol{A} + \tfrac{1}{2}\Delta t^{\,2}\boldsymbol{A}^2 + \tfrac{1}{6}\Delta t^{\,3}\boldsymbol{A}^3 + \tfrac{1}{24}\Delta t^{\,4}\boldsymbol{A}^4\right)(\boldsymbol{x}^n - \boldsymbol{X}) \tag{9.2.28}
$$

which can be rewritten as

$$
\boldsymbol{x}^{n+1} = \boldsymbol{x}^n + \Delta t\ \left(\boldsymbol{I} + \tfrac{1}{2}\Delta t\, \boldsymbol{A} + \tfrac{1}{6}\Delta t^{\,2}\boldsymbol{A}^2 + \tfrac{1}{24}\Delta t^{\,3}\boldsymbol{A}^3\right)(\boldsymbol{A}\boldsymbol{x}^n - \boldsymbol{b}) \tag{9.2.29}
$$

It may readily be verified by straightforward algebraic subsitutions that the same result can be obtained by proceeding in a predictor–corrector sense, making three provisional steps and one final educated step, according to the fourth-order Runge–Kutta algorithm:

$$
\boxed{\begin{array}{ll}
1.\ \text{Compute} & \boldsymbol{f}(\boldsymbol{x}^n) = \boldsymbol{A}\boldsymbol{x}^n - \boldsymbol{b} \\
2.\ \text{Set} & \boldsymbol{x}^{Temp(1)} = \boldsymbol{x}^n + \tfrac{1}{2}\Delta t\, \boldsymbol{f}(\boldsymbol{x}^n) \\
3.\ \text{Compute} & \boldsymbol{f}^{Temp(1)} = \boldsymbol{f}\left(\boldsymbol{x}^{Temp(1)}\right) \\
4.\ \text{Set} & \boldsymbol{x}^{Temp(2)} = \boldsymbol{x}^n + \tfrac{1}{2}\Delta t\, \boldsymbol{f}^{Temp(1)} \\
5.\ \text{Compute} & \boldsymbol{f}^{Temp(2)} = \boldsymbol{f}\left(\boldsymbol{x}^{Temp(2)}\right) \\
6.\ \text{Set} & \boldsymbol{x}^{Temp(3)} = \boldsymbol{x}^n + \Delta t\, \boldsymbol{f}^{Temp(2)} \\
7.\ \text{Compute} & \boldsymbol{f}^{Temp(3)} = \boldsymbol{f}\left(\boldsymbol{x}^{Temp(3)}\right) \\
8.\ \text{Compute} & \boldsymbol{f}^{Final} = \tfrac{1}{6}\left(\boldsymbol{f}(\boldsymbol{x}^n) + 2\boldsymbol{f}^{Temp(1)} + 2\boldsymbol{f}^{itTemp(2)} + \boldsymbol{f}^{Temp(3)}\right) \\
9.\ \text{Set} & \boldsymbol{x}^{n+1} = \boldsymbol{x}^n + \Delta t\, \boldsymbol{f}^{Final}
\end{array}} \tag{9.2.30}
$$

The accumulated error is on the order of Δt^4. In Section 9.4, we shall discuss the generalization of the method to nonlinear and nonautonomous systems.

Time Stepping, Successive Mappings, Numerical Diffusion, and Stability

We developed several numerical methods with varying degrees of accuracy and required computational work. But accuracy is not the only issue involved in making a selection: A nominally accurate method may produce a failed solution.

To explain this seemingly paradoxical statement, we collect equations (9.2.17), (9.2.19), (9.2.22), (9.2.25), and (9.2.28) and recast them into the unified form

$$\boldsymbol{d}(t + \Delta t) = \boldsymbol{P}\boldsymbol{d}(t) \tag{9.2.31}$$

where

$$\boldsymbol{d}(t) \equiv \boldsymbol{x}(t) - \boldsymbol{X} \tag{9.2.32}$$

is the distance of the solution point from the fixed point $\boldsymbol{X}$, and $\boldsymbol{P}$ is a *projection* matrix. The various forms of $\boldsymbol{P}$ corresponding to the methods discussed previously in this section are displayed in Table 9.2.1.

The successive application of equation (9.2.31) yields

$$\boldsymbol{d}(k\Delta t) = \boldsymbol{P}^k\boldsymbol{d}(t = 0) \tag{9.2.33}$$

which shows that to compute the distance $\boldsymbol{d}(k\Delta t)$, we effectively premultiply the initial distance $\boldsymbol{d}(t = 0)$ with the mapping matrix $\boldsymbol{P}$, k times. The exact solution given in equation (9.2.11) yields

$$\boldsymbol{d}(k\Delta t) = \mathbf{exp}(k\Delta t\,\boldsymbol{A})\,\boldsymbol{d}(t = 0) \tag{9.2.34}$$

It is evident now that the behavior of the vector $\boldsymbol{d}(k\Delta t)$ computed from equation (9.2.33) will depend on the spectral radius of the projection matrix $\boldsymbol{P}$, defined as the maximum of the norm of its eigenvalues and denoted by $\rho(\boldsymbol{P})$. If $\rho(\boldsymbol{P})$ is less than unity, the length of $\boldsymbol{d}(k\Delta t)$ will gradually decrease and eventually

Table 9.2.1 **The projection matrix $\boldsymbol{P}$ for the linear autonomous system (9.2.2), and the relation between the eigenvalues of $\boldsymbol{P}$ and those of $\boldsymbol{A}$.**

Method	Projection Matrix	Eigenvalues
Euler explicit	$\boldsymbol{P} = \boldsymbol{I} + \Delta t\,\boldsymbol{A}$	$\lambda_P = 1 + \Delta t\,\lambda_A$
Euler implicit	$\boldsymbol{P} = (\boldsymbol{I} - \Delta t\,\boldsymbol{A})^{-1}$	$\lambda_P = 1/(1 - \Delta t\,\lambda_A)$
Crank–Nicolson implicit	$\boldsymbol{P} = \left(\boldsymbol{I} - \frac{1}{2}\Delta t\,\boldsymbol{A}\right)^{-1}\left(\boldsymbol{I} + \frac{1}{2}\Delta t\,\boldsymbol{A}\right)$	$\lambda_P = \left(1 + \frac{1}{2}\Delta t\,\lambda_A\right)\Big/\left(1 - \frac{1}{2}\Delta t\,\lambda_A\right)$
Modified Euler explicit	$\boldsymbol{P} = \boldsymbol{I} + \Delta t\,\boldsymbol{A} + \frac{1}{2}\Delta t^2\boldsymbol{A}^2$	$\lambda_P = 1 + \Delta t\,\lambda_A + \frac{1}{2}\Delta t^2\lambda_A^2$
Fourth-order Runge–Kutta explicit	$\boldsymbol{P} = \boldsymbol{I} + \Delta t\,\boldsymbol{A} + \frac{1}{2}\Delta t^2\boldsymbol{A}^2$	$\lambda_P = 1 + \Delta t\,\lambda_A + \frac{1}{2}\Delta t^2\lambda_A^2$
	$+\frac{1}{6}\Delta t^3\boldsymbol{A}^3 + \frac{1}{24}\Delta t^4\boldsymbol{A}^4$	$+\frac{1}{6}\Delta t^3\lambda_A^3 + \frac{1}{24}\Delta t^4\lambda_A^4$

tend to zero during the time steppings, whereas if $\rho(\boldsymbol{P})$ is greater than unity, the length of $\boldsymbol{d}(k\Delta t)$ will keep increasing roughly by a factor that is equal to $\rho(\boldsymbol{P})$, while its orientation will remain nearly constant (Section 2.11). The relation between the eigenvalues of $\boldsymbol{A}$, λ_A, and those of $\boldsymbol{P}$, λ_P, can readily be deduced and is shown in the third column of Table 9.2.1.

Numerical diffusion

Let us assume that the real part of an eigenvalue λ_A is positive, which means that the exact solution exhibits an exponential growth. Since $\Delta t > 0$, for Euler's method $|\lambda_P| > 1$ and $\rho(\boldsymbol{P}) > 1$ for any value of Δt; for the other methods, $|\lambda_P|$ may be either larger or lower than unity, depending on the size of Δt. Thus Euler's method produces a solution that grows, whereas other methods may yield an erroneous decay. This discrepancy is attributed to the action of *numerical diffusion* or *dissipation.*

Numerical instability

If, on the other hand, the real part of each eigenvalue λ_A is negative, in which case the exact solution exhibits an exponential decay, $\rho(\boldsymbol{P})$ for the explicit Euler method may be either larger or lower than unity; that is, the numerical solution may yield an erroneous growth described as a *numerical instability.* This observation warns us that the result of a numerical solution will not necessarily be an accurate reflection of a physical behavior of a natural or engineering system as modeled by the dynamical system.

The implicit Euler method and the Crank–Nicolson method reproduce the correct behavior irrespectively of the value of Δt and are thus *unconditionally stable.* This result is in agreement with the rule that *implicit methods are generally unconditionally stable*; the additional work required for solving a linear system of equations at each step may be worth the effort.

The modified Euler and the fourth-order Runge–Kutta methods are stable only when Δt lies within a certain range; that is, they are both *conditionally stable.*

Stiff problems

To be more specific, let us consider the simplest case of one equation, $N = 1$, where the matrix $\boldsymbol{A}$ reduces to a scalar a with corresponding eigenvalue $\lambda_A = a$. If a is real and negative, the exact solution will decay exponentially toward the equilibrium point; the numerical solution will observe this behavior as long as $-1 < \lambda_P < 1$. For the explicit Euler method, this inequality is satisfied when $\Delta t < 2/|a|$. The existence of a restriction of the time step renders the method conditionally stable. The graph on the next page shows the onset of an unstable behavior as the value of Δt approaches and then exceeds the critical threshold.

It is important to note, in particular, that a very large negative exponent a requires a very small time step Δt throughout the computation, even after the solution has virtually reached the equilibrium point following a short initial transient period of decay. As curious as it may seem, the rate of decay has a profound effect on the performance of the numerical method at *all* times. Problems whose solutions exhibit this type of behavior are described as *stiff*; the numerical handling of such problems will be discussed in Section 9.6.

PROBLEMS

9.2.3 *Error in Euler's method.*

Equation (9.2.33) for the explicit Euler method with one equation gives $d(k\Delta t) = (1 + \Delta t\, a)^k d(t = 0)$, whereas the exact solution computed from equation (9.2.34) is $d(k\Delta t) = \exp(k\Delta t\, a) d(t = 0)$. Show that when $a\Delta t \geq -1$, the numerical solution underpredicts the magnitude of $d(k\Delta t)$.

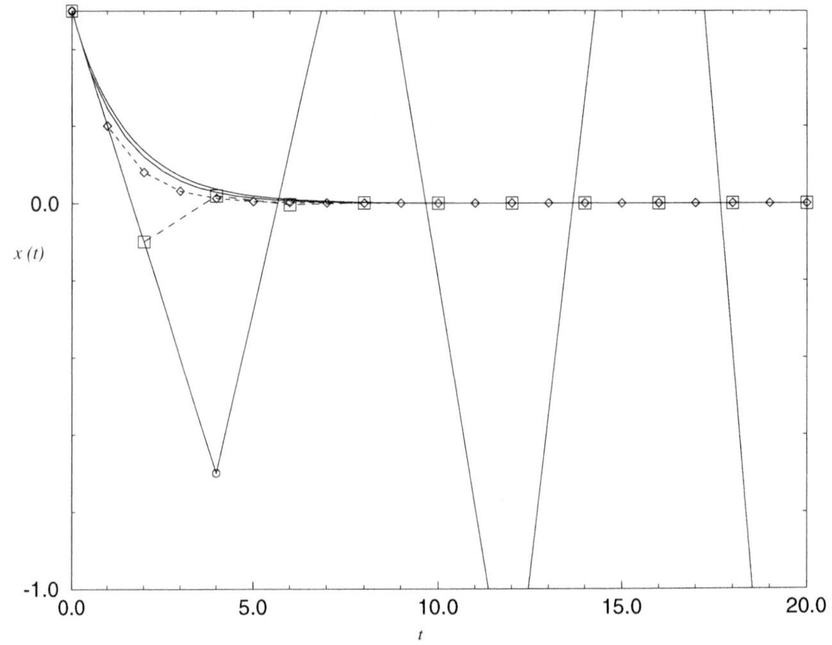

9.2.4 ***Stability and stiffness of a*** 2×2 ***system.***

Consider a 2×2 system with $A_{1,1} = a$, $A_{2,2} = a$, $A_{1,2} = b$, $A_{2,1} = b$. Show that when $a \pm b < 0$, the explicit Euler method reproduces the exponential approach to the fixed point as long as $\Delta t < 2/|a \pm b|$. Discuss the implications of this result on the numerical solution of a stiff system where a, b, or both assume large negative values.

Midpoint Method

Another way to improve the accuracy is to advance the solution using information at more than two time levels; this is an alternative to information gathering by making predictive steps. The idea is that the information does not need to be gained, but can be extracted by interrogating the solution orbit in the phase space. Disadvantages are that the solution at one or more previous time levels must be stored, and the time step must generally be kept constant.

For example, repeating the steps that led us to equation (9.2.25), we find

$$\boldsymbol{x}^{n-1} = \boldsymbol{X} + \left(\boldsymbol{I} - \Delta t\,\boldsymbol{A} + \tfrac{1}{2}\Delta t^2 \boldsymbol{A}^2\right)(\boldsymbol{x}^n - \boldsymbol{X}) \tag{9.2.35}$$

Subtracting corresponding sides of equation (9.2.35) from equation (9.2.25) with the purpose of eliminating the term involving $\boldsymbol{A}^2$, and rearranging the resulting expression, we obtain the midpoint algorithm

$$\boldsymbol{x}^{n+1} = \boldsymbol{x}^{n-1} + 2\Delta t\,(\boldsymbol{A}\boldsymbol{x}^{n} - \boldsymbol{b}) \tag{9.2.36}$$

The first point $\boldsymbol{x}^1$ must be computed using a two-level method, performing a sequence of substeps with a small enough time step so as not to introduce an important source of error. Each subsequent step introduces a numerical error on the order of Δt^3.

Numerical stability

To study the stability of the midpoint method, we recast equation (9.2.36) into the form

$$\boldsymbol{d}^{n+1} = \boldsymbol{d}^{n-1} + 2\Delta t\,\boldsymbol{A}\boldsymbol{d}^{n} \tag{9.2.37}$$

where $\boldsymbol{d}(t) = \boldsymbol{x}(t) - \boldsymbol{X}$, and then rewrite it in the expanded form

$$\boldsymbol{z}^{n+1} = \boldsymbol{P}\boldsymbol{z}^{n} \tag{9.2.38}$$

for $n \geq 1$, where we have introduced the double-length $2N$-dimensional vector $\boldsymbol{z}^{n+1} = (\boldsymbol{d}^{n+1}, \boldsymbol{d}^{n})^T$, and the $2N \times 2N$ extended projection matrix

$$\boldsymbol{P} = \begin{bmatrix} 2\Delta t\,\boldsymbol{A} & \boldsymbol{I} \\ \boldsymbol{I} & \boldsymbol{0} \end{bmatrix} \tag{9.2.39}$$

According to our preceding discussion, the behavior of the numerical solution will depend on the magnitude of the spectral radius of $\boldsymbol{P}$.

It is instructive to consider again the case of one equation, where the matrix $\boldsymbol{A}$ reduces to a scalar a, and the matrix $\boldsymbol{P}$ becomes

$$\boldsymbol{P} = \begin{bmatrix} 2\Delta t\,a & 1 \\ 1 & 0 \end{bmatrix} \tag{9.2.40}$$

The eigenvalues of $\boldsymbol{P}$ satisfy the quadratic equation $\lambda_P^2 - 2a\Delta t\,\lambda_P - 1 = 0$ whose solution is

$$\lambda_P = a\Delta t \pm (a^2\Delta t^2 + 1)^{1/2} \tag{9.2.41}$$

When a is real and negative, corresponding to an exponentially decaying solution, the eigenvalue corresponding to the minus sign is less than -1, and thus the spectral radius of $\boldsymbol{P}$ is larger than unity. The computations will produce an erroneously growing solution as illustrated in the diagram on the next page, which means that the method is numerically unstable.

PROBLEMS

9.2.5 *Theoretical investigation of stability.*

Consider the scalar equation $dx/dt = ax - b$, where a and b are real constants, and investigate the numerical stability of all numerical methods displayed in Table 9.2.1.

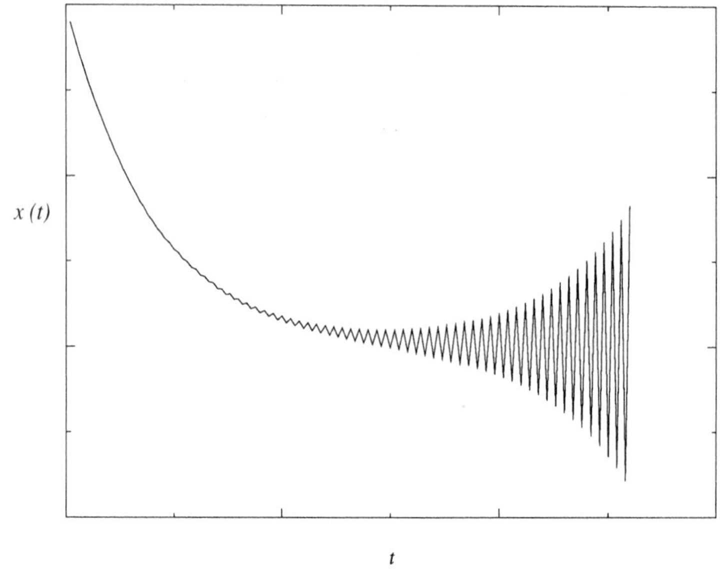

9.2.6 ***Comparison between exact and numerical solutions.***

Consider the system of two scalar differential equations $d\boldsymbol{x}/dt = \mathbf{A}\boldsymbol{x} + (1, 2)^T$, where $A_{1,1} = 0.90$, $A_{1,2} = c$, $A_{2,1} = c$, $A_{2,2} = 0.90$, with initial condition $\boldsymbol{x}(t = 0) = (1, 1)^T$. Compute the solution up to $t = 1.0$ using the explicit Euler method, the implicit Euler method, and the Crank–Nicolson method with $\Delta t = 0.05$, and compare the numerical results with the exact solution for (*a*) $c = 0$ and (*b*) $c = 0.50$.

9.3 Explicit and Implicit Methods for Nonlinear Systems

The numerical methods for solving linear autonomous systems discussed in Section 9.2 can be extended in several ways for solving more general nonlinear, autonomous or nonautonomous systems.

Perhaps the simplest venue for carrying out this extension departs from the integrated form of equation (9.1.3):

$$\boldsymbol{x}^{n+1} = \boldsymbol{x}^n + \int_{t^n}^{t^{n+1}} \boldsymbol{f}(\boldsymbol{x}, t)\, dt \tag{9.3.1}$$

The idea is to approximate the phase-space velocity $\boldsymbol{f}(\boldsymbol{x}, t)$ between the time instants t^n and t^{n+1} using methods of polynomial function interpolation and extrapolation discussed in Chapter 6, and then compute the integral on the right-hand side of equation (9.3.1) by analytical methods. The result is an expression for $\boldsymbol{x}^{n+1}$ in terms of $\boldsymbol{x}^n$ and values of $\boldsymbol{f}$ at a number of past, present, or future time instants.

The use of past and present values, $\ldots, \boldsymbol{f}^{n-1}, \boldsymbol{f}^n$, leads to *explicit* methods, whereas the use of past, present, and one imminent value, $\ldots, \boldsymbol{f}^{n-1}, \boldsymbol{f}^n, \boldsymbol{f}^{n+1}$, leads to *implicit* methods. Explicit methods are much easier to implement and work well for nice problems, but implicit methods are necessary when the system of differential equations is *stiff*, that is, the solution varies over time with two or more *disparate* time scales as was indicated in Section 9.2 and will be discussed in more detail in Section 9.6. Explicit methods are generally only conditionally stable, whereas implicit methods are typically unconditionally stable. The interplay between accuracy and numerical stability is important in making a selection.

Another way of developing numerical methods is to approximate the function $\mathbf{x}(t)$ with a Hermite interpolating polynomial that agrees with $\mathbf{x}(t)$ and its first derivative $\mathbf{x}'(t)$ at a number of data points. In the present context, $\mathbf{x}'(t)$ is the phase-space velocity $\mathbf{f}$. Explicit methods arise from extrapolation, and implicit methods arise from interpolation.

In the remainder of this section, we discuss a class of methods based on the general formula (9.3.1). The methods that have been selected are either illustrative of the procedures or find extensive usage in practice. A comprehensive compilation of a wide variety of additional procedures are presented by Lapidus and Seinfeld (1971).

Euler Method

Euler's method approximates the function $\mathbf{f}(\mathbf{x}, t)$ between the time instants t^n and t^{n+1}, with the constant value $\mathbf{f}^n \equiv \mathbf{f}(\mathbf{x}^n, t^n)$, and thus advances the solution over a small interval Δt by moving in the phase space with the velocity at the beginning of the time step, producing

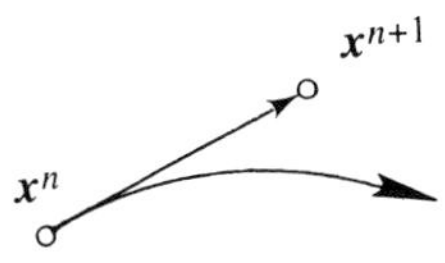

$$\mathbf{x}^{n+1} = \mathbf{x}^n + \Delta t \, \mathbf{f}^n \tag{9.3.2}$$

Note that this equation also arises from the first-order forward-difference discretization of the left-hand side of equation (9.1.3).

Each step produces a numerical error of order Δt^2, and the accumulated error over a time period T is on the order of $T\Delta t$. This amount of error is excessive in all but the most difficult problems where the evaluation of the phase-space velocity $\mathbf{f}$ consumes a large amount of computational time. The single advantage of Euler's method is that it requires only one function evaluation at each time step.

Implicit and Semi-implicit Euler Methods

The implicit Euler method approximates the phase-space velocity $\mathbf{f}(\mathbf{x}, t)$ with the constant value $\mathbf{f}^{n+1} \equiv \mathbf{f}(\mathbf{x}^{n+1}, t^{n+1})$, that is, it advances the solution over a small interval Δt by moving with the phase-space velocity anticipated at the end of the time step, producing

$$\mathbf{x}^{n+1} = \mathbf{x}^n + \Delta t \, \mathbf{f}^{n+1} \tag{9.3.3}$$

Alternatively, this formula may be derived from the first-order backward-difference discretization of equation (9.1.3). As with the explicit Euler method, each step introduces a numerical error on the order of Δt^2.

Because the right-hand side of equation (9.3.3) involves the unknown vector $\mathbf{x}^{n+1}$, carrying out each step requires solving one or a system of nonlinear algebraic equations. In certain cases, this can be done using the method of one-point iterations, that is, by guessing $\mathbf{x}^{n+1}$, computing the right-hand side of equation (9.3.3), setting $\mathbf{x}^{n+1}$ equal to the computed value, and continuing the iterations until convergence has been achieved. More generally, the algebraic system must be solved using Newton's method. The initial guess can be set equal to $\mathbf{x}^n$ or computed by extrapolation.

Semi-implicit Euler method

We observe that the difference between $\boldsymbol{x}^{n+1}$ and $\boldsymbol{x}^n$ is on the order of Δt, and that formula (9.3.3) carries a discretization error on the order of Δt^2, and this allows us to replace $\boldsymbol{f}^{n+1}$ on the right-hand side of equation (9.3.3) with its linearized Taylor series expansion. When the system is autonomous, we obtain

$$\boldsymbol{x}^{n+1} = \boldsymbol{x}^n + \Delta t \, [\boldsymbol{f}^n + \boldsymbol{J}^n (\boldsymbol{x}^{n+1} - \boldsymbol{x}^n)] \tag{9.3.4}$$

where

$$J_{i,j} = \frac{\partial f_i}{\partial x_j} \tag{9.3.5}$$

and the superscript n indicates evaluation at $\boldsymbol{x}^n$ and t^n. Rearranging equation (9.3.4), we derive a system of linear equations for the unknown vector $\boldsymbol{x}^{n+1}$, which is the basis for the *semi-implicit Euler method* expressed by

$$(\boldsymbol{I} - \Delta t \, \boldsymbol{J}^n)\boldsymbol{x}^{n+1} = (\boldsymbol{I} - \Delta t \, \boldsymbol{J}^n)\boldsymbol{x}^n + \Delta t \, \boldsymbol{f}^n \tag{9.3.6}$$

In practice, however, the low-order accuracy of the implicit and semi-implicit Euler methods makes their implementation hardly worth the effort.

Crank–Nicolson Method

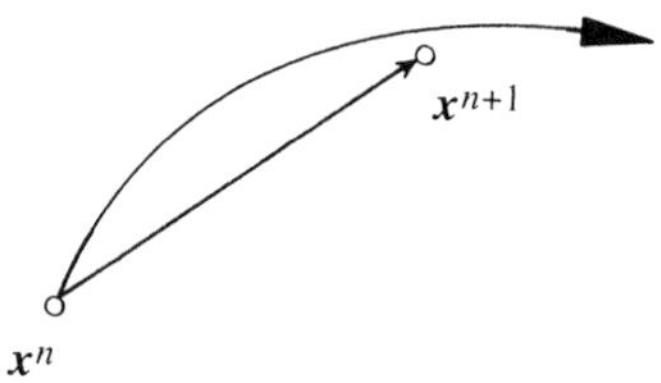

A higher level of accuracy may be achieved by using the Crank–Nicolson method. The idea is to approximate $\boldsymbol{f}(\boldsymbol{x}, t)$ in equation (9.3.1) with the average of the initial and anticipated phase-space velocities, yielding

$$\boldsymbol{x}^{n+1} = \boldsymbol{x}^n + \tfrac{1}{2}\Delta t \, (\boldsymbol{f}^n + \boldsymbol{f}^{n+1}) \tag{9.3.7}$$

Each step introduces a numerical error on the order of Δt^4. The semi-implicit version of the method is analogous to that of the Euler method described in the preceding subsection.

In practice, despite the improved numerical stability, the need to solve a system of nonlinear algebraic equations makes this method inferior to its alternatives discussed in the next section.

Three-point Methods

Another way of achieving second-order accuracy is to work with the counterpart of equation (9.3.1) for the time interval between t^{n-1} and t^{n+1}, yielding

$$\boldsymbol{x}^{n+1} = \boldsymbol{x}^{n-1} + \int_{t^{n-1}}^{t^{n+1}} \boldsymbol{f}(\boldsymbol{x}, t) \, dt \tag{9.3.8}$$

For simplicity, we shall assume that the time step Δt is held constant, that is, $t^{n+1} - t^{n-1} = 2\Delta t$; relaxing this restriction requires only minor and straightforward modifications.

Midpoint method

First, we approximate the function $f(x, t)$ between the points t^{n-1} and t^{n+1}, with the constant value $f(x^n, t^n)$, and obtain the explicit *midpoint algorithm* expressed by

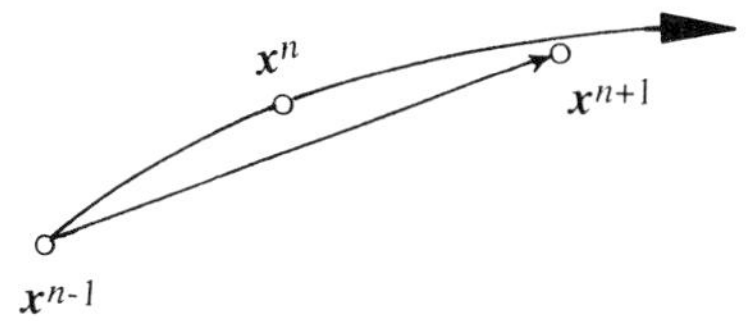

$$x^{n+1} = x^{n-1} + 2\Delta t f^n \tag{9.3.9}$$

The first point $x^{(1)}$ may be found using Euler's method, by executing a sequence of substeps with a time step that is an integral fraction of Δt. Each subsequent step introduces a numerical error on the order of Δt^4.

Unfortunately, this method may suffer from a *weak numerical instability*, which introduces sawtooth type oscillations. Although these oscillations can be filtered out either concurrently or *a posteriori* by repositioning the points x^{n+1} along the orbit in the phase space, resulting in a *modified midpoint method*, the additional work puts the method at a practical disadvantage. The modified midpoint method is used primarily for solving stiff problems, as will be discussed in Section 9.6.

Milne's method

A more accurate method emerges by approximating the function $f(x, t)$ over the interval between t^{n-1} and t^{n+1} with a parabola that passes through the two known points (t^{n-1}, f^{n-1}), (t^n, f^n), and the unknown point (t^{n+1}, f^{n+1}). This is reminiscent of Simpson's rule for computing integrals based on local quadratic interpolation. The result is *Milne's implicit algorithm*

$$x^{n+1} = x^{n-1} + \tfrac{1}{3}\Delta t\,(f^{n-1} + 4f^n + f^{n+1}) \tag{9.3.10}$$

Each step introduces a numerical error on the order of Δt^5. The need for an iterative solution of an algebraic system is a disadvantage, but a slight modification of the method finds usage as a module in a predictor–corrector algorithm to be discussed in Section 9.4.

Adams–Bashforth Methods

Returning to equation (9.3.1), we approximate the phase-space velocity $f(x, t)$ between the time levels t^n and t^{n+1} with a polynomial that interpolates through a specified set of $p + 1$ known points

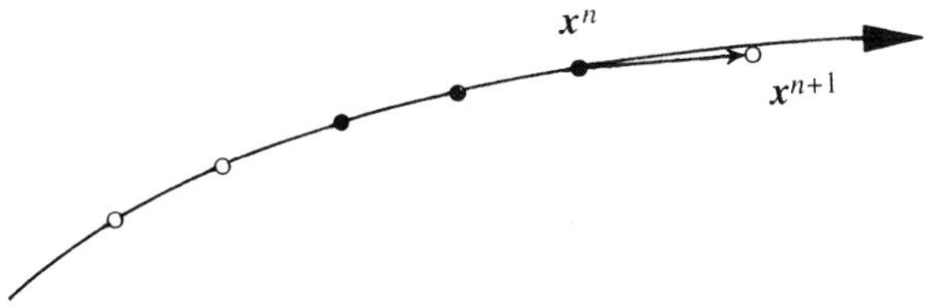

$$(t^{n-p}, f^{n-p}), (t^{n-1}, f^{n-1}), \ldots, (t^n, f^n) \tag{9.3.11}$$

and obtain the family of the Adams–Bashforth methods. The first member of the family corresponding to $p = 0$ is Euler's method, and more advanced methods with $p = 1, 2, 3, 4$ are expressed by

$$\boldsymbol{x}^{n+1} = \boldsymbol{x}^n + \tfrac{1}{2}\Delta t\,(-\boldsymbol{f}^{n-1} + 3\boldsymbol{f}^n) \tag{9.3.12}$$

$$\boldsymbol{x}^{n+1} = \boldsymbol{x}^n + \tfrac{1}{12}\Delta t\,(5\boldsymbol{f}^{n-2} - 16\boldsymbol{f}^{n-1} + 23\boldsymbol{f}^n) \tag{9.3.13}$$

$$\boldsymbol{x}^{n+1} = \boldsymbol{x}^n + \tfrac{1}{24}\Delta t\,(-9\boldsymbol{f}^{n-3} + 37\boldsymbol{f}^{n-2} - 59\boldsymbol{f}^{n-1} + 55\boldsymbol{f}^n) \tag{9.3.14}$$

$$\boldsymbol{x}^{n+1} = \boldsymbol{x}^n + \tfrac{1}{720}\Delta t\,(251\boldsymbol{f}^{n-4} - 1274\boldsymbol{f}^{n-3} + 2616\boldsymbol{f}^{n-2} - 2774\boldsymbol{f}^{n-1} + 1901\boldsymbol{f}^n) \tag{9.3.15}$$

Each step introduces, respectively, a numerical error on the order of Δt^3, Δt^4, Δt^5, and Δt^6.

For future use, we note that the leading-order term of the error of the fourth-order method expressed by equation (9.3.14), corresponding to $p = 3$, is given by

$$\boldsymbol{x}^{n+1} - \hat{\boldsymbol{x}}^{n+1} = -\tfrac{251}{720}\Delta t^5 \left(\frac{d^5\boldsymbol{x}}{dt^5}\right)_{t=t^n} \tag{9.3.16}$$

where a caret denotes the value that would have arisen if we were able to integrate the differential equations exactly from time t^n to time $t^n + \Delta t$.

Adams–Moulton Methods

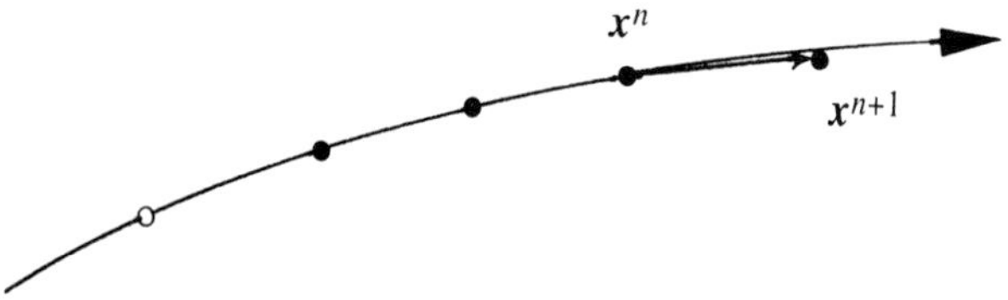

Involving the point $(t^{n+1}, \boldsymbol{f}^{n+1})$ in the polynomial approximation of the function $\boldsymbol{f}(\boldsymbol{x}, t)$ between the time levels t^n and t^{n+1} yields the Adams–Moulton methods. The first member of the family corresponding to $p = 0$ is the implicit Euler method, the second member corresponding to $p = 1$ is the Crank–Nicolson method, and more advanced methods corresponding to $p = 2, 3, 4$ are represented by

$$\boldsymbol{x}^{n+1} = \boldsymbol{x}^n + \tfrac{1}{12}\Delta t\,(-\boldsymbol{f}^{n-1} + 8\boldsymbol{f}^n + 5\boldsymbol{f}^{n+1}) \tag{9.3.17}$$

$$\boldsymbol{x}^{n+1} = \boldsymbol{x}^n + \tfrac{1}{24}\Delta t\,(\boldsymbol{f}^{n-2} - 5\boldsymbol{f}^{n-1} + 19\boldsymbol{f}^n + 9\boldsymbol{f}^{n+1}) \tag{9.3.18}$$

$$\boldsymbol{x}^{n+1} = \boldsymbol{x}^n + \tfrac{1}{720}\Delta t\,(-19\boldsymbol{f}^{n-3} + 106\boldsymbol{f}^{n-2} - 264\boldsymbol{f}^{n-1} + 646\boldsymbol{f}^n + 251\boldsymbol{f}^{n+1}) \tag{9.3.19}$$

Each step introduces, respectively, a numerical error on the order of Δt^4, Δt^5, or Δt^6, but the magnitude of the numerical coefficients that multiply these powers are substantially less than those for the corresponding explicit Adams–Bashforth methods given in equations (9.3.12)–(9.3.15).

For example, the leading-order term of the error of the fourth-order method expressed by equation (9.3.18), corresponding to $p = 3$, is given by

$$\boldsymbol{x}^{n+1} - \hat{\boldsymbol{x}}^{n+1} = \tfrac{19}{720}\Delta t^5 \left(\frac{d^5\boldsymbol{x}}{dt^5}\right)_{t=t^n} \tag{9.3.20}$$

where a caret denotes the value that would have been obtained if we were able to integrate the differential equations exactly from time t^n to time $t^n + \Delta t$.

The generally nonlinear algebraic equations (9.3.17)–(9.3.19) may be solved either by the method of successive substitutions or by linearizing them to derive a system of linear equations in the spirit of the semi-implicit method discussed in a previous subsection.

In practice, the Adams–Moulton methods are used predominantly as modules in predictor–corrector schemes.

PROBLEMS

9.3.1 *Lorenz system.*

The Lorenz system (9.1.24) is typically studied for $k = 10$, $b = \frac{8}{3}$, and for different values of r.

(*a*) Taking $k = 10$, $b = \frac{8}{3}$, calculate the equilibrium points for $r = 0.5, 5.0, 28.0$.

(*b*) Write a computer code that integrates the Lorenz system using the fourth-order Adams–Bashforth method. Perform a series of computations, for $k = 10$, $b = \frac{8}{3}$, $A(0) = T(0) = D(0) = 1.0$, and $r = 0.5, 5.0, 28.0$. Plot and discuss the projections of the solution orbit in the AT and AD planes, and investigate the effect of the time step.

9.3.2 *Evolution of an ecosystem.*

Competition between different species plays an important role in natural selection. It is the basis for Darwin's theory of evolution, originally proposed by Aristotle, and the starting point for the analysis of ecological systems pioneered by the Italian mathematician Vito Volterra in 1920. The dynamics of competing microorganisms in an ecosystem was advanced in the 1970s, partly thanks to the feasibility of obtaining solutions to ordinary differential equations using numerical methods.

Consider an ecological system containing a predator and a prey. Prey multiplies autonomously but is consumed by the predator. The evolution of the prey population x_1 is governed by the differential equation

$$\frac{dx_1}{dt} = ax_1 - \chi x_1 x_2 \tag{9.3.21}$$

where x_2 is the population of the predator, a is a positive birth-rate constant, and χ is another positive constant expressing how often a predator catches a prey. The evolution of the predator, on the other hand, is governed by the differential equation

$$\frac{dx_2}{dt} = -bx_2 + \chi \varepsilon x_1 x_2 \tag{9.3.22}$$

where b is a death-rate constant due to starvation, and ε is a constant that indicates how many individual predators are necessary in order to destroy one individual prey.

The preceding two equations comprise the Lotka–Volterra system. There are two steady-state solutions, denoted by (x_1^*, x_2^*): the trivial one $(x_1^*, x_2^*) = (0, 0)$, and a nontrivial one that can be found by setting the right-hand sides of the differential equations (9.3.21) and (9.3.22) equal to zero, and solving the resulting system of nonlinear algebraic equations for x_1 and x_2.

Consider a system with $\varepsilon = 1$, $a = 0.400$, $b = 0.450$, $\chi = 0.50$. Calculate the nontrivial steady-state concentrations (x_1^*, x_2^*), and integrate the Lotka–Volterra system using the fourth-order Adams–Bashforth method subject to the initial conditions $x_1(t = 0) = 0.10$, $x_2(t = 0) = 0.20$. Carry the integration for a sufficiently long time so that you can assess the asymptotic behavior of the system at long times. Plot and discuss the trajectories of the solution in the phase plane.

9.3.3 *Mathieu's equation.*

Mathieu's equation (5.1.15), resolved into the first-order system (9.1.4), describes the oscillations of the free surface of a liquid in a vertically vibrating container. The nature of the periodic solutions was reviewed in Section 5.1.

(*a*) Use the fourth-order Adams–Bashforth method to compute an even periodic solution corresponding to the solution branch $p = a_0(q)$ for $q = 1.0$, subject to the initial conditions $x(0) = 1.0$ and $x'(0) = 0.0$. Discuss the results of your computations.

(*b*) Repeat part (*a*) for an odd periodic solution corresponding to the solution branch $p = b_1(q)$ for $q = 1.0$, and initial conditions $x(0) = 0.0$ and $x'(0) = 1.0$.

For the evaluation of p, use the asymptotic expansions (Abramowitz and Stegun 1972, Chapter 20):

$$p = a_0(q) = -\tfrac{1}{2}q^2 + \tfrac{7}{128}q^4 - \tfrac{29}{2304}q^6 + \cdots$$
$$p = b_1(q) = 1 - q - \tfrac{1}{8}q^2 + \tfrac{1}{64}q^3 - \tfrac{1}{1536}q^4 + \cdots \tag{9.3.23}$$

Störmer–Verlet Algorithm for Second-order Systems

The mathematical modeling of physical systems involving bodies and particles whose motion is governed by Newton's second law often results in systems of second-order differential equations of the form

$$\frac{d^2\boldsymbol{x}}{dt^2} = \boldsymbol{f}(\boldsymbol{x}, t) \tag{9.3.24}$$

to be solved subject to the initial conditions

$$\boldsymbol{x}(t = 0) = \boldsymbol{x}^{(0)}, \qquad \left(\frac{d\boldsymbol{x}}{dt}\right)_{t=0} = \boldsymbol{v}^{(0)} \tag{9.3.25}$$

where $\boldsymbol{v}$ stands for *velocity*. The absence of the first derivative $d\boldsymbol{x}/dt$ from the right-hand side of equation (9.3.24) reflects the lack of a resistive force associated with energy dissipation. For example, in molecular dynamics, the vector $\boldsymbol{x}$ contains the Cartesian coordinates of a collection of molecules in an ensemble, and the vector $\boldsymbol{f}$ contains the corresponding components of the force due to an intermolecular potential (e.g., Allen and Tildesley 1987). The left-hand side of equation (9.3.24) is then the particles' acceleration.

We can certainly transform the system (9.3.24) into an equivalent larger system of first-order differential equations, but, in this case, it is better to work directly with the second-order system. Evaluating equation (9.3.24) at the nth time level, and approximating the second derivative with the second-order central-difference formula, we obtain the Störmer–Verlet algorithm that is best implemented as follows (Dahlquist and Björck 1974, pp. 352–363).

- To make the initial step, we compute

$$\boldsymbol{a}^{(0)} = \boldsymbol{f}(\boldsymbol{x}^{(0)}, t = 0)$$
$$\boldsymbol{d}^{(0)} = \Delta t\,\left(\boldsymbol{v}^{(0)} + \tfrac{1}{2}\Delta t\,\boldsymbol{a}^{(0)}\right) \tag{9.3.26}$$
$$\boldsymbol{x}^{(1)} = \boldsymbol{x}^{(0)} + \boldsymbol{d}^{(0)}$$

where $\boldsymbol{a}$ and $\boldsymbol{d}$ stand, respectively, for *acceleration* and *displacement*, and where $\boldsymbol{x}^{(1)} \equiv \boldsymbol{x}(\Delta t)$.

- To make the kth step, where $k \geq 1$, we compute

$$\begin{aligned} \boldsymbol{a}^{(k)} &= \boldsymbol{f}\big(\boldsymbol{x}(k\Delta t), k\Delta t\big) \\ \boldsymbol{d}^{(k)} &= \boldsymbol{d}^{(k-1)} + \Delta t^{2}\boldsymbol{a}^{(k)} \\ \boldsymbol{x}^{(k+1)} &= \boldsymbol{x}^{(k)} + \boldsymbol{d}^{(k)} \end{aligned} \tag{9.3.27}$$

where $\boldsymbol{x}^{(k)} = \boldsymbol{x}(k\Delta t)$.

If the velocity is also desired for the purpose of monitoring, for example, the kinetic energy, it can be obtained from the formula

$$\boldsymbol{v}^{(k)} = \frac{1}{\Delta t}\boldsymbol{d}^{(k-1)} + \tfrac{1}{2}\Delta t \boldsymbol{f}\left(\boldsymbol{x}^{(k)}, t^{k}\right) \tag{9.3.28}$$

It can be shown that the error associated with the Störmer–Verlet algorithm can be expressed as a series involving even powers of Δt, beginning with Δt^{4}. Thus applying the Richardson extrapolation raises the order of the method by two units. Refinements and extensions of the method are discussed by Allen and Tildesley (1986, pp. 78–82).

PROBLEM

9.3.4 *Motion of charged point particles.*

Consider N similarly charged particles, initially sitting at the vertices of a regular N-sided polygon. The motion of the jth particle is described by equation (9.3.24), where the force exerted on it is given by

$$\boldsymbol{f}_{(j)} = \sum_{\substack{i=1 \\ i \neq j}}^{N} \frac{\boldsymbol{x}_{(j)} - \boldsymbol{x}_{(i)}}{\left|\boldsymbol{x}_{(j)} - \boldsymbol{x}_{(i)}\right|^{3}} \tag{9.3.29}$$

Compute the trajectories of the particles from the quiescent initial state using the Störmer–Verlet method for $N = 2, 4, 8, 16$, and discuss the physics of the motion.

Integration of ODEs and Nonlinear Algebraic Mappings

The reader must have already noticed that there is an intimate connection between stepping in time and computing a sequence of vectors by a generally nonlinear algebraic mapping of the general form

$$\boldsymbol{x}^{n+1} = \boldsymbol{q}(\boldsymbol{x}^{n+1}, \boldsymbol{x}^{n}, \boldsymbol{x}^{n-1}, \ldots, \boldsymbol{x}^{n-p}) \tag{9.3.30}$$

where $p \geq 0$. The particular form of the iteration function $\boldsymbol{q}$ depends on the size of the time step and the selected numerical method. For explicit methods, the point $\boldsymbol{x}^{n+1}$ does not appear in the argument of $\boldsymbol{q}$. For Euler's method corresponding to $p = 0$, $\boldsymbol{q}$ is a function of $\boldsymbol{x}^{n}$ alone, given by $\boldsymbol{q}(\boldsymbol{x}^{n}) = -\boldsymbol{x}^{n} + \Delta t \boldsymbol{f}(\boldsymbol{x}^{n})$.

A similar type of iteration was used in Chapter 4 for solving the nonlinear system of algebraic equations. If $\boldsymbol{x}^{n}$ settles to a steady value, then that value is a solution of the equation

$$\boldsymbol{h}(\boldsymbol{x}) = \boldsymbol{x} - \boldsymbol{q}(\boldsymbol{x}, \boldsymbol{x}, \ldots, \boldsymbol{x}) = \boldsymbol{0} \tag{9.3.31}$$

In Chapter 4, the objective was to compute the fixed point of the nonlinear iteration function $\boldsymbol{g}(\boldsymbol{x}) \equiv \boldsymbol{q}(\boldsymbol{x}, \boldsymbol{x}, \ldots, \boldsymbol{x})$, which coincides with a root of the function $\boldsymbol{h}(\boldsymbol{x})$.

Euler's method with a constant time step, in particular, is identical to the method of one-point iterations. This observation suggests a way of deriving differential equations from iteration functions and *vice versa*, and thereby benefitting from the complementary perspectives of this dual approach. For example, the differential equation corresponding to the logistic mapping shown in equation (4.3.33) is

$$\frac{dx}{dt} = \alpha x(1 - x) - \frac{1}{\Delta t} x \tag{9.3.32}$$

where $\alpha = \lambda / \Delta t$. Applying Euler's method to equation (9.3.32) recovers equation (4.3.33). When α is held fixed, increasing Δt, which amounts to increasing λ, leads to the behavior shown in Figure 4.3.2. Conversely, applying the Euler method to the system of equations (9.3.21) and (9.3.22) yields the iteration functions

$$g_1(x_1, x_2) = x_1^n(1 + a - \chi x_2^n), \qquad g_2(x_1, x_2) = x_2^n(1 - b + \chi \varepsilon x_1^n) \tag{9.3.33}$$

Such iteration functions are sometimes called *discrete dynamical systems* or *cellular automata.*

PROBLEM

9.3.5 *Lorenz system.*

Derive a discrete dynamical system by applying Euler's method to the Lorenz system (9.1.24). Study and discuss its behavior for $k = 10$, $b = \frac{8}{3}$, and $r = 0.5, 5.0, 28.0$.

9.4 *Predictor–corrector Methods*

Implicit methods require solving a system of nonlinear algebraic equations at each step, which is undesirable. How can we get around this complication without compromising accuracy and jeopardizing stability? Competitive alternatives are provided by the twofold class of predictor–corrector methods.

The idea is to advance the solution over a certain time interval by executing a *sequence* of steps. Certain of these steps are *provisional* or *exploratory*, whereas others are *educated*, in the sense that they sensibly correct rough estimates. The correction is based on knowledge that has been accumulated during the provisional or exploratory steps.

Multipoint and Runge–Kutta Methods

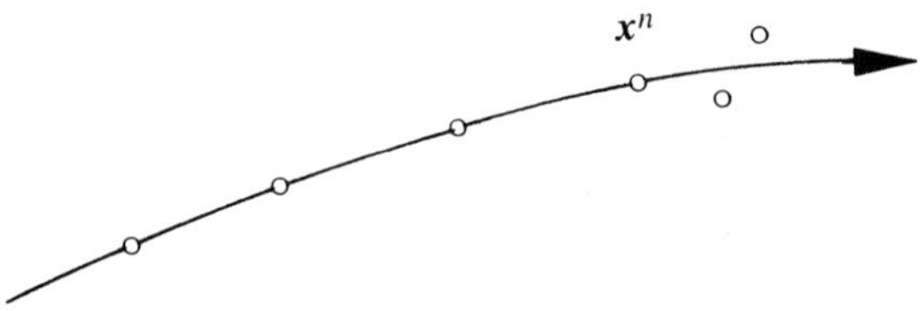

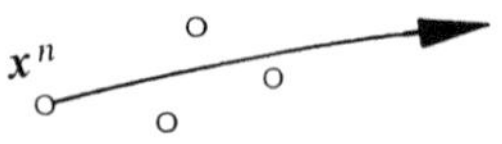

The provisional steps may utilize already available information on the recent history of the motion in the phase space; this results in *multipoint methods.* Or we may choose to abandon all previous knowledge and collect information on the imminent motion by performing exploratory steps; this results in the *Runge–Kutta* methods. The distinction between, and relative merits of, these two approaches will become apparent as we describe the algorithms.

Numerical stability is an important factor in making a selection. Multipoint methods are generally stable but may foster the propagation or even growth of spurious nonphysical oscillations, and are tolerated only when the numerical oscillations grow at a rate that is significantly less than that of the exact solution. Sometimes, these methods produce an accurate solution for a certain initial period of time, but then generate nonphysical behaviors (e.g., Ferziger and Perić 1996). In some instances, these pathologies may be cured with a careful choice of the starting method. A common remedy is to restart the computation when the solution has become unsatisfactory.

Runge–Kutta methods, on the other hand, do not cause spurious oscillations and can be stabilized by using a sufficiently small time step (e.g., Lapidus and Seinfeld 1971, Lambert 1991).

What then are the advantages of the multipoint methods? For the same level of accuracy, they require a lesser number of phase-space velocity evaluations than the Runge–Kutta methods.

Adams–Bashforth–Moulton Multipoint Methods

This family of methods emerges by making a predictor step using an *Adams–Bashforth* method, and then a corrector step using a modified version of the *Adams–Moulton* method.

Fourth-order method

A popular explicit method employs the fourth-order Adams–Bashforth method and a modified fourth-order Adams–Moulton method, according to the algorithm:

$$\begin{array}{ll}
\text{1. Compute} & \mathbf{f}^n = \mathbf{f}(\mathbf{x}^n, t^n) \\
\text{2. Set} & \mathbf{x}^{Temp} = \mathbf{x}^n + \frac{1}{24}\Delta t\,(-9\mathbf{f}^{n-3} + 37\mathbf{f}^{n-2} - 59\mathbf{f}^{n-1} + 55\mathbf{f}^n) \\
\text{3. Compute} & \mathbf{f}^{Temp} = \mathbf{f}(\mathbf{x}^{Temp}, t^n + \Delta t) \\
\text{4. Set} & \mathbf{x}^{n+1} = \mathbf{x}^n + \frac{1}{24}\Delta t\,(\mathbf{f}^{n-2} - 5\mathbf{f}^{n-1} + 19\mathbf{f}^n + 9\mathbf{f}^{Temp})
\end{array} \tag{9.4.1}$$

A complete step requires *two velocity* evaluations and introduces a numerical error on the order of Δt^5.

The computation starts at $n = 3$, at which point we require the values of $\mathbf{x}$ and $\mathbf{f}$ at $n = 0, 1, 2, 3$. The values at the levels $n = 1, 2, 3$ can be obtained by applying a single-step method with a time step that is an integral fraction of Δt that is small enough so that the generated numerical error is tolerable.

One disadvantage of the method is that Δt must be kept constant throughout the course or during parts of the computation. This constraint imposes logistics difficulties in solving problems where a small time step is required during a certain stage of the computation.

Error estimate, step-size control, and accuracy improvement

Our knowledge of the leading error term of the predictor and corrector steps can be used to obtain an estimate of its magnitude, to improve the accuracy of the computation, and to adjust, if necessary, the magnitude of the time step.

Using expressions (9.3.16) and (9.3.20), we write

$$\mathbf{x}^{Temp} - \hat{\mathbf{x}}^{n+1} = -\tfrac{251}{720}\Delta t^{5}\left(\frac{d^5\mathbf{x}}{dt^5}\right)_{t=t^n}$$
$$\mathbf{x}^{n+1} - \hat{\mathbf{x}}^{n+1} = \tfrac{19}{720}\Delta t^{5}\left(\frac{d^5\mathbf{x}}{dt^5}\right)_{t=t^n} \tag{9.4.2}$$

Combining these equations, we derive a computable estimate for the error introduced in the predictor step:

$$\mathbf{x}^{Temp} - \hat{\mathbf{x}}^{n+1} = -\tfrac{251}{270}(\mathbf{x}^{n+1} - \mathbf{x}^{Temp}) \tag{9.4.3}$$

If this is larger that can be tolerated, we decrease the size of the time step and proceed with the computation with a new set of past values obtained by interpolation.

To improve the accuracy, we take into account the error estimates in both the predictor and corrector steps, according to the preceding equations. The resulting algorithm, incorporating mop-up, that is, error extrapolation from the previous time step, is

1. Compute $\mathbf{f}^n = \mathbf{f}(\mathbf{x}^n, t^n)$
2. Set $\mathbf{x}^{P,n+1} = \mathbf{x}^n + \frac{1}{24}\Delta t\,(-9\mathbf{f}^{n-3} + 37\mathbf{f}^{n-2} - 59\mathbf{f}^{n-1} + 55\mathbf{f}^n)$
3. Correct $\mathbf{x}^{Temp} = \mathbf{x}^{P,n+1} + \frac{251}{270}(\mathbf{x}^{C,n} - \mathbf{x}^{P,n})$
4. Compute $\mathbf{f}^{Temp} = \mathbf{f}(\mathbf{x}^{Temp}, t^n + \Delta t)$
5. Set $\mathbf{x}^{C,n+1} = \mathbf{x}^n + \frac{1}{24}\Delta t\,(\mathbf{f}^{n-2} - 5\mathbf{f}^{n-1} + 19\mathbf{f}^n + 9\mathbf{f}^{Temp})$
6. Correct $\mathbf{x}^{n+1} = \mathbf{x}^{C,n+1} - \frac{19}{270}(\mathbf{x}^{C,n+1} - \mathbf{x}^{P,n+1})$

(9.4.4)

When algorithm (9.4.4) is executed for the first time, corresponding to $n = 0$, the vectors $\mathbf{x}^{C,0}$ and $\mathbf{x}^{P,0}$ required in step 3 are not available. To bypass this difficulty, we execute the loops twice, the first time replacing the equation in step 3 with $\mathbf{x}^{Temp} = \mathbf{x}^{P,1}$, and the second time skipping the predictor step and using the values obtained in the first trial to make the correction in step 6.

Runge–Kutta Methods

One undesirable feature of the multistep methods is that Δt must be kept constant. To become more flexible, we abandon information on the history of the motion in the phase space and choose, instead, to collect information on the imminent motion. This choice leads us to the family of Runge–Kutta methods.

A comprehensive discussion of various aspects of the Runge–Kutta methods can be found in the monograph of Lapidus and Seinfeld (1971). In the remainder of this section, we discuss several popular choices.

Second-order Runge–Kutta method, RK2

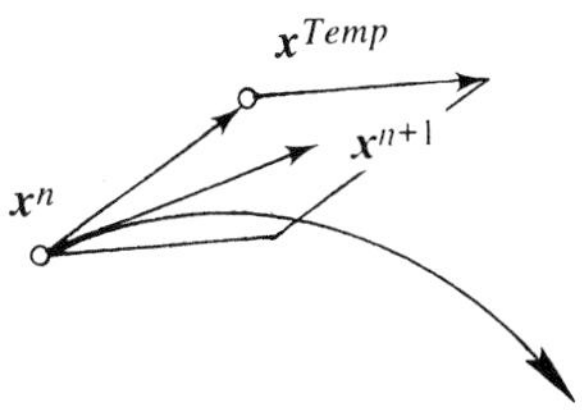

This is a genuine predictor–corrector method involving one provisional step, and requiring two velocity evaluations for each complete step. The main idea is to move in the phase space with a velocity that is equal to a weighted average of the local and current velocity $\boldsymbol{f}^n$, and the velocity $\boldsymbol{f}^{Temp}$ evaluated at a point near the current position, at some later time instant. This leads us to the formula

$$\boldsymbol{x}^{n+1} = \boldsymbol{x}^n + \Delta t\,[(1-\alpha)\boldsymbol{f}^n + \alpha \boldsymbol{f}^{Temp}] \tag{9.4.5}$$

where α is a weighting factor. For lack of a better alternative, we set

$$\boldsymbol{x}^{Temp} = \boldsymbol{x}^n + \kappa \Delta t \boldsymbol{f}(\boldsymbol{x}^n, t^n) \tag{9.4.6}$$

and

$$\boldsymbol{f}^{Temp} = \boldsymbol{f}(\boldsymbol{x}^{Temp}, t^n + \kappa \Delta t) \tag{9.4.7}$$

where κ is a constant. We shall see that these stipulations provide us with sufficient degrees of freedom to achieve our goals.

We want to evaluate α and κ so as to attain the highest possible level of accuracy. For this purpose, we expand $\boldsymbol{x}^{n+1}$ in a Taylor series about the point $\boldsymbol{x}^n$ and obtain

$$\boldsymbol{x}^{n+1} = \boldsymbol{x}^n + \boldsymbol{f}^n \Delta t + \tfrac{1}{2}\left(\frac{d\boldsymbol{f}}{dt}\right)_{t^n} \Delta t^2 + \cdots \tag{9.4.8}$$

Using the chain rule of differentiation, we write

$$\frac{d\boldsymbol{f}}{dt} = \frac{\partial \boldsymbol{f}}{\partial t} + \frac{\partial \boldsymbol{f}}{\partial x_i}\frac{dx_i}{dt} = \frac{\partial \boldsymbol{f}}{\partial t} + \frac{\partial \boldsymbol{f}}{\partial x_i} f_i \tag{9.4.9}$$

where summation is implied over i. Combining this expression with equation (9.4.8), we find

$$\boldsymbol{x}^{n+1} = \boldsymbol{x}^n + \Delta t \boldsymbol{f}^n + \tfrac{1}{2}\Delta t^2 \left(\frac{\partial \boldsymbol{f}}{\partial t} + \frac{\partial \boldsymbol{f}}{\partial x_i} f_i\right)^n + \cdots \tag{9.4.10}$$

Furthermore, we use the chain rule of differentiation once more to write

$$\boldsymbol{f}^{Temp} = \boldsymbol{f}^n + \kappa \Delta t \left(\frac{\partial \boldsymbol{f}}{\partial t} + \frac{\partial \boldsymbol{f}}{\partial x_i} f_i\right)^n + \cdots \tag{9.4.11}$$

Substituting equation (9.4.11) into equation (9.4.5), and requiring that, in the limit as Δt tends to zero, the right-hand side of the resulting equation reproduces the first three terms on the right-hand side of equation (9.4.10), we arrive at the condition

$$2\alpha\kappa = 1 \tag{9.4.12}$$

We have a multitude of choices, and some possibilities are:

- Set $\alpha = \frac{1}{2}$ and $\kappa = 1$ to obtain the standard version of RK2, also called the *modified Euler* method.
- Set $\alpha = 1$ and $\kappa = \frac{1}{2}$ to obtain the *midpoint* RK2.
- Set $\alpha = \frac{3}{4}$ and $\kappa = \frac{2}{3}$ to obtain Heun's method.

The algorithm involves selecting a value for α, computing κ from equation (9.4.12), and then executing the following steps:

$$\begin{array}{ll} 1.\ \text{Compute} & \boldsymbol{f}(\boldsymbol{x}^n, t^n) \\ 2.\ \text{Set} & \boldsymbol{x}^{Temp} = \boldsymbol{x}^n + \kappa \Delta t \boldsymbol{f}(\boldsymbol{x}^n, t^n) \\ 3.\ \text{Compute} & \boldsymbol{f}^{Temp} = \boldsymbol{f}(\boldsymbol{x}^{Temp}, t^n + \kappa \Delta t) \\ 4.\ \text{Compute} & \boldsymbol{f}^{Final} = (1-\alpha)\boldsymbol{f}(\boldsymbol{x}^n, t^n) + \alpha \boldsymbol{f}^{Temp} \\ 5.\ \text{Set} & \boldsymbol{x}^{n+1} = \boldsymbol{x}^n + \Delta t \boldsymbol{f}^{Final} \end{array} \tag{9.4.13}$$

Each complete step introduces a numerical error on the order of Δt^3.

PROBLEMS

9.4.1 ***Exercises with nonautonomous and autonomous systems.***

Use the modified Euler method to integrate the following systems from $t = 0$ to $t = 10.0$ using $\Delta t = 0.10$, with initial conditions $x(0) = 1.0$, $y(0) = 1.0$.

$$\begin{array}{ll} (a)\ \dfrac{dx}{dt} = e^{-x} + y\,(1.0 + t), & \dfrac{dy}{dt} = x\,(y+1) + t \\ (b)\ \dfrac{dx}{dt} = y\,(x+1), & \dfrac{dy}{dt} = x\,(y+1) \end{array} \tag{9.4.14}$$

Plot the trajectory of the solution in the phase space, and verify the overall second-order accuracy of the method.

9.4.2 ***A stiff problem.***

Solve equation (9.6.8) with initial condition $x(0) = 0$ up to $t = 4.0$ using the modified Euler method, with a Δt that is sufficiently small so that the numerical results agree with the exact solution up to the third significant figure. Discuss the behavior of your computations for different values of Δt.

Third-order Runge–Kutta method, RK3

Working in a similar manner, we derive higher-order Runge–Kutta methods. To obtain the third-order method, we set

$$\boldsymbol{x}^{n+1} = \boldsymbol{x}^n + \Delta t\,[(1-\alpha-\beta)\boldsymbol{f}^n + \alpha \boldsymbol{f}^{Temp(1)} + \beta \boldsymbol{f}^{Temp(2)}] \tag{9.4.15}$$

where α and β are two weighting factors. We have introduced the auxiliary velocities

$$\boldsymbol{f}^{Temp(1)} = \boldsymbol{f}(\boldsymbol{x}^n + \kappa\,\Delta t\, \boldsymbol{f}(\boldsymbol{x}^n, t^n), t^n + \kappa\,\Delta t\,) \tag{9.4.16}$$

and

$$\boldsymbol{f}^{Temp(2)} = \boldsymbol{f}(\boldsymbol{x}^n + (\lambda - \mu)\Delta t\, \boldsymbol{f}(\boldsymbol{x}^n, t^n) + \mu\,\Delta t\, \boldsymbol{f}^{Temp(1)}, t^n + \lambda\Delta t\,) \tag{9.4.17}$$

involving the three constants κ, λ, and μ.

To determine the five constants α, β, κ, λ, and μ, we work as outlined in the preceding subsection for the second-order method. Requiring, in particular, that the right-hand side of equation (9.4.15) approximates the exact solution with an error on the order of $\Delta t^{\,4}$, we obtain the algebraic equations

$$\alpha\kappa + \beta\lambda = \tfrac{1}{2}, \qquad \alpha\kappa^2 + \beta\lambda^2 = \tfrac{1}{3}, \qquad \beta\kappa\mu = \tfrac{1}{6} \tag{9.4.18}$$

Two of the five constants may be given arbitrary values, and the other three must be computed so as to satisfy the system (9.4.18).

In the standard version of the RK3 method, we choose $\alpha = \frac{2}{3}$ and $\beta = \frac{1}{6}$, in which case $\kappa = \frac{1}{2}$, $\lambda = 1$, and $\mu = 2$. The algorithm involves the following steps:

$$\begin{array}{lll} 1.\ \text{Compute} & \boldsymbol{f}(\boldsymbol{x}^n, t^n) \\ 2.\ \text{Set} & \boldsymbol{x}^{Temp(1)} = \boldsymbol{x}^n + \tfrac{1}{2}\Delta t\, \boldsymbol{f}(\boldsymbol{x}^n, t^n) \\ 3.\ \text{Compute} & \boldsymbol{f}^{Temp(1)} = \boldsymbol{f}\left(\boldsymbol{x}^{Temp(1)}, t^n + \tfrac{1}{2}\Delta t\right) \\ 4.\ \text{Set} & \boldsymbol{x}^{Temp(2)} = \boldsymbol{x}^n + \Delta t\,(2\boldsymbol{f}^{Temp(1)} - \boldsymbol{f}(\boldsymbol{x}^n, t^n)) \\ 5.\ \text{Compute} & \boldsymbol{f}^{Temp(2)} = \boldsymbol{f}\left(\boldsymbol{x}^{Temp(2)}, t^n + \Delta t\right) \\ 6.\ \text{Compute} & \boldsymbol{f}^{Final} = \tfrac{1}{6}[\boldsymbol{f}(\boldsymbol{x}^n, t^n) + 4\boldsymbol{f}^{Temp(1)} + \boldsymbol{f}^{Temp(2)}] \\ 7.\ \text{Set} & \boldsymbol{x}^{n+1} = \boldsymbol{x}^n + \Delta t\, \boldsymbol{f}^{Final} \end{array} \tag{9.4.19}$$

Each complete step introduces a numerical error on the order of $\Delta t^{\,4}$.

Fourth-order Runge–Kutta method, RK4

When the cost of function evaluations is reasonable, extremely high accuracy is not required, and economizing the CPU time is not a first priority, the RK4 method is considered the best compromise. Better accuracy and higher efficiency may be achieved using the embedding Runge–Kutta methods to be discussed in Section 9.5, but their implementation requires more work.

The fourth-order Runge–Kutta method is developed working as discussed in the preceding subsections for the lower-order methods. A particular selection of the numerical parameters produces the standard version of the RK4 algorithm:

$$\begin{array}{ll}
\text{1. Compute} & \boldsymbol{f}(\boldsymbol{x}^n, t^n) \\
\text{2. Set} & \boldsymbol{x}^{Temp(1)} = \boldsymbol{x}^n + \tfrac{1}{2}\Delta t \boldsymbol{f}(\boldsymbol{x}^n, t^n) \\
\text{3. Compute} & \boldsymbol{f}^{Temp(1)} = \boldsymbol{f}\left(\boldsymbol{x}^{Temp(1)}, t^n + \tfrac{1}{2}\Delta t\right) \\
\text{4. Set} & \boldsymbol{x}^{Temp(2)} = \boldsymbol{x}^n + \tfrac{1}{2}\Delta t \boldsymbol{f}^{Temp(1)} \\
\text{5. Compute} & \boldsymbol{f}^{Temp(2)} = \boldsymbol{f}\left(\boldsymbol{x}^{Temp(2)}, t^n + \tfrac{1}{2}\Delta t\right) \\
\text{6. Set} & \boldsymbol{x}^{Temp(3)} = \boldsymbol{x}^n + \Delta t \boldsymbol{f}^{Temp(2)} \\
\text{7. Compute} & \boldsymbol{f}^{Temp(3)} = \boldsymbol{f}(\boldsymbol{x}^{Temp(3)}, t^n + \Delta t) \\
\text{8. Compute} & \boldsymbol{f}^{Final} = \tfrac{1}{6}[\boldsymbol{f}(\boldsymbol{x}^n, t^n) + 2\boldsymbol{f}^{Temp(1)} + 2\boldsymbol{f}^{Temp(2)} + \boldsymbol{f}^{Temp(3)}] \\
\text{9. Set} & \boldsymbol{x}^{n+1} = \boldsymbol{x}^n + \Delta t \boldsymbol{f}^{Final}
\end{array} \tag{9.4.20}$$

Each complete step introduces a numerical error on the order of Δt^5.

A different selection of the numerical parameters results in the somewhat more accurate algorithm:

$$\begin{array}{ll}
\text{1. Compute} & \boldsymbol{f}(\boldsymbol{x}^n, t^n) \\
\text{2. Set} & \boldsymbol{x}^{Temp(1)} = \boldsymbol{x}^n + \tfrac{1}{3}\Delta t \boldsymbol{f}(\boldsymbol{x}^n, t^n) \\
\text{3. Compute} & \boldsymbol{f}^{Temp(1)} = \boldsymbol{f}\left(\boldsymbol{x}^{Temp(1)}, t^n + \tfrac{1}{3}\Delta t\right) \\
\text{4. Set} & \boldsymbol{x}^{Temp(2)} = \boldsymbol{x}^n - \tfrac{1}{3}\Delta t \boldsymbol{f}(\boldsymbol{x}^n, t^n) + \Delta t \boldsymbol{f}^{Temp(1)} \\
\text{5. Compute} & \boldsymbol{f}^{Temp(2)} = \boldsymbol{f}\left(\boldsymbol{x}^{Temp(2)}, t^n + \tfrac{2}{3}\Delta t\right) \\
\text{6. Set} & \boldsymbol{x}^{Temp(3)} = \boldsymbol{x}^n + \Delta t \boldsymbol{f}(\boldsymbol{x}^n, t^n) - \Delta t \boldsymbol{f}^{Temp(1)} + \Delta t \boldsymbol{f}^{Temp(2)} \\
\text{7. Compute} & \boldsymbol{f}^{Temp(3)} = \boldsymbol{f}(\boldsymbol{x}^{Temp(3)}, t^n + \Delta t) \\
\text{8. Compute} & \boldsymbol{f}^{Final} = \tfrac{1}{8}[\boldsymbol{f}(\boldsymbol{x}^n, t^n) + 3\boldsymbol{f}^{Temp(1)} + 3\boldsymbol{f}^{Temp(2)} + \boldsymbol{f}^{Temp(3)}] \\
\text{9. Set} & \boldsymbol{x}^{n+1} = \boldsymbol{x}^n + \Delta t \boldsymbol{f}^{Final}
\end{array} \tag{9.4.21}$$

Butcher arrays

The numerical coefficients involved in the formulas of the various Runge–Kutta methods can be conveniently displayed in the *Butcher arrays*. The array of the kth order method arises by writing

$$\begin{aligned}
\boldsymbol{x}^{Temp(1)} &= \boldsymbol{x}^n + a_{2,1}\Delta t \boldsymbol{f}(\boldsymbol{x}^n, t^n)\\
\boldsymbol{x}^{Temp(j)} &= \boldsymbol{x}^n + \Delta t \left[a_{j+1,1}\boldsymbol{f}(\boldsymbol{x}^n, t^n) + \sum_{l=1}^{j-1} a_{j+1,l+1}\boldsymbol{f}^{Temp(l)} \right] \qquad \text{for } j = 2, \ldots, k-1\\
\boldsymbol{f}^{Temp(j)} &= \boldsymbol{f}(\boldsymbol{x}^{Temp(j)}, t^n + c_{j+1}\Delta t) \qquad \text{for } j = 1, \ldots, k-1\\
\boldsymbol{f}^{Final} &= w_1 \boldsymbol{f}(\boldsymbol{x}^n, t^n) + \sum_{j=2}^{k} w_j \boldsymbol{f}^{Temp(j-1)}
\end{aligned} \tag{9.4.22}$$

and then constructing the table

$$\begin{array}{c|ccccc}
0 & & & & & \\
c_2 & a_{2,1} & & & & \\
c_3 & a_{3,1} & a_{3,2} & & & \\
c_4 & a_{4,1} & a_{4,2} & a_{4,3} & & \\
\ldots & \ldots & \ldots & \ldots & & \\
\hline
 & w_1 & w_2 & w_3 & \ldots & w_k
\end{array} \tag{9.4.23}$$

For example, the Butcher array of the RK4 algorithm (9.4.20) is

$$\begin{array}{c|cccc}
0 & & & & \\
\frac{1}{2} & \frac{1}{2} & & & \\
\frac{1}{2} & 0 & \frac{1}{2} & & \\
1 & 0 & 0 & 1 & \\
\hline
 & \frac{1}{6} & \frac{2}{6} & \frac{2}{6} & \frac{1}{6}
\end{array} \tag{9.4.24}$$

PROBLEMS

9.4.3 *RK3 and RK4.*

(*a*) Derive the system of equations (9.4.18).
(*b*) Verify that each complete step according to the RK4 method introduces a numerical error of order Δt^5.

9.4.4 *Motion of point vortices.*

Point vortices are mathematical idealizations arising in the limit as compact areas of finite vorticity shrink down to a point, while their strength remains constant (e.g., Pozrikidis 1997, Chapter 11). Consider a collection of N point vortices in an infinite fluid, moving in the complex z plane. The position of the ith point vortex, $z_i(t)$, with strength κ_i, changes according to the ordinary differential equation

$$\frac{dz_i(t)}{dt} = \frac{1}{2\pi I} \sum_{\substack{j=1\\ i\neq j}}^{N} \frac{\kappa_i}{z_i(t) - z_j(t)} \tag{9.4.25}$$

where I is the imaginary unit.

Compute the motion of four point vortices of equal strength from the initial positions

$$z_1(0) = I, \quad z_2(0) = 2 - I, \quad z_3(0) = 1 + 2I, \quad z_4(0) = 1 - 2I \tag{9.4.26}$$

using a method of your choice. Discuss the results of your computations.

9.4.5 *Motion of a vortex patch.*

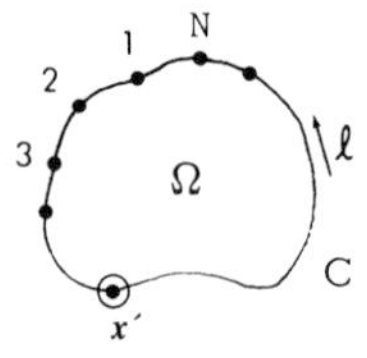

Consider the motion of a region of constant vorticity in an inviscid two-dimensional fluid of infinite expanse (e.g., Pozrikidis 1997, Chapter 11). The velocity at a point $\boldsymbol{x} = (x, y)$ may be expressed as a line integral, around the boundary of the vortex C, in the form

$$\boldsymbol{u}(\boldsymbol{x}) = -\frac{\Omega}{2\pi}\int_C \ln|\boldsymbol{x}-\boldsymbol{x}'|\, dl(\boldsymbol{x}') \tag{9.4.27}$$

where Ω is the constant vorticity of the vortex patch, $l = |d\boldsymbol{x}'|$ is the arc length around C, and the point $\boldsymbol{x}'$ is located on the vortex boundary.

To describe the motion of the vortex boundary, we trace it with N marker points $\boldsymbol{x}_i$ that move with the velocity of the fluid, and approximate the line integral in equation (9.4.27) with the sum of individual integrals over straight segments S_j connecting adjacent marker points, obtaining

$$\frac{d\boldsymbol{x}_i}{dt} = -\frac{\Omega}{2\pi}\sum_{j=1}^{N}\int_{S_j} \ln|\boldsymbol{x}_i-\boldsymbol{x}'|\, dl(\boldsymbol{x}') \tag{9.4.28}$$

for $i = 1, \ldots, N$. Denoting

$$-\frac{\Omega}{2\pi}\int_{S_j} \ln|\boldsymbol{x}_i-\boldsymbol{x}'|\, dl(\boldsymbol{x}') \equiv \boldsymbol{v}(\boldsymbol{x}_i, \boldsymbol{x}_j, \boldsymbol{x}_{j+1}) \tag{9.4.29}$$

we express equation (9.4.28) in the symbolic form

$$\frac{d\boldsymbol{x}_i}{dt} = \sum_{j=1}^{N}\boldsymbol{v}(\boldsymbol{x}_i, \boldsymbol{x}_j, \boldsymbol{x}_{j+1}) \tag{9.4.30}$$

for $i = 1, \ldots, N$, which provides us with a system of ordinary differential equations governing the motion of the marker points.

(*a*) Write a routine that returns $\boldsymbol{v}$ for given $\boldsymbol{x}_i, \boldsymbol{x}_j, \boldsymbol{x}_{j+1}$. The integral in equation (9.4.29) should be computed using an analytical or numerical method, whichever you think is more expedient. Special care will be required when $i = j$ or $i = j + 1$, as the integrand becomes singular.

(*b*) Compute the evolution of an initially elliptical vortex patch with axes ratio equal to 2, 5, or 10, and discuss its behavior. The system (9.4.30) should be integrated using the fourth-order Runge–Kutta method.

(*c*) Repeat part (*b*) for a vortex patch with a nonelliptical initial shape of your choice. A variety of shapes are tabulated by Lawrence (1972).

9.4.6 *Burgers's equation.*

Consider a complex function u of a complex variable z and real time t, evolving according to the generalized Burgers equation

$$\frac{\partial u}{\partial t} + u\frac{\partial u}{\partial z} = \kappa\frac{\partial^2 u}{\partial z^2} \tag{9.4.31}$$

where κ is a positive constant playing the role of diffusivity.

(*a*) Verify that an exact particle-like solution of equation (9.4.31) is given by

$$u(z, t) = -2\kappa\sum_{i=1}^{N}\frac{1}{z - z_i(t)} \tag{9.4.32}$$

where N is an arbitrary number of poles whose positions z_i evolve according to the equation

$$\frac{dz_i(t)}{dt} = -2\kappa \sum_{\substack{j=1 \\ i \neq j}}^{N} \frac{1}{z_i(t) - z_j(t)} \tag{9.4.33}$$

(Choodnovsky and Choodnovsky, 1977). Note that if the poles are initially chosen to form complex-conjugate pairs, then u will remain real at all times for real values of x.
(*b*) Integrate equation (9.4.33) for $N = 4$ using a method of your choice, from the initial positions given in equations (9.4.26), and discuss the results of your computations.

9.5 Error Estimate and Adaptive Step-size Control

Each complete time step according to the mth order Runge–Kutta method introduces a numerical error on the order of $\Delta t^{\,m+1}$. Knowledge of the order of the error allows us to use the method of Richardson extrapolation discussed in Section 1.6, with two objectives:

- Raise the accuracy to the order $\Delta t^{\,m+2}$.
- Decrease the time step to a small enough value so as to achieve a desired level of accuracy; and increase it again when the difficult times have passed.

For example, Richardson extrapolation with the RK2 method can be implemented by carrying one crude step with step size $\Delta t = h$, which requires two velocity evaluations, followed by two refined steps with step size $\Delta t = \frac{1}{2}h$, which requires three additional velocity evaluations; a total of five velocity evaluations. The numerical implementation of this dual procedure is discussed by Gear (1971). For the RK4 method, we have a total of eleven velocity evaluations. This may be a high price to pay.

Embedding Methods

An estimate of the magnitude of the numerical error and an effective control of the time step can be achieved at a much lesser cost with the use of the *embedding methods*. Stoer and Bulirsch (1980) discuss the principles of these methods and develop a composite scheme whose distinguishing features are: Use of the midpoint method to obtain an error that depends on even powers of Δt, and use of Richardson extrapolation with a rational function approximation. The implementation of this method is discussed in detail by Press et al. (1992, p. 718).

In the remainder of this section, we present a popular class of methods that are based on an ingenious modification of the classical Runge–Kutta algorithms invented by Fehlberg. A detailed discussion can be found in the texts of Stoer and Bulirsch (1980) and Lambert (1991).

Runge–Kutta–Fehlberg method of orders 2 and 3, RKF23

Carrying the ideas underlying the development of the classical Runge–Kutta method one step further, and following the exposition of Stoer and Bulirsch (1980, p. 426), we advance the solution in two different ways. First, we use the formula

$$\mathbf{x}^{n+1,1} = \mathbf{x}^n + \Delta t\,[(1-\alpha-\beta)\mathbf{f}^n + \alpha \mathbf{f}^{Temp(1)} + \beta \mathbf{f}^{Temp(2)}] \tag{9.5.1}$$

and then the formula

$$\boldsymbol{x}^{n+1,2} = \boldsymbol{x}^n + \Delta t\,[(1-\gamma-\delta-\varepsilon)\boldsymbol{f}^n + \gamma \boldsymbol{f}^{Temp(1)} + \delta \boldsymbol{f}^{Temp(2)} + \varepsilon \boldsymbol{f}^{Temp(3)}] \tag{9.5.2}$$

The Greek coefficients represent a collection of five yet unspecified numerical parameters. Moreover, we set

$$\begin{aligned}
\boldsymbol{x}^{Temp(1)} &= \boldsymbol{x}^n + \Delta t\, b_{1,0} \boldsymbol{f}^n \\
\boldsymbol{f}^{Temp(1)} &= \boldsymbol{f}(\boldsymbol{x}^{Temp(1)}, t^n + a_1 \Delta t\,) \\
\boldsymbol{x}^{Temp(2)} &= \boldsymbol{x}^n + \Delta t\,(b_{2,0} \boldsymbol{f}^n + b_{2,1} \boldsymbol{f}^{Temp(1)}) \\
\boldsymbol{f}^{Temp(2)} &= \boldsymbol{f}(\boldsymbol{x}^{Temp(2)}, t^n + a_2 \Delta t\,) \\
\boldsymbol{x}^{Temp(3)} &= \boldsymbol{x}^n + \Delta t\,(b_{3,0} \boldsymbol{f}^n + b_{3,1} \boldsymbol{f}^{Temp(1)} + b_{3,2} \boldsymbol{f}^{Temp(2)}) \\
\boldsymbol{f}^{Temp(3)} &= \boldsymbol{f}(\boldsymbol{x}^{Temp(3)}, t^n + a_3 \Delta t\,)
\end{aligned} \tag{9.5.3}$$

where a_i and $b_{i,j}$ is a collection of nine numerical parameters.

Requiring that equations (9.5.1) and (9.5.2) produce solutions with respective one-step errors on the order of $\Delta t^{\,3}$ and $\Delta t^{\,4}$ provides us with six equations that relate the fourteen parameters of the numerical method. Eight additional constraints may be imposed with the objective of optimizing the efficiency of this seemingly cumbersome method.

To this end, one might think that something has gone terribly wrong. Equation (9.5.1) is third-order accurate, while requiring three velocity evaluations, whereas the RK2 method is also third-order accurate, while requiring only two velocity evaluations. An analogous remark can be made about equation (9.5.2). The crux of the RKF method is that the numerical coefficients may be adjusted so that the number of function evaluations on the right-hand sides is reduced from the superficial value of 4 to the actual value of 3. More importantly, the availability of two fromulas of different orders allows us to estimate the magnitude of the error and then use it to adjust the size of the time step in an adaptive fashion.

In the standard RKF23 method, we set $\boldsymbol{x}^{n+1} = \boldsymbol{x}^{n+1,1}$ which is computed from the right-hand side of equation (9.5.1), and require $\boldsymbol{f}(\boldsymbol{x}^{n+1}, t^n + \Delta t\,) = \boldsymbol{f}^{Temp(3)}$. This necessitates that $a_3 = 1$ and $\boldsymbol{x}^{n+1,1} = \boldsymbol{x}^{Temp(3)}$, or

$$(1-\alpha-\beta)\boldsymbol{f}^n + \alpha \boldsymbol{f}^{Temp(1)} + \beta \boldsymbol{f}^{Temp(2)} = b_{3,0} \boldsymbol{f}^n + b_{3,1} \boldsymbol{f}^{Temp(1)} + b_{3,2} \boldsymbol{f}^{Temp(2)} \tag{9.5.4}$$

and, therefore

$$b_{3,0} = 1 - \alpha - \beta, \qquad b_{3,1} = \alpha, \qquad b_{3,2} = \beta \tag{9.5.5}$$

a total of four constraints. Imposing four additional constraints with the objective of minimizing the magnitude of the error leads us to the RKF23 method described in Algorithm 9.5.1. The method includes adaptive step-size control according to a certain error estimate. The quantity ε in step 10 is a preset tolerance. Carrying out a complete step requires only three velocity evaluations, except at the very first step. To reduce the computer storage, the temporary variables designated by *Temp* may be overwritten.

Runge–Kutta–Fehlberg Method of Orders 4 and 5, RKF45

The RKF45 method is developed in an analogous manner. The algorithm involves the steps shown in Algorithm 9.5.2; ε in step 14 is a preset tolerance. Completing a step requires six velocity evaluations.

A more accurate version of the method developed by Cash and Karp (1990) proceeds according to the Algorithm 9.5.3.

ALGORITHM 9.5.1 The Runge–Kutta–Fehlberg RKF23 method.

1. Compute $\boldsymbol{f}(\boldsymbol{x}^n, t^n)$

2. Set $\boldsymbol{x}^{Temp(1)} = \boldsymbol{x}^n + \frac{1}{4}\Delta t \boldsymbol{f}^n$

3. Compute $\boldsymbol{f}^{Temp(1)} = \boldsymbol{f}\left(\boldsymbol{x}^{Temp(1)}, t^n + \frac{1}{4}\Delta t\right)$

4. Set $\boldsymbol{x}^{Temp(2)} = \boldsymbol{x}^n + \frac{1}{800}\Delta t\,(-189\boldsymbol{f}^n + 729\boldsymbol{f}^{Temp(1)})$

5. Compute $\boldsymbol{f}^{Temp(2)} = \boldsymbol{f}\left(\boldsymbol{x}^{Temp(2)}, t^n + \frac{27}{40}\Delta t\right)$

6. Set $\boldsymbol{x}^{Temp(3)} = \boldsymbol{x}^n + \frac{1}{891}\Delta t\,(214\boldsymbol{f}^n + 27\boldsymbol{f}^{Temp(1)} + 650\boldsymbol{f}^{Temp(2)})$

7. Set $\boldsymbol{x}^{n+1} = \boldsymbol{x}^{Temp(3)}$,

Time step adjustment

8. Compute $\boldsymbol{f}^{Temp(3)} = \boldsymbol{f}(\boldsymbol{x}^{Temp(3)}, t^n + \Delta t)$

9. Set $\boldsymbol{x}^{n+1,2} = \boldsymbol{x}^n + \frac{1}{2106}\Delta t\,[533\boldsymbol{f}^n + 1600\boldsymbol{f}^{Temp(2)} - 27\boldsymbol{f}^{Temp(3)}]$

10. Replace Δt by $\Delta t\,(\varepsilon/|\boldsymbol{x}^{n+1} - \boldsymbol{x}^{n+1,2}|)^{1/3}$

11. Set $\boldsymbol{f}(\boldsymbol{x}^{n+1}, t^{n+1}) = \boldsymbol{f}^{Temp(3)}$

12. Return to step 2

ALGORITHM 9.5.2 The Runge–Kutta–Fehlberg RKF45 method.

1. Compute $\boldsymbol{f}(\boldsymbol{x}^n, t^n)$

2. Set $\boldsymbol{x}^{Temp(1)} = \boldsymbol{x}^n + \frac{1}{4}\Delta t \boldsymbol{f}^n$

3. Compute $\boldsymbol{f}^{Temp(1)} = \boldsymbol{f}\left(\boldsymbol{x}^{Temp(1)}, t^n + \frac{1}{4}\Delta t\right)$

4. Set $\boldsymbol{x}^{Temp(2)} = \boldsymbol{x}^n + \frac{1}{32}\Delta t\,(3\boldsymbol{f}^n + 9\boldsymbol{f}^{Temp(1)})$

5. Compute $\boldsymbol{f}^{Temp(2)} = \boldsymbol{f}\left(\boldsymbol{x}^{Temp(2)}, t^n + \frac{3}{8}\Delta t\right)$

6. Set $\boldsymbol{x}^{Temp(3)} = \boldsymbol{x}^n + \frac{1}{2197}\Delta t\,(1932\boldsymbol{f}^n - 7200\boldsymbol{f}^{Temp(1)} + 7296\boldsymbol{f}^{Temp(2)})$

7. Compute $\boldsymbol{f}^{Temp(3)} = \boldsymbol{f}\left(\boldsymbol{x}^{Temp(3)}, t^n + \frac{12}{13}\Delta t\right)$

8. Set $\boldsymbol{x}^{Temp(4)} = \boldsymbol{x}^n + \frac{1}{4104}\Delta t\,(8341\boldsymbol{f}^n - 32832\boldsymbol{f}^{Temp(1)}$

$$+ 28864\boldsymbol{f}^{Temp(2)} - 845\boldsymbol{f}^{Temp(3)})$$

9. Compute $\boldsymbol{f}^{Temp(4)} = \boldsymbol{f}(\boldsymbol{x}^{Temp(4)}, t^n + \Delta t)$

10. Set $\boldsymbol{x}^{n+1,1} = \boldsymbol{x}^n + \Delta t\,\left(\frac{25}{216}\boldsymbol{f}^n + \frac{1408}{2565}\boldsymbol{f}^{Temp(2)} + \frac{2197}{4104}\boldsymbol{f}^{Temp(3)} - \frac{1}{5}\boldsymbol{f}^{Temp(4)}\right)$

11. Set $\boldsymbol{x}^{Temp(5)} = \boldsymbol{x}^n + \Delta t\,\left(-\frac{8}{27}\boldsymbol{f}^n + 2\boldsymbol{f}^{Temp(1)} + \frac{3544}{2565}\boldsymbol{f}^{Temp(2)} + \frac{1859}{4101}\boldsymbol{f}^{Temp(3)} - \frac{11}{40}\boldsymbol{f}^{Temp(4)}\right)$

12. Compute $\boldsymbol{f}^{Temp(5)} = \boldsymbol{f}\left(\boldsymbol{x}^{Temp(5)}, t^n + \frac{1}{2}\Delta t\right)$

13. Set $\boldsymbol{x}^{n+1} = \boldsymbol{x}^n + \Delta t\,\left(\frac{16}{135}\boldsymbol{f}^n + \frac{6656}{12{,}825}\boldsymbol{f}^{Temp(2)} + \frac{28{,}561}{56{,}430}\boldsymbol{f}^{Temp(3)} - \frac{9}{50}\boldsymbol{f}^{Temp(4)} + \frac{2}{55}\boldsymbol{f}^{Temp(5)}\right)$

14. Replace Δt by $0.84\Delta t\,(\varepsilon/|\boldsymbol{x}^{n+1} - \boldsymbol{x}^{n+1,1}|)^{1/5}$

ALGORITHM 9.5.3 The Runge–Kutta–Fehlberg–Cash–Karp RKF45 method.

1. Compute $\boldsymbol{f}(\boldsymbol{x}^n, t^n)$
2. Set $\boldsymbol{x}^{Temp(1)} = \boldsymbol{x}^n + \frac{1}{5}\Delta t \boldsymbol{f}^n$
3. Compute $\boldsymbol{f}^{Temp(1)} = \boldsymbol{f}\left(\boldsymbol{x}^{Temp(1)}, t^n + \frac{1}{5}\Delta t\right)$
4. Set $\boldsymbol{x}^{Temp(2)} = \boldsymbol{x}^n + \frac{1}{40}\Delta t\,(3\boldsymbol{f}^n + 9\boldsymbol{f}^{Temp(1)})$
5. Compute $\boldsymbol{f}^{Temp(2)} = \boldsymbol{f}\left(\boldsymbol{x}^{Temp(2)}, t^n + \frac{3}{10}\Delta t\right)$
6. Set $\boldsymbol{x}^{Temp(3)} = \boldsymbol{x}^n + \frac{1}{10}\Delta t\,(3\boldsymbol{f}^n - 9\boldsymbol{f}^{Temp(1)} + 12\boldsymbol{f}^{Temp(2)})$
7. Compute $\boldsymbol{f}^{Temp(3)} = \boldsymbol{f}\left(\boldsymbol{x}^{Temp(3)}, t^n + \frac{3}{10}\Delta t\right)$
8. Set $\boldsymbol{x}^{Temp(4)} = \boldsymbol{x}^n + \frac{1}{54}\Delta t\,(-11\boldsymbol{f}^n + 135\boldsymbol{f}^{Temp(1)} - 140\boldsymbol{f}^{Temp(2)} + 70\boldsymbol{f}^{Temp(3)})$
9. Compute $\boldsymbol{f}^{Temp(4)} = \boldsymbol{f}(\boldsymbol{x}^{Temp(4)}, t^n + \Delta t)$
10. Set $\boldsymbol{x}^{Temp(5)} = \boldsymbol{x}^n + \Delta t\,\left(\frac{1631}{55{,}296}\boldsymbol{f}^n + \frac{175}{512}\boldsymbol{f}^{Temp(1)} + \frac{575}{12{,}824}\boldsymbol{f}^{Temp(2)} + \frac{44{,}275}{11{,}0592}\boldsymbol{f}^{Temp(3)} + \frac{253}{4096}\boldsymbol{f}^{Temp(4)}\right)$
11. Compute $\boldsymbol{f}^{Temp(5)} = \boldsymbol{f}\left(\boldsymbol{x}^{Temp(5)}, t^n + \frac{7}{8}\Delta t\right)$
12. Set $\boldsymbol{x}^{n+1,1} = \boldsymbol{x}^n + \Delta t\,\left(\frac{2825}{27{,}648}\boldsymbol{f}^n + \frac{18575}{48{,}384}\boldsymbol{f}^{Temp(2)} + \frac{13{,}525}{55{,}296}\boldsymbol{f}^{Temp(3)} + \frac{277}{14{,}336}\boldsymbol{f}^{Temp(4)} + \frac{1}{4}\boldsymbol{f}^{Temp(5)}\right)$

13. Set $\quad \mathbf{x}^{n+1} = \mathbf{x}^n + \Delta t\,\big(\tfrac{37}{378}\mathbf{f}^n + \tfrac{250}{621}\mathbf{f}^{Temp(2)} + \tfrac{125}{594}\mathbf{f}^{Temp(3)} + \tfrac{512}{1771}\mathbf{f}^{Temp(5)}\big)$

14. Replace Δt by $\Delta t\,(\varepsilon/|\mathbf{x}^{n+1} - \mathbf{x}^{n+1,1}|)^{1/5}$

9.6 Stiff Problems

In Section 9.1, we discussed the concept of *inherent instability*, and in Sections 9.2–9.5 we discussed the concept of *numerical instability*. An inherently unstable problem requires an accurate numerical method, for example, a high-order method with a small time step, or computer arithmetic with extended precision. Doubling the size of the time step does not necessarily cause a numerical instability, but it may produce an erroneous behavior, especially at long times. We may say then that it is practically impossible to predict the behavior of the system modeled by the differential equations at long times.

There is another class of problems situated at the other end of the spectrum, called *stiff* problems, where the converse is true; two examples were discussed at the end of Section 9.2. Here, a small time step is necessary not for accuracy, but for securing numerical stability. The distinction between inherently unstable problems and stiff problems can be made more clear by observing that stiff problems are typically *superstable*. This means that the phase-space orbits emanating from two adjacent points tend to converge at an exponential rate with a large negative coefficient multiplying t in the exponent.

An example of a stiff problem is furnished by the seemingly innocuous linear but nonautonomous differential equation

$$\frac{dx}{dt} = -100(x - \sin t) \tag{9.6.1}$$

with initial condition $x(0) = 0$. The solution may be found in closed form and is given by

$$x(t) = \frac{1}{1.0001}\big(\sin t - 0.01\,\cos t + 0.01\,\exp(-100t)\big) \tag{9.6.2}$$

The presence of two disparate time scales on the right-hand side is noticeable. If we attempt to compute the solution using an explicit method, we shall find that a time step that is smaller than roughly $\frac{1}{100}$ is required in order to ensure a stable behavior, even beyond the initial transient period where the last term on the right-hand side ceases to be important. There are even stiffer problems whose solutions require the step size to be so small that it becomes comparable to the round-off error.

Measure of Stiffness

A measure of the stiffness can be devised in terms of two time scales: (a) the short time scale τ, which is the duration of the initial transient evolution—for equation (9.6.1), $\tau = \frac{1}{100}$; and (b) the total travel time in the phase space T. The stiffness of the problem is then quantified by the index $s = T/\tau$; the higher the value of s, the more difficult the problem at hand. Thus if the period of integration of equation (9.6.1) were restricted to the initial transient period, the problem would *not* be stiff.

A more practical measure of stiffness is the computational expense required for solving the problem using a standard method with an adaptive time-stepping module and some sensible error estimator; examples of such methods were discussed in the preceding two sections. More precise definitions are discussed in the monograph of Aiken (1985).

Numerical Methods

In order to keep the size of the time step at a reasonable level, stiff problems must be solved using implicit methods, such as the Adams–Moulton method or an implicit Runge–Kutta method. The development and numerical implementation of such methods are discussed by Gear (1971), and more recent reviews with practical advice are offered by Aiken (1985) and Byrne and Hindmarsh (1987). Specific implementations are described and coded by Press et al. (1992, pp. 727–39).

Gear methods

Gear, in particular, devised a family of implicit methods whose stability is improved notably with respect to that of the Adams–Moulton family of methods. The first-order Gear method is the implicit Euler method, and the second-, third-, and fourth-order methods are expressed by

$$\mathbf{x}^{n+1} = \tfrac{1}{3}(2\Delta t\,\mathbf{f}^{n+1} + 4\mathbf{x}^n - \mathbf{x}^{n-1}) \tag{9.6.3}$$

$$\mathbf{x}^{n+1} = \tfrac{1}{11}(6\Delta t\,\mathbf{f}^{n+1} + 18\mathbf{x}^n - 9\mathbf{x}^{n-1} + 2\mathbf{x}^{n-2}) \tag{9.6.4}$$

$$\mathbf{x}^{n+1} = \tfrac{1}{25}(12\Delta t\,\mathbf{f}^{n+1} + 48\mathbf{x}^n - 36\mathbf{x}^{n-1} + 16\mathbf{x}^{n-2} - 3\mathbf{x}^{n-3}) \tag{9.6.5}$$

The nonlinear algebraic equations are solved either by using an iterative method or by linearization in the spirit of the semi-implicit methods.

Modified midpoint method

Bader and Deuflhard (1983) developed a modification of the midpoint method that results in a semi-implicit method. The first step is to replace the explicit midpoint formula (9.3.9) with the implicit formula

$$\mathbf{x}^{n+1} = \mathbf{x}^{n-1} + 2\Delta t\,\mathbf{f}\left(\tfrac{1}{2}(\mathbf{x}^{n+1} + \mathbf{x}^{n-1}), t^n\right) \tag{9.6.6}$$

Linearizing the right-hand side, as shown in equation (9.3.4), we derive the modified version of the semi-implicit formula (9.3.6),

$$(\mathbf{I} - \Delta t\,\mathbf{J}^n)\mathbf{x}^{n+1} = (\mathbf{I} + \Delta t\,\mathbf{J}^n)\mathbf{x}^{n-1} + 2\Delta t\,(\mathbf{f}^n - \mathbf{J}^n\mathbf{x}^n) \tag{9.6.7}$$

The first step is executed using the semi-implicit Euler method.

PROBLEM

9.6.1 ***A stiff problem.***

Consider the differential equation

$$\frac{dx}{dt} = -1000(x - \mathrm{e}^{-t}) - \mathrm{e}^{-t} \tag{9.6.8}$$

with initial condition $x(0) = 0$ (Byrne and Hindmarsh 1987). (*a*) Compute the solution analytically in closed form and show that it involves two disparate time scales. (*b*) Compute the solution numerically from $t = 0$ up to $t = 4.0$ using the second-order Runge–Kutta method, with a Δt that is sufficiently small so that the numerical results agree with the exact solution up to the third significant figure. Repeat with the second-order Gear method, and discuss the relative efficiency of the two methods.

References

Abramowitz, M., and Stegun, I. A., 1972, *Handbook of Mathematical Functions.* Dover.

Aiken, R. C. (ed.), 1985, *Stiff Computation.* Oxford University Press.

Allen, M. P., and Tildesley, D. J., 1987, *Computer Simulation of Liquids.* Oxford University Press.

Bader, G., and Deuflhard, P., 1983, A semi-implicit mid-point rule for stiff systems of ordinary differential equations. *Numer. Math.* **41**, 373–493.

Byrne, G. D., and Hindmarsh, A. C., 1987, Stiff ODE solvers: a review of current and coming attractions. *J. Comput. Phys.* **70**, 1–62.

Cash, J. R., and Karp, A. H., 1990, *ACM Trans. Math. Software* **16**, 201–22.

Choodnovsky, D. V., and Choodnovsky, G. V., 1977, Pole expansions of nonlinear partial differential equations. *Nuovo Cimento* **B40**, 339–53.

Dahlquist, G., and Björck, Å, 1974, *Numerical Methods.* Prentice-Hall.

Ferziger, J. H., and Perić, M., 1996, *Computational Methods for Fluid Dynamics.* Springer-Verlag.

Gear, C. W., 1971, *Numerical Initial Value Problems in Ordinary Differential Equations.* Prentice-Hall.

Lambert, J. D., 1991, *Numerical Methods for Ordinary Differential Systems: The Initial Value Problem.* Wiley.

Lapidus, L., and Seinfeld, J. H., 1971, *Numerical Solution of Ordinary Differential Equations.* Academic Press.

Lawrence, J. D., 1972, *A Catalog of Special Plane Curves.* Dover.

Lichtenberg, A. J., and Lieberman, M. A., 1983, *Regular and Stochastic Motion.* Springer-Verlag.

Lorenz, E. N., 1963, Deterministic nonperiodic flow. *J. Atmos. Sci.* **20**, 130–41.

Pozrikidis C., 1997, *Introduction to Theoretical and Computational Fluid Dynamics.* Oxford University Press.

Press, W. H., Flannery, B. P., Teukolsky, S. A., and Vetterling, W. T., 1992, *Numerical Recipes*, 2nd ed. Cambridge University Press.

Stoer, J., and Bulirsch, R., 1980, *Introduction to Numerical Analysis.* Springer-Verlag.

Chapter 10

Ordinary Differential Equations; Boundary-Value, Eigenvalue, and Free-Boundary Problems

In Chapter 9, we discussed numerical methods for solving ordinary differential equations subject to an initial condition that establishes the point of departure of the solution orbit in the phase space; these were *initial-value problems.*

In this chapter, we focus our attention on three different but somewhat related classes of problems: *Boundary-value problems, eigenvalue problems,* and *free-boundary problems.* The distinguishing feature of these problems is that the point of departure of the solution orbit, the total travel time, or both are *not* specified at the outset. Instead, the solution is required to satisfy an appropriate number of boundary conditions or global constraints of different sorts.

The problems addressed in this chapter will be tackled using a new class of numerical methods, some of which incorporate as modules the methods for initial-value problems discussed in Chapter 9. Many of these methods can be generalized and implemented to solving a corresponding class of partial differential equations, albeit the extensions are not always straightforward. Thus the introduction of these methods in the present context of ordinary differential equations may serve as a preamble to, and a prerequisite of, a more general subject.

For example, finite-difference methods for ordinary differential equations will be discussed in Section 10.2, and their implementation for solving partial differential equations will be the exclusive subject of Chapter 11.

10.1 Two-point Boundary-value Problems

Consider a system of N first-order ordinary equations $d\mathbf{x}/dt = \mathbf{f}(\mathbf{x}, t)$, as described in Section 9.1, and introduce the corresponding phase space. At the outset, we make two assumptions:

1. The total travel time of the solution orbit, T, is specified; this assumption is a distinguishing feature of a general *boundary-value* problem.

2. The constraints or stipulations that accompany the N differential equations involve only the departure and arrival points, $\boldsymbol{x}(t = 0)$ and $\boldsymbol{x}(t = T)$, but no intermediate points; this assumption is the distinguishing feature of a *two-point boundary-value* problem.

Many-point boundary-value problems are formulated and solved by methods that are generalizations of those discussed in the following few sections for two-point boundary-value problems. Problems where the travel time T is not specified fall under the auspices of free-boundary problems to be discussed at the end of this chapter.

Separated Boundary Conditions

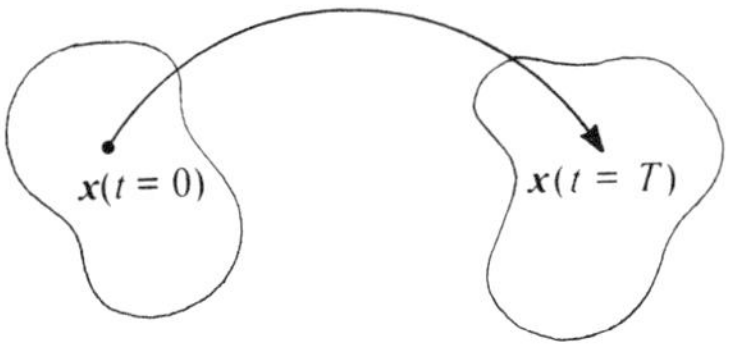

First, we assume that the statement of the problem requires that the departure point $\boldsymbol{x}(t = 0)$ lies on a certain $(N - L)$-dimensional *departure manifold* of the phase space, whereas the point of arrival $\boldsymbol{x}(t = T)$ lies on another L-dimensional *arrival manifold.* These are *separated* boundary conditions, meaning that the scalar equations expressing them involve either the departure point, or the arrival point, but not both.

Two special cases are already familiar:

- When $L = N$, the departure manifold is a point, the arrival manifold is the whole phase space, and the boundary-value problem degenerates to an initial-value problem.
- When $L = 0$, the departure manifold is the whole phase space, and the arrival manifold is a point; making the substitution $t = T - \tau$ produces an initial-value problem with respect to the reversed time τ.

More generally, we obtain a two-point boundary-value problem whose difficulty can be measured by the value of the smallest of the two integers L and $N - L$. The more balanced the number of boundary conditions between the beginning and the end of the solution orbit, the more difficult the problem at hand. In the present context, a problem whose degree of difficulty is equal to zero is an initial-value problem.

Mixed boundary conditions

In their most general form, separated boundary conditions specify the values of N linear or nonlinear combinations of several or all of the components of the departure *or* arrival points.

In particular, *linear* mixed boundary conditions specify the values of L components of the departure point, expressed by the vector

$$\boldsymbol{A}\,\boldsymbol{x}(t = 0) \equiv \boldsymbol{a} \tag{10.1.1}$$

and the values of $N - L$ components of the arrival point, expressed by the vector

$$\boldsymbol{B}\,\boldsymbol{x}(t = T) \equiv \boldsymbol{b} \tag{10.1.2}$$

where $\boldsymbol{A}$ and $\boldsymbol{B}$ are two constant matrices.

For example, for a system of two equations, the statement of the problem may stipulate that

$$x_1(t=0) + \alpha x_2(t=0) = a$$
$$x_1(t=T) + \beta x_2(t=T) = b \qquad (10.1.3)$$

where α, β, a, and b are four constants. In this case, the departure and arrival manifolds of the associated two-dimensional phase plane are straight lines whose slopes are equal, respectively, to $-1/\alpha$ and $-1/\beta$.

When $N = 3$ and $L = 1$, the phase space is three-dimensional, the departure manifold is a surface embedded in it, and the arrival manifold is a line.

PROBLEM

10.1.1 ***Departure and arrival manifolds.***

Write out the equation that describes (a) a spherical departure manifold in the three-dimensional phase space, and (b) an arrival manifold that has the shape of an ellipse in the two-dimensional phase plane.

Pure boundary conditions

When the matrices $\boldsymbol{A}$ and $\boldsymbol{B}$ introduced in equations (10.1.1) and (10.1.2) are diagonal, each scalar component of the boundary conditions involves only one scalar component of the departure or arrival points $\boldsymbol{x}(t=0)$ or $\boldsymbol{x}(t=T)$, and the boundary conditions are pure.

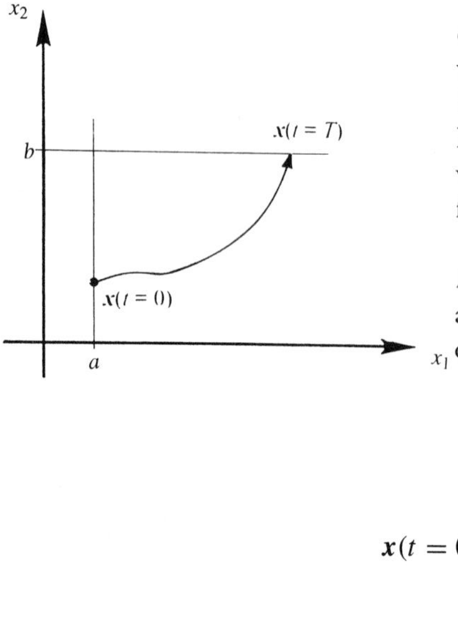

For example, when $N = 2$, the problem may require that the departure point $\boldsymbol{x}(t=0)$ lies on the vertical line $x_1(t=0) = a$, whereas the arrival point $\boldsymbol{x}(t=T)$ lies on the horizontal line $x_2(t=T) = b$, where a and b are two constants, corresponding to $L = 1$. In this case, we must begin traveling from some unknown point on the vertical line, so that after a specified time interval T has elapsed, we shall find ourselves at some point on the horizontal line.

To illustrate the nature of the problem in the more general case of N equations, we designate the specified components of the departure and arrival points with X, and the unspecified components with Ø, and obtain the visual image

$$\boldsymbol{x}(t=0) = \begin{bmatrix} \mathrm{X} \\ \emptyset \\ \mathrm{X} \\ \emptyset \\ \mathrm{X} \\ \cdots \\ \emptyset \\ \mathrm{X} \end{bmatrix}, \qquad \boldsymbol{x}(t=T) = \begin{bmatrix} \emptyset \\ \mathrm{X} \\ \mathrm{X} \\ \emptyset \\ \emptyset \\ \cdots \\ \emptyset \\ \mathrm{X} \end{bmatrix} \qquad (10.1.4)$$

where the total number of X in the first vector is equal to L, and the total number of X in both vectors is equal to N. Two special cases are familiar :

- If all entries of $\boldsymbol{x}(t=0)$ are filled with X, in which case all entries of $\boldsymbol{x}(t=T)$ must be filled with Ø, we obtain an initial-value problem.

- If all entries of $\mathbf{x}(t = T)$ are filled up with X, in which case all entries of $\mathbf{x}(t = 0)$ must be filled up with Ø, we can make the substitution $t = T - \tau$ and obtain a standard initial-value problem with respect to τ.

Electrostatic potential between concentric spheres

As an example, consider the distribution of the electrostatic potential V between two charged spheres with radii R_1 and R_2. Conservation of electrical energy requires that V satisfy Laplace's equation

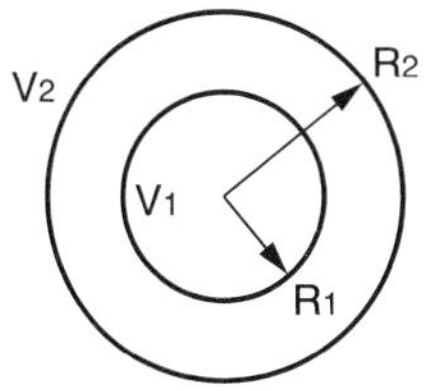

$$\nabla^2 V \equiv \frac{1}{r^2}\frac{\partial}{\partial r}\left(r^2 \frac{\partial V}{\partial r}\right) + \frac{1}{r^2 \sin\theta}\frac{\partial}{\partial \theta}\left(\sin\theta \frac{\partial V}{\partial \theta}\right) + \frac{1}{r^2 \sin^2\theta}\frac{\partial^2 V}{\partial \varphi^2} = 0 \qquad (10.1.5)$$

where (r, θ, φ) are the spherical polar coordinates, ∇^2 is the Laplacian operator, and V is defined in the range $R_1 \leq r \leq R_2$.

If the potential at the surfaces of the two spheres is kept constant at the values V_1 and V_2, then the distribution of V will be independent of the polar and azimuthal angles θ and φ. Laplace's equation (10.1.5) then reduces to the linear second-order ordinary differential equation

$$\frac{d^2 V}{dr^2} + \frac{2}{r}\frac{dV}{dr} = 0 \qquad (10.1.6)$$

with boundary conditions

$$V(R_1) = V_1, \qquad V(R_2) = V_2 \qquad (10.1.7)$$

To derive the equivalent system of first-order equations, we define $x_1 \equiv V$, $x_2 \equiv dV/dr$, and obtain

$$\frac{dx_1}{dr} = x_2, \qquad \frac{dx_2}{dr} = -\frac{2}{r}x_2 \qquad (10.1.8)$$

with the pure boundary conditions

$$x_1(R_1) = V_1, \qquad x_1(R_2) = V_2 \qquad (10.1.9)$$

In this case, $N = 2$ and $L = 1$. The departure and arrival points are given by

$$\mathbf{x}(r = R_1) = \begin{bmatrix} V_1 \\ Ø \end{bmatrix}, \qquad \mathbf{x}(r = R_2) = \begin{bmatrix} V_2 \\ Ø \end{bmatrix} \qquad (10.1.10)$$

where a Ø designates an *a priori* unknown value.

We can now define $t \equiv r - R_1$ and derive the canonical form of a boundary-value problem with respect to t. The total travel time is $T = R_2 - R_1$.

Blasius boundary layer

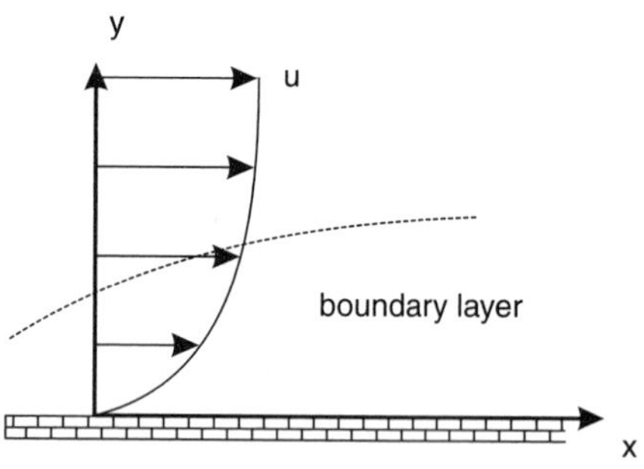

As another example, we consider high-speed flow past a solid surface, such as an airfoil. Far from the airfoil, the motion of fluid parcels is determined by a balance between the inertial and pressure fluid forces. Near the airfoil, a force due to the small but nonzero viscosity of the air is important. In particular, there is a thin boundary layer around the surface of the airfoil whose structure may be described by the simplified boundary layer equations of fluid flow (e.g., Pozrikidis 1997, Chapter 8).

A model of the boundary layer established around an airfoil is provided by the boundary layer developing over a flat plate that is held still parallel to an incident uniform stream. The equations of fluid mechanics allow us to express the streamwise velocity component $u(x, y)$ in terms of a function $f(\eta)$ in the form

$$u = \frac{df}{d\eta} \tag{10.1.11}$$

where

$$\eta \equiv y\sqrt{\frac{U}{\nu x}} \tag{10.1.12}$$

is a *similarity variable*, and ν is a physical constant called the kinematic viscosity of the fluid. The function $f(\eta)$ is computed by solving the *third-order nonlinear* differential equation

$$\tfrac{1}{2}ff'' + f''' = 0 \tag{10.1.13}$$

where a prime denotes a derivative with respect to η. The boundary conditions are

$$f = f' = 0 \quad \text{at } \eta = 0, \qquad \text{and} \qquad f' = 1 \quad \text{at } \eta = \infty \tag{10.1.14}$$

To derive the equivalent system of first-order equations, we define $x_1 \equiv f, x_2 \equiv f', x_3 \equiv f'', t \equiv \eta$, and obtain

$$\begin{aligned} \frac{dx_1}{dt} &= x_2 \\ \frac{dx_2}{dt} &= x_3 \\ \frac{dx_3}{dt} &= -\tfrac{1}{2}x_1x_3 \end{aligned} \tag{10.1.15}$$

with boundary conditions

$$x_1(t = 0) = 0, \qquad x_2(0) = 0, \qquad x_2(\infty) = 1 \tag{10.1.16}$$

Equations (10.1.15) and (10.1.16) provide us with a boundary-value problem with $N = 3$, $L = 2$, and $T = \infty$. The departure and arrival points are

$$\mathbf{x}(t=0) = \begin{bmatrix} 0 \\ 0 \\ \varnothing \end{bmatrix}, \qquad \mathbf{x}(t=\infty) = \begin{bmatrix} \varnothing \\ 1 \\ \varnothing \end{bmatrix} \tag{10.1.17}$$

where $\varnothing$ designates a value that is *a priori* unknown.

In this case, $\mathbf{x}(t = 0)$ is required to lie on the line $x_1(t = 0) = 0$, $x_2(t = 0) = 0$, which is simply the x_3 axis, and the arrival point $\mathbf{x}(t = \infty)$ is required to lie in the plane $x_2(t = \infty) = 1$. This boundary-value problem was first formulated and solved by Blasius in 1908 using a semi-analytical method.

Nonseparated Boundary Conditions

More generally, the boundary conditions specify N linear or nonlinear combinations of several of the components of the arrival *and* departure points, yielding nonseparated boundary conditions.

Linear nonseparated boundary conditions specify all N components of the vector

$$z \equiv \mathbf{A}\,\mathbf{x}(t=0) + \mathbf{B}\,\mathbf{x}(t=T) \tag{10.1.18}$$

where $\mathbf{A}$ and $\mathbf{B}$ are two constant matrices. For example, when $N = 2$, the problem may specify the values of both components of the vector

$$\begin{aligned} z &\equiv \begin{bmatrix} 1 & 1 \\ 1 & 0 \end{bmatrix} \mathbf{x}(t=0) + \begin{bmatrix} -1 & 2 \\ 1 & -1 \end{bmatrix} \mathbf{x}(t=T) \\ &= \begin{bmatrix} x_1(t=0) + x_2(t=0) - x_1(t=T) + 2\,x_2(t=T) \\ x_1(t=0) + x_1(t=T) - x_2(t=T) \end{bmatrix} \end{aligned} \tag{10.1.19}$$

In the following sections, we shall discuss methods for solving two-point boundary-value problems with separated boundary conditions. Problems with nonseparated boundary conditions can be formulated and solved by analogous methods.

10.2 The Shooting Method

Conceptually, this is the simplest method for solving two-point boundary-value problems. Undesired features may include a pronounced sensitivity to the numerical parameters and inflated computational cost.

To illustrate the numerical procedure, we consider a model boundary-value problem with pure boundary conditions. Specifically, we assume that the statement of the problem specifies the travel time T, the values of L coordinates of the departure point $\mathbf{x}(t = 0)$, and the values of $N - L$ coordinates of the arrival point $\mathbf{x}(t = T)$. The departure and arrival vectors are

$$\mathbf{x}(t=0) = \begin{bmatrix} a_1 \\ a_2 \\ \cdots \\ a_{L-1} \\ a_L \\ \varnothing \\ \cdots \\ \varnothing \\ \varnothing \end{bmatrix}, \qquad \mathbf{x}(t=T) = \begin{bmatrix} \varnothing \\ \varnothing \\ \cdots \\ \varnothing \\ \varnothing \\ b_{L+1} \\ \cdots \\ b_{N-1} \\ b_N \end{bmatrix} \tag{10.2.1}$$

where Ø indicates an unspecified value. The nonoverlapping of the initial and final values is not obligatory but was adopted to facilitate the notation.

In the first step, we guess the $N - L$ unspecified coordinates of the departure point,

$$\begin{aligned} &x_{L+1}(t = 0) \\ &x_{L+2}(t = 0) \\ &\dots \\ &x_N(t = 0) \end{aligned} \tag{10.2.2}$$

and solve the associated initial-value problem using one of the numerical methods discussed in Chapter 9. Part of the solution will include the $N - L$ coordinates of the arrival point

$$\begin{aligned} &x_{L+1}(t = T) \\ &x_{L+2}(t = T) \\ &\dots \\ &x_N(t = T) \end{aligned} \tag{10.2.3}$$

which are not necessarily equal to those required by the boundary condition (10.2.1).

The key observation is that the computed values (10.2.3) are generally nonlinear functions of the guessed values (10.2.2). Denoting these functions by G_i, we write

$$x_{L+i}(t = T) = G_i\big(x_{L+1}(t = 0), \dots, x_N(t = 0)\big) \tag{10.2.4}$$

where $i = 1, \dots, N - L$.

The problem has been reduced to finding the unspecified initial values (10.2.2) that satisfy the $N - L$ generally coupled nonlinear algebraic equations

$$R_i\big(x_{L+1}(t = 0), \dots, x_N(t = 0)\big) \equiv G_i\big(x_{L+1}(t = 0), \dots, x_N(t = 0)\big) - b_{L+i} = 0 \tag{10.2.5}$$

for $i = 1, \dots, N - L$. The $N - L$ elements of $\boldsymbol{R}$ provide us with the *residual* or *objective functions.* The solution can be found using the iterative methods discussed in Chapter 4. Examples are the secant method and Newton's method whose implementation will be discussed later in this section.

For reasons that are self-evident, the process of computing the solution in this manner is called *shooting.* The only feature distinguishing the shooting method from a standard method of solving nonlinear equations is that the functions G_i defining the residuals in equations (10.2.5) must be evaluated by solving initial-value problems using the methods discussed in Chapter 9.

In general, an infinite number of shootings will be required before we can satisfy all boundary conditions with infinite precision; some exceptions will be discussed in the following subsections.

To reduce the computational cost, it is helpful to note the following:

- In solving the initial-value problems, which is necessary for evaluating the functions G_i, a high level of accuracy is not necessary, except at the final iterations when the iterations are about to converge. Thus as long as difficulties with numerical instabilities do not arise, the use of a large but successively refined time step is recommended.

- After the iterations have converged, the overall accuracy of the solution will be limited by the error introduced in the numerical evaluation of the functions G_i. Thus the iterations should be terminated when the iteration error has become comparable to the error due to the time integration. The latter can be estimated and improved by the method of Richardson extrapolation (Section 1.6).

Implementation of Newton's Method

The nonlinear algebraic system of equations (10.2.5) can be solved using Newton's method. One way to compute the elements of the Jacobian matrix is by numerical differentiation, setting, for example,

$$\frac{\partial R_i}{\partial x_{L+j}(t=0)} \cong$$

$$\frac{1}{\varepsilon}\Big[G_i\big(x_{L+1}(t=0), \ldots, x_{L+j-1}(t=0), x_{L+j}(t=0)+\varepsilon, x_{L+j+1}(t=0), \ldots, x_N(t=0)\big) \tag{10.2.6}$$

$$-G_i\big(x_{L+1}(t=0), \ldots, x_{L+j-1}(t=0), x_{L+j}(t=0), x_{L+j+1}(t=0), \ldots, x_N(t=0)\big)\Big]$$

for $j = 1, \ldots, N-L$, where ε is a sufficiently small perturbation. Each evaluation of the Jacobian matrix requires solving $N-L+1$ initial-value problems, with initial values that are identified with the current guess or with small perturbations of the current guess, as indicated by the arguments of the first function on the right-hand side of the finite-difference approximation (10.2.6).

A more direct way of evaluating the Jacobian involves differentiating both sides of the original differential equations $dx_i/dt = f_i(\boldsymbol{x}, t)$ with respect to $x_{L+j}(t=0)$, where $j = 1, \ldots, N-L$, thereby producing a new system of ordinary differential equations called the system of the *first variational equations*, of the form

$$\frac{dQ_{i,j}}{dt} = \frac{\partial f_i(\boldsymbol{x})}{\partial x_m} Q_{m,j} \tag{10.2.7}$$

for $i = 1, \ldots, N$, where summation over m is implied on the right-hand side, and where

$$Q_{i,j} \equiv \frac{\partial x_i}{\partial x_{L+j}(t=0)} \tag{10.2.9}$$

Note that the first variational equations are *linear* with respect to $Q_{i,j}$. Differentiating also the components of the departure vector with respect to $x_{L+j}(t=0)$, we derive the initial condition accompanying the system (10.2.7),

$$Q_{i,j}(t=0) = \delta_{i,L+j} \tag{10.2.9}$$

The partial derivatives on the right-hand side of equation (10.2.7) are evaluated using values of $\boldsymbol{x}$ that are already available at discrete points from the numerical solution of the original system of equations $d\boldsymbol{x}/dt = \boldsymbol{f}(\boldsymbol{x}, t)$, with boundary conditions given in equations (10.2.1). Alternatively, the systems $d\boldsymbol{x}/dt = \boldsymbol{f}(\boldsymbol{x}, t)$ and (10.2.7) may be integrated *simultaneously* in a single pass.

The Jacobian matrix follows from

$$\frac{\partial R_i}{\partial x_{L+j}(t=0)} = Q_{L+i,j}(t=T) \tag{10.2.10}$$

In summary, each evaluation of the Jacobian matrix requires solving the initial-value problem expressed by equations (10.2.7) with boundary conditions (10.2.9) $N-L$ times, for $j=1,\ldots,N-L$.

If the original system of differential equations is linear with respect to $\boldsymbol{x}$, but not necessarily linear with respect to t—that is, if the phase-space velocity is given by $\boldsymbol{f}(\boldsymbol{x},t) = \boldsymbol{A}(t)\boldsymbol{x} - \boldsymbol{b}(t)$, where the elements of the matrix $\boldsymbol{A}(t)$ and of the vector $\boldsymbol{x}(t)$ are functions of t—then the solution is found exactly after one iteration, subject to the error introduced by the function evaluations.

Blasius Problem

For example, for the Blasius boundary-value problem expressed by equations (10.1.15) and (10.1.16), $N=3$, $L=2$, and $N-L=1$. The residual or objective function is

$$R\big(x_3(t=0)\big) \equiv G(x_3(t=0)) - 1 \equiv x_2(t=\infty) - 1 \tag{10.2.11}$$

and the first variational equations (10.2.7) take the specific form

$$\begin{aligned}\frac{dQ_{1,1}}{dt} &= Q_{2,1}\\ \frac{dQ_{2,1}}{dt} &= Q_{3,1}\\ \frac{dQ_{3,1}}{dt} &= -\tfrac{1}{2}(Q_{1,1}x_3 + Q_{3,1}x_1)\end{aligned} \tag{10.2.12}$$

which are to be solved subject to the initial condition

$$Q_{1,1}(t=0)=0, \qquad Q_{2,1}(t=0)=0, \qquad Q_{3,1}(t=0)=1 \tag{10.2.13}$$

In this case, the size of the Jacobian matrix is 1×1, and its sole element is equal to $Q_{2,1}(t=\infty)$. The shooting method proceeds according to the following steps:

1. Guess a value for $x_3(t=0)$.
2. Solve the system (10.1.15), treating it as an initial-value problem with $x_1(t=0)=0$ and $x_2(t=0)=0$.
3. Solve the system (10.2.12), treating it as an initial-value problem with initial conditions given in equations (10.2.13). The right-hand side of the third of equations (10.2.12) is evaluated using values of $\boldsymbol{x}$ already available from step 2.
4. Replace the current value $x_3(t=0)$ with the updated value

$$x_3(t=0) - \frac{x_2(t=\infty)-1}{Q_{2,1}(t=\infty)} \tag{10.2.14}$$

PROBLEMS

10.2.1 *Implementation of Newton's method.*

Develop an algorithm based on Newton's method for solving the boundary-value problem describing the distribution of electrostatic potential between two concentric spheres discussed in Section 10.1.

10.2.2 *Secant method.*

Develop an algorithm for solving a two-point boundary-value problem with $N = 2$ and $L = 1$ based on the secant method. Discuss the conditions under which the exact solution will be found after one iteration.

10.2.3 *Shooting but not killing.*

Compute the solution of the *nonlinear* differential equation

$$x'' - xx' = \sin(\pi x/2) \tag{10.2.15}$$

where a prime signifies a derivative with respect to t, within the interval [0, 1], subject to the boundary conditions $x(0) = 1$, $x(1) = -1.2$, using a method of your choice. The solution should produce the value of $x'(0)$, accurate to the second decimal place. *Hint*: confine your search for $x'(0)$ in the range $[-2, -5]$.

Rescaling

We return to examining the Blasius system (10.1.15) with the boundary conditions stated in equations (10.1.16), with the objective to demonstrate that careful inspection may suggest a way of circumventing the shootings.

In this case, we introduce the scaled variable $\xi = \alpha t$ where α is an arbitrary constant, express equations (10.1.15) in the form

$$\begin{aligned} \frac{d}{d\xi}\left(\frac{x_1}{\alpha}\right) &= \frac{x_2}{\alpha^2} \\ \frac{d}{d\xi}\left(\frac{x_2}{\alpha^2}\right) &= \frac{x_3}{\alpha^3} \\ \frac{d}{d\xi}\left(\frac{x_2}{\alpha^3}\right) &= -\frac{1}{2}\frac{x_1}{\alpha}\frac{x_3}{\alpha^3} \end{aligned} \tag{10.2.16}$$

and note that, when regarded as functions of ξ, the scaled functions

$$X_1 \equiv x_1/\alpha, \qquad X_2 \equiv x_2/\alpha^2, \qquad X_3 \equiv x_3/\alpha^3 \tag{10.2.17}$$

satisfy the system (10.1.15), with boundary conditions

$$X_1(\xi = 0) = 0, \qquad X_2(\xi = 0) = 0, \qquad X_2(\xi = \infty) = 1/\alpha^2 \tag{10.2.18}$$

The solution may then be found working in three stages:

1. Integrate the system (10.2.15) for X_1, X_2, X_3 with respect to ξ, using as initial conditions $X_1(\xi = 0) = 0$, $X_2(\xi = 0) = 0$, $X_3(\xi = 0) = a$, where a has an arbitrary value.

2. Set

$$\alpha^2 = 1/X_2(\xi = \infty) \tag{10.2.19}$$

3. Recover

$$x_1\left(t = \frac{\xi}{\alpha}\right) = \alpha X_1(\xi), \qquad x_2\left(t = \frac{\xi}{\alpha}\right) = \alpha^2 X_2(\xi), \qquad x_3\left(t = \frac{\xi}{\alpha}\right) = \alpha^3 X_3(\xi) \tag{10.2.20}$$

What we have done is *rescaling*. We multiplied the independent and dependent variables with a constant, and then fixed the value of the constant so that the solution satisfies the boundary conditions imposed on it. Unfortunately, this simplification is possible only for systems that have special forms.

PROBLEMS

10.2.4 ***Scaling.***

Verify that if the function $f(\eta)$ is a solution of equation (10.1.13), then the function $q(\eta) = \beta f(\beta\eta)$ will also be a solution of this equation for any value of the constant β.

10.2.5 ***Blasius boundary layer.***

Truncate the infinite domain of solution with respect to η at the value $\eta = 10$, and compute the solution of the Blasius boundary-layer equation, accurate to the third significant figure, using a shooting method of your choice. Plot the velocity profile of u against η.

Solution of Linear Systems by Superposing One Particular and a Set of Homogeneous Solutions

Assume now that the phase-space velocity $\boldsymbol{f}$ is a *linear* function of $\boldsymbol{x}$, but not necessarily a linear function of t, that is,

$$\boldsymbol{f} = \boldsymbol{A}(t)\,\boldsymbol{x} - \boldsymbol{b}(t) \tag{10.2.21}$$

where $\boldsymbol{A}(t)$ and $\boldsymbol{b}(t)$ are, respectively, a matrix and a vector function of t. In this case, only a *finite* number of $N + 1 - L$ initial-value problems must be solved before all boundary conditions can be satisfied to computer accuracy. This simplification is consistent with our previous observation for linear systems, that Newton's method produces the solution after only one iteration; constructing the Jacobian matrix requires solving $N - L + 1$ initial-value problems with appropriate initial conditions.

To formalize the method, let $\boldsymbol{x}^P(t)$ be a *particular* solution, and

$$\boldsymbol{x}^{(1)}(t), \quad \boldsymbol{x}^{(2)}(t), \quad \ldots, \quad \boldsymbol{x}^{(N-L)}(t) \tag{10.2.22}$$

be a collection of $N - L$ *homogeneous* solutions of the differential equations, computed with $x_i^{(j)} = 0$, $i = 1, \ldots L$, and different combinations of values for the missing set (10.2.2). Furthermore, the homogeneous solutions are computed by neglecting the term $\boldsymbol{b}(t)$ on the right-hand side of equation (10.2.21). When $\boldsymbol{A}(t)$ and $\boldsymbol{b}(t)$ are constant, both the particular and homogeneous solutions may be found analytically in closed form, as discussed in Section 9.2.

The key observation is that any linear combination of the homogeneous solutions added to the particular solution will also satisfy the linear system. Expressing the desired solution in the form

$$\mathbf{x}(t) \equiv \mathbf{x}^P(t) + c_1\mathbf{x}^{(1)}(t) + \cdots + c_{N-L}\mathbf{x}^{(N-L)}(t) \tag{10.2.23}$$

which is consistent with the first set of the boundary conditions (10.2.1), and requiring the second set of boundary conditions (10.2.1), we demand that

$$\begin{aligned} x_{L+1}(T) &\equiv x^P_{L+1}(T) + c_1 x^{(1)}_{L+1}(T) + \cdots + c_{N-L}x^{(N-L)}_{L+1}(T) = b_{L+1} \\ &\cdots \\ x_N(T) &\equiv x^P_N(T) + c_1 x^{(1)}_N(T) + \cdots + c_{N-L}x^{(N-L)}_N(T) = b_N \end{aligned} \tag{10.2.24}$$

and thereby obtian a system of $N - L$ linear algebraic equations for the $N - L$ coefficients $c_1, \ldots, c_{N-L}$.

For example, for a system of two equations with $N = 2$ and $L = 1$, with boundary conditions

$$\mathbf{x}(t = 0) = \begin{bmatrix} a_1 \\ \varnothing \end{bmatrix}, \qquad \mathbf{x}(t = T) = \begin{bmatrix} \varnothing \\ b_2 \end{bmatrix} \tag{10.2.25}$$

where $\varnothing$ signifies an unspecified value, we compute a particular and a homogeneous solution, the first with $x_1(t = 0) = a_1$ and the second with $x_1(t = 0) = 0$ and a nonzero starting value $x_2(t = 0)$, and set

$$\mathbf{x}(t) \equiv \mathbf{x}^P(t) + c_1\mathbf{x}^{(1)}(t) \tag{10.2.26}$$

where

$$c_1 = \frac{b_2 - x_2^P(T)}{x_2^{(1)}(T)} \tag{10.2.27}$$

Partial Inverse Shooting

It is possible that the shooting method will misbehave, even in the case of linear equations, due to the deleterious effects of the round-off error, the stiffness of the equations, to an unfortunate initial guess, or to a combination of these reasons.

For example, if one or several of the homogeneous solutions discussed in the preceding subsection increase rapidly during the integration from $t = 0$ to $t = T$, then the corresponding final value will dominate the left-hand side of equations (10.2.24), allowing for the loss of significant digits due to cancellation. For the system of two equations discussed in the last subsection, this will occur when the magnitude of $x_2^{(1)}(T)$ is large compared to that of b_2.

A cure for this pathology is provided by the method of *partial inverse shooting*. The procedure involves generating some of the homogeneous solutions by integrating the system of differential equations from the end toward the beginning, using a negative time step.

To illustrate the implementation of the method, consider a system of two equations, with the boundary conditions stated in equations (10.2.25). We define an intermediate matching time T_M and proceed as follows:

1. Guess the value of $x_2(t = 0)$, and integrate the system forward in time from $t = 0$ to $t = T_M$, thus producing the vector $\mathbf{x}^F(t = T_M)$.

2. Guess the value of $x_1(t = T)$, and integrate the system backward in time from $t = T$ to $t = T_M$, thus producing the vector $\boldsymbol{x}^B(t = T_M)$.
3. Revise the guesses $x_2(t = 0)$ and $x_1(t = T)$ to make $\boldsymbol{x}^F(t = T_M) = \boldsymbol{x}^B(t = T_M)$. This can be accomplished by requiring that $\boldsymbol{R}[\boldsymbol{x}^F(t = T_M) - \boldsymbol{x}^B(t = T_M)] = \boldsymbol{0}$, where a root of the objective function $R(z)$ is the null point $\boldsymbol{0}$.

Nonseparated Boundary Conditions

Problems with nonseparated boundary conditions are treated in a similar fashion. The functions of the components of the initial and final points that are prescribed in the statement of the boundary conditions are used to define a system of generally nonlinear algebraic equations, which is then solved using a standard method. At the end, the problem is reduced to finding the roots of a system of properly defined objective functions.

For example, for the linear mixed boundary conditions stated in equation (10.1.18), the objective function is

$$f(\boldsymbol{x}(t = 0)) = z - \boldsymbol{A}\,\boldsymbol{x}(t = 0) - \boldsymbol{B}\,\boldsymbol{x}(t = T) \tag{10.2.28}$$

The roots can be found using, for example, Newton's method.

Further discussion of the shooting method and its variations can be found in specialized monographs, including those by Keller (1968, 1976) and Ascher et al. (1995).

10.3 *Finite-difference and Finite-volume Methods*

Linear boundary-value problems, and to a lesser extent nonlinear problems, may be tackled efficiently using the related classes of finite-difference and finite-volume methods. In practice, these methods are applied directly to the primary differential equations of second or higher order, without prior conversion into an equivalent first-order system, but this is a technical rather than an essential feature.

The accuracy of finite-difference and finite-volume methods is comparable, and often superior, to that of the shooting method discussed in Section 10.2; the latter is sometimes followed by the former in order to improve the accuracy.

Linear Equations

A finite-difference method for solving a linear second-order ordinary differential equation was discussed in Section 3.4. We recall, in particular, that the numerical formulation produced a tridiagonal system of equations that can be solved using the method of Gauss elimination or, even more efficiently, the Thomas algorithm. The finite-volume method, to be discussed later in this section, produces a similar tridiagonal system of equations. This strategy of computation circumvents two undesirable features of the methods discussed in the preceding sections: (*a*) The need for iterations, apart from those that might be involved in solving the system of algebraic equations, and (*b*) the computation and storage of families of homogeneous solutions.

Finite-difference formulation

To further exemplify the implementation of the finite-difference methods, consider the linear second-order equation

$$\frac{d^2x}{dt^2} + (4 + \varepsilon \cos t)x = 0 \tag{10.3.1}$$

where ε is a constant. The solution is to be found over a domain that is subtended between $t = 0$ and $t = T = \pi/2$, with boundary conditions

$$\left(\frac{dx}{dt}\right)_{t=0} = -2, \qquad x\left(t = \frac{\pi}{2}\right) = -1 \tag{10.3.2}$$

A boundary condition for the first derivative is called a *Neumann* or a *natural* boundary condition, whereas a boundary condition for the unknown function itself is called a *Dirichlet* or an *essential* boundary condition. A combination of the two is called a *Robin* or a *mixed* boundary condition.

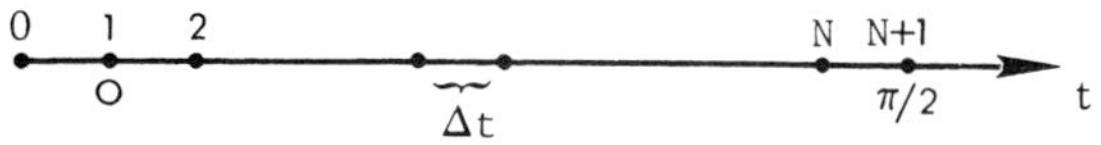

We begin by discretizing the solution domain $(0, \pi/2)$ into N evenly spaced intervals separated by the nodal points $t_i = (i-1)\Delta t$, where $\Delta t = \pi/(2N)$ and $i = 1, \ldots, N+1$. Next, we apply the differential equation (10.3.1) at the ith nodal point and approximate the second derivative using the second-order central finite-difference formula, setting $(d^2x/dt^2)_{x_i} = (x_{i+1} - 2x_i + x_{i-1})/\Delta t^2$, where we have designated $x_i = x(t_i)$. In this manner, we convert the diferential equation to the finite-difference equation

$$x_{i-1} - \big(2 - \Delta t^2(4 + \varepsilon \cos t_i)\big)x_i + x_{i+1} = 0 \tag{10.3.3}$$

The Dirichlet boundary condition at $t = \pi/2$ requires that $t_{N+1} = -1$.

Consider now the numerical implementation of the boundary condition at the left end. Using a forward-difference formula to approximate the first derivative at $x = 0$, we obtain the difference form

$$x_2 - x_1 = -2\Delta t \tag{10.3.4}$$

Appending this equation to the finite-difference equation (10.3.3) written for $i = 2, \ldots, N$, we obtain the linear algebraic system $\boldsymbol{Ax} = \boldsymbol{b}$, where

$$\boldsymbol{x} = (x_1, x_2, \ldots, x_N)^T \tag{10.3.5}$$

$\boldsymbol{A}$ is the $N \times N$ tridiagonal matrix

$$\boldsymbol{A} = \begin{bmatrix} 1 & -1 & 0 & \cdots & 0 \\ 1 & -2+\Delta t^2(4+\varepsilon\cos t_2) & 1 & \cdots & 0 \\ 0 & 1 & -2+\Delta t^2(4+\varepsilon\cos t_3) & \cdots & 0 \\ \cdots & \cdots & \cdots & \cdots & \cdots \\ 0 & 0 & \cdots & -2+\Delta t^2(4+\varepsilon\cos t_{N-1}) & 1 \\ 0 & 0 & \cdots & 1 & -2+\Delta t^2(4+\varepsilon\cos t_N) \end{bmatrix} \tag{10.3.6}$$

and

$$\boldsymbol{b} = (2\Delta t, 0, \ldots, 0, 1)^T \tag{10.3.7}$$

The solution of the linear system may be found efficiently using the Thomas algorithm discussed in Section 3.4.

The finite-difference equation (10.3.3) is second-order accurate in Δt, whereas the finite-difference equation for the boundary condition (10.3.4) is first-order accurate in Δt; this inconsistency limits the overall accuracy of the finite-difference method.

To obtain a truly second-order accurate solution, we extend the domain of solution to the left beyond the natural boundary $t = 0$, introduce the fictitious node $t_0 = -\Delta t$, apply equation (10.3.3) for $i = 1$, and discretize the boundary condition using the second-order central-difference formula to obtain $x_2 - x_0 = -4\Delta t$ or

$$x_0 = x_2 + 4\Delta t \tag{10.3.8}$$

Combing this equation with the finite-difference equation (10.3.3) written for $i = 1$, and appending the result to the rest of equations (10.3.3) corresponding to $i = 2, \ldots, N$, we obtain the linear algebraic system $\boldsymbol{Ax} = \boldsymbol{b}$, where $\boldsymbol{x}$ is given in equation (10.3.5), $\boldsymbol{A}$ is the $N \times N$ tridiagonal matrix

$$\boldsymbol{A} = \begin{bmatrix} -2 + \Delta t^2(4+\varepsilon) & 2 & 0 & \cdots & 0 \\ 1 & -2 + \Delta t^2(4+\varepsilon \cos t_2) & 1 & \cdots & 0 \\ 0 & 1 & -2 + \Delta t^2(4+\varepsilon \cos t_3) & \cdots & 0 \\ \cdots & \cdots & \cdots & \cdots & \cdots \\ 0 & 0 & \cdots & -2 + \Delta t^2(4+\varepsilon \cos t_{N-1}) & 1 \\ 0 & 0 & \cdots & 1 & -2 + \Delta t^2(4+\varepsilon \cos t_N) \end{bmatrix} \tag{10.3.9}$$

and

$$\boldsymbol{b} = (-4\Delta t, 0, \ldots, 0, 1)^T \tag{10.3.10}$$

PROBLEM

10.3.1 ***Solution of a model linear problem.***

(*a*) Compute the solution of equation (10.3.1) with the stated boundary conditions for $\varepsilon = 0$ analytically, in closed form.
(*b*) Solve the tridiagonal systems of equations for $\boldsymbol{x}$ defined in equation (10.3.5) using the Thomas algorithm with $\boldsymbol{A}$ and $\boldsymbol{b}$ given in equations (10.3.6) and (10.3.7), for $\varepsilon = 0$ and 5.0, in each case for $N = 2, 4, 8, 16, 32$. Compare the numerical results for $\varepsilon = 0$ with the exact solution, and assess the order of the method.
(*c*) Repeat part (*b*) with $\boldsymbol{A}$ and $\boldsymbol{b}$ given in eqauations (10.3.9) and (10.3.10).

Finite-volume formulation

The term *finite volume* derives from the practice of discretizing a three-dimensional solution domain into subregions or finite volumes. In the present context of a one-dimensional domain, namely, the t axis, the finite volumes are finite segments or linear elements.

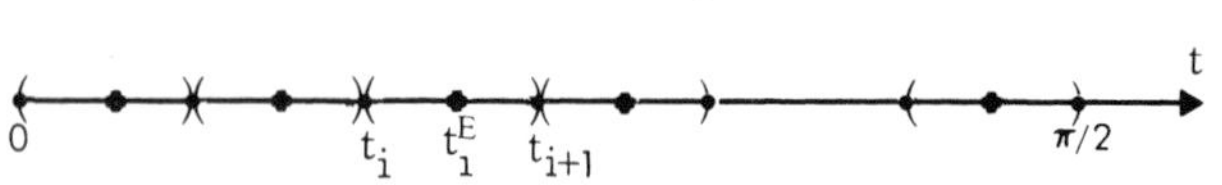

The idea is to divide the solution domain into N elements, as shown in the diagram, integrate both sides of the differential equation over each element, and then manipulate the second derivative d^2x/dt^2 to form the difference between the first derivatives or fluxes dx/dt at the two end-points; in two or three dimensions, the last step is accomplished by using the divergence theorem. Restating the differential equation in terms of the internal fluxes in this manner facilitates the numerical implementation. The advantages for one-dimensional problems presently considered are marginal, but the method is particularly effective in two and three dimensions where the finite volumes may have various shapes to accommodate complex boundary geometries.

For example, with reference to the differential equation (10.3.1), and in terms of the notation defined in the last illustration, we obtain

$$\left(\frac{dx}{dt}\right)_{t=t_{i+1}} - \left(\frac{dx}{dt}\right)_{t=t_i} + \int_{t_i}^{t_{i+1}} (4 + \varepsilon \cos t) x \, dt = 0 \tag{10.3.11}$$

for $i = 1, \ldots, N$. In the standard finite-volume formulation, we introduce the values of x at the middle of each element $t = t_i^E$, denoted as x_i^E, and compute the values and the end-points of the element, $t = t_i$, by linear interpolation.

Assuming that the integrand in equation (10.3.11) is constant over each element, approximating the first derivative using the first-order central-difference formula, and using the first of the boundary conditions (10.3.2), we obtain

$$\frac{x_2^E - x_1^E}{\Delta t} + 2 + (4 + \varepsilon \cos t_1^E) x_1^E \, \Delta t = 0 \tag{10.3.12}$$

and

$$\frac{x_{i+1}^E - x_i^E}{\Delta t} - \frac{x_i^E - x_{i-1}^E}{\Delta t} + (4 + \varepsilon \cos t_i^E) x_i^E \, \Delta t = 0 \tag{10.3.13}$$

for $i = 2, \ldots, N-1$. To complete the finite-volume formulation, we extend the domain of solution beyond the right boundary, introduce the phantom element number $N+1$, apply equation (10.3.13) for $i = N$, and use the second of the boundary conditions (10.3.2) to set $(x_{N+1}^E + x_N^E)/2 = -1$. Collecting the preceding finite-difference equations yields a tridiagonal system of equations for the element values x_i^E.

PROBLEM

10.3.2 ***Solution of a model linear problem.***

Formulate the linear system, and repeat part (*b*) of problem 10.3.1 with the finite-volume method described in the text.

Nonlinear Equations

The second-order nonlinear equation

$$\frac{d^2x}{dt^2} = f\left(\frac{dx}{dt}, x, t\right) \tag{10.3.14}$$

where f is a generally nonlinear function of its arguments, may be tackled in a similar fashion. The numerical implementation of the finite-difference or finite-volume method produces a *nonlinear* algebraic system of equations of the form

$$\mathbf{A}\mathbf{x} = \mathbf{q}(\mathbf{x}) + \mathbf{b} \tag{10.3.15}$$

where $\boldsymbol{A}$ is a constant matrix, $\boldsymbol{x}$ is the solution vector containing the unknown grid values, and $\boldsymbol{q}$ is a generally nonlinear function of its argument.

For example, approximating both the second and the first derivative with second-order central differences, and evaluating equation (10.3.14) at the ith grid point, we find

$$x_{i-1} - 2x_i + x_{i+1} = \Delta t^2 f\left(\frac{x_{i+1} - x_{i-1}}{2\Delta t}, x_i, t_i\right) \tag{10.3.16}$$

The equations corresponding to the first and last grid points are modified to incorporate the boundary conditions.

The resulting system of nonlinear algebraic equations can be solved using one of the iterative methods discussed in Chapter 4, including Newton's method and its variations. Since the Jacobian matrix corresponding to equation (10.3.16) is tridiagonal, the iteration vectors can be obtained economically using the efficient Thomas algorithm, in a process that is sometimes called *relaxation*. Depending, however, on the stiffness of the equations, a good initial guess may be required for a successful solution.

Algorithms for solving the nonlinear algebraic systems that arise from finite-difference formulations are discussed by Deuflhard (1979), Fox (1980), Jain (1984).

PROBLEM

10.3.3 ***A model nonlinear problem.***

Consider the nonlinear equation

$$\frac{d^2x}{dt^2} = -x + \frac{2}{x}\left(\frac{dx}{dt}\right)^2 \tag{10.3.17}$$

to be solved over the interval $-1 \le t \le 1$, with pure and separated boundary conditions $x(-1) = x(1) = 1/(\mathrm{e} + 1/\mathrm{e})$ (Atkinson 1989, p. 439).
(a) Compute the solution analytically in closed form.
(b) Compute the solution numerically based on the formalism of equation (10.3.16), using the standard or a modified Newton method, with number of intervals $N = 2, 4, 8, 16, 32$. The boundary conditions should be implemented so that the results are consistent second-order accurate in Δt.

10.4 Finite-element Methods

Finite-element methods belong to a broader class of *weighted-residual methods* to be discussed in Section 10.5. Rather than introducing them from a more general context, we prefer to first expose their distinguishing features by means of an example.

Consider the second-order linear differential equation

$$x'' + x + t = 0 \tag{10.4.1}$$

where a prime designates a derivative with respect to t, to be solved in the interval between 0 and 1, subject to one Dirichlet and one Neumann boundary condition,

$$x(0) = 0, \qquad x'(1) = 0 \tag{10.4.2}$$

The exact solution is readily found to be

$$x(t) = \frac{\sin t}{\cos 1} - t \tag{10.4.3}$$

Pretending that we are oblivious to this formula, we set out to compute the solution using an entry-level finite-element method.

We begin by subdividing the interval [0, 1] into N segments of equal length $h = 1/N$, separated by $N+1$ grid points, and approximate the unknown function $x(t)$ with a collection of linear local interpolating polynomials $P_1^{(i)}(t)$, as discussed in Section 6.6. In the present context, the segments are called *finite elements*, and the points separating them are called *nodes*.

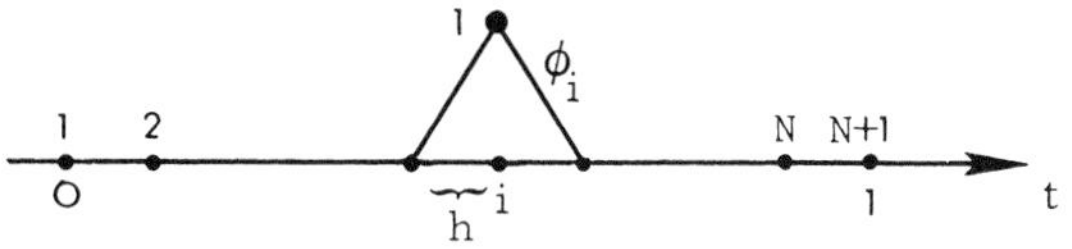

Adopting the linear local interpolation, we represent the function $x(t)$ as an expansion of the tent-like basis functions $\phi_i(t)$ shown in equation (6.6.5), setting

$$x(t) \cong \sum_{i=1}^{N} P_1^{(i)}(t) = \sum_{i=1}^{N+1} x_i \phi_i(t) \tag{10.4.4}$$

where we have denoted $x_i = x(t_i)$.

This is a seemingly absurd proposition: The second derivative of the expansion is clearly equal to zero everywhere except at the data points, where it becomes infinite. How can we possibly satisfy the second-order equation (10.4.1)? Not to worry: Using the ingenious device of the *Galerkin finite-element projection*, we can obtain a *weak solution*, that is, a solution that satisfies a collection of integrated forms of the differential equation involving first-order derivatives.

The main idea is to multiply the differential equation (10.4.1) with each one of the basis functions $\phi_j(t)$, $j = 2, \ldots, N+1$, integrate the product with respect to t between 0 and 1, and then integrate by parts the term involving the second derivative to reduce its order. The exclusion of the first basis function $\phi_1(t)$ is owing to the Dirichlet boundary condition required at the left end. Working in this manner, we obtain

$$(x'\phi_j)_{t=1} - (x'\phi_j)_{t=0} - \int_0^1 x'(t)\phi_j'(t)\,dt + \int_0^1 x(t)\phi_j(t)\,dt + \int_0^1 t\phi_j(t)\,dt = 0 \tag{10.4.5}$$

Next, we substitute the right-hand side of expansion (10.4.4) in place of x within the first and second integral, note that $\phi_j(t = 0) = 0$, for $j = 2, \ldots, N+1$, and use the second of the boundary conditions (10.4.2) to obtain

$$\sum_{i=1}^{N+1} (A_{j,i} - B_{j,i})x_i = \int_0^1 t\phi_j(t)\,dt \tag{10.4.6}$$

for $j = 2, \ldots, N+1$, where $\boldsymbol{A}$ is a symmetric tridiagonal matrix defined as

$$A_{i,j} = \int_0^1 \phi_j'(t)\,\phi_i'(t)\,dt = \begin{cases} \dfrac{1}{h} & \text{if } i = j = 1, N+1 \\ \dfrac{2}{h} & \text{if } i = j \neq 1, N+1 \\ -\dfrac{1}{h} & \text{if } i = j-1 \text{ or } i = j+1 \\ 0 & \text{otherwise} \end{cases} \tag{10.4.7}$$

and where $\boldsymbol{B}$ is another symmetric tridiagonal matrix defined as

$$B_{i,j} = \int_0^1 \phi_j(t)\,\phi_i(t)\,dt = \begin{cases} \frac{1}{3}h & \text{if } i = j = 1, N+1 \\ \frac{2}{3}h & \text{if } i = j \neq 1, N+1 \\ \frac{1}{6}h & \text{if } i = j-1 \text{ or } i = j+1 \\ 0 & \text{otherwise} \end{cases} \tag{10.4.8}$$

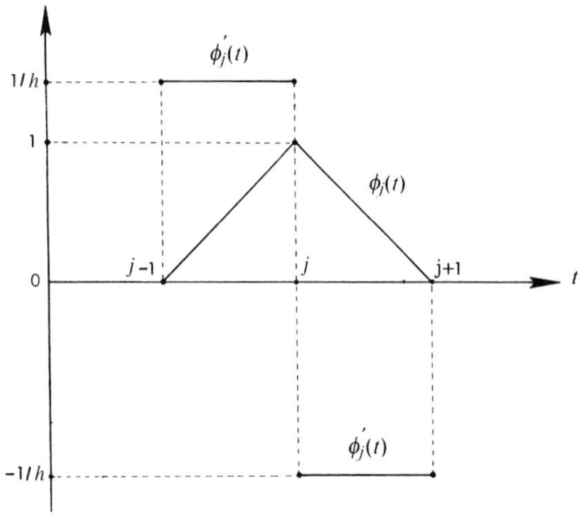

In computing the right-hand side of equation (10.4.7), we have noted that

$$\phi_j'(t) = \begin{cases} \dfrac{1}{h} & \text{if } t_{j-1} < t < t_j \\ -\dfrac{1}{h} & \text{if } t_j < t < t_{j+1} \\ 0 & \text{otherwise} \end{cases} \tag{10.4.9}$$

Applying equation (10.4.6) for $j = 2, \ldots, N+1$, we obtain a system of N equations for the N unknown values x_j, $j = 2, \ldots, N+1$. Using the first of the boundary conditions (10.4.2), we set $x_1 = 0$, and derive the tridiagonal form

$$\begin{bmatrix} a & b & 0 & 0 & 0 & \ldots & \ldots & 0 & 0 \\ b & a & b & 0 & 0 & \ldots & \ldots & 0 & 0 \\ 0 & b & a & b & 0 & \ldots & \ldots & 0 & 0 \\ \ldots & \ldots & \ldots & \ldots & \ldots & \ldots & \ldots & 0 & 0 \\ 0 & 0 & 0 & 0 & 0 & \ldots & b & a & b \\ 0 & 0 & 0 & 0 & 0 & \ldots & 0 & b & \frac{1}{2}a \end{bmatrix} \begin{bmatrix} x_2 \\ x_3 \\ x_4 \\ \ldots \\ x_N \\ x_{N+1} \end{bmatrix} = \begin{bmatrix} c_2 \\ c_3 \\ c_4 \\ \ldots \\ c_N \\ c_{N+1} \end{bmatrix} \tag{10.4.10}$$

where

$$a = \tfrac{2}{3}\left(\frac{3}{h} - h\right), \quad b = -\tfrac{1}{6}\left(\frac{6}{h} + h\right) \tag{10.4.11}$$

and

$$c_j = \int_0^1 t\phi_j(t)\,dt = \begin{cases} ht_j & \text{for } j = 2, \ldots, N \\ \tfrac{1}{2}h\left(1 - \dfrac{h}{3}\right) & \text{for } j = N+1 \end{cases} \tag{10.4.12}$$

Note that, although the boundary condition $x'(1) = 0$ was used in the construction of the linear system (10.4.10), the solution of the system will reproduce this boundary condition only in an approximate sense: The slope of the last segment of the polygonal graph of the solution will not be precisely equal to zero.

Graphs of the exact and numerical solutions for $N = 2, 4, 8, 16$ are shown in Figure 10.4.1. Convergence of the numerical results to the exact values with increasing number of subdivisions N is apparent. Careful inspection reveals that, in fact, the rate of convergence is quadratic with respect to $1/N$.

Relation to the Finite-difference Method

The $i - 1$ equation of the system (10.4.10), corresponding to c_i, where $i = 2, \ldots, N$, can be written in the form

$$\frac{x_{i+1} - 2x_i + x_{i-1}}{h^2} + \frac{x_{i+1} + 4x_i + x_{i-1}}{6} + t_i = 0 \tag{10.4.13}$$

It is illuminating to compare this difference equation with the one that arises by evaluating equation (10.4.1) at the ith node, and approximating the second derivative with the second-order central-difference formula, obtaining

$$\frac{x_{i+1} - 2x_i + x_{i-1}}{h^2} + x_i + t_i = 0 \tag{10.4.14}$$

The second term on the right-hand side of equation (10.4.13) is a *filtered* or *smoothed* version of the corresponding term on the right-hand side of equation (10.4.14). The filtering was effected implicitly by means of the Galerkin projection. This smoothing has the consequence that the numerical error is distributed more uniformly over the domain of solution than that of the finite-difference method.

Features of the Finite-element Method

To this end, we pause to point out certain salient features of the finite-element method just described:

- Function interpolation with an interpolating polynomial whose order can be lower than that of the differential equation.
- Integration of the ordinary differential equation over the domain of solution, after it has been weighted by the basis functions of the interpolating polynomial.
- Integration by parts of the weighed differential equation over its domain of solution, balancing the order of the derivatives of the basis functions and those of the computed solution, as shown in equation (10.4.5).

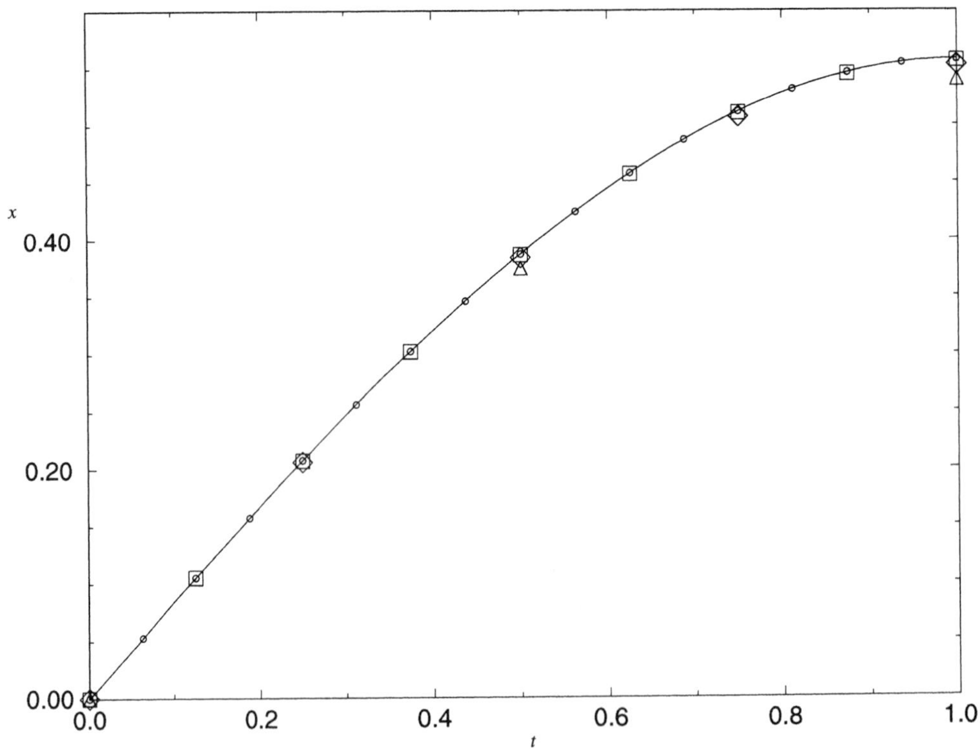

Figure 10.4.1 Graphs of the exact solution of the ordinary differential equation (10.4.1) with boundary conditions (10.4.2), shown with the solid line, and of the finite-element solution dicussed in the text for $N = 2, 4, 8, 16$.

PROBLEMS

10.4.1 *Solution by Thomas's algorithm.*

Solve the system of equations (10.4.10) using Thomas's algorithm for $N = 2, 4, 8, 16$, reproduce Figure 10.4.1, and verify that the error is indeed quadratic in h.

10.4.2 *Finite-element formulation of a linear ODE.*

Derive the system of linear algebraic equations that arises from the finite-element formulation of the equation

$$x'' - x = 0 \tag{10.4.15}$$

over the interval $[0, 2]$ and as described in the text, with boundary conditions $x(0) = 1$ and $x(2) = e^2$, and compare it with the corresponding system that arises from a finite-difference formulation where the second derivative is approximated with the second-order finite difference formula.

Compute the finite-element solution for $N = 2, 4, 8, 16$, reproduce the counterpart of Figure 10.4.1, and show that the error is quadratic in h.

General Formulation of Finite-element Methods

The finite-element method just described can be generalized in several ways with two objectives: Improve the accuracy of the results and enhance the spatial resolution at regions where the solution is expected to show strong variations. To satisfy these requirements, we use finite elements that have different sizes and employ second- or higher-order local interpolating polynomials. The implementation of the method is carried out as follows.

Element nodes

First, we describe the ith element E_i by the two *end-nodes*, and a collection of $M_i - 1$ internal nodes, a total number of $M_i + 1$ nodes, situated at the points

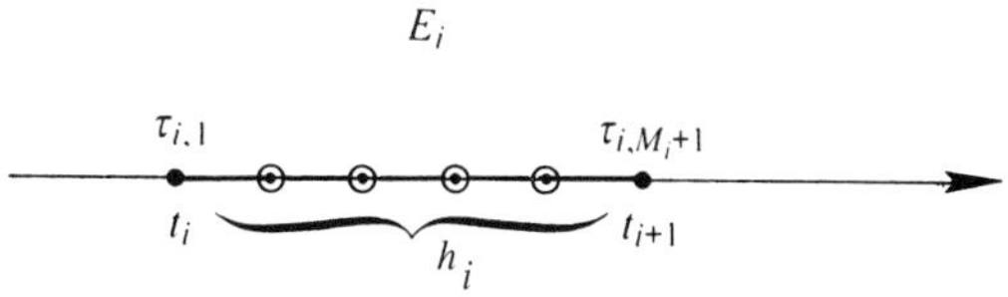

$$t = \tau_{i,j}, \qquad j = 1, 2, \ldots, M_i + 1$$

Observing that the last node of the ith element E_i is the first node of the next element E_{i+1}, we write

$$\tau_{i,1} = t_i, \qquad \tau_{i,M_i+1} = t_{i+1}, \qquad \text{for } i = 1, \ldots, N \tag{10.4.16}$$

The particular element–node distribution depicted in the last drawing was adopted for the purpose of illustration, and is not obligatory. Other element–node selections, including distributions corresponding to the zeros of orthogonal polynomials of a certain family, generally lead to more accurate solutions. For example, if we insist on maintaining the end-nodes, we can position the internal nodes at the zeros of the Lobatto polynomials displayed in Table B.7.

Local interpolation

In the next step, we approximate the unknown function $x(t)$ over the element E_i with an M_i-degree interpolating polynomial. Two examples with $M_i = 1$ or 2, corresponding to a *linear* or *quadratic* local interpolating polynomial, are shown in the diagram. For brevity, the qualifiers *linear* and *quadratic* are sometimes ascribed to the elements; a linear element is an element that supports a linear interpolating polynomial.

Denoting the value of the function unknown $x(t)$ at $t = \tau_{i,j}$ by $X_{i,j}$, and recalling that the last node of the ith element E_i is the first node of the next element E_{i+1}, we write

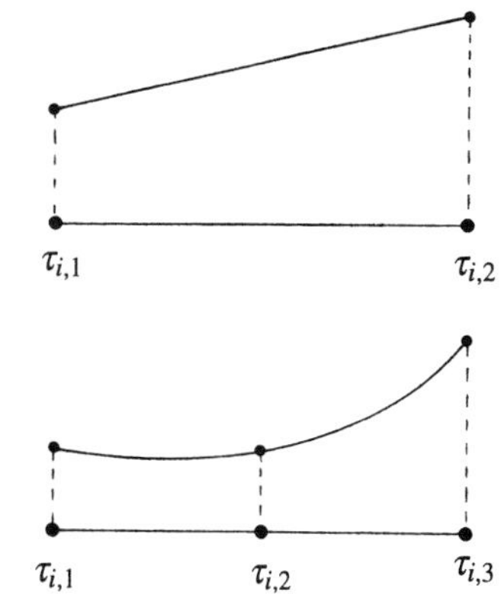

$$X_{i,1} = x_i, \qquad X_{i,M_i+1} = x_{i+1}, \qquad \text{for } i = 1, \ldots, N \tag{10.4.17}$$

Local interpolating polynomials

Over the ith element E_i of length h_i, we describe the local interpolating polynomial in the form

$$P_{M_i}^{(i)}(t) = \sum_{j=1}^{M_i+1} X_{i,j} \psi_{i,j}\left(\frac{t - t_i}{h_i}\right) \tag{10.4.18}$$

where $\psi_{i,j}(\xi)$ are *local interpolating polynomials* defined in the range $0 \le \xi \le 1$, satisfying the cardinal property

$$\psi_{i,j}\left(\frac{\tau_{i,k} - t_i}{h_i}\right) = \delta_{k,j} \tag{10.4.19}$$

and where $\delta_{k,j}$ is Kronecker's delta.

For a *linear* element with two end-nodes and no middle nodes, the local interpolating polynomials are given by

$$\psi_{i,j}(\xi) = 1 - \xi$$
$$\psi_{1,2}(\xi) = \xi \tag{10.4.20}$$

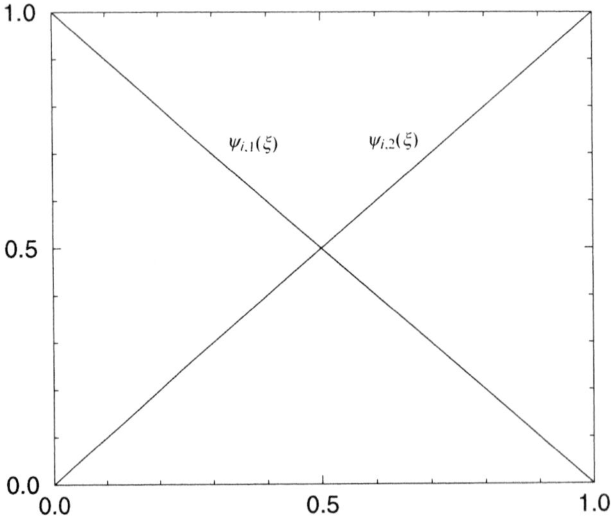

For a *quadratic* element with two end-nodes and one middle node located midway between the end-nodes,

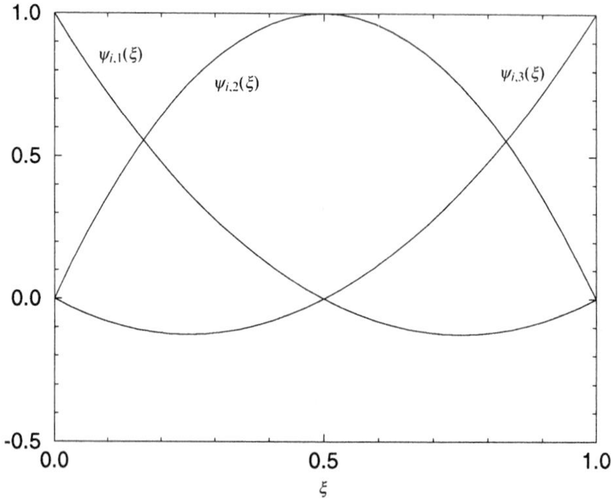

$$\psi_{i,1}(\xi) = 1 - 3\xi + 2\xi^2$$
$$\psi_{i,2}(\xi) = 4\xi(1 - \xi) \tag{10.4.21}$$
$$\psi_{i,3}(\xi) = \xi(2\xi - 1)$$

It will prove useful to define the extensions of the local interpolating polynomials outside their host element, by requiring that they vanish beyond the end-nodes.

Global interpolating polynomials

Continuing with the implementation, we express the global interpolating polynomial in terms of the *extended* local interpolating polynomials as

$$x(t) \cong \sum_{i=1}^{N} P_{M_i}^{(i)}(t) = \sum_{i=1}^{N} \sum_{j=1}^{M_i+1} X_{i,j}\psi_{i,j}\left(\frac{t - \tau_{i,1}}{h_i}\right) \tag{10.4.22}$$

Partial consolidation of the double sum can be made with the help of equation (10.4.17).

For example, if all elements are linear, which means that there are no middle nodes, the consolidation produces expansion (10.4.4) in terms of the basis functions ψ_i given by

$$\phi_1(t) = \psi_{1,1}\left(\frac{t - t_1}{h_i}\right)$$
$$\phi_i(t) = \psi_{i-1,2}\left(\frac{t - t_{i-1}}{h_{i-1}}\right) + \psi_{i,1}\left(\frac{t - t_i}{h_i}\right), \qquad \text{for } i = 2, \ldots, N \tag{10.4.23}$$
$$\phi_{N+1}(t) = \psi_{N,2}\left(\frac{t - t_N}{h_N}\right)$$

If all elements are quadratic, we introduce the $(2N + 1)$-dimensional vector z that contains that values of x at sequentially positioned end-node and middle node, and express the right-hand side of equation (10.4.22) as a single sum in the form

$$z = \begin{bmatrix} X_{1,1} \\ X_{1,2} \\ X_{1,3} \\ X_{2,2} \\ X_{2,3} \\ \cdot \\ \cdot \\ \cdot \\ X_{N,2} \\ X_{N,3} \end{bmatrix}$$

$$x(t) \cong \sum_{i=1}^{2N+1} z_i\phi_i(t) \tag{10.4.24}$$

where ϕ_i are expressible in terms of the local interpolating polynomials as

$$\phi_1(t) = \psi_{1,1}\left(\frac{t - t_1}{h_i}\right)$$
$$\phi_i(t) = \psi_{i/2,2}\left(\frac{t - t_{i/2}}{h_i}\right), \qquad i = 2, 4, \ldots, 2N$$
$$\phi_i(t) = \psi_{(i-1)/2,3}\left(\frac{t - t_{(i-1)/2}}{h_{(i-1)/2}}\right) + \psi_{(i+1)/2,1}\left(\frac{t - t_{(i+1)/2}}{h_{(i+1)/2}}\right), \qquad i = 3, \ldots, 2N - 1 \tag{10.4.25}$$
$$\phi_{2N+1}(t) = \psi_{N,3}\left(\frac{t - t_N}{h_N}\right)$$

Galerkin projection

The finite-element solution proceeds by substituting expansion (10.4.24) into the differential equation, multiplying the result by the global interpolating functions ϕ_j, integrating the product over the whole of the domain of solution, integrating by parts the term with the highest derivative to reduce its order, and thus deriving a system of linear equations for the nodal values z_i.

In practice, the final integral is computed in terms of the sum of the individual projections of the local interpolating polynomials; the latter are placed in the *element stiffness matrix.* For example, with reference to the first integral on the left-hand side of equation (10.4.5), we write

$$\int_0^1 x'(t)\phi_j'(t)\,dt = \sum_{i=1}^{N} \int_{E_i} x'(t)\phi_j'(t)\,dt$$
$$= \sum_{i=1}^{N}\sum_{l=1}^{M_i+1} X_{i,l} \int_{E_i} \psi_{i,l}'(t - t_i)/h_i)\phi_j'(t)\,dt \tag{10.4.26}$$

where, again, partial consolidation of the double sum can be made with the help of equations (10.4.17). The integral on the right-hand side of equation (10.4.26) is part of the element stiffness matrix; additional contributions to this matrix are made by the rest of the integrals in equation (10.4.5).

PROBLEM

10.4.3 ***Solution with quadratic elements.***

Show that the finite-element solution of equation (10.4.1) with boundary conditions stated in (10.4.2), implemented with quadratic elements of equal length h, and internal nodes located in the middle of each element, leads to the finite-difference equation (Chung 1978, p. 259)

$$\frac{-x_{i+1} + 8x_{i+1/2} - 14x_i + 8x_{i-1/2} - x_{i-1}}{h^2} + \frac{-x_{i+1} + 2x_{i+1/2} + 8x_i + 2x_{i-1/2} - x_{i-1}}{10}$$
$$+ \frac{-t_{i+1} + 2t_{i+1/2} + 8t_i + 2t_{i-1/2} - t_{i-1}}{10} = 0 \tag{10.4.27}$$

10.5 *Weighted-residual Methods*

The finite-element method described in Section 10.4 falls into the more general class of methods of weighted residual. To illustrate the common thread linking these methods and expose their distinguishing features, we consider the Kth-order scalar ordinary differential equation

$$f(x^{(K)}, x^{(K-1)}, \ldots, x, t) = 0 \tag{10.5.1}$$

for the function $x(t)$, where f is generally a nonlinear function of its arguments and the superscript (i) denotes the ith derivative. The independent variable t ranges in the interval $[a, b]$, and separated boundary conditions are prescribed at one or both ends $t = a$ and $t = b$, as discussed in Section 10.1.

The idea is to look for a solution $x(t)$ that does not necessarily satisfy the primary differential equation (10.5.1) but instead makes a set of weighted residuals vanish. That is, we require that

$$\int_a^b f(x^{(K)}, x^{(K-1)}, \ldots, x, t)\psi_j(t)\, dt = 0 \tag{10.5.2}$$

where $\psi_j(t)$, $j = 1, \ldots, M$, is a collection of properly selected *weighting functions*. In this context, the left-hand side of equation (10.5.1) evaluated for an arbitrary function $x(t)$ is a residual; if the residual vanishes for any value of t, and the function $x(t)$ satisfies the prescribed boundary conditions at $t = a$ and b, then we have recovered the exact solution.

The integration on the left-hand side of equation (10.5.2) effectively filters or smooths the residual in a manner that depends on the choice of the weighting functions $\psi_j(t)$.

In the practical implementation of the method, the function $x(t)$ is expanded in a series of appropriate *basis functions* $\phi_i(t)$ that are usually, but not always, chosen to individually satisfy the boundary conditions, writing

$$x(t) = \sum_{i=1}^{N+1} a_i \phi_i(t) \tag{10.5.3}$$

Different choices for the weighting and for the basis functions lead to different methods, some of which are already familiar. In the remainder of this section, we outline the implementation of methods used most frequently in applications. Extensive discussions can be found in the monographs of Finlayson (1972), Gottlieb and Orszag (1977), and Fornberg (1996).

Galerkin or Tau Method for Basis Functions that Satisfy the Boundary Conditions

Assume first that the basis functions $\phi_i(t)$ satisfy all boundary conditions. In the Galerkin or Tau method developed by Lanczos, we identity $\psi_i(t)$ with $\phi_i(t)$, where $i = 1, \ldots, N + 1$, and use equation (10.5.2) to generate a set of $N + 1$ algebraic equations for the $N + 1$ unknown coefficients a_i. The linear system is then solved using a standard numerical method.

Consider, for example, the differential equation (10.4.1), repeated here for ready reference,

$$f(x'', x, t) \equiv x'' + x + t = 0 \tag{10.5.4}$$

where a prime designates a derivative with respect to t, with the homogeneous Dirichlet boundary conditions

$$x(0) = 0 \qquad \text{and} \qquad x(1) = 0 \tag{10.5.5}$$

The exact solution is readily found to be

$$x(t) = \frac{\sin t}{\sin 1} - t \tag{10.5.6}$$

Pretending that we are unaware of this formula, we introduce the two-term expansion

$$x(t) = a_1 t(1-t) + a_2 t^2(1-t) \tag{10.5.7}$$

corresponding to $N = 1$, involving the two basis functions

$$\phi_1(t) = t(1-t), \qquad \phi_2(t) = t^2(1-t) \tag{10.5.8}$$

which have been designed to satisfy both boundary conditions (10.5.5).

Substituting the right-hand side of equation (10.5.7) into the middle side of equation (10.5.4), and the result into equation (10.5.2), and setting $\psi_1(t) = \phi_1(t)$ and $\psi_2(t) = \phi_2(t)$, we find

$$\int_0^1 [a_1(-2+t-t^2) + a_2(2-6t+t^2-t^3) + t]\,(1-t)\,t^i\,dt = 0 \tag{10.5.9}$$

where $i = 1, 2$. Carrying out the integrations, solving the emerging system of two equations for a_1 and a_2, and substituting the result back into expansion (10.5.7), we obtain

$$x(t) = t(1-t)\left(\tfrac{71}{369} + \tfrac{7}{41}t\right) \tag{10.5.10}$$

The numerical solution, shown with the thick dashed line in Figure 10.5.1, is in excellent agreement with the exact solution plotted with the solid line.

Galerkin Method for Basis Functions that Do Not Satisfy the Boundary Conditions

A dichotomy between the Galerkin and the Tau method arises when the basis functions do *not* satisfy the boundary conditions, as will be demonstrated in the following examples.

Considering first the Galerkin method with continued reference to equation (10.5.4) subject to the boundary conditions given in equations (10.5.5), we introduce the two-term expansion

$$x(t) = a_1 t + a_2 t^2 \tag{10.5.11}$$

involving the two basis functions

$$\phi_1(t) = t, \qquad \phi_2(t) = t^2 \tag{10.5.12}$$

which satisfy the boundary condition at the left end but not the boundary condition at the right end. Enforcing the right-end boundary condition, we find $a_2 = -a_1$, which we then substitute into equation (10.5.11) to find the one-term expansion

$$x(t) = a_1 t(1-t) \tag{10.5.13}$$

involving the new basis function $\Phi_1(t) = t(1-t)$. Note that $\Phi_1(t)$ is identical to the first basis function of the expansion (10.5.7). The approximate solution will inevitably be symmetric with respect to the midpoint $t = 0.50$. Orthogonalizing the residual with respect to $\Phi_1(t)$, we write

$$\int_0^1 [a_1(-2+t-t^2)+t]\,t(1-t)\,dt = 0 \tag{10.5.14}$$

which is satisfied when $a_1 = \frac{5}{18}$. The results, plotted with the thin long–short dashed line in Figure 10.5.1, are in fair agreement with the exact solution.

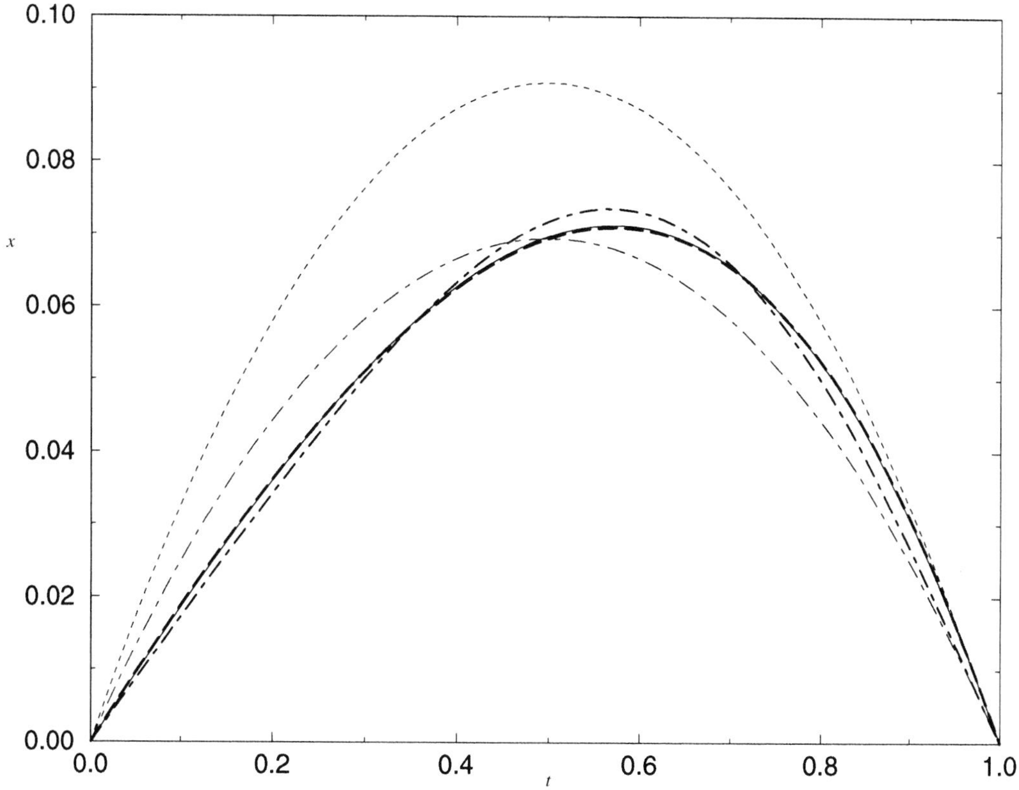

Figure 10.5.1 Performance of several weighted-residual methods for solving equation (10.5.4) with boundary conditions given in equations (10.5.5). The solid line represents the exact solution, and the broken lines represent various numerical approximations.

Tau Method for Basis Functions that Do Not Satisfy the Boundary Conditions

Maintaining the expansion (10.5.11), we proceed in a different manner. First, we orthogonalize the residual with respect to the basis function $\phi_1(t)$, setting

$$\int_0^1 [a_1 t + a_2(2 + t^2) + t]\, t\, dt = 0 \tag{10.5.15}$$

Second, we enforce the right-end boundary condition to find $a_2 = -a_1$, which together with equation (10.5.15) yields expansion (10.5.13) with $a_1 = \frac{4}{11}$. The results, shown with the thin dashed line in Figure 10.5.1, are in fair agreement with the exact solution.

Why Tau?

Tau, Greek spelling τ, is the nineteenth letter of the modern Greek alphabet. To explain why the present method was coined as the *Tau method*, we temporarily generalize our discussion.

Consider the linear ordinary differential equation

$$L(x(t)) = 0 \tag{10.5.16}$$

to be solved for the function $x(t)$ over the interval $a \le t \le b$, subject to the N_B boundary conditions

$$(B_i(x))_{t=a,b} = 0 \tag{10.5.17}$$

for $i = 1, \ldots, N_B$, where L and B_i are linear differential operators. In the Tau method, we introduce the expansion (10.5.3) and proceed in three stages:

- We substitute expansion (10.5.3) into the left-hand side of equation (10.5.16), multiply the result with each one of the test functions $\phi_i(t)$, $i = 1, \ldots, N - N_B + 1$, integrate the products with respect to t from a to b, and thus derive a system of $N - N_B + 1$ linear algebraic equations involving the coefficients a_i.
- We substitute expansion (10.5.3) into the left-hand side of the boundary conditions (10.5.17), and thus derive a system of N_B algebraic equations involving the coefficients a_i.
- We solve the system of $N + 1$ equations, derived in the two previous steps, for a_i.

Lanczos (1956) showed that the solution computed in this manner satisfies *exactly* the modified differential equation

$$L(x(t)) = \sum_{l=1}^{N_B} \tau_l\, \phi_{N-N_B+l+1}(t) \tag{10.5.18}$$

with the boundary conditions (10.5.17). The right-hand side of equation (10.5.18) involves the N_B coefficients τ_l whose values are given by

$$\tau_l = \frac{\int_a^b L(x(t))\phi_{N-N_B+l+1}(t)\, dt}{\int_a^b \phi^2_{N-N_B+l+1}\, dt} \tag{10.5.19}$$

Finite-element Galerkin Method

The finite-element method emerges by identifying $\phi_i(t)$ with the global interpolating functions introduced in Section 10.4. Integrating, in particular, the right-hand side of equation (10.5.2) by parts, to reduce the order of the derivatives, permits us to use global interpolating functions whose order is lower than that of the differential equation. Details on the implementation of the method are given in texts on finite-element methods (e.g., Chung 1978, Lapidus and Pinder 1982).

PROBLEM

10.5.1 ***Galerkin method.***

Consider equation (10.4.1) with boundary conditions stated in equations (10.4.2). Express the solution in terms of the two-term expansion (10.5.11), and use Galerkin's method with the second derivative integrated by parts to compute the coefficients a_1 and a_2. Does the computed solution satisfy the required boundary conditions?

Spectral Methods

In this class of methods, the basis functions employed in the expansion (10.5.3) are selected from a set of functions $\phi_i(t)$ that are mutually orthogonal with respect to a certain weighting function $w(t)$ over a certain interval. For problems with homogeneous or periodic boundary conditions, expansions in sine and cosine functions are appropriate. More generally, expansions in Chebyshev polynomials lead to more accurate solutions (Fox and Parker 1968, Gottlieb and Orszag 1977). The method proceeds by identifying $\psi_i(t)$ with $w(t)\phi_i(t)$, and then using the orthogonality condition to derive a sparse set of algebraic equations for the coefficients a_i.

Consider, for example, equation (10.5.4) with boundary conditions stated in equations (10.5.5), and introduce the two-term Fourier expansion

$$x(t) = a_1\phi_1(t) + a_2\phi_2(t) \tag{10.5.20}$$

involving the two basis functions

$$\phi_1(t) = \sin \pi t, \qquad \phi_2(t) = \sin 2\pi t \tag{10.5.21}$$

that are mutually orthogonal over the domain of solution [0, 1] with respect to the uniform weighting function $w(t) = 1$.

Substituting the right-hand side of equation (10.5.20) into equation (10.5.4), and the result into equation (10.5.2), and using the orthogonality property

$$\int_0^1 \sin(n\pi t)\, \sin(m\pi t)\, dt = \tfrac{1}{2}\delta_{n,m} \tag{10.5.22}$$

we obtain two *uncoupled* algebraic equations for a_1 and a_2, whose solution is

$$a_1 = \frac{2}{\pi}\frac{1}{\pi^2 - 1}, \qquad a_2 = \frac{1}{\pi}\frac{1}{1 - 4\pi^2} \tag{10.5.23}$$

With these values, the approximation (10.5.20) represented by the thick long–short dashed line in Figure 10.5.1 compares favorably with the exact solution.

Spectral methods are most efficient for linear equations with constant coefficients. Under more general circumstances, the *pseudospectral* methods described later in this section are preferred.

Spectral-element method

When the computed function is expected to show significant variations over certain parts of the solution domain requiring a high-order expansion, it is appropriate to divide the domain into a set of evenly or unevenly distributed elements, and then expand the unknown function over each element in a series of properly scaled orthogonal functions that belong to a particular set. This methodology provides us with the class of spectral-element methods.

The main advantage of the spectral-element methods is that a desired level of resolution can be achieved throughout the domain of solution using only low-order expansions. The coefficients of the local expansions are computed either by exploiting the orthogonality property described previously in this subsection or by using the collocation method to be described in the next subsection. In the second case, we obtain the family of *pseudospectral-element* methods.

Collocation Methods

In these methods, we expand the solution $x(t)$ in a set of basis functions as shown in equation (10.5.3), carry out the differentiations involved in equation (10.5.1), and identify the weighting functions ψ_i with a collection of one-dimensional Dirac delta functions $\delta(t - \xi_i)$, $i = 1, \ldots, M$. The collocation points ξ_i are distributed in some manner over the domain of solution, $a \leq \xi_i \leq b$. Thus

$$\psi_i(t) = \delta(t - \xi_i) \tag{10.5.24}$$

The distinguishing property of the Dirac delta function is that if $g(t)$ is an arbitrary function, then

$$\int_a^b g(t)\delta(t - \xi_i)\, dt = g(\xi_i) \tag{10.5.25}$$

Using this property, we transform equations (10.5.2) to

$$f\left(x^{(K)}(\xi_i), x^{(K-1)}(\xi_i), \ldots, x(\xi_i), \xi_i\right) = 0 \tag{10.5.26}$$

Appending to these equations the boundary conditions, we obtain a system of L algebraic equations for the $N + 1$ unknown coefficients a_i. If the basis functions satisfy the boundary conditions, in which case $L = M$, the boundary conditions do not participate in the final system of algebraic equations. When $L > N + 1$, the system is overdetermined and must be solved using a least-squares or a minimization method.

As an example, we consider the differential equation (10.5.4) with boundary conditions given in equations (10.5.5). Substituting the two-term expansion (10.5.7) into equation (10.5.4), we obtain

$$a_1\big(-2 + t(1 - t)\big) + a_2\big(2 - 6t + t^2(1 - t)\big) + t = 0 \tag{10.5.27}$$

Selecting the two collocation points $\xi_1 = 0$ and $\xi_2 = 1$, we find $a_1 = a_2 = \frac{1}{6}$. The numerical solution,

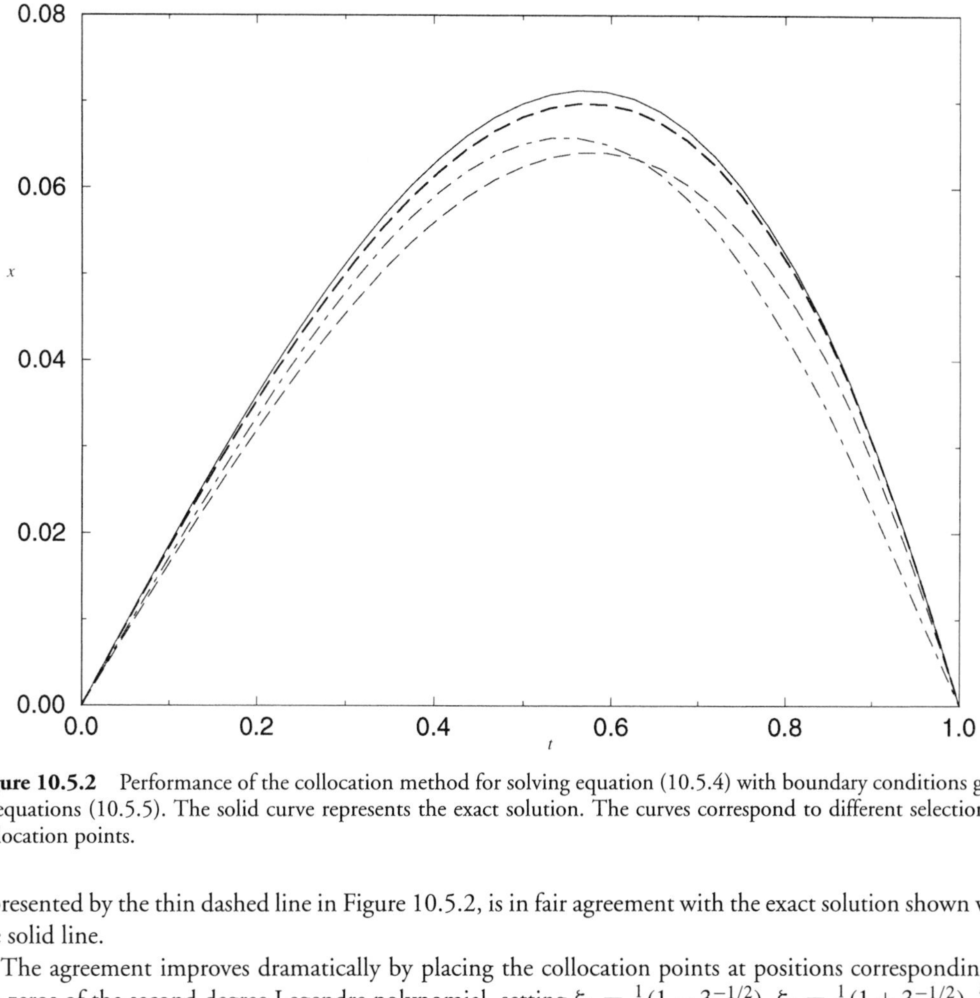

Figure 10.5.2 Performance of the collocation method for solving equation (10.5.4) with boundary conditions given in equations (10.5.5). The solid curve represents the exact solution. The curves correspond to different selections of collocation points.

represented by the thin dashed line in Figure 10.5.2, is in fair agreement with the exact solution shown with the solid line.

The agreement improves dramatically by placing the collocation points at positions corresponding to the zeros of the second degree Legendre polynomial, setting $\xi_1 = \frac{1}{2}(1 - 3^{-1/2})$, $\xi_2 = \frac{1}{2}(1 + 3^{-1/2})$. The corresponding numerical solution is represented by the thick dashed line in Figure 10.5.2.

Orthogonal collocation

It is evident that the accuracy of the collocation method will depend on the number and positioning of the collocation points. The positioning is especially important when a low-order expansion is used, that is, when N is small.

In the orthogonal collocation method advocated by Lanczos (1956), the collocation points correspond to the zeros of a selected family of orthogonal polynomials, such as the Legendre or Chebyshev polynomials. The implementation and performance of the method are discussed by Finlayson (1972, Chapter 5), de Boor and Swartz (1973), and Ascher (1986). The search for an optimal solution strategy leads us to the more advanced class of pseudospectral methods.

Pseudospectral Methods

In this variation of the collocation method, the basis functions are selected from an *orthogonal set,* and the solution is computed over a scaled interval corresponding to the domain of definition of the orthogonal set. The implementation and performance of the method are discussed by Villadsen and Strewart (1967), Gottlieb and Orszag (1977), and Fornberg (1996). In the remainder of this section, we discuss several choices.

Sine and cosine basis functions provide us with a trigonometric orthogonal set. When these functions respect the boundary conditions at both ends, the best results are obtained by distributing the collocation points evenly over the domain of solution. For example, with reference to the differential equation (10.5.4) and boundary conditions stated in equations (10.5.5), we introduce the two-term expansion (10.5.20) and select the collocation points $\xi_1 = \frac{1}{3}$ and $\xi_2 = \frac{2}{3}$. Equation (10.5.26) becomes

$$a_1(\pi^2 - 1) \sin \pi\xi_i + a_2(4\pi^2 - 1) \sin 2\pi\xi_i = \xi_i \tag{10.5.28}$$

for $i = 1, 2$. Solving for the expansion coefficients, we obtain

$$a_1 = \frac{1}{\sqrt{3}} \frac{1}{\pi^2 - 1}, \qquad a_2 = \frac{1}{3\sqrt{3}} \frac{1}{1 - 4\pi^2} \tag{10.5.29}$$

The numerical results represented by the long–short dashed line in Figure 10.5.2 show a good agreement with the exact solution.

When using Chebyshev polynomial expansions that respect the boundary conditions at both ends, the best results are obtained by placing the collocation points at the zeros of the $(N + 1)$ degree Chebyshev polynomials listed in Table B.3.

When the orthogonal expansion does not satisfy the boundary conditions at either end, we identify the collocation points with the zeros of the $(N - 1)$ degree Lobatto polynomials (Table B.7, Appendix B) and enforce the boundary conditions at both ends to obtain two additional equations. Explicit expressions and the roots of the first few Lobatto polynomials are shown in Table 7.3.3.

For example, with reference to the differential equation (10.5.4) accompanied by the boundary conditions (10.5.5), we express the solution as a linear combination of the first three Chebyshev polynomials, corresponding to $N = 2$. This amounts to approximating the solution with the quadratic function

$$x(t) = a_0 + a_1 t + a_2 t^2 \tag{10.5.30}$$

The boundary conditions require that $a_0 = 0, a_0 + a_1 + a_2 = 0$. Substituting expansion (10.5.30) into equation (10.5.4), and requiring the satisfaction of the resulting expression for $t = \frac{1}{2}$, we find $2a_2 + a_0 + \frac{1}{2}a_1 + \frac{1}{4}a_2 + \frac{1}{2} = 0$. Solving the system of the preceding two equations produces the approximation $x(t) = \frac{2}{7}t(1 - t)$.

When the orthogonal expansion satisfies the boundary conditions only at one end, we identify the collocation points with the zeros of the N degree Radau polynomials (Table B.6, Appendix B), and enforce the boundary condition at the other end to obtain one additional equation. The procedure is similar to that illustrated in the preceding example.

10.6 Eigenvalue Problems

In this section, we turn our attention to a different class of problems that are similar in some respects to the two-point boundary-value problem discussed earlier in this chapter, and also to the algebraic eigenvalue problem; that is, the problem of finding the eigenvalues of a matrix discussed in Chapter 5. In their simplest version, these differential-equation eigenvalue problems are described as follows:

- Both the departure point $\boldsymbol{x}(t = 0)$ and arrival point $\boldsymbol{x}(t = T)$ of the phase-space orbit are required to lie on the same *manifold* of the N-dimensional phase space, where T is a specified travel time.
- There is a particular point $\boldsymbol{x}^*$ on this manifold representing a so-called trivial solution or contrived phase-space orbit; that is, $\boldsymbol{x}(t) = \boldsymbol{x}^*$ for all t.
- The phase-space velocity $\boldsymbol{f}$ involves a scalar parameter κ, and nontrivial phase-space orbits exist only for certain values of κ, which are the eigenvalues of the differential equations subject to the stated boundary conditons.

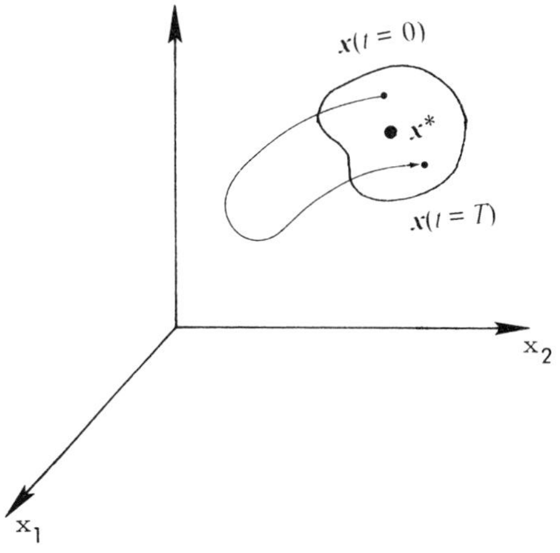

Consider, for example, the second-order linear differential equation

$$\frac{d^2x_1}{dt^2} + \kappa^2\pi^2 x_1 = 0 \tag{10.6.1}$$

where κ is an unspecified parameter, with boundary conditions $x_1(0) = 0$ and $x_1(1) = 0$, discussed earlier in Section 5.1, and rewrite it in the form of a system of two first-order equations as

$$\begin{aligned} \frac{dx_1}{dt} &= x_2 \\ \frac{dx_2}{dt} &= -\kappa^2\pi^2 x_1 \end{aligned} \tag{10.6.2}$$

In this case $N = 2$, the total travel time is $T = 1$, and the eigenvalue manifold is the vertical line $x_1 = 0$ of the x_1x_2 phase plane.

An elementary computation reveals the existence of a trivial solution $x_1^* = 0$ and $x_2^* = 0$ for any value of κ, and a family of nontrivial solutions described by

$$x_1 = c\ \sin(\kappa\pi t), \qquad x_2 = c\kappa\pi\ \cos(\kappa\pi t) \tag{10.6.3}$$

when κ is an integer, where c is an arbitrary constant. Thus $\kappa = m$, where m is an integer, are the eigenvalues of the differential equation (10.6.1) or its equivalent system (10.6.2). The first few eigensolutions are shown in the following diagram for $c = 1$.

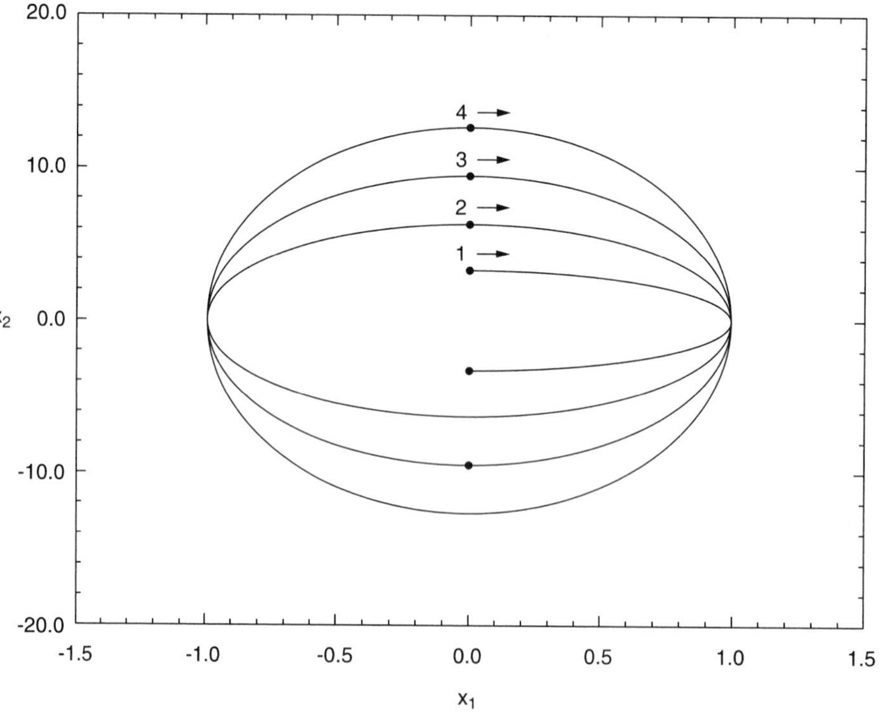

Reduction to a Boundary-value Problem

The differential-equation eigenvalue problems just described may be transformed to two-point boundary-value problems simply by relabeling $\kappa = x_{N+1}$, regarding x_{N+1} as an additional unknown, and appending to the original system of N differential equations the $N + 1$ equation $dx_{N+1}/dt = 0$. The solution may then be found using one of the methods discussed in the preceding section, but this is not necessarily the most expedient approach. Our expertise in solving algebraic eigenvalue problems suggests better methods.

Reduction to an Algebraic Eigenvalue Problem

Linear eigenvalue problems, and to a lesser extent nonlinear eigenvalue problems, may be solved efficiently by first reducing them to standard or generalized, linear or nonlinear *algebraic* eigenvalue problems discussed in Chapter 5, using finite-difference methods, finite-volume methods, or methods of weighted residuals. The reduction of the problem expressed by equation (10.6.1) with the accompanying boundary conditions was already discussed in Section 5.1. In the remainder of this section, we present additional examples that exemplify the numerical implementation. Tau methods are discussed by Gardner et al. (1989).

Vibration of a circular membrane: to beat a drum with a finite-difference method

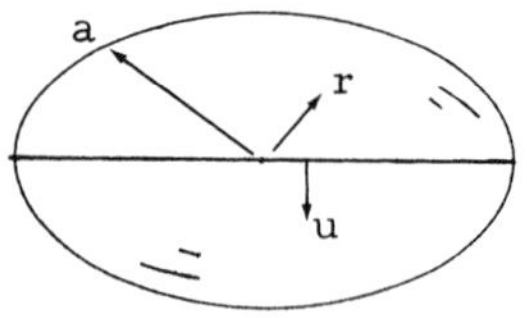

Consider the vibrations of a membrane stretched across the open mouth of a circular ring of radius a. Assuming that the shape of the membrane remains axisymmetric at all times, we regard its displacement from the flat position, u, as a function of the radial distance r and time t, writing $u = u(r, t)$. Requiring that the motion of an elementary membrane patch observes Newton's second law of motion and neglecting damping, we find that u satisfies the wave equation

$$\frac{\partial^2 u}{\partial t^2} = c^2 \frac{1}{r} \frac{\partial}{\partial r}\left(r \frac{\partial u}{\partial r}\right) \tag{10.6.4}$$

subject to (*a*) the boundary condition $u = 0$ at $r = a$, which ensures that the membrane is attached to the rim at all times, and (*b*) the regularity condition that all partial derivatives of u with respect to its arguments are finite everywhere on the disk. The square of the wave speed c^2 is equal to T/ρ, where T is the uniform tension of the stretched membrane, and ρ is the density of the membrane material.

To describe periodic vibrations with angular frequency ω, we set $u(r, t) = f(r) \cos(\omega t)$ and find that the function $f(r)$ satisfies the second-order ordinary differential equation

$$\frac{d}{dr}\left(r \frac{df}{dr}\right) + \kappa^2 r f = 0 \tag{10.6.5}$$

where $\kappa \equiv \omega/c$.

It is clear that equation (10.6.5) with the aforementioned boundary conditions has the trivial solution $f(r) = 0$, but we are interested in nontrivial eigensolutions expressing different modes of vibration. These are possible only for particular values of the parameter κ, that is, for specific frequencies of vibration.

It is convenient to introduce the dimensionless radius $R = r/a$ and recast equation (10.6.5) into the dimensionless form

$$\frac{d}{dR}\left(R \frac{df}{dR}\right) + \mu^2 R f = 0 \tag{10.6.6}$$

where $\mu \equiv \kappa a = \omega a/c$ is a dimensionless characteristic frequency of vibration. Expanding out the derivative on the left-hand side, we obtain

$$\frac{d^2 f}{dR^2} + \frac{1}{R} \frac{df}{dR} + \mu^2 f = 0 \tag{10.6.7}$$

The requirement that the membrane be attached to the ring demands that

$$f(R = 1) = 0 \tag{10.6.8}$$

A second condition emerges by requiring that all derivatives of f be finite at $R = 0$. Expanding the second term in equation (10.6.7) in a Taylor series about $R = 0$, and then taking the limit as R tends to zero, we find

$$\left(2 \frac{d^2 f}{dR^2} + \mu^2 f\right)_{R=0} = 0 \tag{10.6.9}$$

Although this equation was derived from the differential equation, the intermediate assumption of regular behavior at $R = 0$ allows it to play the role of a boundary condition.

The exact solution of the eigenvalue problem just posed is given by

$$f(R) = b J_0(\mu R) \tag{10.6.10}$$

where b is an arbitrary constant, and J_0 is the zeroth-order Bessel function of the first kind (e.g., Hildebrand 1974, Sections 4.8, 9.9). The requisite eigenvalues μ are the zeros of J_0; that is, $J_0(\mu) = 0$.

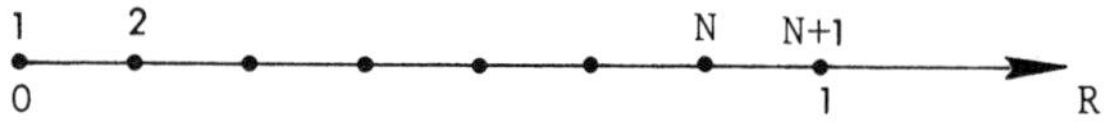

Pretending that we have no knowledge of the Bessel functions, we proceed to compute the eigenvalues using a finite-difference method. We begin by discretizing the solution domain [0, 1] into N evenly spaced intervals demarcated by the $N+1$ nodes $R_i = (i-1)\Delta R$, where $\Delta R = 1/N$ and $i = 1, \ldots, N+1$. Next, we apply the differential equation (10.6.7) at the ith node, approximate the first and second derivatives using second-order central differences, that is, we set

$$\left(\frac{df}{dR}\right)_{R_i} = \frac{(f_{i+1} - f_{i-1})}{2\Delta R}$$
$$\left(\frac{d^2 f}{dR^2}\right)_{R_i} = \frac{(f_{i+1} - 2f_i + f_{i-1})}{\Delta R^2} \tag{10.6.11}$$

where we have designated $f_i = f(x_i)$, and thus obtain the finite-difference equation

$$\left(-1 + \frac{1}{2}\frac{\Delta R}{R_i}\right) f_{i-1} + 2 f_i - \left(1 + \frac{1}{2}\frac{\Delta R}{R_i}\right) f_{i+1} = \Delta R^2 \mu^2 f_i \tag{10.6.12}$$

for $i = 2, \ldots, N$. The boundary condition (10.6.8) requires that $f_{N+1} = 0$. An analogous treatment of the boundary condition (10.6.9) yields the finite-difference equation

$$4 f_1 - 4 f_2 = \Delta R^2 \mu^2 f_1 \tag{10.6.13}$$

Putting equation (10.6.13) at the top of equation (10.6.12), we obtain a system of equations for the N unknowns f_i,

$$\boldsymbol{A f} = \Delta R^2 \mu^2 \boldsymbol{f} \tag{10.6.14}$$

where $\boldsymbol{f} = (f_1, f_2, \ldots, f_N)^T$, and $\boldsymbol{A}$ is an $N \times N$ tridiagonal matrix whose structure may be inferred readily from the preceding equations. It is clear that $\lambda = \Delta R^2 \mu^2$ is an eigenvalue of the matrix $\boldsymbol{A}$, and

$$\mu = \frac{\sqrt{\lambda}}{\Delta R} \tag{10.6.15}$$

For example, for the coarsest possible discretization with $N = 2$ corresponding to $\Delta R = \frac{1}{2}$, the matrix $\boldsymbol{A}$ takes the form

$$\boldsymbol{A} = \begin{bmatrix} 4 & -4 \\ -\frac{1}{2} & 2 \end{bmatrix} \tag{10.6.16}$$

and its eigenvalues are $\lambda = 1.268, 4.732$. The corresponding values of μ, computed from equation (10.6.15), are 2.252 and 4.351. Considering the crudeness of the method, these are reasonable approximations to the first two zeros of the zeroth-order Bessel function, which are equal, respectively, to 2.405 and 5.520 (e.g., Hildebrand 1974, p. 230).

Vibration of a circular membrane: to beat a drum with a Galerkin method

Next, we tackle the same problem using a Galerkin method. We begin by expanding the solution of equation (10.6.7) into a series of trigonometric functions, each one of them satisfying the boundary condition (10.6.8) and the regularity condition $(df/dR)_{R=0} = 0$, setting

$$f(R) = \sum_{i=1}^{N+1} a_i \phi_i(R) \tag{10.6.17}$$

where

$$\phi_i(R) = \cos\left((i - \tfrac{1}{2})\pi R\right) \tag{10.6.18}$$

Substituting expansion (10.6.17) into equation (10.6.7) and multiplying the left-hand side by R, we find

$$\begin{aligned}\sum_{i=1}^{N+1} a_i \big[&R(i - \tfrac{1}{2})^2\pi^2 \cos\left((i - \tfrac{1}{2})\pi R\right) \\ &+(i - \tfrac{1}{2})\pi \sin\left((i - \tfrac{1}{2})\pi R\right) - R\mu^2 \cos\left((i - \tfrac{1}{2})\pi R\right)\big] = 0\end{aligned} \tag{10.6.19}$$

To apply the Galerkin projection, we multiply both sides of equation (10.6.19) by $\phi_j(R)$, integrate the product with respect to R from 0 to 1, and obtain a homogeneous system of linear equations for a_i. Setting the determinant of the coefficient matrix equal to zero provides us with a secular equation involving μ.

For example, when $N = 0$, equation (10.6.19) for $i = 1$ yields

$$\begin{aligned}a_1 \Bigg[&\frac{\pi^2}{4}\int_0^1 R \cos^2\left(\frac{\pi R}{2}\right) dR + \frac{\pi}{2}\int_0^1 \cos\left(\frac{\pi R}{2}\right) \sin\left(\frac{\pi R}{2}\right) dR \\ &-\mu^2 \int_0^1 R \cos^2\left(\frac{\pi R}{2}\right) dR\Bigg] = 0\end{aligned} \tag{10.6.20}$$

Requiring that the term enclosed by the square brackets vanish, we obtain

$$\mu = \frac{\pi}{2}\left(\frac{\pi^2 + 4}{\pi^2 - 4}\right)^{1/2} = 2.415 \tag{10.6.21}$$

which is in remarkable agreement with the exact value 2.405.

PROBLEMS

10.6.1 *Vibrations of a circular membrane.*

Consider the matrix $\mathbf{A}$ defined in equation (10.6.14). Compute the eigenvalue with the smallest magnitude and the corresponding value of μ for $N = 4, 8, 16, 32$ using one of the methods described in Chapter 5. Compare the numerical results with the exact value, and discuss their convergence with respect to N.

10.6.2 *Vibrations of a circular membrane with Galerkin's method.*

With reference to the Galerkin method discussed in the text, compute and discuss the accuracy of (a) the two eigenvalues corresponding to $N = 1$, and (b) of the three eigenvalues corresponding to $N = 2$.

10.6.3 *Vibrations of a square membrane.*

Present the numerical formulation of the algebraic eigenvalue problem for a flat membrane stretched and vibrating over a square frame.

Hydrodynamic stability

An important class of eigenvalue problems emerge from the linear stability analyses of a broad range of physical systems. The objective is to examine the behavior of small perturbations introduced in a certain steady or time-periodic state. An example from the field of hydrodynamics will serve to illustrate the methods.

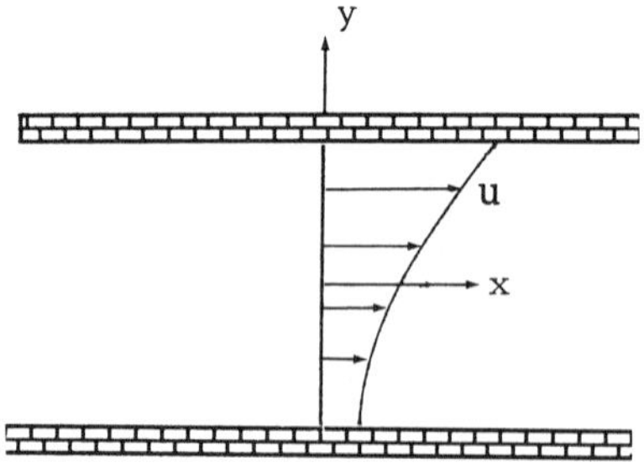

Consider the steady unidirectional flow of a fluid through a channel that is bounded by two walls, where the streamwise component of the velocity, u, is directed along the x axis and is a function of y, $u = U(y)$. To investigage the physical relevance of the flow, that is, to examine its ability to dampen disturbances of sufficiently small amplitude, we carry out a linear stability analysis.

Specifically, we consider the behavior of *normal mode* disturbances in the xy plane. The associated velocity field and other flow variables are assumed to depend on the downstream position x and time t in an exponential or sinusoidal fashion. For example, the cross-stream component of the perturbation velocity along the y axis is assumed to take the functional form

$$v(x, y, t) = f(y) \exp[Ik(x - ct)] \tag{10.6.22}$$

where $f(y)$ is an unknown complex function to be determined as part of the solution, k is the wave number, c is the *complex phase velocity*, and I is the imaginary unit. Both the real and imaginary parts of v represent physically acceptable solutions.

Stipulating that k is real, we obtain spatially periodic perturbations that evolve in time, corresponding to the *temporal stability problem.* If the imaginary part of c is positive, the perturbation will grow in time at an exponential rate, and the disturbance will be unstable; if the imaginary part of c is negative, the disturbance will be stable; and if it is equal to zero, the disturbance will be neutrally stable.

It can be shown that, if the effects of the fluid viscosity are neglected, the function $f(y)$ must satisfy the Rayleigh equation

$$U\frac{d^2 f}{dy^2} - \left(Uk^2 + \frac{d^2 U}{dy^2}\right) f = c\left(\frac{d^2 f}{dy^2} - k^2 f\right) \tag{10.6.23}$$

which is to be solved subject to the no-penetration condition requiring that $f(y)$ vanish at the channel walls (e.g., Pozrikidis 1997, Chapter 9). To study the temporal stability subject to perturbations of a specified wavelength, we specify the value of the real wave number k, and compute the complex phase velocity c by requiring the existence of a nontrivial solution for the homogeneous equation (10.6.23).

To make matters more specific, let us consider flow in a channel that is confined between two impermeable walls located at $y = -A$ and $y = B$. Adopting a finite-difference method, we introduce a one-dimensional uniform grid of nodes located at $y_i, i = 0, \ldots, N+1$, where $y_0 = -A$ and $y_{N+1} = B$, and approximate $d^2 f/dy^2$ at the ith node using the second-order central-difference formula. This discretization replaces equation (10.6.23) with the finite-difference equation

$$U_i f_{i-1} - [2U_i + \Delta y^2(U_i k^2 + U_i'')] f_i + U_i f_{i+1} = c[f_{i-1} - (2 + k^2\Delta y^2) f_i + f_{i+1}] \tag{10.6.24}$$

where $\Delta y = (A + B)/(N + 1)$, and we have designated $f_i = f(y_i)$. To satisfy the no-penetration condition, we require $f_0 = 0$ and $f_{N+1} = 0$. Finally, we collect the values of $f_i, i = 1, \ldots, N$, into the N-dimensional vector $\boldsymbol{f} = (f_1, f_2, \ldots, f_{N-1}, f_N)$ and form the linear system of equations

$$\boldsymbol{A f} = c\boldsymbol{B f} \tag{10.6.25}$$

where $\boldsymbol{A}$ and $\boldsymbol{B}$ are two $N \times N$ tridiagonal matrices. The elements of $\boldsymbol{A}$ and $\boldsymbol{B}$ may readily be inferred from the finite-difference equation (10.6.24). Two points are worth noting: The structure of the matrices depends on the finite-difference method chosen to approximate the derivatives; and the matrix $\boldsymbol{B}$ is independent of the particular shape of the velocity profile $U(y)$.

Thus we have reduced the problem of computing the eigenvalues of Rayleigh's equation, c, to the solving a generalized algebraic eigenvalue problem expressed by equation (10.6.25). Solving the algebraic problem produces N complex eigenvalues, the most important of which is the eigenvalue with the maximum growth rate $\sigma_I = \text{Im}(kc)$. As N is increased, the computed σ_I tends to a limit corresponding to the continuous problem expressed by equation (10.6.23).

One way to solve the generalized eigenvalue problem is to rewrite equation (10.6.25) in the form of a homogeneous equation $\boldsymbol{Ef} = \boldsymbol{0}$. The structure of the matrix $\boldsymbol{E}$ follows readily by recasting equation (10.6.24) into the form

$$f_{i-1} - \left[2 + \Delta y^2 \left(k^2 + \frac{U_i''}{U_I - c}\right)\right] f_i + f_{i+1} = 0 \tag{10.6.26}$$

where a prime designates a derivative with respect to y. The eigenvalues c are the roots of the nonlinear algebraic equation $\text{Det}(\boldsymbol{E}) = 0$. Having specified the value of k, we may now compute the eigenvalues according to the following steps:

1. Guess a complex value for c.
2. Compute $\text{Det}(\boldsymbol{E})$.
3. Correct the value of c to make both the real and imaginary parts of $\text{Det}(\boldsymbol{E})$ vanish.

The correction in step 3 may be done using, for example, Newton's method, setting

$$c^{New} = c^{Old} - \frac{\text{Det}[\boldsymbol{E}(c^{Old})]}{\left(\dfrac{d\text{Det}[\boldsymbol{E}(c)]}{dc}\right)_{c=c^{Old}}} \tag{10.6.27}$$

where $\text{Det}(\boldsymbol{E})$ is computed using the efficient method discussed in Section 2.4.

Since $\text{Det}[\boldsymbol{E}(c)]$ is an analytic function of the complex variable c, the derivative $d\text{Det}[\boldsymbol{E}(c)]/dc$ may be approximated using a finite-difference method, setting, for example,

$$\frac{d\text{Det}[\boldsymbol{E}(c)]}{dc} \cong \frac{\text{Det}[\boldsymbol{E}(c+\varepsilon)] - \text{Det}[\boldsymbol{E}(c)]}{\varepsilon} \tag{10.6.28}$$

where ε is a complex number with a sufficiently small magnitude.

Results of computations performed using this method for a shear flow with velocity profile given by $U(y) = U_0 \tanh(y/b)$, where U_0 and b are two constants, are shown in Figure 10.6.1. The depicted family of curves correspond to a sequence of channel widths with $A = B$. The dashed lines show the dimensionless imaginary part of the *phase velocity* c_I/U_0, and the solid lines show the scaled *growth rate* $4b\sigma_I/U_0$ of unstable perturbations. It is evident that the presence of the walls reduces the growth rate of the perturbations.

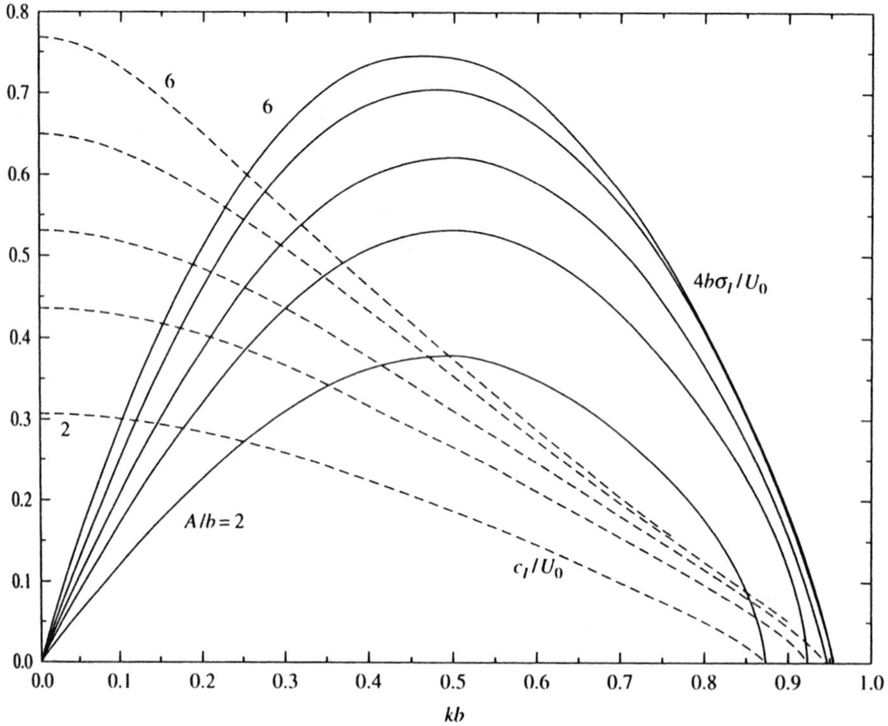

Figure 10.6.1 Phase velocity (dashed lines) and growth rate (solid lines) of perturbations, for flow in a channel with velocity profile $U(y) = U_0 \tan\text{h}(y/b)$, where U_0 and b are two constants, for several values of the channel semiwidth $A/b = 2.0, 2.5, 3.0, 4.0, 6.0$. From C. Pozrikidis, 1997, *Introduction to Theoretical and Computational Fluid Dynamics*, Oxford University Press.

Other methods of computing the eigenvalues are reviewed by Pozrikidis (1997, Chapter 7).

PROBLEM

10.6.4 ***Linear stability of the Bickley jet.***

Construct the counterpart of Figure 10.6.1 for a flow with velocity profile $U(y) = U_0\text{sech}^2(y/b)$, where U_0 and b are two constants.

10.7 Free-boundary Problems

In closing this chapter, we turn our attention to a slightly more subtle class of problems, described as free-boundary or moving-front problems.

The statement of a free-boundary problem includes a set of N scalar initial or, more generally, boundary conditions for the N unknown functions involved in the system of differential equations $d\mathbf{x}/dt = \mathbf{f}(\mathbf{x}, t)$, as discussed in Section 10.1. But the phase-space travel time T is *not* specified; in its place, we have an additional boundary condition or scalar constraint that can be written in the symbolic form

$$G(\boldsymbol{x}(\tau), \tau) = 0 \tag{10.7.1}$$

The expression on the left-hand side may involve the values of the solution at the initial instant $\tau = 0$, at the final instant $\tau = T$, or at any or all intermediate instants $0 \le \tau \le T$.

Free-boundary problems arise in physical applications involving evolving interfaces and advancing fronts. An example is the moving interface between a gas and a liquid in the process of condensation; another example is the spreading of a liquid droplet over a solid wall. The qualifier *free* indicates that the location of the boundary is *a priori* unknown and must be computed as part of the solution along with the rest of the unknowns. In the second of the aforementioned examples, the portion of the wall occupied by the drop must be computed simultaneously with the drop surface.

Rescaling

When equation (10.7.1) takes the form of a boundary condition, a proper rescaling of the independent variable t reduces the free-boundary problem to a standard boundary-value problem. We simply introduce the new time-like variable $\tau = t/T$, where $0 \le \tau \le 1$, substitute $t = \tau T$ into the system of differential equations whose right-hand side now involves T, relabel $T = x_{N+1}$, and enhance the original system of N equations with the $N+1$ equation $dx_{N+1}/d\tau = 0$.

When equation (10.7.1) has a more general form than that of a boundary condition—for example, it expresses a global constraint—a more creative approach may be required; an example follows.

Mass Transport from a Dissolving Particle

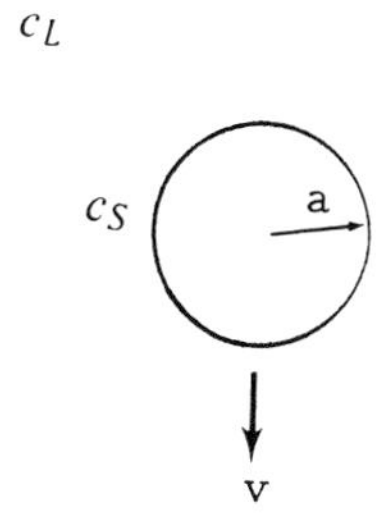

Consider mass transport of a chemical species from a dissolving spherical particle settling down a stationary column of a liquid of length L. The concentration of the dissolving species at the particle surface is constant, equal to c_S, whereas the concentration at the bulk of the liquid has the constant value c_L, where $c_L < c_S$. Because of the dissolution, the radius of the particle decreases in time, and we write $a = a(t)$. We want to find the size of the largest particle that will dissolve by the end of its travel through the column.

Proceeding with the mathematical modeling, we note that the vertical velocity of the particle, v, changes according to Newton's second law. Setting the rate of change of momentum of the particle equal to the sum of the gravitational force, the buoyancy force, and the force due to the fluid expressed by Stokes's law $F = 6\pi\mu a v$, where μ is the viscosity of the fluid, we find

$$\frac{dv}{dt} = -\frac{9\mu}{2a^2\rho_P} v + \frac{\rho_P - \rho_L}{\rho_P} g \tag{10.7.2}$$

where ρ_P, and ρ_L are, respectively, the density of the particle and the density of the fluid, and g is the magnitude of the acceleration of gravity (e.g., Pozrikidis 1997).

The rate of mass transfer from the particle, Q, can be expressed in the form

$$Q = Sh(c_S - c_L) \tag{10.7.3}$$

where S is the instantaneous particle surface, $S = 4\pi a^2$, and h is a *mass-transfer coefficient* whose value depends on the instantaneous value of the dimensionless Péclet number $Pe = av/D$; D is the dissolving species diffusivity.

Setting the left-hand side of equation (10.7.3) equal to the rate of change of particle mass, we obtain

$$\rho_P \frac{d}{dt}(\tfrac{4}{3}\pi a^3) = 4\pi a^2 h(c_S - c_L) \tag{10.7.4}$$

which may be rearranged to give

$$\frac{da}{dt} = \frac{h}{\rho_P}(c_S - c_L) \tag{10.7.5}$$

Equations (10.7.2) and (10.7.5) provide us with a system of ordinary differential equations for v and a, with initial condition

$$v(t = 0) = 0 \tag{10.7.6}$$

expressing the initial state of rest, and final condition

$$a(t = T) = 0 \tag{10.7.7}$$

The travel time T is *a priori* unknown. But the requirement that the particle travels a total distance equal to L yields the constraint

$$L = \int_0^T v\,dt \tag{10.7.8}$$

which is a specific form of equation (10.7.1).

The problem may be solved using the shooting method described in Section 10.2. The only necessary modification is that, simultaneously with solving the system of equations (10.7.2) and (10.7.5), we must also integrate the equation

$$\frac{dx}{dt} = v \tag{10.7.9}$$

with the initial condition $x(t = 0) = 0$, where x is the position of the particle down the column. The integration is stopped when $x = L$, whereupon $t = T$.

PROBLEM

10.7.1 *Reduction to a boundary-value problem.*

Recast the problem of the dissolving particle described in the text into a standard boundary-value problem with a fixed travel time involving the four unknowns a, v, x, and T.

References

Atkinson, K. E., 1989, *An Introduction to Numerical Analysis.* Wiley.
Ascher, U. M., 1986, Collocation for two-point boundary-value problems revisited. *SIAM J. Numer. Anal.* **23**, 596–609.

ASCHER, U. M., MATTHEIJ, M. M., and RUSSELL, R. D., 1995, *Numerical Solution of Boundary Value Problems for Ordinary Differential Equations.* SIAM, Philadephia.

DE BOOR, C., and SWARTZ, B., 1973, Collocation and Gaussian points. *SIAM J. Numer. Anal.* **10**, 582–606.

CHUNG, T. J., 1978, *Finite Element Analysis in Fluid Dynamics.* McGraw-Hill.

DEUFLHARD, P., 1979, Nonlinear equation solvers in boundary-value problem codes. In *Codes for Boundary-Value Problems in Ordinary Differential Equations.* B. Childs et al. (eds.), Lecture Notes in Computer Science **76**. Springer-Verlag.

FINLAYSON, B., 1972, *The Method of Weighted Residuals and Variational Principles.* Academic Press.

FORNBERG, B., 1996, *A Practical Guide to Pseudospectral Methods.* Cambridge University Press.

FOX, L., and PARKER, I. B., 1968, *Chebyshev Polynomials in Numerical Analysis.* Oxford University Press.

FOX, L., 1980, Numerical methods for boundary-value problems. In *Computational Techniques for Ordinary Differential Equations,* I. Gladwell and D. Sayers, (eds.). Academic Press.

GARDNER, D. R., TROGDON, S. A., and DOUGLASS, R. W., 1989, A modified Tau spectral method that eliminates spurious eigenvalues. *J. Comput. Phys.* **80**, 137–67.

GOTTLIEB, D., and ORSZAG, S. A., 1977, *Numerical Analysis of Spectral Methods: Theory and Applications.* SIAM, Philadelphia.

HILDEBRAND, F. B., 1974, *Advanced Calculus for Applications.* Prentice-Hall.

JAIN, M. K., 1984, *Numerical Solution of Differential Equations.* Halsted Press.

KELLER, H. B., 1968, *Numerical Methods for Two-Point Boundary-Value Problems.* Blaisdell.

KELLER, H. B., 1976, *Numerical Solution of Two-Point Boundary-Value Problems. SIAM Regional Conf. Ser. Appl. Math.* **24**, Philadelphia.

LANCZOS, C., 1956, *Applied Analysis.* Reprinted by Dover, 1988.

LAPIDUS, L., and PINDER, G. F., 1982, *Numerical Solution of Partial-Differential Equations in Science and Engineering.* Wiley.

POZRIKIDIS, C., 1996, *Introduction to Theoretical and Computational Fluid Dynamics.* Oxford University Press.

STOER, J., and BULIRSCH, R., 1980, *Introduction to Numerical Analysis.* Springer-Verlag.

VILLADSEN, J. V., and STEWART, W. E., 1967, Solution of boundary-value problems by orthogonal collocation. *Chem. Eng. Sci.* **22**, 1483–501.

Chapter 11

Finite-difference Methods for Partial Differential Equations

11.1 Introduction and Procedures

Finite-difference methods provide us with powerful tools for generating numerical solutions to the partial differential equations of engineering and mathematical physics in a broad range of applications. Evidence for their efficiency was presented in Chapters 3, 5, and 10, in the context of ordinary differential equations. Rather than developing these methods from a general and abstract perspective, we prefer to illustrate their implementation with reference to several prototypical but pervasive classes of equations: The *unsteady diffusion equation, Laplace's equation,* their parental *convection–diffusion equation*—sometimes called the *advection–diffusion* equation—and the *Poisson equation.*

We begin by presenting these equations and then proceed to outline the basic principles involved in their finite-difference formulation. Specific implementations will be the subject of later sections.

Convection–Diffusion Equation

We seek to compute a scalar function of position $\boldsymbol{x} = (x, y, z)$ and time t, $f(\boldsymbol{x}, t)$, that satisfies the scalar convection–diffusion equation

$$\frac{\partial f}{\partial t} + U_x \frac{\partial f}{\partial x} + U_y \frac{\partial f}{\partial y} + U_z \frac{\partial f}{\partial z} = \kappa \left(\frac{\partial^2 f}{\partial x^2} + \frac{\partial^2 f}{\partial y^2} + \frac{\partial^2 f}{\partial z^2} \right) \tag{11.1.1}$$

where:

- The convection velocity vector $\boldsymbol{U} = (U_x, U_y, U_z)$ is a known function of position $\boldsymbol{x}$ and time t in two ways: Explicitly, and implicitly through its dependence on f. Thus, in general, the three scalar components of $\boldsymbol{U}$ are functions of $\boldsymbol{x}$, t, and f.
- When $\boldsymbol{U}$ is constant or an explicit function of $\boldsymbol{x}$ and t, we obtain a linear convection–diffusion problem, whereas when $\boldsymbol{U}$ depends on f we obtain a nonlinear convection–diffusion problem.
- κ is a nonnegative scalar constant called the *diffusivity*.

For simplicity, the diffusivity will be assumed to be constant in time and uniform throughout the domain of solution. Revoking this assumption requires only straightforward modifications in the numerical formulation.

To simplify the notation, we introduce the *del* or *gradient* operator and the *Laplacian* operator, respectively defined as

$$\nabla \equiv \left(\frac{\partial}{\partial x}, \frac{\partial}{\partial y}, \frac{\partial}{\partial z} \right), \qquad \nabla^2 \equiv \frac{\partial^2}{\partial x^2} + \frac{\partial^2}{\partial y^2} + \frac{\partial^2}{\partial z^2} \tag{11.1.2}$$

and rewrite equation (11.1.1) in the compact form

$$\frac{\partial f}{\partial t} + \boldsymbol{U} \cdot \nabla f = \kappa \nabla^2 f \tag{11.1.3}$$

where

$$\boldsymbol{U} \cdot \nabla f \equiv U_x \frac{\partial f}{\partial x} + U_y \frac{\partial f}{\partial y} + U_z \frac{\partial f}{\partial z} \tag{11.1.4}$$

Special forms of equation (11.1.3) arise in a variety of applications and in many different contexts: From heat and mass transport, to momentum transport in fluid mechanics, to the statistical mechanics of random systems.

Unsteady diffusion equation

When all three components of $\boldsymbol{U}$ are equal to zero, the convection–diffusion equation (11.1.1) simplifies to the unsteady diffusion equation

$$\frac{\partial f}{\partial t} = \kappa \left(\frac{\partial^2 f}{\partial x^2} + \frac{\partial^2 f}{\partial y^2} + \frac{\partial^2 f}{\partial z^2} \right) \tag{11.1.5}$$

which is a prototypical *parabolic* differential equation in the time–space domain.

Laplace's equation

If, in addition, f is independent of t, that is, the physical system described by the differential equation is at steady state, the unsteady diffusion equation reduces to Laplace's equation

$$\frac{\partial^2 f}{\partial x^2} + \frac{\partial^2 f}{\partial y^2} + \frac{\partial^2 f}{\partial z^2} = 0 \tag{11.1.6}$$

which is a prototypical *elliptic* differential equation in the spatial domain.

Convection equation

When κ is equal to zero, equation (11.1.1) reduces to the convection equation

$$\frac{\partial f}{\partial t} + U_x \frac{\partial f}{\partial x} + U_y \frac{\partial f}{\partial x} + U_z \frac{\partial f}{\partial z} = 0 \tag{11.1.7}$$

which is a prototypical *hyperbolic* differential equation in the time–space domain.

We shall see later in this chapter that the seemingly academic classification of the last three equations into the parabolic, elliptic, and hyperbolic categories has important consequences on the effectiveness of the various finite-difference methods.

Vectorial Convection–Diffusion Equation

In the vector version of the convection–diffusion problem, we want to compute a vector function $\boldsymbol{f}$, where the individual scalar functions comprising $\boldsymbol{f}$ depend on position $\boldsymbol{x} = (x, y, z)$ and time t, that satisfies the convection–diffusion equation

$$\frac{\partial \boldsymbol{f}}{\partial t} + \boldsymbol{U} \cdot \nabla \boldsymbol{f} = \kappa \nabla^2 \boldsymbol{f} \tag{11.1.8}$$

In general, the components of $\boldsymbol{U}$ depend on $\boldsymbol{x}$ and t, as well as on the components of $\boldsymbol{f}$.

When the convection velocity field is *solenoidal*, that is,

$$\nabla \cdot \boldsymbol{U} \equiv \frac{\partial U_x}{\partial x} + \frac{\partial U_y}{\partial y} + \frac{\partial U_z}{\partial z} = 0 \tag{11.1.9}$$

equation (11.1.8) can be recast into the *conservative form*

$$\frac{\partial f_i}{\partial t} + \frac{\partial (U_j f_i)}{\partial x_j} = \kappa \nabla^2 f_i \tag{11.1.10}$$

where summation is implied over the repeated index j on the left-hand side. In contrast, the primary equation (11.1.8) expresses the *nonconservative form.*

This qualifier *conservative* indicates the possible telescoping of discrete values of the matrix $\boldsymbol{Uf}$ at the grid points. Sometimes, the telescoping conserves certain invariants of the solution. The conservative form is usually preferred over the nonconservative form for reasons of enhanced accuracy and numerical stability.

Initial and Boundary Conditions

The solution of the generally vectorial convection–diffusion equation must be found subject to a specified initial condition

$$\boldsymbol{f}(\boldsymbol{x}, t = 0) \equiv \boldsymbol{F}(\boldsymbol{x}) \tag{11.1.11}$$

where $\boldsymbol{F}$ is a known function. Requiring a proper number of boundary conditions completes the statement of the problem.

When the diffusivity κ is nonzero, the convection–diffusion equation is a second-order partial differential equation. In this case, we must supply a number of boundary conditions, involving the function $\boldsymbol{f}$ or its first spatial derivatives, that are equal to the dimension of $\boldsymbol{f}$. Thus if $\boldsymbol{f}$ is a three-dimensional vector, we require three scalar conditions, one for each component of $\boldsymbol{f}$ over each boundary. If the diffusivity is zero, the boundary conditions must be chosen more carefully according to the geometry of the domain of solution (e.g., Strikwerda 1989).

Heat conduction through a rod

As an example, we consider a cylindrical aluminum rod of length $L = 30$ cm and radius $a = 1$ cm supporting a pipe that carries a hot fluid. One end of the rod is attached to the pipe, and the other to the wall. We want to compute the temperature distribution along the rod and the consequent heat loss.

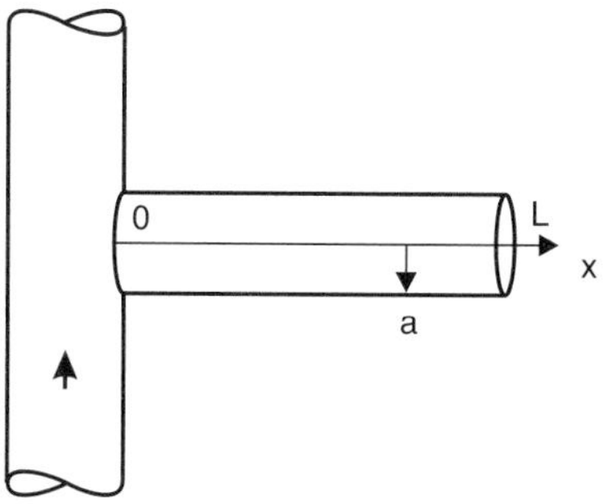

To proceed with the mathematical formulation, we introduce the x axis along the rod, with origin at the pipe. The temperature of the rod T is a function of x and time t. Conservation of thermal energy requires that the temperature field evolve according to the unsteady heat-conduction equation

$$\frac{\partial T}{\partial t} = \kappa \frac{\partial^2 T}{\partial x^2} \tag{11.1.12}$$

where $\kappa = k/(\rho c_p)$ is the *thermal diffusivity*, k is the *thermal conductivity*, ρ is the density of the rod, and c_p is its specific heat at constant pressure.

In one particular case, the temperature of the rod is initially uniform, equal to the wall temperature $T_W = 25$ °C. At the origin of time, hot fluid starts flowing through the pipe, and the temperature of the pipe starts rising, as shown in the diagram. Measurements by means of a thermocouple have shown that the temperature at the hot end of the rod, at $x = 0$, increases in time according to the equation

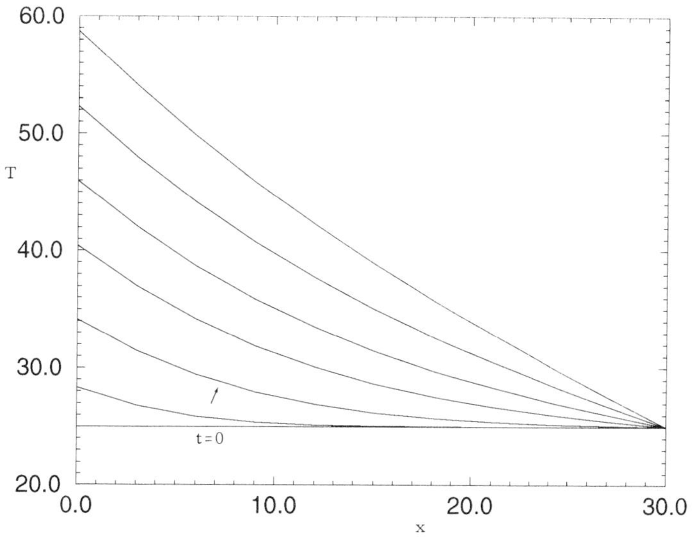

$$T(x = 0) = 25 \exp(-0.003t) + 80[1 - \exp(-0.002t)] \tag{11.1.13}$$

where all temperatures are measured in degrees Celsius, and time is measured in seconds. The temperature at the wall-end of the rod, at $x = L$, remains constant at all times, equal to 25 °C. The statement of the problem is now complete, and the solution must be found using a numerical method.

The heat loss from the pipe is given by

$$q_{lost} = -k \left(\frac{\partial T}{\partial x} \right)_{x=0} A \tag{11.1.14}$$

where A is the cross-sectional area of the rod. To compute the right-hand side, we must first produce the temperature distributions subject to the aforementioned initial and boundary conditions.

PROBLEMS

11.1.1 ***Heat conduction through a radiating rod.***

Consider heat conduction through a rod of length L. One end of the rod is insulated—there is no heat loss through it—and the other end is maintained at a constant temperature T_L. Heat radiates from the rod to the ambient medium, so that the distribution of the absolute temperature T is described by the equation

$$\frac{\partial T}{\partial t} = \kappa \frac{\partial^2 T}{\partial x^2} - \beta T^4 \tag{11.1.15}$$

where κ is the thermal diffusivity, and β is a positive constant.
(a) What are the units of κ and β?
(b) Do you expect that the temperature field will tend to a steady state at long times? If yes, can you compute the asymptotic distribution in analytical form?

11.1.2 ***Flux-conservative form.***

An evolution equation for a certain vector function $\boldsymbol{f}$ of position x and time t is said to be in a flux-conservative form when it can be expressed as

$$\frac{\partial \boldsymbol{f}}{\partial x} + \frac{\partial \boldsymbol{Q}(x, t, \boldsymbol{f})}{\partial x} = \boldsymbol{0} \tag{11.1.16}$$

where $\boldsymbol{Q}$ is a function of its arguments. Express the one-dimensional wave equation

$$\frac{\partial^2 u}{\partial t^2} = c^2 \frac{\partial^2 u}{\partial x^2} \tag{11.1.17}$$

where c is the constant wave speed, in the flux-conservative form.

Poisson's Equation

Next, we look for a scalar function of position $f(\boldsymbol{x})$, where $\boldsymbol{x} = (x, y, z)$, that satisfies Poisson's equation

$$\nabla^2 f = g(\boldsymbol{x}) \tag{11.1.18}$$

within a specified one-dimensional, two-dimensional, or three-dimensional domain, where g is a known forcing function; the Laplacian operator ∇^2 was defined in equations (11.1.2). When $g = 0$, Poisson's equation reduces to Laplace's equation $\nabla^2 f = 0$.

To complete the statement of the problem, we must specify one scalar boundary condition for f over the boundary of the domain of solution. In practice, this boundary condition may come in three flavors:

- The *Dirichlet* boundary condition specifies the boundary value of f.

- The *Neumann* boundary condition specifies the value of the normal derivative

$$\boldsymbol{n} \cdot \nabla f \equiv n_x \frac{\partial f}{\partial x} + n_y \frac{\partial f}{\partial y} + n_z \frac{\partial f}{\partial z} \tag{11.1.19}$$

where $\boldsymbol{n}$ is the unit vector normal to the boundary, pointing either inward or outward from the domain of solution.

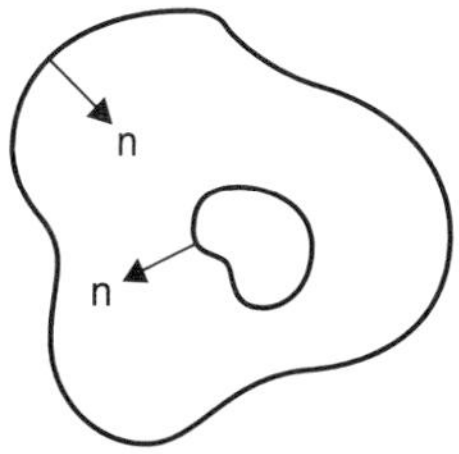

- The *mixed* or *Robin* boundary condition, provides us with the value of the linear combination

$$\alpha f + \beta \boldsymbol{n} \cdot \nabla f \equiv \alpha f + \beta \left(n_x \frac{\partial f}{\partial x} + n_y \frac{\partial f}{\partial y} + n_z \frac{\partial f}{\partial z} \right) \tag{11.1.20}$$

where α and β are two scalar constants.

A problem whose mathematical formulation led us to a Poisson equation with different types of boundary conditions was discussed in Section 3.1.

Finite-difference Procedures

The central goal of a finite-difference method is to generate the values of the unknown scalar function f or vector function $\boldsymbol{f}$ at the nodes of a *grid* that covers the domain of solution, at a sequence of discrete time levels that are separated by the constant or variable *time step* Δt. If the unknown function does not depend on t, the computation is done at one time corresponding to the steady state.

The finite-difference grid may be defined in Cartesian coordinates (x, y, z) or other orthogonal or nonorthogonal *curvilinear* coordinates (ξ, η, ζ), as illustrated in Figure 11.1.1. The choice of coordinates is guided by the geometry of the domain of solution and is made with the goal of facilitating the implementation of the boundary conditions.

For example, when the domain of solution is the exterior or interior of a sphere, the boundary conditions are naturally described in terms of spherical polar coordinates (r, θ, φ) with origin at the center of the sphere,

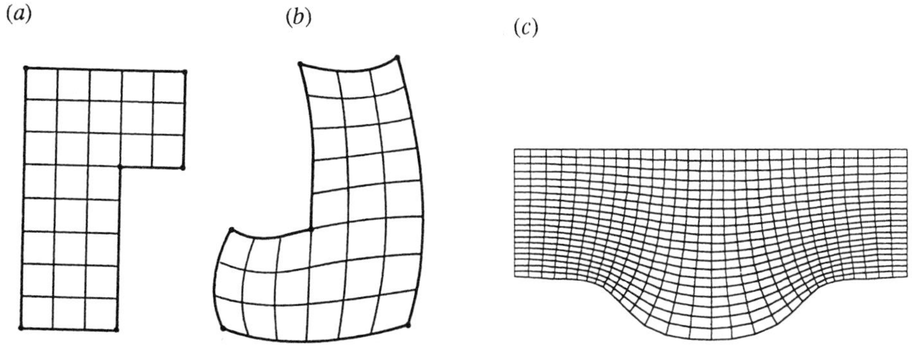

Figure 11.1.1 Finite-difference grids in two dimensions: (*a*) Cartesian, (*b*) non-orthogonal curvilinear, and (*c*) orthogonal curvilinear.

and the governing equation is solved for the spherical polar coordinates of the unknown vector function. The use of orthogonal coordinates is desirable for algebraic simplicity and improved numerical stability.

In Cartesian coordinates, the finite-difference grid is comprised of an array of straight lines that run parallel to the x, y, and z axes, with grid spacings Δx, Δy, and Δz that may vary across the domain of solution in order to allow for enhanced spatial resolution at regions where the solution is expected to exhibit sharp variations. The grid becomes finer as the grid spacings become smaller.

Once the discrete finite-difference solution has been computed, the values of f between grid points and time levels are obtained by applying standard methods of function interpolation, extrapolation, or approximation discussed in Chapters 6 and 8.

Finite-difference Discretizations

The distinguishing feature of a finite-difference method is the approximation of the temporal and spatial partial derivatives in the governing equation with finite differences relating the values of the unknown functions at a set of neighboring grid points at various time levels. This approximation replaces the partial differential equation, denoted in short by PDE, with a finite-difference equation, denoted in short by FDE. The process of replacing the partial derivatives with algebraic differences is called the *finite-difference approximation* or *discretization* of the differential equation.

Applying the finite-difference equation sequentially at the internal nodes of the finite-difference grid yields a system of linear or nonlinear algebraic equations that relate the values of the unknown function at the nodes. We have seen in previous sections that, in certain cases, it is beneficial to extend the domain of solution beyond the natural boundaries of the physical problem, and to apply the finite-difference equation at boundary nodes with the objective of improving the accuracy of the difference equations expressing the boundary conditions.

Consistency

The accuracy of a numerical computation based on a finite-difference method depends on the size of the grid spacings and time step, which are the control parameters of the numerical method. If, in the limit as both the grid spacings and time step are reduced simultaneously but in an arbitrary manner that allows them to have different orders of magnitude, the finite-difference equation approximates the partial differential equation with increasing accuracy, then the finite-difference method is consistent.

The consistency of a finite-difference equation that arises by applying well-established finite-difference formulas to approximate the temporal and spatial derivatives of the partial differential equation is guaranteed. The consistency of a finite-difference equation that arises by making heuristic or ad hoc modifications to well-established finite-difference approximations, however, must be confirmed, as will be discussed in subsequent sections.

The consistency of a finite-difference method can be assessed by pretending that all variables in the finite-difference equation are continuous functions of space and time, and then expanding them in a Taylor series around a selected grid point, at a certain time instant. In this manner, the finite-difference equation yields a new differential equation, called the *modified differential equation*, denoted in short by MDE (Warming and Hyett 1974). If, in the limit as the size of the time step and grid spacings are reduced simultaneously but independently, the MDE reduces to the original PDE, then the finite-difference method is consistent. Phrased differently, if the finite-difference method is consistent, the difference between the MDE and the PDE involves terms that are proportional to powers of the grid sizes and time step. The exponents of these powers define the order of the numerical error or the order of the finite-difference method.

We shall see in subsequent sections that certain finite-difference equations emerge by applying the differential equation at a particular grid point, and then replacing it with a combination of values of the unknown function f at a group of neighboring grid points. In these cases, the coefficients that multiply the values of the function are computed by imposing certain restrictions including consistency with the differential equation, and a desired degree of accuracy in the approximation of the partial derivatives.

Stability

Let us assume that the exact solution of the convection–diffusion equation, or some other partial differential equation, subject to an initial condition and a proper number of boundary conditions, does not grow in time but either stays constant or decays at every point. It is not unreasonable to demand that the finite-difference solution reproduces this behavior, that is, it provides us with a bounded solution that is free of artificial oscillations. If it does, the finite-difference method is stable, otherwise it is unstable. When the exact solution of the differential equation grows in time, the finite-difference method is considered stable if it provides us with a numerical solution that grows at a rate that is equal to, or lower than, that of the exact solution. An alternative definition of stability concerns the behavior of the numerical solution at a fixed time, in the limit as the spatial and temporal steps become increasingly smaller. The issue is discussed further in monographs on finite-difference methods cited at the end of this section.

The stability of relatively simple finite-difference methods for linear partial differential equations may be assessed by several methods including the *von Neumann stability method*, the *projection matrix method*, and the *discrete-perturbation method*. The first method is easiest to carry out but does not generally account for the effect of the boundary conditions. The stability of involved finite-difference methods is difficult to investigate and, in practice, it is often warranted by the absence of noticeable spatial or temporal oscillations in the results of a computation.

The stability of finite-difference methods for nonlinear differential equations is typically examined by linearizing the differential equation about a particular grid point, and then studying the performance of the finite-difference method with reference to the linearized equation, as discussed in Chapter 9. Experience has shown that the local stability criteria obtained in this manner provide us with a reliable characterization of the overall performance of the numerical method.

Convergence

Stability imposes a modest requirement on the performance of a numerical method. Before we can claim that the numerical results bear any degree of physical relevance to the physical problem described by the partial differential equation, we must ensure that, as the size of the grid and time step are made finer, the numerical solution converges to the exact solution.

Lax's equivalence theorem guarantees that if a numerical solution of a *linear* partial differential equation obtained using a consistent finite-difference approximation is stable, then in the limit as the grid spacings and time step tend to zero the numerical solution will converge to the exact solution (Lax and Richtmyer 1956, Richtmyer and Morton 1967). Thus *consistency and stability ensure convergence and vice versa.*

The convergence of finite-difference methods for nonlinear differential equations is more difficult to assess, but experience has shown that if the numerical method is consistent and locally stable, then the finite-difference solution will converge to the exact solution as the grid spacings and size of the time step are refined.

Where to Look for More

The subject of finite-difference methods is broad and diverse, and we must necessarily confine our attention to discussing the fundamental principles and procedures, and to presenting a selected class of methods that either illustrate the methodology or find extensive applications in practice.

Extensive discussion can be found in specialized monographs and texts on numerical methods for partial differential equations including those by Richtmyer and Morton (1967), Mitchell (1969), Ames (1977), Ferziger (1981), Lapidus and Pinder (1982), Sod (1985), Strickwerda (1989), and Hoffman (1992).

11.2 One-dimensional Unsteady Diffusion

We begin by presenting finite-difference methods for the one-dimensional unsteady diffusion equation

$$\frac{\partial f}{\partial t} = \kappa \frac{\partial^2 f}{\partial x^2} \tag{11.2.1}$$

which is a simplified form of the multidimensional unsteady diffusion equation (11.1.5). The solution is to be found subject to the initial condition $f(x, 0) = F(x)$, where $F(x)$ is a known function. The diffusivity κ is assumed to be a positive constant.

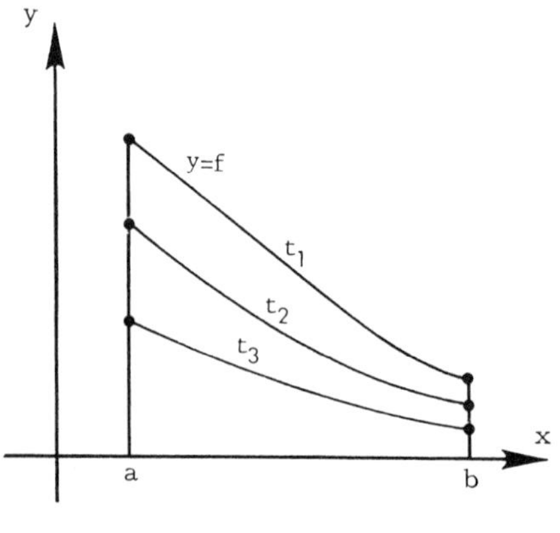

For simplicity, we assume that the domain of solution extends over the whole x axis and require the homogeneous far-field conditions $f(x = \pm\infty, t) = 0$. If the domain of solution were bounded, for example $a < x < b$ as shown in the diagram, we would have to require one boundary condition for f, $\partial f/\partial x$, or their linear combination, at both ends $x = a$ and $x = b$, or boundary conditions for both f, $\partial f/\partial x$, or their linear combination, at one end.

Equation (11.2.1) is a *parabolic* partial differential equation in time, with a well-known analytical solution given by

$$f(x, t) = \frac{1}{\sqrt{4\kappa\pi t}} \int_{-\infty}^{\infty} F(x + u) \exp\left(-\frac{u^2}{4\kappa t}\right) du \tag{11.2.2}$$

(e.g., Carslaw and Jaeger 1959, p. 53). Our present objective is to generate the discrete version of this solution using a finite-difference method.

Discretization

As a first step toward developing a finite-difference method, we assume that f remains infinitesimal outside a certain computational domain $a < x < b$, during a certain initial period of evolution, and introduce a two-dimensional grid that covers the semi-infinite strip $a < x < b, 0 < t < \infty$ in the time–space plane, as depicted in Figure 11.2.1. Our objective is to compute the values of the function f_i^n, at the grid points $x_i, i = 1, \ldots, K+1$, at a sequence of successive time levels t^n, beginning from the initial time level $t^0 = 0$, and subject to the boundary conditions $f_1^n = 0$ and $f_{K+1}^n = 0$.

Explicit FTCS Method

Applying equation (11.2.1) at the grid point x_i at the time instant t^n, and approximating the time derivative with a first-order forward difference and the space derivative with a second-order centered difference, we derive the FTCS finite-difference equation

$$\frac{f_i^{n+1} - f_i^n}{\Delta t} + O(\Delta t) = \kappa \frac{f_{i+1}^n - 2f_i^n + f_{i-1}^n}{\Delta x^2} + O(\Delta x^2) \qquad (11.2.3)$$

which is first-order accurate in time and second-order accurate in space. The FTCS differentiation stencil is indicated in Figure 11.2.1 with hollow circles. Solving equation (11.2.3) for f_i^{n+1}, we obtain

$$f_i^{n+1} = \alpha f_{i-1}^n + (1 - 2\alpha) f_i^n + \alpha f_{i+1}^n \qquad (11.2.4)$$

where

$$\alpha = \frac{\kappa \Delta t}{\Delta x^2} \qquad (11.2.5)$$

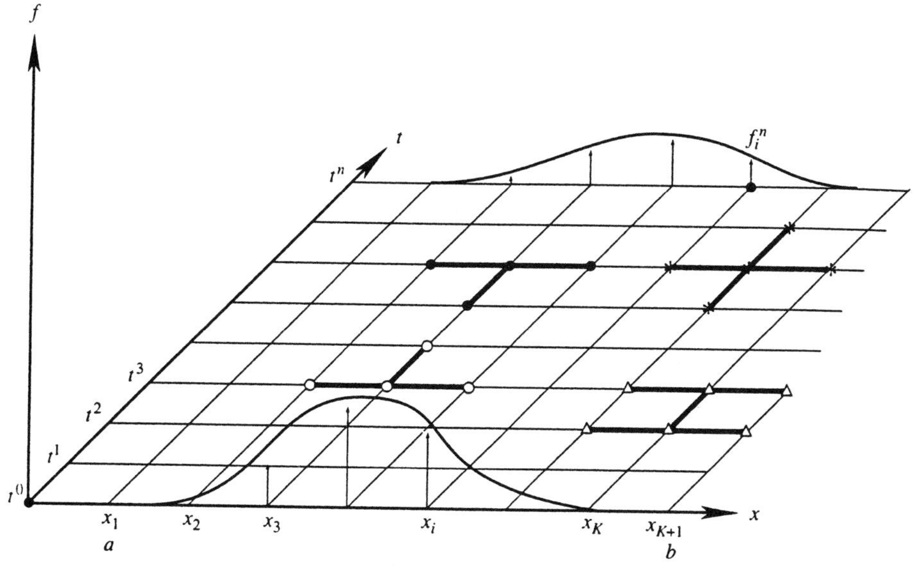

Figure 11.2.1 Discretization of the space–time domain for solving the one-dimensional unsteady diffusion equation. The initial distribution diffuses and tends to occupy an increasingly wider interval along the x axis. The hollow circles, asterisks, solid circles, and triangles indicate, respectively, the finite-difference stencils for the FTCS, CTCS, BTCS, and Crank–Nicolson method. From *Introduction to Theoretical and Computational Fluid Dynamics* by C. Pozrikidis. Copyright © 1997 Oxford University Press. Used by permission of Oxford University Press, Inc.

is a positive dimensionless constant called the *diffusion number.* Equations (11.2.3) and (11.2.4) can be applied at the *internal* grid points $i = 2, \ldots, K$ but not at the boundary points $i = 1, K + 1$.

Consistency and accuracy

To confirm the consistency of equation (11.2.4), we pretend that all discrete variables are continuous functions of space and time, expand them in a Taylor series about the point (x_i, t^n), and simplify to obtain the associated *modified differential equation*

$$f_t + \tfrac{1}{2} f_{tt} \Delta t + O(\Delta t^2) = \kappa f_{xx} + \tfrac{1}{12} \kappa f_{xxxx} \Delta x^2 + O(\Delta x^4) \tag{11.2.6}$$

where subscripts denote partial derivatives with respect to corresponding variables, and all variables are evaluated at x_i and t^n. Since, in the limit as Δt and Δx tend to zero, equation (11.2.6) reduces to equation (11.2.1), the FTCS difference equation is consistent with the original differential equation.

Differentiating equation (11.2.1) once with respect to t and twice with respect to x, and combining the resulting equations, we derive the fourth-order equation $f_{tt} = \kappa^2 f_{xxxx}$. Eliminating f_{xxxx} on the right-hand side of equation (11.2.6) in favor of f_{tt}, and combining the resulting expression with the second term on the left-hand side shows that, when $\alpha = \frac{1}{6}$, the accuracy of the FTCS method becomes of second order in time.

Classification and algorithm

Equation (11.2.4) provides us with a straightforward algorithm for computing the value of f at a grid point at the time level $n + 1$ in terms of the values of f at three grid points at the previous time level n. Since the method does not require inverting a matrix or solving a system of algebraic equations, it is classified as *explicit.*

In summary, the FTCS discretization provides us with an *explicit two-level method* with accuracy that is first-order in time and second-order in space.

Successive mappings and numerical stability

$$f^{n+1} = \begin{bmatrix} 1\text{-}2\alpha & \alpha & 0 & \cdots & 0 & 0 \\ \alpha & 1\text{-}2\alpha & \alpha & 0 & \cdots & 0 \\ 0 & \alpha & 1\text{-}2\alpha & \alpha & 0 & \cdots \\ 0 & \ddots & \ddots & \ddots & & 0 \\ \cdots & 0 & \alpha & 1\text{-}2\alpha & \alpha & 0 \\ 0 & \cdots & 0 & \alpha & 1\text{-}2\alpha & \alpha \\ 0 & 0 & \cdots & 0 & \alpha & 1\text{-}2\alpha \end{bmatrix} f^n$$

To formalize the action of the FTCS method, we collect the values of f_i^n at the grid points $i = 2, \ldots, K$ into the vector $\boldsymbol{f}^n$, and use equation (11.2.4) in conjunction with the boundary conditions $f_1^n = 0$ and $f_{K+1}^n = 0$, to obtain $\boldsymbol{f}^{n+1} = \boldsymbol{B}\boldsymbol{f}^n$, where $\boldsymbol{B}$ is a $(K-1) \times (K-1)$ tridiagonal matrix with superdiagonal, diagonal, and subdiagonal elements equal to α, $1 - 2\alpha$, and α, respectively. This matrix form shows that the solution vector $\boldsymbol{f}^{n+1}$ derives by premultiplying the vector $\boldsymbol{f}^n$ with the matrix $\boldsymbol{B}$, thereby establishing a relation between *time stepping* and *successive mapping.*

The behavior of the vector $\boldsymbol{f}^n$ with respect to n will depend on the properties of the *projection matrix* $\boldsymbol{B}$ and, in particular, on the spectral radius of $\boldsymbol{B}$ denoted by $\rho(\boldsymbol{B})$ and defined as the maximum of the norm of the eigenvalues of $\boldsymbol{B}$. The theory of matrix calculus discussed in Section 2.11 shows that if $\rho(\boldsymbol{B})$ is equal to unity, less than unity, or larger than unity, the length of $\boldsymbol{f}^n$ will stay roughly constant, decrease, or increase during the successive mappings. Since, according to the exact solution, the amplitude of f diminshes due to diffusion, we tolerate the first behavior, accept the second behavior, but dismiss the third behavior as being numerically unstable.

Using the results of Section 5.3, we find that the eigenvalues of $\boldsymbol{B}$ are given by

$$\lambda_m = 1 - 4\alpha \sin^2\left(\frac{m\pi}{2K}\right) \tag{11.2.7}$$

for $m = 1, \ldots, K-1$. We observe that the spectral radius of $\boldsymbol{B}$ is less than unity for any value of K only when $\alpha < \frac{1}{2}$, and this shows that the FTCS method is only *conditionally stable.*

Effect of the boundary conditions

If boundary conditions other than the homogeneous Dirichlet conditions $f_1^n = 0$ and $f_{K+1}^n = 0$ were specified on one or both ends of the computational domain, the structure of the projection matrix would be different. The performance of the finite-difference method, however, would still be determined by the spectral radius of the corresponding projection matrix.

Consider, for example, the time-independent Neumann boundary condition $\partial f/\partial x = g$ at the left end, at $x = a$, and maintain the homogeneous Dirichlet condition at the right end, at $x = b$. To implement the first boundary condition with second-order accuracy, we extend the domain of solution beyond the physical boundary at $x = a$, introduce the fictitious node $x_0 = x_1 - \Delta x$, and use central differences to obtain $f_2^n - f_0^n = 2g\Delta x$. Having extended the domain, we may apply the differential equation at the first node x_1 and write (11.2.4) with $i = 1$.

To derive the corresponding mapping matrix $\boldsymbol{B}$, we collect the values of f_i^n at the grid points $i = 1, \ldots, K$ into the vector $\boldsymbol{f}^n$ and use equation (11.2.4) along with the aforementioned boundary conditions to obtain $\boldsymbol{f}^{n+1} = \boldsymbol{B}\boldsymbol{f}^n + \boldsymbol{b}$, where $\boldsymbol{B}$ is a $K \times K$ tridiagonal matrix with superdiagonal, diagonal, and subdiagonal elements equal to α, $1 - 2\alpha$, and α, respectively, except that the second entry in the first row is equal to 2α. All entries of the vector $\boldsymbol{b}$ are equal to zero except for the first one which is equal to $-2g\Delta x$.

$$f^{n+1} = \begin{bmatrix} 1-2\alpha & 2\alpha & 0 & \cdots & 0 & 0 \\ \alpha & 1-2\alpha & \alpha & 0 & \cdots & 0 \\ 0 & \alpha & 1-2\alpha & \alpha & 0 & \cdots \\ 0 & \ddots & \ddots & \ddots & & 0 \\ \cdots & 0 & \alpha & 1-2\alpha & \alpha & 0 \\ 0 & \cdots & 0 & \alpha & 1-2\alpha & \alpha \\ 0 & 0 & \cdots & 0 & \alpha & 1-2\alpha \end{bmatrix} f^n + \begin{bmatrix} -2g\,\Delta x \\ 0 \\ 0 \\ \cdot \\ \cdot \\ 0 \\ 0 \end{bmatrix}$$

The presence of the constant vector $\boldsymbol{b}$ does not affect the significance of the projection matrix $\boldsymbol{B}$ regarding the behavior of the solution discussed in a previous subsection. When $\rho(\boldsymbol{B})$ is equal to unity, less than unity, or larger than unity, the length of $\boldsymbol{f}^n$ will stay roughly constant, decrease, or increase during the successive mappings. Unfortunately, we can no longer compute the eigenvalues of $\boldsymbol{B}$ in closed form.

von Neumann stability analysis

The simple structure of the projection matrix associated with the FTCS method, subject to homogeneous Dirichlet boundary conditions at both ends, allowed us to compute its eigenvalues and spectral radius exactly in closed form and thereby assess the stability of the numerical method. Unfortunately, closed-form solutions are not readily available for more advanced finite-difference discretizations and more general types of boundary conditions. An alternative would be to compute the eigenvalues using the numerical methods discussed in Chapter 5, but this is computationally arduous task, to say the least.

Another way to assess the stability of the numerical method is to perform the von Neumann stability analysis of the finite-difference equation neglecting the boundary conditions. The idea is to examine the behavior of the numerical solution subject to a harmonic initial condition of a certain wavelength L. Motivated by the linearity of the governing equation, we separate the temporal from the spatial dependence, writing

$$f_i^n = A^n \exp(Ii\theta) \tag{11.2.8}$$

where $I = \sqrt{-1}$ is the imaginary unit, $\theta = 2\pi\,\Delta x/L$ is the *phase angle*, and A^n is a coefficient whose value depends on the time level n. Substituting equation (11.2.8) into equation (11.2.4) and simplifying, we find

$$\frac{A^{n+1}}{A^n} \equiv G = 1 + 2\alpha(\cos\theta - 1) = 1 - 4\alpha\,\sin^2\frac{\theta}{2} \tag{11.2.9}$$

where G is the *growth factor, gain,* or *amplification factor.*

When $\alpha > \frac{1}{2}$, the magnitude of the right-hand side of equation (11.2.9) is greater than unity for a certain range of values of θ, and the numerical method is unstable. When $\alpha < \frac{1}{2}$, the magnitude of the right-hand side of equation (11.2.9) is less than unity for any value of θ, and the numerical method is stable. These results are consistent with our previous conclusions based on the spectral radius of the projection matrix $\boldsymbol{B}$.

The efficiency of the von Neumann stability analysis is now evident. One limitation of the method is that, in its simple form described above, it does not incorporate the inhomogeneous boundary conditions, which may have a destabilizing effect on the finite-difference method.

Assessment of the FTCS method

Since α is proportional to the temporal step Δt and inversely proportional to the square of the spatial step Δx, the stability constraint $\alpha < \frac{1}{2}$ of the FTCS method requires a time step that is excessively small and may demand a prohibitive amount of computational time. The low-order accuracy of the method, combined with the conditional stability, renders the method less attractive compared to its alternatives.

Explicit CTCS or Leapfrog Method

One way to achieve second-order accuracy in both time and space is to use centered differences for both independent variables. Applying equation (11.2.1) at the point x_i at the time instant t^n, and using centered differences in both time and space, we obtain the CTCS difference equation

$$\frac{f_i^{n+1} - f_i^{n-1}}{2\Delta t} + O(\Delta t^2) = \kappa\frac{f_{i+1}^n - 2f_i^n + f_{i-1}^n}{\Delta x^2} + O(\Delta x^2) \tag{11.2.10}$$

The corresponding finite-difference stencil is illustrated with asterisks in Figure 11.2.1. Rearranging, we derive the three-time-level explicit algorithm

$$f_i^{n+1} = f_i^{n-1} + 2\alpha f_{i-1}^n - 4\alpha f_i^n + 2\alpha f_{i+1}^n \tag{11.2.11}$$

where α was defined in equation (11.2.5). The solution at the first time level $n = 1$ must be computed using a two-level method, such as the FTCS method discussed previously, with a time step that is an integral fraction of Δt and that is small enough so as to prevent the onset of deleterious oscillations.

To examine the stability of the method, we substitute equation (11.2.8) into equation (11.2.11), set $A^{n+1}/A^n \equiv A^n/A^{n-1} \equiv G$, and obtain the quadratic equation for the gain, $G^2 + \beta G - 1 = 0$, where $\beta = 2\alpha(1 - \cos\theta) = 4\alpha \sin^2(\frac{1}{2}\theta)$ is a real nonnegative parameter. The two roots of the quadratic equation are $G = \frac{1}{2}[-\beta \pm (\beta^2 + 4)^{1/2}]$. The magnitude of the root corresponding to the minus sign is higher than unity, and this means that the CTCS method is *unconditionally unstable* and thus of no practical value.

The Du Fort and Frankel Explicit Method

Du Fort and Frankel (1953) proposed a modification of the CTCS discretization with the objective of maintaining the second-order accuracy but improving the numerical stability. Their method proceeds by replacing the middle term in the numerator on the right-hand side of equation (11.2.10) with an average value, yielding

$$\frac{f_i^{n+1} - f_i^{n-1}}{2\Delta t} + O(\Delta t^2) = \kappa \frac{f_{i+1}^n - 2[\frac{1}{2}(f_i^{n+1} + f_i^{n-1})] + f_{i-1}^n}{\Delta x^2} + O(\Delta x^2) \tag{11.2.12}$$

Rearranging, we obtain the three-level explicit algorithm

$$f_i^{n+1} = \frac{1-2\alpha}{1+2\alpha} f_i^{n-1} + \frac{2\alpha}{1+2\alpha}(f_{i-1}^n + f_{i+1}^n) \tag{11.2.13}$$

The computations must be started using a two-level method.

Performing the von Neumann stability analysis, we find that the amplification factor satisfies the quadratic equation

$$(1 + 2\alpha)G^2 - 4G\alpha \cos\theta - 1 + 2\alpha = 0 \tag{11.2.14}$$

Examining the roots, we find that $|G| < 1$ for any value of α, and this ensures that the Du Fort–Frankel method is *unconditionally stable.*

Since, however, equation (11.2.12) has been derived on the basis of an ad hoc modification of the well-founded CTCS discretization, its consistency must be examined by comparing the associated modified differential equation with the original differential equation (11.2.1). To derive the former, we regard all discrete variables in equation (11.2.12) as continuous functions of space and time, expand them in a Taylor series about the point (x_i, t^n), and simplify to obtain

$$f_t = \kappa f_{xx} - \kappa \left(\frac{\Delta t}{\Delta x}\right)^2 f_{tt} \tag{11.2.15}$$

Equation (11.2.15) will be an accurate approximation of equation (11.2.1) only when $\kappa(\Delta t/\Delta x)^2$ is sufficiently small. No matter how small Δt and Δx are, if $\kappa(\Delta t/\Delta x)^2$ has a nonzero value, the Du Fort–Frankel method will produce the solution of a fictitious problem described by equation (11.2.15), instead of the diffusion problem described by equation (11.2.1). Thus the method cannot be said to be consistent in general.

Observing that equation (11.2.15) is a hyperbolic differential equation in time, owing to the presence of the second derivative with respect to time, whereas equation (11.2.1) is a parabolic differential equation, suggests that *adding a term with a wave-like character has a stabilizing influence.*

Because of its desirable properties regarding stability, the Du Fort–Frankel method has enjoyed extensive applications in practice. When using it, however, care must be taken that the product $\kappa(\Delta t/\Delta x)^2 = \alpha \Delta t$ is sufficiently small, otherwise the results will not be physically meaningful.

Implicit BTCS or Laasonen Method

Thus far, we have considered explicit methods where the solution at a particular time level is computed directly from the solution at one or two previous time levels, without solving any systems of equations. We turn now to considering implicit discretizations that require solving systems of equations, in the hope of gaining unconditional stability while maintaining consistency, and thus relaxing the restriction on Δt.

Applying equation (11.2.1) at the point x_i at the time instant t^{n+1}, and approximating the temporal derivative with a backward difference and the spatial derivative with a central difference, we obtain the BTCS difference equation

$$\frac{f_i^{n+1} - f_i^n}{\Delta t} + O(\Delta t) = \kappa \frac{f_{i+1}^{n+1} - 2f_i^{n+1} + f_{i-1}^{n+1}}{\Delta x^2} + O(\Delta x^2) \tag{11.2.16}$$

The corresponding finite-difference stencil is shown with solid circles in Figure 11.2.1. Rearranging equation (11.2.16), we derive the two-level implicit algorithm

$$-\alpha f_{i-1}^{n+1} + (1+2\alpha) f_i^{n+1} - \alpha f_{i+1}^{n+1} = f_i^n \tag{11.2.17}$$

Algorithm

Recasting equation (11.2.17) into a matrix form, and implementing the homogeneous Dirichlet boundary conditions, we obtain the linear system of equations $\boldsymbol{A}\boldsymbol{f}^{n+1} = \boldsymbol{f}^n$, where $\boldsymbol{A}$ is a tridiagonal matrix with superdiagonal, diagonal, and subdiagonal elements equal to $-\alpha$, $1+2\alpha$, and $-\alpha$, respectively. Solving for $\boldsymbol{f}^{n+1}$ yields $\boldsymbol{f}^{n+1} = \boldsymbol{A}^{-1}\boldsymbol{f}^n$, where $\boldsymbol{A}^{-1}$ is the inverse of $\boldsymbol{A}$, which shows that stepping in time is equivalent to successively mapping the initial vector $\boldsymbol{f}^0$ with the projection matrix $\boldsymbol{A}^{-1}$.

$$f^n = \begin{bmatrix} 1+2\alpha & -\alpha & 0 & \cdots & 0 & 0 \\ -\alpha & 1+2\alpha & -\alpha & 0 & \cdots & 0 \\ 0 & -\alpha & 1+2\alpha & -\alpha & 0 & \cdots \\ 0 & \ddots & \ddots & \ddots & & 0 \\ \cdots & 0 & -\alpha & 1+2\alpha & -\alpha & 0 \\ 0 & \cdots & 0 & -\alpha & 1+2\alpha & -\alpha \\ 0 & 0 & \cdots & 0 & -\alpha & 1+2\alpha \end{bmatrix} f^{n+1}$$

In practice, in order to compute the solution at the $n+1$ time level, we solve the system of linear algebraic equations $\boldsymbol{A}\boldsymbol{f}^{n+1} = \boldsymbol{f}^n$, which renders the BTCS method implicit.

Since the matrix $\mathbf{A}$ is tridiagonal and diagonally dominant, the linear system may be solved using the Thomas algorithm or Jacobi's method discussed in Chapter 3.

Stability

To study the stability of the method, we consider the eigenvalues of the projection matrix $\mathbf{A}^{-1}$. Using the results of Section 5.3, and remembering that the eigenvalues of $\mathbf{A}^{-1}$ are equal to the inverses of the eigenvalues of $\mathbf{A}$, we obtain

$$\lambda_m = \left(1 + 4\alpha \cos^2 \frac{m\pi}{2K}\right)^{-1} \tag{11.2.18}$$

where $m = 1, \ldots, K-1$. Since the spectral radius of the projection matrix is less than unity, the BTCS method is *unconditionally stable.*

An independent way of arriving at the same result is to perform the von Neumann stability analysis, obtaining the amplification factor

$$G = \frac{1}{1 + 2\alpha(1 - \cos\theta)} = \frac{1}{1 + 4\alpha \sin^2 \frac{\theta}{2}} \tag{11.2.19}$$

whose magnitude is less than unity for any value of α or θ.

The main limitation of the BTCS method is its low-order temporal accuracy, which places a restriction on the maximum size of the time step for an accurate solution.

The Crank–Nicolson Method

Continuing our search for an efficient method, we target an algorithm that is second-order accurate in both time and space and unconditionally stable. To this end, we recall that the explicit FTCS method emerged by applying equation (11.2.1) at the point x_i and at the time level t^n, whereas the implicit BTCS method emerged by applying equation (11.2.1) at the point x_i at the time level t^{n+1}.

Being adventurous, the distinguishing attribute of the reader of this book, we apply equation (11.2.1) at the intermediate grid point $(x_i, t^{n+1/2})$ that is located halfway between the grid points (x_i, t^n) and (x_i, t^{n+1}), set the spatial derivative at the $t^{n+1/2}$ level equal to the average value of the spatial derivatives at the t^n and t^{n+1} levels, and arrive at the finite-difference equation

$$\frac{f_i^{n+1} - f_i^n}{\Delta t} = \kappa \frac{1}{2}\left(\frac{f_{i+1}^{n+1} - 2f_i^{n+1} + f_{i-1}^{n+1}}{\Delta x^2} + \frac{f_{i+1}^n - 2f_i^n + f_{i-1}^n}{\Delta x^2}\right) \tag{11.2.20}$$

(Crank and Nicolson 1947). The corresponding finite-difference stencil is shown with triangles in Figure 11.2.1. Rearranging equation (11.2.20), we obtain a tridiagonal system of equations

$$-\alpha f_{i-1}^{n+1} + 2(1+\alpha) f_i^{n+1} - \alpha f_{i+1}^{n+1} = \alpha f_{i-1}^n + 2(1-\alpha) f_i^n + \alpha f_{i+1}^n \tag{11.2.21}$$

Deriving and examining the corresponding modified differential equation shows that the Crank–Nicolson method is consistent and *second-order accurate in both time and space.*

Stability

Recasting equation (11.2.21) into a matrix form, we obtain the system of linear equations $\boldsymbol{A}\boldsymbol{f}^{n+1} = \boldsymbol{B}\boldsymbol{f}^{n}$, where $\boldsymbol{A}$ and $\boldsymbol{B}$ are tridiagonal matrices with superdiagonal, diagonal, and subdiagonal elements equal to $-\alpha, 2(1+\alpha), -\alpha$, and $\alpha, 2(1-\alpha), \alpha$, respectively. Solving for $\boldsymbol{f}^{n+1}$ yields $\boldsymbol{f}^{n+1} = \boldsymbol{A}^{-1}\boldsymbol{B}\boldsymbol{f}^{n}$, which shows that stepping in time is equivalent to mapping with the projection matrix $\boldsymbol{A}^{-1}\boldsymbol{B}$. The spectral radius of the projection matrix may be shown to be less than unity, and this ensures that the Crank–Nicolson method is *unconditionally stable* (Problem 11.2.4).

$$
\begin{bmatrix}
2(1+\alpha) & -\alpha & 0 & \cdots & 0 & 0 \\
-\alpha & 2(1+\alpha) & -\alpha & 0 & \cdots & 0 \\
0 & -\alpha & 2(1+\alpha) & -\alpha & 0 & \cdots \\
0 & \ddots & \ddots & \ddots & & 0 \\
\cdots & 0 & -\alpha & 2(1+\alpha) & -\alpha & 0 \\
0 & \cdots & 0 & -\alpha & 2(1+\alpha) & -\alpha \\
0 & 0 & \cdots & 0 & -\alpha & 2(1+\alpha)
\end{bmatrix}
f^{n+1} =
\begin{bmatrix}
2(1-\alpha) & \alpha & 0 & \cdots & 0 & 0 \\
\alpha & 2(1-\alpha) & \alpha & 0 & \cdots & 0 \\
0 & \alpha & 2(1-\alpha) & \alpha & 0 & \cdots \\
0 & \ddots & \ddots & \ddots & & 0 \\
\cdots & 0 & \alpha & 2(1-\alpha) & \alpha & 0 \\
0 & \cdots & 0 & \alpha & 2(1-\alpha) & \alpha \\
0 & 0 & \cdots & 0 & \alpha & 2(1-\alpha)
\end{bmatrix}
f^{n}
$$

Carrying out the von Neumann stability analysis yields the amplification factor

$$
G = \frac{1-\alpha(1-\cos\theta)}{1+\alpha(1-\cos\theta)} = \frac{1-2\alpha\,\sin^2\dfrac{\theta}{2}}{1+2\alpha\,\sin^2\dfrac{\theta}{2}} \qquad (11.2.22)
$$

which is always less than unity, confirming that the method is unconditionally stable.

Because of its qualities with respect to both accuracy and stability, the Crank–Nicolson method has become a standard choice in practice.

Multiple substeps

It is instructive to notice that the Crank–Nicolson method may be regarded as the result of the sequential action of a two-step method, where each substep lasts for a time interval of duration $\Delta t/2$. The first substep is carried out using the explicit FTCS method, and the second substep is carried out using the implicit BTCS method according to the finite-difference equations

$$
\frac{f_i^{n+1/2} - f_i^n}{\Delta t/2} = \kappa\frac{f_{i+1}^n - 2f_i^n + f_{i-1}^n}{\Delta x^2}, \qquad \frac{f_i^{n+1} - f_i^{n+1/2}}{\Delta t/2} = \kappa\frac{f_{i+1}^{n+1} - 2f_i^{n+1} + f_{i-1}^{n+1}}{\Delta x^2} \qquad (11.2.23)
$$

Adding the two equations (11.2.23) to eliminate the intermediate variable $f^{n+1/2}$ recovers equation (11.2.20). The unconditional stability of the second substep prevails over the conditional stability of the first substep and renders the method overall unconditionally stable.

Three-level Implicit Methods

Richtmyer and Morton (1967, p. 189) survey three-level implicit methods. The general form of a three-level, five-point method with a T-shaped finite-difference stencil is

$$(1+\beta)\frac{f_i^{n+1}-f_i^n}{\Delta t}-\beta\frac{f_i^n-f_i^{n-1}}{\Delta t}+O(\Delta t)=\kappa\frac{f_{i+1}^{n+1}-2f_i^{n+1}+f_{i-1}^{n+1}}{\Delta x^2}+O(\Delta x^2) \qquad (11.2.24)$$

where β is a positive constant. It can be shown that the numerical algorithm is unconditionally stable for any choice of β.

The particular choice $\beta=\frac{1}{2}$ provides us with a method that is second-order accurate in time and proved to suppress small-scale oscillations. These features render equation (11.2.24) with $\beta=\frac{1}{2}$ preferable over the Crank–Nicolson method when the solution exhibits sharp spatial variations. The choice $\beta=\frac{1}{2}[1-1/(6\alpha)]$ provides us with a method that is second-order accurate in time and fourth-order accurate in space.

PROBLEMS

11.2.1 *Heat conduction through an aluminum rod.*

Consider the problem of heat condution through an aluminum rod discussed in Section 11.1. Compute the evolution of the temperature distribution using the explicit FTCS finite-difference method. Carry out the calculations for a long enough period of time, so that you can assess the long-time behavior.

Specifically, perform three sets of computations with three different diffusion numbers α. First, try $\alpha=0.20$, then $\alpha=0.50$, and finally $\alpha=1.00$. Comment on the numerical stability in each case. Plot the temperature distribution along the rod at several times for a set of computations of your choice, as well as the heat loss given by equation (11.1.14) as a function of time. (The thermal conductivity of aluminum is $k=0.5$ cal/(s.cm.°C), and the thermal of diffusivity is $\kappa=0.86$ cm^2/s).

11.2.2 *Heat conduction through a radiating rod.*

With reference to Problem 11.1.1, compute the evolution of the temperature distribution along the rod from the initial value $T(x,t=0)=T_L$ using the Crank–Nicolson method for different radiation intensities. Discuss the behavior of the solution at long times.

11.2.3 *CTCS and Du Fort–Frankel.*

(*a*) Express equation (11.2.11) in vector notation in terms of the solution vectors $\boldsymbol{f}^{n+1}, \boldsymbol{f}^n$, and $\boldsymbol{f}^{n-1}$.
(*b*) Perform a consistency analysis of the Du Fort–Frankel method and derive the modified differential equation (11.2.15).

11.2.4 *Spectral radius of the projection matrix of the Crank–Nicolson method.*

Verify that the eigenvalues of the projection matrix corresponding to the Crank–Nicolson method are given by

$$\lambda_m(\boldsymbol{A}^{-1}\boldsymbol{B})=\frac{1-2\alpha\,\sin^2[m\pi/(2K)]}{1+2\alpha\,\sin^2[m\pi/(2K)]} \qquad (11.2.25)$$

where $m=1,\ldots,K-1$, and then show that the spectral radius of the projection matrix is less than unity and therefore the method is unconditionally stable.

11.2.5 *Generalized Crank–Nicolson method.*

A more general form of equation (11.2.20) emerges by using a weighted average to approximate the spatial derivative, obtaining

$$\frac{f_i^{n+1} - f_i^n}{\Delta t} = \kappa \left[\beta \frac{f_{i+1}^{n+1} - 2f_i^{n+1} + f_{i-1}^{n+1}}{\Delta x^2} + (1-\beta) \frac{f_{i+1}^n - 2f_i^n + f_{i-1}^n}{\Delta x^2} \right] \tag{11.2.26}$$

where β is a positive parameter taking values in the range [0, 1]. When $\beta = 0$ or $\frac{1}{2}$ we obtain, respectively, the explicit FTCS and the implicit Crank–Nicolson method.
(*a*) Show that, in general, the method is first-order accurate in time and second-order accurate in space.
(*b*) Show that when $\beta = \frac{1}{2}[1 - 1/(6\alpha)]$, the accuracy of the method is raised to second-order in time and fourth-order in space.
(*c*) Discuss the interpretation of equation (11.2.26) in terms of a sequence of two elementary substeps.
(*d*) Perform the von Neumann stability analysis of equation (11.2.26), derive the amplification factor

$$G = \frac{1 - 4(1-\beta)\alpha \sin^2 \frac{\theta}{2}}{1 + 4\beta\alpha \sin^2 \frac{\theta}{2}} \tag{11.2.27}$$

and show that the method is unconditionally stable when $\frac{1}{2} < \beta < 1$ and conditionally stable when $0 < \beta < \frac{1}{2}$. Show that, in the second case, numerical stability is guaranteed when $\alpha < 1/(2 - 4\beta)$ (Richtmyer and Morton 1967, p. 189).

11.2.6. *Exponential equation method.*

Derive the following explicit finite-difference equation and discuss the underlying assumptions (Bhattacharya, 1985):

$$f_i^{n+1} = f_i^n \exp\left(\alpha \frac{f_{i-1}^n - 2f_i^n + f_{i+1}^n}{f_i^n}\right) \tag{11.2.28}$$

11.2.7 *Dispersion.*

Consider the dispersion equation in one dimension

$$\frac{\partial f}{\partial t} = \beta \frac{\partial^3 f}{\partial x^3} \tag{11.2.29}$$

where the constant β is the *dispersion coefficient.*
(*a*) Investigate analytically the evolution of harmonic waves. More specifically, show that

$$f(x, t) = \exp[Ik(x - k^2\beta t)] \tag{11.2.30}$$

is an exact soluion, where I is the imaginary unit, k is a real wave number, and $k^2\beta$ is the phase velocity of traveling harmonic waves. Since the phase velocity depends on the wavelength, *the waves are dispersive*; the dispersion equation propagates an arbitrary initial distribution comprised of a finite or infinite superposition of sinusoidal waves while modifying its shape.
(*b*) The FTCS discretization leads to the explicit five-point two-level finite-difference equation

$$f_i^{n+1} = -\tfrac{1}{2}\gamma f_{i-2}^n + \gamma f_{i-1}^n + f_i^n - \gamma f_{i+1}^n + \tfrac{1}{2}\gamma f_{i+2}^n \tag{11.2.31}$$

where $\gamma = \beta\Delta t/\Delta x^3$, which is first-order accurate in time and second-order accurate in space. Carry out the von Neumann stability analysis and derive the amplification factor

$$G = 1 - I2\gamma \ \sin\theta(1 - \cos\theta) \tag{11.2.32}$$

and thereby show that the method is *unconditionally unstable.*
(*c*) Show that the implicit five-point two-level BTCS discretization leads to the finite-difference equation

$$\tfrac{1}{2}\gamma f_{i-2}^{n+1} - \gamma f_{i-1}^{n+1} + f_i^{n+1} + \gamma f_{i+1}^{n+1} - \tfrac{1}{2}\gamma f_{i+2}^{n+1} = f_i^n \tag{11.2.33}$$

Derive the associated amplification factor

$$G = [1 + I2\gamma \ \sin\theta(1 - \cos\theta)]^{-1} \tag{11.2.34}$$

and thus show that the method is *unconditionally stable.*
(*d*) Repeat part (*b*) for the counterpart of the Crank–Nicolson method.

11.2.8 *Fourth-order diffusion.*

Consider the fourth-order diffusion equation in one dimension

$$\frac{\partial f}{\partial t} = -\nu \frac{\partial^4 f}{\partial x^4} \tag{11.2.35}$$

where the positive constant ν is called the *fourth-order diffusivity.*
(*a*) The FTCS discretization leads to the explicit five-point difference equation

$$f_i^{n+1} = \varepsilon f_{i-2}^n - 4\varepsilon f_{i-1}^n + (1 + 6\varepsilon) f_i^n - 4\varepsilon f_{i+1}^n + \varepsilon f_{i+2}^n \tag{11.2.36}$$

where $\varepsilon = \nu\Delta t/\Delta x^4$, which is first-order accurate in time and second-order accurate in space. Carry out the von Neumann stability analysis, compute the amplification factor, and assess the stability of the method.
(*b*) Repeat part (*a*) for the implicit five-point two-level BTCS method.
(*c*) Develop the counterpart of the Crank–Nicolson method.

11.3 Unsteady Diffusion in Two and Three Dimensions

Expanding the scope of our discussion, we consider the unsteady diffusion equation in two and three dimensions

$$\frac{\partial f}{\partial t} = \kappa \left(\frac{\partial^2 f}{\partial x^2} + \frac{\partial^2 f}{\partial y^2} \right), \qquad \frac{\partial f}{\partial t} = \kappa \left(\frac{\partial^2 f}{\partial x^2} + \frac{\partial^2 f}{\partial y^2} + \frac{\partial^2 f}{\partial z^2} \right) \tag{11.3.1}$$

subject to the initial condition $f(\mathbf{x}, t = 0) = F(\mathbf{x})$, where F is a known function. For simplicity, we assume that the domain of solution extends over the whole two-dimensional plane or three-dimensional space, and the function f vanishes at infinity.

The majority of the finite-difference methods discussed in Section 11.2 can be extended in a straightforward manner to handle a second or a third dimension; several examples will be discussed in the remainder of this section. We shall see, however, that the need for computationally efficient algorithms necessitates the development of new procedures.

Finite-difference Procedures

In three dimensions, the goal of the finite-difference method is to generate the values of the function f at the nodes of a three-dimensional grid that is parametrized by the three indices i, j, k. In two dimensions, the grid is parametrized by two indices i, j. For simplicity, we assume that the grid spacings Δx, Δy, and Δz are uniform but not necessarily identical throughout the domain of solution.

To carry out the von Neumann stability analysis, we set

$$f^n_{i,j,k} = A^n \exp[I(i\theta_x + j\theta_y + k\theta_z)] \tag{11.3.2}$$

where $I = \sqrt{-1}$ is the imaginary unit, and $\theta_x = 2\pi\Delta x/L_x, \theta_y = 2\pi\Delta y/L_y, \theta_z = 2\pi\Delta z/L_z$ are phase angles with corresponding wavelengths L_x, L_y, L_z, and then study the magnitude of the amplification factor $G \equiv A^{n+1}/A^n$.

In general, the stability criteria for two-dimensional and three-dimensional diffusion turn out to be substantially more restrictive than those for one-dimensional diffusion, and this renders the conditionally stable methods prohibitively expensive.

Explicit FTCS Method

With the preceding considerations in mind, we proceed to develop the FTCS discretization for the two-dimensional problem. Approximating the two-dimensional Laplacian operator with the five-point formula discussed in Section 6.12, we derive the difference equation

$$\frac{f^{n+1}_{i,j} - f^n_{i,j}}{\Delta t} = \kappa\left(\frac{f^n_{i+1,j} - 2f^n_{i,j} + f^n_{i-1,j}}{\Delta x^2} + \frac{f^n_{i,j+1} - 2f^n_{i,j} + f^n_{i,j-1}}{\Delta y^2}\right) \tag{11.3.3}$$

Rearranging, we derive

$$f^{n+1}_{i,j} = f^n_{i,j} + \alpha_x(f^n_{i+1,j} + f^n_{i-1,j}) + [1 - 2(\alpha_x + \alpha_y)]f^n_{i,j} + \alpha_y(f^n_{i,j+1} + f^n_{i,j-1}) \tag{11.3.4}$$

where

$$\alpha_x = \frac{\kappa\Delta t}{\Delta x^2}, \qquad \alpha_y = \frac{\kappa\Delta t}{\Delta y^2} \tag{11.3.5}$$

are the diffusivities for the x and y directions.

Carrying out the von Neumann stability analysis, we find that the method is stable provided that $\alpha_x + \alpha_y < \frac{1}{2}$. This constraint imposes a strong restriction on the size of the time step and renders the explicit discretization inefficient. Similar but more severe difficulties are encountered in three dimensions.

Implicit BTCS Method

To achieve unconditional stability, we resort to an implicit method. The BTCS discretization yields the two-level algorithm

$$-\alpha_x(f^{n+1}_{i-1,j} + f^{n+1}_{i+1,j}) + (1 + 2\alpha_x + 2\alpha_y)f^{n+1}_{i,j} - \alpha_y(f^{n+1}_{i,j-1} + f^{n+1}_{i,j+1}) = f^n_{i,j} \tag{11.3.6}$$

with first-order accuracy in time and second-order accuracy in space, which may be shown to be *unconditionally stable.*

Unfortunately, the numerical implementation of equation (11.3.6) results in a pentadiagonal system of linear algebraic equations whose solution can no longer be carried out using specialized methods such as the Thomas algorithm.

ADI Method in Two Dimensions

To reduce the computational burden of the implicit BTCS method for two-dimensional diffusion, Peaceman and Rachford (1955) and Douglas (1955) proposed splitting each time step into two substeps of equal duration $\Delta t/2$, and approximating the spatial derivatives in a partially implicit manner, while working sequentially and alternatingly in the x and y directions. The computations proceed according to the finite-difference equations

$$\frac{f_{i,j}^{n+1/2} - f_{i,j}^{n}}{\Delta t/2} = \kappa\left[\frac{f_{i+1,j}^{n+1/2} - 2f_{i,j}^{n+1/2} + f_{i-1,j}^{n+1/2}}{\Delta x^2}\right] + \kappa\frac{f_{i,j+1}^{n} - 2f_{i,j}^{n} + f_{i,j-1}^{n}}{\Delta y^2} \tag{11.3.7}$$

and

$$\frac{f_{i,j}^{n+1} - f_{i,j}^{n+1/2}}{\Delta t/2} = \kappa\frac{f_{i+1,j}^{n+1/2} - 2f_{i,j}^{n+1/2} + f_{i-1,j}^{n+1/2}}{\Delta x^2} + \kappa\left[\frac{f_{i,j+1}^{n+1} - 2f_{i,j}^{n+1} + f_{i,j-1}^{n+1}}{\Delta y^2}\right] \tag{11.3.8}$$

where $n+\frac{1}{2}$ is an intermediate time level, and the square brackets enclose implicit discretizations. The first substep is carried out according to the implicit BTCS method for the x direction, and the second substep is carried out according to the implicit BTCS method for the y direction. To eliminate the bias associated with this particular arrangement, we alternate this order after the completion of each step. The overall accuracy of the method is of second order in both space and time.

Rearranging equations (11.3.7) and (11.3.8), we obtain a two-step implicit algorithm representing the x and y sweeps,

$$\alpha_x f_{i-1,j}^{n+1/2} - 2(1+\alpha_x)f_{i,j}^{n+1/2} + \alpha_x f_{i+1,j}^{n+1/2} = -\alpha_y f_{i,j-1}^{n} - 2(1-\alpha_y)f_{i,j}^{n} - \alpha_y f_{i,j+1}^{n} \tag{11.3.9}$$

and

$$\alpha_y f_{i,j-1}^{n+1} - 2(1+\alpha_y)f_{i,j}^{n+1} + \alpha_y f_{i,j+1}^{n+1} = -\alpha_x f_{i-1,j}^{n+1/2} - 2(1-\alpha_x)f_{i,j}^{n+1/2} - \alpha_x f_{i+1,j}^{n+1/2} \tag{11.3.10}$$

Completing one time step requires solving two systems of tridiagonal equations, but this can be done efficiently using the Thomas algorithm.

Carrying out the von Neumann stability analysis yields the amplification factor

$$G = \frac{[1+\alpha_x(\cos\theta_x - 1)][1+\alpha_y(\cos\theta_y - 1)]}{[1-\alpha_x(\cos\theta_x - 1)][1-\alpha_y(\cos\theta_y - 1)]} = \frac{\left(1-2\alpha_x\sin^2\frac{\theta_x}{2}\right)\left(1-2\alpha_y\sin^2\frac{\theta_y}{2}\right)}{\left(1+2\alpha_x\sin^2\frac{\theta_x}{2}\right)\left(1+2\alpha_y\sin^2\frac{\theta_y}{2}\right)} \tag{11.3.11}$$

Close examination shows that $|G| < 1$ under any conditions, which ensures that the ADI method is *unconditionally stable.* The second-order accuracy combined with the unconditional stability have made the ADI method an attractive choice.

Crank–Nicolson and Approximate Factorization

The ADI method allows us to advance the solution over one time step by solving two pseudo-one-dimensional implicit problems, which amounts to decoupling the diffusion processes in the two spatial directions. To show this clearly, we introduce the centered-difference operators

$$\Delta_x^2\langle f_{i,j}^k\rangle = f_{i-1,j}^k - 2f_{i,j}^k + f_{i+1,j}^k, \qquad \Delta_y^2\langle f_{i,j}^k\rangle = f_{i,j-1}^k - 2f_{i,j}^k + f_{i,j+1}^k \tag{11.3.12}$$

and rewrite equations (11.3.9) and (11.3.10) in the forms

$$(2-\alpha_x\Delta_x^2)f_{i,j}^{n+1/2} = (2+\alpha_y\Delta_y^2)f_{i,j}^n, \qquad (2-\alpha_y\Delta_y^2)f_{i,j}^{n+1} = (2+\alpha_x\Delta_x^2)f_{i,j}^{n+1/2} \tag{11.3.13}$$

Combining these equations to eliminate the intermediate solution at the $n+\frac{1}{2}$ level, we obtain

$$(2-\alpha_x\Delta_x^2)(2-\alpha_y\Delta_y^2)f_{i,j}^{n+1} = (2+\alpha_x\Delta_x^2)(2+\alpha_y\Delta_y^2)f_{i,j}^n \tag{11.3.14}$$

The aforementioned decoupling is manifested in the factorized nature of the difference operators on either side of equation (11.3.14).

Now, the fully implicit Crank–Nicolson discretization of the two-dimensional diffusion equation can be expressed in the symbolic form

$$(2-\alpha_x\Delta_x^2-\alpha_y\Delta_y^2)f_{i,j}^{n+1} = (2+\alpha_x\Delta_x^2+\alpha_y\Delta_y^2)f_{i,j}^n \tag{11.3.15}$$

which can be restated as

$$(2-\alpha_x\Delta_x^2)(2-\alpha_y\Delta_y^2)f_{i,j}^{n+1} = (2+\alpha_x\Delta_x^2)(2+\alpha_y\Delta_y^2)f_{i,j}^n + \alpha_x\alpha_y\Delta_x^2\Delta_y^2(f_{i,j}^{n+1} - f_{i,j}^n) \tag{11.3.16}$$

The method can be shown to be unconditionally stable (Problem 11.3.1). The ADI equation (11.3.14) derives from equation (11.3.16) by neglecting the last term on the right-hand side. This simplification is permissible, for the neglected term is of fourth-order in the spatial step, whereas equation (11.3.14) is meant to be accurate only up to second order. In this light, the ADI method emerges as the result of the approximate factorization of the difference operators on either side of equation (11.3.15).

ADI Method in Three Dimensions

The standard ADI method in three dimensions is carried out in three substeps of equal duration $\Delta t/3$, where one spatial dimension is treated implicitly, while the other two dimensions are treated explicitly within each step, in the spirit of equations (11.3.7) and (11.3.8). The method is first-order accurate in time and second-order accurate in space. Unfortunately, the partial BTCS discretizations result in an algorithm that is stable only when $a_x + a_y + a_z < \frac{3}{2}$, which places an unaffordable restriction on the size of the time step.

Douglas (1962) developed an ADI method that is second-order accurate in both time and space, and unconditionally stable. The method proceeds in a *predictor–corrector* sense, in three substeps: the first substep produces a predicted solution using the Crank–Nicolson method for the x direction, while treating the y and z directions explicitly; the second substep produces a predicted solution using the x discretization of the first substep and the Crank–Nicolson method for the y direction, while treating the z direction explicitly; and the third substep advances the solution using the x and y discretizations of the second substep, while using the Crank–Nicolson method for the z direction. The finite-difference equations are:

$$\frac{f_{i,j,k}^{n+1/3} - f_{i,j,k}^{n}}{\Delta t} = \kappa \left\{ \frac{1}{2}\left[\left(\frac{\partial^2 f}{\partial x^2}\right)^{n+1/3} + \left(\frac{\partial^2 f}{\partial x^2}\right)^{n} \right] + \left(\frac{\partial^2 f}{\partial y^2}\right)^{n} + \left(\frac{\partial^2 f}{\partial z^2}\right)^{n} \right\}$$

$$\frac{f_{i,j,k}^{n+2/3} - f_{i,j,k}^{n}}{\Delta t} = \kappa \left\{ \frac{1}{2}\left[\left(\frac{\partial^2 f}{\partial x^2}\right)^{n+1/3} + \left(\frac{\partial^2 f}{\partial x^2}\right)^{n} + \left(\frac{\partial^2 f}{\partial y^2}\right)^{n+2/3} + \left(\frac{\partial^2 f}{\partial y^2}\right)^{n} \right] + \left(\frac{\partial^2 f}{\partial z^2}\right)^{n} \right\} \tag{11.3.17}$$

$$\frac{f_{i,j,k}^{n+1} - f_{i,j,k}^{n}}{\Delta t} = \kappa \frac{1}{2}\left[\left(\frac{\partial^2 f}{\partial x^2}\right)^{n+1/3} + \left(\frac{\partial^2 f}{\partial x^2}\right)^{n} + \left(\frac{\partial^2 f}{\partial y^2}\right)^{n+2/3} + \left(\frac{\partial^2 f}{\partial y^2}\right)^{n} + \left(\frac{\partial^2 f}{\partial z^2}\right)^{n+1} + \left(\frac{\partial^2 f}{\partial z^2}\right)^{n} \right]$$

where the second-order partial derivatives are discretized using centered differences. To eliminate the spatial bias associated with this particular arrangement, we alternate the sequence of the three substeps in a cyclic manner.

Operator Splitting and Fractional Steps

Another way to preserve the tridiagonal nature of the one-dimensional implicit discretization is to replace the diffusion equation in three dimensions, or its two-dimensional counterpart, with a set of three or two one-dimensional evolution equations that operate successively in fractional steps (Yanenko 1970). In two dimensions, we obtain

$$\frac{\partial f}{\partial t} = \kappa \frac{\partial^2 f}{\partial x^2}, \qquad \frac{\partial f}{\partial t} = \kappa \frac{\partial^2 f}{\partial y^2} \tag{11.3.18}$$

each for $t^n < t < t^n + \Delta t$, which amounts to allowing diffusion to operate sequentially in the two directions, each time neglecting the other dimension. Each fractional step proceeds for the full time interval of Δt, and the time is set back to the initial value at the end of the first fractional step. To preserve the spatial isotropy of the Laplacian operator, the order of equations (11.3.18) is switched after the completion of a full time step.

Carrying out the fractional steps may be done using different methods for the component diffusion equations, as discussed previously in Section 11.2. The stability restrictions of the overall method are comprised of the collection of the restrictions imposed on the individual fractional steps.

Approximate factorization implicit method, AFI

This method emerges by applying the implicit BTCS discretization to each fractional step in equations (11.3.18). Each fractional step requires solving a tridiagonal system of equations, which can be done using the efficient Thomas algorithm. It can be shown that the AFI method derives from the approximate factorization of the fully implicitly BTCS discretization of the second of equations (11.3.1) (Problem 11.3.2).

Using the Crank–Nicolson method for each fractional step, we obtain an algorithm that is second-order accurate in time and space, and unconditionally stable. The component equations are

$$-\alpha_x f_{i-1,j}^{*} + 2(1+\alpha_x) f_{i,j}^{*} - \alpha_x f_{i+1,j}^{*} = \alpha_x f_{i-1,j}^{n} + 2(1-\alpha_x) f_{i,j}^{n} + \alpha_x f_{i+1,j}^{n} \tag{11.3.19}$$

and

$$-\alpha_y f_{i,j-1}^{n+1} + 2(1+\alpha_y) f_{i,j}^{n+1} - \alpha_y f_{i,j+1}^{n+1} = \alpha_y f_{i,j-1}^{*} + 2(1-\alpha_y) f_{i,j}^{*} + \alpha_y f_{i,j+1}^{*} \tag{11.3.20}$$

where an asterisk designates the solution at the end of the first fractional step. The scheme remains unconditionally stable in three dimensions.

PROBLEMS

11.3.1 ***Generalized fully implicit Crank–Nicolson method in two dimensions.***

A generalized version of equation (11.3.15) is

$$[1-\gamma(\alpha_x\Delta_x^2+\alpha_y\Delta_y^2)]f_{i,j}^{n+1}=[1+(1-\gamma)(\alpha_x\Delta_x^2+\alpha_y\Delta_y^2)]f_{i,j}^{n} \tag{11.3.21}$$

where γ is a numerical parameter; the standard Crank–Nicolson method derives by setting $\gamma=\frac{1}{2}$. Perform the von Neumann stability analysis, derive the amplification factor

$$G=\frac{1-4(1-\gamma)\left(\alpha_x^2\sin^2\frac{\theta_x}{2}+\alpha_y^2\sin^2\frac{\theta_y}{2}\right)}{1+4\gamma\left(\alpha_x^2\sin^2\frac{\theta_x}{2}+\alpha_y^2\sin^2\frac{\theta_y}{2}\right)} \tag{11.3.22}$$

and show that the method is stable when $(1-2\gamma)(\alpha_x^2+\alpha_y^2)<\frac{1}{2}$. Explain why the standard Crank–Nicolson method corresponding to $\gamma=\frac{1}{2}$ is unconditionally stable.

11.3.2 *AFI.*

Show that the AFI method may be regarded as the result of the approximate factorization of the fully implicit BTCS discretization of equations (11.3.1).

11.3.3 ***Exponential equation method.***

Derive the counterpart of equation (11.2.28) in three dimensions (Bhattacharya, 1985).

11.4 Poisson and Laplace Equations

In Section 3.1, we developed a finite-difference method for solving the Poisson equation (3.1.3) in a two-dimensional rectangular domain with Dirichlet boundary conditions along the left, bottom, and top sides, and Neumann boundary condition along the right side. Approximating the partial derivatives with finite differences provided us with the system of linear algebraic equations (3.1.9) and (3.1.12), which we wrote in the symbolic form $\boldsymbol{As}=\boldsymbol{b}$. The vector $\boldsymbol{s}$ contains the various values $f_{i,j}$. In the example,

$$\begin{aligned} \boldsymbol{s}=(&f_{2,2},\ f_{3,2},\ \ldots,\ f_{N+1,2},\\ &f_{2,3},\ f_{3,3},\ \ldots,\ f_{N+1,3},\\ &\ldots,\\ &f_{2,M},\ f_{3,M},\ \ldots,\ f_{N+1,M})^T \end{aligned} \tag{11.4.1}$$

The solution of the linear system can be found by a standard or a custom-made iterative method, as will be discussed later in this section.

Programmable Construction of the Linear System

The manual construction of the coefficient matrix $\boldsymbol{A}$ and constant vector $\boldsymbol{b}$ arising from a finite-difference formulation can be taxing. Fortunately, the work can be automated by introducing the residual function $\boldsymbol{R}(\boldsymbol{s})$, rewriting the equation $\boldsymbol{A}\boldsymbol{s} = \boldsymbol{b}$ in the form

$$\boldsymbol{R} = \boldsymbol{A}\boldsymbol{s} - \boldsymbol{b} = \boldsymbol{0} \tag{11.4.2}$$

and making two key observations:

- The matrix $\boldsymbol{A}$ contains the constant first partial derivatives of the function $\boldsymbol{R}$ with respect to the components $f_{i,j}$ of $\boldsymbol{s}$.
- The constant vector $\boldsymbol{b}$ can be evaluated as $\boldsymbol{b} = -\boldsymbol{R}(\boldsymbol{s} = \boldsymbol{0})$.

Thus for a $K \times K$ matrix $\boldsymbol{A}$, we set

$$\begin{aligned} A_{i,j} = \frac{\partial R_i}{\partial s_j} &= R(s_1 = 0, \ldots, s_{j-1} = 0, s_j = 1, s_{j+1} = 0, \ldots, s_K = 0) \\ &\quad - R(s_1 = 0, \ldots, s_{j-1} = 0, s_j = 0, s_{j+1} = 0, \ldots, s_K = 0) \end{aligned} \tag{11.4.3}$$

To compute $\boldsymbol{A}$ and $\boldsymbol{b}$, all we need is a subroutine that receives the components of the vector $\boldsymbol{s}$ and produces the components of the vector $\boldsymbol{R}$. For example, for the aforementioned problem where the vector $\boldsymbol{s}$ is defined in equation (11.4.1), the matrix $\boldsymbol{A}$ and the constant vector $\boldsymbol{b}$ may be found using Algorithm 11.4.1.

ALGORITHM 11.4.1 Algorithm for computing the coefficient matrix $\boldsymbol{A}$ and constant vector $\boldsymbol{b}$ for the Poisson equation (3.1.3) using a finite-difference method. The Laplacian is approximated with the five-point formula. Modifications can be made to avoid the numerous multiplications by zeros and prevent idle operations. The implementation of more accurate finite-difference approximations proceeds along similar lines.

Do $j = 2, M$	Initialize all unknowns to zero
Do $i = 2, N+1$	
$f_{i,j} = 0$	
END DO	
END DO	
Do $j = 2, M$	Dirichlet boundary condition at left side
$f_{1,j} = w_j$	
END DO	
Do $i = 2, N+1$	Dirichlet boundary conditions up and down
$f_{i,1} = z_i$	
$f_{i,M+1} = v_i$	
END DO	
	First, will compute $\boldsymbol{b}$
Do $j = 2, M$	Newmann boundary condition at right side
$f_{N+2,j} = 2\Delta x q_j$	Fantom-node values
END DO	

```
p = 0                                                   Scan grid points row-by-row to compute b
Do j = 2, M
  Do i = 2, N + 1
  p ← p + 1
  R = f_{i+1,j} − 2(1 + β) f_{i,j} + f_{i−1,j} + β f_{i,j+1} + β f_{i,j−1} − Δx^2 g_{i,j}
  b(p) = −R
  END DO
END DO

                                                        Proceed to compute A
t = 0                                                   Scan grid points row-by-row to compute A
Do s = 2, M
  Do l = 2, N + 1
  f_{l,s} = 1.0
  t ← t + 1
    Do k = 2, M                                         Compute values at fantom nodes at right side
    f_{N+2,k} = f_{N,k} + 2Δx q_k
    END DO
    p = 0
    Do j = 2, M
      Do i = 2, N + 1
      p ← p + 1
      R = f_{i+1,j} − 2(1 + β) f_{i,j} + f_{i−1,j} + β f_{i,j+1} + β f_{i,j−1} − Δx^2 g_{i,j}
      A(p, t) = R + b(p)
      END DO
    END DO
  f_{l,s} = 0.0
  END DO
END DO
```

Algorithms of this kind are particularly useful when more accurate approximations, involving more than five grid points, are used to represent the partial derivatives in the differential equation and the boundary conditions. Straightforward modifications yield the corresponding forms of $\boldsymbol{A}$ and $\boldsymbol{b}$ for different types of boundary conditions.

PROBLEM

11.4.1 *Linear system for Poisson's equation with Neumann boundary conditions.*

(*a*) Derive a system of $(N+1)(M+1)$ linear equations with second-order accuracy for Poisson's equation with Neumann boundary conditions that specify the normal derivative of f around all boundaries of a rectangular domain. Develop an alorithm for the automated evaluation of $\boldsymbol{A}$ and $\boldsymbol{b}$, and implement it in a computer program. (*b*) Run the program and verify that a vector with constant elements falls in the null space of $\boldsymbol{A}$. This means that the equation $\boldsymbol{A}\boldsymbol{s} = \boldsymbol{b}$ has either no or a multiplicity of solutions. The physical explanation is that a steady-state solution will exist only if the rate of production of heat represented by the function g balances the outward flux of f across the boundaries. And when it does, the values of f may be shifted by an arbitrary constant, which means that s is a constant vector with equal elements is an eigenvector.

Iterative Solution of Laplace and Poisson Equations

The finite-difference methods for solving the unsteady diffusion equations (11.3.1) provide us with efficient iterative procedures for solving the Laplace and Poisson equations in two and three dimensions.

Consider, for example, the Poisson equation (3.1.3), and introduce a fictitious unsteady diffusion problem in the presence of a sink term, governed by the unsteady reaction–diffusion equation

$$\frac{\partial f}{\partial t} = \kappa \left(\frac{\partial^2 f}{\partial x^2} + \frac{\partial^2 f}{\partial y^2} - g(x, y) \right) \tag{11.4.4}$$

subject to a certain initial condition. The asymptotic solution of equation (11.4.4) at long times is identical to the solution of the Poisson equation (3.1.3), and this suggests that advancing the solution of the diffusion–reaction problem in time amounts to iterating on the solution of Poisson's equation with a projection matrix that arises from the finite-difference method.

An algorithm that derives from the explicit FTCS method on a uniform two-dimensional grid, and its modified versions corresponding to the Gauss–Seidel and successive over-relaxation methods, are collected in Table 11.4.1. All methods are second-order accurate in Δx and Δy.

The explicit point–Gauss–Seidel scheme given in the second entry of Table 11.4.1 derives from the FTCS scheme given in the first entry by setting $1/\alpha_x = 2(1+\beta)$, which satisfies the stability criterion $\alpha_x + \alpha_y < \frac{1}{2}$ in a marginal way. The relaxation parameter ω for the PSOR and LSOR schemes varies between 1 and 2; when $\omega = 1$, the SOR methods reduce to the corresponding Gauss–Seidel methods. The implicit LGS and LSOR methods require solving tridiagonal systems of equations for each grid line parallel to the x axis, which can be done using the efficient Thomas algorithm.

To develop the ADI method, we recast the ADI equations (11.3.9) and (11.3.10) into the form shown in the sixth entry of Table 11.4.1, where we have introduced the new parameter $\rho = 2/\alpha_x$. Since the ADI method is unconditionally stable, it might appear that the fastest approach to the steady state will be achieved by using a large value for Δt or small value for ρ. Careful analysis, however, shows that a minimum number of iterations for a specified level of accuracy are achieved with a certain repetitive sequence of values of ρ. Unless this sequence is known, the ADI method competes in efficiency with the successive over-relaxation method (Ames 1977, Hoffman 1992, p. 446). A generalized form of the ADI method involving the relaxation parameter ω is given in the last entry of Table 11.4.1 (Hoffmann and Chiang 1993, vol. II, p. 9).

PROBLEMS

11.4.2 *ADI method for the Poisson equation in two dimensions.*

Write a computer program that solves the Poisson equation in a two dimensional rectangular domain that has been covered with an $N \times M$ grid defined in terms of $N + 1$ and $M + 1$ grid lines, using the ADI method given in Table 11.4.1, with Dirichlet boundary conditions all around.

Run the program and compute the solution within a square box with a forcing function and Dirichlet boundary conditions of your choice. Examine the rate of convergence of your results with respect to ρ, and test the reliability of your solution by comparing the results against a known analytical solution of your choice.

11.4.3 *ADI method for Poisson's equation in three dimensions.*

Repeat Problem 11.4.2 for the Poisson equation in three dimensions within a rectangular box that has been covered with an $N \times M \times L$ grid, based on the ADI method.

Table 11.4.1 Iterative methods for solving Poisson's equation $\nabla^2 f = g$ on a uniform rectangular grid with $\beta = (\Delta x/\Delta y)^2$. All methods are second-order accurate in Δx and Δy. The relaxation constant w ranges between 1 and 2.

***Explicit* FTCS**

$$f_{i,j}^{k+1} = \alpha_x \left\{ f_{i+1,j}^k + f_{i-1,j}^k + \left[\frac{1}{\alpha_x} - 2(1+\beta)\right] f_{i,j}^k + \beta(f_{i,j+1}^k + f_{i,j-1}^k) - \Delta x^2 g_{i,j} \right\}$$

***Explicit point–Gauss–Seidel,* PGS**

$$f_{i,j}^{k+1} = \frac{1}{2(1+\beta)}[f_{i+1,j}^k + f_{i-1,j}^k + \beta(f_{i,j+1}^k + f_{i,j-1}^k) - \Delta x^2 g_{i,j}]$$

***Explicit point–successive over-relaxation,* PSOR**

$$f_{i,j}^{k+1} = (1-\omega) f_{i,j}^k + \frac{\omega}{2(1+\beta)}[f_{i+1,j}^k + f_{i-1,j}^k + \beta(f_{i,j+1}^k + f_{i,j-1}^k) - \Delta x^2 g_{i,j}]$$

***Implicit line–Gauss–Seidel for the x direction,* LGS**

$$f_{i-1,j}^{k+1} - 2(1+\beta) f_{i,j}^{k+1} + f_{i+1,j}^{k+1} = -\beta(f_{i,j+1}^k + f_{i,j-1}^k) + \Delta x^2 g_{i,j}$$

***Implicit line successive over-relaxation for the x direction,* LSOR**

$$\omega f_{i-1,j}^{k+1} - 2(1+\beta) f_{i,j}^{k+1} + \omega f_{i+1,j}^{k+1} = 2(\omega - 1)(1+\beta) f_{i,j}^k - \omega\beta(f_{i,j+1}^k + f_{i,j-1}^k) + \Delta x^2 g_{i,j}$$

ADI

$$f_{i-1,j}^{k+1/2} - (2+\rho) f_{i,j}^{k+1/2} + f_{i+1,j}^{k+1/2} = -\beta\left[f_{i,j-1}^k - \left(2 - \frac{\rho}{\beta}\right) f_{i,j}^k + f_{i,j+1}^k\right] + \Delta x^2 g_{i,j}$$

$$f_{i,j-1}^{k+1} - \left(2 + \frac{\rho}{\beta}\right) f_{i,j}^{k+1} + f_{i,j+1}^{k+1} = -\frac{1}{\beta}[f_{i-1,j}^{k+1/2} - (2-\rho) f_{i,j}^{k+1/2} + f_{i+1,j}^{k+1/2}] + \Delta y^2 g_{i,j}$$

SOR-ADI

$$\omega f_{i-1,j}^{k+1/2} - 2(1+\beta) f_{i,j}^{k+1/2} + \omega f_{i+1,j}^{k+1/2} = 2(\omega - 1)(1+\beta) f_{i,j}^k - \omega\beta(f_{i,j-1}^{k+1/2} + f_{i,j+1}^k) + \omega\Delta x^2 g_{i,j}$$

$$\omega\beta f_{i,j-1}^{k+1} - 2(1+\beta) f_{i,j}^{k+1} + \omega\beta f_{i,j+1}^{k+1} = 2(\omega - 1)(1+\beta) f_{i,j}^{k+1/2} - \omega(f_{i-1,j}^{k+1} + f_{i+1,j}^{k+1/2}) + \omega\Delta y^2 g_{i,j}$$

11.4.4 *Steady temperature field in a porous medium.*

The temperature distribution in a porous rock in the presence of groundwater flow directed along the x axis is described by the equation

$$U\frac{\partial T}{\partial x} = \kappa \nabla^2 T \qquad (11.4.5)$$

where U is the constant water velocity, and κ is the effective thermal diffusivity. The domain of solution is the rectangle confined by $0 < xU/\kappa < 30$, $0 < yU/\kappa < 10$.
(a) Compute the temperature distribution using a finite-difference method of your choice with boundary conditions $T = 65$ °C at $x = 0$, $T = 25$ °C at $x = 30\kappa/U$, $T = 25$ °C at $y = 10\kappa/U$, $\partial T/\partial y = 0$ at $y = 0$.
(b) Repeat the solution, but this time impose a periodicity condition along the x axis, $T(x) = T(x + 30\kappa/U)$.

Cyclic Reduction and Multigrid Methods

These somewhat related methods provide us with efficient algorithms for solving systems of linear algebraic equations that arise from the finite-difference discretization of Laplace's, Poisson's, and similar partial differential equations.

Cyclic reduction methods

The principle behind the method was discussed in Section 3.4 in the context of tridiagonal systems. The basic idea is to use a grid whose size is a power of 2 and successively eliminate unknowns from equations until a simple form has been obtained (Swartzrauber 1977).

Multigrid methods

The idea is to tackle the problem on a coarse grid, by iterating on a modified set of equations that arise by combining the finite-difference equations on the fine grid, and then interpolate the correction onto an increasingly finer grid. Interrogating the projection matrices shows that iterating on the coarse grid is much more effective than iterating on the fine grid (e.g., Hackbush 1985; Press et al. 1992, pp. 862–880; Ferziger and Perić 1996). For example, the Gauss–Seidel method for Laplace's equation in two or three dimensions converges, respectively, four or eight times as fast on a coarse grid as on a fine grid with twice as large dimensions.

The basic steps of a *two-grid* method are the following:

1. Introduce a *fine* grid, and perform k iterations on the associated finite-difference system $\mathbf{A s} = \mathbf{b}$ using an iterative method, until the error vector $\mathbf{e}^{(k)} = \mathbf{s}^{(k)} - \mathbf{S}$ defines a smoothly varying function; $\mathbf{s}^{(k)}$ is the kth iterate, and $\mathbf{S}$ satisfies $\mathbf{A S} = \mathbf{b}$.
2. Compute the residual vector $\mathbf{R}$ defined in equation (11.4.2). If a certain norm of $\mathbf{R}$ is less than a preestablished threshold, stop.
3. Introduce a *coarse* grid, and combine the finite-difference equations on the fine grid to derive a modified set of finite-difference equations $\mathbf{A}'\mathbf{e} = \mathbf{b}'$ for the error vector $\mathbf{e} = \mathbf{s} - \mathbf{S}$ on the coarse grid; the right-hand side $\mathbf{b}'$ involves the components $\mathbf{R}$ defined in step 2. Perform iterations on the coarse grid to compute $\mathbf{e}$ with a specified accuracy.
4. Interpolate $\mathbf{e}$ onto the fine grid, compute the fine-grid solution $\mathbf{s}^{(k)} - \mathbf{e}$, and return to step 2.

In a multigrid method, we continue the procedure until the coarsest possible grid has been considered. The computation involves a cascade of interpolations. Clever applications of the multigrid method allow for substantial savings, sometimes by a factor as large as 100.

PROBLEM

11.4.5 ***Equations for a two-grid method.***

Consider the Poisson equation in one dimension

$$\frac{d^2 f}{dx^2} = g(x) \tag{11.4.6}$$

Introduce a uniform fine grid with spacing Δx and the underlying coarse grid with spacing $2\Delta x$, and derive the linear systems $\mathbf{A s} = \mathbf{b}$ and $\mathbf{A}'\mathbf{e} = \mathbf{b}'$ involved in steps 1 and 3 of the two-grid method discussed in the text (Ferziger and Perić 1996, pp. 107–108).

11.5 One-dimensional Convection

In this and the next sections, we discuss finite-difference methods for solving the convection equation (11.1.7). We begin by considering the one-dimensional equation

$$\frac{\partial f}{\partial t} + U\frac{\partial f}{\partial x} = 0 \tag{11.5.1}$$

where U is a positive or negative constant convection velocity, which is to be solved subject to the initial condition $f(x, 0) = F(x)$, where $F(x)$ is a known function. For simplicity, we assume that the domain of solution extends over the whole x axis and require the far-field conditions $f(\pm\infty, t) = 0$. If the domain of solution were bounded, we would have to specify one boundary condition only at one end.

One may readily verify that the exact solution to equation (11.5.1) is given by $f(x, t) = F(x - Ut)$, which states that the initial distribution F travels along the x axis with velocity U as illustrated in Figure 11.5.1. If U is negative, the motion is directed toward the negative direction of the x axis. The value of f remains constant along the characteristic line $x = X + Ut$ in the xt plane, shown with the dashed line in Figure 11.5.1, where X is an arbitrary point.

As a preliminary, we differentiate equation (11.5.1) twice, once with respect to t and the second time with respect to x, and combine the resulting expressions to obtain the *wave equation*

$$\frac{\partial^2 f}{\partial t^2} = U^2\frac{\partial^2 f}{\partial x^2} \tag{11.5.2}$$

which is a prototypical second-order hyperbolic partial differential equation. The hyperbolic nature of the first-order equation (11.5.1) thus becomes apparent.

Explicit FTCS Method

Applying equation (11.5.1) at the point x_i at the time instant t^n, and approximating $\partial f/\partial t$ with a forward difference and $\partial f/\partial x$ with a central difference, we obtain the finite-difference equation

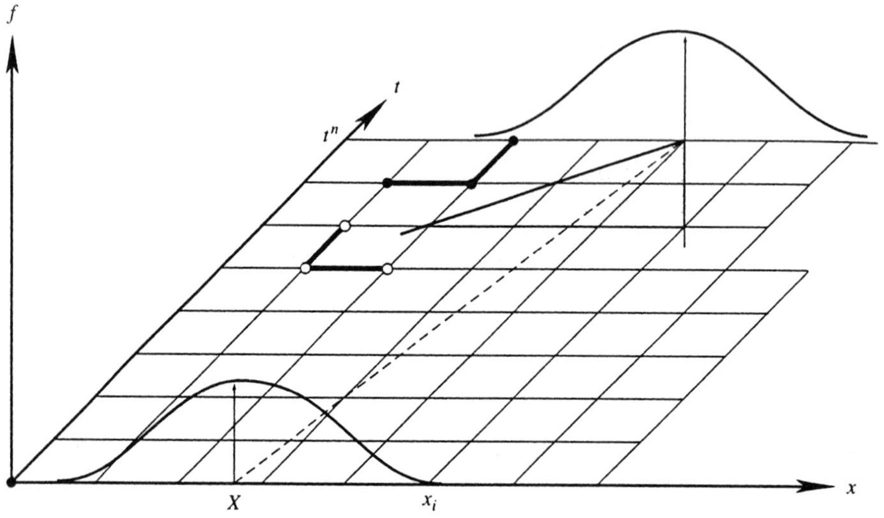

Figure 11.5.1 Discretization of the space–time plane for solving the one-dimensional convection equation. The initial distribution travels along the x axis with velocity U. The dashed line is a characteristic line along which the value of the solution remains constant. The solid and hollow circles indicate the FTBS and FTFS discretization stencils. The numerical cone of influence is shown as a dotted area. From *Introduction to Theoretical and Computational Fluid Dynamics*, by C. Pozrikidis. Copyright © 1997 by Oxford University Press. Used by permission of Oxford University Press, Inc.

$$\frac{f_i^{n+1} - f_i^n}{\Delta t} + U\frac{f_{i+1}^n - f_{i-1}^n}{2\Delta x} + O(\Delta t) + O(\Delta x^2) = 0 \tag{11.5.3}$$

The associated finite-difference stencil is shown with hollow circles in Figure 11.2.1. Since equation (11.5.3) arose by applying standard finite-difference approximations, its consistency is guaranteed and does not need to be examined. Solving equation (11.5.3) for f_i^{n+1}, we obtain an explicit method expressed by

$$f_i^{n+1} = \frac{c}{2} f_{i-1}^n + f_i^n - \frac{c}{2} f_{i+1}^n \tag{11.5.4}$$

where

$$c \equiv \frac{U\,\Delta t}{\Delta x} \tag{11.5.5}$$

is the *convection number*, also called the *Courant number.*

Successive mappings and numerical stability

According to equation (11.5.4), the solution vector $\boldsymbol{f}^{n+1}$ emerges by premultiplying $\boldsymbol{f}^n$ with a tridiagonal matrix $\boldsymbol{B}$ whose subdiagonal, diagonal, and superdiagonal elements are equal to $c/2$, 1, and $-c/2$, respectively. When the spectral radius of $\boldsymbol{B}$, denoted by $\rho(\boldsymbol{B})$, is equal to, less than, or larger than unity, the length of $\boldsymbol{f}^n$ will stay roughly constant, decrease, or increase during successive projections. Remembering that, according to the exact solution, the initial distribution is convected without change in shape, we accept the first behavior, tolerate the second behavior, and dismiss the third behavior as numerically unstable. The eigenvalues and hence the spectral radius of $\boldsymbol{B}$ can be computed analytically using the results of Section 5.3.

$$f^{n+1} = \begin{bmatrix} 1 & -c/2 & 0 & \cdots & 0 & 0 \\ c/2 & 1 & -c/2 & 0 & \cdots & 0 \\ 0 & c/2 & 1 & -c/2 & 0 & \cdots \\ 0 & \ddots & \ddots & \ddots & & 0 \\ \cdots & 0 & c/2 & 1 & -c/2 & 0 \\ 0 & \cdots & 0 & c/2 & 1 & -c/2 \\ 0 & 0 & \cdots & 0 & c/2 & 1 \end{bmatrix} f^n$$

An equivalent way of assessing the stability of the method is to perform the von Neumann analysis, obtaining the amplification factor

$$\frac{A^{n+1}}{A^n} \equiv G = 1 - Ic\,\sin\theta \tag{11.5.6}$$

Since the magnitude of the right-hand side of equation (11.5.6) is greater than unity for any value of c and θ, the magnitude of $\boldsymbol{f}^n$ will amplify for any value of the wavelength L. Thus the FTCS method is *unconditionally unstable* and must be abandoned. Recall that the FTCS method for the unsteady diffusion equation is conditionally stable.

FTBS Method

We proceed to explore a different type of discretization in the hope of improving numerical stability. Now, we apply equation (11.5.1) at the point x_i at the time instant t^n, and approximate $\partial f/\partial t$ with a forward difference and $\partial f/\partial x$ with a backward difference, to obtain

$$\frac{f_i^{n+1} - f_i^n}{\Delta t} + U\frac{f_i^n - f_{i-1}^n}{\Delta x} + O(\Delta t) + O(\Delta x) = 0 \tag{11.5.7}$$

The corresponding finite-difference stencil is shown with solid circles in Figure 11.5.1. Rearranging equation (11.5.7), we derive the two-level explicit method expressed by

$$f_i^{n+1} = cf_{i-1}^n + (1-c)f_i^n \tag{11.5.8}$$

It is interesting to note that when $c = 1$, we obtain $f_i^{n+1} = f_{i-1}^n$, which reproduces the exact solution.

Carrying out the von Neumann stability analysis, we derive the amplification factor

$$G = 1 - c[1 - \exp(-I\theta)] \tag{11.5.9}$$

which reveals that G is located on a circle in the complex plane. The center of the circle is at the point $(1 - c, 0)$, and its radius is equal to c. To ensure stability we require $|G| \le 1$ and derive the restriction $0 \le c \le 1$. Clearly, if the convection velocity U is negative, the method will be unstable.

FTFS Method

Returning to equation (11.5.1), we use forward finite differencing for both the time and space derivatives, and obtain an explicit method expressed by

$$f_i^{n+1} = (1+c)f_i^n - cf_{i+1}^n \tag{11.5.10}$$

The corresponding finite-difference stencil is shown with hollow circles in Figure 11.5.1. When $c = -1$, we recover the exact analytical solution.

Carrying out the von Neumann stability analysis, we derive the amplification factor

$$G = 1 + c[1 - \exp(-I\theta)] \tag{11.5.11}$$

which shows that $|G| \le 1$, and the method is stable, provided that $-1 \le c \le 0$. When the convection velocity U is positive, the method is unstable.

Upwind Differencing, Numerical Diffusivity, and the CFL Condition

The complementary successes of FTBS and FTFS discretizations for positive and negative values of the convection velocities, respectively, suggest the method of *upwind differencing*: use FTBS when $U > 0$, use FTFS when $U < 0$, and always maintain $|c| \le 1$. The restriction $|c| \le 1$ is known as the *Courant–Friedrichs–Lewy* or CFL *stability criterion.*

Upwind differencing is particularly effective for nonlinear problems where the convection velocity U is not constant but varies in time and space over the domain of solution. In physical terms, upwind differencing carries information on the structure of the solution forward from the direction of a traveling wave, and thus suppresses the growth of unwanted perturbations propagated from other directions.

Numerical diffusivity

To further explain the conditional stability of the one-sided FTBS and FTFS difference methods, which may be contrasted with the unconditional instability of the FTCS method, we perform a consistency analysis.

Considering first equation (11.5.8), we pretend that the discrete variables in the difference equation are continuous functions of space and time, and we expand them in Taylor series about the grid point (x_i, t^n) to obtain

$$f_i^n + \left(\frac{\partial f}{\partial t}\right)_i^n \Delta t + \tfrac{1}{2}\left(\frac{\partial^2 f}{\partial t^2}\right)_i^n \Delta t^2 + O(\Delta t^3) = c\left[f_i^n - \left(\frac{\partial f}{\partial x}\right)_i^n \Delta x + \tfrac{1}{2}\left(\frac{\partial^2 f}{\partial x^2}\right)_i^n \Delta x^2\right] + (1-c) f_i^n + O(\Delta x^3) \tag{11.5.12}$$

Rearranging, and using equation (11.5.2), we find

$$\left(\frac{\partial f}{\partial t} + U\frac{\partial f}{\partial x}\right)_i^n = \frac{1-c}{2} U\Delta x \left(\frac{\partial^2 f}{\partial x^2}\right)_i^n + O(\Delta t) + \frac{1}{\Delta t} O(\Delta x^3) \tag{11.5.13}$$

As expected, in the limit as Δx tends to zero, equation (11.5.13) reduces to the original differential equation (11.5.1), thereby confirming the consistency of the FTBS discretization. The key observation for present purposes is that the right-hand side of equation (11.5.13) also involves a small diffusive term with an artificial or numerical diffusivity that is equal to $\frac{1}{2}(1-c)U\Delta x$, which is nonnegative when $0 \le c \le 1$.

Working similarly with the FTFS method, we derive a modified differential equation that is identical to equation (11.5.13) except that the numerical diffusivity is equal to $-\frac{1}{2}(1-c)U\Delta x$, which is nonnegative when $-1 \le c \le 0$. Combining these results with those of the von Neumann stability analysis suggests that *positive numerical diffusivity is instrumental in maintaining numerical stability.*

What went wrong with the FTCS method? Carrying out a consistency analysis, we derive an equation that is identical to equation (11.5.13), except that the numerical diffusivity is now equal to $-\frac{1}{2}cU\Delta x = -\frac{1}{2}U^2\Delta t$ which is negative for any value of c or Δt. The negative numerical diffusivity is the source of the numerical instability.

Numerical cone of influence

To obtain a graphical interpretation of the CFL condition, we recall that the exact solution at the grid point (x_i, t^n) may be found by traveling backward in time along the characteristic line emanating from that point, shown with the dashed line in Figure 11.5.1, until we reach the time level $t = 0$ where $x = X$; then, $f_i^n = F(X)$, where $F(x)$ is the initial distribution. According to equations (11.5.8) and (11.5.10), in order to compute the solution at the grid point (x_i, t^n), we use information at all grid points within a planar angle with vertex at the point (x_i, t^n) shown as a dotted area in Figure 11.5.1, called the *numerical cone of influence.* The CFL condition requires that the characteristic line lie within the numerical cone of influence.

In general, the requirement that the characteristic line emanating from a certain point lie within the numerical cone of influence is necessary but not sufficient for numerical stability. Thus, if the numerical cone of influence does not include the characteristic line that passes through the vertex, the numerical method will certainly be unstable.

Lax's Modification of the FTCS Method

Lax (1954) proposed a modification of the unconditionally unstable FTCS method that introduces a certain stabilizing diffusive action. This is done by replacing f_i^n in the temporal finite-difference approximation of equation (11.5.3) with the average value $\frac{1}{2}(f_{i+1}^n + f_{i-1}^n)$, yielding

$$\frac{f_i^{n+1} - \frac{1}{2}(f_{i+1}^n + f_{i-1}^n)}{\Delta t} + U\frac{f_{i+1}^n + f_{i-1}^n}{2\Delta x} = 0 \tag{11.5.14}$$

Rearranging, we obtain the explicit two-level algorithm

$$f_i^{n+1} = \tfrac{1}{2}(1 + c)f_{i-1}^n + \tfrac{1}{2}(1 - c)f_{i+1}^n \tag{11.5.15}$$

Since equation (11.5.14) was derived from a rather *ad hoc* modification of the FTCS discretization, its consistency must be examined. Expanding all variables in Taylor series about the point (x_i, t^n), we derive the modified differential equation

$$\left(\frac{\partial f}{\partial t} + U\frac{\partial f}{\partial x}\right)_i^n = -\tfrac{1}{2}\Delta t\left(\frac{\partial^2 f}{\partial t^2}\right)_i^n + \tfrac{1}{2}\frac{\Delta x^2}{\Delta t}\left(\frac{\partial^2 f}{\partial x^2}\right)_i^n + O(\Delta x) + O(\Delta t) \tag{11.5.16}$$

which shows that Lax's method is consistent only when Δx and Δt are reduced simultaneously so that the ratio $\Delta x^2/\Delta t$ tends to zero. Maintaining, in particular, the ratio $\Delta x/\Delta t$ at a constant value yields a method that is consistent and first-order accurate in both time and space.

Carrying out the von Neumann stability analysis, we find the amplification factor

$$G = \cos\theta - Ic\sin\theta \tag{11.5.17}$$

which shows that Lax's method is stable provided that the CFL criterion is fulfilled, $|c| < 1$. To explain this behavior, we substitute the right-hand side of equation (11.5.2) in place of the first term on the right-hand side of equation (11.5.16), and group the first with the second term to obtain a diffusion term with numerical diffusivity equal to $\frac{1}{2}U\Delta x(1 - c^2)/2c$, which is positive when $|c| < 1$. The magnitude of the numerical diffusivity becomes excessively large as c tends to zero, and this places a serious restriction on the effectiveness of Lax's method: the improvement in stability is penalized by a loss of accuracy.

The Explicit Lax–Wendroff Method

Upwind differencing offers numerical stability but suffers from low-order spatial accuracy. FTFS differencing, on the other hand, offers second-order spatial accuracy but suffers from unconditional instability. Is it possible to devise a two-level explicit method that combines accuracy and stability, that is, a method that is second-order accurate in both time and space and unconditionally stable?

To answer this question, we express f_i^{n+1} as a linear combination of f_{i-1}^n, f_i^n, and f_{i+1}^n, writing

$$f_i^{n+1} = a_{-1}f_{i-1}^n + a_0 f_i^n + a_1 f_{i+1}^n \tag{11.5.18}$$

where a_{-1}, a_0, and a_1 are three constant coefficients. It will be noted that equation (11.5.18) is a generalization of all two-level methods considered previously in this section.

To ensure the consistency of equation (11.5.18), we expand all variables in a Taylor series about the point (x_i, t^n) and derive the modified differential equation

$$(1 - a_{-1} - a_0 - a_1) f_i^n + \Delta t \left(\frac{\partial f}{\partial t} + \frac{U}{c}(a_{-1} - a_1)\frac{\partial f}{\partial x} \right)_i^n + \tfrac{1}{2}\Delta t^2 \left(\frac{\partial^2 f}{\partial t^2} \right)_i^n$$
$$= \tfrac{1}{2}\Delta x^2 (a_{-1} + a_1) \left(\frac{\partial^2 f}{\partial x^2} \right)_i^n \tag{11.5.19}$$

Substituting the left-hand side of equation (11.5.2) into the last term on the left-hand side of equation (11.5.19), requiring that in the limit as Δx and Δt tend to zero equation (11.5.19) reduces to equation (11.5.1), and stipulating that the second-order temporal error cancels the second-order spatial error, we obtain the following system of algebraic equations for the three coefficients a_{-1}, a_0, and a_1:

$$\begin{bmatrix} 1 & 1 & 1 \\ 1 & 0 & -1 \\ 1 & 0 & 1 \end{bmatrix} \begin{bmatrix} a_{-1} \\ a_0 \\ a_1 \end{bmatrix} = \begin{bmatrix} 1 \\ c \\ c^2 \end{bmatrix} \tag{11.5.20}$$

whose solution is

$$a_{-1} = \tfrac{1}{2}c(c+1), \qquad a_0 = 1 - c^2, \qquad a_1 = \tfrac{1}{2}c(c-1) \tag{11.5.21}$$

Substituting these values into equation (11.5.18), we obtain the Lax–Wendroff method (1960).

Carrying out the standard von Neumann stability analysis, we find the amplification factor

$$G = 1 - c^2 + c^2 \cos\theta - Ic \sin\theta \tag{11.5.22}$$

which shows that G traces an ellipse in the complex plane with center at $(1 - c^2, 0)$ and semi-axes equal to c and c^2. Geometrical arguments reveal that $|G| < 1$ when $|c| < 1$, and this shows that the method is conditionally stable.

The Lax–Wendroff method provides us with an efficient algorithm: it is explicit, second-order accurate, and conditionally stable.

Explicit CTCS or Leapfrog Method

One way to guarantee second-order accuracy in both time and space is to use central differences for both the temporal and spatial derivatives $\partial f/\partial t$ and $\partial f/\partial x$, obtaining the CTCS difference equation

$$\frac{f_i^{n+1} - f_i^{n-1}}{2\Delta t} + U\frac{f_{i+1}^n - f_{i-1}^n}{2\Delta x} + O(\Delta t^2) + O(\Delta x^2) = 0 \tag{11.5.23}$$

Rearranging, we derive a three-level explicit method expressed by

$$f_i^{n+1} = f_i^{n-1} + cf_{i-1}^n - cf_{i+1}^n \tag{11.5.24}$$

The solution at the first time level $n = 1$ must be computed using a two-level method with a sufficiently small time step in order to prevent oscillations.

Performing the von Neumann stability analysis, we find that the amplification factor satisfies the quadratic equation $G^2 + 2I\delta G - 1 = 0$, where $\delta = c \sin\theta$ is a real parameter. The roots of this equation are

$$G = -I\delta \pm \sqrt{1-\delta^2} \tag{11.5.25}$$

To assess the magnitude of the amplification factor, we distinguish two cases: (*a*) if $\delta^2 > 1$, then $G = I[-\delta \pm (\delta^2 - 1)^{1/2}]$, and the magnitude of the root corresponding to the minus sign is greater than unity; (*b*) if $\delta^2 \le 1$ then $|G| = 1$, which shows that the magnitude of A^n stays constant during the successive mappings in agreement with the exact solution. We note that $\delta^2 \le 1$ when $|c| \le 1$ for any value of θ and find that the CTCS method will be stable provided that the CFL condition is fulfilled. To this end, we recall that the CTCS discretization for the diffusion equation was found to be unconditionally unstable, and suggests that the presence of diffusivity does not necessarily promote the numerical stability.

In practice, the efficiency of the CTCS method is hindered by increased memory requirements associated with the storing of information at three time levels, and the occurrence of dual numerical error that grows independently at every other time step. This behavior is known as *even–odd coupling*.

Implicit BTCS Method

We turn next to considering implicit discretizations in the hope of achieving unconditional stability and thus relaxing the restriction on the time step Δt. The BTCS discretization yields the difference equation

$$\frac{f_i^{n+1} - f_i^n}{\Delta t} + U\frac{f_{i+1}^{n+1} - f_{i-1}^{n+1}}{2\Delta x} + O(\Delta t) + O(\Delta x^2) = 0 \tag{11.5.26}$$

which may be rearranged to give

$$-cf_{i-1}^{n+1} + 2f_i^{n+1} + cf_{i+1}^{n+1} = 2f_i^n \tag{11.5.27}$$

$$f^n = \begin{bmatrix} 2 & c & 0 & \cdots & 0 & 0 \\ -c & 2 & c & 0 & \cdots & 0 \\ 0 & -c & 2 & c & 0 & \cdots \\ 0 & \ddots & \ddots & \ddots & & 0 \\ \cdots & 0 & -c & 2 & c & 0 \\ 0 & \cdots & 0 & -c & 2 & c \\ 0 & 0 & \cdots & 0 & -c & 2 \end{bmatrix} f^{n+1}$$

In the matrix notation, we obtain $\boldsymbol{B}\boldsymbol{f}^{n+1} = \boldsymbol{f}^n$, where $\boldsymbol{B}$ is a tridiagonal matrix with subdiagonal, diagonal, and superdiagonal elements equal to $-c/2$, 1, and $c/2$, respectively. Note that the matrix $\boldsymbol{B}$ is proportional to the transpose of the matrix $\boldsymbol{B}$ corresponding to the FTCS method.

Carrying out the von Neumann stability analysis, we find the amplification factor

$$G = \frac{1}{1 + Ic \sin\theta} \tag{11.5.28}$$

Since $|G|$ is always less than unity, the BTCS method is unconditionally stable. The increase in computational effort required for solving the linear system (11.5.27) is compensated with unconditional stability. The reason is that the boundaries of the numerical cone of influence include all points at the $n+1$ time level: The numerical cone of influence reduces to a rectangular strip that is guaranteed to contain the characteristic line passing through (x_i, t^n).

Crank–Nicolson Method

As in the case of the unsteady diffusion equation, the Crank–Nicolson method proceeds by applying the differential equation midway between the time levels t^n and t^{n+1}, and approximating the spatial derivative with the average of the two derivatives at the two time levels, obtaining

$$\frac{f_i^{n+1} - f_i^n}{\Delta t} + U\frac{1}{2}\left(\frac{f_{i+1}^{n+1} - f_{i-1}^{n+1}}{2\Delta x} + \frac{f_{i+1}^n - f_{i-1}^n}{2\Delta x}\right) + O(\Delta t^2) + O(\Delta x^2) = 0 \quad (11.5.29)$$

Rearranging, we derive the implicit difference equation

$$-cf_{i-1}^{n+1} + 4f_i^{n+1} + cf_{i+1}^{n+1} = cf_{i-1}^n + 4f_i^n - cf_{i+1}^n \quad (11.5.30)$$

which produces tridiagonal matrices. The amplification factor is equal to the ratio of two complex conjugate numbers,

$$G = \frac{2 - Ic\,\sin\theta}{2 + Ic\,\sin\theta} \quad (11.5.31)$$

Since $|G| = 1$, the method is unconditionally stable for any value of c.

Comparison of the Methods

The restrictions on the time step of the conditionally stable explicit methods for the convection equation are typically not prohibitive. Implicit methods allow the use of larger time steps, but the associated numerical error may erode the accuracy and therefore the physical relevance of the solution. Thus explicit methods are a standard choice.

Modified Dynamics and Explicit Numerical Diffusion

We saw that the finite-difference discretizations introduce some type of numerical error whose leading-order term may be proportional to the second, the third, or the fourth spatial derivative of the solution. The presence of these terms introduces, respectively, *numerical diffusion*, *dispersion*, and *fourth-order diffusion.*

Numerical diffusion is indispensable for maintaining numerical stability and dampening small-scale irregularities yielding *monotone schemes*, that is, schemes that produce solutions that are free of artificial oscillations. Unfortunately, in practice, a prohibitively fine grid may be required in order to reduce the artificial smearing of sharp gradients. A remedy would be to use a higher-order method, but the associated modified differential equation typically contains a dispersive term that causes local oscillations. The magnitude of the oscillations may be reduced by enhancing the original differential equation with explicit diffusion-like terms, expressing regular or fourth-order diffusion. In one dimension, these are expressed, respectively, by the terms

$$\beta(x)\frac{\partial^2 f}{\partial x^2}, \qquad -\gamma(x)\frac{\partial^4 f}{\partial x^4} \tag{11.5.32}$$

The positive coefficients β and γ are allowed to vary in space and time according to the expected structure of the solution. The optimal values of these coefficients depend on the particular problem under consideration and must be found by numerical experimentation.

One justification for introducing explicit numerical diffusion is that the approximations that lead us to the convection equation have neglected certain physical diffusion-like processes, and it is possible that introducing them back into the differential equation will not affect the physical relevance of the solution.

Explicit Multistep Methods

One way to improve the accuracy of explicit methods is to advance the solution over each time step using a number of elementary substeps that are carried out using different numerical methods. Some of these substeps are predictive, in the sense that they seek to estimate the solution at the next time level using a crude method, while others correct the predictions using a more sophisticated method.

The overall action of a multistep method for *linear* convection can be reduced to that of a single-step explicit method discussed previously in this section. Their discussion in the present context serves as a point of departure for developing methods for problems with nonconstant convection velocity, which will be pursued later in this section.

Richtmyer's method

Richtmyer (1963) suggested carrying out a complete step in two substeps of equal duration $\Delta t/2$. The first substep is executed using Lax's method according to equation (11.5.15), and the second substep is executed using the CTCS method according to equation (11.5.24). The finite-difference equations are

$$f_i^{n+1/2} = \tfrac{1}{2}(1+\tfrac{1}{2}c)f_{i-1}^n + \tfrac{1}{2}(1-\tfrac{1}{2}c)f_{i+1}^n, \qquad f_i^{n+1} = f_i^n - \tfrac{1}{2}c(f_{i+1}^{n+1/2} - f_{i-1}^{n+1/2}) \tag{11.5.33}$$

where $c = U\Delta t/\Delta x$. Eliminating the intermediate solution from the second equation, we find

$$f_i^{n+1} = \frac{c}{4}\left(\frac{c}{2}+1\right)f_{i-2}^n + \left(1-\frac{c^2}{4}\right)f_i^n + \frac{c}{4}\left(\frac{c}{2}-1\right)f_{i+2}^n \tag{11.5.34}$$

which is the Lax–Wendroff formula (11.5.18) with coefficients given in equation (11.5.21) and grid spacing equal to $2\Delta x$. This shows that Richtmyer's method is second-order accurate in both time and space and stable as long as $|c| \le 2$.

Multistep Lax–Wendroff

Burstein (1967) suggested a modification of Richtmyer's method according to which, in the first substep, the convection equation is applied at the intermediate grid point $i+\frac{1}{2}$, whereas in the second substep, the equation is applied at the regular grid point i, both times with a spatial step equal to $\Delta x/2$. The counterparts of equations (11.5.33) are

$$f_{i+1/2}^{n+1/2} = \tfrac{1}{2}(1+c)f_i^n + \tfrac{1}{2}(1-c)f_{i+1}^n, \qquad f_i^{n+1} = f_i^n - c(f_{i+1/2}^{n+1/2} - f_{i-1/2}^{n+1/2}) \tag{11.5.35}$$

where $c = U\Delta t/\Delta x$. Eliminating the intermediate solution from the second equation yields the Lax–Wendroff formula (11.5.18) with the coefficients given in equations (11.5.21); this ensures that the method is second-order accurate in both time and space, and stable as long as $|c| \le 1$.

MacCormack method

MacCormack (1969) developed a genuine predictor–corrector method that has become popular in engineering practice. The predictor step provides us with an approximation to f_i^{n+1} denoted by f_i^*, computed using the explicit FTFS discretization according to equation (11.5.10), as

$$f_i^* = (1+c)f_i^n - cf_{i+1}^n \tag{11.5.36}$$

The second step uses the explicit forward time approximation and a hybrid forward/backward space approximation that involves the predicted values, according to the difference equation

$$\frac{f_i^{n+1} - f_i^n}{\Delta t} + U\frac{1}{2}\left(\frac{f_{i+1}^n - f_i^n}{\Delta x} + \frac{f_i^* - f_{i-1}^*}{\Delta x}\right) = 0 \tag{11.5.37}$$

Rearranging equation (11.5.37) and using equation (11.5.36) to eliminate f_{i+1}^n in favor of f_i^*, we derive the explicit formula

$$f_i^{n+1} = \tfrac{1}{2}[f_i^n + f_i^* - c(f_i^* - f_{i-1}^*)] \tag{11.5.38}$$

Eliminating the intermediate variable f_i^* from equations (11.5.36) and (11.5.38), we obtain the Lax–Wendroff method; this ensures that MacCormack's method is second-order accurate in both time and step and stable for $|c| \le 1$.

The bias in the solution due to the one-sided differencing involved in equations (11.5.36) and (11.5.38) may be prevented by alternating the direction of the one-sided differences after completion of one time step. Thus, in the next step, we use FTBS and FTBS-FS discretizations to obtain the equations

$$f_i^* = cf_{i-1}^n + (1-c)f_i^n, \qquad f_i^{n+1} = \tfrac{1}{2}[f_i^n + f_i^* - c(f_{i+1}^* - f_i^*)] \tag{11.5.39}$$

We shall see later in this section that MacCormack's method is particularly effective for problems with variable convection velocity.

Flux-Corrected Transport

We saw earlier in this section that explicit numerical diffusion may be necessary in order to enhance the performance of a first-order method. The idea behind the flux-corrected-transport method is to use a predictor–corrector method where the predictor step involves an artificial dampening term, while the corrector step removes the excessive dissipation by introducing antidiffusion, that is, diffusion with a negative diffusivity. The method is developed and discussed in three parts by Boris and Book (1973), Book et al. (1975), and Boris and Book (1976).

Nonlinear Convection

Most of the methods for linear convection discussed earlier in this section may be extended to nonlinear convection, where U is no longer a constant. The main difference is that the convection velocity must be evaluated at the grid point where the differential equation is applied to yield the finite-difference equation. When a finite-difference method is conditionally stable, the size of the time step must be kept sufficiently small in order to satisfy the stability criteria derived for linear convection with the maximum value of the velocity over the domain of the solution.

Implicit methods lead to systems of nonlinear algebraic equations that must be solved using iterative procedures. This complication often introduces pragmatic difficulties associated with excessive computational cost and renders the implicit methods a mere academic alternative.

Multistep methods are efficient and easy to implement, and offer high-order accuracy. Richtmyer's and MacCormack's two-step methods are two popular choices.

Inviscid Burgers Equation

To illustrate the implementation of finite-difference methods for nonlinear problems, we consider the inviscid Burgers equation whose nonconservative and conservative forms are

$$\frac{\partial f}{\partial t} + f\frac{\partial f}{\partial x} = 0 \qquad \text{and} \qquad \frac{\partial f}{\partial t} + \frac{\partial E}{\partial x} = 0 \tag{11.5.40}$$

where $E \equiv \frac{1}{2}f^2$. Physically, Burgers' equation describes the propagation of wave fronts with a local convection velocity that is equal to the local amplitude of the wave. The dependence of the velocity on the amplitude may cause the formation of discontinuous fronts or shocks from smooth initial distributions.

Lax's method

Lax's modification of the FTCS discretization yields an explicit two-level algorithm expressed by

$$f_i^{n+1} = \tfrac{1}{2}(f_{i-1}^n + rE_{i-1}^n) + \tfrac{1}{2}(f_{i+1}^n - rE_{i-1}^n) \tag{11.5.41}$$

where $r = \Delta t/\Delta x$, which is first-order accurate in time second-order accurate in space, and stable as long as $|c_{Max}| = |rf_{Max}| < 1$.

Lax–Wendroff method

Proceeding as in the case of linear convection with the objective of achieving second-order accuracy, we develop the nonlinear Lax–Wendroff method expressed by

$$\begin{aligned} f_i^{n+1} &= f_i^n - \tfrac{1}{2}r(E_{i+1}^n - E_{i-1}^n) \\ &\quad + \tfrac{1}{4}r^2[-f_{i-1}^n(E_i^n - E_{i-1}^n) + f_i^n(E_{i+1}^n - 2E_i^n + E_{i-1}^n) + f_{i+1}^n(E_{i+1}^n - E_i^n)] \end{aligned} \tag{11.5.42}$$

which is second-order accurate in both space and time, and stable as long as $|c_{Max}| < 1$ (e.g., Hoffmann and Chiang 1993, Vol. I, p. 207).

Implicit BTCS method

The implicit BTCS method results in the quadratically nonlinear system of equations

$$c(f_{i+1}^{n+1})^2 + 4f_i^{n+1} - c(f_{i-1}^{n+1})^2 = 4f_i^n \tag{11.5.43}$$

which must be solved using an iterative method. A suitable initial guess is the converged solution at the previous time step. In practice, increased computational demands discourage the selection of this method.

MacCormack's method

The two-step MacCormack method proceeds by predicting a solution, designated with an asterisk, and then correcting it according to the finite-difference equations

$$f_i^* = f_i^n - r(E_{i+1}^n - E_i^n), \qquad f_i^{n+1} = \tfrac{1}{2}[f_i^n + f_i^* - r(E_i^* - E_{i-1}^n)] \tag{11.5.44}$$

The method is stable as long as $|c_{Max}| = |r f_{Max}| < 1$. It will be noted that equations (11.5.44) are straightforward extensions of equations (11.5.36) and (11.5.38). MacCormack's method has become a popular choice in practice.

PROBLEMS

11.5.1 ***Numerical diffusivity.***

Carry out a consistency analysis of the FTCS and Lax's methods, and derive the corresponding numerical diffusivities.

11.5.2 ***Lax–Wendroff method for the inviscid Burgers equation.***

Derive equation (11.5.42) working by analogy with the case of linear convection.

11.5.3 ***Burgers equation with MacCormack's method.***

Use MacCormack's method to solve the inviscid Burgers equation in an infinite domain with initial conditions: (*a*) the Heaviside step function $F(x) = 1$ for $x < 0$ and $F(x) = 0$ for $x > 0$; (*b*) $F(x) = \exp(-x^2)$. Discuss the behavior of the numerical solution in each case.

11.5.4 ***Burgers equation with an implicit method.***

Repeat Problem 11.5.3 with the implicit BTCS method, and discuss the performance of the method and required computational cost.

11.6 *Convection in Two and Three Dimensions*

Generalizing the discussion of Section 11.5, we consider linear convection in three dimensions described by equation (11.1.7). The convection velocity vector $\boldsymbol{U} = (U, V, W)$ is assumed to be constant in time and uniform in space. Two-dimensional convection arises by setting $W = 0$. The solution is to be found subject to an appropriate initial condition $f(\boldsymbol{x}, t = 0) = F(\boldsymbol{x})$.

Assuming, for simplicity, that the domain of solution extends over the whole three-dimensional space or two-dimensional plane, we find that the exact solution is given by $F(\boldsymbol{x} - \boldsymbol{U}t)$, which states that the initial distribution F travels with the constant velocity $\boldsymbol{U}$. The characteristic lines along which the function f remains constant are described by the equation $\boldsymbol{x} - \boldsymbol{U}t = \boldsymbol{X}$, where $\boldsymbol{X}$ is an arbitrary point.

Finite-difference methods for two-dimensional and three-dimensional convection arise as direct and straightforward extensions of the methods for one-dimensional convection discussed in the preceding section. In the remainder of this section, we illustrate several characteristic examples.

Lax's Method

Lax's method in three dimensions emerges by replacing equation (11.1.7) with the finite-difference equation

$$\frac{1}{\Delta t}[f_{i,j,k}^{n+1} - \tfrac{1}{6}(f_{i+1,j,k}^n + f_{i-1,j,k}^n + f_{i,j+1,k}^n + f_{i,j-1,k}^n + f_{i,j,k+1}^n + f_{i,j,k-1}^n)]$$
$$+\frac{U}{2\Delta x}(f_{i+1,j,k}^n - f_{i-1,j,k}^n) + \frac{V}{2\Delta y}(f_{i,j+1,k}^n - f_{i,j-1,k}^n) + \frac{W}{2\Delta z}(f_{i,j,k+1}^n - f_{i,j,k-1}^n) = 0 \tag{11.6.1}$$

Rearranging the various terms, we obtain an *explicit* two-level algorithm expressed by

$$f_i^{n+1} = \tfrac{1}{2}(\tfrac{1}{3} + c_x)f_{i-1,j,k}^n + \tfrac{1}{2}(\tfrac{1}{3} - c_x)f_{i+1,j,k}^n + \tfrac{1}{2}(\tfrac{1}{3} + c_y)f_{i,j-1,k}^n + \tfrac{1}{2}(\tfrac{1}{3} - c_y)f_{i,j+1,k}^n$$
$$+\tfrac{1}{2}(\tfrac{1}{3} + c_z)f_{i,j,k-1}^n + \tfrac{1}{2}(\tfrac{1}{3} - c_z)f_{i,j,k+1}^n \tag{11.6.2}$$

where

$$c_x = \frac{U\,\Delta t}{\Delta x}, \qquad c_y = \frac{V\,\Delta t}{\Delta y}, \qquad c_z = \frac{W\,\Delta t}{\Delta z} \tag{11.6.3}$$

Performing the von Neumann stability analysis, we find that the method is stable provided that

$$c_x^2 + c_y^2 + c_z^2 < \tfrac{1}{3} \tag{11.6.4}$$

This condition imposes a strong restriction on the size of the time step. For convection in two dimensions, we find the analogous condition $c_x^2 + c_y^2 < \frac{1}{2}$, which is still a stringent constraint.

Implicit Methods

To achieve unconditional stability, one may resort to a fully implicit method. Unfortunately, the finite-difference equations result in a system of algebraic equations that is sparse but not tridiagonal, and its solution may require a prohibitive computational effort. One remedy is to use an ADI method that requires solving a one-dimensional problem within each substep. ADI methods for hyperbolic equations are developed and discussed by Lee (1962), Douglas and Gunn (1964), and Hirsch (1988, Vol. I, p. 442). The standard implementation of the ADI method leads to an algorithm that is unconditionally stable in two dimensions, but unconditionally unstable in three dimensions.

Operator Splitting

Following the general idea of operator splitting and fractional steps, we replace equation (11.1.7) with a system of three equations that operate in a sequential manner, each for the full time interval Δt,

$$\frac{\partial f}{\partial t} + U\frac{\partial f}{\partial x} = 0, \qquad \frac{\partial f}{\partial t} + V\frac{\partial f}{\partial y} = 0, \qquad \frac{\partial f}{\partial t} + W\frac{\partial f}{\partial z} = 0 \tag{11.6.5}$$

all for $t^n < t < t^n + \Delta t$. Adopting the implicit BTCS discretization, we advance the solution over each fractional step by solving the tridiagonal systems of equations

$$(1 + c_x\Delta_x)f_{i,j,k}^* = f_{i,j,k}^n, \qquad (1 + c_y\Delta_y)f_{i,j,k}^{**} = f_{i,j,k}^*, \qquad (1 + c_z\Delta_z)f_{i,j,k}^{n+1} = f_{i,j,k}^{**} \tag{11.6.6}$$

where the variables with single and double asterisks designate the solutions after the first and second fractional step, Δ_x is the first-difference operator defined as

$$\Delta_x \left\langle f_{i,j,k} \right\rangle = f_{i+1,j,k} - f_{i-1,j,k} \tag{11.6.7}$$

and Δ_y and Δ_z are defined in a similar manner. Combining the three equations (11.6.6), we obtain the overall finite-difference scheme

$$(1 + c_x \Delta_x)(1 + c_y \Delta_y)(1 + c_z \Delta_z) f_{i,j,k}^{n+1} = f_{i,j,k}^{n} \tag{11.6.8}$$

involving a factorized implicit operator on the left-hand side.

Now, the finite-difference equation corresponding to the fully implicit BTCS discretization is

$$(1 + c_x \Delta_x + c_y \Delta_y + c_z \Delta_z) f_{i,j,k}^{n+1} = f_{i,j,k}^{n} \tag{11.6.9}$$

The left-hand sides of equations (11.6.8) and (11.6.9) are identical up to first order in the spatial intervals, and this means that the fractional-step method may be regarded as the result of the approximate factorization of the explicit BTCS discretization.

PROBLEMS

11.6.1 *Lax's method.*

(*a*) Perform the von Neumann stability analysis to derive the stability criterion displayed in equation (11.6.4), and derive the associated numerical diffusivity.
(*b*) Analyze the stability of the method in two dimensions, and derive the stability constraint given in the text.

11.6.2 *ADI method.*

Devise an ADI method for linear two-dimensional convection, and study its consistency and stability. Then indicate how the method can be extended to three dimensions.

11.7 *Convection–diffusion in One Dimension*

The methods developed in the preceding sections for the modular cases of pure diffusion and pure convection may be extended in a straightforward fashion to tackle combined convection–diffusion. In this section, we discuss extensions for the simplest case of linear one-dimensional convection–diffusion described by the *parabolic* differential equation

$$\frac{\partial f}{\partial t} + U \frac{\partial f}{\partial x} = \kappa \frac{\partial^2 f}{\partial x^2} \tag{11.7.1}$$

subject to the initial condition $f(x, t = 0) = F(x)$. Both the convection velocity U and diffusivity κ are assumed to be constant.

Explicit FTCS and Lax's Method

The FTCS discretization yields the difference equation

$$\frac{f_i^{n+1} - f_i^n}{\Delta t} + U\frac{f_{i+1}^n - f_{i-1}^n}{2\Delta x} = \kappa\frac{f_{i+1}^n - 2f_i^n + f_{i-1}^n}{\Delta x^2} + O(\Delta t) + O(\Delta x^2) \tag{11.7.2}$$

Rearranging, we derive the two-level explicit algorithm

$$f_i^{n+1} = f_i^n - c\tfrac{1}{2}(f_{i+1}^n - f_{i-1}^n) + \alpha(f_{i+1}^n - 2f_i^n + f_{i-1}^n) \tag{11.7.3}$$

where we recall that $c = U\Delta t/\Delta x$ is the Courant number and $\alpha = \kappa\Delta t/\Delta x^2$ is the diffusion number. The ratio between these two numbers

$$Re_c \equiv \frac{c}{\alpha} = \frac{U\Delta x}{\kappa} \tag{11.7.4}$$

is the *cell Reynolds number* or *cell Péclet number*. Physically, Re_c expresses the relative strengths of the convective and diffusive contributions to equation (11.7.1).

In previous sections, we found that the explicit FTCS discretization is conditionally stable for pure unsteady diffusion, and unconditionally unstable for pure linear convection. Performing the von Neumann stability analysis for the present case of mixed convection–diffusion, we derive the amplification factor

$$G = 1 - 2\alpha + 2\alpha\cos\theta - Ic\sin\theta \tag{11.7.5}$$

which shows that G traces an ellipse that passes through the point $(1, 0)$ in the complex plane. The center of the ellipse is located at the point $(1 - 2\alpha, 0)$, and its semi-axes are equal to 2α and c. To guarantee stability, we must ensure that the ellipse resides within the unit disk. Requiring that the lengths of the semi-axes are less than unity yields the restrictions $\alpha < \frac{1}{2}$, $c < 1$. A third restriction emerges by requiring that the curvature of the ellipse at the point $(1, 0)$ be higher than that of the unit circle, and this demands that $c^2 < 2\alpha$. Combining these three restrictions, we obtain

$$c^2 < 2\alpha < 1 \tag{11.7.6}$$

Recalling that the FTCS discretization for convection alone leads to an unconditionally unstable method suggests that *the presence of the diffusion term has a stabilizing influence.*

The modified differential equation corresponding to the finite-difference equation (11.7.3) is

$$\frac{\partial f}{\partial t} + U\frac{\partial f}{\partial x} = (\kappa - \tfrac{1}{2}U^2\Delta t)\frac{\partial^2 f}{\partial x^2} \tag{11.7.7}$$

involving the effective diffusivity $\kappa - \frac{1}{2}U^2\Delta t = \kappa(1 - \frac{1}{2}cRe_c)$. The stability criterion shown in equation (11.7.6) thus states that a necessary condition for the method to be stable is that the positive physical diffusivity be larger than the negative numerical diffusivity in absolute value.

An equivalent form of equation (11.7.3) is

$$f_i^{n+1} = \tfrac{1}{2}\alpha(2 + Re_c)f_{i-1}^n + (1 - 2\alpha)f_i^n + \tfrac{1}{2}\alpha(2 - Re_c)f_{i+1}^n \tag{11.7.8}$$

Consider an initial condition where the function f is equal to zero at all grid points except for one grid point where it has a positive value. Since the initial distribution will be convected and diffuse, we expect that the value of f should be positive at all grid points, at all subsequent time levels. The observance of this behavior, however, necessitates that all coefficients on the right-hand side of equation (11.7.8) be positive, and this suggests the *physical* restriction

$$Re_c < 2 \tag{11.7.9}$$

Violation of inequality (11.7.9) leads to unphysical overshooting.

The stability restriction (11.7.6) requires that the size of the time step be excessively small, which renders the FTCS discretization uneconomical. Lax's modification discussed in Section 11.5 leads to an unconditionally unstable method. Consequently, the FTCS method and its variations are of limited practical value.

Upwind-differencing Methods

One might argue that the stability properties of the FTCS method will improve by using upwind differencing for the convective derivative. Assuming that U is positive, let us use a forward difference for the time derivative, a backward difference for the first spatial derivative, and a centered difference for the second spatial derivative, to obtain

$$\frac{f_i^{n+1} - f_i^n}{\Delta t} + U\frac{f_i^n - f_{i-1}^n}{\Delta x} = \kappa\frac{f_{i+1}^n - 2f_i^n + f_{i-1}^n}{\Delta x^2} + O(\Delta t) + O(\Delta x) \tag{11.7.10}$$

Rearranging, we find

$$f_i^{n+1} = f_i^n - c(f_i^n - f_{i-1}^n) + \alpha(f_{i+1}^n - 2f_i^n + f_{i-1}^n) \tag{11.7.11}$$

Carrying out a consistency analysis, we find that the corresponding modified differential equation is the convection–diffusion equation with an effective diffusivity that is equal to $\kappa[1 + \frac{1}{2}(1 - c)Re_c]$. Since Re_c vanishes as Δx tends to zero, the method is confirmed to be consistent.

Performing the von Neumann stability analysis, we find the amplification factor

$$G = 1 - c - 2\alpha + (c + 2\alpha)\cos\theta - Ic\sin\theta \tag{11.7.12}$$

which shows that G traces an ellipse passing through the point $(1, 0)$ in the complex plane. The center of the ellipse is located at the point $(1-c-2\alpha, 0)$, and its semi-axes are equal to $c+2\alpha$ and c. To guarantee stability, we must ensure that the ellipse resides within the unit disk, and this provides us with the stability criterion

$$c^2 < c + 2\alpha < 1 \tag{11.7.13}$$

When U is negative, we use a forward difference for the convective spatial derivative and work similarly to find that the method will be stable provided that $c^2 < |c|+2\alpha < 1$. In practice, the numerical diffusivity associated with the upwind method may be substantial. This feature, combined with the first-order accuracy and the conditional stability, renders the method inferior to its alternatives. A generalization of the method will be discussed in Problem 11.7.5.

Higher-order Methods

To improve the accuracy and reduce the numerical diffusivity of the first-order upwind method, Leonard et al. (1978) proposed approximating the first spatial derivative using the third-order backward difference involving four points, while maintaining the central difference for the second spatial derivative. When U is positive, the finite-difference equation is

$$\frac{f_i^{n+1} - f_i^n}{\Delta t} + U\frac{2f_{i+1}^n + 3f_i^n - 6f_{i-1}^n + f_{i-2}^n}{6\Delta x} = \kappa\frac{f_{i+1}^n - 2f_i^n + f_{i-1}^n}{\Delta x^2} + O(\Delta t) + O(\Delta x^2) \quad (11.7.14)$$

Rearranging, we obtain

$$f_i^{n+1} = f_i^n - \tfrac{1}{6}c(2f_{i+1}^n + 3f_i^n - 6f_{i-1}^n + f_{i-2}^n) + \alpha(f_{i+1}^n - 2f_i^n + f_{i-1}^n) \quad (11.7.15)$$

Examining the corresponding modified differential equation, and carrying out the von Neumann stability analysis, we find that the method is consistent and conditionally stable. More importantly, the numerical diffusivity, which is identical to that of the FTCS method, and the stability criteria are milder than those of the first-order upwind method (e.g., Hoffman 1992, p. 679).

Explicit CTCS and the Du Fort–Frankel Method

In previous sections, we found that the CTCS discretization is unconditionally unstable for the unsteady diffusion equation and conditionally stable for the convection equation. Does adding convection to diffusion have a stabilizing influence? Surprisingly, we find that the answer is negative; the CTCS discretization for the convection–diffusion equation leads to an unconditionally unstable method.

The Du Fort–Frankel method discussed in Section 11.3 is a variation of the CTCS discretization that proceeds according to the difference equation

$$\frac{f_i^{n+1} - f_i^{n-1}}{2\Delta t} + U\frac{f_{i+1}^n - f_{i-1}^n}{2\Delta x} = \kappa\frac{f_{i+1}^n - 2[\frac{1}{2}(f_i^{n+1} + f_i^{n-1})] + f_{i-1}^n}{\Delta x^2} \quad (11.7.16)$$

which represents the CTCS discretization, except that the middle term in the numerator on the right-hand side has been replaced by an average value. Rearranging equation (11.7.16), we obtain the three-level explicit algorithm

$$f_i^{n+1} = \frac{c + 2\alpha}{1 + 2\alpha}f_{i-1}^n + \frac{1 - 2\alpha}{1 + 2\alpha}f_i^{n-1} - \frac{c - 2\alpha}{1 + 2\alpha}f_{i+1}^n \quad (11.7.17)$$

A consistency analysis shows that the Du Fort–Frankel method produces reliable results only when the ratio $(\Delta t/\Delta x)^2$ is sufficiently small. Carrying out the von Neumann stability shows that the amplification factor satisfies the quadratic equation

$$(1 + 2\alpha)G^2 - 2(2\alpha\cos\theta - Ic\sin\theta)G - 1 + 2\alpha = 0 \quad (11.7.18)$$

Upon detailed examination, we find that $|G| < 1$ as long as $|c| < 1$, which means that the Du Fort–Frankel method is stable as long as the CFL condition is fulfilled.

The absence of a stability restriction on α allows the use of time steps of large size but, in practice, a small time step is required in order to obtain a solution that is sufficiently accurate and consistent with the original differential equation.

Implicit Methods

Implicit methods were found to be unconditionally stable for pure convection and pure diffusion, and remain unconditionally stable for mixed convection–diffusion.

BTCS

Implementing a backward-difference approximation in time and centered-difference approximations for both the convective and diffusive spatial derivatives, we obtain the fully implicit BTCS difference equation

$$\frac{f_i^{n+1} - f_i^n}{\Delta t} + U\frac{f_{i+1}^{n+1} - f_{i-1}^{n+1}}{2\Delta x} = \kappa\frac{f_{i+1}^{n+1} - 2f_i^{n+1} + f_{i-1}^{n+1}}{\Delta x^2} + O(\Delta t) + O(\Delta x^2) \quad (11.7.19)$$

Rearranging, we derive the tridiagonal form

$$-(c + 2\alpha)f_{i-1}^{n+1} + 2(1 + 2\alpha)f_i^{n+1} + (c - 2\alpha)f_{i+1}^{n+1} = 2f_i^n \quad (11.2.20)$$

The corresponding amplification factor is found to be

$$G = \frac{1}{1 + 2\alpha(1 - \cos\theta) + Ic\sin\theta} \quad (11.7.21)$$

One may show that $|G| < 1$ for any value of α and c, which reveals that the method is unconditionally stable. The physical restriction $Re_c < 2$ must, however, be observed for the results to be physically meaningful.

Crank–Nicolson

To improve the temporal accuracy of the BTCS method, we implement the fully implicit Crank–Nicolson method according to the difference equation

$$\begin{aligned}\frac{f_i^{n+1} - f_i^n}{\Delta t} &+ U\frac{1}{2}\left(\frac{f_{i+1}^{n+1} - f_{i-1}^{n+1}}{2\Delta x} + \frac{f_{i+1}^n - f_{i-1}^n}{2\Delta x}\right) \\ &= \kappa\frac{1}{2}\left(\frac{f_{i+1}^{n+1} - 2f_i^{n+1} + f_{i-1}^{n+1}}{\Delta x^2} + \frac{f_{i+1}^n - 2f_i^n + f_{i-1}^n}{\Delta x^2}\right)\end{aligned} \quad (11.7.22)$$

The accuracy of the method is of second order in both time and space. Rearranging, we obtain the tridiagonal form

$$\begin{aligned}&-(c + 2\alpha)f_{i-1}^{n+1} + 4(1 + \alpha)f_i^{n+1} + (c - 2\alpha)f_{i+1}^{n+1} \\ &= (c + 2\alpha)f_{i-1}^n + 4(1 - \alpha)f_i^n - (c - 2\alpha)f_{i+1}^n\end{aligned} \quad (11.7.23)$$

The amplification factor is given by

$$G = \frac{2 - 2\alpha(1 - \cos\theta) - Ic\sin\theta}{2 + 2\alpha(1 - \cos\theta) - Ic\sin\theta} \quad (11.7.24)$$

One may show that $|G| < 1$ for any value of α and c, and thus that the method is unconditionally stable. The physical restriction $Re_c < 2$ must be fulfilled.

Three-level fully implicit method

Another way of achieving second-order accuracy in time is to use a three-level method for the approximation of the temporal derivative at the $n+1$ time level, while maintaining the fully implicit spatial discretizations, yielding the difference equation

$$\frac{3f_i^{n+1} - 4f_i^n + f_i^{n-1}}{2\Delta t} + U\frac{f_{i+1}^{n+1} - f_{i-1}^{n+1}}{2\Delta x} = \kappa\frac{f_{i+1}^{n+1} - 2f_i^{n+1} + f_{i-1}^{n+1}}{\Delta x^2} + O(\Delta t^2) + O(\Delta x^2) \quad (11.7.25)$$

The method is unconditionally stable and capable of dampening small-amplitude oscillations. Thus it is preferable over the Crank–Nicolson method when the solution exhibits sharp variations. The physical restriction $Re_c < 2$ must be fulfilled.

Multistep Methods

The multistep methods discussed in Section 11.5 for pure convection can be extended in a straightforward manner to handle combined convection–diffusion.

MacCormack's explicit method

This is a genuine predictor–corrector method. The predictor step is an extension of equation (11.5.36):

$$f_i^* = (1+c)f_i^n - cf_{i+1}^n + \alpha(f_{i+1}^n - 2f_i^n + f_{i-1}^n) \quad (11.7.26)$$

and the corrector step is an extension of equation (11.5.38):

$$f_i^{n+1} = \tfrac{1}{2}[f_i^n + f_i^* - c(f_i^* - f_{i-1}^*) + \alpha(f_{i+1}^* - 2f_i^* + f_{i-1}^*)] \quad (11.7.27)$$

The method is second-order accurate in both time and space, and stable as long as $c < 0.90$ and $\alpha < 0.50$ (e.g., Hoffman 1992, p. 692).

MacCormack's implicit method

This method derives in a manner that is completely analogous to that which led us to equations (11.7.26) and (11.7.27), except that the diffusion term is treated implicitly in both the predictor and corrector steps. The derivation of the difference equations is left as an exercise for the reader in Problem 11.7.2 (Hoffmann and Chiang 1993, Vol. I, p. 263).

Operator Splitting and Fractional Steps

In the previous sections, we saw that certain types of discretization work well for the convection equation, while others work well for the diffusion equation. This realization suggests the use of a fractional-step method, where the convective and diffusive parts are treated independently by different methods according to the component equations

$$\frac{\partial f}{\partial t} + U\frac{\partial f}{\partial x} = 0, \qquad \frac{\partial f}{\partial t} = \kappa\frac{\partial^2 f}{\partial x^2} \tag{11.7.28}$$

both for $t^n < t < t^n + \Delta t$. The time is reset back to the initial value t^n at the end of the first fractional step. In general, the overall stability criteria of a fractional-step method consist of the union of the criteria of the individual methods used for handling the convective and diffusive steps.

Hopscotch

The idea behind the hopscotch method, named after a children's game, is that the solution at different grid points can be advanced using different methods, and a judicious combination of these methods leads to high efficiency and improved accuracy (Gourlay 1970; Mitchell and Griffiths 1980, p.77).

The method proceeds by using the explicit FTCS discretization to advance the solution at the odd-numbered grid points x_{2j+1}, and then the implicit BTCS discretization to advance the solution at the even-numbered grid points x_{2j}, where j is an integer. The order is reversed after the completion of each time step. The crucial advantage is that, since the solution at every other grid point at the new time level is known, the implicit step does not have to be done through matrix inversion or by solving a system of equations, and the method is effectively explicit.

The hopscotch method is first-order accurate in time, second-order accurate in space, and stable as long as $|c| < 1$. There is no stability restriction imposed on the diffusion number α. The efficiency of the method can be further improved by replacing the FTCS difference equation with the equivalent equation $f_i^{n+2} = 2f_i^{n+1} - f_i^n$ after the first step.

Nonlinearities

The significance and implications of nonlinearities on the development of finite-difference methods were discussed in Section 11.5 in the context of the convection equation; the discussion carries over to the present case of combined convection–diffusion.

Burgers equation

A prototypical equation for studying the performance of finite-difference methods is Burgers convection–diffusion equation whose nonconservative and conservative forms are

$$\frac{\partial f}{\partial t} + f\frac{\partial f}{\partial x} = \kappa\frac{\partial^2 f}{\partial x^2} \quad \text{and} \quad \frac{\partial f}{\partial t} + \frac{\partial E}{\partial x} = \kappa\frac{\partial^2 f}{\partial x^2} \tag{11.7.29}$$

where $E = \frac{1}{2}f^2$. Remarkably, the solution in an unbounded domain subject to an arbitrary initial condition may be found analytically using the Cole–Hopf transformation $f = -(2\kappa/u)du/dx$ (Benton and Platzman 1972). Note that this transformation fails when κ is equal to zero, in which case we obtain the inviscid form. It can be shown that the function u satisfies the linear unsteady diffusion equation (11.2.1) whose solution was given in closed form in equation (11.2.2).

As an example, the explicit MacCormack method arises by straightforward modifications of equations (11.7.26) and (11.7.27). The predictor step is

$$f_i^* = (1+r)E_i^n - rE_{i+1}^n + \alpha(f_{i+1}^n - 2f_i^n + f_{i-1}^n) \tag{11.7.30}$$

and the corrector step is

$$f_i^{n+1} = \tfrac{1}{2}[E_i^n + E_i^* - r(E_i^* - E_{i-1}^*) + \alpha(f_{i+1}^* - 2f_i^* + f_{i-1}^n)] \qquad (11.7.31)$$

where $r = \Delta t/\Delta x$.

PROBLEMS

11.7.1 *Du Fort–Frankel method.*

Perform a consistency analysis of the Du Fort–Frankel method and show that the corresponding partial differential equation is given by equation (11.2.15), enhanced with the convection term on the left-hand side.

11.7.2 *Implicit MacCormack method.*

Write out the finite-difference equations for the implicit MacCormack method.

11.7.3 *The method of undetermined coefficients for a two-level method.*

The general form of an implicit method for the one-dimensional convection–diffusion equation, involving three grid points and two time levels, is

$$b_{-1} f_{i-1}^{n+1} + b_0 f_i^{n+1} + b_1 f_{i+1}^{n+1} = a_{-1} f_{i-1}^n + a_0 f_i^n + a_1 f_{i+1}^n \qquad (11.7.32)$$

where a_i and b_i are six constant coefficients. Requiring that equation (11.7.32) is consistent with equation (11.7.1), show that $a_{-1} + a_0 + a_1 = b_{-1} + b_0 + b_1 = 1$, where the last equality represents an arbitrary normalization. Then carry out the von Neumann stability analysis to derive the amplification factor

$$G = \frac{1 - a_1 - a_{-1} + (a_1 + a_{-1})\cos\theta + I(a_1 + a_{-1})\sin\theta}{1 - b_1 - b_{-1} + (b_1 + b_{-1})\cos\theta + I(b_1 + b_{-1})\sin\theta} \qquad (11.7.33)$$

(Peyret and Taylor 1983, p. 39). Verify that equation (11.7.33) is a generalization of equation (11.7.24).

11.7.4 *Implicit* BTBC-CS *method.*

Write out the difference equation for the BTBC-CS method, using BS differencing for the convection term and CS differencing for the diffusion term, and discuss its stability.

11.7.5 *Generalized explicit upwind differencing.*

A generalized form of the upwind differencing method is expressed by the finite-difference equation

$$\frac{f_i^{n+1} - f_i^n}{\Delta t} + \tfrac{1}{2}U\left[(1-\beta)\frac{f_{i+1}^n - f_i^n}{\Delta x} + (1+\beta)\frac{f_i^n - f_{i-1}^n}{\Delta x}\right] = \kappa\frac{f_{i+1}^n - 2f_i^n + f_{i-1}^n}{\Delta x^2} \qquad (11.7.34)$$

where β is an arbitrary constant. Setting $\beta = 1$ when $U > 0$, and $\beta = -1$ when $U < 0$, yields the first-order upwind differencing method; setting $\beta = 0$, we recover the fully explicit FTCS method.

(*a*) Derive the modified differential equation corresponding to equation (11.7.34), show that the method is consistent, and compute the effective diffusivity.

(*b*) The amplification factor corresponding to equation (11.7.34) may be deduced from equation (11.7.33). Show that the method is stable provided that

$$c^2 < 2\alpha + \beta c < 1 \qquad (11.7.35)$$

(Peyret and Taylor 1983, p. 43). Verify that these stability criteria encompass those shown in equations (11.7.6) and (11.7.13).

11.7.6 *Burgers equation.*

An exact solution to the Burgers equation is

$$f = -2\frac{\kappa}{L}\frac{\cosh(x/L)}{\sinh(x/L) + \exp(-\kappa t/L^2)} \tag{11.7.36}$$

where L is an arbitrary length (Benton and Platzman 1972). Compute the evolution of the solution from $t = 0$ using (*a*) the FTCS method, (*b*) the Du Fort–Frankel method, and (*c*) MacCormack's explicit method. Compare the exact with the numerical solutions. *Warning*: Note that the right-hand side of equation (11.7.36) becomes singular at a point.

11.7.7 *Korteweg–de Vries equation.*

A standard form of the Korteweg–de Vries equation is

$$\frac{\partial f}{\partial t} + \varepsilon f\frac{\partial f}{\partial x} + \mu\frac{\partial^3 f}{\partial x^3} = 0 \tag{11.7.37}$$

where ε and μ are positive constants.
(*a*) Develop an explicit finite-difference method.
(*b*) An exact solution to equation (11.7.37) in an unbounded domain, expressing the propagation of a solitary wave, is given by

$$f = 3c\,\mathrm{sech}^2\left(\sqrt{\frac{\varepsilon c}{4\mu}}(x - \varepsilon ct - d)\right) \tag{11.7.38}$$

where c and d are two arbitrary constants, and $c \geq 0$ (Greig and Morris 1976). Compute the evolution of the solution from the initial state using a finite-difference method of your choice, and discuss the accuracy of your results.

11.8 Convection–diffusion in Two and Three Dimensions

Finite-difference methods for the three-dimensional linear convection–diffusion equation

$$\frac{\partial f}{\partial t} + U\frac{\partial f}{\partial x} + V\frac{\partial f}{\partial y} + W\frac{\partial f}{\partial z} = \kappa\left(\frac{\partial^2 f}{\partial x^2} + \frac{\partial^2 f}{\partial y^2} + \frac{\partial^2 f}{\partial z^2}\right) \tag{11.8.1}$$

and its reduced form in two dimensions emerge as straightforward extensions of the methods for the one-dimensional case discussed in the preceding section. In the remainder of this concluding section, we discuss selected examples.

FTCS

The fully explicit FTCS method is first-order accurate in time, and second-order accurate in space. The stability restrictions in three dimensions are

$$\alpha_x + \alpha_y + \alpha_z < 2 \qquad \text{and} \qquad \frac{c_x^2}{\alpha_x} + \frac{c_y^2}{\alpha_y} + \frac{c_z^2}{\alpha_z} < \tfrac{1}{2} \tag{11.8.2}$$

(Hindmarsh et al. 1984). In two dimensions, the sums on the left-hand sides are over x and y.

Upwind Differencing

The first-order upwind-differencing method applied to each convective spatial derivative, combined with a central-differencing method for the diffusive derivatives, leads to a consistent method. The stability constraint in two dimensions with $\Delta x = \Delta y$ is (Peyret and Taylor 1983, p. 66)

$$4\alpha_x + |c_x| + |c_y| < 1 \tag{11.8.3}$$

Hopscotch Method

The implementation of the hopscotch method is similar to that for the one-dimensional case outlined in Section 11.7 (Gourlay 1970). In two dimensions, we first use the explicit FTCS method to advance the solution at the grid points $x_{i,j}$, where $i + j$ is an odd integer, and then use the implicit BTCS method to advance the solution at the grid points $x_{i,j}$, where $i + j$ is an even integer. The order is reversed after the completion of a time step. After the first time step, the FTCS difference equation is replaced with the equivalent equation $f_{i,j}^{n+2} = 2f_{i,j}^{n+1} - f_{i,j}^n$. The method is overall explicit, first-order accurate in time, second-order accurate in space, and stable as long as $|c| < 1$.

ADI in two dimensions

Implicit methods are preferred because of their unconditional stability. The ADI method in two dimensions proceeds according to the finite-difference equations

$$\frac{f_{i,j}^{n+1/2} - f_{i,j}^n}{\Delta t/2} + U\left[\frac{f_{i+1,j}^{n+1/2} - f_{i-1,j}^{n+1/2}}{2\Delta x}\right] + V\frac{f_{i,j+1}^n - f_{i,j-1}^n}{2\Delta y}$$
$$= \kappa\left[\frac{f_{i+1,j}^{n+1/2} - 2f_{i,j}^{n+1/2} + f_{i-1,j}^{n+1/2}}{\Delta x^2}\right] + \kappa\frac{f_{i,j+1}^n - 2f_{i,j}^n + f_{i,j-1}^n}{\Delta y^2} \tag{11.8.4}$$

and

$$\frac{f_{i,j}^{n+1} - f_{i,j}^{n+1/2}}{\Delta t/2} + U\frac{f_{i+1,j}^{n+1/2} - f_{i-1,j}^{n+1/2}}{2\Delta x} + V\left[\frac{f_{i,j+1}^{n+1} - f_{i,j-1}^{n+1}}{2\Delta y}\right]$$
$$= \kappa\frac{f_{i+1,j}^{n+1/2} - 2f_{i,j}^{n+1/2} + f_{i-1,j}^{n+1/2}}{\Delta x^2} + \kappa\left[\frac{f_{i,j+1}^{n+1} - 2f_{i,j}^{n+1} + f_{i,j-1}^{n+1}}{\Delta y^2}\right] \tag{11.8.5}$$

The terms in the square brackets designate implicit discretizations. The ADI method is second-order accurate in both time and space, and unconditionally stable (Peyret and Taylor 1983, p. 66).

ADI in Two Dimensions with Time-dependent Velocities

When the convection velocities U and V are not constant but change in time, the ADI method is first-order accurate in the temporal step. To maintain second-order accuracy, we replace the constants U and V in equation (11.8.4) with the weighted averages

$$U = a_1 U^{n+1} + (1 - a_1 - a_2)U^n + a_2 U^{n-1}, \qquad V = b_1 V^{n+1} + (1 - b_1 - b_2)V^n + b_2 V^{n-1} \quad (11.8.6)$$

and the constants U and V in equation (11.8.5) with the weighted averages

$$U = (1 - a_1 + a_2 + a_3)U^{n+1} + (a_1 - a_2 - 2a_3)U^n + a_3 U^{n-1},$$
$$V = (1 - b_1 + b_2 + b_3)V^{n+1} + (b_1 - b_2 - 2b_3)V^n + b_3 V^{n-1} \quad (11.8.7)$$

where a_i, b_i are six arbitrary constants (Peyret and Taylor 1983, p. 66).

Fractional Steps

One fractional-step method in three dimensions treats convection–diffusion in each dimension individually and separately through a sequence of three one-dimensional steps of equal duration Δt, according to the one-dimensional equations

$$\frac{\partial f}{\partial t} + U\frac{\partial f}{\partial x} = \kappa\frac{\partial^2 f}{\partial x^2}, \qquad \frac{\partial f}{\partial t} + V\frac{\partial f}{\partial y} = \kappa\frac{\partial^2 f}{\partial y^2}, \qquad \frac{\partial f}{\partial t} + W\frac{\partial f}{\partial z} = \kappa\frac{\partial^2 f}{\partial z^2} \quad (11.8.8)$$

all for $t^n < t < t^n + \Delta t$. The time is reset back to the initial value after completion of the first and second fractional steps. Each step may be carried out using an unconditionally stable implicit method.

PROBLEMS

11.8.1 ***Hopscotch method for linear three-dimensional convection–diffusion.***

Develop a hopscotch algorithm for linear convection–diffusion in three dimensions.

11.8.2 ***Fractional-step method.***

Write out the finite-difference equations corresponding to the Crank–Nicolson discretization of the three equations in equation (11.8.8).

References

AMES, W. F., 1977, *Numerical Methods for Partial Differential Equations.* Academic Press.

BHATTACHARYA, M. C., 1985, An explicit conditionally stable finite-difference equation for heat conduction problems. *Int. J. Numer. Meth. Eng.* **21**, 239–65.

BENTON, E. R., and PLATZMAN, G. W., 1972, A table of solutions of the one-dimensional Burgers equation. *Quart. Appl. Math.* **30**, 195–212.

BURSTEIN, S. Z., 1967, Finite-difference calculations for hydrodynamic flows containing discontinuities. *J. Comput. Phys.* **2**, 198–222.

BOOK, D. L., BORIS, J. P., and HAIN, K., 1975, Flux-corrected transport II: generalizations of the method. *J. Comput. Phys.* **18**, 248–83.

BORIS, J. P., and BOOK, D. L., 1973, Flux-corrected transport. I. SHASTA, a fluid transport algorithm that works. *J. Comput. Phys.* **11**, 38–69.

BORIS, J. P., and BOOK, D. L., 1976, Flux-corrected transport. III. Minimal-error FCT algorithms. *J. Comput. Phys.* **20**, 397–431.

CARSLAW, H. S., and JAEGER, J. C., 1959, *Conduction of Heat in Solids.* Oxford University Press.

CRANK, J., and NICOLSON, P., 1947, A practical method for numerical evaluation of solutions of partial differential equations of the heat-conduction type. *Proc. Camb. Philos. Soc.* **43**, 50–67.

DOUGLAS, J., 1955, On the numerical solution of $\partial^2 u/\partial x^2 + \partial^2 u/\partial y^2 = \partial u/\partial t$ by implicit methods. *J. Soc. Indust. Appl. Math.* **3**, 42–65.

DOUGLAS, J., 1962, Alternating direction methods for three space variables. *Numer. Math.* **4**, 41–63.

DOUGLAS, J., and GUNN, J. E., 1964, A general formulation of alternating direction methods. *Numer. Math.* **6**, 428–53.

DU FORT, E. C., and FRANKEL, S. P., 1953, Stability conditions in the numerical treatment of parabolic differential equations. *Math. Tables Other Aids to Comput.* **7**, 135–52.

FERZIGER, J. H., 1981, *Numerical Methods for Engineering Application.* Wiley.

FERZIGER, J. H., and PERIĆ, M., 1996, *Computational Methods for Fluid Dynamics.* Springer-Verlag.

GOURLAY, A. R., 1970, Hopscotch: a fast second-order partial differential equation solver. *J. Inst. Math. Applic.* **6**, 375–90.

GREIG, I. S., and MORRIS, J. L., 1976, A hopscotch method for the Korteweg–de Vries equation. *J. Comput. Phys.* **20**, 64–80.

HACKBUSCH, W., 1985, *Multi-Grid Methods and Applications.* Springer-Verlag.

HINDMARSH, A. C., GRESHO, P. M., and GRIFFITHS, D. F., 1984, The stability of explicit Euler time-integration for certain finite-difference approximations of the multi-dimensional advection–diffusion equation. *Int. J. Numer. Meth. Fluids* **4**, 853–97.

HIRSCH, C., 1988, *Numerical Computation of Internal and External Flows.* 2 volumes. Wiley.

HOFFMAN, J. D., 1992, *Numerical Methods for Engineers and Scientists.* McGraw-Hill.

HOFFMANN, K. A., and CHIANG, S. T., 1993, *Computational Fuid Dynamics for Engineers, Vols. I and II.* Engineering Education System, Wichita, Kansas, 67208-1078, USA.

LAPIDUS, L., and PINDER, G. F., 1982, *Numerical Solution of Partial Differential Equations in Science and Engineering.* Wiley.

LAX, P. D., 1954, Weak solutions of nonlinear hyperbolic equations and their computation. *Comm. Pure Appl. Math.* **7**, 159–93.

LAX, P. D., and RICHTMYER, R. D., 1956, Survey of the stability of linear finite difference equations. *Comm. Pure Appl. Math.* **9**, 267–93.

LAX, P., and WENDROFF, B., 1960, Systems of conservation laws. *Comm. Pure Appl. Math.* **13**, 217–37.

LEE, M., 1962, Alternating direction methods for hyperbolic differential equations. *J. Soc. Indust. Appl. Math.* **10**, 611–16.

LEONARD, B. P., LESCHZINER, M. A., and MCGUIRK, J., 1978, Third-order finite-difference method for steady two-dimensional convection. *Numer. Meth. Laminar and Turbulent Flow.* 807–19.

MACCORMACK, R. W., 1969, *The effect of viscosity in hypervelocity impact cratering.* AIAA paper No. 69–354.

MITCHELL, A. R., 1969, *Computational Methods in Partial Differential Equations.* Wiley.

MITCHELL, A. R., and GRIFFITHS, D. F., 1980, *The Finite-Difference Method in Partial Differential Equations.* Wiley.

PEACEMAN, D. W., and RACHFORD, H. H., 1955, The numerical solution of parabolic and elliptic differential equations. *J. Soc. Indust. Appl. Math.* **3**, 28–41.

PEYRET, R., and TAYLOR, T. D., 1983, *Computational Methods for Fluid Flow.* Springer-Verlag.

PRESS, W. H., FLANNERY, B. P., TEUKOLSKY, S. A., and VETTERLING, W. T., 1992, *Numerical Recipes*, 2nd ed. Cambridge University Press.

RICHTMYER, R. D., 1963, *A Survey of Difference Methods for Non-steady Fluid Dynamics.* NCAR Technical Notes 63–2.

RICHTMYER, R. D., and MORTON, K. W., 1967, *Difference Methods for Initial-Value Problems.* Interscience.

SOD, G. A., 1985, *Numerical Methods in Fluid Fynamics. Initial and Initial Boundary-Value Problems.* Cambridge University Press.

STRIKWERDA, J. C., 1989, *Finite Difference Schemes and Partial Differential Equations.* Brooks/Cole.

SWARTZRAUBER, P. N., 1977, The methods of cyclic reduction, Fourier analysis, and the FACR algorithm for the discrete solution of Poisson's equation on a rectangle. *SIAM Rev.* **19**, 490–501.

WARMING, R. F., and HYETT, B. J., 1974, The modified equation approach to the stability and accuracy analysis of finite-difference methods. *J. Comput. Phys.* **14**, 159–79.

YANENKO, N. N., 1970, *The Method of Fractional Steps.* Springer-Verlag.

Appendix A

Calculus Refresher

A.1 Functions of a Real Variable

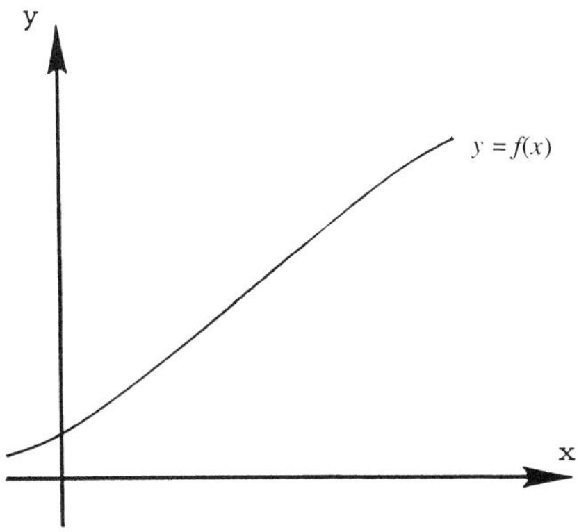

A function of a real variable is an engine that receives a real number x and produces another real number $y = f(x)$. We can also say that the function maps the point x to the point $f(x)$. Roughly speaking, if the graph of the function does not have any jumps within a certain interval $a \le x \le b$, then $f(x)$ is continuous over that interval.

More rigorous definitions and further concepts of differential and integral calculus may be found in standard texts of calculus (e.g., Boas 1966). In the remainder of this appendix, we summarize certain salient results that are used frequently in the development and analysis of numerical methods.

Intermediate-value Theorem

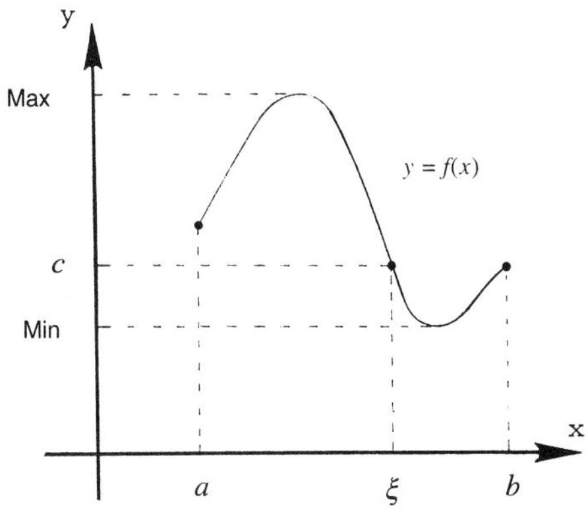

Let the function $f(x)$ be continuous over the interval $a \le x \le b$, let $Min \equiv$ Infimum$(f(x))$ be the minimum value of $f(x)$, and $Max \equiv$ Supreme$(f(x))$ be the maximum value of $f(x)$ over that interval. The intermediate-value theorem asserts, and geometrical intuition confirms, that for any number c where $Min \le c \le Max$, there is at least one value ξ, located in the interval $a \le \xi \le b$, so that $f(\xi) = c$.

Mean-value Theorem for the Derivative

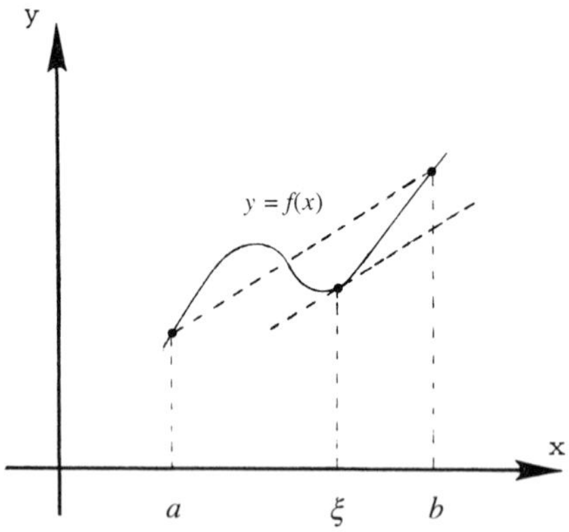

Let the function $f(x)$ be continuous over the closed interval $a \le x \le b$, and its derivative $f'(x)$ be finite at every point over the open interval $a < x < b$. Then there exists at least one point ξ, located in the interval $a < \xi < b$, where the first derivative of $f(x)$ is equal to the slope of the secant line that is subtended between the two end points, that is,

$$f'(\xi) = \frac{f(b) - f(a)}{b - a} \tag{A.1.1}$$

A consequence of this mean-value theorem is *Rolle's theorem*: If $f(a) = 0$ and $f(b) = 0$, then there is at least one point ξ, located in the interval $a < \xi < b$, where $f'(\xi) = 0$.

Mean-value Theorem for the Integral

Let the function $f(x)$ be continuous over the closed interval $a \le x \le b$, and let $w(x)$ be a nonnegative function whose definite integral over that interval is finite. Then there exists at least one point ξ, located in the interval $a \le \xi \le b$, with the property

$$\int_a^b f(x)\, w(x)\, dx = f(\xi) \int_a^b w(x)\, dx \tag{A.1.2}$$

Taylor Series Expansion

If the point x is close to the point x_0, if the function $f(x)$ is continuous in a neighborhood of x_0, and if the graph of $f(x)$ does not tend to become vertical at x_0, then it is reasonable to expect that the difference between $f(x)$ and $f(x_0)$ will be proportional to the distance between x and x_0. The Taylor series provides us with a polynomial expression for $f(x) - f(x_0)$ in terms of the difference $x - x_0$.

Specifically, if the first $N + 1$ derivatives of the function $f(x)$ are continuous at every point between and including the points x and x_0, then

$$f(x) - f(x_0) = \sum_{m=1}^{N} \left[\frac{1}{m!} f^{(m)}(x_0)\, (x - x_0)^m \right] + R_N(x, x_0) \tag{A.1.3}$$

where the superscript (m) signifies the mth derivative. The last term of the right-hand side, called the *remainder*, is given by

$$R_N(x, x_0) = \frac{1}{(N+1)!} f^{(N+1)}(\xi)\, (x - x_0)^{N+1} = \frac{1}{N!} \int_{x_0}^{x} f^{(N+1)}(x')\, (x - x')^N\, dx' \tag{A.1.4}$$

If $x < x_0$, $x \le \xi \le x_0$; and if $x_0 < x$, $x_0 \le \xi \le x$.

As N tends to infinity, we obtain an infinite series representation. It is important to note that the remainder does not necessarily vanish as N tends to infinity but may oscillate or grow without a bound. Thus the infinite Taylor series without the remainder does not necessarily converge to the function $f(x)$. The condition for convergence is discussed in Section 8.1 in the context of function approximation.

If x_0 is set at the origin, $x_0 = 0$, the infinite Taylor series without the remainder reduces to the *Maclaurin series*. Table A.1.1 presents Maclaurin and Taylor series expansions of a number of common functions.

A.2 Functions of Two Variables

A function of two variables is an engine that receives a pair of numbers x and y and produces another number $z = f(x, y)$. Thus the function $f(x, y)$ maps the pair (x, y) to the point $f(x, y)$. Functions of a higher number of variables are defined in a similar manner.

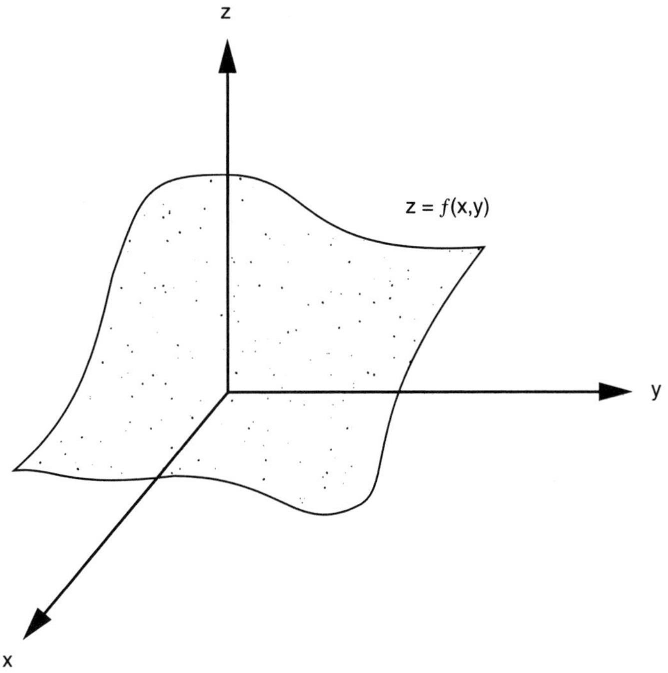

Taylor Series Expansion

The Taylor series expansion of a function of two variables about the point (x_0, y_0) is

$$f(x, y) - f(x_0, y_0) = \sum_{m=1}^{N} \left[\frac{1}{m!} \left((x - x_0)\frac{\partial}{\partial x} + (y - y_0)\frac{\partial}{\partial y} \right)^m_{x_0, y_0} f(x, y) \right] + R_N(x, y, x_0, y_0) \quad \text{(A.2.1)}$$

The remainder is given by

$$R_N(x, y, x_0, y_0) = \left[\frac{1}{(N+1)!} \left((x - x_0)\frac{\partial}{\partial x} + (y - y_0)\frac{\partial}{\partial y} \right)^{N+1}_{\xi_x, \xi_y} f(x, y) \right] \quad \text{(A.2.2)}$$

If $x < x_0$, $x \le \xi_x \le x_0$; and if $x_0 < x$, $x_0 \le \xi_x \le x$. Similarly, for ξ_y.

Table A.1.1 Maclaurin series, Taylor series, asymptotic series, and product expansions of several common functions adapted from the CRC *Standard Mathematical Tables* (Beyer 1987).

Exponential	
$e = 1 + \frac{1}{1!} + \frac{1}{2!} + \frac{1}{3!} + \frac{1}{4!} + \cdots$	
$e^x = 1 + x + \frac{x^2}{2!} + \frac{x^3}{3!} + \frac{x^4}{4!} + \cdots$	(all real values of x)
$a^x = 1 + x \log_e a + \frac{(x \log_e a)^2}{2!} + \frac{(x \log_e a)^3}{+} \cdots$	
$e^x = e^a \left[1 + (x-a) + \frac{(x-a)^2}{2!} + \frac{(x-a)^3}{3!} + \cdots\right]$	
Logarithmic	
$\log_e x = \frac{x-1}{x} + \frac{1}{2}\left(\frac{x-1}{x}\right)^2 + \frac{1}{3}\left(\frac{x-1}{x}\right)^3 + \cdots$	$\left(x > \frac{1}{2}\right)$
$\log_e x = (x-1) - \frac{1}{2}(x-1)^2 + \frac{1}{3}(x-1)^3 - \cdots$	$(2 \geq x > 0)$
$\log_e x = 2\left[\frac{x-1}{x+1} + \frac{1}{3}\left(\frac{x-1}{x+1}\right)^3 + \frac{1}{5}\left(\frac{x-1}{x+1}\right)^5 + \cdots\right]$	$(x > 0)$
$\log_e(1+x) = x - \frac{1}{2}x^2 + \frac{1}{3}x^3 - \frac{1}{4}x^4 + \cdots$	$(-1 < x \leq 1)$
$\log_e(n+1) - \log_e(n-1) = 2\left[\frac{1}{n} + \frac{1}{3n^3} + \frac{1}{5n^5} + \cdots\right]$	
$\log_e(a+x) = \log_e a + 2\left[\frac{x}{2a+x} + \frac{1}{3}\left(\frac{x}{2a+x}\right)^3 + \frac{1}{5}\left(\frac{x}{2a+x}\right)^5 + \cdots\right]$	$(a > 0, -a < x < +\infty)$
$\log_e \frac{1+x}{1-x} = 2\left[x + \frac{x^3}{3} + \frac{x^5}{5} + \cdots + \frac{x^{2n-1}}{2n-1} + \cdots\right]$	$(-1 < x < 1)$
$\log_e x = \log_e a + \frac{(x-a)}{a} - \frac{(x-a)^2}{2a^2} + \frac{(x-a)^3}{3a^3} - + \cdots$	$(0 < x \leq 2a)$
Trigonometric	
$\sin x = x - \frac{x^3}{3!} + \frac{x^5}{5!} - \frac{x^7}{7!} + \cdots$	(all real values of x)
$\cos x = 1 - \frac{x^2}{2!} + \frac{x^4}{4!} - \frac{x^6}{6!} + \cdots$	(all real values of x)
$\tan x = x + \frac{x^3}{3} + \frac{2x^5}{15} + \frac{17x^7}{315} + \frac{62x^9}{2835} + \cdots + \frac{(-1)^{n-1}2^{2n}(2^{2n}-1)B_{2n}}{(2n)!}x^{2n-1} + \cdots$	$\left(x^3 < \frac{\pi^3}{4}\right.$, and B_n represents the nth Bernoulli number$\left.\right)$
$\cot x = \frac{1}{x} - \frac{x}{3} - \frac{x^3}{45} - \frac{2x^5}{945} - \frac{x^7}{4725} - \cdots - \frac{(-1)^{n+1}2^{2n}}{(2n)!}B_{2n}x^{2n-1} - \cdots$	($x^2 < \pi^2$, and B_n represents the nth Bernoulli number)
$\sec x = 1 + \frac{x^2}{2} + \frac{5}{24}x^4 + \frac{61}{720}x^6 + \frac{277}{8064}x^8 + \cdots + \frac{(-1)^n}{(2n)!}E_{2n}x^{2n} + \cdots$	$\left(x^2 < \frac{\pi^2}{4}\right.$, and E_n represents the nth Euler number$\left.\right)$
$\csc x = \frac{1}{x} + \frac{x}{6} + \frac{7}{360}x^3 + \frac{31}{15{,}120}x^5 + \frac{127}{604{,}800}x^7 + \cdots + \frac{(-1)^{n+1}2(2^{2n-1}-1)}{(2n)!}B_{2n}x^{2n-1} + \cdots$	($x^2 < \pi^2$, and B_n represents nth Bernoulli number)
$\sin x = x\left(1 - \frac{x^2}{\pi^2}\right)\left(1 - \frac{x^2}{2^2\pi^2}\right)\left(1 - \frac{x^2}{3^2\pi^2}\right)\cdots$	$(x^2 < \infty)$
$\cos x = \left(1 - \frac{4x^2}{\pi^2}\right)\left(1 - \frac{4x^2}{3^2\pi^2}\right)\left(1 - \frac{4x^2}{5^2\pi^2}\right)\cdots$	$(x^2 < \infty)$
$\sin^{-1} x = x + \frac{x^3}{2 \cdot 3} + \frac{1 \cdot 3}{2 \cdot 4 \cdot 5}x^5 + \frac{1 \cdot 3 \cdot 5}{2 \cdot 4 \cdot 6 \cdot 7}x^7 + \cdots$	$\left(x^2 < 1, -\frac{\pi}{2} < \sin^{-1} x < \frac{\pi}{2}\right)$
$\cos^{-1} x = \frac{\pi}{2} - \left(x + \frac{x^3}{2 \cdot 3} + \frac{1 \cdot 3}{2 \cdot 4 \cdot 5}x^5 + \frac{1 \cdot 3 \cdot 5x^7}{2 \cdot 4 \cdot 6 \cdot 7} + \cdots\right)$	$\left(x^2 < 1, 0 < \cos^{-1} x < \pi\right)$
$\tan^{-1} x = x - \frac{x^3}{3} + \frac{x^5}{5} - \frac{x^7}{7} + \cdots$	$(x^2 < 1)$
$\tan^{-1} x = \frac{\pi}{2} - \frac{1}{x} + \frac{1}{3x^3} - \frac{1}{5x^5} + \frac{1}{7x^7} - \cdots$	$(x > 1)$
$\tan^{-1} x = -\frac{\pi}{2} - \frac{1}{x} + \frac{1}{3x^3} - \frac{1}{5x^5} + \frac{1}{7x^7} - \cdots$	$(x < -1)$
$\cot^{-1} x = \frac{\pi}{2} - x + \frac{x^3}{3} - \frac{x^5}{5} + \frac{x^7}{7} - \cdots$	$(x^2 < 1)$
$\log_e \sin x = \log_e x - \frac{x^2}{6} - \frac{x^4}{180} - \frac{x^6}{2835} - \cdots$	$(x^2 < \pi^2)$

Continued

Table A.1.1 Continued

$\log_e \cos x = -\frac{x^2}{2} - \frac{x^4}{12} - \frac{x^6}{45} - \frac{17x^8}{2520} - \cdots$	$\left(x^2 < \frac{\pi^2}{4}\right)$
$\log_e \tan x = \log_e x = \frac{x^2}{3} + \frac{7x^4}{90} + \frac{62x^6}{2835} + \cdots$	$\left(x^2 < \frac{\pi^2}{4}\right)$
$e^{\sin x} = 1 + x + \frac{x^2}{2!} - \frac{3x^4}{4!} - \frac{8x^5}{5!} - \frac{3x^6}{6!} + \frac{56x^7}{7!} + \cdots$	
$e^{\cos x} = e\left(1 - \frac{x^2}{2!} + \frac{4x^4}{4!} - \frac{31x^6}{6!} + \cdots\right)$	
$e^{\tan x} = 1 + x + \frac{x^2}{2!} + \frac{3x^3}{3!} + \frac{9x^4}{4!} + \frac{37x^5}{5!} + \cdots$	$\left(x^2 < \frac{\pi^2}{4}\right)$
$\sin x = \sin a + (x-a)\cos a - \frac{(x-a)^2}{2!}\sin a - \frac{(x-a)^3}{3!}\cos a + \frac{(x-a)^4}{4!}\sin a + \cdots$	

The operators on the right-hand sides of equations (A.2.1) and (A.2.2) are defined in terms of the binomial expansion as

$$\left((x - x_0)\frac{\partial}{\partial x} + (y - y_0)\frac{\partial}{\partial y}\right)^m_{x_0, y_0} f(x, y)$$
$$= \sum_{m_x=0}^{m} \binom{m}{m_x} \left(\frac{\partial^m f(x, y)}{\partial x^{m_x}\, \partial y^{m-m_x}}\right)_{x_0, y_0} (x - x_0)^{m_x} (y - y_0)^{m-m_x} \tag{A.2.3}$$

The first set of parentheses on the right-hand side designate the combinatorial, defined in equation (1.4.6).

References

Beyer, W. H., 1987, *Standard Mathematical Tables.* CRC Press.
Boas, M. L., 1966, *Mathematical Methods in the Physical Sciences.* Wiley.

Appendix B

Orthogonal Polynomials

Consider a triangular family of polynomials $p_i(t)$, $i = 0, 1, 2, \ldots$, where $p_i(t)$ is an ith degree polynomial of the independent variable t defined over a certain interval $[c, d]$, with leading-order coefficient equal to A_i. Explicitly,

$$\begin{aligned} p_0(t) &= A_0 \\ p_1(t) &= A_1 t + A_1' \\ &\cdots \\ p_i(t) &= A_i t^i + A_i' t^{i-1} + \cdots \end{aligned} \tag{B.1}$$

where A_i, A_i', $\cdots$ are constant coefficients. If the mutual weighted projection of any pair of these polynomials satisfies the orthogonality condition

$$(p_i, p_j) \equiv \int_c^d p_i(t)\, p_j(t)\, w(t)\, dt = \begin{cases} h_i & \text{if } i = j \\ 0 & \text{if } i = j \end{cases} \tag{B.2}$$

where $w(t)$ is a certain weighting function and h_i is a constant, then the triangular family is *orthogonal.* If $h_i = 1$ for any value of i, then the family is *orthonormal.*

It can be shown that for every weighting function $w(t)$ that takes *nonnegative* values for any value of t within the interval $[c, d]$, there is a unique corresponding triangular family of orthonormal polynomials defined over that interval. The members of the family can be computed using the Gram–Schmidt orthogonalization method discussed in Problem 6.4.1.

Extensive discussion of the properties of orthogonal polynomials can be found in the texts of Sansone (1959), Szegö (1959), Krylov (1962), Stroud and Secrest (1966), Fox and Parker (1968), Luke (1969), Arfken (1970), Hamming (1973), Hildebrand (1974), Rivlin (1974), and Dahlquist and Björck (1974). Certain important theorems are:

- Any polynomial $Q_m(t)$ of degree m or less can be expressed as a linear combination of any group of $m+1$ orthogonal polynomials $p_0(t), p_1(t), \ldots, p_m(t)$ that belong to a certain family. This implies that $p_i(t)$ is orthogonal to any such polynomial $Q_m(t)$ with $m < i$; that is,

$$(p_i, Q_m) \equiv \int_c^d p_i(t)\, Q_m(t)\, w(t)\, dt = 0 \tag{B.3}$$

  This property provides us with a method of computing the coefficients multiplying the monomials of $p_{m+1}(t)$: we set the value of one of the coefficients at an arbitrary value—for example, the value of the leading-order coefficient A_{m+1}; apply equation (B.3) for a collection of m linearly independent polynomials $Q_m(t)$; and solve the system of equations for the rest of the coefficients.

- An mth degree orthogonal polynomial has m real distinct roots that lie within the domain $[c, d]$. Thus the graph of the polynomial does not cross the t axis outside the interval $[c, d]$.

- The roots of an mth degree orthogonal polynomial interleave those of the $(m-1)$ degree polynomial in the same family. That is, there is exactly one root of the former between two consecutive roots of the latter.

- The $p_l(t)$ orthogonal polynomial has a set of l points $t_i, i = 1, \ldots, l$, so that the following discrete orthogonality condition holds:

$$\sum_{i=1}^{l} p_n(t_i) p_m(t_i) = 0 \qquad \text{if } n \neq m \tag{B.4}$$

  where $n < l$ and $m < l$.

Weighting with Delta Functions and Discrete Orthogonality

We can allow the weigting function $w(t)$ to be a sum of delta functions centered at $N+1$ points $t_l, l = 1, \ldots, N+1$, that are distributed is some manner over $[c, d]$. That is, we can set

$$w(t) = \sum_{l=1}^{N+1} w_l \delta(t - t_l) \tag{B.5}$$

where w_l are weighting coefficients and δ is the one-dimensional delta function. The orthogonality condition (B.2) becomes

$$(p_i, p_j) = \sum_{l=1}^{N+1} p_i(t_l)\, p_j(t_l)\, w_l = \begin{cases} h_i & \text{if } i = j \\ 0 & \text{if } i \neq j \end{cases} \tag{B.6}$$

It can be shown that the corresponding family contains only $N+1$ polynomials, $p_0(t), p_1(t), \ldots, p_N(t)$; higher-order polynomials are equal to zero. An example is the family of *Gram polynomials* shown in Table B.10.

Recursion Relation

Any set of orthogonal polynomials satisfies the recursion relation:

$$\begin{aligned} p_0(t) &= A_0 \\ p_1(t) &= \alpha_0\,(t-\beta_0)\,p_0(t) \\ p_{i+1}(t) &= \alpha_i\,(t-\beta_i)\,p_i(t) - \gamma_i\,p_{i-1}(t) \qquad \text{for } i = 1, 2, \ldots \end{aligned} \tag{B.7}$$

The coefficients α_i, β_i, and γ_i are generally functions of i whose precise form depends on the particular family under consideration. It can be shown, in particular, that, for any family,

$$\begin{aligned} \alpha_i &= \frac{A_{i+1}}{A_i} \qquad \text{for } i = 0, 1, \ldots \\ \beta_i &= \frac{(tp_i, p_i)}{(p_i, p_i)} \qquad \text{for } i = 0, 1, \ldots \\ \gamma_i &= \alpha_i \frac{(tp_i, p_{i-1})}{(p_{i-1}, p_{i-1})} = \frac{\alpha_i}{\alpha_{i-1}} \frac{(p_i, p_i)}{(p_{i-1}, p_{i-1})} \qquad \text{for } i = 1, 2, \ldots \end{aligned} \tag{B.8}$$

The projection operator represented by the parentheses was defined in equation (B.2). If the graph of $w(t)$ is symmetric about the point $t = \beta$, then $\beta_i = \beta$ for all i.

Matrix formulation

Comparing the recursion relation (B.7) with that shown in algorithm (2.4.3) for computing the determinant of a tridiagonal matrix, we find

$$p_{n+1}(t) = A_0 \operatorname{Det}(\boldsymbol{T}(t)) \tag{B.9}$$

where $\boldsymbol{T}$ is a $(n+1) \times (n+1)$ tridiagonal matrix whose elements satisfy the relations

$$\begin{aligned} T_{i,i} &= \alpha_{i-1}(t-\beta_{i-1}) \qquad \text{for } i = 1, 2, \ldots, n+1 \\ T_{i,i-1}T_{i-1,i} &= \gamma_{i-1} \qquad \text{for } i = 2, 3, \ldots, n+1 \end{aligned} \tag{B.10}$$

Setting, in particular, the superdiagonal elements $T_{i-1,i}$ equal to 1, we obtain the matrix

$$\boldsymbol{T}(t) = \begin{bmatrix} \alpha_0(t-\beta_0) & 1 & 0 & 0 & 0 & 0 & \ldots \\ \gamma_1 & \alpha_1(t-\beta_1) & 1 & 0 & 0 & 0 & \ldots \\ 0 & \gamma_2 & \alpha_2(t-\beta_2) & 1 & 0 & 0 & \ldots \\ 0 & 0 & \gamma_3 & \alpha_3(t-\beta_3) & 1 & 0 & \ldots \\ \ldots & \ldots & \ldots & \ldots & \ldots & \ldots & \ldots \\ 0 & 0 & 0 & 0 & \gamma_{n-1} & \alpha_{n-1}(t-\beta_{n-1}) & 1 \\ 0 & 0 & 0 & 0 & 0 & \gamma_n & \alpha_n(t-\beta_n) \end{bmatrix} \tag{B.11}$$

Based on this matrix formulation, we can compute the roots of families of orthogonal polynomials by using methods for computing the eigenvalues of tridiagonal matrices discussed in Chapter 5.

Clenshaw's Method of Computing an Expansion of Orthogonal Polynomials

The fastest way to compute the polynomial expansion

$$P_N(t) = \sum_{i=0}^{N} c_i p_i(t) \tag{B.12}$$

where c_i are given coefficients, is by Clenshaw's method. With the coefficients α_i, β_i, and γ_i introduced in equations (B.7), we exercise the algorithm:

$$\begin{array}{l} d_N = c_N \\ d_{N-1} = (t - \beta_{N-1})\, d_N + c_{N-1} \\ \text{Do } k = N-2, 0, -1 \\ \quad d_k = \alpha_k (t - \beta_k)\, d_{k+1} - \gamma_{k+1}\, d_{k+2} + c_{N-1} \\ \text{End Do} \\ P_N(t) = A_0\, d_0 \end{array} \tag{B.13}$$

Standard Families of Orthogonal Polynomials

Several standard families of orthogonal polynomials are summarized in Table B.1, and their properties are listed in Tables B.2–B.10. Further information can be found in the references cited at the end of this appendix.

Table B.1 Standard families of orthogonal polynomials

Name	Notation	Domain of definition	$w(t)$	Standard normalization	Normalization coefficient defined in eq. (B2)
Legendre	$L_i(t)$	$[-1, 1]$	1	$L_i(1) = 1$	$h_i = \frac{2}{2i+1}$
Chebyshev	$T_i(t)$	$[-1, 1]$	$(1-t^2)^{-1/2}$	$T_i(1) = 1$	$h_0 = \pi$, $h_i = \frac{1}{2}\pi$ for $i = 1, 2, \ldots$
Chebyshev of the second kind	$U_i(t)$	$[-1, 1]$	$(1-t^2)^{1/2}$	$U_i(1) = i + 1$	$h_i = \frac{\pi}{2}$
Jacobi	$P_i^{(\alpha,\beta)}(t)$	$[-1, 1]$	$(1-t)^{\alpha}(1+t)^{\beta}$	$P_i^{(\alpha,\beta)}(1) = \left(\frac{\Gamma(\alpha+i+1)}{\Gamma(\alpha+1)\Gamma(i+1)}\right)$	
	$\alpha > -1, \beta > -1$				$h_i = \frac{2^{\alpha+\beta+1}}{i!} \frac{\Gamma(\alpha+i+1)\,\Gamma(\beta+i+1)}{(\alpha+\beta+1+2i)\,\Gamma(\alpha+\beta+i+1)}$
			Γ is the Gamma function: if m is an integer, $\Gamma(m+1) = m!$		
Radau	$R_i(t)$	$[-1, 1]$	$1+t$	$R_i(1) = 1$	$h_i = \frac{2}{i+1}$
Lobatto	$Lo_i(t)$	$[-1, 1]$	$1-t^2$	$Lo_i = L'_{i+1}$	$h_i = \frac{2(i+1)\,(i+2)}{2i+3}$
Laguerre	$\mathcal{L}_i(t)$	$[0, \infty)$	$\exp(-t)$	$A_i = (-1)^i$	$h_i = (i!)^2$
Hermite	$H_i(t)$	$(-\infty, +\infty)$	$\exp(-t^2)$	$A_i = 2^i$	$h_i = \sqrt{\pi}2^i(i!)$
Gram			See Table B.10		

Table B.2 *Legendre* polynomials and their properties.

Domain of definition: $[-1, 1]$

$c = -1, d = 1$

Weighting function: $w(t) = 1$

Members

$$L_0 = 1$$
$$L_1 = t$$
$$L_2 = \tfrac{1}{2}(3\,t^2 - 1)$$
$$L_3 = \tfrac{1}{2}(5\,t^2 - 3)t$$
$$L_4 = \tfrac{1}{8}(35\,t^4 - 30t^2 + 3)$$
$$L_5 = \tfrac{1}{8}(63\,t^4 - 70t^2 + 15)t$$
$$\dots$$
$$L_i(t) = \frac{1}{2^i i!}\frac{d^i}{dt^i}(t^2 - 1)^i = \frac{(2i)!}{2^i (i!)^2} t^i - \cdots$$

Standard normalization

$$L_i(1) = 1$$

Orthogonality property

$$\int_{-1}^{1} L_i(t)\, L_j(t)\, dt = \frac{2}{2i + 1}\delta_{i,j}$$

Recursion relation

$$L_{i+1}(t) = \frac{2i+1}{i+1} t\, L_i(t) - \frac{i}{i+1} L_{i-1}(t)$$

with reference to equations (B.7),

$$\alpha_i = \frac{2i+1}{i+1}, \qquad i = 0, 1, \dots$$
$$\beta_i = 0, \qquad i = 0, 1, \dots$$
$$\gamma_i = \frac{i}{i+1}, \qquad i = 1, 2, \dots$$

Recurrence relations for the first derivative

$$(1 - t^2)L'_{i+1}(t) = (i + 1)(-tL_{i+1}(t) + L_i(t))$$
$$= (i + 2)(-L_{i+2}(t) + tL_{i+1}(t))$$
$$L'_{i+1} - L'_{i-1} = (2i + 1)L_i$$

Table B.2 (*Continued*) *Legendre* polynomials and their properties.

Zero-mean property

Applying the orthogonality property for $j = 0$ yields

$$\int_{-1}^{1} L_i(t)\,dt = 0 \qquad i \geq 1$$

Range of variation

$$|L_i(t)| \leq 1 \qquad \text{for } -1 \leq t \leq 1$$

Differential equation

$$(1 - t^2)L_i''(t) - 2tL_i'(t) + i(i+1)L_i(t) = 0$$

Gauss–Legendre *m*-point quadrature weights

$$w_i = \frac{2}{(1 - t_i^2)}\frac{1}{L_m'^2(t_i)} = -\frac{2}{m+1}\frac{1}{L_m'(t_i)\,L_{m+1}(t_i)}$$

$$= \frac{2}{m}\frac{1}{L_m'(t_i)\,L_{m-1}(t_i)} = \frac{2}{(m+1)^2}\frac{(1 - t_i^2)}{L_{m+1}^2(t_i)}$$

where t_i are the zeros of $L_m(t)$, $i = 1, \ldots, m$

Table B.3 *Chebyshev* polynomials and their properties.

Domain of definition: $[-1, 1]$

$c = -1, d = 1$

Weighting function: $w(t) = (1 - t^2)^{-1/2}$

Members

$$T_0 = 1$$
$$T_1 = t$$
$$T_2 = 2\,t^2 - 1$$
$$T_3 = 4\,t^3 - 3t$$
$$T_4 = 8\,t^4 - 8\,t^2 + 1$$
$$T_5 = 16\,t^5 - 20\,t^3 + 5t$$
$$\ldots$$
$$T_i = \cos(i\theta) = 2^{i-1}\,t^i - i\,2^{i-3}\,t^{i-2} + \cdots = \frac{1}{2}(z^i + z^{-i}),$$

where $t = \cos\theta$,

$$z^2 - 2\,xz + 1 = 0$$

Standard normalization

$$T_i(1) = 1$$

Orthogonality property

$$\int_{-1}^{1} \frac{T_i(t)\,T_j(t)}{\sqrt{1-t^2}}dt = \begin{cases} \pi & \text{if } i = j = 0 \\ \frac{1}{2}\pi & \text{if } i = j = 0 \neq 0 \\ 0 & \text{Otherwise} \end{cases}$$

Recursion relation

$$T_{i+1}(t) = 2\,t\,T_i(t) - T_{i-1}(t)$$

with reference to equations (B.7),

$$\alpha_0 = 1, \alpha_i = 2, \qquad i = 1, 2, \ldots$$
$$\beta_i = 0, \qquad i = 0, 1, \ldots$$
$$\gamma_i = 1, \qquad i = 1, 2, \ldots$$

Recurrence relations for the first derivative

$$(1 - t^2)T'_{i+1}(t) = (i+1)(-t\,T_{i+1}(t) + T_i(t))$$
$$\frac{T'_{i+1}}{i+1} - \frac{T'_{i-1}}{i-1} = 2\,T_i$$

Zeros

$T_n(t)$ has n zeros in the interval $[-1, 1]$ given by

$$t_i = \cos\left[\left(i - \tfrac{1}{2}\right)\pi/n\right], i = 1, 2, \ldots, n \text{ for } n > 0$$

Table B.3 (*Continued*) *Chebyshev* polynomials and their properties.

Discrete orthogonality property

$$\sum_{i=1}^{l} T_n(t_i)\, T_m(t_i) = \begin{cases} 0 & \text{if } n \neq m \\ l & \text{if } n = m = 0 \\ \frac{1}{2}l & \text{if } n = m \neq 0 \end{cases}$$

where $n < l, m < l,$ and

$t_i = \cos\left[\left(i - \frac{1}{2}\right)\pi/l\right], \quad i = 1, \ldots, l$ are the zeros of $T_l(t)$

Minimax property

Of all nth degree polynomials with leading coefficient equal to 2^{n-1}, T_n has the least maximum value in the interval $[-1, 1]$

Extrema

T_n has $n + 1$ extrema in the interval $[-1, 1]$ given by $t_i = \cos(i\pi/n), i = 0, 2, \ldots, n.$

Range of variation

$$|T_i(t)| \leq 1, \qquad \text{for } -1 \leq t \leq 1$$

Differential equation

$$(1 - t^2)T_i''(t) - t\, T_i'(t) + i^2\, T_i(t) = 0$$

Gauss–Chebyshev *m*-point quadrature weights

$$w_i = \frac{\pi}{m}$$

Product to sum resolution

$$2\, T_n(t)\, T_m(t) = T_{n+m}(t) + T_{n-m}(t), \qquad \text{for } n \geq m$$

Expansion of a function and its derivative

$$\text{If } f(t) = \sum_{n=0}^{N} a_n\, T_n(t), \qquad \text{then} \quad f'(t) = \sum_{n=0}^{N} b_n\, T_n(t)$$

$$\text{where:} \quad b_n = \frac{2}{\alpha_n} \sum_{\substack{p=n+1 \\ p+n \text{ is odd}}}^{N} p a_p, \qquad \alpha_0 = 2, \alpha_i = 1 \quad \text{for } i = 1, 2, \ldots$$

Table B.3 (*Continued*) *Chebyshev* polynomials and their properties.

Polynomial economization relations

$$1 = T_0$$

$$t = T_1$$

$$t^2 = \tfrac{1}{2}(T_0 + T_2)$$

$$t^3 = \tfrac{1}{4}(3T_1 + T_3)$$

$$t^4 = \tfrac{1}{8}(3T_0 + 4T_2 + T_4)$$

$$t^5 = \tfrac{1}{16}(10T_1 + 5T_3 + T_5)$$

$$\cdots$$

If k is even:

$$t^k = \frac{1}{2^{k-1}}\left[T_k(x) + \binom{k}{1}\ T_{k-2}(x) + \binom{k}{2}\ T_{k-4}(x) + \cdots + \frac{1}{2}\binom{k}{k/2}\ T_0(x)\right]$$

If k is odd:

$$t^k = \frac{1}{2^{k-1}}\left[T_k(x) + \binom{k}{1}\ T_{k-2}(x) + \binom{k}{2}\ T_{k-4}(x) + \cdots + \binom{k}{(k-1)/2}\ T_1(x)\right]$$

Table B.4 ***Chebyshev*** **polynomials of the** ***second kind.***

Domain of definition: $[-1, 1]$

$$c = -1, d = 1$$

Weighting function: $w(t) = (1 - t^2)^{1/2}$

Members

$$U_0 = 1$$
$$U_1 = 2t$$
$$U_2 = 4\,t^2 - 1$$
$$U_3 = 8\,t^3 - 4t$$
$$\dots$$
$$U_i(t) = \frac{\sin[(i+1)\theta]}{\sin\,\theta} = \frac{1}{i+1}\,T'_{i+1}(t) = 2^i\,t^i + \cdots$$

where $t = \cos\theta$

Standard normalization

$$U_i(1) = i + 1$$

Orthogonality property

$$\int_{-1}^{1} U_i(t)\,U_j(t)\,\sqrt{1-t^2}\,dt = \frac{\pi}{2}\delta_{i,j}$$

Recursion relation

$$U_{i+1}(t) = 2\,tU_i(t) - U_{i-1}(t)$$

with reference to equations (B.7),

$$\alpha_i = 2, \qquad i = 0, 1, \dots$$
$$\beta_i = 0, \qquad i = 0, 1, \dots$$
$$\gamma_i = 1, \qquad i = 1, 2, \dots$$

Zeros

U_l has l roots in the interval $[-1, 1]$ given by

$t_i = \cos[i\pi/(l+1)], i = 1, 2, \dots, l.$

Range of variation

$$|U_i(t)| \le i + 1, \qquad \text{for } -1 \le t \le 1$$

Differential equation

$$(1 - t^2)U_i''(t) - 3t\,U_i'(t) + i(i+2)\,U_i = 0$$

Second orthogonality property

$$\int_{-1}^{1} |U_n(t)|\,\mathrm{Sign}(T_m(t))\,dt = 0 \qquad \text{for } n < m$$

Of all nth degree polynomials with leading coefficient equal to 2^n, U^n gives the minimum value of the integral of the absolute value over the interval $[-1, 1]$

Table B.5 ***Jacobi*** **polynomials and their properties.**

Domain of definition: $[-1, 1]$

$c = -1, d = 1$

Weighting function: $w(t) = (1-t)^{\alpha}\,(1+t)^{\beta}$

where $\alpha$and $\beta$are constants, both greater than -1.

$\alpha = 0$ and $\beta = 0$ yield the *Legendre* polynomials.

$\alpha = -\frac{1}{2}$ and $\beta = -\frac{1}{2}$ yield the *Chebyshev* polynomials.

$\alpha = \frac{1}{2}$ and $\beta = \frac{1}{2}$ yield the *Chebyshev polynomials of the second kind.*

Members

$$P_0^{(\alpha,\beta)} = 1$$

$$P_1^{(\alpha,\beta)} = \tfrac{1}{2}(\alpha+\beta+2)t + \tfrac{1}{2}(\alpha-\beta)$$

$$\begin{aligned} P_2^{(\alpha,\beta)} &= \tfrac{1}{8}(\alpha+\beta+3)(\alpha+\beta+4)t^2 \\ &+ \tfrac{1}{4}(\alpha-\beta)(\alpha+\beta+3)t \\ &+ \tfrac{1}{8}[(\alpha-\beta)^2-(\alpha+\beta+4)] \end{aligned}$$

$$\cdots$$

$$P_i^{(\alpha,\beta)} = \frac{1}{2^i}\sum_{m=0}^{i}\binom{i+\alpha}{m}\binom{i+\beta}{i-m}(t-1)^{i-m}(t+1)^m$$

Standard normalization

$$P_i^{(\alpha,\beta)}(1) = \left(\frac{\Gamma(\alpha+i+1)}{\Gamma(\alpha+1)\Gamma(i+1)}\right)$$

where Γ the the *Gamma* function.

Orthogonality property

$$\int_{-1}^{1} P_i^{(\alpha,\beta)}(t)\,P_j^{(\alpha,\beta)}(t)\,w(t)\,dt = \frac{2^{\alpha+\beta+1}}{i!}\,\frac{\Gamma(\alpha+i+1)\,\Gamma(\beta+i+1)}{(\alpha+\beta+1+2i)\Gamma(\alpha+\beta+i+1)}\,\delta_{i,j}$$

Recursion relation

$$\begin{aligned} P_{i+1}^{(\alpha,\beta)}(t) &= \frac{(2i+\alpha+\beta+1)(2i+\alpha+\beta+2)}{2(i+1)(i+\alpha+\beta+1)}\left(t - \frac{\beta^2-\alpha^2}{(2i+\alpha+\beta)(2i+\alpha+\beta+2)}\right)P_i^{(\alpha,\beta)}(t) \\ &- \frac{(i+\alpha)(i+\beta)(2i+\alpha+\beta+2)}{(i+1)(i+\alpha+\beta+1)(2i+\alpha+\beta)}P_{i-1}^{(\alpha,\beta)}(t) \end{aligned}$$

Table B.5 (*Continued*) *Jacobi* polynomials and their properties.

with reference to equations (B.7),

$$\alpha_i = \frac{(2i+\alpha+\beta+1)(2i+\alpha+\beta+2)}{2(i+1)(i+\alpha+\beta+1)}, \qquad i = 0, 1, \ldots$$

$$\beta_i = \frac{\beta^2-\alpha^2}{(2i+\alpha+\beta)(2i+\alpha+\beta+2)}, \qquad i = 0, 1, \ldots$$

$$\gamma_i = \frac{(i+\alpha)(i+\beta)(2i+\alpha+\beta+2)}{(i+1)(i+\alpha+\beta+1)(2i+\alpha+\beta)}, \qquad i = 1, 2, \ldots$$

Differential equation

$$(1-t^2)P_i''(t) - [\alpha-\beta+(\alpha+\beta+2)t]\ P_i'(t) + i(i+\alpha+\beta+1)\ P_i(t) = 0$$

Table B.6 ***Radau*** **polynomials and their properties.**

Domain of definition: $[-1, 1]$

$c = -1, d = 1$

Weighting function: $w(t) = 1 + t$

Members

$$R_0(t) = 1$$
$$R_1(t) = \tfrac{1}{2}(3t - 1)$$
$$R_2(t) = \tfrac{1}{2}(5\,t^2 - 2t - 1)$$
$$R_3(t) = \tfrac{1}{8}(35t^3 - 15t^2 - 15t + 3)$$
$$\ldots$$

Standard normalization

$$R_i(1) = 1$$

Orthogonality property

$$\int_{-1}^{1} R_i(t)\; R_j(t)\; (1 + t)\; dt = \frac{2}{i+1}\,\delta_{i,j}$$

Relation to the Legendre polynomials

$$R_i(t) = \frac{L_i(t) + L_{i+1}(t)}{1 + t}$$

Recursion relation

$$R_{i+1}(t) = \frac{(2i+1)(2i+3)t - 1}{(i+2)(2i+1)}\,R_i(t) - \frac{i(2i+3)}{(i+2)(2i+1)}\,R_{i-1}(t)$$

Gauss–Radau $(m+1)$-point quadrature weights

$$w_i = \frac{1}{(m+1)^2}\,\frac{1 - t_i}{L_m^2(t_i)} = \frac{1}{1 - t_i}\,\frac{1}{L_m'^2(t_i)}$$

where t_i are the zeros of $R_m(t)$, $i = 1, \ldots, m$

Table B.7 ***Lobatto*** **polynomials and their properties.**

Domain of definition: $[-1, 1]$

$c = -1, d = 1$

Weighting function: $w(t) = 1 - t^2$

Members

$$Lo_0(t) = 1$$
$$Lo_1(t) = 3t$$
$$Lo_2(t) = \tfrac{3}{2}(5\,t^2 - 1)$$
$$Lo_3(t) = \tfrac{1}{2}(35t^2 - 15)t$$
$$\ldots$$

Orthogonality property

$$\int_{-1}^{1} Lo_i(t)\,Lo_j(t)\,(1 - t^2)\,dt = \frac{2(i+1)(i+2)}{2i+3}\,\delta_{i,j}$$

Relation to the Legendre polynomials

$$Lo_i(t) = L'_{i+1}(t)$$

Gauss–Lobatto $(m+1)$-point quadrature weights

$$w_i = \frac{2}{m(m-1)}\,\frac{1}{L^2_{m-1}(t_i)}$$

where t_i are the zeros of $Lo_{m-1}(t)$, $i = 1, \ldots, m-1$

Table B.8 *Laguerre* **polynomials and their properties.**

Domain of definition: $[0, \infty)$

$c = 0, d = \infty$

Weighting function: $w(t) = \exp(-t)$

Members

$$\mathcal{L}_0 = 1$$
$$\mathcal{L}_1 = -t + 1$$
$$\mathcal{L}_2 = t^2 - 4t + 2$$
$$\mathcal{L}_3 = -t^3 + 9t^2 - 18t + 6$$
$$\mathcal{L}_4 = t^4 - 16t^3 + 72t^2 - 96t + 24$$
$$\mathcal{L}_4 = -t^5 + 25t^4 - 200t^3 + 600t^2 - 600t + 120$$
$$\cdots$$
$$\mathcal{L}_i(t) = e^t \frac{d^i}{dt^i}(t^i e^{-t}) = (-1)^i t^i + \cdots$$

Standard normalization

Leading-power coefficient $A_i = (-1)^i$

Orthogonality property

$$\int_0^\infty \mathcal{L}_i(t)\, \mathcal{L}_j(t)\, e^{-t}\, dt = (i!)^2\, \delta_{i,j}$$

Recursion relation

$$\mathcal{L}_{i+1}(t) = (1 + 2i - t)\, \mathcal{L}_i(t) - i^2\, \mathcal{L}_{i-1}(t)$$

Recursion relation for the first derivative

$$t\, \mathcal{L}'_{i+1}(t) = (i+1)\, (\mathcal{L}_{i+1}(t) - (i+1)\, \mathcal{L}_i(t))$$
$$= \mathcal{L}_{i+2}(t) - (t - i - 2)\, \mathcal{L}_{i+1}(t)$$

Differential equation

$$t\mathcal{L}''_i(t) + (1 - t)\, \mathcal{L}'_i(t) + i\mathcal{L}_i(t) = 0$$

Gauss–Laguerre m-point quadrature weights

$$w_i = \frac{(m!)^2}{\mathcal{L}'_m(t_i)\, \mathcal{L}_{m+1}(t_i)} = -\frac{((m-1)!)^2}{\mathcal{L}'_m(t_i)\, \mathcal{L}_{m-1}(t_i)} = \frac{(m!)^2 t_i}{\mathcal{L}^2_{m+1}(t_i)}$$

where t_i are the zeros of $\mathcal{L}_m(t)$, $i = 1, \ldots, m$.

Table B.9 ***Hermite*** **polynomials and their properties.**

Domain of definition: $(-\infty, +\infty)$

$$c = -\infty, d = +\infty$$

Weighting function: $w(t) = \exp(-t^2)$

Members

$$H_0 = 1$$
$$H_1 = 2t$$
$$H_2 = 4t^2 - 2$$
$$H_3 = 8t^3 - 12t$$
$$H_4 = 16t^4 - 48t^2 + 12$$
$$H_5 = 32t^5 - 160t^3 + 120t$$
$$\cdots$$
$$H_i(t) = (-1)^i e^{t^2} \frac{d^i e^{-t^2}}{dt^i} = 2^i t^i + \cdots$$

Standard normalization

Leading-power coefficient $A_i = 2^i$

Orthogonality property

$$\int_{-\infty}^{\infty} H_i(t)\ H_j(t)\ e^{-t^2}\ dt = \sqrt{\pi}\ 2^i (i!)\ \delta_{i,j}$$

Recursion relation

$$H_{i+1}(t) = 2\ t\ H_i(t) - 2i\ H_{i-1}(t)$$

Recursion relation for the first derivative

$$H'_{i+1}(t) = 2(i+1)\ H_i(t) = 2t\ H_{i+1}(t) - H_{i+2}(t)$$

Differential equation

$$H_i''(t) - 2t H_i'(t) + 2i H_i(t) = 0$$

Gauss–Hermite m-point quadrature weights

$$w_i = \sqrt{\pi}\frac{2^{m+1}m!}{H'_m(t_i)\ H_{m+1}(t_i)} = \sqrt{\pi}\frac{2^m(m-1)!}{H'_m(t_i)\ H_{m-1}(t_i)} = \sqrt{\pi}\frac{2^{m+1}m!}{H'^2_m(t_i)} = \sqrt{\pi}\frac{2^{m+1}m!}{H^2_{m+1}(t_i)}$$

where t_i are the zeros of $H_m(t)$, $i = 1, \ldots, m$

Table B.10 *Gram* polynomials with $N+1$ base points and their properties.

Domain of definition: $[-1, 1]$

$$c = -1, d = +1$$

Weighting function:

$$w(t) = \sum_{i=1}^{N+1} \delta(t - t_i), \qquad t_i = -1 + (i-1)\frac{2}{N}$$

Members

$$G_0 = \frac{1}{\sqrt{N+1}}$$

$$G_1 = \frac{N}{\sqrt{N+1}} \left(\frac{3}{(N+1)^2 - 1} \right)^{1/2} t$$

$$\cdots$$

Orthogonality property

$$\sum_{i=1}^{N+1} G_l(t_i)\, G_m(t_i) = \delta_{l,m}$$

Recursion relation

$$G_{i+1}(t) = \alpha_i\, t\, G_i(t) - \gamma_i\, G_{i-1}(t)$$

where

$$\alpha_i = \frac{N}{i+1} \left(\frac{4(i+1)^2 - 1}{(N+1)^2 - (i+1)^2} \right)^{1/2}$$

$$\gamma_i = \frac{\alpha_i}{\alpha_{i-1}}$$

References

ARFKEN, G., 1970, *Mathematical Methods for Physicists.* Academic Press.

DAHLQUIST, G., and BJÖRCK, Å., 1974, *Numerical Methods.* Prentice-Hall.

FOX, L., and PARKER, I. B., 1968, *Chebyshev Polynomials in Numerical Analysis.* Oxford University Press.

HAMMING, R. W., 1973, *Numerical Methods for Scientists and Engineers.* Reprinted by Dover, 1986.

HILDEBRAND, F. B., 1974, *Introduction to Numerical Analysis.* Reprinted by Dover, 1987.

KRYLOV, V. I., 1962, *Approximate Calculation of Integrals.* MacMillan.

LUKE, Y. L., 1969, *The Special Functions and their Approximations.* Academic Press.

RIVLIN, T., 1974, *The Chebyshev Polynomials.* Wiley.

SANSONE, G., 1959, *Orthogonal Functions.* Reprinted by Dover, 1991.

STROUD, A. H., and SECREST, D., 1966, *Gaussian Quadrature Formulas.* Prentice-Hall.

SZEGÖ, G., 1959, *Orthogonal Polynomials.* American Mathematical Society, Providence, RI.

Appendix C

UNIX Primer

The following summary was adapted from the document *Survival UNIX*, a publication of the San Diego Supercomputer Center.

Summary of commonly used UNIX shell commands

What	*UNIX command*
Display a file	cat
Page through a file	more
Remove a file	rm
Copy a file	cp
Rename a file	mv
Move a file	mv
Make a directory	mkdir
Remove a directory	rmdir
Change working directory	cd
Print working directory	pwd
List files	ls
List files in detail	ls -l
Print a file	lpr
Visual editor	vi
Type out manual pages	man
List who is logged in	who
Show what jobs are running	ps
Send mail	mail
Log out	logout

The UNIX C-Shell

The UNIX shell is the program that is running on your terminal or in your window after you log in. It reads your command line input that you type and executes it. There are several shell programs available such as the *sh* (bourne), *ksh* (korn), *bash*, *tcsh*, and *csh* (pronounced C-shell).

There are a lot of niceties in *csh*. To find them all, look at the *csh* man page in the manuals, or look at the information on-line by typing:

```
man csh
```

The following are some of the *csh* highlights.

When *csh* starts up, it looks for a file named *.cshrc* in your home directory. If found, it takes the lines in *.cshrc* as command lines to set up a few things specifically for you.

csh: redirection

One nice feature of *csh* is the ability to redirect your program's input and output. If your program would normally have printed to the terminal via the C *printf* call, you can redirect that output into a file with the > ("goes to") symbol:

```
project1 > outputfile
```

Likewise, if your program would have read from the terminal via the C *getchar* or *scanf* calls, you can read from a file instead by using the < ("comes from") symbol:

```
project1 < inputfile
```

You can also combine these:

```
project1 < inputfile > outputfile
```

One program's output can also be used as another program's input by using the | ("pipes to") symbol:

```
project1 | lpr
```

csh: wildcards

The *csh* has the ability to expand wildcards. For example, if you wanted to remove all of your C source files, you could list them individually:

```
rm proj1.c proj2.c proj3.c proj4.c proj5.c
```

or you could use the * ("everything") wildcard:

```
rm proj*.c
```

This is obviously very dangerous and should be used with caution! If you are about to expand something with wildcards, it is sometimes nice to see what files it will expand the wildcard into:

```
echo proj*.c
```

There are other wildcards that let you get a little more specific. If you want to delete just the first three project's source files, type

```
rm proj [1–3].c
```

Again, be careful with wildcards. They can lead to disaster at very inopportune times. A common mistake is to remove all projects' executable files. You could type

```
rm proj1 proj2 proj3 proj4 proj5
```

but, instead you type

```
rm proj*
```

which promptly wipes out all of your proj*.c files as well. *Warning*: Don't let your cleverness get the better of you!

csh: command history

If you ask it to, the *csh* will keep a limited history of the previous commands you have typed. To tell *csh* to start doing this, put the following line in your *.cshrc*:

```
set history = 100
```

This will cause it to record the last 100 commands you have entered. You can see the full command record by typing

```
history
```

Each command that you type is given a command number. This is shown in the history list and typically in the *csh* prompt. You can re-execute a previous command by typing an exclamation point (pronounced "bang") and the command number. To re-execute command 57, type

```
!57
```

You can also re-excute commands with the bang and the first letters from the previous command line. To execute the *vi* editor on the same file that you edited before and then run the C compiler as you did before, type

```
!vi
!cc
```

Be careful with this! The intent of the following is to run the ranlib program on all archive files in the current directory and on all archive files in the *Projects* directory. What will really happen? (You might have to read on the UNIX filesystem first and then come back to this frightening thought exercise.)

```
ranlib *.a
cd tmp
rm *
cd ../Projects
!r
```

csh: aliasing

The *csh* will allow you to alias command names. Suppose that, being raised on VMS, you think that *dir* is easier to type than *l's* when you want to list the files in a directory. An easy way to accomplish this is to say

```
alias dir   ls -l
```

You can also substitute listed arguments into an aliased string. The symbol for "listed arguments" is \!*, so that one could create a print alias with

```
alias print   "pr \!* | lpr"
```

From then on, you just need to type

```
print file1 file2 file3 ...
```

Typically, users define their aliases in their *.cshrc* file so that they get created upon each log in.

csh: environment variables

Many UNIX programs require pieces of information that do not change much from one invocation of the program to the next. Rather than always having to pass this information to the program explicitly, some programs look for the information in environment variables first. For example, many programs need to know what type of terminal you are on so that they can do the proper type of screen erase, scrolling, and so on. To specify that you are on a VT100 terminal or terminal emulator, type

```
setenv TERM vt100
```

Typically, users define their environment variables in their *.cshrc* file so that they get created upon each log in.

csh: terminal type

And speaking of setenv term, the following lists what setenv commands need to be used for different terminals/workstations:

Terminal/Workstation	*setenv TERM*
SGI IRIS	iris-ansi
Sun	sun
vt100	vt100
X windows	xterm

The UNIX filesystem

The UNIX filesystem is arranged in a *hierarchy.* When you log in, you are placed in your *home directory.* You can create directories "underneath" your home directory by typing, for example,

```
mkdir Project1
```

This only creates the directory but does not actually put you there.

We recommend that you create a subdirectory for each project. It seems to be a good way to organize your workstation life.

To **c**hange to a sub**d**irectory, type for example,

```
cd Project1
```

You can move several levels at once by placing a slash between directory names:

```
cd Project/Data
```

To change directory to your own home directory, just type

```
cd
```

The "directory" named ..(two periods) is an abbreviation for "the directory right above here." To go one level up in the hierarchy, type

```
cd ..
```

This can also be chained into a several-directory leap. To move from one project directory to another, you might type

```
cd Project1
cd ../Project2
```

As you can guess, it is easy to get lost in all of this. To find out where you are in the directory structure, you will want to **p**rint your **w**orking **d**irectory. Type

```
pwd
```

There are some other special cases. The directory named . (one period) specifies the directory you are in right now. The directory named ~*username* specifies that user's home directory.

These directory names then become part of the names of files. To copy a file from the *ames293* data directory into your own, you might type

```
cp ~ames293/Data/teapot ./Project1/Data/teapot
```

The *vi* editor

The UNIX *vi* editor is the one most commonly used to edit files. It is entered by typing

```
vi filename
```

vi is a screen editor: it visually shows you what has been done. It is a character editor, that is, you can move the cursor into the middle of a line and do operations on individual characters.

One very important thing to remember about *vi* is that it is a *mode editor*. It has two modes: edit and enter. When you invoke *vi*, you are automatically in the edit mode. You go into the enter mode by appending, inserting, or changing. You go from the enter mode back to edit mode by hitting the *Escape* key.

The following is a survival summary of the most common *vi* commands:

Moving Around in the File

Go to the first line	1G
Go to the last line	G
Go to line #n	nG
Move up	↑, k
Move down	↓, j, Carriage
Move left	←, h
Move right	→, l, Spacebar
Move right to the next word	w
Move right to the end of the next word	e
Move left to the previous word	b
Scroll down one line	Control-y
Scroll down half a screen	Control-d
Scroll down a full screen	Control-f
Scroll up one line	Control-e
Scroll up half a screen	Control-u
Scroll up a full screen	Control-b
Mark this line with symbolic letter c	mc
Use that letter as a line number later	'c

Appending (after)

Append after current character	a
Append at end of current line	A
Open a new line after the current one	o

Inserting (before)

Insert before current character	i
Insert before current line	I
Open a new line before the current one	O

Deleting (and copy to the paste buffer)

Delete current character	x
Delete n characters	nx
Delete a word	dw
Delete n words	ndw
Delete to end of current line	D
Delete current line	dd
Delete n lines	ndd
Delete lines m–n	:m,nd

Yanking (copy to the paste buffer, but don't delete)

Yank to end of current line	Y
Yank current line	yy
Yank n lines	nyy

Pasting

Paste before	P
Paste after	p

Changing

Replace current character with c	rc
Substitute for a character	s
Substitute for n characters	ns
Change word	cw
Change n word	ncw
Change to end of current line	C
Change current line	cc
Change n lines	ncc

Global Changes

Global change	:g/string/s/ /string2/g
Global anything	:g/string/command

Searching

Search forward for a string	/string
Search backward for a string	?string
Repeat search for same string	n
Repeat search backward	N

Reading and Writing

Write the file	:w
Write to a different file	:w filename
Read another file into this one after line n	:nr filename
Edit a new filename	:e filename

Miscellaneous

Toggle upper/lower case	~
Print information on file	Control-G
Write and quit	:wq
Quit	:q!

Compiling

The following is typical of a C compile and link line:

```
cc -o name name.c subs1.o subs2.o -lgl_s -lm
```

where

cc	is the name of the compiler
-o name	indicates that the executable output is to go into a file called *name* (otherwise, it goes into *a.out*)
name.c	the name of the program to compile
subs1.o	the name of the subroutine object modules to link in
-lgl_s	link in the gl shared library
-lm	link in the math library

Commonly used object modules can be created by

```
cc -c -o subs1.o subs1.c
```

where things are as they were before except

-c	indicates that this module is just to be compiled, not linked

The *make* program

Remembering all of this compile line stuff can be a pain. We recommend that you get comfortable with the *make* program. To use *make*, you need to create a file named *Makefile*. This file will contain a list of targets to make, the files that are necessary to make them, and what rules to follow. For example, a *Makefile* might contain:

```
name:       name.o subs1.o
            cc -o name name.o subs1.o -lgl_s -lm

name.o:     name.c
            cc -c -o name.c

subs1.o:    subs1.c
            cc -c -o subs1.o subs1.c

clean:
            rm name.o subs1.o
```

When you type

```
make name
```

the *make* program will follow the targets (name.o subs1.o) and produce whatever is necessary. A nice additional feature is that *make* will compare the modify dates of the target and its source file to see if it actually has to make the target at all. You can (and should) have a separate *Makefile* in each project directory.

Appendix D

FORTRAN Primer

D.1 FORTRAN 77 statements

Despite pessimistic prognoses, FORTRAN remains a competitive computer language for scientific and engineering computation. A summary of FORTRAN 77 statements borrowed from Borse (1991) follows.

FORTRAN statements are entered in columns 7–72. Blanks are ignored. An * or a C in column 1 identifies a comment line. A nonzero, nonblank character in column 6 indicates that the line is a continuation of a previous line. Some statements may be given an identifying statement number in columns 1–5. In this summary, brackets [ ] around an item indicate an optional part of a statement and parentheses around an item indicate the item is to be supplied by the user.

Procedure Statements	
PROGRAM (name)	Optional
SUBROUTINE (name) [(argument list)]	Referenced by a CALL (name) in another program unit.
[type] FUNCTION (name) (argument list)	Referenced by (name) (argument list) in another program unit.
Statement function: F(A,B,X)=	User-defined single statement computation.
Specification Statements (Nonexecutable)	
Type Declaration Statements	
INTEGER (name list)	
REAL (name list)	
CHARACTER [*sl] (name list)	
COMPLEX (name list)	
LOGICAL (name list)	
IMPLICIT (type) ($a_1 - a_2$)	a_1, a_2 are single letters specifying the range of default variable typing.
PARAMETER (name = constant)	Defines named constants.
DATA (name list)/(value list)/	Initialize variables during compilation.
DIMENSION (array name) (subscript bounds) or (type) (array name) (subscript bounds)	The size of an array is specified by the limits on its subscripts.

EXTERNAL (procedure name)	Defines a name as a user-written subprogram.
INTRINSIC (procedure name)	Identifies a name as a library subprogram.
COMMON/(block name)/(name list)/	Identifies global variables.
SAVE (name list)	Used to retain values of local variables between calls.
BLOCK/DATA (name)	Subprogram procedure to assign values of variables, particularly those in labeled COMMON, at compilation time.

Assignment and Program Control Statements (Executable)

Assignment statement (name) = (expression)	The expression may be any of the FORTRAN data-types depending on the type of (name).
STOP [tag]	Terminates program execution and displays the integer or character [tag].
PAUSE [tag]	Interrupts program execution and displays the integer or character [tag].

Flow-Control Statements

RETURN [statement number]	Causes a return to the calling program or subprogram. The optional [statement number] is for alternate returns from a subroutine.
CALL (subroutine name) (argument list)	Transfer of control to subroutine (name).
CONTINUE	Loop terminator or target of a GO TO.
GO TO (statement number)	Unconditional GO TO.
GO TO (statement number list), (arithmetic expressions)	Computed GO TO.
IF (arithmetic expression) sl_1, sl_2, sl_3	Arithmetic IF statement.
IF (logical expression) (executable statement)	Logical IF statement.
IF (logical expression) THEN	Logical block IF statement.
ELSE	Optional ELSE block.
ELSE IF (logical expression) THEN	Optional conditional block.
END IF	Block IF terminator.
DO (statement number) (name) = e_1, e_2, [e_3]	DO loop. (name) is the name of the loop index; e_1, e_2, e_3, are expressions specifying the index bounds.

File-Directive Statements

OPEN (unit number, FILE = '(name)', [options])	Connects a file to an I/O unit.
CLOSE (unit number)	Disconnects a file.
REWIND (unit number)	Positions a sequential file at the beginning.
BACKSPACE (unit number)	Backspace a sequential file one record.
ENDFILE (argument list)	Write an End-Of-File mark on a sequential file.

INPUT/OUTPUT Statements

READ ((unit number), (format), [options]), (in-list)	
READ (format), (in-list)	
READ*, (in-list)	List-directed READ.
READ (*,*) (in-list)	List-directed READ.
WRITE ((unit number), (format), [options]), (out-list)	
PRINT (format), (out-list)	
PRINT*, (out-list)	List-directed PRINT.
WRITE (*,*) (out-list)	List-directed WRITE.
FORMAT (format specification list)	I/O editing specifications.

The Order of FORTRAN Statements

<table>
<tr><td colspan="3">PROGRAM/SUBROUTINE/FUNCTION</td><td rowspan="6">Comments</td></tr>
<tr><td>IMPLICIT</td><td rowspan="3">PARAMETER</td><td rowspan="5">FORMAT</td></tr>
<tr><td>Type Specifications (REAL, INTEGER, etc.)</td></tr>
<tr><td>Other Specifications
COMMON, EXTERNAL, INTRINSIC, etc.</td></tr>
<tr><td>Statement functions</td><td rowspan="2">DATA</td></tr>
<tr><td>ALL executable statements</td></tr>
<tr><td colspan="4">END</td></tr>
</table>

D.2 FORTRAN Intrinsic Functions

Several commonly used FORTRAN 77 intrinsic functions, borrowed from Borse (1991), are displayed in the following tables.

In FORTRAN 77 and more advanced versions, the type of the argument matches the type of the result: If the argument is in single precision, the result will also be in single precision; if the argument is in double precision, the result will also be in double precision. In older versions of FORTRAN, the prefix D or C had to be introduced in front of the generic function name to obtain double precision or complex arithmetic. For example, the square root of the real and positive number X in double precision is DSQRT(X). FORTRAN 77 is compatible with these older conventions.

FORTRAN mathematical intrinsic functions

		Permissible Arguments			
Generic Function Name	Description	Real	Integer	Double Precision	Complex
SQRT (X)	$\sqrt{\pi}$	Yes	No	Yes	Yes
EXP (X)	e^x, exponential	Yes	No	Yes	Yes
LOG (X)	$\ln(x)$, natural logarithm	Yes	No	Yes	Yes
LOG10 (X)	$\log(x)$, base-10 logarithm	Yes	No	Yes	No
ABS (X)	$\|x\|$ absolute value, for complex argument returns the norm	Yes	Yes	Yes	Yes
SIN (X)	Trigonometric sine, argument is in radians	Yes	No	Yes	Yes
COS (X)	Trigonometric cosine, argument is in radians	Yes	No	Yes	Yes
TAN (X)	Trigonometric tangent, argument is in radians	Yes	No	Yes	No
ASIN (X)	Inverse trigonometric functions, result is in	Yes	No	Yes	No
ACOS (X)	radians; if $x = \tan(\theta)$, then $\theta = \text{ATAN(X)}$	Yes	No	Yes	No
ATAN (X)		Yes	No	Yes	No
SINH (X)	Hyperbolic functions $\sinh(x) = \frac{1}{2}(e^x - e^{-x})$	Yes	No	Yes	No
COSH (X)	$\cosh(x) = \frac{1}{2}(e^x + e^{-x})$	Yes	No	Yes	No
TANH (X)	$\tanh(x) = \sinh(x)/\cosh(x)$	Yes	No	Yes	No
MOD (X,Y)	Remainder of division of x by y	Yes	Yes	Yes	No
MAX (X1,X2,...)	Maximum element in list	Yes	Yes	Yes	No
MIN (X1,X2,...)	Minimum element in list	Yes	Yes	Yes	No

FORTRAN intrinsic functions for converting data types

Function Name	Description	Permissible Arguments: Real	Integer	Double Precision	Complex	Type of Result
REAL (X)	Converts argument to real	Y	Y	Y	Y	Real
INT (X)	Truncates argument to integer	Y	Y	Y	Y	Integer
CMPLX (X) or	Converts argument to complex (X,0)	Y	Y	Y	Y	Complex
CMPLX (X1,X2)	Converts argument to complex (X1,X2)					
NINT (X)	Rounds argument to nearest integer	N	Y	Y	N	Same as argument
LEN (CH)	Returns length of character string	Single-character string				Integer
INDEX (CH1,CH2)	Returns position of string CH2 within string CH1	Two-character strings				Integer
CHAR (I)	Returns *i*th character in system-defined collating sequence	N	Y	N	N	A single character
ICHAR (CH)	Returns the position of the single character CH in system-collating sequence	A character string of length 1				Integer

Reference

Borse, G. J., 1991, FORTRAN *77 and Numerical Methods for Engineers.* PWS-Kent.

Appendix E

FORTRAN Programs

A library of FORTRAN 77 programs and subroutines were written by *Steve Yon* and the author to accompany this book. The notation and algorithms follow those of the text. The level of difficulty varies over a broad range; some programs are elementary and are included for the purpose of instruction and warming up; others are more advanced and should be useful in science and engineering research. Every effort was made to debug the codes, but an error-free product cannot be claimed. When a choice had to be made, clarity was preferred over efficiency.

The source codes are freely available through the internet, and can be accessed on the worldwide web at http://www.ames.ucsd.edu/research/pozrikidis/ncse or at http://www.oup-usa.org.

Alternatively, they can be transferred through anonymous ftp. A typical session is:

ftp stokes.ucsd.edu	Type to begin the session
login: anonymous	The computer prompts you with *login:*
password: type your e-mail address	The computer prompts you with *password:*
cd /pub	Type to access the public subdirectory
mget *filename*	Substitute *filename* with the name of the file you want
bye	Type to end the session

The programs are arranged in 12 directories. A table of contents is provided in each directory. Stand-alone subroutines are accompanied by drivers; if *xian.f* is the file name of a subroutine, then *xian_dr.f* is the name of the driver. The first 11 directories correspond to the same-numbered chapters of the text. Compilation and linking should be done as indicated in the *makefile* of the host directory (see appendix C). For convenience of transport, all directories have been condensed into the *tar* file *ncse.tar.* To unravel the directories type: *tar xvf ncse.tar*

Following is a *partial* list of the directory contents in alphabetical order; the names of the subroutine drivers are not listed. The FORTRAN qualifier *.f* is appended at the end of each file name.

Directory: *01_num_comp*

General numerical computation

Programs and subroutines

binary	Converts a positive number to the corresponding binary.

bits	Greatest integer that can be described with a specified number of bits.
combinatorial	Computes the combinatorial by two methods.
euclid	Finds the greatest common divisor of two positive integers.
fast_sum	Sums an infinite series whose terms decay like $1/n^2$.
mapping	Multiplies a vector by a matrix many times.
pie	Computes π by considering the perimeter of a regular polygon.
prime	Ascertains whether a specified integer is prime.
quadratic	Produces the roots of a quadratic equation.
ran_bsort	Generates a string of random numbers and ranks them using bubble-sort.
ran_sort	Generates a string of random numbers, indexes and ranks them.
richardson	Richardson extrapolation.
von_koch	Generates the von Koch line.

Directory: *02_lin_calc*

Linear algebra and calculus

Programs and subroutines

chol	Cholesky decomposition of a symmetric and positive-definite matrix by column or row.
det_hess	Determinant of a lower or upper Hessenberg matrix.
det_33c	Determinant of a 3×3 complex matrix.
gram_sc	Gram-Schmidt orhogonalization of a set of vectors.
inv_l	Inverse of a lower-triangular matrix.
inv_u	Inverse of an upper-triangular matrix.
ldu	***LDU*** decomposition of a symmetric matrix.
lu_c	Crout ***LU*** decomposition, determinant and inverse of a matrix.
lu_d	Doolittle ***LU*** decomposition, determinant and inverse of a matrix.
lu_d_t	Doolittle ***LU*** decomposition of a tridiagonal matrix.
qr_dec	***QR*** decomposition of an arbitrary matrix by three methods: Gram-Schmidt, rotation, and reflection.

Directory: *03_lin_eq*

Systems of linear algebraic equations

Programs and subroutines

bcg	Biconjugate gradients method for arbitrary systems.
cg	Conjugate gradients method for symmetric and positive-definite systems.
cramer_33	Solution of a 3×3 system by Cramer's rule.
cycl_red	Cyclic reduction of a tridiagonal matrix with constant diagonal elements.
gel	Gauss elimination with the option of pivoting and operation-reduction for a symmetric matrix; ***LU*** decomposition.
gel_mrhs	Gauss elimination with multiple right-hand sides; inverse of a matrix.
penta	Solution of a pentadiagonal system.
solve_chol_c	Solution of a real, symmetric, and positive definite system by the Cholesky decomposition.
thomas	Solution of a tridiagonal system.

Directory: *04_nl_eq*

Systems of nonlinear algebraic equations

Programs and subroutines

bairstow	Bairstow's method for polynomial root finding.
fpi1	One-point iterations for one equation.
fpi2	One-point iterations for two equations.
non_lin	Solves a system of nonlinear equations by the Newton or Broyden method.
non_lin1	Solves one nonlinear equation by several methods.

Directory: *05_eigen*

Eigenvalues and eigenvectors of matrices

Programs and subroutines

char_poly	Generates the coefficients of the characteristic polynomial by two methods.
cond_number	$\| \|_2$ condition number of a real matrix.
house_sym	Householder method for making a symmetric matrix tridiagonal.
jacobi	Jacobi method for computing all eigenvalues of a real symmetric matrix.
lanczos	Lanczos method for making a matrix tridiagonal.
lanczos_sym	Lanczos method for making a symmetric matrix tridiagonal.
power	Real eigenvalues by the power method with shifting and deflation, and with option for inverse iteration.
power_sym	Real eigenvalues of a symmetric matrix by the power method.

Directory: *06_interp_diff*

Function interpolation and differentiation

Programs and subroutines

arc_3d	Computes the circular arc that passes through three specified points in space.
chebyshev	Applications of Chebyshev interpolation and approximation.
hermite	Hermite interpolation of a function of one variable by two methods.
int_1d	Interpolation of a function of one variable by several methods.
lagrange	Applications of Lagrange interpolation.
lagrange_es	Lagrange interpolation with evenly-spaced points.
laplacian_2d	Finite-difference formulae for the Laplacian of a function of two variables.
splines_cb	Cubic spline interpolation with several options.
trgl_octa	Triangulation of the surface of a sphere beginning with a regular octahedron.

Directory: *07_integration*

Integration of a function of one or two variables

Programs and subroutines

integral	Computes an integral of a function of one variable by several methods.
integral_lag	Computes an integral by the Gauss-Laguerre method.
integral_log	Computes an integral with a logarithmic singularity using a quadrature.
filon	Filon rule for an oscillatory integrand.
gauss_lag	Base points and weights of the Gauss-Laguerre quadrature.
gauss_leg	Base points and weights of the Gauss-Legendre quadrature.
gauss_log	Base points and weights of the Gauss quadrature with a log singularity.
gauss_sp	Base points and weights of an integration quadrature over a sphere.
gauss_trgl	Base points and weights of an integration quadrature over a planar triangle.

Directory: *08_approximation*

Approximation of a function of one variables

Programs and subroutines

bernstein	Approximates a function with a Bernstein polynomial.
b_spl_un	Computes and displays uniform B-splines.
b_splines	Approximates a function with B-splines.
fft	Fast Fourier transform of a function of one variable.
ft_1d	Fourier transform of a function of one variable by the conventional method.
ft_1d_nu	Complex Fourier transform of a periodic function of one variable for non-uniformly spaced points, by the conventional method.
ft_2d	Fourier transform of a function of two variables over a rectangle.

Directory: *09_ode_ivp*

Ordinary differential equations; initial-value problems

Programs and subroutines

ode	Solves a system of odes by several methods.

Directory: *10_ode_bvp*

Ordinary differential equations; boundary-value problems

Programs and subroutines

fdm	Finite-difference method for an ode.
fem	Finite-element method for an ode.

Directory: *11_pde*

Partial differential equations

Programs and subroutines

poiss_fds_NNNN	Generates the finite-difference system for Poisson equation in a rectangle with Neumann boundary conditions on the four sides. Uses second-order centered differences for the equation and the boundary conditions.
poiss_fds_DNDD	Same but with Dirichet and Neumann boundary conditions.
poiss_fds1_NNNN	Generates the finite-difference system for Poisson equation in a rectangle with Neumann boundary conditions on the four sides. Uses second-order centered differences for the equation and one-sided second-order differences for the boundary conditions.
poiss_NNNN	Solves the Poisson equation in a rectangle with Neumann boundary conditions on the four sides, with various discretization options and regularization of the source term.
rod	Unsteady heat conduction through a rod.

Directory: *12_spec_fnc*

Special functions

Programs and subroutines

bessel_j0	Evaluates the Bessel function J_0.
bessel_j1	Evaluates the Bessel function J_1.
bessel_mod	Evaluates the modified Bessel functions.
ell_int	Evaluates the complete elliptic integrals of the first and second kind.
error_f	Error, complementary error, and inverse error function.
exp_int	Evaluates the exponential integral
gso_poly	Produces the coefficients of several families of orthogonal polynomials by the Gram-Schmidt method.

Index